ASSISTIVE ROBOTICS

ASSISTIVE ROBOTICS

Editors

Hongye Su
Zhejiang University, China

Tianmiao Wang
Beijing University of Aeronautics and Astronautics, China

Mohammad O Tokhi
University of Sheffield, UK

Gurvinder S Virk
University of Gävle, Sweden

World Scientific

NEW JERSEY • LONDON • SINGAPORE • BEIJING • SHANGHAI • HONG KONG • TAIPEI • CHENNAI • TOKYO

Published by

World Scientific Publishing Co. Pte. Ltd.
5 Toh Tuck Link, Singapore 596224
USA office: 27 Warren Street, Suite 401-402, Hackensack, NJ 07601
UK office: 57 Shelton Street, Covent Garden, London WC2H 9HE

British Library Cataloguing-in-Publication Data
A catalogue record for this book is available from the British Library.

The front cover image has been provided by:
Professor Seungbin Moon, Sejong University, South Korea.

ASSISTIVE ROBOTICS
Proceedings of the 18th International Conference on CLAWAR 2015

ISBN 978-981-4725-23-1

In-house Editor: Rhaimie Wahap

Printed in Singapore

PREFACE

It is a pleasure to welcome everybody to beautiful HangZhou, China for the 2015 CLAWAR (Climbing and Walking Robots) conference, 18th in the series. The CLAWAR conference series has acted as a key platform for dissemination, debate and emergence of new research developments and findings in the area of mobile robotics, and CLAWAR 2015 is built upon the same mission.

The CLAWAR 2015 proceedings present state of the art research findings around the theme of assistive robotics in 89 technical articles by authors from 16 countries from throughout the five continents. The works presented are geared towards provision of assistance with certain degree of autonomy in various applications. As usual with previous issues of the conference there is a significant number of articles tackling the challenging issues of robot locomotion either legged, wheeled or hybrid. However assistive and autonomous robots are featured more dominantly in this issue. This is mainly due to the current drive and motivation to the need for helping the elderly and the disabled in different environments. A further strong theme in the conference is designs based on biological inspirations, looking at various mechanisms and designs that may be realised resembling features of biological species.

The editors would like to thank members of the International Program Committee, International Advisory Committee and National Organising Committee for their efforts in reviewing the submitted articles, and the authors in addressing the comments and suggestions of the reviewers in their final submissions. It is intended that the CLAWAR 2015 proceedings will be a valuable source of reference for research and development in rapidly growing area of mobile service robotics.

H. Su, T. Wang, M. O. Tokhi, G. S. Virk

CONFERENCE ORGANISERS

Zhejiang University
HangZhou, China

HangZhou Association for Science and Technology
HangZhou, China

HangZhou Automation Society
HangZhou, China

CLAWAR Association

A charitable non-profit making membership based professional organization serving the robotics community
www.clawar.org

CONFERENCE COMMITTEES AND CHAIRS

Conference Chairs

of Eighteenth International Conference on Climbing and Walking Robots and the Support Technologies for Mobile Machines

H. Su (General co-Chair)	– Zhejiang University, China
T. Wang (General co-Chair)	– Beijing University of Aeronautics and Astronautics, China
M. O. Tokhi (Program Chair)	– University of Sheffield, UK
G. S. Virk (IAC Chair)	– University of Gävle, Sweden

International Advisory Committee (IAC)

of Eighteenth International Conference on Climbing and Walking Robots and the Support Technologies for Mobile Machines

H. Levent Akin	– Turkey
Manuel Armada	– Spain
A. K. M. Azad	– USA
Karsten Berns	– Germany
Philippe Bidaud	– France
Bryan Bridge	– UK
Anibal T de Almeida	– Portugal
H. Fujimoto	– Japan
Qiang Huang	– China
Krzysztof Kozlowski	– Poland
Giovanni Muscato	– Italy
Kenneth J Waldron	– USA
Ming Xie	– Singapore
Jie Zhao	– China

International Programme Committee (IPC)

of Eighteenth International Conference on Climbing and Walking Robots and the Support Technologies for Mobile Machines

Ahmad S.	– Malaysia
Alam M. S.	– Bangladesh
Aldbrez F. M.	– UK
Aloimonos Y.	– USA
Arkin R.	– USA
Belter D.	– Poland
Ben Amar F.	– France
Bhattacharya B.	– India
Billingsley J.	– Australia
Bonsignorio F.	– Italy
Bostelman R. V.	– USA
Bouazza-Marouf K.	– UK
Cameron S.	– UK
Chen W.	– China
Cheng H.	– China
Chevallerau C.	– France
Chew C.-M.	– Singapore
Chugo D.	– Japan
Cruse H.	– Germany
Cubero S.	– Australia
Dehghani A.	– UK
Ding Y.	– USA
Dodd T. J.	– UK
Dogramadzi S.	– UK
Dutra M. S.	– Brazil
Dutta A.	– India
Fortuna L.	– Italy
Fukuda T.	– Japan
Garcia E.	– Spain
Gatsoulis Y.	– UK
Goher K. M.	– Oman
Goldman D.	– USA
González de Santos	– Spain
Gradetsky V.	– Russia
Grand C.	– France
Guan Y.	– China
Gupta S. K.	– USA
Tenreiro Machado J.	– Portugal
Heyes N.	– UK
Howard D.	– UK
Hwang K.-S.	– Taiwan
Ion I.	– Romania
Johnson A. M.	– USA
Kadar E. E.	– UK
Kiriazov P.	– Bulgaria
Shabgahi R. L.	– Iran
Lefeber D.	– Belgium
Leon-Rodriguez H.	– Columbia
Lewis M. A.	– USA
Li B.	– China
Li C.	– USA
Liu D.	– Australia
Ma H.	– China
Marques L.	– Portugal
Massoud R.	– Syria
Mohamed Z.	– Malaysia
Molfino R.	– Italy
Moon S.	– Korea
Nakamura T.	– Japan
Penders J.	– UK
Pratihar D. K.	– India
Prescott T. J.	– UK
Preucil L.	– Czech Republic
Qian J.	– China
Rachkov M.	– Russia
Ribeiro M.	– Portugal
Santos V.	– Portugal
Sattar T. P.	– UK
Schmiedeler J.	– USA
Sharma S. K.	– UK
Silva F.	– Portugal
Singh S. P. N.	– Australia
Skibniewski M.	– USA
Steinicke L.	– Belgium
Tan M.	– China
Yigit A.	– Kuwait

National Organising Committee

of Eighteenth International Conference on Climbing and Walking Robots and the Support Technologies for Mobile Machines

CONFERENCE SPONSORS AND CO-SPONSORS

State Key Laboratory of Industrial Control Technology

Zhejiang Association of Automation

GUOZI Company

TABLE OF CONTENTS

Section-4: Biologically-inspired systems and solutions

Section-10: Modelling and simulation of CLAWAR

Section-11: Perception, localization and rescue operations

Section-12: Planning and control

Section-13: Underwater and sea robotics

SECTION–1

PLENARY PRESENTATIONS

INFRASTRUCTURE ROBOTICS: OPPORTUNITIES AND CHALLENGES

GAMINI DISSANAYAKE

Mechanical and Mechatronic Engineering
University of Technology, Sydney, Australia

Current applications of robotics is distinguished from more traditional automation by the focus on machines that operate in relatively unstructured, difficult and often hazardous environments. The past decade has seen the deployment of a number of robotic systems in highly challenging application domains such as mining and agriculture. In particular, the potential safety, cost and health impacts from the use of robotics aids for periodic inspection and maintenance of civil infrastructure has led to a significant expansion of research activity in "infrastructure robotics". Two of the prerequisites for deploying a robot in the field are the ability to acquire and maintain a representation of the environment, and efficient motion planning. In scenarios where a machine and a human must collaborate to perform a task, an intuitive user interface, a joint understanding of abilities and joint management of task execution are also essential. This talk will summarise current research on these key competencies that underpin robot deployments in unstructured environments, with the focus on simultaneous localisation and mapping, and human robot collaboration. Future potential and key challenges in infrastructure robotics will also be presented together with a review of the research activities in this field at the Centre for Autonomous Systems, University of Technology Sydney.

COGNITION-INSPIRED ROBOT LEARNING AND CONTROL

JIANWEI ZHANG

Institute of Technical Aspects of Mulitimodal Systems
University of Hamburg, Germany

In a dynamic and changing world, a robust and effective robot system must have adaptive behaviours, incrementally learnable skills and a high-level conceptual understanding of the world it inhabits, as well as planning capabilities for autonomous operations. Future robot systems will benefit from the recent research on neurocognitive models in processing multisensory data, exploiting synergy, integrating high-level knowledge and learning, etc. I will first introduce multisensory integration methods for intelligent control of robots. Then I will present our investigation and experiments on synergy technique which uses fewer parameters to govern the high DOF of multifinger robot movement. The third part of my talk will demonstrate how an intelligent system like a robot can evolve its model as a result of learning from experiences; and how such a model allows a robot to better understand new situations by integration of knowledge, planning and learning. I will show some integrated results of operational mobile robot platforms with grasping facilities in a restaurant service scenario.

BIOLOGICALLY-INSPIRED MINIATURE JUMPING ROBOT: FROM DESIGN TO CONTROL

NING XI

Department of Electrical and Computer Engineering
Michigan State University, East Lansing, Michigan, USA

Robots will transform our daily lives in the near future by moving from controlled industrial lines to unstructured environments such as home, offices, or outdoors with various applications from healthcare, service, to defence. Nevertheless, two fundamental problems remain unsolved for robots to work in such environments. On one hand, how to equip robots, especially meso-scale ones with sizes of a few centimetres, with multiple locomotion abilities to deal with the unstructured environment is still a daunting task. On the other hand, how to control such robots to dynamically interact with the uncertain environment for agile and robust locomotion also requires tremendous efforts. In this talk, I will present my research efforts to tackle these two problems in the framework of biologically inspired robotics. First, I will show how to use biologically principles found in nature to build efficient meso-scale robots with various locomotion abilities such as jumping, rolling, and aerial manoeuvring. Second, I will present a novel non-vector space control method for meso-scale robots which have limited computation power. The research in these two thrusts will pave the way for next generation bio-inspired, low cost, and agile robots.

UNDERSTANDING ANIMAL LOCOMOTION USING BIO-INSPIRED ROBOTICS AND SOFT ROBOTICS

TIANMIAO WANG

School of Mechanical Engineering and Automation
Beijing University of Aeronautics and Astronautics, China

The current studies of fish-like aquatic propulsion are quite limited in their ability to control for physical determinates such as body shape, kinetic motions, flexural stiffness, skin surface scale and Mucus properties etc. Keeping one of these determinates constant while altering others in a controllable manner is impossible for the swimming live fishes. Bio-inspired robotic models offered the ability to manipulate and control individual physical determinate that affect the aquatic propulsive performance.

The research group at Beijing University of Aeronautics and Astronautics has developed bio-inspired robotics as well as multi-material prototypes to experimentally investigate several new topics of aquatic propulsion. A key aspect of the experimental setup is that the bio-inspired robotics and multi-material prototypes are self-propelled (thus to satisfy Newtonian equation of balance) and can swim against the flow in the lab water tank. While self-propelled swimming, measurement of speed, external forces and torques, internal power consumptions, cost-of-transport (COT) of the bio-inspired robot and the multi-material prototypes are synchronized with the motion program and high-speed video of the wake flow. Finally, pneumatically actuated soft robotics, that can efficiently grasp objects with different sizes and shapes, will be introduced in this presentation. This robotic device is completely soft and made of low-cost lightweight material. The grasping performance can be realized by very simple control without any feedback.

SECTION–2

ASSISTIVE ROBOTS

A BEHAVIOR ADAPTATION METHOD BASED ON HIERARCHICAL POMDPS *

YONG TAO†

School of Mechanical Engineering and Automation, Beihang University, Beijing Beijing, 100191, China

YOUDONG CHEN

School of Mechanical Engineering and Automation, Beihang University, Beijing Beijing, 100191, China

DONG XU

School of Automation Science and Electrical Engineering, Beihang University, Beijing Beijing, 100191, China

JIAQI ZHENG

School of Machinery and Automation, Wuhan University of Science and Technology, Hubei, Wuhan, 430081,China

This paper proposes a behavior adaptation method for companion robot based on the hierarchical partially observable Markov decision processes (H-POMDPs). The H-POMDPs have recently emerged as a powerful method for optimizing human-robot interaction (HRI) decision making in partially observable environments. The model is used for representing and learning HRI models for indoor elderly companion robot. This approach proposes an approximate solution by employing the H-POMDPs in the uncertain environments. The HRI states such as daily dialogues, news and whether broadcasting, motion speed and navigation-assist are used throgh the H-POMDP. The user's speech command, touch-screen input, head position and body posture are detected as subconscious signals that indicate a user's interaction preference. The experimental results show the effectiveness and feasibility of designing the H-POMDP behavior adaptation method. The H-POMDPs can train significantly faster than the original MDP. The H-POMDPs requires much less data, and can easily be extended to variables reducing the time and sample complexity.

* This work is supported by the Youth Science Fund Project NO.61305116 of the National Natural Science Foundation.

† Corresponding author： Yong Tao, phone: 086-10-8233-8271; fax: 086-10-8233-8271; e-mail: taoy@buaa.edu.cn).

1. Introduction

In the aging of the society, limited mobility and high dependence on others affects people's self-confidence negatively. The current services for the elderly are insufficient. We need to find alternative ways of providing care to the elderly population, which will not only lower the costs, but also increase the comfort of living with the level of dignity that our elderly deserve [1].

For improving the living environment, life style and care model of the elderly through high tech equipment, many intelligent assistive devices such as intelligent wheelchair, smart nursing bed and companion robot, have been developed in the last decades. Companion robots are expected to communicate with non-experts in a natural and intuitive way. A companion robot will act in an unstructured environment, such as a private home or a rest home for old people, with people roaming around. Currently, there are no robots that are able to perform a combination of these tasks efficiently, accurately and robustly.

The interaction between service robot and human is a research focus and challenge. Especially, finding optimal strategies for human robot interaction operating in an uncertain environment is an active field of research in service robotics [2-5]. One of the major goals of HRI is designing as: robot systems that perceive their environment and execute actions. In particular, robot should be able to plan their interactive actions with human, so as to fulfill their task as accurate as possible.

The service robots are required to accomplish their interaction tasks based on perceptions about the environment and the user's intention collected via the sensors. The decision making in the observable static environments has been solved effectively via the well-known Markov decision process (MDP) [6].The MDP solutions setup applies to a single adaptive interactive actions in a certain environment at discrete time steps. However, in most real-world domains, the robots are not able to discern the exact environment state and the user's intention as the sensors are noisy and limited. The faulty and inaccurate observations are given with the sensor noise, and the robot is unable to observe the difference between states that cannot be detected by the sensor.

The companion robot sensor reading might require different action choices. In this paper, we consider the problem of how robot can cooperate to jointly accomplish a task, while facing the sensing limitations such as the capability to observe the state of the interaction. The case is studied wherein robots have to learn how to plan the actions, given uncertainty in the actions and sensor readings error. The need for HRI learning and planning is particularly pressing, given that smart device such as smart phone, Pad, watch and glasses, are becoming ubiquitous in society.

Partially observable Markov decision processes (POMDP) is a method for calculating optimal control actions under uncertainty. It is fit for a wide range of real-world domains where joint planning and tracking is necessary. It has been successfully applied to problems of location, path planning and human-robot interaction. Other types of applications that have been considered include visual planning for mobile robot teams using hierarchical POMDPs [7], mobile robots localizing, path planning and navigation with POMDPs [8-12], parameter estimation via real-time re-planning of continuous Gaussian POMDPs [13], as well as multi-Robot cooperation problems [14-15]. Manipulation planning problem for Grasping via POMDPs [16-17].

A POMDP Framework for modelling human interaction with assistive robot [18-19]. A Bayesian-game-based fuzzy reinforcement learning (RL) controller for decentralized partially observable Markov decision processes (Dec-POMDPs) is proposed [20]. Representing hierarchical POMDPs as DBNs for multi-scale map learning is given [21].

In this paper, we have formalized the decision-making problem in HRI of elderly companion robot variations of the POMDPs framework. Combined with the large amount of uncertainty in such real-world environments, decision-theoretic frameworks of elderly companion robot are more than appropriate. A behavior adaptation method based on H-POMDP is proposed. Based on the speech command, touch-screen input, head position and body posture from the user, the reward function is described through the observation. It focuses on the behavioral parameters in order to fit the user's interaction preference that maximize the reward. A reward method is proposed which consists of the user's info query and navigation-assist rate in HRI of elderly companion robot.

We use the progressive solution to the H-POMDP and effectively generalize over the continuous global belief space. The method is a crucial step toward both an efficient approximation technique and an algorithm for the H-POMDP domains. The rest of the paper is organized as follows. Section II briefly sheds define the H-POMDPs model: the hardware design of the companion robot and POMDPs model are introduced In Section IV, the H-POMDPs for elderly companion robot is presented. In Sections V, the proposed behavior adaptation method based on H-POMDPs is proposed. Section VI presents empirical results and the experiments in the indoor uncertain environment with different operate mode. The information inqury and navigation, which demonstrate the superiority of H-POMDPs over flat POMDPs, both in terms of accuracy and speed of learning. Finally, Section VI concludes the paper with a discussion on the future scope of the proposed approach.

2. Background and Related Work

2.1. *The POMDPs model*

POMDPs provide a mathematic model that permits to characterize this type of systems. The POMDPs model represent the dynamics of the environment, such as the probabilistic outcomes of the actions, the transition function T, the reward function R and the probabilistic relationships between the agents observations and the states of the environment. The POMDPs framework is a systematic method that uses belief states to represent memory of past actions and observations [22].

The POMDPs model is defined as $\langle S, A, T, R, Z, \gamma, O \rangle$, a seven tuple.

The real state is not known in the POMDPs based on the uncertain observation. As shown in Figure 1, the control stage of the POMDPs is divided into two blocks.

1) State estimator: the module receives observations from the environment, the last action taken and the previous belief state and produces an updated belief state.

2) Politics: the input of this block is the current belief state and its output is the action to perform. This block obtains the optimal action to perform in the next execution step to maximizing the recompense R .

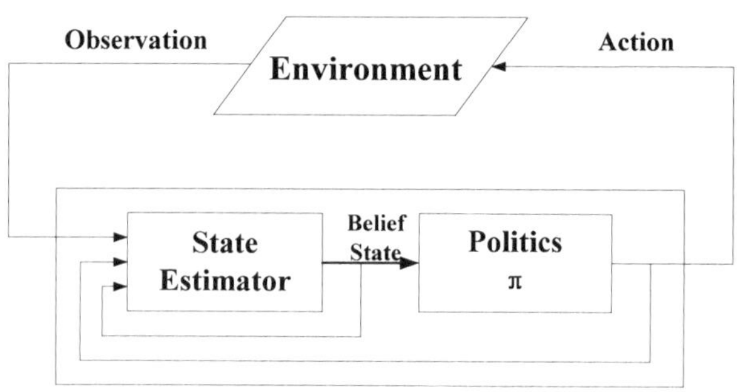

Figure 1. The POMDP model.

The Belief state $B(S)$ would be updated whenever a new action or observation is carried out. When an action a is executed and a new observation o is taken, the new probabilities become:

$$B_t(s') = \eta \times p(o \mid s') \times \sum_{s \in S} p(s' \mid s, a) \times B_{t-1}(s), \quad \forall s' \in S \tag{2}$$

2.2. *Introduction of the elderly companion robot*

As shown in Figure 2, the robot is equipped with a URG-04lx-ug01 laser range finder, capable of measuring distances at an angular resolution of one degree and a spatial resolution of 5 cm, within a planar perceptual field that covers a 180 degree range. The robot is with 8 ultrasonic sensors and 8 infrared sensors.

There are two on-board PCs, connected to the Internet via a 2Mbit/sec wireless Ethernet link. The robot is also equipped with a bright, touch-sensitive color display conveniently at approximate eye height for sitting people.

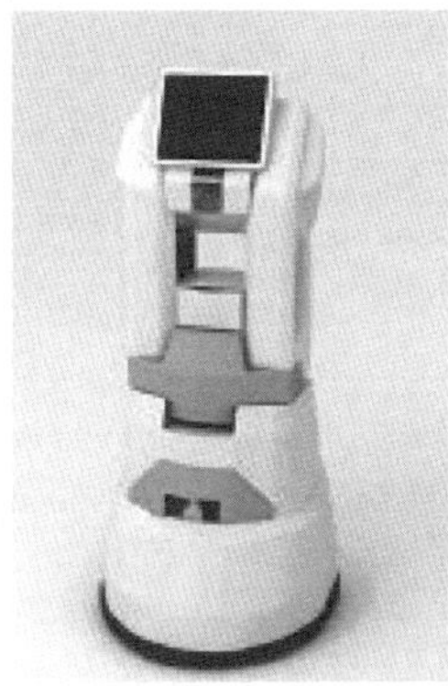

Figure 2. The elderly companion robot.

There is a CCD camera with an approximate aperture angle of 100 degrees. The camera is connected to a frame grabber and JPEG encoder for image processing and high-bandwidth communication. There are a speaker system and a microphone array, for recording and synthesizing speech. The battery lifetime is approximately 60 minutes.

3. The Hierarchical POMDPS

The action space is decomposed into smaller chunks in the hierarchical POMDPs. It's assumed that the maps are available and the map representation is coded and programmed. The basic concept of the action hierarchy is illustrated in Figure 3.

The action hierarchy is a tree, where each leaf is labeled by an action from the action set of POMDPs. Each action $a \in A_0$ must be attached to at least one leaf, A_0 is the primitive action. In each internal node, there is an abstract action [16].

In the H-POMDPs model, the hierarchical action is set to replace the full POMDPs by a set of smaller POMDPs. It will reduce the computing complexity. Through the hierarchical action set, we achieve a collection of POMDPs that are smaller than the original POMDPs, yet collectively define a complete policy.

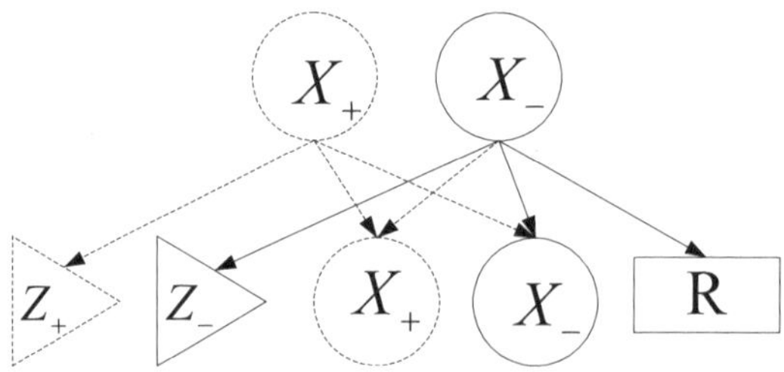

Figure 3. Dynamic belief network.

Each internal node $\overline{a}_i$ in the action hierarchy is corresponding to a subtask P_i. Given that the action hierarchy spans a collection of separate POMDPs subtasks, the independent policy independently optimize for each subtask, such that we obtain a collection of corresponding *local policies*.

The state and observation abstraction are related to the parameters of the hierarchical POMDPs model. With a subtask P_i, action set A_i, state set S_i and observation set O_i, the state space is corresponding to state features $X_i = \{X_1, X_2, ..., X_m\}$ and the observation space is corresponding to features $Z_i = \{Z_1, Z_2, ..., Z_n\}$. The state features are divided into two different sets, X_+ and X_-, and similarly observation features are divided into two disjoint sets, Z_+ and Z_-. The state features X_+ and observation features Z_+ are irrelevant to subtask P_i if $\forall a_j \in A_i$:

$$R(X_+, X_-, a_j) = R(X_-, a_j) \tag{3}$$

The constraints are illustrated in the form of a dynamic belief network. The state features in set X_+ have no effect on the reward function, and furthermore provide no transition or observation information regarding those features (namely X_-) that do have an effect on the reward function. The state features X_+ and observation features Z_+ can have no effect on the value function, and therefore can be safely ignored. The feature irrelevance condition should hold for all actions in a given subtask.

In the H-POMDPs model, abstraction can be applied independently to each subtask, thus reducing $|S|$ and $|O|$, without influencing the policy optimization any further than what is attributed to the action decomposition.

The state and observation abstraction is a natural consequence of the action decomposition. The action decomposition and state/observation abstraction are carried out alternatively. The four parameters are adapted by the system. These were speech command, touch-screen input, head position and body posture.

The reward function is based on the information providing by the elderly and the navigation-assist rate in the home environment. An analysis of user's command of speech, touch-screen and user body position in HRI that the evaluation from users had a positive correlation with the information query times and navigation-assist acceptance.

The head's positions and body position of the elderly people are measured by a camera capture system. The speech command is collected by microphone array in front of the user. The navigation directions are measured by the touch-screen input. The news and weather information query and entrainment are interacted through the speech command.The contributions of the factors were of equal size as a result.

4. A Behavior Adaptation Method Based on H-POMDPs

The H-POMDPs is presented by representing the state-space at multiple levels of abstraction, scale much better to larger environments. The H-POMDPs simplify the planning and learning, the states have lower entropy and the number of free parameters is reduced. The parameters of H-POMDP require an inference algorithm, since the true state of the world is not observed. The multi-scale method learns much larger models much faster, without requiring manual segmentation of the training data.

4.1. *HRI behaviors of H-POMDPs*

There are six interaction behaviors, *Dialogue, News-broadcasting*, *Tell-weather*, *GotoKitchen*, *GotoBathroom* and *GotoTV.room*. Each behavior takes 25s. During the interaction, the robot tries to provide the *information and assist navigation* of the category to which the behavior belongs. The *information query and navigation-assist are* measured through the interaction interface, user's head position and body posture.

The interaction behaviors are classified into three interaction categories: *information-query, navigation-assist,* and *Other*.

It also meets the human's head position in a cyclic manner, where the companion robot meets and averts the human's head position in cycles that last average 6s through the camera. The parameter *head-front-ratio* is the portion of each cycle spent meeting the user's head.

The *information-query* controls the information displayed on the screen or voice introduction through the speaker. When it performs behaviors from (a) *dialogue* to (b) *news broadcasting* that require different commands on the part of the user, the companion robot starts the actions after it makes an *information-query* processing.

The *navigation-assist* controls the movement of the robot. If the *GotoBathroom* is given, the motion is carried out based on the *location*, *destination* and *path-planning*.

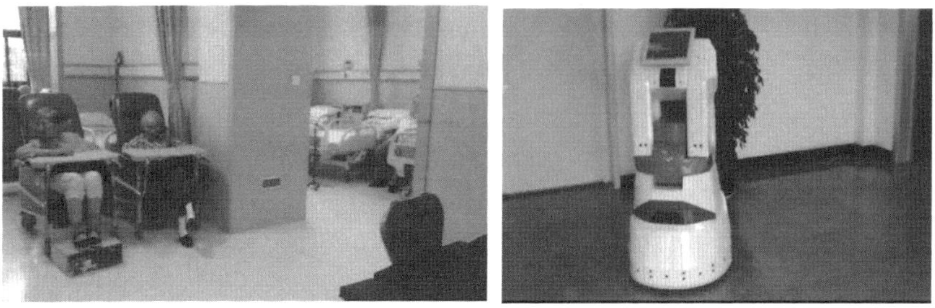

Figure 4. The elderly companion robot in a living room.

4.2. *Behavior adaptation*

During the interaction, the H-POMDP behavior adaptation method was running on the robot. For the duration of each interaction behavior, or the test of a policy, the robot keeps the interaction observation and other parameters according to π_{Pi}. The reward function R was calculated for each executed interactive action of the robot using the command input and user head position for the duration of the behavior.

The duration starts from just after the robot selects the behavior and it ends at the end of the behavior. A total of ten different parameter combinations were tried before optional policy π_i was calculated and the parameter values updated. The subject did not notice the update of the policy during interaction.

5. Experiments

To investigate the advantages of behavior adaption method, we performed experiments using an indoor environment. A total of 8 subjects are tesetd. They were not familiar with the use of the robot, and had no prior experience of interaction with the robot. We exposed 8 subjects to the behaviors, and let them choose what kind of interaction they were comfortable with for each of these. The robot did not give information or navigation service in the pre-study. There are variations of preferences for each subject within the same class.

The information inqury experiments are implemented using the speech interaciton and touchscree input-output. During these experiments, the robot gave 69 information answer about the weather, news and time. The *Tell-time* converged in the acceptable range for 6 of 8 subjects. The *News-broadcasting* and *Tell-weather* converged in acceptable range for 7 and 8 subjects, respectively. The learned *information-query* is calculated as the average parameter value during the last 15 min of each study run since the H-POMDP algorithm keeps searching for the optimum value.

The robot autonomously led 15 full transportation service for 8 different testers. The *GotoKitchen*, *GotoBathroom*, *GotoTV.room*, converged to the values near to selected values for 6, 7 and 8 of 8 subjects, respectively.

It is different in the success rate between different parameters. It is a typical character for the behavior adaptation that parameters having a greater impact on the reward function are adjusted faster. The behaviors of the subject did not display any difference for policies π_i if the current parameter was too far from the preferred values.

6. Conclusion

This paper presents a novel hierarchical POMDPs scheme for elderly companion. We show how H-POMDPs approximation can be used as a principled method to effectively combat the time and space complexity of H-POMDPs. We elucidate the feasibility of the proposed approach with empirical results on the elderly companion robot. The proposed hierarchical POMDPs method successfully discovers a high-quality optimal policy solution in a reasonable number of trials and running time. The method uses detecting the user's speech command, touch-screen input, head position and body posture as subconscious signals that indicate a user's interaction preference. The experimental results and comparative evaluation show the effectiveness and feasibility of employing the H-POMDP behavior adaptation method.

Acknowledgments

This research is supported by the Youth Science Fund Project NO.61305116 of the National Natural Science Foundation.

References

1. G. Baltus, D. Fox, F. Gemperle, et al. Towards Personal Service Robots for the Elderly. *Workshop on Interactive Robots and Entertainment.* (2000).
2. M. Chang, M.He, and X. Luo. AstonCAT-Plus: An efficient specialist for the TAC market design tournament. *IJCAI*, pp. 146-151 (2011).
3. M. Chang, M. He, and X. Luo. Designing a successful adaptive agent for TAC ad auction. *Proc. Eur. Artif. Intell.*, pp. 587-592 (2010).
4. F.A.Oliehoek. Decentralized POMDPs, *Reinforcement Learning: State of the Art, M. Wiering and M. van Otterlo, Eds.* (2012).
5. M. Spaan, F. Oliehoek, and C. Amato. Scaling up optimal heuristic search in Dec-POMDPs via incremental expansion. *IJCAI,* pp. 2027-2032 (2011).
6. M. L. Puterman. *Markov Decision Processes—Discrete Stochastic Dynamic Programming.* New York: Wiley (1994).

7. Shiqi Zhang, Mohan Sridharan, and Christian Washington. Active Visual Planning for Mobile Robot Teams Using Hierarchical POMDPs. *IEEE Transaction on Rpbotics*, Vol. 29, No. 4, pp.975-985 (2013).
8. Salvatore Candido and Seth Hutchinson. Minimum Uncertainty Robot Path Planning using a POMDP Approach. *IROS*, pp.1408-1413 (2010).
9. H. Kurniawati, T. Bandyopadhyay, and N. M. Patrikalakis. Global Motion Planning under Uncertain Motion,Sensing, and Environment Map. *MIT Press eBook Chapters: Robotics:Science and Systems VII*, pp. 161- 168 (2012).
10. W. Adiprawita, A. Suwandi Ahmad and Jaka Sembiring. Reinforcement learning with Heuristic to solve POMDP problem in Mobile Robot Path Planning. *ICEEI*, pp. 1- 5 (2011).
11. Candido, S., Hutchinson, S. Minimum uncertainty robot navigation using information-guided POMDP planning. *ICRA,* pp.6102 – 6108 (2011).
12. Becerra, I., Valentin-Coronado, L.M. , Murrieta-Cid, R. ,et al. Appearance-based Motion Strategies for Object Detection. *ICRA*, pp. 6455- 6461 (2014).
13. Webb, D.J. ,Crandall, K.L. Online parameter estimation via real-time replanning of continuous Gaussian POMDPs. *ICRA*, pp. 5998 - 6005 (2014).
14. Capitan, J.,Spaan, M.T.J.,Merino, L., et al.Decentralized Multi-Robot Cooperation with Auctioned POMDPs. *ICRA*, pp.3323-3328 (2012).
15. Eker, B. , Ozkucur, E. , Mericli, C. , et al. A Finite Horizon DEC-POMDP Approach to Multi-robot Task Learning. *AICT*, pp.1-5 (2011).
16. Joni Pajarinen and Ville Kyrki. Robotic manipulation in object composition space. *IEEE/RSJ International IROS*, pp.1-6 (2014).
17. Matanya Horowitz and Joel Burdick. Interactive Non-Prehensile Manipulation for Grasping via POMDPs. *ICRA*, pp.3257-3264 (2013).
18. Tarek Taha, Jaime Valls Mir´o. POMDP Framework for Modelling Human Interaction with Assistive Robots. *ICRA,* pp.544-549 (2011).
19. Frank Broz, Illah Nourbakhsh and Reid Simmons. Designing POMDP Models of Socially Situated Tasks. *ISRHIC,* pp.39-46 (2011).
20. R. Sharma and M. Spaan. Bayesian-Game-Based Fuzzy Reinforcement Learning Control for Decentralized POMDPs. *IEEE Transactions on Computational Intelligence and AI in Games*, pp.309-328 (2012).
21. Theocharous, G. , Murphy, K. . Representing hierarchical POMDPs as DBNs for multi-scale robot localization. *ICRA,* pp. 1045 - 1051 (2004).
22. Kaelbling, L. P., Littman, M. L.. Planning and acting in partially observable stochastic domains. *Artificial Intelligence*, pp. 99-134 (1998).

ROBOTIC TRANSTIBIAL PROSTHESIS WITH DAMPING CONTROL IMPROVES AMPUTEE'S LEVEL-GROUND WALKING AT DIFFERENT SPEEDS

YANGGANG FENG[a,b], KEBIN YUAN[a,b], BAOJUN CHEN[a,b], and QINING WANG[a,b,*]

a. The Robotics Research Group, College of Engineering, Peking University, Beijing 100871, China.

b. Beijing Engineering Research Center of Intelligent Rehabilitation Engineering, Beijing 100871, China

**E-mail: qiningwang@pku.edu.cn*

Robotic transtibial prostheses which can intelligently adjust ankle angle are gaining popularity. In this paper, based on the light-weight robotic transtibial prosthesis with damping control, we further study the improvement on amputee's walking performance at different speeds on level-ground. Experiments on one transtibial amputee subject indicate that, compared with the passive prosthesis, the proposed robotic transtibial prosthesis has a larger range of ankle movement, which is closer to that of the able-bodied, and can improve amputee's gait symmetry and walking stability at different speeds.

Keywords: Robotic prosthesis, damping control, speed adaptation, level-ground walking, transtibial amputee.

1. Introduction

The main way to regenerate bipedal gaits for a walking amputee is the lower-extremity prosthesis.[1] Most existing commercial transtibial prostheses are passive without actuation whose ankle joints are either rigid or rotatable in a limited range, which causes asymmetrical gait pattern and more metabolic power consumption during walking than the able-bodied. As a result, robotic transtibial prostheses which can intelligently adjust ankle angle and torque are gaining popularity in the research field.

According to the joint actuation types, existing robotic transtibial prostheses can be roughly divided into two main categories. The first one is the pneumatic actuator based transtibial prosthesis. Klute *et al.* and Versluys *et al.* have developed different transtibial prostheses powered by pneumatic actuators, respectively.[2,3] Although these pneumatic actuators are capable of providing enough force to propel amputees forward, their deficiencies such as low position accuracy and large dimensions resulting from the high pressure air pump have limited the development of pneumatic prostheses. The other category is the motor-driven prosthesis.[4–9] Herr *et al.* developed a motor-driven robotic powered prosthesis, which uses a combination of a parallel spring and a series elastic actuator.[4] SPARKy prosthesis is another motor-driven prosthesis.[5] Though these two powered prostheses can output sufficient power and thus improve the amputees' energetic efficiency, their weight are more than $2kg$ due to their high power motors and complex mechanical structures. A heavy prosthesis tends to increase the knee extension load and cause larger interaction force between the adaptor and the residual limb. To solve the problems, we have developed a light-weight robotic transtibial prosthesis,[9] which focuses on environmental adaptation instead of providing net positive work. The prosthesis is named as PKU-RoboTPro, short for ROBOtic Transtibial PROsthesis, Peking University.

In our previous studies, we have evaluated the performance of the PKU-RoboTPro prosthesis for walking on different terrains. Wang *et al.* proposed a damping control strategy, which improved the amputees' performance on different terrains including level ground, stairs, and ramps.[9] Chen *et al.* designed a myoelectric controller for adaptive slope walking with this prosthesis.[10] Experimental results show that the prosthesis can help amputees to adaptively walk on the ground with varied inclination angles in daily life. However, the performance of the proposed prosthesis on speed adaptation has not been investigated. In this paper, we further study the improvements by the robotic prosthesis on amputee's performances at different walking speeds on level-ground.

2. Robotic Transtibial Prosthesis PKU-RoboTPro

The prototype of PKU-RoboTPro is shown in Fig. 1(a). Current version is an integrated one that takes all the modules in the transtibial prosthesis including mechanical structure, control circuits, sensors and battery.[9] As shown in Fig. 1(b), the model of the ankle joint can be simplified as a three-bar mechanism which comprises three bars a, b, and c, and three hinges, A, B, and C. To visualize the model, a can be seen as the foot, b as the shank, and C as the ankle joint. c is a customized bar, which is made up of a motor-driven ball screw transmission. The screw pitch is $2mm$. The motor system uses a $50W$ DC brushless motor from Maxon (EC $45 - 50W$), equipped with a $5.8:1$ reduction gearbox. The angle range of the ankle joint is from 25° dorsiflexion to 25° plantar flexion. The total weight of the proposed prosthesis (excluding the rechargeable Li-ion battery) is $1.3kg$.

Three kinds of sensors are installed on the prosthesis including one load cell, one angle sensor and two inertial measurement units (IMUs), as shown in Fig. 1(a). The single-axis load cell (Interface LBS) has a measurement range of $0-$

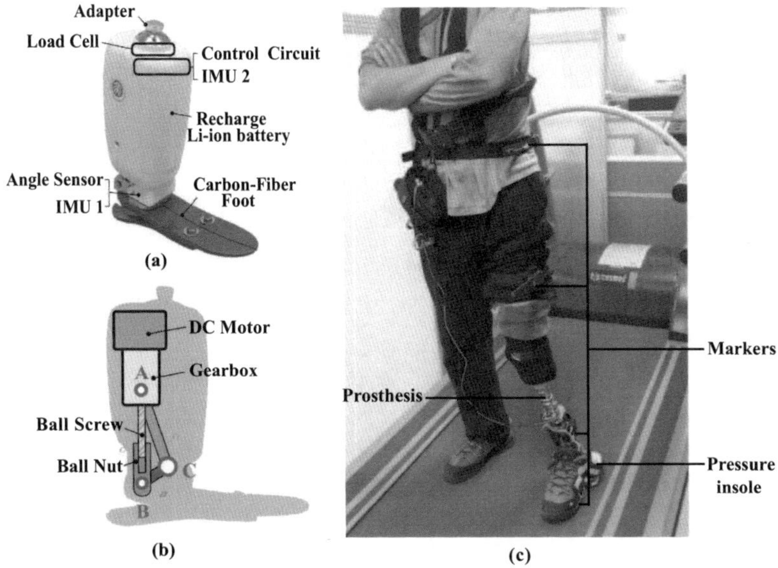

Fig. 1. (a) The prototype of the proposed prosthesis (PKU-RoboTPro), including an active ankle joint and a carbon-fiber foot. (b) Ankle model, simplified as a three-bar mechanism which comprises three bars *a*, *b*, and *c*, and three hinges, *A*, *B*, and *C*. *a* can be seen as the foot, *b* as the shank, and *C* as the ankle joint. (c) Experiment setup.

$250lbf$ and is used to detect the interaction force between the residual limb of the amputee and the prosthesis. The absolute angle sensor (Angtron-RE-25) is used to measure the ankle angle with a 0-360° range and 12-bit resolution. Two IMUs are used to measure the inclination angle and other inertial information such as the acceleration and the rotation rates. One IMU is installed on the upper surface of the foot, and the other is installed on the shank of the prosthesis. Each IMU has an embedded tri-axis gyroscope and a tri-axis accelerometer. The gyroscope has a full-scale range of $2000^o/s$ and a resolution of $0.06^o/s$ while the accelerometer has a full-scale range of $157m/s^2$ and a resolution of $0.005m/s^2$.

3. Motor-Driven Nonlinear Damping

As the ankle joint has different kinematic and kinetic properties during stance phase and swing phase, control strategies for two phases are different. During stance phase, we proposed a damping control strategy, while during swing phase, we developed position control.

3.1. *Realization of damping control*

Here we implement motor-driven damping control in the ankle joint.[9] When the motor rotates, it behaves as a generator and generates an induced voltage E_a, which is defined as

$$E_a = C_e \phi n, \tag{1}$$

where C_e is the electromotive constant, ϕ is the magnetic field intensity, and n is the motor speed.

A current I_a will be generated by the induced voltage E_a when the stator windings of the brushless motor are shorted. And it produces a braking torque τ_b to prevent the motor from rotating

$$\tau_b = C_T \phi I_a = C_T \phi \cdot \frac{C_e \phi n}{R_a} = \frac{C_T C_e \phi^2}{R_a} n, \tag{2}$$

where C_T is the motor torque constant, and R_a is the motor armature resistance.

A pulse width modulation (PWM) signal is used to switch on/off of the motor-winding-short, and the effect of braking torque on ankle joint rotation is determined by the duty cycle (D) of the PWM signal. The equivalent braking torque τ_{eb} can be approximately estimated as

$$\tau_{eb} = D\tau_b = \frac{C_T C_e \phi^2}{R_a} Dn. \tag{3}$$

As C_T, C_e, ϕ and R_a are constants, the damping property of the ankle joint is only affected by duty cycle D, which is controlled to adjust damping coefficient of the prosthetic ankle in this study.

3.2. *Damping control in stance phase*

In this study, we propose a variable damping control strategy. The damping consists of two parts: one is the angle related damping, the other is ground reaction force (GRF) related damping.

The duty cycle expressing GRF related damping is determined by

$$D_\theta = \begin{cases} \dfrac{D_1}{1 + e^{a_1(\theta - b_1)}}, & \theta \geq 0, \text{ and } \dot{\theta} \geq 0, & \text{(4a)} \\ k_1 \cdot \theta, & \theta < 0, \text{ and } \dot{\theta} \geq 0, & \text{(4b)} \\ \dfrac{D_2}{1 + e^{a_2(\theta - b_2)}}, & \theta \geq 0, \text{ and } \dot{\theta} < 0, & \text{(4c)} \end{cases}$$

otherwise where D_1, a_1, b_1, $k_1 (\geq 0)$, D_2, a_2 and b_2 are parameters to describe the ankle angle related damping, and θ (positive value of ankle angle denotes dorsiflexion while negative value denotes plantar flexion) is the currently measured ankle angle. D_1, a_1, b_1, D_2, a_2 and b_2 are not changed for different ground slopes.

The duty cycle expressing GRF related damping is determined by

$$D_{grf} = D_3 \cdot (\frac{F - F_0}{F_{max} - F_0} + d), \tag{5}$$

where D_3 is the maximum duty cycle of GRF related damping, F is the currently measured GRF, F_{max} is the maximum GRF value that can be measured during walking, F_0 is the GRF threshold value for foot-contact detection, and d is a positive constant. In this study, we set $d = 0.05$ and $F_{max} = F_0 + 200$ to avoid the measured GRF exceeding F_{max}.

The total duty cycle D for damping control is

$$D = D_\theta + D_{grf}. \tag{6}$$

3.3. *Position control in swing phase*

As for the swing phase, a Proportional-Derivative (PD) position controller is used.[9] The ankle joint is controlled to a predefined ankle angle θ_0.

4. Experimental Methods

To evaluate the performance of the PKU-RoboTPro at different speeds, a male unilateral transtibial amputee subject (age: 29 years; height: $170cm$; mass: $68kg$) volunteered to participate in the study and provided written consent prior to testing. The amputee subject has been amputated (left leg) for seven years.

The experiments consisted of the training session and the test session. During the training session, the subject walked on the treadmill (see Fig. 1(c)) at different speeds ($0.5m/s$, $0.7m/s$, and $0.9m/s$) with the passive prosthesis and the robotic prosthesis, respectively, at least 30 minutes. According to the amputee's comfort feedback, nine parameters in Table 1 were tuned at different speeds. During the testing session, the participant wearing the passive prosthesis walked on the treadmill at three different speeds. Each testing lasted at least 10 minutes. Then the robotic prosthesis was set to the proposed damping control mode and repeated the same locomotion.

Table 1. CONTROL PARAMETERS AT DIFFERENT SPEEDS

speeds(m/s)	a_1	b_1	D_1	k_1	a_2	b_2	D_2	$\theta_0(^\circ)$	D_3
0.5	-0.4	20	40	0	0.2	-20	20	0	40
0.7	-0.4	20	40	0	0.2	-20	20	0	70
0.9	-0.4	20	40	0	0.2	-20	20	0	100

We use the ankle angle trajectories and three gait symmetry indicators to evaluate the performance improvements. The ankle angle and medial-lateral (ML) dis-

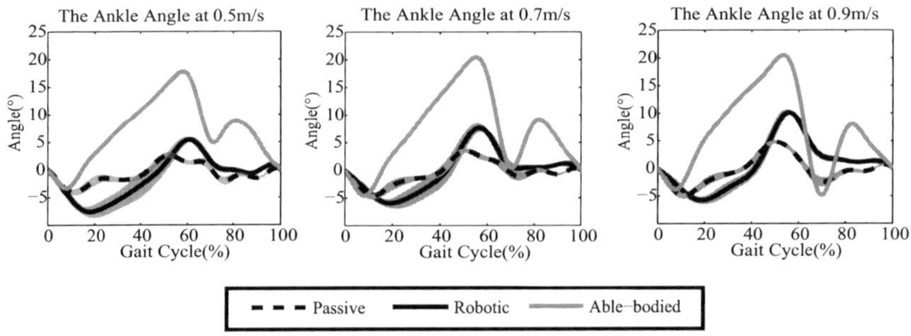

Fig. 2. The ankle angle trajectories of the able-bodied and amputee with the passive prosthesis and the robotic one, respectively,at different speeds($0.5m/s$, $0.7m/s$, $0.9m/s$). The red line refers to the reference angle trajectory of the able-bodied, the dashed black line refers to the angle trajectory of the passive prosthesis, while the solid black line refers to the angle trajectory of the robotic prosthesis. The grey shade area depicts the standard error of mean.

placement were measured by a 3-D motion tracking device (Codamotion, Charnwood Dynamics Ltd.). The three gait symmetry indicators are calculated based on the ground reaction force, which were measured by a wireless flexible sensorized insole.[12] The insole with 64 pressure-sensitive elements integrates these features including low temperature drift, no needed repeated calibrations during long-duration and high sampling frequency ($100Hz$). The data measured by insoles was sent to the PC by the bluetooth wireless transmitter.

Three indicators related to the gait symmetry and walking stability are as follows.

- *Normalized peak GRF*. Peak GRF refers to the maximal GRF during stance. It is associated with the large impact on the residual limb and may lead to unstable or asymmetric gaits. To evaluate the performance of the PKU-RoboTPro, the peak GRF is normalized.
- *Normalized impulse in the whole foot*. Impulse in the whole foot refers to the sum of GRF in one gait cycle. To derive normalized impulse, the left impulse was divided by the right one in the whole foot.
- *Maximal shift in medial-lateral (ML)*. Maximal shift in ML refers to maximal side displacement when walking.

5. Experimental Results

5.1. *Ankle angle trajectories*

The ankle angle trajectories at different speeds, averaged over more than 50 steps from the amputee, are shown in Fig. 2. The gait cycle starts with heel strike.

The faster the able-bodied walk, the larger is the max plantar flexion angle, which illustrates that the faster the able-bodied walk, the more is the work generated. It is necessary that the ankle angle trajectories of the able-bodied compares with that of the amputee at the same speed.

From Fig. 2, compared with the passive prosthesis, the robotic prosthesis has a larger range of ankle movement, which is much closer to that of the able-bodied. As the proposed robotic transtibial prosthesis PKU-RoboTPro cannot provide large assistive torque, the prosthetic ankle is unable to reach an obvious plantar flexion position at toe-off.

5.2. *Gait indicators*

The calculated indications of the proposed robotic prosthesis at different speeds are shown in Fig. 3. At each speed, the mean value of normalized peak GRF of PKU-RoboTPro is much closer to 1 than that of the passive prosthesis. Moreover, the standard deviation of normalized peak GRF is smaller. The Normalized impulse in the whole foot shows that similar results. This shows that the PKU-RoboTPro has a smaller impact on the residual limb and produce a more symmetric gait pattern than the passive one. In addition, the mean of maximal shift in ML of the PKU-RoboTPro is smaller than that of the passive one,which indicates that the gravity shift is smaller. The results show that the PKU-RoboTPro has improved the amputee's symmetry and walking stability.

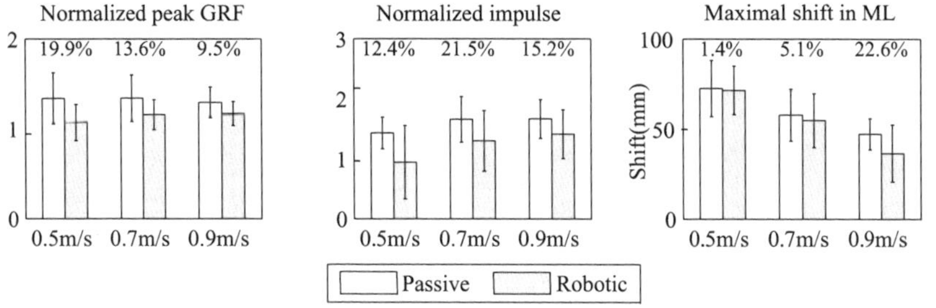

Fig. 3. Comparison of different indicators between the passive and robotic prosthesis. The numbers on the bar express the variation quantity of the robotic prosthesis compared with the passive one. The red number means that the indicator has been improved.

6. Conclusion

In this paper, we have evaluated the performance of a light-weight robotic transtibial prosthesis with damping control at different speeds. Preliminary results of one transtibial amputee subject with robotic prosthesis at different speeds indicate that the range of ankle angle is much closer to that of the able-bodied, and the robotic prosthesis improves amputee's walking performance at different speeds. The robotic prothesis is promising for the future clinical application. In the future, more efforts will be made on control strategy of speed self-adaption.

Acknowledgment

This work was supported by the 2011 R&D Project of the Beijing Disabled Persons' Federation, the PKU-Biomedical Engineering Joint Seed Grant 2014, and the Beijing Nova Program (No. Z141101001814001).

References

1. M. Goldfarb, B. E. Lawson, and A. H. Shultz, "Realizing the promise of robotic leg prostheses," *Sci. Tranl. Med.*, vol. 5, no. 210, p. 210ps15, 2013.
2. G. K. Klute, J. Czerniecki, and B. Hannaford, "Development of powered prosthetic lower limb," *The 1st Natl. Meeting, Veterans Affairs Rehabil. R and D service*, 1998.
3. R. Versluys, A. Desomer, G. Lenaerts, O. Pareit, B. Vanderborght, G. Perre, L. Peeraer, and D. Lefeber, "A biomechatronical transtibial prosthesis powered by pleated pneumatic artificial muscles," *Int. J. Modell. Identif. Control*, vol. 4, pp. 394-405, 2008.
4. S. Au, J. Weber, and H. Herr, "Powered ankle-foot prosthesis improves walking metabolic economy," *IEEE Trans. Robot.*, vol. 25, no. 1, pp. 51-66, 2009.
5. J. Hitt, T. Sugar, M. Holgate, R. Bellmann, and K. Hollander, "Robotic transtibial prosthesis with biomechanical energy regeneration," *Industrial Robot*, vol. 36, no. 5, pp. 441-447, 2009.
6. A. H. Shultz, J. E. Mitchell, D. Truex, B. E. Lawson, and M. Goldfarb, "Preliminary evaluation of a walking controller for a powered ankle prosthesis," *Proc. of the Int. Conf. Robotics and Automation*, pp. 4838-4843, 2013.
7. J. Zhu, Q. Wang, and L. Wang, "On the design of a powered transtibial prosthesis with stiffness adaptable ankle and toe joints," *IEEE Trans. Ind. Electron.*, vol. 61, no. 9, pp. 4797-4807, 2014.
8. P. Cherelle, V. Grosu, A. Matthys, B. Vanderborght, and D. Lefeber, "Design and validation of the ankle mimicking prosthetic (AMP-) foot 2.0," *IEEE Trans. Neural. Syst. Rehabil. Eng.*, vol. 22, no. 1, pp. 138-148, 2014.
9. Q. Wang, K. Yuan, J. Zhu, and L. Wang, "A robotic transtibial prosthesis with nonlinear damping behaviors for terrain adaptation," *IEEE Robotics and Automation Magazine*, 2015. (accepted)
10. B. Chen, Q. Wang, and L. Wang, "Adaptive slope walking with a robotic transtibial prosthesis based on volitional EMG control," *IEEE/ASME Trans. Mechatronics*, 2014. (online first)
11. C. Kendell, E. D. Lemaire, N. L. Dudek, and J. Kofman, "Indicators of dynamic stability in transtibial prosthesis users," *Gait & Posture*, vol. 31, pp. 375-379, 2010.
12. S. Crea, M. Donati, S. M. M. De Rossi, C. M. Oddo, and N. Vitiello, "A wireless flexible sensorized insole for gait analysis," *Sensors*, pp. 1073-1093, 2014.

IMPROVING RUNNING SMOOTHNESS OF WHEG-DRIVEN SYSTEMS WHILE MAINTAINING THE HIGH VALUED OBSTACLE PERFORMANCE*

SEBASTIAN KÖHRING, MAX FREMEREY, OMAR NASSAR, MANUEL SCHÖNE, KARL WEINMEISTER and HARTMUT WITTE

Group of Biomechatronics, Faculty of Mechanical Engineering, Technische Universität Ilmenau, Haus F, Max Planck Ring 12, D-98693 Ilmenau, Germany

ĐORĐEVIĆ, GORAN

Control Engineering Department and Robotic Lab, University of Niŝ, Faculty of Electronic Engineering, Aleksandra Medvedeva, 14, 18000 Nis, Serbia

Following common mechatronic design methodology VDI 2206, this paper introduces three different variants of wheg design. During construction focus was placed on a smooth locomotion of all wheg-mechanisms over flat terrain, which means low alternation of CoM in vertical direction y (in the following addressed as 'quietness' of a wheg) as well as the ability to overcome obstacles. Subject to the number of spokes, quietness and theoretically manageable obstacle height were calculated. The results were taken as reference for experimental evaluation of each variant and a successive comparison. The phase shifting double wheg, is described in further detail.

1. Introduction

Stable walking over unstructured and unknown terrain in a robust manner is still a challenging task for an autonomously acting machine or robot. Considering the fact that the technical system has to encounter different kinds of obstacles, the implementation of bio-inspired legs for locomotion purposes is a possible approach.

BIGDOG [11], CHEETAH-CUB Robot [14] or TEKKEN II [5] give excellent examples for successful bio-inspired walking machines. It is quite interesting that (while neglecting other design criteria) robustness can be achieved either by

* This work is supported by NATO grant EAP.SFP 984560 and the 'Thüringer Innovationszentrum Mobilität (ThIMo)' in line with the project 'Silvermobility - Nahfeldmobilitätskonzepte für die Altersgruppe 50+' (no. 2011FGR0127) funded by the European Social Fond (ESF), and the Thuringian ministry of economy, work and technology TMWAT via 'Thüringer Aufbaubank' (TAB)

software algorithms (BIGDOG), embodiment (well-designed, even compliant mechanics like in CHEETAH-CUB), or a balanced mixture of both.
However, legged robots still have some disadvantages due to their kind of locomotion. One negative issue is the high number of actuators used. BIGDOG requires 16 [11], and even CHEETAH-CUB featuring a bio-inspired pantograph-like mechanism (c.f. Witte & Fischer [3]) requires two actuators per leg, and thus eight in sum [14]. In addition, the proper synchronization of the legs still needs effort in control.

Thus the authors want to point on the other possible, yet established approach for robust robot locomotion: the concept of whegs (like shown in chapter 2), and their improvement. Fig.1 illustrates the principle of a wheg (wheel + leg) and its derivation from a wheel.

Figure 1: Illustration of a wheel and two whegs consisting of a hub and a different number n of spokes fixed to it.

Whegs as a combination of the principles of wheel and leg provide advantages of both. They achieve a high horizontal velocity due to the possibility of using a continuously rotating actuator, while keeping control effort low. Aside the spokes allow the wheg to overcome obstacles in a leg-like manner: 1) a stance phase where the spoke is in contact with the substrate, and 2) the swing phase where the spoke is not in contact with the substrate.
By controlling the wheg's pace and phasing relative to other whegs, a wheg-equipped robot simply adapts to different environments. Therefore whegs impose efficient dynamic stability and robustness in interacting with unexpected obstacles and unknown terrain.

2. State-of-the-Art of Wheg-Driven Robots

The use of whegs, or rimless wheels respectively, is no new idea. A brief study of a rimless spoked wheel in 2D with a single degree of freedom according to [2] was carried out by McGeer. In 1997 Coleman et al. [2] carried out a 3D mathematical analysis including 3D stability of a rimless wheel with rigid spokes rolling on an inclined ramp, considering the effect of gravity.
One of the first robots that used a rimless wheel (or wheg) was PROLERO (PROtotype of LEgged ROver) by Alvarez et al. in 1996 [1]. Featuring six

simple spokes driven by a single actuator each, PROLERO preceded the WHEGS™ series as well as RHEX [15]. RHEX, a robot having six rimless wheels with one passive compliant spoke each, was developed in 2001 in Martin Bühler's lab at McGill University by Saranli et al. [12]. Each wheg was driven by a separate actuator allowing a change of rotational speed and of phase of each wheg individually. Being redesigned several times and equipped with differently shaped whegs, RHEX is able to overcome rough terrain in a stable manner and by even human-sized steps with up to 42% slope [7].

An alternative approach was introduced in 2002 by Quinn et al. [9]. They developed WHEGS™ I, a cockroach inspired robot featuring six whegs, each wheg having three instead of one spokes. Aside the driving torque required is delivered by a single drive and separated towards the whegs by gear mechanisms. Therewith the drive motor is able to run at constant speed instead of accelerating and decelerating like in RHEX during each walking cycle. In addition WHEGS™ I has compliant axles which enable the robot to overcome a large variety of obstacles, stairs and barriers without changing parts of the design. Based on WHEGS™ I Quinn and coworkers developed several other wheg driven robots like WHEGS II, DAGSI WHEGS, MINI WHEGS, CLIMBING MINI WHEGS, or SEADOG [15].

Another small sized wheg-driven robot is the modular WARMOR robot (Fremerey et al. [4]). Equipped with only four whegs each featuring four compliant spokes (Material: FR4, length: 75 mm, thickness 0.3 mm) the robot is able to deal with unstructured environments covered e.g. with gravel or debris. Due to a balanced mechanical design the robot uses the energy stored in the spokes during Touch down for reducing the amount of energy required during next Lift off. For robustness purposes the robot also is able to flip without any lack in performance.

3. Advantages and Disadvantages of Wheg-Driven Systems

Whegs are an applicable variant if the environment, the robot/technical system has to deal with, consists of planes with different orientations and obstacles unknown in size and orientation. Therefore these obstacles might be of random type like stones or debris as well as structured like steps or platforms.

However due to the discontinuous spoke-to-ground contact of the wheg the center of mass (CoM) of the wheg (and thus the robot) is forced to an alternating movement perpendicular to the substrate. This alternation has to be considered in comparison to the quietness of the wheg-driven system. It is best when the number of spokes n is infinite; the wheg becomes a wheel. The ability to overcome obstacles is poor. The quietness is worst having whegs with n = 1, which results in a high vertical movement of the center of mass, but is the best way to deal with obstacles (compare Fig. 3). However all previous wheg-driven systems show limitations either concerning the quietness or the range of feasible obstacle height. Therefore modifications of hitherto common wheg robot design are necessary.

For overcoming higher obstacles than a common wheg does, Hong et al. in 2006 developed a new concept called IMPASS (Intelligent Mobility Platform with Active spoke system) [6]. Here the focus was placed onto individual foot placement to overcome even extreme irregular terrain. Therefore the length of each of the three spokes may be actively changed (independently of each other). To improve the quietness of wheg-driven robots without losing the ability to overcome obstacles Shen et al. in 2009 introduced WHEEL-LEG HYBRID ROBOT [13], a design where an additional actuator changes the shape of the wheg between a circle and a semi-circle (by two c-shaped spokes). So the WHEEL-LEG HYBRID ROBOT is able to use a wheel-like configuration while traveling over flat terrain and a wheg-like configuration when dealing with unstructured environment. This principle, but designed as a triple wheel, also was used by Quaglia et al. in 2013 with their EPI.Q robot family [10].

4. Influence of Number of Spokes on Wheg Kinematics

Finding a proper ratio between an acceptable quietness (means less alternation of CoM in y-direction) and good obstacle dealing seems obvious for an adequate wheg design. Therefore the calculation of both design criteria is successively introduced. Fig. 2 illustrates all relevant elements of a single wheg like spoke length *r*, the number of spokes *n*, the angle between two spokes α (= 360°/n), the height y_{min} using to calculate the alternation *y* and the (theoretically) manageable obstacle height *h*. Due to fig. 2 y_{min} is used to calculate the alternation *y*. It is described by:

$$y_{min} = r \cdot \cos(\alpha) \qquad \text{eq. 1}$$

$$y = r \cdot y_{min} \qquad \text{eq. 2}$$

$$\alpha = \frac{360°}{(2 \cdot n)} \qquad \text{eq. 3}$$

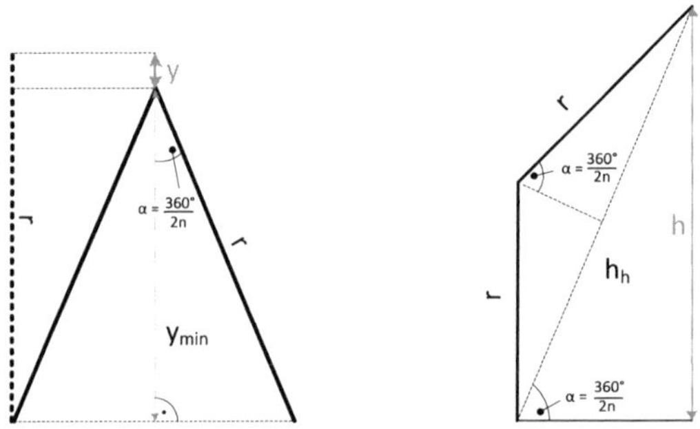

Figure 2: Calculation of alteration (= quietness) in y-direction using y_{min} (left) and theoretically manageable obstacle height h subject to number of spokes n (right).

For the theoretically manageable obstacle height h eq. 4 has to be used:

$$h = \sin(\alpha) \cdot \text{sqrt}\,(2r^{\dagger} \cdot (1-\cos(\alpha)) \qquad \text{eq. 4}$$

The results of the alternation of the CoM in y-direction as well as calculation of theoretically manageable obstacle height subject only to number of spokes n is shown in Fig. 3. Here the percental results eliminating the need for spoke length r are shown.

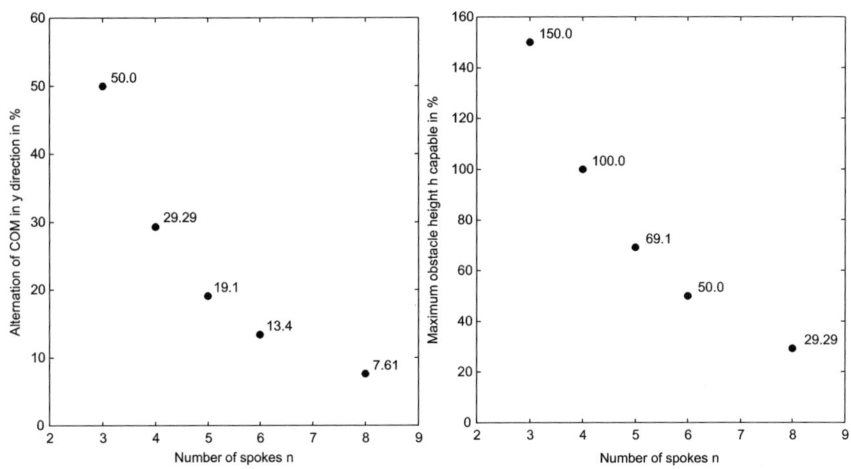

Figure 3: Results of percental alteration (= deviation from a quietness *Q of 100 %*) in y-direction (left) and theoretically manageable obstacle height h both subject to number of spokes n (right).

With increasing number of spokes the alternation of CoM of the observed single wheg decreases, which results in an improved quietness Q and a smooth locomotion. But in addition the ability to overcome obstacles decreases, too. For example, having six whegs, the CoM only alternates 13.4 % of spoke length in y-direction. However the manageable obstacle height only amounts 50 % of spoke length.

5. Design Approaches for Improving the Performance of Wheg-Driven Robots

Following eq. 1 and eq. 2 an acceptable and maybe even online adaptable ratio between quietness and possible obstacle height is favored. Therefore the authors introduce two different design variants. A successive evaluation will be done to show advantages and disadvantages of each variant.

5.1. *A phase shifting double wheg-module*

The phase shifting double wheg consists of a common wheg with four spokes; which is rigidly attached to a shaft (outer wheg). Around this first shaft a hallow axle turns a second wheg (inner wheg) which also features four spokes. Each shaft is driven by a distinct actuator. Due to the use of two motors the two whegs are able to turn independently of each other. Using PI control and two magnetic rotary encoders enable a smooth setting of a phase difference between inner and outer wheg. Therefore the amount of spokes of the double wheg-module is changeable: It features eight spokes when travelling over flat terrain (45° phase shift) and four spokes when dealing with obstacles (0° phase shift). Fig. 4 shows the technical principle and the resulting construction.

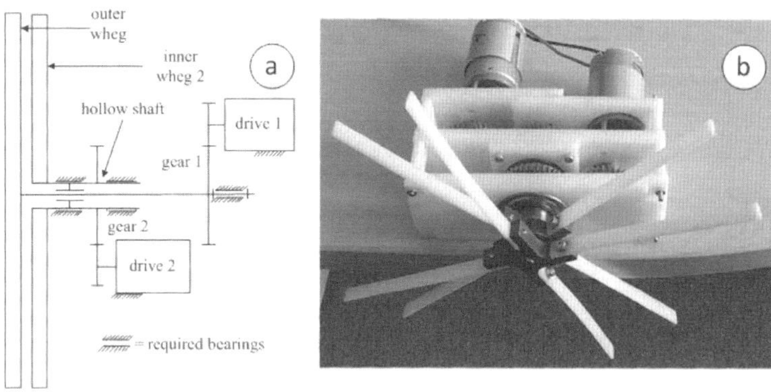

Figure 4: a) Technical principle and b) design of a phase shifting wheg-module.

5.2. *A wheg-axle-tracking mechanism*

If transportation of persons or shock-sensitive objects frames the major requirement during design, the compensation of the alternation of the wheg mechanism gains importance. For this reason the next approach aims at the maximally possible elimination of alternation and a maximum of quietness. A possible solution is to use a wheg-axle-tracking mechanism in phase to the whegs rotation angle, which operates according to eq. (5) and eq. (6), managed via an appropriate actuator. It keeps the body of the mobile platform in constant distance to the ground. In case of integration of a separate impellent, aspects of available space, energy supply, control and timing have to be considered.

Pfefferle [7] in 2004 introduced a mechanism which is capable to use the existing impellent for locomotion as well as for the compensation of the alternation. Condition for this purpose is the application of a gear train having a transmission ratio of $i = n$ superposed to a crankshaft as shown in Fig. 5.

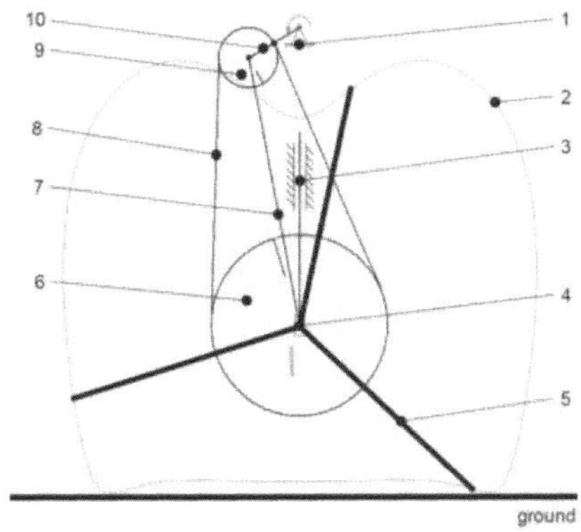

1 = crankshaft bearing
2 = path of the contact point
3 = linear bearing
4 = axle bearing
5 = wheg
6 = timing belt pulley at the output of the gearbox with diameter $d_2 = n \cdot d_1$
7 = couple
8 = timing belt
9 = timing belt pulley at the input of the gearbox with diameter d_1
10 = crank

9 and 10 in rigid coupling

Figure 5: Optimized mechanism for the compensation of the alternation via only one impellent for both locomotion and for compensation of the alternation.

This mechanism is traced and enhanced, and with focus on the path generated by the contact point of the wheg in the 120° stance phase, plotted in Fig. 5. It is detectable that the major part of the alternation is eliminated, but as well that a component not negligible remains. Currently this residual alternation amounts to an optimum value of 8.4 %, with five extreme values.

$$h(a) = H\left|\sin(\frac{n}{2}\alpha)\right| \quad \text{eq. 5}$$

$$k(a) = H - h(a) \quad \text{eq. 6}$$

with α = angular position of the wheg, h = offset depending on the angular position of the wheg, H = maximum offset, k = compensation offset depending on the angular position of the wheg.

5.3. *Additional elasticity*

Both the phase shifting double-wheg module in the illustrated version and the wheg-axle-tracking mechanism leave a residual alternation of approx. 10 %. So the idea was to create a combined solution based on one of the processed methods and elasticity. While the phase shifting double-wheg module has implemented elasticity within the spokes, the wheg-axle-tracking mechanism was mounted on a wishbone capable of cushing the disturbing effects of incomplete compensated alternation. As shown in Fig. 6, a hybrid form of MacPherson strut and double wishbone suspension is applied. While the lower wishbone and the linear bearing are engaged for guiding the up and down movement, the upper wishbone acts as shock absorbing mounting of the whole mechanical compensation module. This exemplary realization involves an adjustable spring-damper-unit for experimental extracting the best parameters for antagonizing vibrations caused by remaining residual alternation.

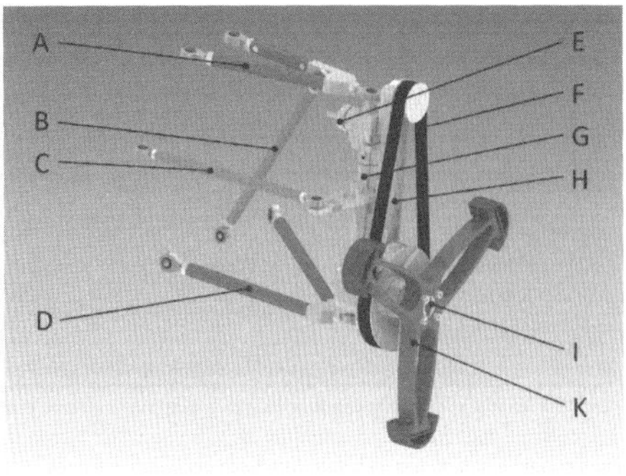

A = upper wishbone
B = pull rod (transferring vibrations to spring-damper-unit)
C = tie rod
D = lower wishbone
E = crank (impellent engages)
F = timing belt gearbox
G = linear bearing
H = couple
I = svivel
K = wheg

Figure 6: Technical realization of the compensation mechanism as a hybrid form of MacPherson strut and double wishbone suspension.

6. Conclusion

Both solutions compensating the main part of alternation are suited when handling obstacles is desired as well as a smooth locomotion over flat terrain. Due to the use of only a single drive, the wheg-axle-tracking mechanism succeeds regarding efficiency of energy consumption. Aside the continuous turning motor reduces control effort which lowers the degree of complexity. In Combination with elasticities both variants can provide a high compensation quality for transportation tasks with a higher demand on running smoothness.

References

[1] M. Alvarez, W. De Peuter, J. Hillebrand, P. Putz, A. Matthyssen and J. F. de Weerd, WAC '96, Montpellier, France, (1996)
[2] M. Coleman, A. Chatterjee and A. Ruina, Dynamics and Stability of Systems, 12, 139–159, (1997)
[3] M. S. Fischer and H. Witte, Phil. Trans. R. Soc. A, pp. 185–198, (2007)
[4] M. Fremerey, G. Djordjevic and H. Witte, Biomimetic and Biohybrid Systems. pp. 345– 346, (2012)
[5] Y. Fukuoka, H. Kimura and A. H. Cohen. Int. J. of Robotics Research, 3–4, 187–202, (2003)
[6] D. Hong and D.Laney. Conf. on Science, Technology and Entrepreneurship, (2006)
[7] E. Z. Moore, D. Campbell, F. Grimminger and M. Buehler, IEEE Int. Conf. Robotics and Automation, (2002)
[8] H. Pfefferle, Gebrauchsmusterschrift DE 20 2004 010 935 U1 – Gehvorrichtung, (2004)
[9] R.D. Quinn, J. T. Offi, D. A. Kingsley and R. E. Ritzmann, IEEE/RSJ, vol. 3, pp. 2652-2657, (2002)
[10] G. Quaglia, R. Oderio, L. Bruzzone and R. Razzoli, Int. J. of Advanced Robotic Systems, 10, (2013)
[11] M. Raibert, K. Blankespoor, G. Nelson and R. Playter, Proc. of the 17th World Congress , pp. 10823-10825, (2008)
[12] U. Saranli, M. Buehler and D. E. Koditschek, Int. J. of Robotics Research 20(7), 616– 631, (2001)
[13] S. Shen, C. Li, C. Cheng, J. Lu, S. Wang and P. Lin, Proc. of the IEEE/RSJ Int. Conf. on Intelligent Robots and Systems, pp. 4682–4687, (2009)
[14] S. Spröwitz, A. Tuleu, M. Vespignani, M. Ajallooeian, E. Badri and A. J. Ijspeert, The Int. J. of Robotics Research, 0(0), 1–19, (2013)
[15] Biologically Inspired Robotics, Case Western Reserve University, http://biorobots.case.edu/projects/whegs/

ABOVE KNEE PROSTHESIS FOR ASCENDING/DESCENDING STAIRS WITH NO EXTERNAL ENERGY SOURCE

SREY SOPHYN and KOICHI KOGANEZAWA

Department of Mechanical Engineering, Tokai University, 4-1-1 Kitakaname, Hiratsuka, Kanagawa 259-1292, Japan E-mail: kogane@keyaki.cc.u-tokai.ac.jp

This paper proposes a design of above knee prosthesis that allows above knee amputees to walk on ground level and ascend/descend stairs by using only hydraulic circulation system (HCS) as self-energized actuator. The mechanism utilizes amputee thigh motion to drive oil flow in the HCS to enable the prosthesis to make extension, flexion, dorsiflexion and plantar flexion. Oil flow is closed for ascending stairs and adjusted for level walking and stairs descending by a flow control valve (FCL).

1. Mechanism of Prosthesis

There are three main incidences for humans to become low limb amputees. First, they have abnormal conditions which cause their limbs not to grow normally since their birth. Second, they get severe injuries in wars or car accidents. Third, they get malfunctions in vascular system such as diabetes [1]. It is found that in Japan more than ten thousand people lose their lower limbs every year by the third problems [2].To cope with the problems, some bionic technologies have been used in above knee prostheses to restore disable legs to be able to walk just the same as normal persons. Unfortunately, their commercially available products are very expensive. BiOM, a biomic leg designed by Hugh Herr, is able to approach natural gate due to its ankle technology [3]. It embeds two microprocessors and twelve sensors with the prosthesis. Genium knee by Ottobock is another bionic prosthetic system having microrocessor-controlled knee joint that works with complex sensory system and sophisticated rule sets to mimic natural gait [4]. Commercially available prostheses with no electric devices such as 3R80 by Ottobock, a hydraulic modular knee, are unable to ascend stairs because they lack of some functions on the foot necessary for ascending stairs [5]. This paper proposes a novel design for an above knee prosthesis that allows to ascend stairs by using a hydraulic circulation system (HCS) as self-energized actuator. The proposed prosthesis utilizes human thigh motion to drive oil flow in the HCS to produce extension moment about knee joint, to support body weight during stance phase and swing phase of the human gait cycle. It also shows the experiments of level walking and stairs ascending/descending by using the developed prosthesis with thigh socket made for normal subject to test prosthetic walking.

2. Mechanism of Prosthesis

2.1. *Mechanism*

Fig. 1 illustrates the proposed prosthesis. It consists of four main parts, knee, HCS, frames and ankle. Knee consists of an adapter used to connect the prosthesis with a thigh socket. It is connected with upper piston of the HCS and the lateral frame by rotary joints. Stoppers are attached to the knee to prevent its over flexion and extension. The frame is used to connect the other three parts by rotary joints and to support body weight. The ankle is connected to the frame by an ankle joint and to lower piston of HCS by a rotating link. Stoppers are attached at the lower part of the frame to the ankle to prevent its over dorsiflexion and plantar flexion. The weight without the cosmetic foot is 1.90 kg.

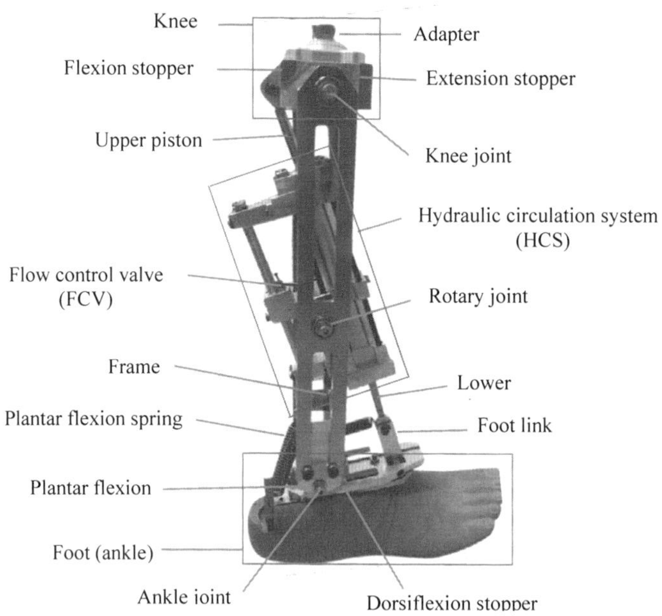

Fig. 1 Appearance of the developed A/K prosthesis.

2.2. *Hydraulic circulation system (HCS)*

As illustrated in Fig. 2, the HCS consists of one cylinder, four check valves, one flow control valve (FCV), and one upper and lower pistons. The check valves are used to make the oil flow in one direction. The FCV is used to adjust oil flow rate from chamber to the chamber of the HCS. The spring in cylinder is used to suppress excessive increase of oil pressure. An elastic rubber bar is embedded in the cylinder to adjust the volume of the oil in chamber when its volume varies during the walking cycle. The HCS is designed for allowing the oil flow to provide a normal cooperative movement of the knee and the ankle during the ground level walking and the ascending/descending stairs. Moving up and down of the upper piston causes the knee to extend and flex respectively. Moving up and down of the lower piston causes the ankle to make dorsiflexion and plantar flexion respectively. The HCS is connected to the frame by a rotary joint as shown in Fig. 1. This joint not only allows HCS to rotate while both pistons to move up and down according to the oil flow but also makes the knee and the foot rotate according to the movement of the amputee.

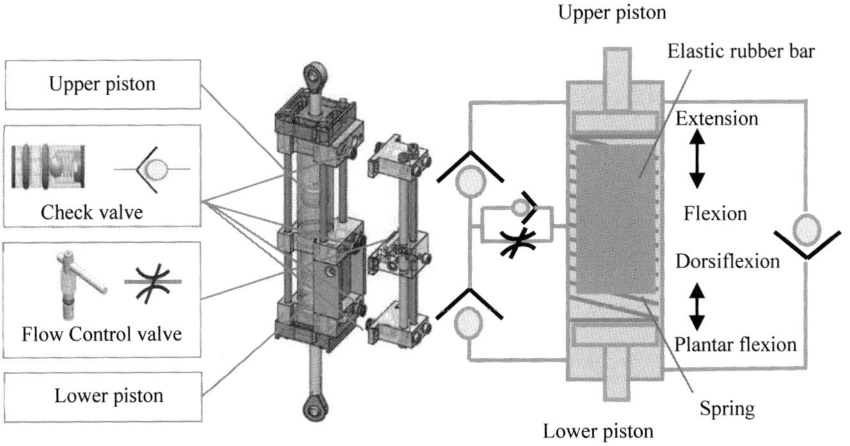

Fig. 2 Hydraulic circulation system (HCS).

3. Oil Flow During Walking

In order that the knee joint extends and flexes and the ankle makes dorsiflexion and plantar flexion for walking, oil flow is divided into two cases by adjusting the FCV. It should be completely closed for ascending stairs or walking on incline surface because this flow provides large extension moment about knee joint to support and shift the body weight and provides enough force to prosthetic foot to keep balance of amputee from falling down during the stance phase [6]. In the other case, the FCV should be appropriately adjusted for descending the stairs, walking on flat and decline surface because it allows the knee to make enough flexion and extension for those walking.

3.1. *Oil flow during level walking*

Fig. 3 illustrates the oil flow during a level walking. (a): the upper piston is at highest position and the lower piston is at lowest position by the spring in the chamber. (b): while body weight is supported by the intact leg, the prosthetic leg starts to take off the ground and the knee flexes. (c): the upper piston goes downward against the direction of the lower piston that tends to go upward in the cylinder. Thus it becomes a good dumper for the prosthesis to support the body weight. (d): the intact leg takes off forward to next step. This causes the upper piston to downward more and the lower piston to go upward. The prosthetic leg supports and shifts body weight to the next step. (e): the intact leg strikes on the ground while the prosthetic leg starts to take off the ground. (f): the body weight is supported only by the intact leg and the prosthetic leg starts to take off the ground that cause the upper piston start to go upward. Then the action starts to repeat from (b).

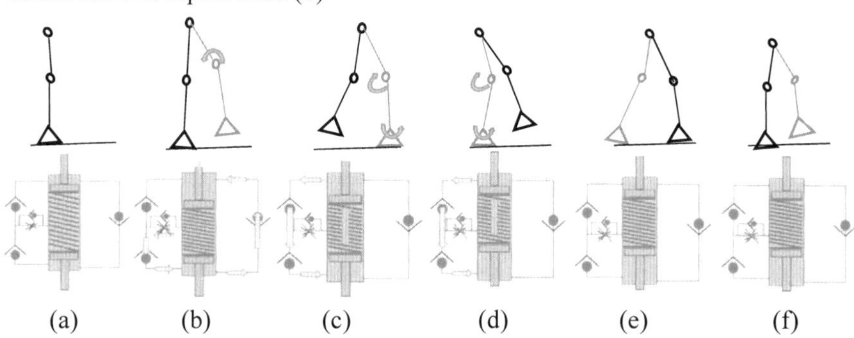

Fig. 3 Oil flow during level walking.

3.2. *Oil flow during stairs ascending*

Fig. 4 illustrates the oil flow during a stairs ascending. (b): the body weight is supported by the intact leg. The knee starts to flex while the prosthetic leg places on the first stair. Then the amputee manually closes the FCV by rotating the shaft (see Fig. 2). (c): the upper piston and the lower piston go downward simultaneously because there is no oil flowing from the middle chamber of the cylinder due to the FCV been closed. (d): the intact leg takes off and goes forward to second step. This causes the upper piston and the lower piston to go upward simultaneously. The prosthetic leg supports and shifts body weight to the second step. (e): the intact leg starts to strike on the second step while the prosthetic leg starts to take off the second step. The upper piston fully extend.

(f): the body weight is supported only by the intact leg on the second step. The prosthetic leg starts to take off the first step and this causes the upper piston to start to go upward. Then the action starts to repeat from (b) for the other steps.

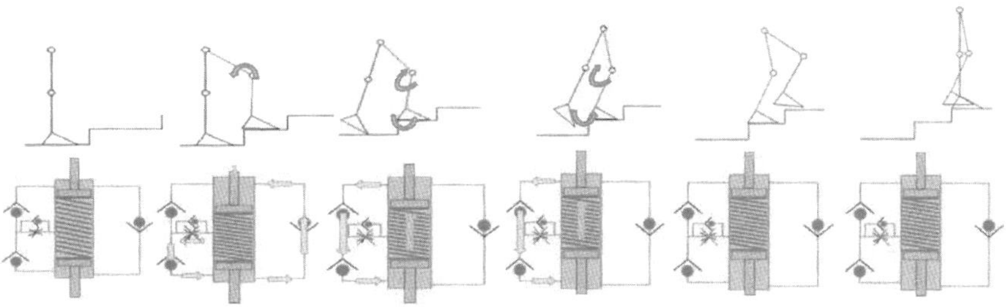

(a) (b) (c) (d) (e) (f)

Fig. 4 Oil flow during stairs ascending.

3.3. *Oil flow during stairs descending*

Fig. 5 illustrates the oil flow during a stairs descending. (b): amputee adjusts the FCV. The prosthetic leg supports the body weight. The prosthetic foot starts to make dorsiflexion, and it causes the lower piston to go upward, which works to prevent an abrupt flexion of the knee. In that moment, the intact leg takes off forward to the next second step. The knee flexes and causes the upper piston to go down forward against direction of the lower piston in the cylinder. This becomes good dumper for supporting body weight. (c): the intact leg strikes the second step. The upper piston keep moving downward again the direction of the lower piston which also moving upward. (d): after the intact leg placed on the second step fully, the prosthetic leg starts to take off the first step, and it causes the upper piston goes upward. (e): while supporting body weight, the intact leg bends forward quickly. This movement causes the upper piston to go downward and the prosthetic leg takes off completely from the step. (f): the upper piston goes downward, and pushes the spring inside the cylinder to make the lower piston goes downward too. The prosthetic ankle makes plantar flexion and strikes the second step. Next the intact leg starts to take off the second step and the prosthetic leg starts to repeat action from (b).

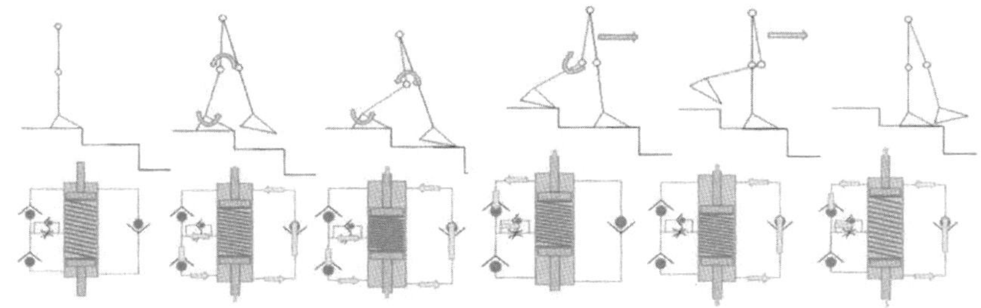

(a) (b) (c) (d) (e) (f)

Fig. 5 Oil flow during stairs descending.

4. Experiment

Stairs with height of 170mm and HAS-500 high speed camera by DITECT [7] were used to measure the variation of knee and ankle angles for the walking experiments as illustrated in Fig. 6. One normal subject (22 years old male) wore the thigh socket specifically fabricated for him to experience prosthetic walking (see Fig. 7). It was observed that the assembled prosthesis allowed the subject to walk stably on ground level, stairs ascending/ descending without any assistance by using handrail to keep balance. The subject also tried to ascend and descend the stairs without using handrail. The experiments of prosthetic walking and normal walking were conducted by the same subject to compare the variation of knee and ankle angles on the three kinds of walking.

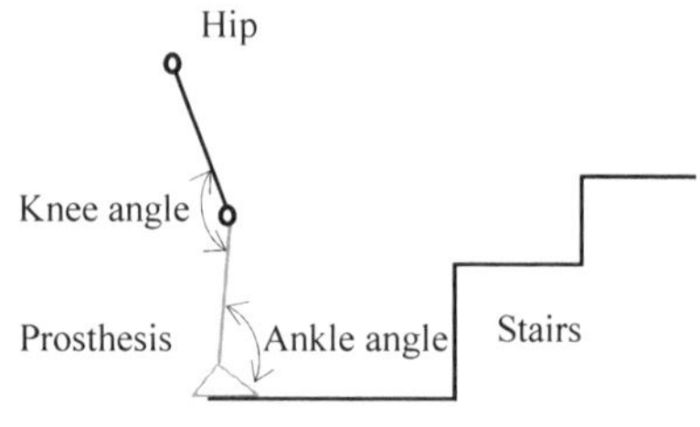

Fig. 6 Angle and stairs definition.

4.1. *Level walking*

Fig. 7 illustrates the level walking experiment. (a): the subject could stand on ground comfortably without feeling to lose balance before he propelled the prosthesis leg forward. (b): the knee flexes a little before the prosthesis leg started to shift the body. In this moment, the upper piston started going downward and the knee angle started flexing. (c): the prosthesis supported and shifted the body. Next the intact leg started to strike the ground. The gate was good and the subject did not feel to fall down. It was observed that the knee prosthesis flexed higher than normal leg with almost the same period but prosthesis ankle flexion was slower than the normal leg with almost the same angle.

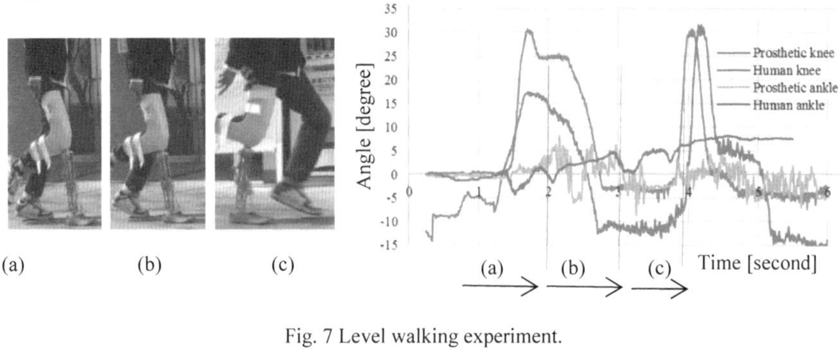

Fig. 7 Level walking experiment.

4.2. *Stairs ascending*

Fig. 8 illustrates the stairs ascending experiment. (a): the subject placed the prosthesis leg on the first step and completely closed the FCV by hand. This made prosthetic ascending experiment slower than the normal one. (b): he started to shift his body to second step. In this moment, it was observed that the upper piston started going down until its stopper and the knee started flexing to maximum value but it did less value than that of the normal one with the same pattern. Lower piston went up a little and the prosthetic foot started making dorsiflexion about 6 degree. Accordingly, the prosthetic knee joint started to extend rapidly in the middle of the period (b). (c): the intact leg reached the second step and the prosthetic leg was placed on the third step. It was observed that the subject was able to ascend the stair stably without noticing any difficulty. It was also confirmed the knee extended while sustaining the body weight by the proposed HCS with no external energy source.

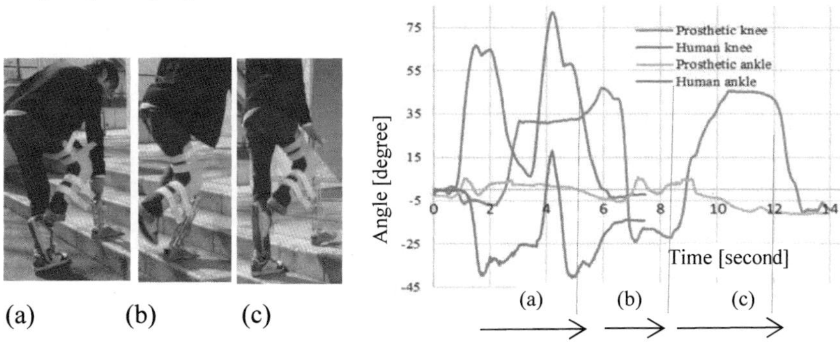

Fig. 8: Stairs ascending experiment.

4.3. *Stairs descending*

Fig. 9 illustrates the stairs descending experiment. (a): after adjusting the flow control valve, the subject hang the prosthetic leg to first step. (b): the prosthetic leg supported and shifted body. Then the knee gradually flexed. (c): just after the intact leg stroke on the next step, the prosthetic leg was propelled forward to strike the next step. Since the subject adjusted the flow control valve, which allowing oil from the chamber, the knee flexed smoothly. The prosthetic descending experiment was almost as the same period as the normal one and it flexed almost as the same as that normal one with the same pattern. Unfortunately, the prosthetic ankle flexed slower than that of normal one. This can be considered due to the lower piston being gone down so little, which made plantar flexion slower on the swing phase because of undue bending of the foot link (see Fig. 1). In this case, it was observed that some traction force is necessary to restore the foot link.

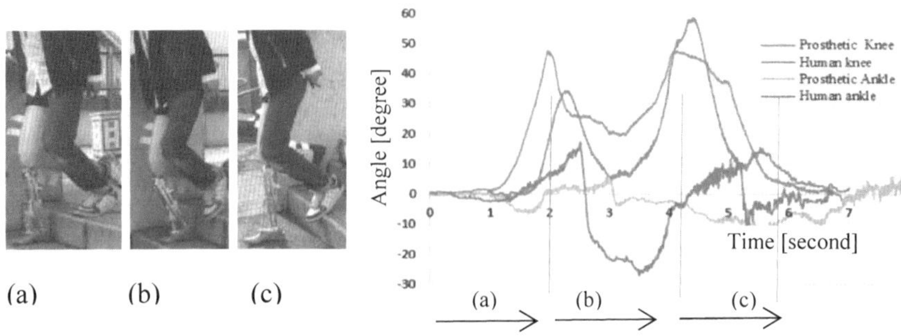

Fig. 9 Stairs descending experiment.

5. Conclusion

We have designed and made the prosthesis according to our analyses. The developed above/knee prosthesis allows amputees to walk on a level ground, ascend and descend stairs with no external energy source. The experiments revealed that flexion of ankle was very slow on stairs ascending as well as stairs descending during the swing phase. An appropriate spring should be used to between back of the foot part and lower part of the frame to pull back the foot part after it make dorsiflexion (help the ankle to make plantar flexion during swing phase).The FCV should be adjusted automatically when ascending the stairs or walking on incline surface by using a motor controlled FCV. Now we are refining these problems and planning to achieve walking experiments again.

References

1. Minerva V. Pillai, H.Kazerooni and Andrew Hurwich: “Design of a Semi-Active Knee-Ankle Prosthesis” International Conference on Robotic and Automation, 2012
2. Japan society of limb salvage and podiatric medicine, 2013 http://www.dm-net.co.jp/calendar/2013/019706.php (Access on 13 April 2015)
3. BiOM by Hugh Herr, http://www.biom.com/ (Access on 13 April 2015)
4. Genium by Ottobock, http://professionals.ottobockus.com/cps/rde/xchg/ob_us_en/hs.xsl/53013.html (Access on 13 April 2015)
5. 3R80 by Ottobock http://professionals.ottobockus.com/cps/rde/xchg/ob_us_en/hs.xsl/609.html?id=14966#t14966 (Access on 13 April 2015)
6. Koh Inoue, Takahiro Wada and Ryuchi Harada: “Novel Knee Joint Mechanism of Transfemoral Prosthesis for Stair Ascend” International Conference on Rehabilitation Robotic, 2013
7. Digital Image Technology, http://www.ditect.co.jp/ (Access on 13 April 2015)

ADAPTIVE GAIT TRAJECTORY BASED ON ITERATIVE LEARNING CONTROL FOR LOWER EXTREMITY REHABILITATION EXOSKELETON*

HAITAO ZHOU, LINING SUN

State Key Laboratory of Robotics and System, Harbin Institute of Technology, Harbin 150001, China

JUAN LI†, WEIDA LI, XIAOWEI CAI, LONGHAI LU

School of Mechanical and Electric Engineering, Key Laboratory of Robotics and System of Jiangsu Province, Soochow University, Suzhou 215021, China

The design of a controller is one of the key tasks and major difficulties in the development of rehabilitation exoskeletons. An algorithm about an adaptive gait trajectory based on the iterative learning control for lower extremity rehabilitation exoskeleton is proposed in this paper. First of all, dynamic model is built up based on the Lagrange equations for the lower extremity rehabilitation exoskeleton. Secondly, an adaptive gait trajectory based on the iterative learning controller is put forward to achieve the active mode of patients. Finally, a simulation experiment is conducted in MATLAB based on the standard gait data which were collected by an optical motion capture system. The simulation results show that the control algorithm can achieve the desired adaptive tracking for joint trajectory and enable patients' active participation in rehabilitation.

1. Introduction

Rehabilitation training could improve muscular strength and movement coordination of patients with neurological or orthopedic lesions such as stroke [1] and spinal cord injuries [2]. Conventionally, rehabilitation training demands for therapists. However, this is limited by personnel shortage and fatigue of therapists. In order to overcome these deficiencies, several rehabilitation exoskeletons have been developed [3]-[5], namely, LOKOMAT [6], lower extremity powered exoskeleton(LOPES) [7], active leg exoskeleton (ALEX) [8], an intrinsically compliant robotic orthosis for treadmill training [9], etc.

---

* This work is supported by National Science Foundation of China(No.51475314 and No.61203367)

† Corresponding author: lijuan@suda.edu.cn

Meanwhile, the design of the corresponding rehabilitation exoskeleton's controller has been studied by many researchers. For example, impedance control and adaptive control were applied in LOKOMAT [6]. Subsequently, adaptive impedance control [10] and iterative learning impedance control have been proposed [11].The control structure of Human-machine collaborative active training usually is composed of an inner loop controller and an outer loop controller. There are abundant researches about inner loop controllers in the traditional robot control systems. Nevertheless, the application of the inner loop controller lacks sufficient attentions in lower extremity rehabilitation exoskeletons. As far as we know, PD control served as an inner loop controller in LOKOMAT, which is simple but un-preferable in tracking performance. The boundary layer augmented sliding mode control was used as a basic trajectory tracking controller in order to guide the subject's limbs on reference physiological trajectories [12], which may causes chattering degraded system instability. As we well known, inner loop controller directly affects the accuracy and feasibility of the control algorithm. It is difficult to establish accurate dynamics model in the lower extremity rehabilitation exoskeleton as it is a nonlinear, strong coupling system.

In this paper, an iterative learning controller is proposed as the inner loop controller. It could achieve favorable trajectory tracking performance and strong robustness because it is not strictly depending on the precise mathematical model. Moreover, we propose a gait adaptive control algorithm on the basis of iterative learning controller to make patients participate actively in rehabilitation training. This paper is organized as follows. Section 2 builds dynamic model. Section 3 explains the adaptive gait trajectory based on an iterative learning algorithm. Section 4 explains the simulation results of the proposed control structure. Section 5 presents the conclusions of the described work.

2. Dynamic Modeling

The lower extremity rehabilitation exoskeleton is characterized to be symmetric and wearable. The simplified model of the exoskeleton is shown in Fig 1. Each leg of it has three active DOFs in the sagittal plane, namely, hip, knee and ankle, which is modeled based on the Lagrange equations. $G(q)$, $C(q,\dot{q})$ and $M(q)$ are calculated as equation (1), (2) and (3), respectively.

$$\mathrm{G(q)} = \begin{bmatrix} (m_1gd_1 + m_2gl_1 + m_3gl_1)\sin \mathrm{q}_1 + (m_2gd_2 + m_3gl_2)\sin(\mathrm{q}_1 - \mathrm{q}_2) + m_3gd_3\sin(\mathrm{q}_1 - \mathrm{q}_2 + \mathrm{q}_3 + pi/2) \\ -(m_2gd_2 + m_3gl_2)\sin(\mathrm{q}_1 - \mathrm{q}_2) - m_3gd_3\sin(\mathrm{q}_1 - \mathrm{q}_2 + \mathrm{q}_3 + pi/2) \\ m_3gd_3\sin(\mathrm{q}_1 - \mathrm{q}_2 + \mathrm{q}_3 + pi/2) \end{bmatrix}. \quad (1)$$

$$C(q,\dot{q}) = \begin{bmatrix} \begin{matrix}[-2m_2l_1d_2\sin q_2 - 2m_3l_1l_2\sin q_2 \\ +2m_3l_1d_3\cos(q_2-q_3)]\dot{q}_2\end{matrix} & \begin{matrix}[m_2l_1d_2\sin q_2 + m_3l_1l_2\sin q_2 \\ -m_3l_1d_3\cos(q_2-q_3)]\dot{q}_2 \\ +[2m_3l_1d_3\cos(q_2-q_3) \\ +2m_3l_2d_3\cos q_3]\dot{q}_3\end{matrix} & \begin{matrix}[-m_3l_1d_3\cos(q_2-q_3) \\ -m_3l_2d_3\cos q_3]\dot{q}_3 \\ +[-2m_3l_1d_3\cos(q_2-q_3) \\ -2m_3l_2d_3\cos q_3]\dot{q}_1\end{matrix} \\ \begin{matrix}[m_2l_1d_2\sin q_2 + m_3l_1l_2\sin q_2 \\ -m_3l_1d_3\cos(q_2-q_3)]\dot{q}_1\end{matrix} & [-2m_3l_2d_3\cos q_3]\dot{q}_3 & \begin{matrix}[m_3l_2d_3\cos q_3]\dot{q}_3 \\ +[2m_3l_2d_3\cos q_3]\dot{q}_1\end{matrix} \\ \begin{matrix}[m_3l_1d_3\cos(q_2-q_3) \\ +m_3l_2d_3\cos q_3]\dot{q}_1 \\ -[2m_3l_2d_3\cos q_3]\dot{q}_2\end{matrix} & [m_3l_2d_3\cos q_3]\dot{q}_2 & 0 \end{bmatrix}. \quad (2)$$

$$M(q) = \begin{bmatrix} \begin{matrix}[m_1d_1^2 + m_2l_1^2 + m_2d_2^2 + m_3d_3^2 \\ +2m_2l_1d_2\cos q_2 + m_3l_1^2 + m_3l_2^2 \\ +2m_3l_1l_2\cos q_2 - 2m_3l_2d_3\sin q_3 \\ +2m_3l_1d_3\sin(q_2-q_3)]\end{matrix} & \begin{matrix}[-m_2d_2^2 - m_2l_1d_2\cos q_2 \\ -m_3d_3^2 - m_3l_1l_2\cos q_2 \\ -m_3l_1d_3\sin(q_2-q_3) - m_3l_2^2 \\ +2m_3l_2d_3\sin q_3]\end{matrix} & \begin{matrix}[m_3d_3^2 + m_3l_1d_3\sin(q_2-q_3) \\ -m_3l_2d_3\sin q_3]\end{matrix} \\ \begin{matrix}[-m_2d_2^2 - m_2l_1d_2\cos q_2 \\ -m_3l_2^2 - m_3d_3^2 - m_3l_1l_2\cos q_2 \\ -m_3l_1d_3\sin(q_2-q_3) \\ +2m_3l_2d_3\sin q_3]\end{matrix} & \begin{matrix}[m_2d_2^2 + m_3l_2^2 + m_3d_3^2 \\ -2m_3l_2d_3\sin q_3]\end{matrix} & [-m_3d_3^2 + m_3l_2d_3\sin q_3] \\ \begin{matrix}[m_3d_3^2 + m_3l_1d_3\sin(q_2-q_3) \\ -m_3l_2d_3\sin q_3]\end{matrix} & [-m_3d_3^2 + m_3l_2d_3\sin q_3] & [m_3d_3^2] \end{bmatrix}. \quad (3)$$

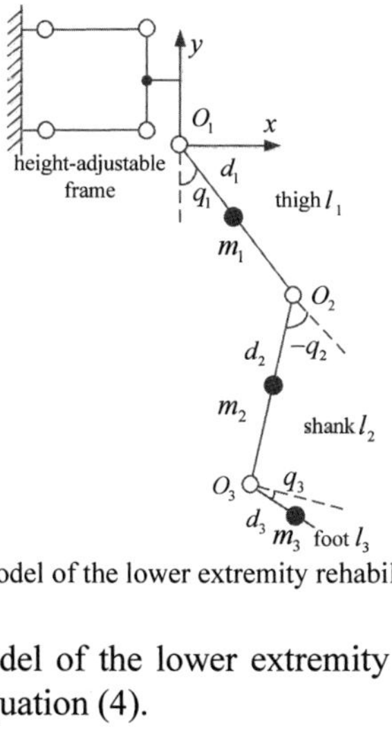

Fig 1. Simplified model of the lower extremity rehabilitation exoskeleton.

The mathematical model of the lower extremity rehabilitation exoskeleton could be described with equation (4).

$$M(q)\ddot{q} + C(q,\dot{q})\dot{q} + G(q) = \tau - \tau_d. \tag{4}$$

Where $q \in \mathbb{R}^3$ is the generalized position vector which comprises the hip, knee and ankle angle. $M(q) \in \mathbb{R}^{3\times3}$ is a symmetric bounded positive definite inertia matrix. $C(q,\dot{q}) \in \mathbb{R}^{3\times3}$ denotes the Coriolis and Centrifugal forces. $G(q) \in \mathbb{R}^3$ is the vector resulting from the gravitational forces. $\tau \in \mathbb{R}^3$ denotes the control input vector containing the torques to be applied at each joint. $\tau_d \in \mathbb{R}^3$ is the vector containing the external torques and un-modeled disturbances.

3. Adaptive Gait Trajectory Based on Iterative Learning Controller

An adaptive gait trajectory based on an iterative learning algorithm (ILC) for the lower extremity rehabilitation exoskeleton is shown in Fig. 2 and the essence of ILC could be seen in Fig. 3. The equation (5) shows the control law of PD-type ILC.

$$u_{k+1}(t) = u_k(t) + K_p e_{k+1}(t) + K_d \dot{e}_{k+1}(t). \tag{5}$$

Where u is the control input; k is the iteration index; K_p and K_d are the proportional and derivative control gain matrices, respectively. e_{k+1} is the system output error at the (k+1)th trail of the proposed learning controller, which given by $e_{k+1}(t)=q_d(t)-q_{k+1}(t)$.

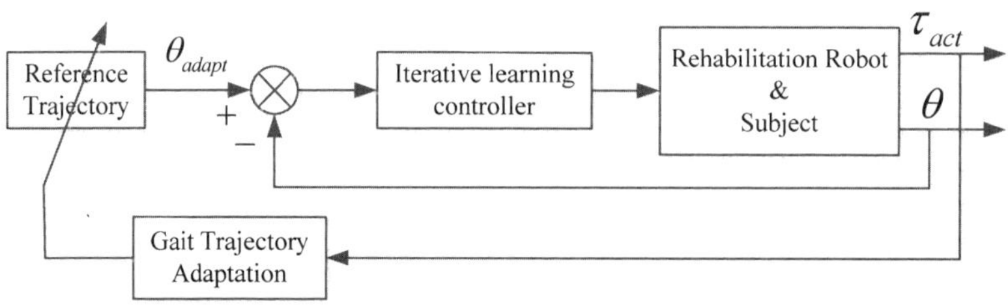

Fig. 2. Adaptive gait trajectory based on iterative learning controller diagram.

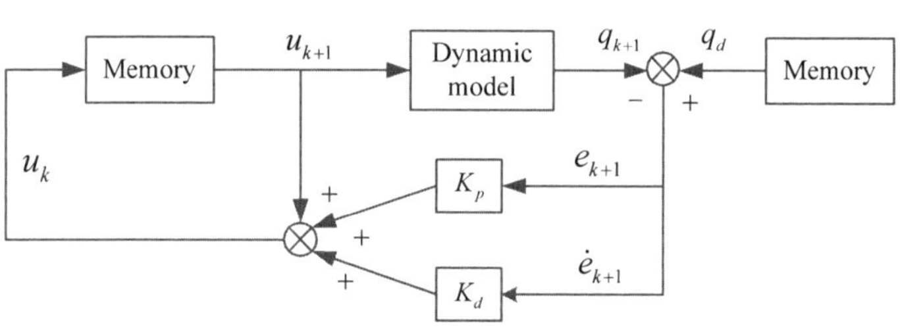

Fig. 3. The block diagram of the PD-type iterative learning control.

The appropriate gait rehabilitation may be supposed to restore the normal gait trajectory. Then the iterative learning control could achieve perfect trajectory tracking. But the normal gait trajectory isn't the best one for every patient and the whole process of rehabilitation. The main disadvantage of the iterative learning control strategies is that it is based on a fixed reference trajectory, which makes it difficult to adjust the trajectory for the individual patient. However, the adaptive gait trajectory control strategy could overcome this disadvantage. It has fully considered the patient's motion intention which is derived from the interaction forces measured by force sensors. The gait reference trajectory parameters can be adjusted automatically according to different patients or the change of interaction forces. The new reference trajectory q_{adapt} are obtained by parameterizing the standard gait trajectory q_{nom}, which could be expressed as equation (6).

$$q_{adapt} = a \cdot q_{nom}(b \cdot t) + c. \tag{6}$$

Where the parameter a represents the amplitude of q_{nom}. b stretches the time and influences the period of gait trajectory. c indicates the joint angle offset. q_{adapt} represents the standard gait trajectory when a=1, b=1, c=0. In order to ensure the adapted gait trajectory remain physiological, parameters of a and b are limited to 20% of the normal values. Besides, the variation range of parameter c is limited to $\pm 4°$ [6]. The new reference trajectory generated by interaction forces could be expressed as q_{nom} and Δq with equation (7).

$$q_{new} = q_{nom} + \Gamma \cdot \Delta q. \tag{7}$$

Where Γ serves as a scaling factor to change the strength of adaptation. Position deviation Δq of the trajectory could be obtained with equation (8).

$$\Delta q = \frac{1}{K_p + K_d s} \tau_{act}. \tag{8}$$

Where τ_{act} represent the interaction torques between subject and exoskeleton. The new joint reference trajectory is fitted with an objective function which is derived from the best square approximation method of the Euclidean norm of the joint trajectory deviation (9).

$$M = min \int_0^T \left\| q_{new} - q_{adapt} \right\|_2^2 dt. \tag{9}$$

4. Simulation Results

To verify the feasibility and effectiveness of the algorithm, simulation experiments are carried out based on the dynamics model of lower extremity rehabilitation exoskeleton.

Parameters used in the simulation are as follows: the weights of thigh, shank and foot of the exoskeleton are separately 10kg, 8kg and 3kg; the lengths of thigh, shank and foot of the exoskeleton are separately 0.45m, 0.44m, 0.1m; the distances between the mass centers of thigh, shank, foot and the rotation centers of hip, knee, ankle are 0.225 m, 0.22 m and 0.05m, respectively.

The un-modeled disturbance torques are added in the form of

$$\tau_d = \text{rand}(1) * [\sin(t); \sin(t); \sin(t)]. \tag{10}$$

Where *rand(1)* is a random signal with amplitude 1.

The sampling time is set to be 10^{-4} sec and the controller parameters are chosen to be $K_p = K_d = diag[250,200,100]$.The standard gait trajectories are collected by the optical motion capture system and fitted by Fourier series (11).

$$\begin{aligned}
q_{hip} &= 0.5207 * \sin(3.133 * t + 0.6142) + 0.5939 * \sin(4.568 * t + 3.134) + \\
&0.08901 * \sin(11.97 * t - 1.365) + 0.0418 * \sin(19.03 * t + 0.01385) + \\
&0.01254 * \sin(33.07 * t - 2.035) \\
q_{knee} &= 1.014 * \sin(3.192 * t + 0.7442) + 0.3184 * \sin(12.32 * t - 0.3614) + \\
&0.0831 * \sin(18.65 * t + 0.9845) + 0.9633 * \sin(4.427 * t + 3.866) \\
q_{ankle} &= 0.0685 * \sin(22.86 * t - 8.746) + 0.1731 * \sin(11.94 * t - 2.664) + \\
&0.1202 * \sin(3.178 * t + 2.077) + 0.02366 * \sin(32.56 * t - 2.094).
\end{aligned} \tag{11}$$

The designed controller is simulated and analyzed in MATLAB. The simulation experiments of standard gait trajectory tracking are shown in Figs. 6 and 7. The results indicate that the accurate tracking of position and speed is achieved by using 10 times iterative learning. As presented in Figs 10 and 11, based on the patient's active forces constantly adjust gait trajectory, the designed controller achieves ideal tracking performance with 10 times iterative learning when adaptive trajectory parameters a=1.1, b=1.2, c=+4°. The standard gait trajectory and adaptive gait trajectory determined from iterative learning control algorithm are shown in Figs. 4, 5, 8 and 9. The results show that the error of position and velocity is converged to zero.

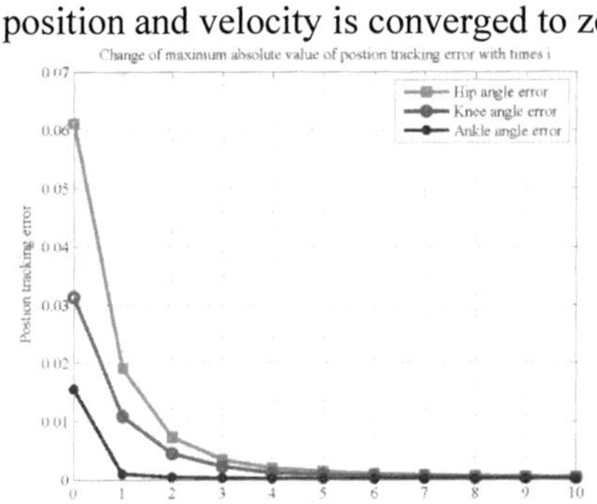

Fig. 4. Position tracking error with 10 times ILC of standard gait trajectory.

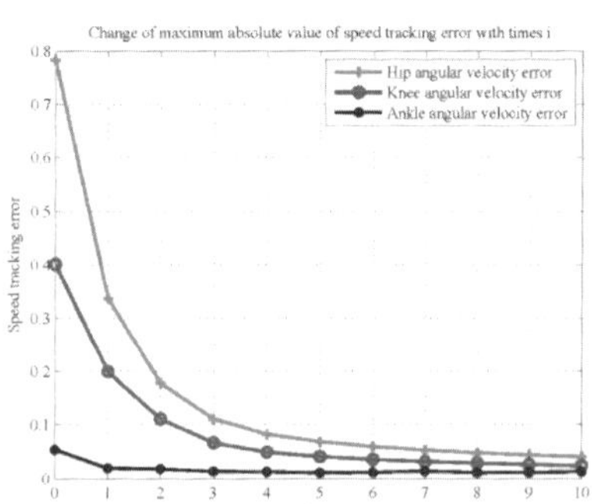

Fig. 5. Speed tracking error with 10 times ILC of standard gait trajectory.

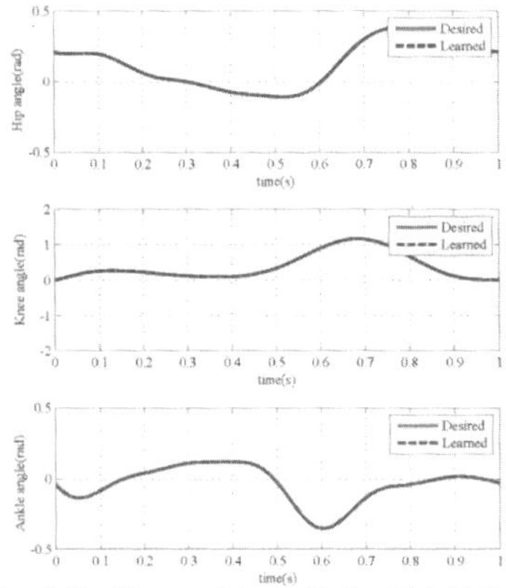

Fig. 6. Position tracking with the 10th ILC of standard gait trajectory.

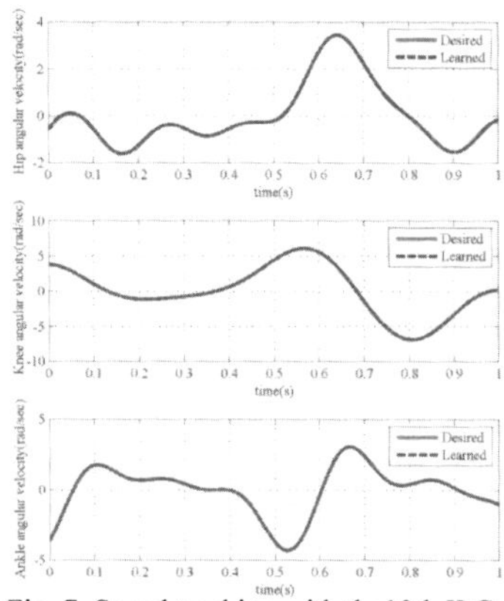

Fig. 7. Speed tracking with the10th ILC of standard gait trajectory.

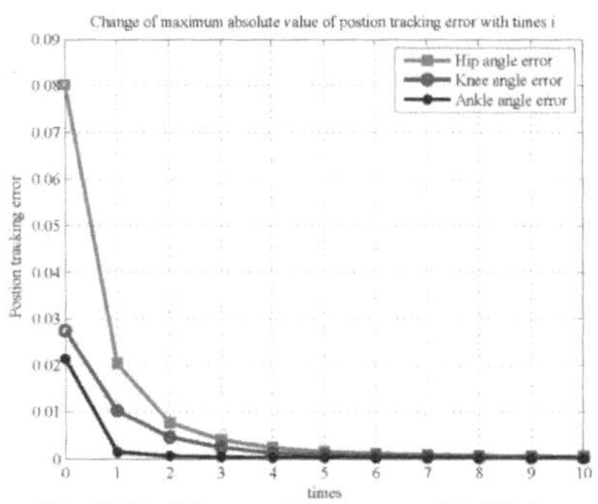

Fig. 8. Position tracking error with 10 times ILC of adaptive gait trajectory.

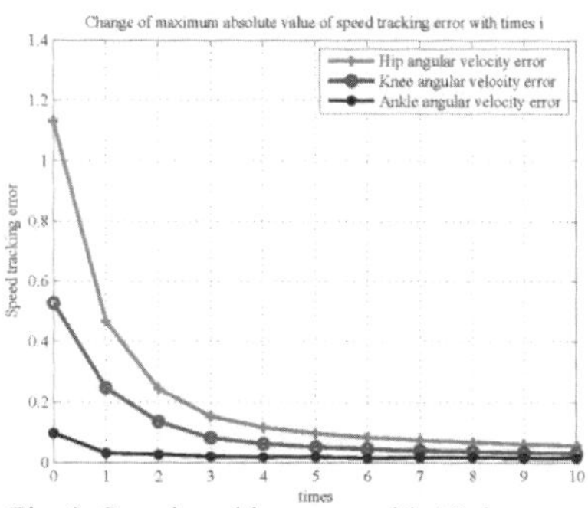

Fig. 9. Speed tracking error with 10 times ILC of adaptive gait trajectory.

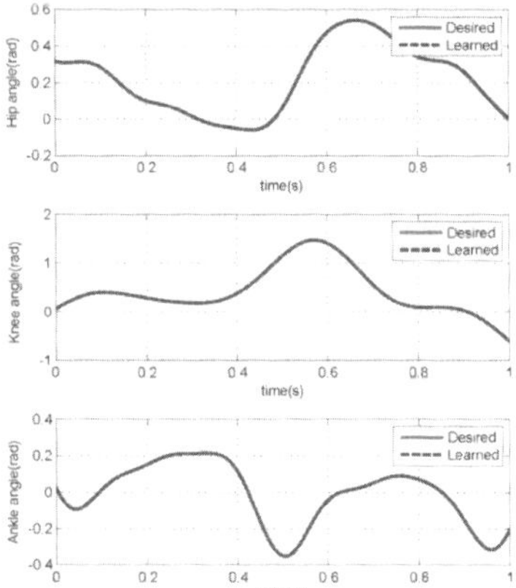

Fig. 10. Position tracking with the 10th ILC of adaptive gait trajectory.

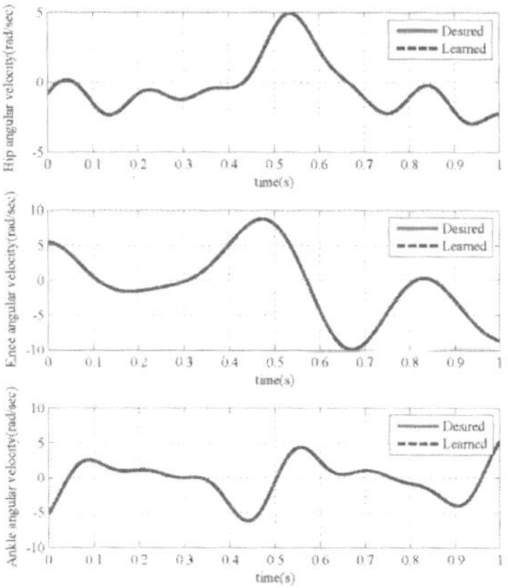

Fig. 11. Speed tracking with the 10th ILC of adaptive gait trajectory.

5. Conclusion

A novel algorithm consisting of gait trajectory adaptive control and iterative learning control for lower extremity rehabilitation exoskeleton is put forward in this paper. Firstly, the dynamic model is built up based on the Lagrange equations for the lower extremity rehabilitation exoskeleton. Then, the adaptive gait trajectory control strategy is proposed to achieve the active mode of patients. Furthermore, the designed controller is simulated and analyzed in MATLAB. The simulation experiment's results show that the control algorithm could achieve a desired adaptive tracking for joint trajectory and the error of position and velocity could be converged to zero.

References

1. B. Husemann, F. Müller, C. Krewer, S. Heller, E. Koenig. *Stroke* **38**, 349 (2007).
2. B. Dobkin, H. Barbeau, D. Deforge, J. Ditunno, R. Elashoff, D. Apple, M. Basso, A. Behrman, L. Fugate, S. Harkema. *Neurorehabilitation and neural repair* **21**, 25(2007).
3. G. Chen, C. K. Chan, Z. Guo, H. Yu. *Critical Reviews™ in Biomedical Engineering* **41**, (2013).
4. S. Hussain, S. Q. Xie, G. Liu. *Medical engineering & physics* **33**, 527 (2011).
5. S. Viteckova, P. Kutilek, M. Jirina. *Biocybernetics and Biomedical Engineering* **33**, 96 (2013).
6. R. Riener, L. Lunenburger, S. Jezernik, M. Anderschitz, G. Colombo, V. Dietz. *Neural Systems and Rehabilitation Engineering, IEEE Transactions on* **13**, 380 (2005).
7. J. F. Veneman, R. Kruidhof, E. E. Hekman, R. Ekkelenkamp, E. H. Van Asseldonk, H. Van Der Kooij. *Neural Systems and Rehabilitation Engineering, IEEE Transactions on* **15**, 379 (2007).
8. S. K. Banala, S. H. Kim, S. K. Agrawal, J. P. Scholz. *Neural Systems and Rehabilitation Engineering, IEEE Transactions on* **17**, 2 (2009).
9. S. Hussain, S. Q. Xie, P. K. Jamwal, J. Parsons. *Medical engineering & physics* **34**, 1448 (2012).
10. S. Hussain, S. Q. Xie, P. K. Jamwal. *Cybernetics, IEEE Transactions on* **43**, 1025 (2013).
11. J. He, X. Zhuge, Z. Luo, G. Li, Y. zhang. *TELKOMNIKA Indonesian Journal of Electrical Engineering* **12**, 1300 (2014).
12. S. Hussain, S. Q. Xie, P. K. Jamwal. *Robotics and Autonomous Systems* **61**, 911 (2013).

AN EDITORIAL PLATFORM FOR SCREENPLAY OF INTERACTIVE ROBOTIC PUPPET SHOWS

KAO-SHING HWANG

Electrical Engineering, National Sun Yat-sen University, 70 Lienhai Rd., Kaohsiung 80424, Taiwan, R.O.C.

YU-CHI KUO

Electrical Engineering, National Sun Yat-sen University, 70 Lienhai Rd., Kaohsiung 80424, Taiwan, R.O.C.

WEI-CHENG JIANG

Electrical Engineering, National Sun Yat-sen University, 70 Lienhai Rd., Kaohsiung 80424, Taiwan, R.O.C.

In recent years, due to the development of computers and the Internet, the robotics industry has steadily grown. Besides academic applications, entertainment robots can also be programed to bring us unprecedented performances that many people can enjoy. Robotic puppet shows are an important part of the entertainment robot field. The robot can integrate the basic ability to speak, make poses and human like expressions. Based on the considerations above, this paper describes an Editorial Platform for Screenplay of Interactive Robotic Puppet Shows (EPFS). EPFS lets users in different age groups control robots and allows them to write the screenplays for drama performances in a simple and fast way. In the platform users have the ability to use game design concepts that allow them to make diverse screenplays and a multitude of different storylines. Finally, the platform controls and records the story-making process by constructing a timeline. In doing so, problems caused by robotic expression and interaction can be solved. In addition, the arrangement of screenplays will be more flexible and accurate. According to the results of the experiment the platform proves to be flexible and easy to use. A video presentation of the paper has been posted on YouTube; please search the title below "An Editorial Platform for Screenplay of Interactive Robotic".

1. Introduction

In today's world robots are replacing humans in many jobs that are too trivial or dangerous for people to do. The use of robotics can be applied to academic and industrial applications as well as many other fields. Service and entertainment robots have gradually made their entrance into our daily lives. Dramas and games are both forms of art that can also be enjoyed as recreational activities.

The biggest difference between drama and games is that a drama is passive, and game is interactive. The content of drama is directly presented to the audience, in the form of television or movies. In dramas the audience only can watch as the story unfolds. In contrast, games allow users to take part in the story, which as a result could make the ending of the story different. Today's robotic puppet shows usually use event-driven story arrangements [1][2] that are too simple to be used in situations where sophisticated functions are needed. In this paper, the time stamps within timelines are implemented in order to arrange performance elements in the shows. The concept of EPFS is by editing the screenplay users can change the robots posture and speech to make their expressions more humanlike. The joining of sound effects and music can also make the show more dramatic. After finishing a screenplay, the development and ending of a drama are fixed. A new feature called "branching" has been added to the platform which allows the user to pick different paths as the story progresses.

2. Background

2.1. *MVC architecture*

MVC is a design pattern for software engineering [3]. It divides an application into three parts: Model, View and Controller. Three parts within the pattern, only the model can access the database directly. The view can develop visual outputs for users, which is used for data display, such as a webpage. The controller is responsible for collecting the input data from the view in order to translate it into commands for updating the state of models or changing the corresponding view's appearance.

3. System Analysis and Design

In the robotic puppet show, the developer looks forward to making the robotic motion more vivid and human-like. But editing robotic motion was usually done by adjusting the parameters of the robot's actuators. Using this process is repetitive and inefficient. In [4], the motion sensing devices based on optics such as Microsoft Kinect that provides users with a cost-effective and state of the art functions to capture human motions.

3.1. *System introduction*

This paper develops a platform to solve the previously mentioned problem and prove it by using a robotic puppet show. The core design concept is to allow users of different ages to easily edit their own screenplays and to make different kinds of performance robots act in the edited screenplays. Since web service has become an essential part of human life. Developing a system to become a web application, restrictions such as distance and time can be broken and less storage space is needed on the client. The wireless technology, Wi-Fi, not only has a high transport speed but also has commonly used in daily life. In addition, the robotic puppets have a built in Wi-Fi function or can be equipped with a Wi-Fi module. Therefore, Wi-Fi is chosen as the main communication type in this system to allow users to have the ability to access the website and control robotic puppets. The concept of system is described in Figure 1.

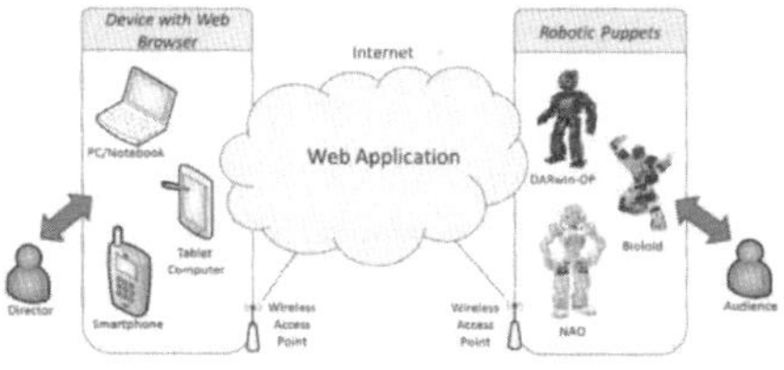

Figure 1. The concept of EPFS.

3.2. *Motion generation*

This platform also allows users to utilize Kinect to capture human motions for extending the motion library applied to robotic puppet shows. By doing so, motion generation can become more intuitive and efficient than the common ways used to control robots. Below are the methods proposed by [4] to implement motion emulation and representation by using Kinect. When the human's image is successfully captured by Kinect, a set of lines that mimic a skeleton will then appear on the top of the body. Then, the user can receive important data such as joint position and orientation. Inverse kinematics refers to the use of a robots kinematics equations to determine the joint parameters that provide the desired position of the end-effector. The joint orientation represents current posture. Solving the inverse kinematics can determine the current joints change in angle.

3.3. *Screenplay editing*

1. Scene Tree

A rooted n-ary tree structure is used to construct a whole screenplay. By using this kind of tree, the users make the plot of story diverge and as a result, the story will become more diverse. There are three different crucial nodes in the tree: root node, decision nodes, and leaf nodes. Each node stands for a single scene, which was depicted in Figure 2. The root node represents the start of the story. Any node that has more than two child nodes is called a decision node. Branching will occur when the first node has finished performing, and then the user can choose one of scenes at the next level of the tree to make the story diverge and proceed. As a result, the leaf node means the end scene of the screenplay. By doing so, even the user play the same screenplay, the development and ending of the story can be distinctive according to the user's choice.

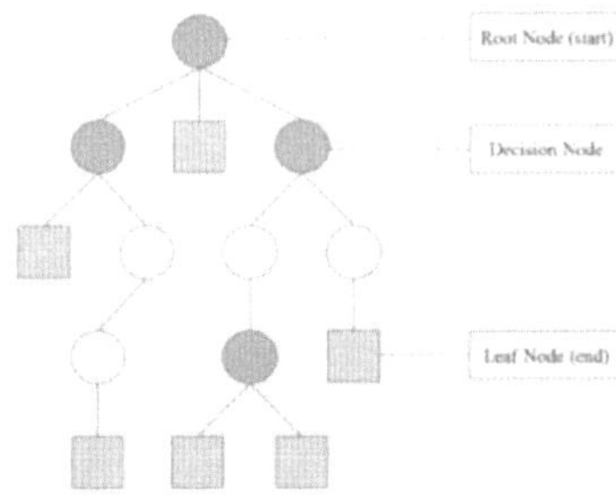

Figure 2. Structure of scene tree.

2. Interactive Screenplay Authoring Platform (ISAP)

ISAP contains two parts: the website and the Node.js. The main function of this website is to provide users with screenplay display, editing, and storage. The implementation of website has followed MVC pattern. Users can access the website by using any device with a web browser to edit their own screenplays. All of the details will be automatically stored in web server within database.

3. Mobile Interpreter (MI)

An Android application has been developed as a mobile interpreter (MI). Based on the existing MVC framework, this interpreter can be regarded as a new View constructed with the original Model and Controller. Compared to ISAP, the MI is playing the role that is similar to the combination of devices within a web browser and Node.js as show in Figure 3. The major difference between ISAP and MI is that the latter one only provides basic functions including screenplay display and performance.

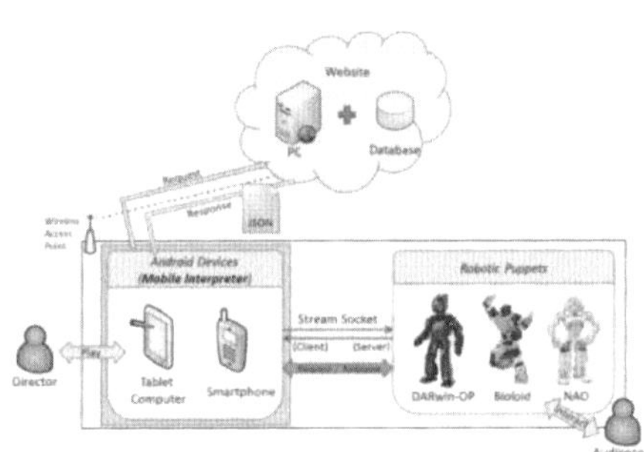

Figure 3. MI architecture.

4. Software Architecture

4.1. *DARwIn-OP API*

The functions of the DARwIn-OP robot can be divided into two modes, multi-threading and single-threading. The muliti-threading mode includes motion, walking, Kinect-based representation and speech functions. On the other hand, color detection is developed in single-threading. In multi-threading mode, the Pthread library allows for four functions to be run simultaneously. Those threads will release its resources and destroy itself when the processes have been completed. On the contrary, the color detection function is developed in a single-threading, in this way the color detection results can be sent back to the client. Below are the specifications of the previously mentioned functions.

4.2. *NAO API*

Since NAO is regarded as a new type actor in EPFS. The command stored in the packet will be extracted based on the identifier in order to execute a python motion file or run a speech function with a line. Moreover, both of them are executed using multi-threading in order to express the two different emotions at the same time.

5. Implementation

5.1. *User interface of ISAP*

1. Screenplay Creation

In Figure 4, users can add personal screenplays. The screenplay title and description need to be filled in by the user. In addition, the user also can pick a corresponding picture as the cover of the screenplay, and then select how many robots of each type are desired.

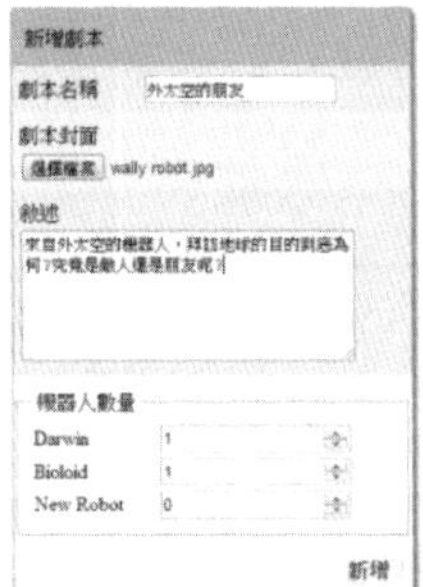

Figure 4. Screenplay creation.

2. Scene Establishment

ISAP allows users to add all of the scenes that are needed in the screenplay, which include the initial, middle and last scene. In Figure 5, when adding a new scene, the scene's title and description need to be filled in first, and the scene selector is used to choose whether the automation or color card mode. Automation mode means the scene is linearly played in an automatic fashion. In other words, the story can never be diversified. On the contrary, the color-card mode allows audiences to use a color card which can be recognized by DARwIn-OP and also can be used to pick different paths as the story progresses.

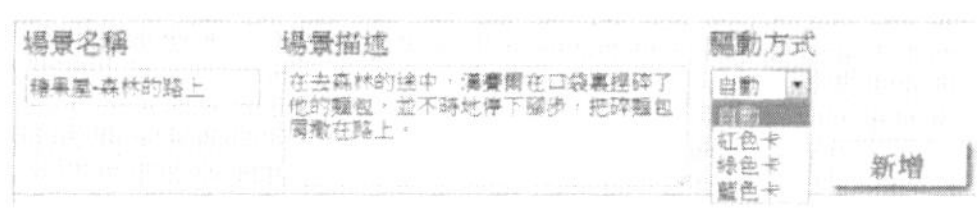

Figure 5. Adding a new scene.

3. Scene Editing

After creating a new scene, the user can double click on the scene to edit its content. The scene-editing page is divided into three major divisions: the motion library, music library, and the timeline, as shown in Figure 6. The motion library provides many predefined postures and motions, which are classified as either robots or Kinect motions. Secondly, the music library has a diverse collection of sound effects and background music.

Figure 6. Scene-editing page.

In Figure 7, the timeline consists of the sound effects, background music as well as the motion and speech of each robot. Each scene only has one local timeline based on seconds. The files in the motion and music library can be dragged into a specific track in order to add new event in the timeline. Fig. 6 shows the user picking up a motion file and dragging it into a track. In the next step the motion file needs to be placed at the desired spot on the timeline. Finally, the new action appears with its name on the timeline along with a box that represents the starting and ending time of the action. Moreover, a user can cancel the action by double clicking on the box that represents the action on the timeline.

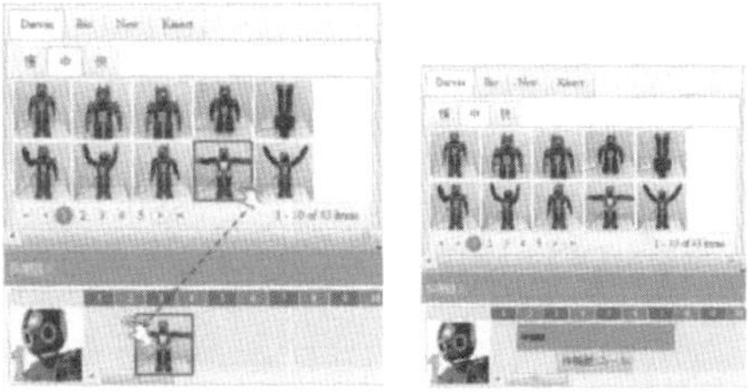

Figure 7. Adding a new motion into a track.

4. The Relationship of Scenes

After finishing the content of single scene, the linking tool is used to establish the relationships between all of the scenes in the screenplay. In Figure 8, the user can drag the linking tools from the source scene to the target; after this action is completed an arrow will appear between the two boxes signifying the completion of the linking process. If the user decides to delete the arrow he/she can do so by double clicking on it.

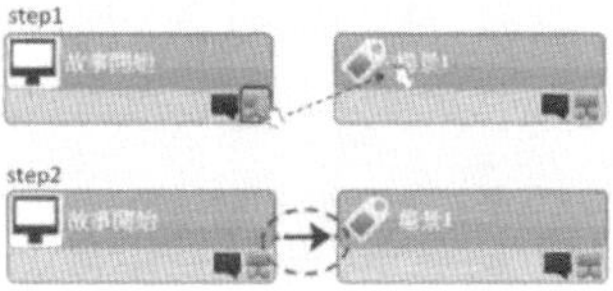

Figure 8. Linking tool of scene.

5. Motion Management

In Figure 9, EPFS has provided the administrator of the website with the ability to add or modify existing DARwIn-OP and Bioloid motions as well as any type of performance robots with motion functions like NAO. The GIF and JPEG are both supported as the format of images in this platform to differentiate whether this command is a motion or a posture.

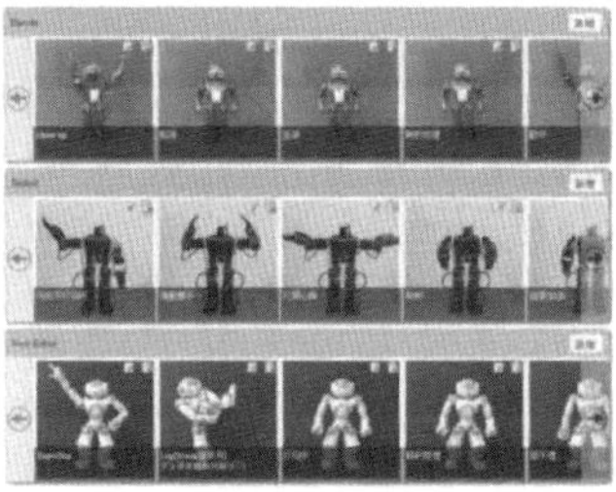

Figure. 9 Motion management page.

6. Kinect-based Motion Uploading for DARwIn-OP

First the user needs to install the Kinect SDK. After the imitation program is executed, the user's movement can be captured and recorded by Kinect. By uploading the imitation texture file named "kinect.txt", the user can add the motion into the robotic puppet show as shown in Figure 10.

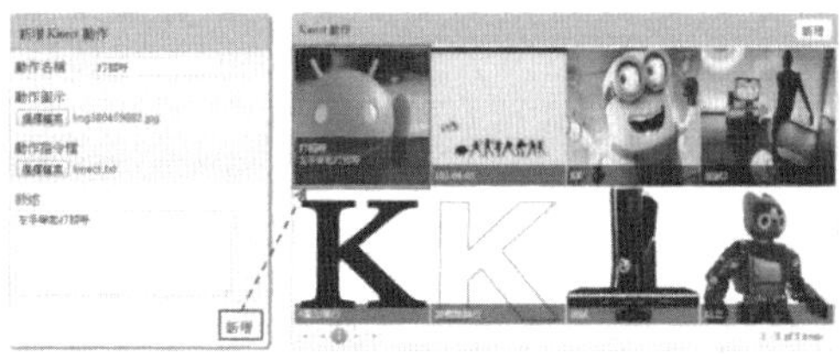

Figure. 10 Uploading a Kinect-based motion file.

6. Conclusion

This paper provide user with a GUI to edit personal screenplays rather than a redundant and complicated way to control the robotic puppets. The screenplays created by the director can be performed with not only robotic puppets but also entertainment elements such as sound effects and background music. The addition of new robot types and the extension of motion library features make this platform more flexible and functional. The audience can interact with robotic puppets by using dynamic screenplays which gives users the opportunity to not only be a viewer but also to participate in the shows. Thanks to the web based architecture, EPFS can be accessed by any devices with a web browser or smart devices connected to Wi-Fi for breaking the physical restrictions. As a result, the robotic puppet show can be presented any place any time and will surely entertain any audience.

References

1. J. J. Siao, "The Graphic Authoring Platform of Screenplays for Robotic Puppet Shows," National Sun Yat-sen University, 2012.
2. W. C. Wu, "The Development of Screenplay Interpreter for Multi-morphic Robots," National Sun Yat-sen University, 2012.
3. "The Model-View-Controller (MVC) Design Pattern for PHP," [Online]. Available: http://www.tonymarston.net/php-mysql/model-view-controller.html. [Accessed 03 12 2012].
4. Y. A. Zheng, "Humanoid Robot Behavior Emulation and Representation," National Sun Yat-sen University, 2012.

DESIGN OF EXO-LEGS EXOSKELETONS

G.S. VIRK, U. HAIDER, I. NYOMAN, N. MASUD
University of Gävle, SE-801 76 Gävle, SWEDEN

I. MAMAEV, P. HOPFGARTEN, B. HEIN
Karlsruhe Institute of Technology, Karlsruhe, Germany

Abstract: The paper describes the design details for realising the EXO-LEGS assistive exoskeletons for Ambient Assisted Living (AAL) applications based on modelling and simulation studies performed for key mobility functionalities in activities for daily living such as stable standing in open space and straight walking. The results provide the basis for selecting sensors and actuators to develop the needed assistive exoskeletons to help the elderly to stay active and independent for as long as possible.

1. Introduction

The EXO-LEGS project focuses on defining the key mobility functionalities which are needed for active daily living (ADL) and identifying the design requirements so that appropriate assistive non-medical exoskeletons can be designed. A survey of the EXO-LEGS end user group comprising 109 members has been carried out to identify main mobility functionalities. In total 67 completed questionnaires (48 EEUG members and 19 Control Group members) have been completed to determine the most important mobility functionalities and how they should be included in the various exoskeleton models considered within the project. The results from the primary, secondary, tertiary end users and the control group all show remarkable similarity showing that all are aligned and all the completed questionnaires can be considered collectively in ranking the overall mobility functionality requirements. The mobility functionalities for normal daily living have been ranked in the following order:

1. Standing
2. Sit-to-stand transfers
3. Straight 2D walking
4. Walking/turning on flat ground
5. Bending down (to pick up item)
6. Crouching with support (e.g. pushing rollator)
7. Walking up/down stairs

8. Stepping over objects
9. Crouching without support (e.g. to avoid low ceiling)
10. Speed walking
11. Walking/running on uneven ground
12. Walking on slippery ground
13. Exercising
14. Using leg to push door
15. Running, skipping
16. Crossing country skiing
17. Jumping, hopping

These ranked motion functions and other suggestions made have been used to define the functionalities to include in the Basic, Standard and Deluxe exoskeleton models in a logical manner to define the core functionalities in the basic model and then build upon them for the more sophisticated designs. It was decided to have a single degree of freedom (DOF) for each leg joint in the Basic exoskeleton design so that it was able to provide support for motions in the sagittal plane. In addition, a low power exoskeleton (as defined in ISO 13482:2014 [6]) providing 30% physical assistance decided upon to ensure it would be affordable. This allowed the top three and the sixth mobility functionalities to be included but not motion 4 (additional DOF needed at hip for turning) or motion 5 (high torques needed for getting up).

2. Modelling and Simulation

2.1. *Stable standing in open space*

There are several daily activities where the human body must reside in a stable position. For studying the balancing issues within EXO-LEGS, MATLAB/Simulink was chosen, as it provides a wide range of tools for modelling and allows fast prototyping as well as easy integration with hardware (Hardware-in-the-Loop). SimMechanics Toolbox provides a multi-body simulation environment for 3D mechanical systems, such as the exoskeletons to be desgined. Models from CAD systems, including mass, inertia, joint constraints and 3D geometry, can be imported into SimMechanics. The exoskeleton studied consists of 7 rigid bodies and 6 revolute joints (see Figure 1a). In order to generate stable trajectories for the exoskeleton it is necessary to have a model and a stability criteria. The following stability criterions were implemented: Center of Mass (CoM), Zero Moment Point (ZMP), Center of Pressure (CoP).

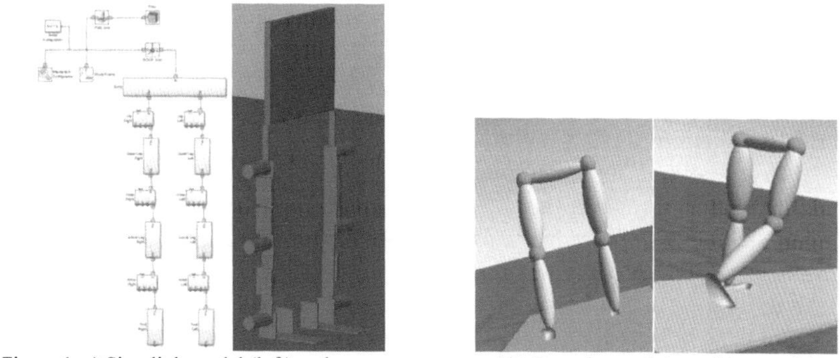

Figure 1. a) Simulink model (left) and SimMechanics output (right)

1b) Opensim straight walking model

The CoM only includes the position of the exoskeleton and ignores motion and so it only ensures stability, if the exoskeleton is motionless. If N is the number of links of the exoskeleton, P_i the position of link i and m_i the mass of link i, then: $CoM = \frac{\sum_{i=1}^{N} P_i \cdot m_i}{\sum_{i=1}^{N} m_i}$ represents the CoM. If the projection of the CoM on to the ground lies within the support polygon, the position of the exoskeleton is stable. The projection can be established by $CoM_y = 0$.

The CoP does not use the position and motion to determine stability. Instead, it uses the pressure of each contact point. Therefore, additional pressure sensors are needed for the feet. If N is the number of contact points, P_i the position of contact point i and F_i the pressure in contact point i, then

$$CoP = \frac{\sum_{i=1}^{N} P_i \cdot F_i}{\sum_{i=1}^{N} F_i}$$

is the CoP and stability is achieved if this lies on or within the support polygon.

The ZMP was defined by Vukobratovic et al. [5]. In addition to the position, like the CoM, it takes into account resulting forces and moments of motions. If A is the point on the ground where the force F_A and the moment M_A of the exoskeleton acts, then P with force F_P and moment M_P is the point to keep the exoskeleton in balance. This point P is defined by the static equilibrium equations:

$$F_A + F_P = 0$$

$$F_A \times p_{OA} + M_A + F_P \times p_{OP} + M_P = 0$$

p_{OA} and p_{OP} are the position vectors of A and P. This point P is the ZMP, hence this is where its name originates from. Point P is the point where all moments are eliminated. If the ZMP is inside the support polygon without its edge, the exoskeleton is in a stable state.

$$M_{Px} = 0, \; M_{Py} = 0$$

$$ZMP_x = \frac{\sum_{i=1}^{N} m_i p_{ix}(a_{iz} + g) - \sum_{i=1}^{N} m_i p_{iz} a_{ix} - \sum_{i=1}^{N} I_{iy}\omega_{iy}}{\sum_{i=1}^{N} m_i (a_{iz} + g)}$$

Thus, in order to have a statically stable position, the exoskeleton should have the field projection of the CoM inside its support polygon. But as soon as the exoskeleton starts to make fast movements, dynamic forces come into play. Therefore, for dynamically stable movement it is necessary to use the ZMP criterion. CoP, on the other hand, presents a good possibility to evaluate control algorithms using pressure sensors.

For the simulation following blocks are necessary:

- CoP observer: to simulate pressure sensors on the foot. The output is the distribution of the pressure relative to the foot centre.
- Position generator with CoM: to generate position with given CoM and 2 of 3 joint angles
- Position generator with ZMP: to generate position with given ZMP and 2 of 3 joint angles
- Point-to-Point (PTP) trajectory generator
- Interpolated trajectory generator

Inverse dynamic control is based on the Newton-Euler approach and calculates the required forces for the target position and velocity.

2.2. *Straight walking*

Straight walking activity is investigated by limiting hip, knee and ankle joints of both legs to 1 degree of freedom (DOF) with human body segmented into thighs(2), shanks(2), feet(2), while the HAT (Head, Arms and Trunk) is considered as 1 segment. A lower body stick model available in Opensim was enhanced to fit the requirements with the thighs(2), shanks(2), feet(2) and pelvis for studying the EXO-LEGS mobility functionality; this is shown in Figure 1b. The mass and length of each body segment is scalable and modified according to the user data.

An exoskeleton was used for collecting straight walking data comprising hip, knee and ankle joints [1] to study human walking pattern. An algorithm was developed to split the data in to walking cycles so that the individual joint angles over the steps could be normalised in time and magnitude. These average values are obtained to yield the average walking trajectory for the individual joints. The mean trajectories for each joint for a single step were then used at each time step in an Opensim inverse dynamical model to obtain the torques needed to drive the walking pattern. To simulate the ground reactions forces (GRFs), pin constraints and contact forces are used between the foot and the ground at estimated times of the heel strike, the single stance phase, the toe off and over the double stance phase [2]. These constraints provide approximations to the GRFs. The overall algorithms are presented in Figure 2.

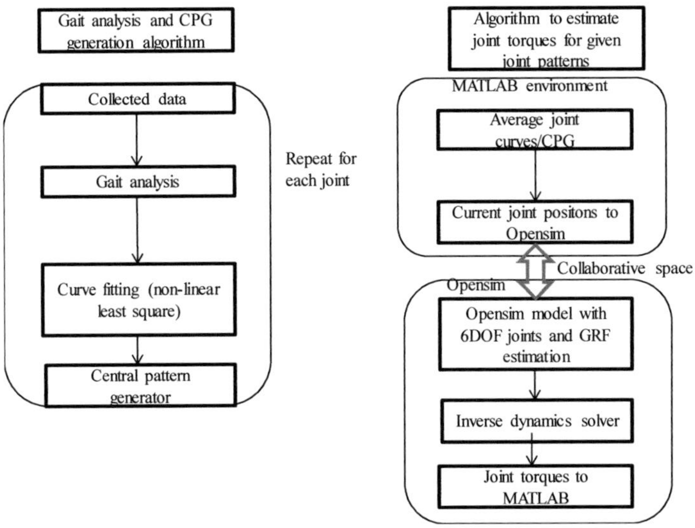

Figure 2. Gait analysis and pattern generation to estimate joint torques for walking.

Single support and double support dynamic models were derived for the 6DOF sagittal plane walking activity. The model equations are given in Eq. (1)

$$\tau = E^{-1}(A(\theta)\ddot{\theta} + D(\theta)\dot{\theta} + G(\theta) - K\dot{\theta} - J^T(\theta)F) \tag{1}$$

where A is the mass inertia matrix, D is the centrifugal force matrix, G is the gravitational force vector, J is the constraint Jacobian vector for the change in the single and double support and F is the row vector representing GRF. The estimated GRFs from the Opensim simulations were used to analyse and validate the mathematical models.

3. Simulations

To validate the developed models several simulations were conducted. For the balance studies, the position generator for the CoM was tested. The model calculates the third joint values based on the positions of the other 2 and the desired CoM. Using the CoP observer, the reached CoM position and joint positions were evaluated. Table 1 shows the simulation data, where empty cell correspond to the calculated joint. The maximum error between desired CoM and observed CoP is 0.0015*. The same joint positions were used for the PTP trajectory tests, where CoP were observed during the movement. Figure 3 shows the CoP plot using the ankle joint.

* Defining the CoM/CoP range as [-1;1], where 1 denotes the frontal boundary of the support polygon and -1 represents the rear boundary.

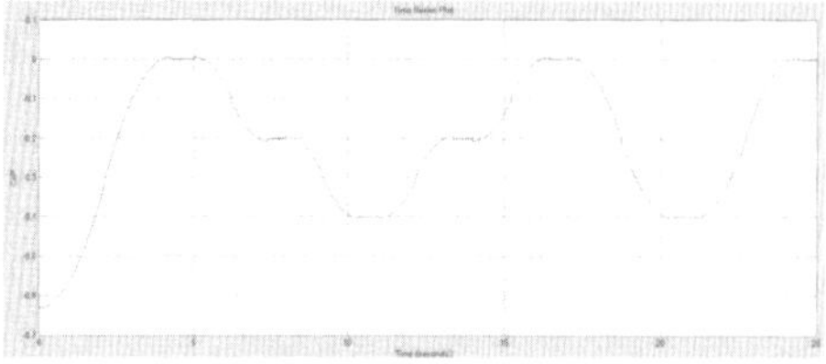

Figure 3: CoP plot for ankle motion at 0.05 rad/s^2

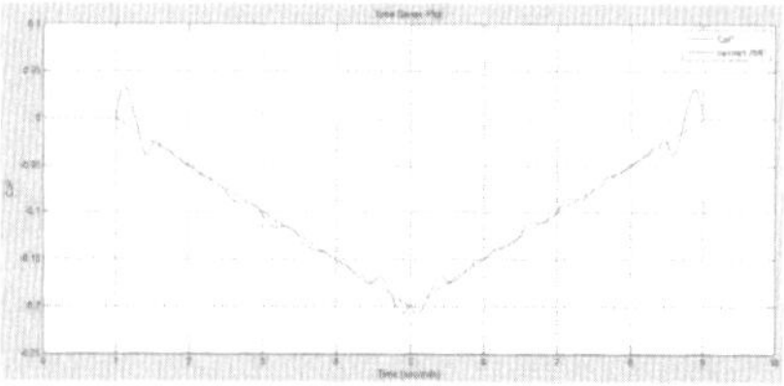

Figure 4: CoP plot for ZMP = -0.2 at the middle

The interpolated trajectory tests were also conducted to examine the adaptation of trajectory to a desired ZMP path using the hip joint. Each trajectory starts and ends in the same position, where the ZMP error equals 0. A 6th order polynomial is used to calculate the trajectories. The trajectories are discretized with $\delta t = 0.5$ for the ZMP adjustment.

Table 1. Target CoM and simulated CoP.

Hip	Knee	Ankle	CoM	CoP
0°	0°		0	0.0000
0°	0°		−0.2	−0.2017
0°	0°		−0.4	−0.4003
−30°	60°		0	0.0006
−30°	60°		−0.2	−0.2003
−30°	60°		−0.4	−0.4001
0°		−30°	0	0.0004
0°		−30°	−0.2	−0.1985
0°		−30°	−0.4	−0.4000
−30		−30°	0	0.0000
−30°		−30°	−0.2	−0.2000
−30°		−30°	−0.4	−0.3999
	20°	−20°	0	−0.0015
	20°	−20°	0.1	0.0999
	50°	−30°	0	−0.0001
	50°	−30°	−0.1	−0.1000

Figure 4 depicts the observed CoP for a trajectory with the ZMP = −0.2 at the middle. The oscillations are visible due to discretization, which is used for velocity and acceleration derivatives. It can be seen though, that those oscillations are marginal and should not cause any problems. This allows the target ZMP to be set at any given time.

For the straight walking simulations, a data set for straight walking from a healthy adult subject with weight 56kg, height 1.67m was collected with cadence of 80 steps/min. The cadence is lower than the average cadence of adults. Then gait analysis was performed to get mean trajectories, from the collected data and the resulting torque joint estimates are presented in Figure 4. As the cadence is low the estimated joint torques are also low but these will increase with increases in cadence.

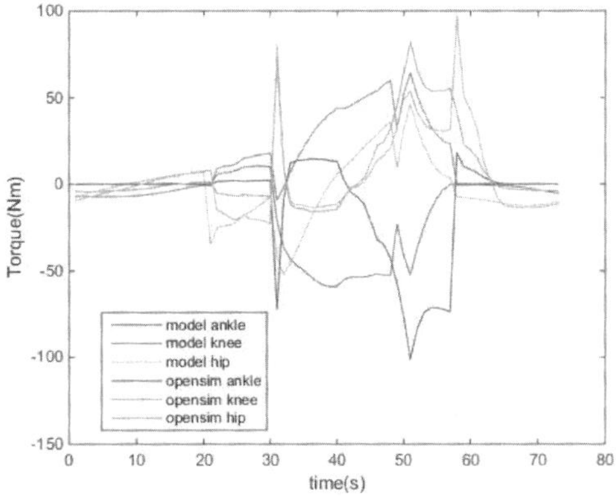

Figure 4. (a) Estimated joint torques from mean curves using inverse dynamics with 6DOF human lower body model scaled to subject

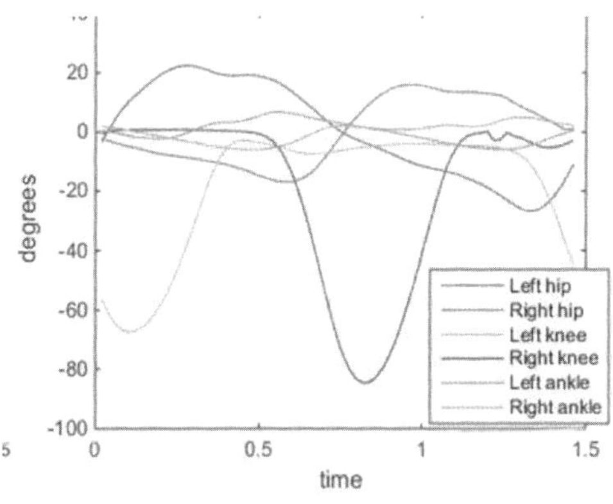

4(b) Estimate mean curves from collected walking data.

The estimated torques are used to select the actuator systems for the exoskeleton such that it remains a low power physical assistant robot as defined in [6]. The mathematical model in Eq. (1) was simulated using the GRFs from Opensim giving the results shown in Figure 5. The torque values from the mathematical model are different at some points than the Opensim model. This can be because the Opensim model is formulated differently as compared to the model produced here. The model produced here considers ground constraints when the foot is in contact with the ground both in single and double support and as well as in double support both legs in toe off and heel strike phase and this reduces the torque required in the toe off phase as seen in figure 6 from time 0.4-0.5sec.

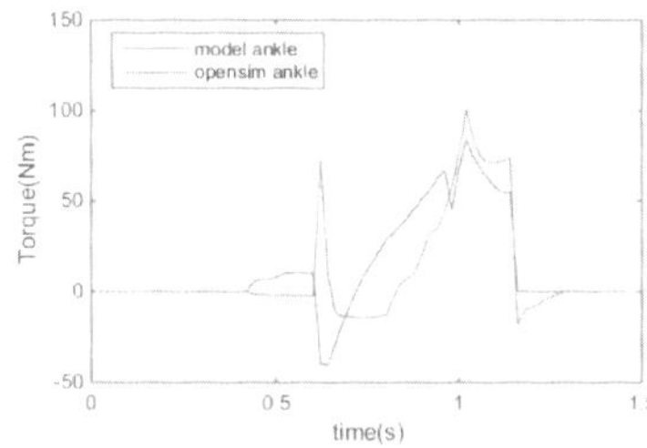

Figure 5. Ankle torque estimated from Opensim model and model in equation (1).

4. Conclusions

A number of simulation environments for modelling and simulating the scenarios important for the EXO-LEGS project were explored. With MATLAB/Simulink and SimMechanics several balance criteria were implemented and tested. The CoM criterion has been used to assess position stability and ZMP was proven to be effective for assessing dynamic stability.

Opensim and MATLAB has been use in the modelling and simulation of a 6DOF human model to estimate the joint torques in straight walking. The results of the simulations were used for the design requirements for sizing of the hardware such as the actuators needed. In future control techniques will also be studied using these simulation tools.

Acknowledgments

Research funded under the AAL Call 4 project EXO-LEGS; in Sweden and Germany the funding is provided by Vinnova and BMBF respectively.

References

1. G. S. Virk, U. Haider, I Nyoman, R. Krishnan and N. Masud, EXO-LEGS for elderly persons, *CLAWAR'14 Conference*, 2014.
2. H. Samuel, R. Hamner, Ajay Seth, K. M. Steele, and S. L. Delp, A rolling constraint reproduces ground reaction forces and moments in dynamic simulations of walking, running, and crouch gait, *Journal of Biomechanics*, 46, 1772-1776, 2013
3. V. Baran, M. Colonna, M. Di Toro and A. B. Larionov, *Nucl. Phys.* **A632**, 287, 1998.
4. V. Baran, M. Colonna, M. Di Toro and V. Greco, *Phys. Rev. Lett.* **86**, 4492, 2001.
5. Vukobratović, Borovac, Zero-Moment Point - Thirty Five Years of Its Life, Int. J. Humanoid Robotics, Vol. 1 No. 1, 2004, p. 161,162.
6. ISO 13482, Robots and robotic devices — Safety requirements for personal care robots, 2014.

DESIGN AND CONTROL OF EXOSKELETON FOR ELDERLY MOBILITY

GHASAQ AL REZAGE, M.O.TOKHI and SITI KHADIJAH ALI
Automatic Control and Systems Engineering, University of Sheffield, Sheffield.UK
gsmousa1@sheffield.ac.uk

Abstract—This paper presents investigations for development of an assistive exoskeleton device for elderly mobility. This exoskeleton is designed to enhance the lower limb and provide support torque in order to augment the torque of knee and hip during the walking cycle. PID Control is designed and implemented in this work. Due to the complexity in identifying the lower limb musculoskeletal system with traditional mathematical approaches, the visual Nastran 4D software is used for development of simulation model of the exoskeleton and a humanoid. Simulation results demonstrating the performance of the adopted approach are presented and discussed.

Keywords—robotics; exoskeleton; walking.

1. Introduction

Many parts of the world have been witnessing a global aging population. These days, much research is needed to address the problems of aging. The United Nations reports that the number of old people above 60 was 8% in 1950, 10% in 2000 and is estimated to rise to 21% by 2050 [1]. Walking efficiency will decrease because of the loss of muscle strength due to reduced motor neuron, muscle fiber and aerobic capacity [2]. Walking is one of the most common features of human mobility. Normally, the elderly cannot keep their step length and their leg muscle power; most people will experience mobility problems when they become old. For example, elderly people ambulate more slowly than young people, and also in terms of falling, older people suffer higher risk [3]. This situation is becoming more serious, especially in developed countries. Many types of instrument or equipment such as the classic crutch or wheelchair have been invented to help the elderly. These are of limited benefit. A key development in addressing the problem of mobility could be the exoskeleton system. This paper is organized as follow Section 2 contains a historical of the exoskeleton. Section 3 illustrates the humanoid and exoskeleton model designed using solid works and 4D visual Nastran softwares as well the control mechanism used for walking. Section 4 describes the PID controller, validation and results, and Section 5 includes the conclusions and future work.

2. Exoskeleton: Historical

Exoskeletons have emerged as assistive technology within the robotic sector. Significant number of researchers have engaged in development of assistive devices for empowering or rehabilitation purpose. The aim is to augment reduced power of the joint and carry part of the body weight or use as aid and substitute the human limb [4, 5]. The exoskeleton used for rehabilitation could be classified in two types: the immobile robot exoskeleton and mobile exoskeleton. Earlier exoskeletons used for rehabilitation were treadmill based gait devices. A system for body weight support was used in treadmill, predetermined pattern for walking was used, normally variation not allowed within the gait pattern. One of the widely used and studied devices in this field is Lokomat developed by the Swiss company Hocoma, Lokomat is the most widely adopted exoskeleton for rehabilitation around the world. The first version was limited on position controller. But it followed by many developments to increase the user motivation and participation [5, 6]. It was driven by dc motor and ball screw. In addition the device has hip and knee joint for each leg. Force sensor was used to estimate the interaction force between the exoskeleton and lower limb. Another prominent treadmill based exoskeleton is ALEX (Active Leg Exoskeleton) designed at the university of Delaware (USA). ALEX has 4 DOF, two for pelvic and two for hip and knee joints, which are actuated using linear actuators , controlled by force-field controller assisting the walk [7]. The modified version of ALEXII, uses rotary motor in order to control the hip and knee by replacing the linear actuator [8]. This rotary motor can address the range limitation problem, and the joint can achieve a wider range of motion. Hybrid Assistive Limb (HAL-3), as a wearable robot has been designed for different functions [9]. The main goal is to help the disorder and elderly people in their walking. HAL consists of three main parts: skeleton and actuator, controller and sensor. Aluminum alloy and steel have been used in the consideration of light weight. DC actuator used to generate proper torque for both hip and knee. Measuring unit, power supply and the controller are contain in the backpack. Force sensor is used in the sole of the foot to measure reaction force, while EMG is attached to skin surface of leg to estimate the muscle activity [9].

3. Lower Extremity

3.1. *Human body*

The major function of humanoid model is to present human model like the physical reality in terms of length and height [10]. The human model segment parameter was built according to anthropometric data presented by Winter (2009). As shown in figure 1 it illustrates the length of each body segment for whole human, a humanoid model of 75 kg weight and 1.7m length is used in this work. The humanoid design in solid works and export to 4D visual Nastran[11].

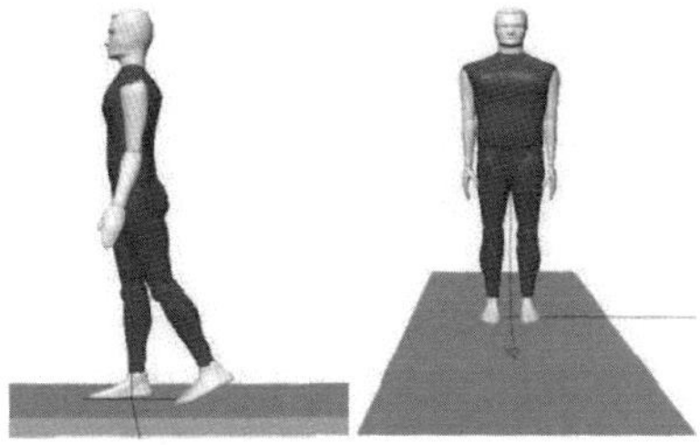

Figure 1. The humanoid model.

3.2. *Exoskeleton*

The exoskeleton can perform many functions. The main purpose of it in this work is assistive mobility for elderly. The second purpose is to provide assistive torque to reduce the torque of the humanoid joint during the walking cycle. The mechanical design should be the first consideration such as light weight and low cost material; a glass fiber reinforced polymer (GFRP) is used because of its character and low cost. The exoskeleton was designed in solid works software. It provides support for back and consists of waist part, upper legs parts, lower legs parts, soles and hip and knee joint as in figure 2. Then the final designed is incorporated into 3-D humanoid model (Figure 3).

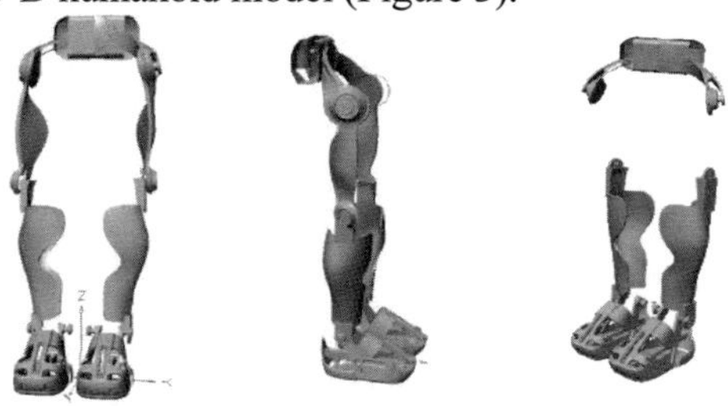

Figure 2. The exoskeleton final design.

Figure 3. The Humanoid and the exoskeleton combination.

3.3. *Actuator selection*

The main function of the actuator is transferring the action generated by the controller to functional movement; usually it is attached to the exoskeleton frame. Various types of actuations are used such as pneumatic, hydraulic, electrical, and electro active Polymers[5]. Many of these types are not suitable to use in exoskeleton because it is required that the actuator provides high torque while working at high speed. This makes the actuator selection difficult because the space is very limited around the joint and the weight should be taken into consideration [12]. Therefore many selection criteria are used for the actuator in exoskeleton such as: velocity, torque-mass ratio, motion range and construability. The comparisons made by Hollerbach for the actuation used in robotic application shows that hydraulic and electrical actuation have the high torque-mass ratio [13].Some works suggest that hydraulic actuation is usually bulky and hard to control while electrical motor shows reduction in power consumption during the gait cycle [14]. Several advantages are offered by brushless DC motor for wearable exoskeleton: high efficiency, high torque, increased reliability, reduced noise, long lifetime and electromagnetic reduction.

3.4. *Controller investigation*

It is evident from the review of previous work, that the main issue of elderly mobility is body weight, so support should be provided by the exoskeleton system to the user all the time. The fundamental goal of the control method is to measure the force required at each lower limb joints during the walking cycle, then the exoskeleton system should be controlled in a way to provide an assistive force to the user. Many exoskeleton system designs do not take into consideration self-balancing control; instead, the user will control the exoskeleton balancing. Moreover, the measured real force generated by human muscle varies with different people. There is little data from previous research for human movement to complete a walking cycle, depending on the Clinical Gait Analysis (CGA) data in this research for both hip and knee joints. According to that, the controller will control both humanoid and the exoskeleton. Two separate controllers are devised, one for the humanoid model and one for the exoskeleton providing 60% and 40% of the overall torque respectively, i.e. 60% of power will drive humanoid, and the other 40% of the power will drive the exoskeleton. The orientation will be controlled for the humanoid lower limb to follow the predefined reference trajectory. The torque of each joint is calculated during the walking cycle. 4 PID controllers are used to control the 2 joints in each leg.

4. PID Controller

Proportional Integrated Derivative control is the most widely used in controlling industrial systems over many years due to its simplicity in implementation and reliability as well as ease to tune [15]. The PID is trying to correct the error calculated between the measured and the desired angle and then adjust the system by calculating the correct action. Three parameters are combined in the PID controller; proportional, integral and derivative. The transfer function of PID is given as:-

$$u(t) = kp\, e(t) + ki \int_0^t e(t)dt + kd\, \frac{de}{dt} \tag{1}$$

Where e (t) is the actual error between actual and reference displacement, kp, ki and kd are the proportional, integral and derivative gains of the controller respectively. Figure 4. Shows the reference trajectory used from Clinical Gait Analysis (CGA) for normal healthy walking data for hip and knee joints.

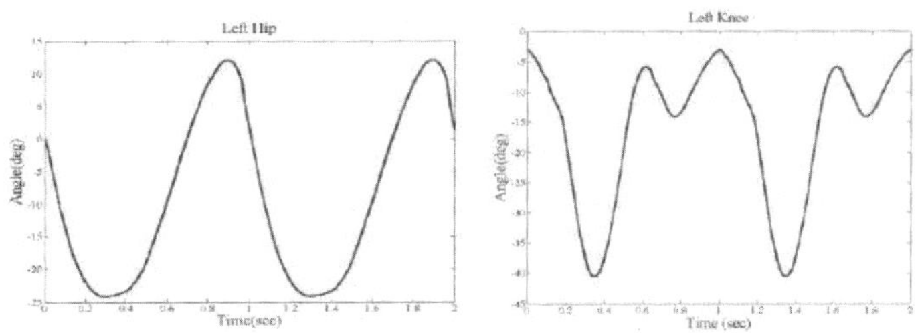

Figure 4. Hip and Knee reference trajectory.

1. Controlling the humanoid

Figure 5 shows a Simulink block diagram of PID control structure utilised. In order to control the knee and hip trajectory the error to the PID is the difference between the hip and knee reference and the actual hip and knee which are measured from the visual Nastran model. The output of the controller is the hip and knee torque, which is fed to hip and knee of the humanoid.

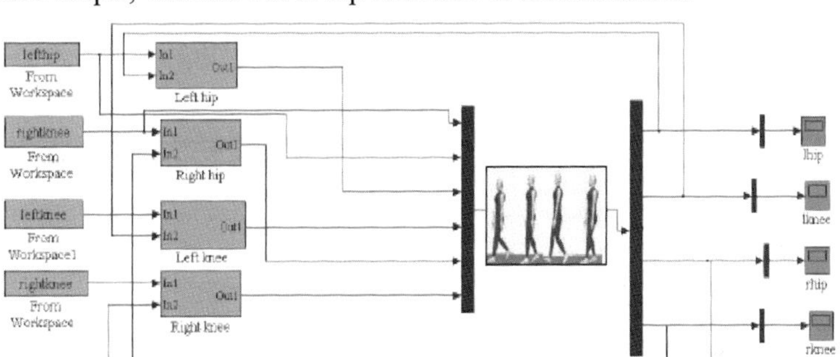

Figure 5. PID controller design using Matlab / Simulink

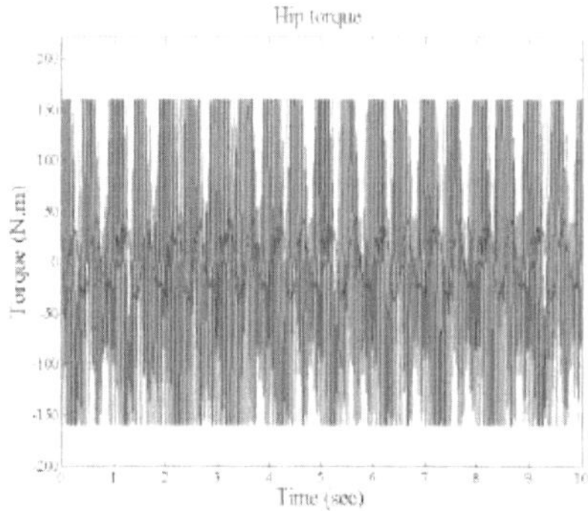

Figure 6. Torque at the Hip joint

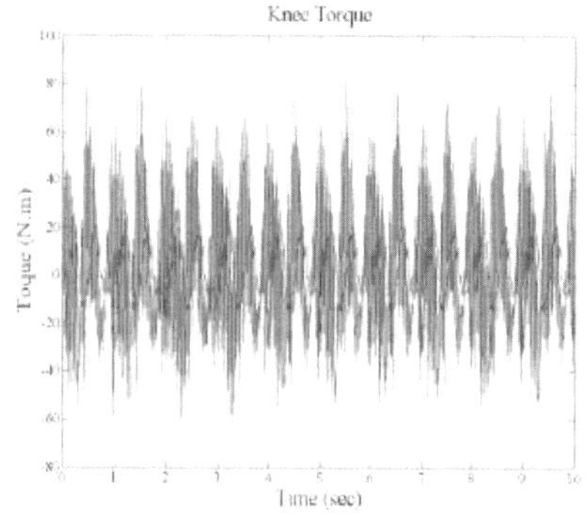

Figure 7. Torque at knee joint.

Figure 6 shows the torque measured at the Hip joint of the humanoid for left and right, which was below 160 N.m through the walking cycle. Figure7 shows the knee torque for left and right, which was below 80 N.m through the walking cycle. Figure8 shows the humanoid with hip and knee joints controlled by PID controllers through the walking cycle.

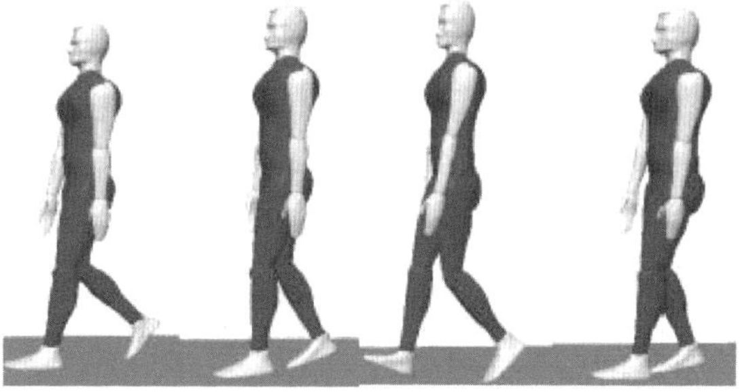

Figure 8 Visual Nastran captured image of humanoid walking cycle.

2. Controlling the humanoid with an exoskeleton

In this section the humanoid and the exoskeleton orientation are controlled to follow predefined reference trajectory for the lower extremity. The torque is calculated during the walking cycle, PID controller is used for both the two joints of the humanoid and the exoskeleton for each leg. Figure 9 shows a Simulink block diagram of the control structure. Figures 10 and 11 show the torque profile of the hip and knee joints respectively for the humanoid and exoskeleton. The corresponding Visual Nastran captured image of the humanoid with exoskeleton during walking cycle is shown in Figure 12.

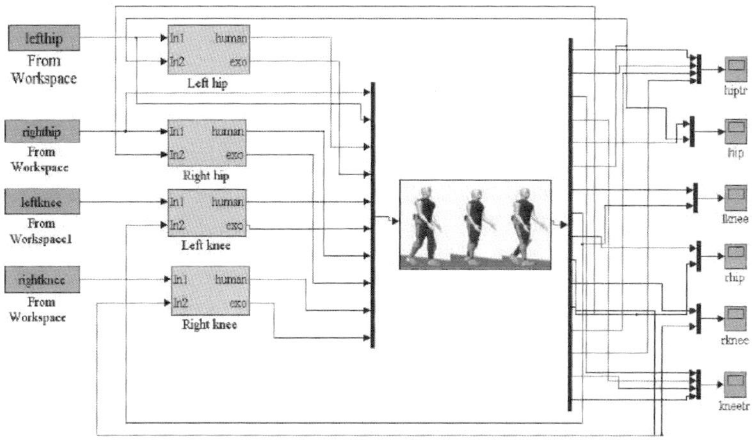

Figure 9. The control structure for walking cycle.

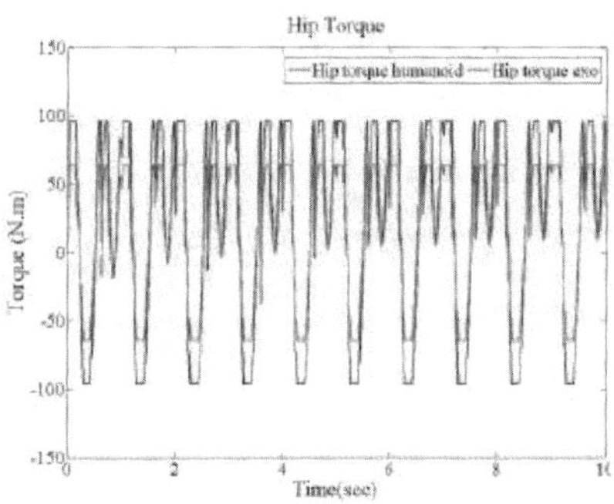

Figure 10. Torque at the hip joint

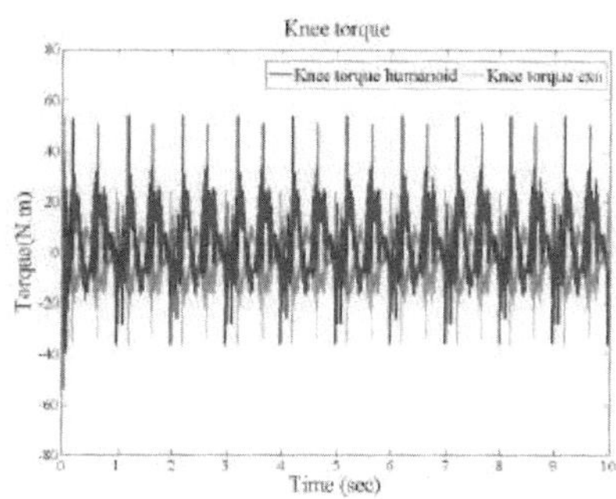

Figure 11. Torque at the knee joint.

From figures above show the humanoid torque required to complete a walking cycle has been augmented, which is meant the exoskeleton has provided a support torque approximately 25 N.m for knee joint and 60 N.m for hip joint.

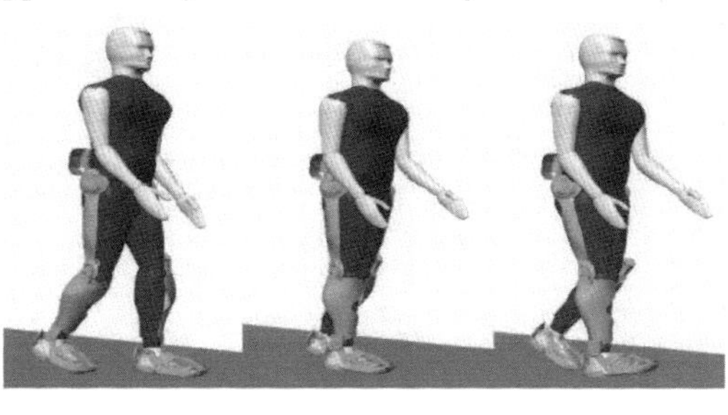

Figure 12. Visual Nastran captured image of humanoid with exoskeleton during walking cycle

5. Conclusion

Lower extremity exoskeleton has been developed to support the elderly mobility. A support torque 40% of the total torque required for the walking cycle is provided by the exoskeleton. This has been achieved for PID controller for the knee and hip joints. Future work will investigate sophisticated control algorithms such as fuzzy logic, fuzzy/PID with optimization algorithm for enhanced control performance.

References

[1] D. o. Economic and S. A. U. Nations, *World population ageing: 1950-2050*: UN, 2002.

[2] F. Prince, H. Corriveau, R. Hébert, and D. A. Winter, "Gait in the elderly," *Gait Posture,* vol. 5, pp. 128-135, 1997.

[3] Y. Barak, R. C. Wagenaar, and K. G. Holt, "Gait characteristics of elderly people with a history of falls: a dynamic approach," *Phys Ther,* vol. 86, pp. 1501-10, Nov 2006.

[4] J. L. Pons, *Wearable robots: biomechatronic exoskeletons* vol. 338: Wiley Online Library, 2008.

[5] Y. W. Hong, Y.-J. King, W.-H. Yeo, C.-H. Ting, Y.-D. Chuah, J.-V. Lee, and E.-T. Chok, "Lower Extremity Exoskeleton: Review and Challenges Surrounding the Technology and its Role in Rehabilitation of Lower Limbs," *Australian Journal of Basic and Applied Sciences,* vol. 7, pp. 520-524, 2013.

[6] G. Colombo, M. Joerg, R. Schreier, and V. Dietz, "Treadmill training of paraplegic patients using a robotic orthosis," *J Rehabil Res Dev,* vol. 37, pp. 693-700, 2000.

[7] S. K. Banala, S. K. Agrawal, and J. P. Scholz, "Active Leg Exoskeleton (ALEX) for gait rehabilitation of motor-impaired patients," in *Rehabilitation Robotics, 2007. ICORR 2007. IEEE 10th International Conference on*, 2007, pp. 401-407.

[8] K. N. Winfree, P. Stegall, and S. K. Agrawal, "Design of a minimally constraining, passively supported gait training exoskeleton: ALEX II," in *Rehabilitation Robotics (ICORR), 2011 IEEE International Conference on*, 2011, pp. 1-6.

[9] H. Kawamoto and Y. Sankai, "Power assist system HAL-3 for gait disorder person," in *Computers helping people with special needs*, ed: Springer, 2002, pp. 196-203.

[10] D. A. Winter, *Biomechanics and motor control of human movement*: John Wiley & Sons, 2009.

[11] S. Wang, "Software review: Motion simulation with working model 2D and MSC visual Nastran 4D," *Working Model 2D—Reviews and Awards,* 2007.

[12] U. Onen, F. M. Botsali, M. Kalyoncu, M. Tinkir, N. Yilmaz, and Y. Sahin, "Design and actuator selection of a lower extremity exoskeleton," *Mechatronics, IEEE/ASME Transactions on,* vol. 19, pp. 623-632, 2014.

[13] I. W. Hunter, J. M. Hollerbach, and J. Ballantyne, "A comparative analysis of actuator technologies for robotics," *Robotics Review,* vol. 2, 1991.

[14] A. Zoss and H. Kazerooni, "Design of an electrically actuated lower extremity exoskeleton," *Advanced Robotics,* vol. 20, pp. 967-988, 2006.

[15] J. Niu, Q. Song, and X. Wang, "Fuzzy PID control for passive lower extremity exoskeleton in swing phase," in *Electronics Information and Emergency Communication (ICEIEC), 2013 IEEE 4th International Conference on*, 2013, pp. 185-188.

BALANCE ANALYSIS AND OPTIMAL POSTURE ESTIMATION DURING ASSISTED WALKING

LEI LI

School of Mechanical and Aerospace Engineering, Nanyang Technological University, Singapore Email: lli8@e.ntu.edu.sg

K. H. HOON

School of Mechanical and Aerospace Engineering, Nanyang Technological University, Singapore Email: mkhhoon@ntu.edu.sg

K. H. LOW

School of Mechanical and Aerospace Engineering, Nanyang Technological University, Singapore Email: mkhlow@ntu.edu.sg

Crutch and walking frames are widely used together with exoskeleton assistive devices to provide body weight support and balance to the user, However, it is unclear what is an optimal posture for the user to choose when using these devices, In this paper, it is found out that human posture, the force from the upper body and the support points are the three key effectors that affect the stability during assisted walking. And it is proposed to evaluate the effect of human posture through a global index which is widely used in robotics field: the manipulability. A three degree of freedom arm model is used and an optimization framework is established to find out the optimal posture during assistive walking.

1. Introduction

The recent development of robotic orthosis or exoskeleton aims at the possibility of home-based rehabilitation together with assistive ability. Some representative devices include *eLegs* [1], *ReWalk* [2], *HAL* [3], *Mina* [4], and *Vanderbilt exoskeleton* [5]. These deviccs require the user to balance themselves through the using of crutches.

The problem identified is the under-actuation of most of the exoskeleton devices (usually with only hip and knee flexion motion are active controlled). Thus such devices require the user to use the upper body to provide propulsion force and increase the stability of the system. As a result, there is a need to use external force to stabilize the gait to achieve a stable assisted walking by using crutch or walking frames.

In this paper, different balance criteria are explored in assisted walking including the concept of Zero Moment Point [6], Generalized Zero Moment Point [7]. For a given posture, the force manipulability [8] can quantify the ability of change of force in all the Cartesian directions. The manipulability is represented by an ellipsoid through which unit torque can be converted into the largest end point force along the major axis of the ellipsoid. This concept is further extended to quantify the ability of change of acceleration of Center of Mass (COM) based on Linear Inverted Pendulum Model (LIPM). At last, an optimization framework is established to provide the optimal posture for assisted walking. Crutch is used in this study for its popularity in daily assisted walking.

2. Stability Analysis based on Zero Moment Point

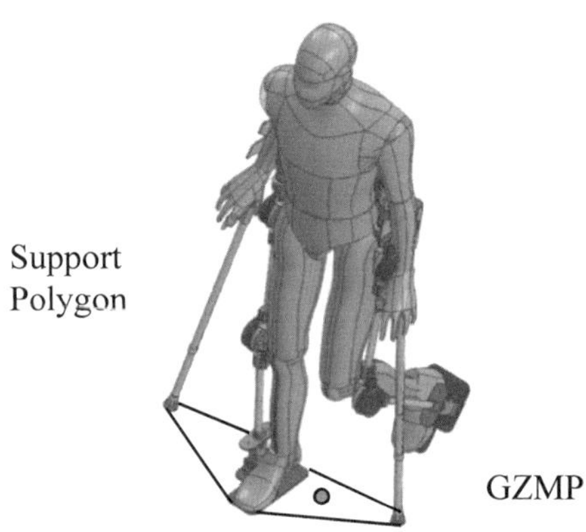

Figure 1 ZMP points and support polygon in assisted walking

Definition 1 Generalized Zero Moment Point (GZMP)

Zero-moment point (ZMP) is defined as the point where the sum of moments of the GRF is zero in the horizontal direction. It has been widely used in defining stability of walking robots where the robot motion is stable as long as the ZMP point is within the support polygon. The Generalized zero-moment point (GZMP) shown in Figure 1 is used for the stability analysis of bipedal walking robots when robots' arm interacts with the environment. Then the support polygon will include the contact points between the hands and the environment as shown in Figure 1, not just the foot contact area. As a result, the balance can be kept even when the foot has only edge contact or point contact with the floor. The generalizer zero moment point can be calculated as:

$$x_{zmp} = \frac{\sum_{i=1}^{n} m_i \{x_i(\ddot{z}_i + g) - \ddot{x}_i z_i\} - I_{iy} w_{iy}}{\sum_{i=1}^{n} m_i(\ddot{z}_i + g)} \tag{1}$$

$$y_{zmp} = \frac{\sum_{i=1}^{n} m_i \{y_i(\ddot{z}_i + g) - \ddot{y}_i z_i\} - I_{ix} w_{ix}}{\sum_{i=1}^{n} m_i(\ddot{z}_i + g)} \tag{2}$$

where n represents the number of links of the robot;

m_i represents the mass of link i;

I_{ix}, I_{iy}, I_{iz} are the components of the inertia tensor of link i

x_i, y_i, z_i represent link i's centre-of-mass position *f*

ω_{ix} ω_{iy} ω_{iz} represents link i's angular velocity

If the GZMP is on the edge of an area defined in Figure 1, the robot will fall down by the moment around the edge of the convex hull. However, the GZMP does not show explicitly how the crutch force can influence the balance of the system. Hence, the following analysis will use the ZMP with external analysis.

Definition 2 ZMP with external force

The position of ZMP with external can be calculated using the equation below:

$$x_{zmp} = \frac{\sum_{i=1}^{n} m_i \{x_i(\ddot{z}_i + g) - \ddot{x}_i z_i\} - I_{iy} w_{iy} + \sum_{i=1}^{n} (s^i \times F^i)}{\sum_{i=1}^{n} m_i(\ddot{z}_i + g) - \sum_{i=1}^{n} F_z^i} \tag{3}$$

$$y_{zmp} = \frac{\sum_{i=1}^{n} m_i \{y_i(\ddot{z}_i + g) - \ddot{y}_i z_i\} - I_{ix} w_{ix} + \sum_{i=1}^{n} (s^i \times F^i)}{\sum_{i=1}^{n} m_i(\ddot{z}_i + g) - \sum_{i=1}^{n} F_z^i} \tag{4}$$

where $F^i[F^i_x, F^i_y, F^i_z]$ represents i's contact force

$S^i[S^i_x, S^i_y, S^i_z]$ represents i's contact point

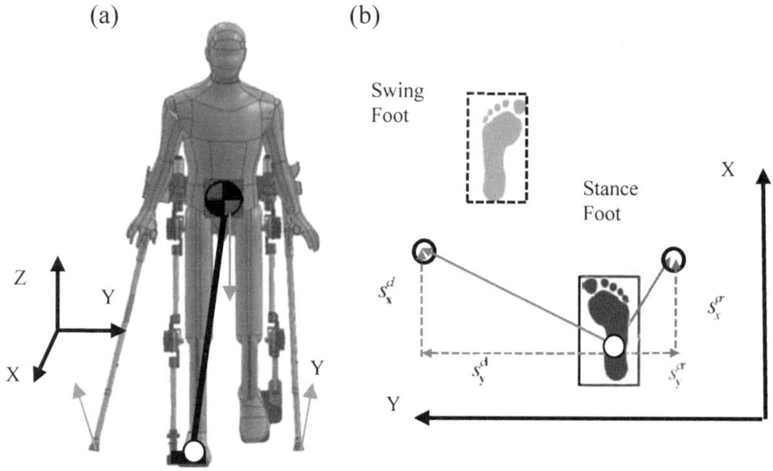

Figure 2 Lumped mass model of assisted walking

In order to get more intuition of how the external force will influence on the ZMP trajectory, the lumped mass model is used as shown in Figure 2 where the whole body mass concentrates at one point. If we make further assumption that the variation of vertical acceleration and the net moment about the COM is small, and express the crutch vertical force in terms of body weight, then the equation can be simplified further as:

$$x_{zmp} = x_{com} + \frac{\ddot{x} \bullet z_0}{g} + \frac{\tau_x}{mg} \tag{5}$$

$$y_{zmp} = y_{com} + \frac{\ddot{x} \bullet z_0}{g} + \frac{\tau_y}{mg} \tag{6}$$

where τ_x and τ_y are the moment about the ankle created by the crutch force and can be calculated by:

$$\tau_x = F_z^r \bullet s_x^{cr} + F_z^l \bullet s_x^{cl} \tag{7}$$

$$\tau_y = F_z^r \bullet s_y^{cr} + F_z^l \bullet s_y^{cl} \tag{8}$$

where s_x^{cr} , s_x^{cr} , s_x^{cr} , s_x^{cr} are the x and y distance for both left and right crutch support point with respect to the stance ankle position.

It can be seen from the equation that, the ZMP is influenced by the magnitude of the force F_z in the vertical direction and also the position of the contact points with respect to the ankle joint $S_{x/y}^{cl/cr}$.

3. Force Manipulability Ellipsoid

When human arm is performing a task, either the transmission of end effect velocity from joint velocity or force from joint torque is preferred. For example, high transmission of end-effect velocity from joint velocity is preferred when the arm is moving to reach certain position. On the other hand, high transmission of end-effect force from joint torque is preferred when the arm is holding heavy load.

In the case of crutch assisted walking, the transmission of end-effect force is preferred since the force dominate is the one which the stability of walking while the speed of movement is relatively small. The index which is widely used in robotics field to quantify these is: the force manipulability. For a given posture, this index quantifies the ability of change of force in all the Cartesian directions. The manipulability is represented by an ellipsoid through which unit torque can be converted into the largest end point force along the major axis of the ellipsoid.

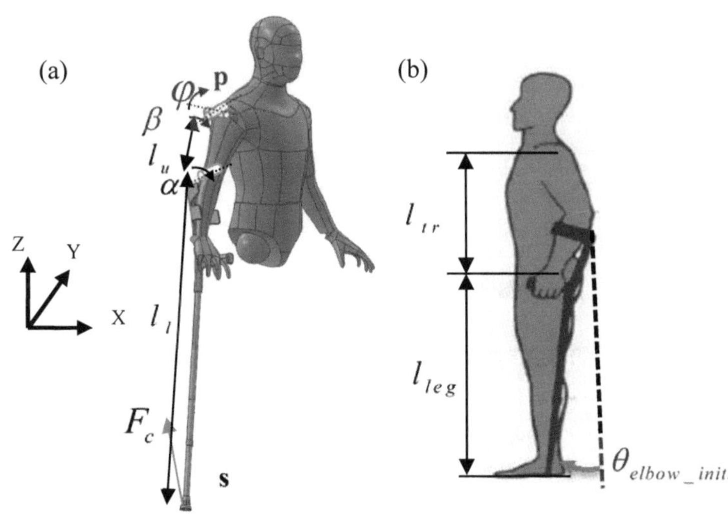

Figure 3 (a) Upper body model of assistive walking (b) Posture to define crutch length

In order to investigate how individual's arm posture affects the vertical force F_z during crutch-assisted walking, the concept of force manipulability is used. The human arm is model as a 3 DOF serial link as shown in Figure 3. The dynamics of the arm can be expressed as:

$$\boldsymbol{\tau} = \mathbf{M(q)\ddot{q}} + \mathbf{h(q,\dot{q})} + \mathbf{g(q)} + \mathbf{J^T(q)F_c} \tag{9}$$

The dynamics of the generalized coordinates q are given by the mass matrix M, the force vector h (which includes the gravitational, Coriolis, and centrifugal forces), the gravitational force g and the contact force F_c. If we neglect the dynamic effect on the arm, the equation can be simplified as:

$$\boldsymbol{\tau} = \mathbf{J^T(q)F_c} \tag{10}$$

The set of joint torques are represented as:

$$\|\boldsymbol{\tau}\|^2 = \tau_1^2 + \tau_2^2 \cdots \tau_n^2 \le 1 \tag{11}$$

By combining equations above we get:

$$\mathbf{F}_c^T \mathbf{J(q)J^T(q)F}_c \le 1 \tag{12}$$

This equation describes a three dimensional ellipsoid and the longer the radius of the ellipsoid the larger force is applied along that direction. However, the above equation does not consider the differences of maximum joint torque inhuman arm. In order to include the difference of joint toque due to different subject and different joints, one simple way to do this is to measure the human joint torque and normalize it by using equation:

$$\tilde{\mathbf{J}} = \mathbf{JT}_\tau (\mathbf{T}_\tau = diag[1/(\tau_{i\max})]) \tag{13}$$

where τ_{imax} is the minimum torque at the ith joint.

Then the manipulating force ellipsoid is re-expressed as:

$$\mathbf{F}_c \hat{\mathbf{J}}\mathbf{(q)}\hat{\mathbf{J}}^T\mathbf{(q)F}_c \le 1 \tag{14}$$

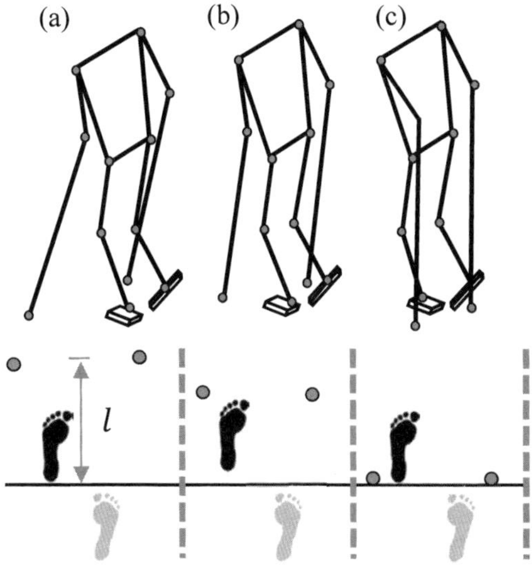

Figure 4 Different arm configurations due to different support point

Besides the crutch force, the shape and orientation of force ellipsoid depends on the configuration of the arm which is determined by the shoulder position and the crutch support position. Thus the parameters which affect the position of the shoulder position and the crutch support position will affect the ellipsoid accordingly. Different arm configuration due to different support point is shown in Figure 4.

Define a frame with its origin located at the should joint position $P_s(p_{sx}, p_{sy}, p_{sz})$ and orientation same as the global frame shown in Figure 3(a). The kinematic relationship between the shoulder joint position $p(p_x, p_y, p_z)$ and the support point $s(s_x, s_y, s_z)$ shown in Figure 3(a) can be found by equations as shown below:

$$p_x - s_x = \sin\varphi(-l_l \sin\alpha \sin\beta + l_l \cos\alpha \cos\beta + l_u \cos\beta) \tag{15}$$

$$p_y - s_y = l_l(-\cos\beta \sin\alpha - \sin\beta \cos\alpha) - l_u \sin\beta \tag{16}$$

$$p_z - s_z = \cos\varphi(-l_l \sin\alpha \sin\beta + l_l \cos\alpha \cos\beta + l_u \cos\beta) \tag{17}$$

where α, β, φ are the arm angle as shown in Figure 3(a).

Given the upper arm length l_u and the fore arm together with crutch length l_l, the angles of the arm joints can be calculated as:

$$\alpha = a\cos\frac{(p_x - s_x)^2 + (p_y - s_y)^2 + (p_z - s_z)^2 - l_u^2 - l_l^2}{2l_u l_l} \tag{18}$$

$$\beta = a\cos\frac{\sqrt{((p_y - s_y)^2 + (p_z - s_z)^2)}(l_u + l_l\cos\alpha) + (p_x - s_x)l_l\sin\alpha}{(l_u)^2 + (l_l)^2 + 2l_l l_u\cos\alpha} \tag{19}$$

$$\phi = a\cos\frac{\sqrt{(p_z - s_z)^2}}{\sqrt{((p_x - s_x)^2 + (p_y - s_y)^2)}} \tag{20}$$

The fore arm together with crutch length l_l is calculated based on the initial elbow bending angle shown in Figure 3(b). The angle is determined as: the user stand erect and put the end of the crutch at the point coinciding with ground projection point of ankle joint. The values of the arm length and leg length are shown in
Table 1.

Table 1 Mean (±SD) data of subject

Trunk Length(From hip to shoulder) l_u	515.5
Leg Length (From Heel to Hip) l_{leg}	965.5
Upper Arm length l_u	281.7
Forearm Length l_f	268.9

$$l_l = l_u\cos(\theta_{elbow\_initial}) + \sqrt{-l_u^2\sin^2(\theta_{elbow\_initial}) + (l_{leg} + l_{tr})^2} \tag{21}$$

Then the relationship between the end-effector velocity and the joint velocity can be defined as follows:

$$\dot{P}_s = J_a\dot{\theta}$$

$$= \begin{bmatrix} 0 & -l_l\cos(\alpha+\beta) - l_u\cos(\beta) & -l_l\cos(\alpha+\beta) \\ \cos(\phi)(l_l\cos(\alpha+\beta)+l_u\cos(\beta)) & -\sin(\phi)(l_l\sin(\alpha+\beta)+l_u\sin(\beta)) & -l_l\sin(\alpha+\beta)\sin(\phi) \\ \sin(\phi)(l_l\cos(\alpha+\beta)+l_u\cos(\beta)) & \cos(\phi)(l_l\sin(\alpha+\beta)+l_u\sin(\beta)) & l_l\sin(\alpha+\beta)\sin(\phi) \end{bmatrix}\begin{bmatrix} \dot{\phi} \\ \dot{\beta} \\ \dot{\alpha} \end{bmatrix} \tag{22}$$

where J_a is the Jacobian of the human arm.

The Jacobian matrix J_a can be represented in the form UDV^T by using singular value decomposition, where $U = [u_1\ u_2\ u_3]$, $V = [v_1\ v_2\ v_3]$ and $D = \mathrm{diag}\ [\sigma_1\ \sigma_2\ \sigma_3]$. The vector U indicates one of the directions of the three principle axes of the manipulability ellipsoid and values of σ_i in D indicates the magnitude of these axes. The figure below shows the dynamic manipulability ellipsoid and different dynamic ellipsoid due to different support posture.

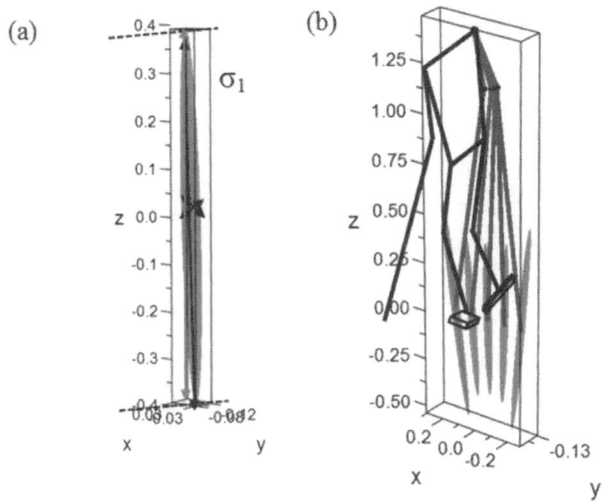

Figure 5 (a) Dynamic Force Manipulability Ellipsoid (b) Different Dynamic Force Manipulability Ellipsoid due to different arm postureOptimal Posture in the sense of CoM manipulability Ellipsoid

It has been shown that, we are able to find the manipulability ellipsoid in a given posture. However, it would be more useful to provide the optimal posture which can provide maximum manipulability. At the same time, as indicated in previous section, the force is not the only factor which affects the stability of walking. In addition to the optimization problems related to maximizing the force manipulability, the COM manipulability will be used as the measure of the optimal posture.

Note that the analytical expression of u_i and σ_i can be found based on the fact that they are the eigenvectors and square root of the non-zero eigenvalues of $J_a \cdot J_a^T$. Since we are only interested on the magnitude of major axis which contributes most in the z direction which is σ_1, the analytical solution of the magnitude of σ_1 and their direction $v_1$can be expressed as:

$$\sigma_1 = \frac{1}{\sqrt{\frac{l_u^{\ 2}}{2} - \xi_1 + l_f^2 + \xi_2}} \tag{23}$$

$$v_1 = \begin{bmatrix} v_{1x} \\ v_{1y} \\ v_{1z} \end{bmatrix} = \begin{bmatrix} -\dfrac{l_l^2\cos(2\beta) + 2l_u^2\cos(2\beta+2\alpha) - \sqrt{\xi_1} + 2l_u l_f\cos(2\beta+\alpha)}{\cos(\phi)(\sin(2\beta)l_u^2 + 2\sin(2\beta+\alpha)l_u l_f + 2\sin(2\beta+2\alpha)l_l^2)} \\ -\dfrac{\sin(\phi)}{\cos(\phi)} \\ 1 \end{bmatrix} \tag{24}$$

where

$$\xi_1 = \frac{\sqrt{l_u^4 + 4l_u^3 l_f\cos(\beta) + 8l_u^2 l_f^2\cos^2(\beta) + 8l_u l_f^3\cos(\beta) + 4l_f^2}}{2}$$

$$\xi_2 = l_u l_f\cos(\beta)$$

Then the optimal posture estimation problem can be described as a general constrained optimization as follows. Find an optimal posture which minimizes the function:

$$CoM^x_{manipulability} = \frac{\sigma_1\|v_{1z}\|}{\sqrt{v_{1x} + v_{1y} + v_{1z}}}\|p_{sy} - y_{ankle}\| \tag{25}$$

$$CoM^y_{manipulability} = \frac{\sigma_1\|v_{1z}\|}{\sqrt{v_{1x} + v_{1y} + v_{1z}}}\|p_{sx} - x_{ankle}\| \tag{26}$$

subject to:

$$g_i(\theta) = 0, i = 1, 2, ..., m$$
$$g_j(\theta) \le 0, j = 1, 2, ..., n$$
$$\theta_l \le \theta \le_u$$

where θ is an joint angle vector of the arm and g_i and g_j are the equalities and inequalities constraints. And the vectors θ_u and θ_l are the upper and lower limits of joint angles respectively.

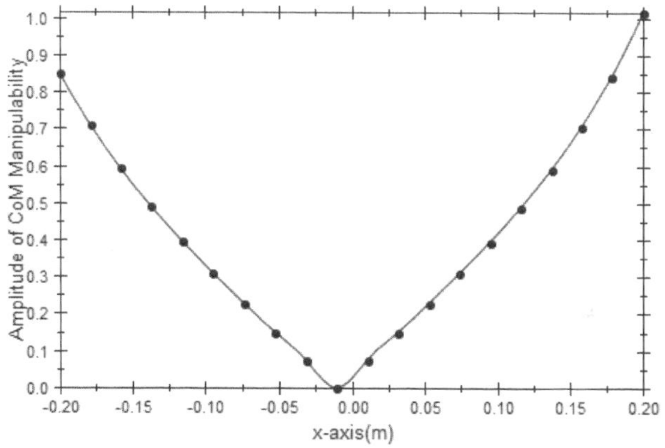

Figure 6 Amplitude of CoM Manipulability in the walking direction

As shown in the figure above, the amplitude of CoM manipulability in the walking direction is increasing as the support point is further apart from the ankle joint. Based on the analysis above, it is suggested that we should extended the crutch as much as possible to have a better control of the motion of COM.

4. Conclusions

From this study, it is concluded that: (1) Crutch force is one of the key effect which affect the stability of walking and the ellipsoid concept reflects well the patient's ability to generate force in crutch assisted walking; (2) If maximal the ability of change of acceleration of COM is an important parameter, then it can be served as the optimized parameter to establish the framework of estimating the optimal posture for assisted walking

Acknowledgments

The authors would like to thank the Tan Tock Seng Hospital (TTSH) and the Robotics Research Center (RRC) of the Nanyang Technological University (NTU) for the support provided to this work. The financial support provided by Lee Foundation and National Healthcare group is acknowledged.

References

1. K. A. Strausser and H. Kazerooni, "The Development and Testing of a Human Machine Interface for a Mobile Medical Exoskeleton," in 2011 IEEE/RSJ International Conference on Intelligent Robots and Systems (IROS 2011), 25-30 Sept. 2011, Piscataway, NJ, USA, 2011, pp. 4911-16.

2. M. Talaty, A. Esquenazi, and J. E. Briceno, "Differentiating ability in users of the ReWalkTM powered exoskeleton: An analysis of walking kinematics," in *2013 IEEE 13th International Conference on Rehabilitation Robotics, ICORR 2013, June 24, 2013 - June 26, 2013*, Seattle, WA, United states, 2013, p. NSF Engineering Research Center for; Sensorimotor Neural Engineering (CSNE); at the University of Washington; IEEE EMB; IEEE Robotics and Automation Society.
3. H. Kawamoto and Y. Sankai, "Power assist method based on Phase Sequence and muscle force condition for HAL," Advanced Robotics, vol. 19, pp. 717-734, 2005.
4. P. D. Neuhaus, J. H. Noorden, T. J. Craig, T. Torres, J. Kirschbaum, and J. E. Pratt, "Design and evaluation of Mina: A robotic orthosis for paraplegics," in Rehab Week Zurich 2011 - 2011 IEEE International Conference on Rehabilitation Robotics, ICORR 2011, June 27, 2011 - July 1, 2011, Zurich, Switzerland, 2011.
5. R. J. Farris, H. A. Quintero, and M. Goldfarb, "Performance evaluation of a lower limb exoskeleton for stair ascent and descent with Paraplegia," in 2012 34th Annual International Conference of the IEEE Engineering in Medicine and Biology Society (EMBC), 28 Aug.-1 Sept. 2012, Piscataway, NJ, USA, 2012, pp. 1908-11.
6. VUKOBRATOVIĆ, M. and B. BOROVAC, ZERO-MOMENT POINT — THIRTY FIVE YEARS OF ITS LIFE. International Journal of Humanoid Robotics, 2004. 01(01): p. 157-173.
7. Harada, K., et al. ZMP analysis for arm/leg coordination. in Intelligent Robots and Systems, 2003. (IROS 2003). Proceedings. 2003 IEEE/RSJ International Conference on. 2003.
8. Chiacchio, P., Y. Bouffard-Vercelli, and F. Pierrot, Force polytope and force ellipsoid for redundant manipulators. Journal of Robotic Systems, 1997. 14(8): p. 613-620.

PID AND ADAPTIVE SPIRAL DYNAMIC ALGORITHM IN CONTROLLING HUMAN ARM MOVEMENTS*

SITI KHADIJAH ALI[1], M. O. TOKHI[1], ASNOR JURAIZA ISHAK[2] and GHASAQ AL REZAGE[1]

[1]*Automatic Control & System Engineering, University of Sheffield, Sheffield, UK*
skali1@sheffield.ac.uk
[2]*Department of Electrical and Electronic Engineering, Universiti Putra Malaysia*

This work presented in this paper focuses on development of assistive robotic control approach for the upper extremities. A set-point tracking position control structure with proportional, integral and derivative (PID) controller is considered. A spiral dynamic optimization algorithm is utilized for tuning of the PID gains. The control strategy is tested and evaluated within simulation model of upper extremities. The results show that good position tracking performance is achieved with the developed control approach.

Keyword: Exoskeleton

1. Introduction

1.1. *Upper-extremities structure*

Upper extremities comprise three categories: bones, joints and muscles each of which play their own role. Bones of the upper extremity are arm, forearm, hand, scapula and clavicle of the pectoral girdle. The bones of the human are as shown in Figure 1. Joints are needed in order for human to move. Generally, for human arm, there are three types of joint: glenohumeral, elbow and wrist joint. In order for joints to move, they need power to generate the movement.

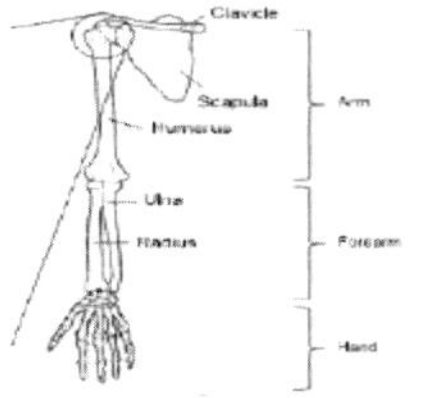

Figure 1: Bones of the upper extremity [1]

* This work is supported by Ministry of Education Malaysia and Universiti Putra Malaysia.

1.2. *Exoskeleton*

Most of the existing exoskeletons are developed for helping stroke patients (rehabilitation purposes) such as ARMin III and NTUH-II (Figure 2 and Figure 3). The target users of ARMin III are chronic stroke patients and NTUH-II's aim is frozen shoulder patients [2][3].

Both can be categorized as active exoskeletons. The actuators used in ARMin III are Maxon Re 35, brushed DC motors whereas NTUH-II use electrical actuators. In terms of sensors, NTUH-II incorporates 2 ATI 6-axis force/torque sensors to get information on users' intention of movement. In NTUH-II, two parameters are proposed: stiffness and control authority. These two parameters are used to control the dynamic torque gain. A combination of admittance and impedance control is implemented in ARMin-III and the control strategy is applied to support the subject.

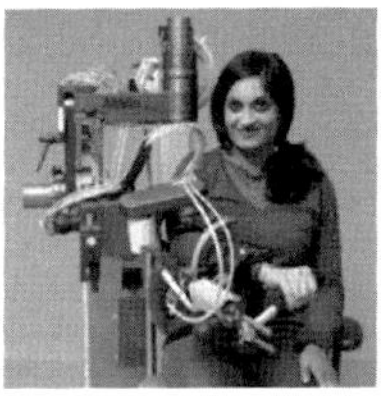

Figure 2: ARMin III with healthy people [2]

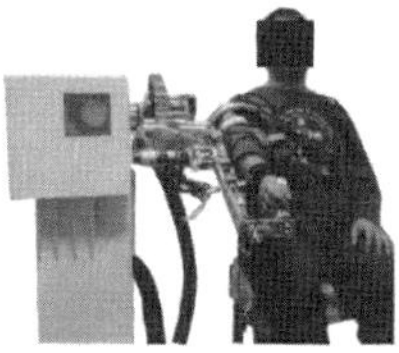

Figure 3: NTUH-II in training [3]

1.3. *PID controller*

PID control is widely used in the industry, as it is simple to operate, is realiable and easy to tune. In this work, PID is used to control the error in positioning the arm so that, the arm will follow and move to desired positions. A PID controller is represented with the differential equation:

$$u(t) = kp\, e(t) + Ki \int_0^t e(t)dt + kd\, \frac{de}{dt} \tag{1}$$

where $e(t)$ and $u(t)$ represent the input and output of the controller respectively, and kp, ki and kd are the proportional, integral and derivative gains of the controller.

1.4. *Spiral dynamic algorithm*

Spiral dynamic algorithm is known as a relatively fast computing algorithm due to its simple structure and stable convergence response when approaching the optimum point. A recently proposed adaptive spiral dynamic algorithm (ADA) [4], which has been shown to have improved convergence speed, accuracy and total computing time is utilised in this work for optimisation of the PID controller gains.

The objective of this paper is to investigate development of a control approach with ASDA optimization algorithm for application to exoskeleton support in upper extremity maneuvers.

2. Methodology

2.1. *Kinematic*

Kinematics is the study of motion without considering the forces/torques or moments that responsible for the motion. In this work, the Denavit-Hartenberg (DH) convention is used to compose coordinate transformation into one homogenous transformation matrix.

Figure 4(a) shows an initial position of the humanoid. Solidworks is used to draw the humanoid. Figure 4(b) shows the mechanical structure of two-link arm. From the mechanical structure, it can be seen that there are three joints and for this experiment, the movements of these joints are confined to the means each joint has 1 degree of freedom (DOF).

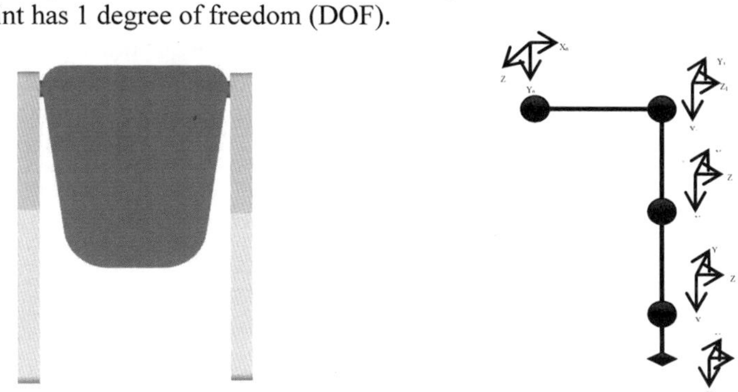

Figure 4 (a) Humanoid (Initial position) (b) The mechanical structure arm

Figure 5(a) shows the position of the arm when the shoulder is moved from initial position to 90°. At this position, the shoulder has extended the arm or can be referred to vertical flexion. When the shoulder joint is fixed and the elbow is bent at 90°, the elbow joint flexes the lower arm or vertically flexed.

Figure 5 (a): Shoulder joint (vertically flexion) (b) Elbow joint (flexion)

The humanoid is designed according to [5]. All the parameters are shown in Table 1.

Table 1 Parameters for the humanoid

Segments	Length (cm)	Mass (kg)	Inertia I (kg m^2)
Upper arm	31.32	2.17	$\begin{bmatrix} 5.13e^{-3} & 0 & 0 \\ 0 & 1.77e^{-3} & 0 \\ 0 & 0 & 5.75e^{-3} \end{bmatrix}$
Forearm / Lower arm	28.08	1.30	$\begin{bmatrix} 1.50e^{-3} & 0 & 0 \\ 0 & 1.63e^{-3} & 0 \\ 0 & 0 & 3.13e^{-3} \end{bmatrix}$
Hand	19.42	0.49	$\begin{bmatrix} 0.13e^{-3} & 0 & 0 \\ 0 & 0.82e^{-3} & 0 \\ 0 & 0 & 0.20e^{-3} \end{bmatrix}$

2.2. *Geometric parameters of a link*

As mentioned above, the advantage of using DH convention is its ability to compose or merge the coordinate transformation to homogenous transformation. In the DH convention, there are four important parameters: a_i, α_i, d_i and θ_i.
All these parameters will be associated with the link and joint i. Let H_i be a homogenous transformation:

$$H_i = Rot(z,\theta_i)Trans(z,d_i)Trans(x,a_i)Rot(x,\alpha_i)$$

$$= \begin{bmatrix} c_{\theta_i} & -s_{\theta_i} & 0 & 0 \\ s_{\theta_i} & c_{\theta_i} & 0 & 0 \\ 0 & 0 & 1 & 0 \\ 0 & 0 & 0 & 1 \end{bmatrix} \begin{bmatrix} 1 & 0 & 0 & 0 \\ 0 & 1 & 0 & 0 \\ 0 & 0 & 1 & d_i \\ 0 & 0 & 0 & 1 \end{bmatrix} \begin{bmatrix} 1 & 0 & 0 & a_i \\ 0 & 1 & 0 & 0 \\ 0 & 0 & 1 & 0 \\ 0 & 0 & 0 & 1 \end{bmatrix} \begin{bmatrix} 1 & 0 & 0 & 0 \\ 0 & c_{\alpha_i} & -s_{\alpha_i} & 0 \\ 0 & s_{\alpha_i} & c_{\alpha_i} & 0 \\ 0 & 0 & 0 & 1 \end{bmatrix}$$

$$= \begin{bmatrix} c_{\theta_i} & -s_{\theta_i}c_{\alpha_i} & s_{\theta_i}s_{\alpha_i} & a_i c_{\theta_i} \\ s_{\theta_i} & c_{\theta_i}c_{\alpha_i} & -c_{\theta_i}s_{\alpha_i} & a_i s_{\theta_i} \\ 0 & s_{\alpha_i} & c_{\alpha_i} & d_i \\ 0 & 0 & 0 & 1 \end{bmatrix}$$

where, $c_{\theta_i} = \cos\theta_i$, $c_{\alpha_i} = \cos\theta_i$ and
$s_{\theta_i} = \sin\theta_i$, $s_{\alpha_i} = \sin\alpha_i$

Table 2 shows the DH parameters for Figure 4(b). This table is representing the position of the human arm for 1 DOF of each joint.

Table 2: Denavit-Hartenberg parameters

Link	a_i	α_i	d_i	θ_i
1	0	0	Length of shoulder	0
2	Length of upper arm	0	0	θ_2
3	Length of lower arm	0	0	θ_3
4	Length of hand	0	0	θ_4

The angles θ_i refer to shoulder joint, elbow joint and wrist joint. Table 3 shows values of a_i and d_i.

Table 3: The values of a_i and d_i

Link	a_i (cm)	α_i	d_i(cm)	θ_i
1	0	0	21.27	0
2	31.32	0	0	θ_2
3	28.08	0	0	θ_3
4	19.42	0	0	θ_4

For the initial configuration in Figure 4(a), the values of θ_i are 0°. Table 4 shows the values and Figure 6 shows the 3D trajectory view for the initial position of the human arm.

Table 4: Values of theta' representing the initial position of the human arm

Initial position	Values of thetas'
θ_2	0
θ_3	0
θ_4	0

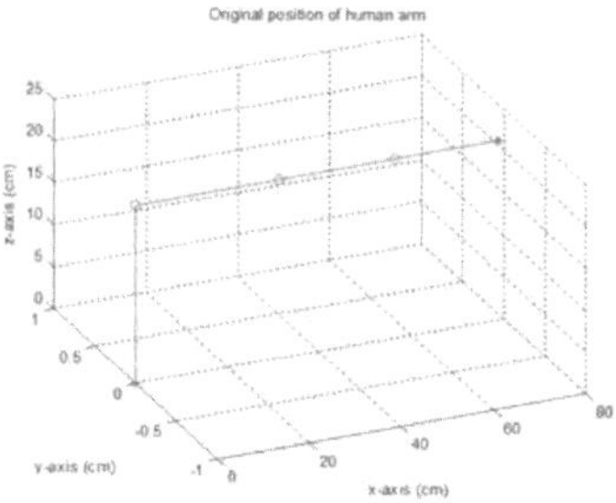

Figure 6: A 3D trajectory view for initial position

Table 5 represents two movements of elbow and shoulder from initial position. The movements are: shoulder vertical flexion and elbow flexion. The value of θ_4 is set to be 0.The combined movements are shown in 3D in Figure 7 and Figure 8 shows the new position in CAD.

Table 5: Position for two movements for shoulder and elbow

Movements	θ_2	θ_3	θ_4
Initial position	0	0	0
End position	50	30	10

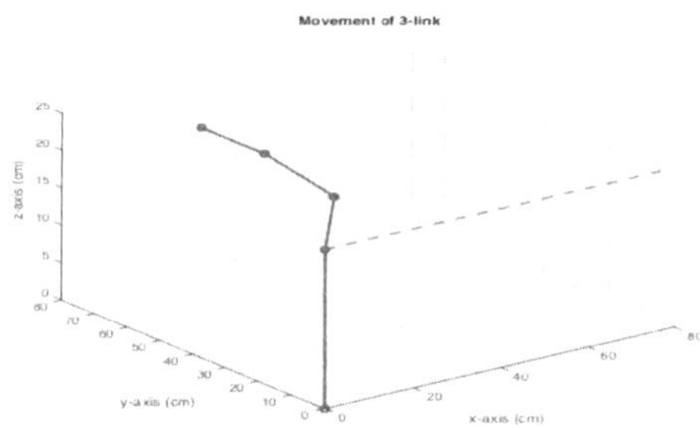

Figure 7: A movement from initial position (dot line) to the new position (solid with bubble line)

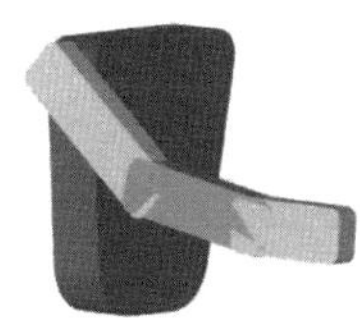

Figure 8: Humanoid's movement

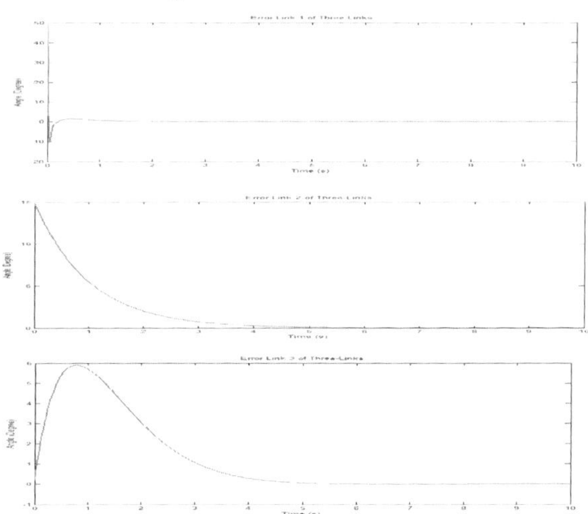

Figure 9: Error for each link.

Figure 9 shows the error for each link of these three links. For link 1, at less than 1 second, the error reach at the steady-state error, for link 2, the system reach at the steady-state error at sec 4.5 and link 3 achieves steady-state error at second 4.3. The error in link 1 is high because this joint needs to carry the weight of the first link, second link and third link. From Figure 9, the errors are reducing for second link and third link since the number of links that need to carry by these joints are reducing.

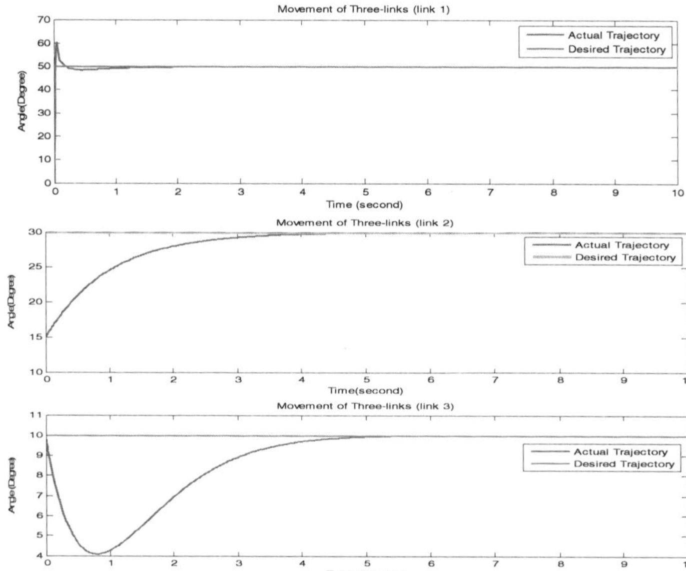

Figure 10: The actual and desired movement of three-links for link-3

Figure 10 shows that the control system are able to construct the links to follow the desired trajectory. With this figure and Figure 9, it is show that the system is stable.

As mentioned, PID controller is used to ensure the arm follows a desired trajectory and [5] is chosen to compare the performance of the PID with [5]. The control system applied in [5] is model-based approach (dynamic). Thus, the desired trajectory (Figure 11) is taken from [5] and the torque obtained is compared.

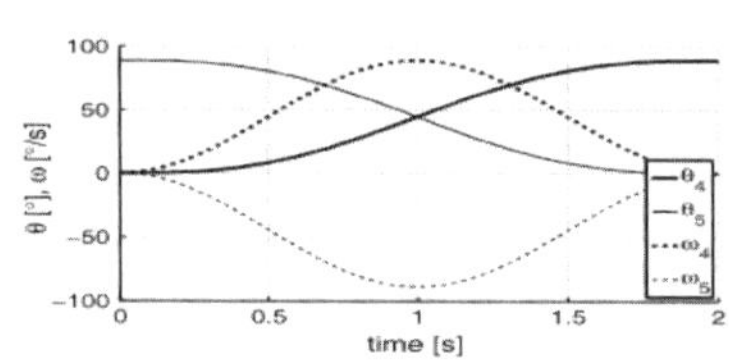

Figure 11: Joint angles and angular velocity [5]

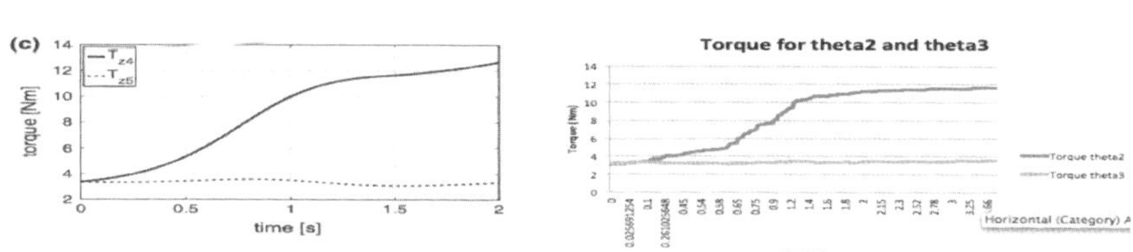

Figure 12: (a) Torque from [5]

(b) Torque from current work

Figure 12 show the torque obtained from [5] and this work. It is shown that the range of the torque in this work and [5] is between 2-4 Nm for elbow flexion. However, the range of the torque for shoulder flexion for the current work is lower compared to [5] which is, between 10-12 Nm. In [5], the maximum torque needed is 12.5 Nm whereas in current work, it is 11.9 Nm.

The maximum torque achieved is equal to the maximum torque of joint z_4 in [5] when the exoskeleton is applied and as stated in [5], this solution can be used in medical training or rehabilitation for stroke people or spine injury. Therefore, in future work, it is aimed to develop an exoskeleton with this algorithm, which will be used for rehabilitation or for supporting upper extremities of in maintenance kind of work.

3. Conclusion

A PID control mechanism with ASDA has been adopted for position control of human arm. The results show that the approach may be used for developing exoskeleton for rehabilitation purposes. Future work will investigate application of the control approach with gravity compensation on upper extremity exoskeleton.

Acknowledgments

Siti Khadijah Ali is funded by Ministry of Malaysia Education and Universiti Putra Malaysia.

References

[1] J. Ambrósio, C. Quental, B. Pilarczyk, J. Folgado, and J. Monteiro, "Multibody biomechanical models of the upper limb," *Procedia IUTAM*, vol. 2, pp. 4–17, 2011.

[2] T. Nef, M. Guidali, and R. Riener, "ARMin III – arm therapy exoskeleton with an ergonomic shoulder actuation," *Appl. Bionics Biomech.*, vol. 6, no. 2, pp. 127–142, 2009.

[3] C.-H. Lin, W.-M. Lien, W.-W. Wang, S.-H. Chen, C.-H. Lo, S.-Y. Lin, L.-C. Fu, and J.-S. Lai, "NTUH - II Robot Arm with Dynamic Torque Gain Adjustment Method for Frozen Shoulder Rehabilitation," *2014 IEEE/RSJ Int. Conf. Intell. Robot. Syst. (IROS 2014)*, no. Iros, pp. 3555–3560, 2014.

[4] a. N. K. Nasir, M. O. Tokhi, N. M. Abd Ghani, and R. M. T. Raja Ismail, "Novel adaptive spiral dynamics algorithms for global optimization," *2012 IEEE 11th Int. Conf. Cybern. Intell. Syst.*, pp. 99–104, Aug. 2012.

[5] S. Głowiński, T. Krzyżyński, S. Pecolt, and I. Maciejewski, "Design of motion trajectory of an arm exoskeleton," *Arch. Appl. Mech.*, Aug. 2014.

MODELLING AND SIMULATION OF ASSISTIVE EXOSKELETON FOR ELDERLY MOBILITY*

D. MIRANDA-LINARES[1], J. LOPEZ-CORONADO[2], C. A. DIAZ-HERNÁNDEZ[2], A. J. ISHAK[1] AND M. O. TOKHI[1,2]

[1] *Department of Automatic Control and Systems Engineering, The University of Sheffield, United Kingdom*
d.miranda-linares@sheffield.ac.uk

[2] *Department of Systems Engineering and Automation, Polytechnical University of Cartagena, Cartagena, 30202, Spain*
jl.coronado@upct.es, carlos.diaz@upct.es

Abstract: This paper presents research on the modelling and simulation of assistive exoskeletons for elderly mobility. The exoskeleton is designed to augment physical capabilities of the elderly, so that the combined natural joint torques, and torques provided by the exoskeleton, result in overall torque equivalent to that of a physically able person. The main mobility functions considered are standing-up, sitting-down, upright balance and level walking. The current paper focuses on standing-up and sitting-down scenarios. Torque profiles required for achieving desired joint displacement are investigated to establish upper limits for control actuation applied at the exoskeleton joints to ensure smooth manoeuvres.

1. Introduction

According to the World Population Ageing 2013 report, the global share of older people, aged 60 years or over, increased by 2.5% from 1990, and will grow 9.4% more by 2050. This means that the number of elderly people will double in the next 35 years, summing more than 2 billion by 2050. They will represent 21.1% of the total population. Moreover, the older population is itself ageing. People over 80 years represented 14% of the older population in 2013, and by 2050, will grow to 19%, meaning there will be three times the number of people 80 years or over in the world. The issue which emerges is that those additional years are not always lived in good health, but threatened by diseases such as memory loss, urinary incontinence, depression, falls or immobility. [1]

* The work is part of the EXO-LEGS project funded by the European Commission under AAL Call 4; grant number 2012-03255

When immobility is not attended, it can start a chain reaction which generates, or greatly boost, additional health risks. Assistive technology devices like crutches, orthoses, wheelchairs and scooters intend to maintain functional capabilities of older people with mobility impairments [2]. However restoration to their original condition has not been possible until the emergence of exoskeleton devices. Exoskeletons for elderly people are thought as an assistive technology with the benefits of the mentioned devices; provide support while standing, enable displacement, usable in limited spaces, and energy efficient. Overall, they will restore and, furthermore, increase previously lost mobility capabilities.

Various studies have been done regarding exoskeletons in recent years. EXPOS is a light exoskeletal device which, combined with a smart caster walker, provides the additional support that older people need to walk [3]. This device is still at the research stage, such as a prototype from Saga University that uses EMG-based control methods to activate the exoskeleton [4]. Honda has been developing and testing a walking assistive device which provides support at hip joint while lifting legs when walking, thus lengthening stride to cover longer distances at a higher speed [5]. Cyberdyne launched into the market the Hybrid Assistive Limb (HAL) living support for lower limbs. HAL is a full robotic lower limb exoskeleton capable of assisting the wearer to walk, stand up and sit down, by sensing bioelectric signals to follow user's motion [6]. It is now available for rent to institutions for an initial installation cost of around £3000 and a monthly rental fee of £1000. Other companies as Argo Medical Technologies, Rex Bionics and Berkeley Bionics have already commercialized exoskeletons to restore walking and other functions in paraplegic patients although the cost is higher than £45,000.

Considering that, nowadays, about two thirds of the world's older persons live in developing countries, and by 2050, 8 in 10 of them will live in under-developed countries, there is an eminent need to produce an exoskeleton device complex enough to safely assist older people in everyday activities, but simple enough to be affordable and manageable by the general public [1]. The EXO-LEGS project for the elderly, focuses on a set of tasks which this exoskeleton need to be able to assist with [7]. Tasks were defined by its end users group (EEUG), standing-up and sitting-down motions are part of those.

This paper describes the modelling and simulation stage of these two motions, and studies the system torque and force requirements under different conditions. Simulation is used due to the complexity of mathematically modelling the whole lower limb musculoskeletal system and exoskeleton device. Moreover, simulation allows testing of controllers under different conditions avoiding risks for the user and equipment, while minimising costs. Experimental verification will be done in future stages of the project.

A number of studies related to standing-up and sitting-down manoeuvres have been carried out. An example is the work of Karlj and Bajd [8], that characterised the standing-up motion in normal subjects into four phases, namely initiation, seat uploading, ascending and stabilisation, lasting around 1.5 s. Orientation of hip, knee and ankle joints for standing-up and sitting-down motions are shown in Figure1. These are used as reference input in this work.

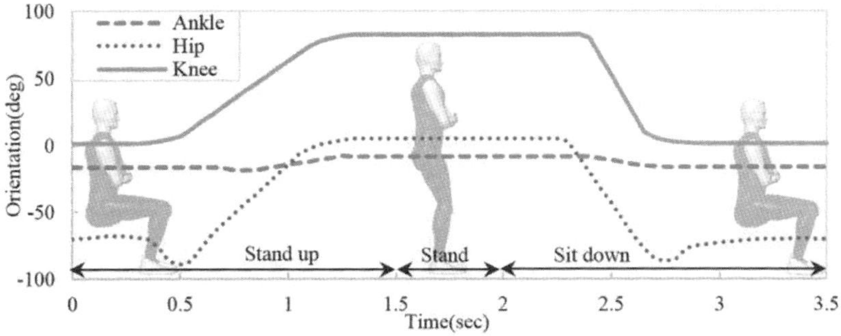

Figure 1. Joint's orientation during standing-up and sitting-down manoeuvres [8].

2. Material

The MSC Visual Nastran 4D (VN4D) software is used for the modelling and simulation of the standing-up and sitting-down manoeuvres. This is a virtual environment created for design and engineering of 3D parts and assemblies [9]. It is used in this project due to its capability to communicate with MATLAB Simulink. VN4D environment allows for the generation of a humanoid model which reacts in a realistic way to external forces such as gravity and weight. In this project, VN4D is used for three main purposes, firstly, to assemble the human like model from segments created in Solid Works and couple them with appropriate kinematic characteristics, secondly, to measure torques, orientation and position of the joints and centroids of links, that are later sent to MATLAB to be used as inputs to control the exoskeleton; and lastly, to visualise the motion of the humanoid and exoskeleton interacting with external forces and disturbances in a realistic manner.

3. Methods

The different segments or parts of a humanoid model of 80 kg and 1.8 m in weight and height respectively, were created in Solid Works, based on dimensions given by Winter [10, 11]. After the design in Solid Works, each part was exported to and assembled in VN4D, considering the anatomical position of joints. For simplification purposes, the humanoid model created in this project has only 3 actuated (A) joints at each leg, 3 (P) passive joints at each arm and 1 passive (P) joint at neck, all in the sagittal or frontal planes, Y and X axes respectively, while the transversal plane, Z axis, is locked.

As mentioned before, an exoskeleton can perform different tasks; in this project, its main purpose is to generate and stop motion in the sagittal plane, thus compensating for muscle incapacity and creating postures needed to accomplish the standing-up and sitting-down manoeuvres. The exoskeleton model used was designed by PCM, Proyecto Control Montaje, S.L. as part of the EXO LEGS project. The CAD model was modified using Solid Works and simplified to have 16 parts per leg. These were later exported to VN4D and linked by 3 actuated (A) and 3 passive (P) revolute joints, as well as 2 prismatic (Pr) joints used to adapt the exoskeleton to humanoid dimensions. Each leg weights 15 kilograms and is attached to the humanoid at waist, thigh, calf and feet. Figure 2 shows the frontal and lateral views of the humanoid model and the exoskeleton.

The final assembly is able to guide the lower body motion during the standing-up and sitting-down motion scenarios once joints are actuated, or follow the motion when joints are passive. Data in Figure 1 was used as reference input to generate standing-up and sitting-down manoeuvres for both humanoid and exoskeleton. Joints can be actuated either by orientation or torque. The first option ensures that the reference orientation will be followed with a minimum error but using unrestricted torque. The second option requires a torque signal to be sent to the joint, torque needs to be controlled to achieve the intended motion. In this research, ankle joints are actuated by orientation while hip and knee joints are connected to controllers and actuated by torque. Ankle joint will be controlled in further stages. Sensors are added to get feedback from the system, orientation and torque at hip, knee and ankle joints are monitored regularly. Additionally, force reactions with ground and seat were measured.

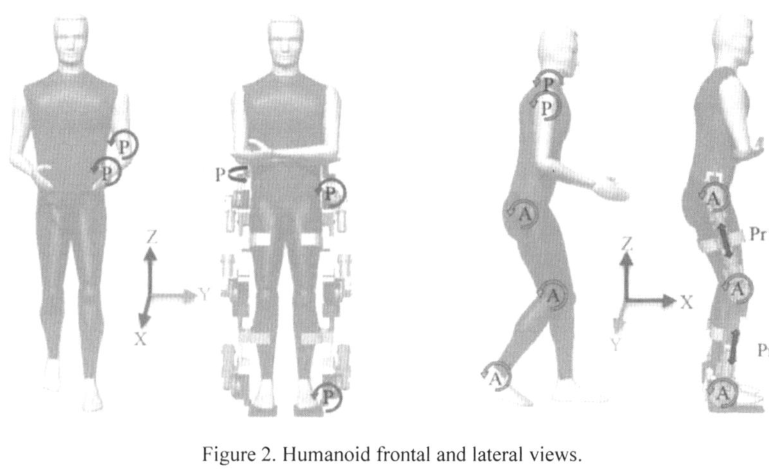

Figure 2. Humanoid frontal and lateral views.

Regarding the controller, a closed-loop set-point tracking PID control was designed for each hip and knee joint. A saturation block was added at the output of the PID controllers to ensure that the value was not above admissible limits of 210 Nm at knee and 160 Nm at hip joints [8]. Various saturation values and 3 different speeds were tested to analyse the impact on the behaviour of the system. As mentioned before, the purpose of this exoskeleton is to aid elderly people to complete the standing-up and sitting-down manoeuvres. For that purpose, an assumption was made that an elderly person is able to generate the 70% of the required torque, while the exoskeleton provide the remaining 30%. Figure 3 shows the implementation of the control scheme in Simulink.

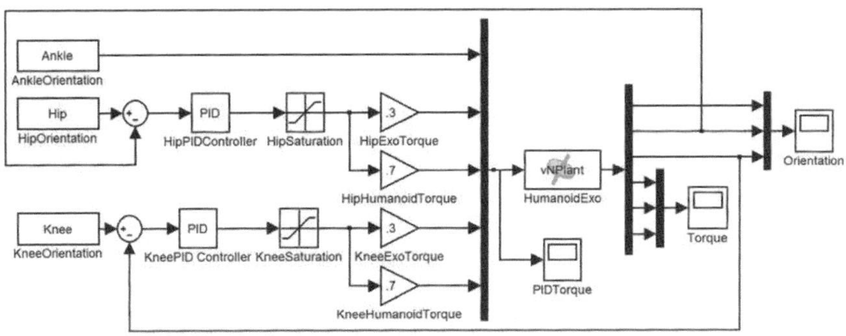

Figure 3. Controller in Simulink.

4. Results

4.1. *Humanoid and exoskeleton sitting-down and standing-up motions*

Several combinations of P, I and D gains were used focusing on minimising MSE. The best tuning was obtained with gains of 14, 1, 1.4 and 14, 7, 1.4 for hip and knee joints, respectively. The system was also tested with different saturation levels to find out if the system was capable of performing with less torque. Although results were satisfactory using lower torques, minimum error was achieved with higher torque values. Figure 4 shows results with saturation limits of 160 Nm for hip and 210 Nm for knee. Orientations shown were achieved with mean-squared errors of 7.25 and 13.96 for hip and knee joints respectively. It is important to note that controllers were tuned focusing on minimising MSE. In order to smoothen torque profiles, a multi-objective optimization algorithm could be used in future stages.

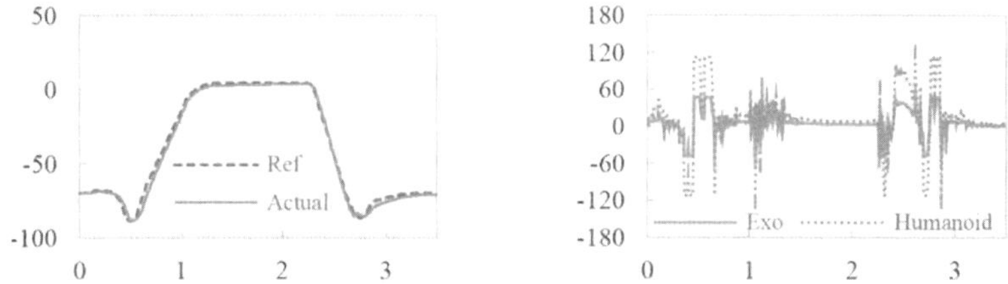

Orientation (deg) vs Time (sec) Torque (Nm) vs Time (sec)

(a) Hip joint with 160 Nm saturation

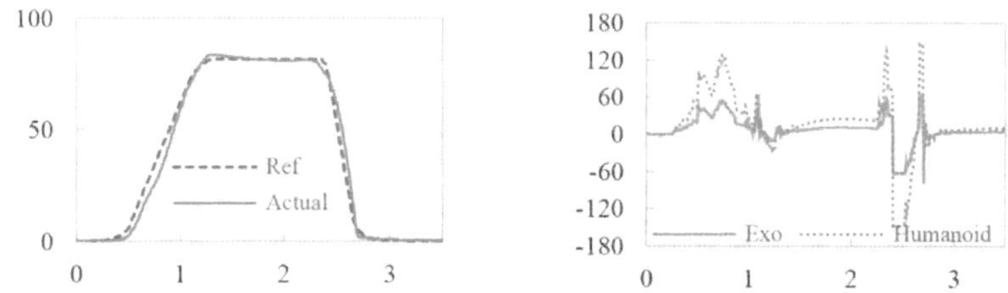

Orientation (deg) vs Time (sec) Torque (Nm) vs Time (sec)

(b) Knee joint with 210 Nm saturation

Figure 4. Hip and knee joint trajectory and torque profiles.

4.2. *Variable speed test*

In order to test the variation of torque and MSE at hip and knee while changing the time to complete the motion, a number of simulations were run. Three motion speeds namely middle (normal), high and low, which complete the manoeuvre over 3.5 sec, 1.75 sec and 5.25 sec respectively, were considered. PID controllers with the saturation level mentioned in the previous subsection were used for hip and knee joints. MSE and average torque required were calculated after each run. In general, these increased at high speed and decreased at low speed. Results obtained with the actuated exoskeleton were satisfactory, with a MSE between 3.37 and 48.32 for hip and 11.59 and 77.8 for knee. Regarding torque, humanoid provided around 45 Nm for hip and 38 Nm for knee, while the exoskeleton added 19 Nm for hip and 16 Nm for knee. Without the use of the exoskeleton device, the humanoid needed about the same torque for hip, but around 15% more for knee, to perform the same motions. Moreover, MSE doubled under this circumstance. Finally, when the exoskeleton needed to be carried by the user, torque and MSE increased for both joints as expected. Figure 5 shows results for the mentioned above cases, normal speed is presented in blue while high and low speed are at the extremes of the bars.

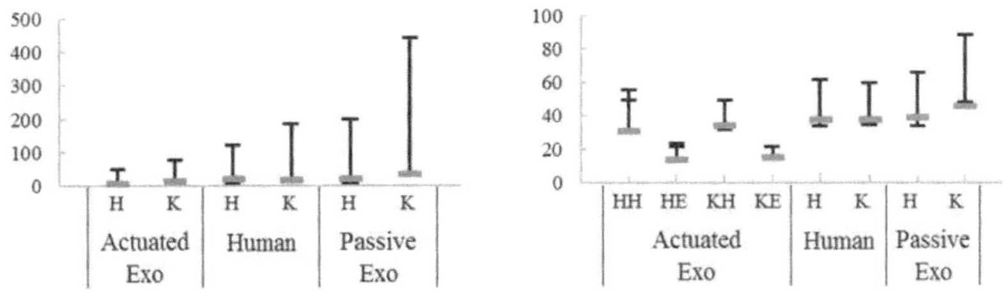

a) MSE b) Average Required Torque (Nm)

Figure 5. Hip and knee joint trajectory and torque profiles.

4.3. *Reaction force with seat and ground*

Once humanoid and exoskeleton were integrated, reaction forces with seat and ground were measured during standing-up and sitting-down manoeuvres at normal speed. In order to get realistic measurements of reaction with the ground, balance had to be guaranteed during the whole length of the motion, slight adjustments had to be made to hip trajectory to ensure this.

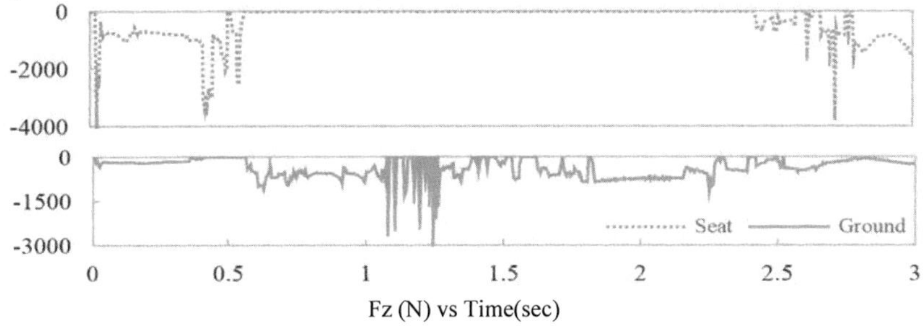

Figure 6. Reaction force with seat and ground during standing-up and sitting-down motions

Figure 6 shows the resulting reaction forces with ground and seat in the vertical axis. Although the behaviour is similar to previous studies, [8] lack of friction with the ground is still an issue that needs to be solved to get clearer and more significant results.

5. Conclusion

Humanoid and exoskeleton models have been developed within a virtual environment for theoretical investigations, through simulation, of levels of mobility assistance for the elderly. Feasibility investigations of required torque profiles for actuating lower limbs' joints motors to achieve proper displacement during standing-up and sitting-down tasks, have been carried out. It has been demonstrated that the joint torques can be contained within allowable limits with suitable control actions, even at different motion speeds. Initial estimations of ground and seat reaction forces have been obtained and will be used as parameters for the following control stages.

Future work will investigate development of soft-computing control approaches with suitable optimisation algorithms to minimise MSE and smoothen torque profiles for sit to stand and stand to sit motions. Different combinations of exoskeleton support will be assessed before proceeding to experimental verification.

References

1. United Nations, Department of Economics and Social Affairs, Population Division (2013).World Population Ageing 2013. ST/ESA/SER.A/348
2. World Health Organization (2011). World report on disability 2011. ISBN 978 92 4 068521 5
3. Kong K., Jeon D. (2006) Design and control of an exoskeleton for the elderly and patients. IEEE/ASME Transaction in Mechatronics 2006.
4. Kiguchi K., Imada Y. (2009) EMG-based control for lower-limb power-assist exoskeletons. IEEE Workshop on Robotic Intelligence in Informationally Structured Space. pp. 19-24
5. Honda (2015) Walking assist device with stride management system. (retrieved on March 1st, 2015)
6. Cyberdyne (2015) HAL for living support (lower limb type). (retrieved on March 1st, 2015)
7. G.S.Virk, U. Haider, I. Nyoman, R.Krishnan and N. Masaud (2014). EXO-LEGS for elderly persons. CLAWAR 2014 conference.
8. Kralj, A. R., & Bajd, T. (1989). Functional electrical stimulation: standing and walking after spinal cord injury. Boca Raton, Fla. : CRC Press
9. Wang, S. (2001). Motion simulation with Working Model 2D and MSC.visualNastran 4D. J. Comput. Inf. Sci. Eng. 1(2), 193-196.
10. Winter, D. A. (1990). Biomechanics and motor control of human movement. Hoboken, N.J. : Wiley ; Chichester : John Wiley
11. Miranda-Linares, D. & Tokhi, M.O. (2014). Finite state control of exoskeleton and wheel walker for gait restoration. CLAWAR 2014 conference.

ASSESSING HUMAN ROBOT INTERACTION: THE ROLE OF LONG-RUN EXPERIMENTS*

ISABEL FERREIRA
Centro de Filosofia da Universidade de Lisboa
Lisbon, Portugal

JOAO SILVA SEQUEIRA
Instituto Superior Técnico/ Institute for Systems and Robotics
Lisbon, Portugal

The present paper aims to contribute to the definition of the structure and substance of long-run experiences indispensable to assess the nature and quality of Human-Robot Interaction (HRI) by describing the procedures involved in the definition of those experiments in the context of the MOnarCH project.

1. Introduction

Personal service robots assist or entertain people in varied existential contexts from domestic to recreational settings. As the application domains where robots interact socially with humans increase, there is an increasingly urge for the creation of interfaces not only easily acceptable by people but also engaging, motivating and safe. In order to ensure that the behavior of the robot carrying out assistive tasks does not have a negative impact on the everyday life of its user, the robot should be user friendly, comfortable to use as well as predictable and safe to interact with, [3].

All this points to the choice of a user-centered design perspective in that the social and situational context of use is described; the individual and social requirements are defined; the efficiency of the system's performance is assessed according to well-established parameters that give salience to the nature of the user experience. This assessment has to develop not just in the preliminary discrete lab experiments, but fundamentally observing the system's long-run impact and performance in the real scenarios where human robot interaction

* This work was supported by projects FP7-ICT-2011-9-601033-MOnarCH and FCT [UID/EEA/50009/2013].

takes place. To be accurate these observations should have a continuous nature that respects the way interactions develop through time. The present paper describes the structure and substance of the tests that have been taking place, since October 2014, in order to assess the quality of HRI in the context of the MOnarCH project. By their durative nature, these extensive tests are contributing to the perfecting of the overall system, constituting an on-going process of assessment that extends through the project's lifetime.

2. Assessing Human Robot Interaction in the MOnarCH Project

2.1. *Physical, social and situational context*

Most of this interaction occurs in the context of edutainment activities taking place in the Pediatrics ward of the IPOL hospital (in Lisbon, Portugal). The space selected for the robots to interact was the main corridor, the playroom, and the classroom (see Figure 1).

Figure 1: Blueprint of the IPOL Pediatrics ward (the areas used by MOnarCH are labeled).

From a social robotics viewpoint, having a robot circulating in a room does not just pose the common navigational problems of moving around in space identifying and avoiding obstacles in order to successfully circulate. Navigating in a human scenario, be it a domestic or public one, demands the adequacy of the robot's movement to the existence of the entities that populate that environment- static or moving physical objects and people, differently moving- as well as adequacy and compliance to the accepted conventions that rule that space. Having a robot interacting in human scenarios demands the understanding the physical and the social dynamics of its context of use, [8].

2.2. *Setting up a long-run experiment*

2.2.1. Goals

The current experiments involving the MOnarCH robots target the:

1. Demonstration of the robustness of the basic functionalities that guarantee the robot's complex behaviors,
2. Verification of the robot's navigational adequacy within that specific physical and social space,
3. Verification of basic user experience features: reaction to the robot's presence.

2.3. *The testbed: technical components*

The experimental set up at IPOL is currently composed of (i) video cameras, (ii) image acquisition system, (iii) wireless network system, and (iv) overhead, static, video projectors.
Figure 2 shows three snapshots of the cameras used to assess the long-run experiments. The areas in the Pediatrics Ward of interest for MOnarCH are those directly involved in the storyboards considered in the project, namely (see also Figure 1), (i) Classroom, in the Teaching Assistant Storyboard, (ii) Playroom, in the Interactive Game, (iii) Main corridor in the Joyful Warden Storyboard.

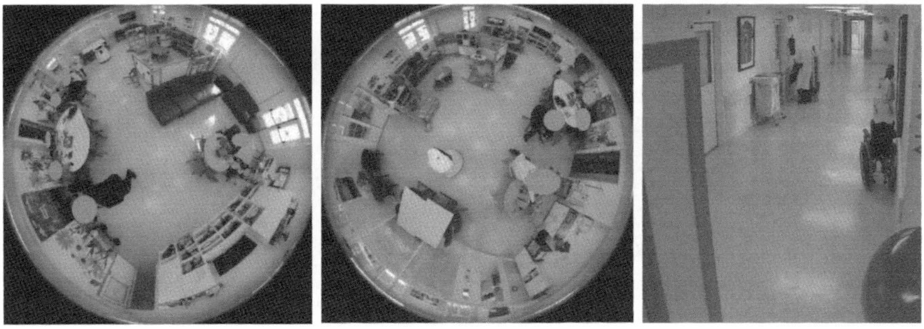

Figure 2: IPOL Pediatrics ward snapshots from the playroom (left and middle), and main corridor (right).

A long-run experiment in MOnarCH involves (i) having a robot interacting with minimal/none engineering assistance by the MOnarCH team, (ii) observing, in a non-intrusive way, the behaviors of people towards the robot, during a large interval of time, and (iii) classifying and discussing the observed behaviors.

Each experiment is evaluated using multiple metrics, in general combining simple objective and subjective metrics.
From a practical point of view both the transient and stationary values of the evaluation metrics are of interest. The former may give indications about how volatile the perception of the observers is in the short term. The later may give indications about what are the important HRI features in the long term.

2.2.2. Defining and assessing the experiments

The experiments are being designed according to (i) the current stage of technological development of the robots and (ii) the ever improving knowledge about the social environment (see Table 1). The structure follows approaches in the literature, that is, first testing isolated components and combinations of them afterwards (see [5]).

Table 1: Categorization of the long-run experiments at IPOL (currently only G1-G4 were addressed)

Experiment groups	**Objective**
G1 - Basic navigation features	Assess robustness of navigation
G2 - Shell Design	Assess the reaction of people to a still non-intentionally interactive robot.
G3 - Basic HRI features	Assess the response of people to the interaction provided by each of the HRI interfaces onboard the robot and by simple combinations of them
G4 - Combined navigation and basic HRI features	Assess the response of people to complex combinations of the HRI interfaces onboard the robots and the features of its movement.
G5 - Full MOnarCH system	Assess the reaction of people at IPOL to each of the storyboards

Each of these groups may be further subdivided in multiple experiments and each may be executed according to two modes, namely (i) in fully automated mode, with the robot moving and interacting without human supervision, and (ii) in semi-autonomous mode, with some of the features of the robot being remotely controlled by a human supervisor.
These experiments aim at (G1) verifying if it can withstand with fast dynamics periods of the environment, and the variety of disturbances induced by the social environment, (G2) how people react towards the presence of the robot and how its physical presence is integrated or not in the environment, (G3) assessing the effectiveness of each of the HRI interfaces onboard the robot, (G4) assessing the combination of navigation and HRI interfaces, and (G5) demonstrating the full system. Currently the first four groups of experiments have been addressed.

2.2.3. Evaluation metrics

Different metrics to assess the performance of social robots have been created and used by several authors. Even though the social context and subjective aspects may somewhat influence the design of the metrics, the combination of multiple metrics may yield the best results.
Metrics directly related to functionalities of typical robot systems were used in [10], namely navigation, perception, management, e.g., deviation from planned route, and obstacles successfully avoided.
Specific metrics such as "Neglect Tolerance", "Robot Attention Demand", "Fan-Out", and "Interaction Effort" were proposed in [4,12]. The introduction of models of human trust and reliability in these metrics was discussed in [9].
Segmentation of experiments in micro-behaviors was used in [2], e.g., move away and touch, and determined the duration of each of them. The segmentation was performed non-automatically by trained observers.
A survey of metrics for proxemics analysis, suitable for automated measurement is presented in [7], involving factors such as, body stance pose, torso pose (Schegloff's metric), left/right arm pose (McNeill's metric), straight-ahead interpersonal distance, relative body orientation (Mehrabian's metric), and interpersonal distance, voice loudness, touch (Hall's metric).
Dependability issues were discussed in [1] where the influence of sensors, control architectures, and strategies to handle contingency situations are recognized to play major roles, though without objective metrics.
The fluency of actions in human-robot interactions was addressed in [6] in the context of collaboration between humans and robots. Subjective metrics include questionnaire based evaluation of multiple indicators such as (i) how well the robot contributed to overall performance, and (ii) how trustworthy was the robot. Objective metrics included (i) the robot idle time, and (ii) time during which concurrent activities were taking place.
The metrics considered for the MOnarCH project target (Metric 1) the robustness of the robots while acting autonomously, (Metric 2) the compliance of the navigation with the physical and social constraints, and (Metric 3) the effectiveness of the human-robot interaction interfaces.

2.2.4. Experimental results

Recent test campaigns at IPOL were scheduled to the static and dynamic display of the robot, and to exhibit low/medium complexity HRI. A total of around 375 minutes of data was obtained for annotation purposes. The total time the robot was on display amounts to 11.5 hours.
Table 2 lists (i) significant behaviors and (ii) the relevant social conditions used for annotation purposes. Results show the average activation rate for each item.

The next to the right hand column was obtained, when no robot is present in the environment, across a period of 3.5 hours in a normal day (no emergency events occurred). In a sense, it represents the baseline for the social environment. The right hand column shows the results when a robot is present.

Table 2: Activation rates for social behaviors/conditions observed at IPOL in the absence/presence of robots in the environment (rates are per minute).

Id	Observed behaviors	Rate (no robot)	Rate (robot)
0	Corridor empty	0,41	0
1	People seated and ignoring events in the ward	0,11	0,11
2	People seated often deviating attention to events in the ward	0,038	0
3	People talking to each other ignoring events in the ward	0,15	0
4	People talking to each other deviating attention at events in the ward	0,075	0,075
5	People often deviating attention to events in the ward	0,3	0,19
6	People wandering through the Pediatric ward (with no clear purpose)	1,088	0,3
7	People entering/leaving a room	1,13	0,23
8	People waiting at a door (of an inpatient room)	0	0
9	People moving between the playroom and the main corridor and opposite	0,94	0,6
10	People moving and stopping to talk to someone	0	0
11	People talking to someone far (at a distance bigger than the maximal public proxemics distance)	0,15	0
12	People talking to someone near (inside the maximal public proxemics distance)	0,38	0,15
13	Children wandering alone in the ward unrestricted	0,11	0,075
14	Children wandering alone in the ward restricted by support equipment (e.g., medical, wheelchair)	0,11	0,34
15	Children wandering in the ward accompanied by adult(s) restricted by support equipment (e.g., medical, wheelchair)	0,075	0,038
16	Child riding a bike through the ward	0	0
17	Children riding bikes (competing with each other) through the ward	0	0
18	Children playing in group (eventually with toys on the floor)	0,038	0
19	Children talking to each other	0	0,11
20	Child(ren) following adult(s)	0,15	0
21	Adult(s) following children	0,038	0
22	Adults talking to each other	0	0
23	Child playing alone	0,075	0
24	Child playing with adult	0,11	0,11
25	Child seat in the classroom	0,038	0
26	Child standing up in the classroom	0	0
27	Teacher seat in the classroom	0,038	0
28	Teacher standing up in the classroom	0,038	0
29	Staff entering/leaving classroom	0	0

30	Staff handling medical equipment	0,038	0,038
31	Child entering/leaving the classroom	0	0
32	Adult doing some activity inside the playroom	0,11	0,075
33	Adult wandering inside the playroom	0,075	0,038
34	Playroom empty	0	0
35	People wandering through the corridor	0	0

Some of the behaviors in Table 2 are marked with 0 activation rates. In some cases such behaviors have been observed during tests not registered, e.g., behaviors 16 and 17. In other cases they are close to another behavior in the list, e.g., behavior 22 is close to either behavior 11 or 12, and the alternative behavior was counted instead. As for behaviors associated to the classroom, e.g., 29, 31, the duration of the test samples was too short for observation. Table 3 lists additional behaviors involving some form of HRI.

Table 3: Observed HRI specific behaviors (rates are per minute).

Id	HRI specific behaviors	Rate
1	Child looking at the robot	0,4875
2	Child talking to the robot	0,2625
3	Child touching the robot	0,6375
4	Child blocking the robot	0
5	Child deviating from the robot	0,1125
6	Child being afraid of the robot	0,3
7	Child approaching the robot	0,8625
8	Child pushing the robot	0,5625
9	Child walking along (in loose formation) with the robot	0,225
10	Staff looking at the robot	0,975
11	Staff talking to/about the robot	0
12	Staff touching the robot	0,3375
13	Staff blocking the robot	0,1125
14	Parent/adult looking at the robot	0,825
15	Parent/adult talking to/about the robot	0
16	Parent/adult touching the robot	0,0375
17	Parent/adult blocking the robot	0
18	Staff not paying attention to the robot	0,1125
19	Child not paying attention to the robot	0,0375
20	Adult/parent not paying attention to the robot	0,2625
21	Staff deviating from the robot	0,075
22	Adult/parent deviating from the robot	0,1125

Table 4 lists the metrics observed from the experiments at IPOL. The definitions in the previous section were adapted with normalizing and averaging steps.
Metric 1 is the "minimum time between interventions of anyone of the MOnarCH team during the experiments". No interventions were registered

during the experiments so this value is identical to the total time of 3.5h of the experiment.
Metric 2 is the "number of direct and indirect interactions between child and robot". This includes both positive and negative interactions. Metric 2 is defined as the average of the all the activation rates of behaviors in Table 2 with the exception of 18, 19 and 20. The success condition can be defined by comparing the average global rate of the direct/indirection interactions with the average global rate of no interaction at all. Higher values for the "direct/indirect interactions" than for "no interaction" may be interpreted as the robot being a device that changes the social environment.
Metric 3a is the "number of clearly positive reactions from a child towards the robot". Expressions of interest such as changing attention (turning the head) to the robot are not counted as positive reactions. Metric 3a is thus the average sum of the activation rates of behaviors 2, 3, 7, 8, 9 in Table 3.
The success condition can be defined by comparing the average global rate for the positive interactions with the average global rate of no interaction or non-interesting interactions, i.e., behaviors 1, 5, 6, and 19. Higher values for the positive interactions than for other interactions may be interpreted as the robot being a device that pleases the children.
Metric 3b is the "number of positive reactions involving HRI interfaces" from a child towards the robot and is defined the average global activation rates of behaviors 2, 11, and 15. This metric aims at decoupling possible effects of acceptance/rejection due only to the general aspect of the robot, i.e., the shape of the outer shell, from Metric 3a. Success condition can be defined by comparing the positive interactions with all the other interactions. A higher value for the average rate in positive interactions using HRI interfaces than the average rate of all the other positive interactions maybe interpreted as good effectiveness by the HRI interfaces.
The additional Metric 4 aims at determining if a robot in a social environment causes significant changes in the behaviors of people. Metric 4 is defined as the average balance between the activation rates of the behaviors in Table 2. Values in the neighborhood of 0 indicate small or no differences between the scenarios with and without a robot.

Table 4: Estimated values for the MOnarCH metrics

Evaluation					
	Metric 1	Metric 2	Metric 3a	Metric 3b	Metric 4
Estimated	3.5h	0.33	0.42	0.088	-0.084
Threshold	168h	0.038	0.23	0.41	0

Metric 1 already suggests a significant robustness, though far from project objective of 168h (1 week) without any human supervision.

The estimated value for Metric 2 suggests that a robot in the environment potentiates the interactions with people.

Metric 3a clearly indicates that a robot pleases the children at IPOL. As for Metric 3b, only 1 HRI interface could be assessed effectively (non-verbal speech) in these experiments and hence the reduced effectiveness shown.

Metric 4 is close to the threshold 0 and hence, though there are slight changes in way people behave in the presence of a robot, relative to the absence, it may not be considered significant. In a sense this means that a robot at IPOL is not likely to disturb the social environment.

3. Conclusions

The lists of behaviors in Table 2 and Table 3 are not exhaustive. New behaviors can be added and some of the listed ones may be decomposed into simpler ones.

During the last test campaign the age range of the inpatient children was 4-7 years old. Additional testing campaigns will target different age ranges.

The physical/medical condition of the children clearly influences their behaviors and hence the metrics. Empirical observations suggest that children with more severe physical problems are less proactive interacting with the robot. During the tests no children with clearly visible physical/medical problems were active at the IPOL premises (corridor, playroom, and classroom). The values obtained suggest that

- The robot is well accepted by children, which is a prime objective of the project, and
- The introduction of a robot in a social environment does not imply significant social changes, which in a sense is re-assuring that the Pediatric ward can continue operating as before.

The long-run tests campaigns will continue until the end of the project, following the guidelines used so far. Future tests will assess other metrics and the behaviors list will be extended to account for each of the specific HRI interfaces.

References

1. "Safe and Dependable Physical Human-Robot Interaction in Anthropic Domains. State of the Art and Challenges". R. Alami, A.Abu-Schaeffer, A. Bicchi, R. Bischoff, R. Chatila, A. De Luca, A. De Santis, G. Giralt, J. Guiochet, G. Hirzinger, F. Ingrand, V. Lippiello, R. Mattone, D. Powell, S. Sen, B. Siciliano, L. Villani. Procs. of the *IEEE/RSJ Int. Conf. on Intelligent Robots and Systems (IROS'06),* Beijing, China(2006).

2. "A Quantitative Technique for Analysing Robot-Human Interactions". K. Dautenhahn, I. Werry. Procs.of the *2002 IEEE/RSJ Int. Conf. on Intelligent Robots and Systems*, Zurich, Switzerland, (2002).
3. "Social Roles and Baseline Proxemic Preferences for a Domestic Service Robot". K. Dautenhahn , K. Koay, D. Syrdal, M. Ashgari, M. Walters. *International journal of Social Robotics*,6(4):469-488, November (2014).
4. "Seven Principles of Efficient Human Robot Interaction". Michael A. Goodrich, Dan R. Olsen, Jr. Procs. of the *IEEE Int. Conf. on Systems Man and Cybernetics*, pp.3943-3948, Washington, USA (2003).
5. "Multimodal Metric Study for Human-Robot Collaboration". Scott A. Green, Scott M. Richardson, Randy J. Stiles, Mark Billinghurst, J. Geoffrey Chase. Procs. of the *1st Int. Conf. on Advances in Computer-Human Interaction*, pp. 218-233, Sainte-Luce, Martinique, (2008).
6. "Evaluating Fluency in Human-Robot Collaboration". Guy Hoffman. Procs.of the *RSS'13 Workshop on Human–Robot Collaboration*, Berlin, Germany (2013).
7. "Proxemic Feature Recognition for Interactive Robots: Automating Metrics from the Social Sciences". Ross Mead, Amin Atrash, Maja Mataric. Procs of the *Third Int. Conf. on Social Robotics* (ICSR'11), pp. 52-61, Amsterdam, The Netherlands, (2011).
8. "Designing a minimalist socially aware robotic agent for the home". L.M. Rocha, I. Wood, S. Sabanovic, M. Francisco. *Journal of Social Robotics*, Springer, June (2014).
9. "Towards Generalized Performance Metrics for Human-Robot Interaction". Jamil Abou Saleh, Fakhreddine Karray. Procs. of the *Int. Conf. on Autonomous and Intelligent Systems (AIS)*, pp. 1-6, Póvoa de Varzim, Portugal, (2010).
10. "Common Metrics for Human-Robot Interaction". A. Steinfeld, T. Fong, D. Kaber, M. Lewis, J. Sholtz, A. Schultz, M. Goodrich. Procs of the *HRI'06*, Salt Lake City, Utah, USA, (2006).
11. "Towards a Framework for Human Robot Interaction". S. Thrun. Journal of Human-Computer Interaction 19(1):9-24, June (2004).
12. "Metrics for evaluating human-robot interactions". D. Olsen, M. Goodrich. Procs. of the Performance Metrics for Intelligent Systems Workshop (PerMIS'03), Gaithersburg, USA,(2003).

SECTION–3

AUTONOMOUS ROBOTS

WALL CLIMBING ROBOT MOTION SIMULATION IN NON-DETERMINISTIC AREA WITH EXISTING MOVING OBJECTS

V. G. GRADETSKY, M. M. KNYAZKOV, A. M. NUNUPAROV, E. A. SEMYONOV and A. N. SUKHANOV

Institute for Problems in Mechanics of the Russian Academy of Sciences101-1, Prospect Vernadskogo, 119526 Moscow, Russia

This paper describes an artificial potential field algorithm for the path-finding problem solving in dynamically changing environments for a wall-climbing robot. Suitability of solving path-finding problem with artificial potential field approach for the wall climbing robot with sliding seal has been analyzed. Simulation demonstrated that this approach provides real-time creation of accurate and smooth trajectories.

Keywords: Mobile robots, artificial potential field, simulation, wall-climbing robot, sliding seal.

The research is supported by Russian Science Foundation, Grant №14-11-00298

1 Introduction

The essential part of designing an autonomous robot is the path-finding algorithm. The path-finding problem solving involves finding the collision-free trajectory from the starting location to the predefined destination point. There are two different groups of path-planning problems. The first one assumes that robots have complete information about the environment, the second one assumes that robots have no any useful information about the environment or the information is incomplete [1-4].The article is devoted to research of the problem with no information about the environment. The artificial potential field approach (APF) was chosen to solve this task. APF is a classical method, which can be schematically described as follows: the mobile robot moves in the field of forces. The goal location to be reached is an attractive pole for mobile robot and obstacles are repulsive surfaces for mobile robot [5-8]. The paper studies the application of APF approach for wall climbing robots with sliding seal.

The overview of the wall-climbing robot (WCR) with a sliding seal and its design schematic are shown in Fig. 1. The mechanical system design of the robot consists of the hull (1) round shaped, the locomotion unit that consist of two DC-motors (2) and two wheels (3), a sliding seal (4) attached to the housing around its bottom edge, the air pump (5), and the low-pressure chamber in the inner area of the hull. The air pump (a fan) creates the required vacuum level in the low-pressure area [5]. The design of this robot allows it to overcome minor obstacles such as cracks and holes on a surface, uneven areas on a surface, and other obstacles robot can bypass. To bypass obstacles the robot has been equipped with laser locator mounted on active rotating axis.

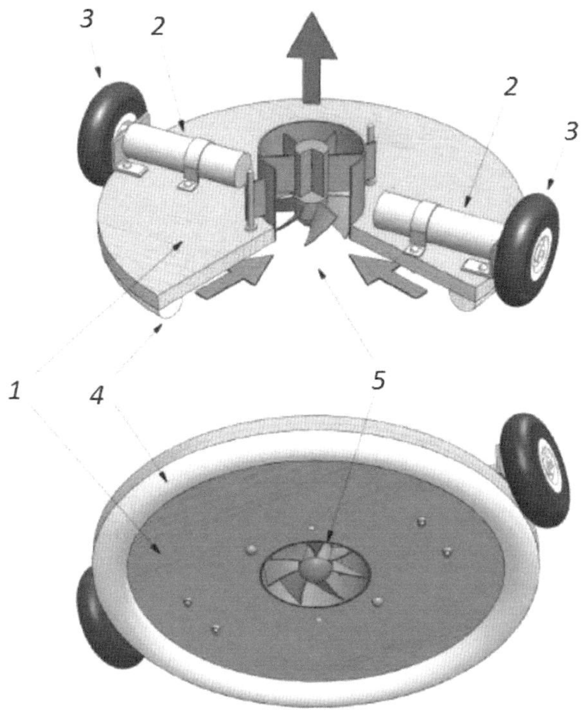

Fig. 1. Design of a wheeled robot with a sliding seal

2 Modeling of the Artificial Potential Field

The fundamental feature of the APF approach is a definition of the potential field U(x):

$$U(x) = U_{attract}(x) + U_{repulse}(x) \tag{1}$$

where $U_{attract}(x)$ is attractive potential field of the goal point, $U_{repulse}(x)$ is the repulsive potential field of nearest obstacles. The definition of $U_{attract}(x)$ and $U_{repulse}(x)$ functions determine the shape of the trajectory. The trajectory can be found with the descent of the potential field to its minimum. A simple way to get to the minimum of a potential field is using the gradient descent technique. The Fig. 2 illustrates the example gradient of the artificial potential field.

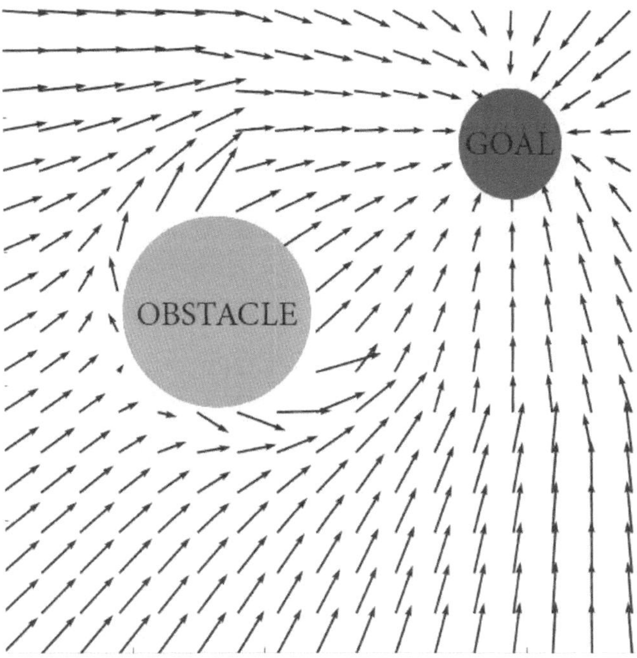

Fig. 2. The gradient of artificial potential field

The attractive and repulsive potential define as following:

$$U_{attract}(x) = \frac{D(x_{goal}, x)}{D(x_{goal}, x_{start})} \tag{2}$$

$$U_{repulse}(x) = e^{-\beta D(x_{obstacle}, x)} \tag{3}$$

Where x_{goal}, x_{start} are coordinates of starting and goal locations, $x_{obstacle}$ is the coordinate of the closest obstacle point to the x point, $D(x_1, x_2)$ is the distance between x_1 and x_2, parameter β determines the distances to the obstacle, when the robot start bypassing.

3 Simulation WCR Motion in Non-deterministic area with Existing Moving Obstacles

For the initial testing of the suitability of APF approach the robot was modeled as a moving oriented point. It had the x_{robot} coordinate, the θ_{robot} as orientation and the R_{robot} as the dimension. The equation of the motion is

$$\dot{x}_{robot} = -k \text{ grad } U(x_{robot}) \tag{4}$$

where k is the gain for gradient descent.

Robot was equipped with the rotating laser locator with working area of $R_{rf\_max} > R_{robot}$. It is assumed that laser locator accomplishes the complete rotation instantly, which means that it builds full distance map for the obstacles at each moment at once. The resulting distances are always adjusted by a robot dimension R_{robot}.

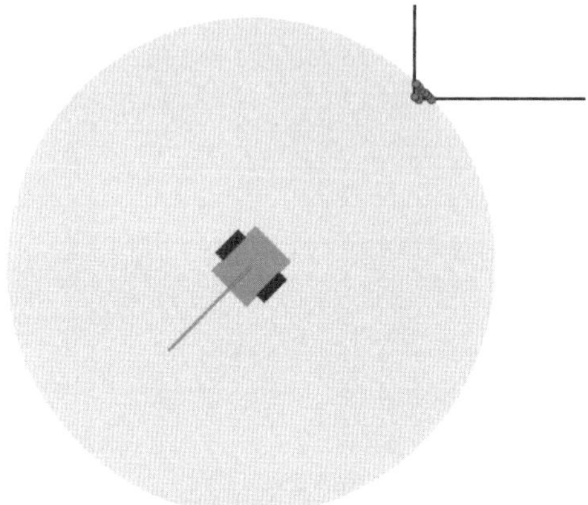

Fig. 3. The robot model with illustrated laser locator working area

Figure 3 illustrates the designed robot model with laser locator working area. The circle describes the laser locator working area, the robot is located at the center of the circle, markers on the obstacle denote found obstacle points, the line following the robot is the robot's path.

It is assumed that velocity vectors of obstacles are constant and their values are smaller than robot's maximum velocity. At the same time, last restriction on the velocity could be weakened by changing the working area of rangefinder R_{rf_max} and repulsion function $U_{repulse}(x)$. In addition, it is possible to weaken the velocity restriction by predicting a future trace of a sensed moving obstacle using its moving history, and calculating repulsive potential field considering several virtual fixed obstacles on this trace [6]. Simulation of the robot motion has been made under specified model, the starting location was set at the left bottom corner of the screen, the goal location was set at the right top corner. The map consists of 2 round obstacles; left obstacle was not moving, right obstacle was slowly moving down with constant velocity. The simulation illustrated chronology in Figs. 4(a)–4(e). At the beginning the robot bypassed the left obstacle on a minimum permitted distance to the obstacle (determined by parameter β in (3), Fig. 4(a), 4(b) until it reached the minimum permitted distance to the second obstacle (Fig. 4(c)). Than the robot turned around and started slowly to avoid the moving obstacle, until the distance to the moving obstacle became more than the minimum permitted (Fig. 4(d)), than the robot turned around and started moving to destination (Fig. 4(e)).

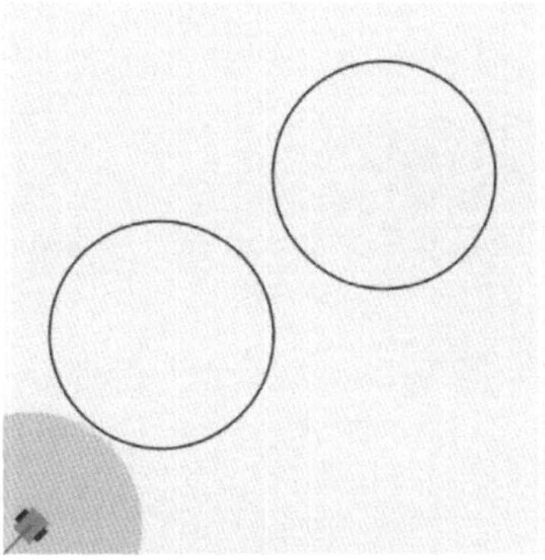

Fig. 4(a). Start

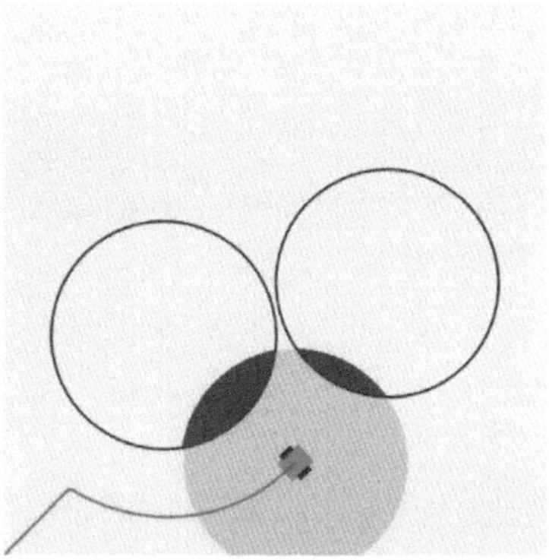

Fig. 4(b). Bypassing stationary obstacle

Fig. 4(c). Getting away from the moving obstacle

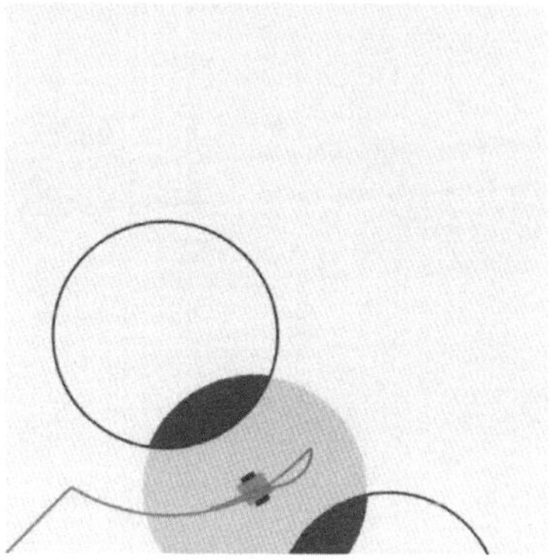

Fig. 4(d). Turned around and continue moving toward goal location

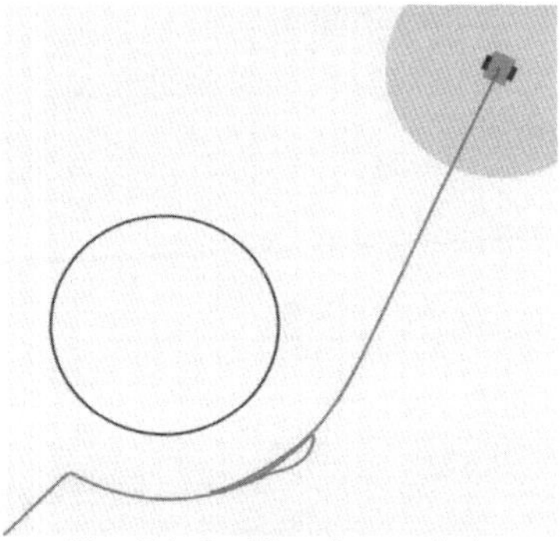

Fig. 4(e). Goal location reached

Figure 4 illustrates that the robot started movement between obstacles with the distance to both of them sufficiently more than it is needed to the collision-free movement. To increase the maneuverability of the robot the coefficient β became a function of relative dislocation of the obstacle to the robot. It is proposed to set the smallest β in front side and then gradually increase it toward the back side, where β has the highest value. Simulation of changed parameter β is shown in Fig. 5.

Implementation of the motion control algorithm is illustrated at Fig. 6.

Fig. 5. Movement between obstacles with decreased β

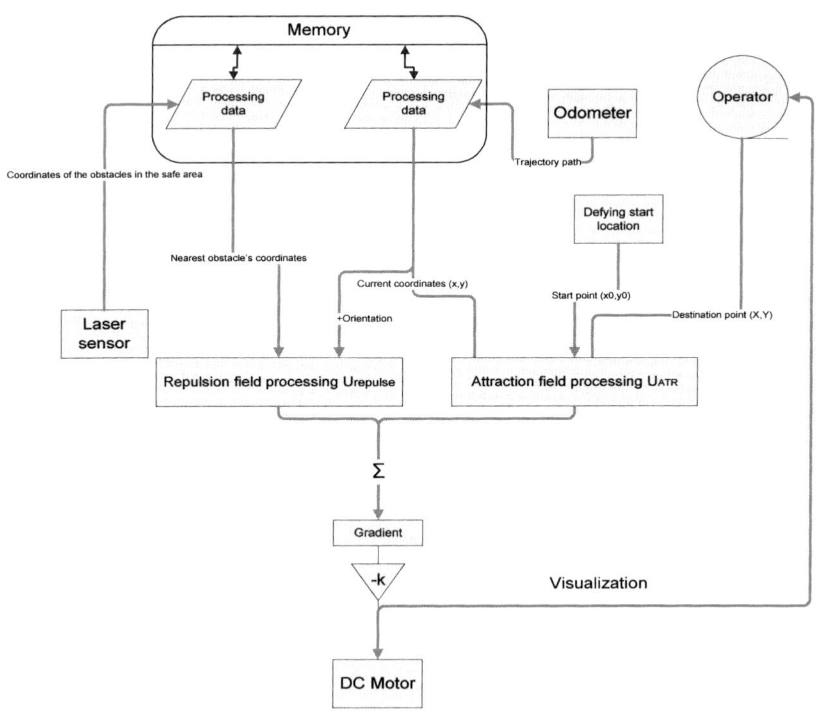

Fig. 6. Scheme of the motion control algorithm

4 Conclusion

Wall climbing robot motion in non-deterministic dynamically changing environment with obstacles' velocity restrictions and onboard sensor's imitation has been considered. As a result of simulating motion, robot reached the goal location without collisions, which illustrates sufficiency of the chosen path-finding approach. The future work is to simulate WCR's motion taking its dynamical parameters into account.

References

1. A. Atyabi, D.M.W. Powers, "Review of classical and heuristic-based navigation and path planning approaches", International Journal of Advancements in Computing Technology (IJACT), 2013, 5(14), pp.1-14.
2. O. Khatib, "Real-time obstacle avoidance for manipulators and mobile robots", International Journal of Robotics Research, vol.5 issue 1, spring, 1986, 90-98.
3. P. Vadakkepat, Kay Chen Tan, Wang Ming-Lang, "Evolutionary Artificial Field and Their Application in Real Time Robot Path Planning", Evolutionary Computation, 2000. Proceedings of the 2000 Congress on vol. 1, 2000, 256-263 vol. 1.
4. A.K. Platonov, I.I. Karpov, A.A. Kirlichenko, "Solving path-finding problem using method of potentials", Keldysh Institute of Applied Mathematics Preprint, AS USSR, 1947, #124, 27p.
5. V Gradetsky, M. Knyazkov, E. Semyonov, A. Sukhanov, "Mobile robot's movement on vertical, horizontal surfaces and slopes under external disturbances and existing moving objects", Mechatronics, Automatization and Control, 2015, Vol.16, №3, pp. 166-173.
6. Y. Lim, K. Jung, "Decentralized Control for Intelligent Robot System to Avoid Moving Obstacles", Intelligent Systems Modelling & Simulation (ISMS), 2013 4th International Conference on, 2013, pp.61-66.
7. Xiuqing Wang, Zeng-Guang Hou, Anmin Zou, Min Tan, Long Cheng, "A behavior controller based on spiking neural networks for mobile robots", Vocational & Technical Institute, Hebei Normal University, Shijiazhuang 050031, China, Neurocomputing (Impact Factor: 2.01). 01/2008; DOI: 10.1016/j.neucom.2007.08.025
8. H.M. Wang, Z-G. Hou, L. Cheng, M. Tan, "Online mapping with a mobile robot in dynamic and unknown environments", Int. J. of Modelling, Identification and Control, 2008 Vol.4, No.4 > pp.415 - 423

INTELLIGENT SENSOR USED FOR *MERO* MODULAR WALKING ROBOTS MOVING ON UNDEVELOPED TERRAIN

ION ION[1], ADRIAN CURAJ[2], CORNEL DINU[3] and CRISTIAN SIMION[1]

[1]*Manufacturing Technology Department*

[2]*Automatic Control and Industrial Informatics Department*

[3]*Aerospatial Department*

POLITEHNICA University of Bucharest, Romania

One of the central problems of motion control of walking robots is the distribution of force between legs and the organization of robot motion within margins of static stability. Support reactions need to be controlled during movements on undeveloped terrain The force/torque sensor is an important component of the measuring system and force control and sliding contact with the ground uneven leg of the modular walking robot MERO. Determination of the real forces distribution in the shifting mechanisms of a walking locomotion system which moves in undeveloped terrain at low speed is necessary for the analysis of stability. The paper will analyze an intelligent sensor developed by the authors with an important role in the complex system control slipping when moving walking robot MERO on undeveloped terrain.

Keywords: Modular walking robot, mechanism, control slipping, stability, force/torque sensor.

1. Introduction

The modular walking robots MERO (old and new model) Fig. 1 [3], [7], [10] represent a special category of robots, characterized by having the power source and technological equipments embarked on the platform which can be transported on undeveloped terrain (exp. agriculture and forestry).

Figure.1 MERO-The modular walking robot (old and new model)

In agriculture and forestry, due to the specific characteristic conditions created by vegetation and state of the ground, as well as in order to protect the environment, caterpillar or wheeled machines used nowadays for performing the activities in these domains have a limited mobility and usually destroy and/or damage the environment, vegetation and young trees in their movements from one place to another. [3], [7].

The Walking robots MERO are recommended for applications in In agriculture and forestry and other, because better protect the environment, as their contact to the ground is discrete, which considerably diminishes the area underwent to crushing; the robot's weight can be optimally distributed all over the leaning surface, by controlling the forces. Altering the distance to the ground, the robot can pass over young trees or other vegetation, growing in the passage area.

Avoiding hurdles such as logs or tree trunks is a considerable advantage, against the other types of vehicles.[7], [3].

The artificial compaction is the consequence of an exaggerated, unreasonable traffic in the field for farming, transport, hydrological and other similar activities. The negative effects of the compaction, irrespective of its nature and causes, are multiple. Thus, the water retention capacity and especially permeability are decreased, worsening the ground water state and, generally, the soil humidity. Aeration is reduced and at the same time, it often increases considerably the resistance to penetration of the tools used for tilling, digging, excavating etc.

The presence of the obstacles on the support surface may lead to changing the direction, the height of the body's movement and its orientation in space as well as the appropriate reorganization of the sequel of watching the regime of the legs' movement.

For safe movement a walking robots, in the conditions described above, is required determination of the real forces, distribution in walking locomotion system, when moved in undeveloped land at low speed, is necessary an analysis of stability.

The stability, besides other factors [1], [8], [9], [10] is influenced by the values of friction coefficients between terrain and feet, the stiffness of terrain, the shape of terrain surface, situation becomes critical when stepping modular robot carrying a technological equipment on a sloping or undeveloped land along the slope.

To improve the robot's stability when it moves along a slope (Fig. 2(b)), there are three strategies such as [1], [2], [4], [5], [6], [8], [10], [11].

- its body's height is diminished and its position adjusted;
- the step's length is also curbed;
- activation of intelligent lock sliding

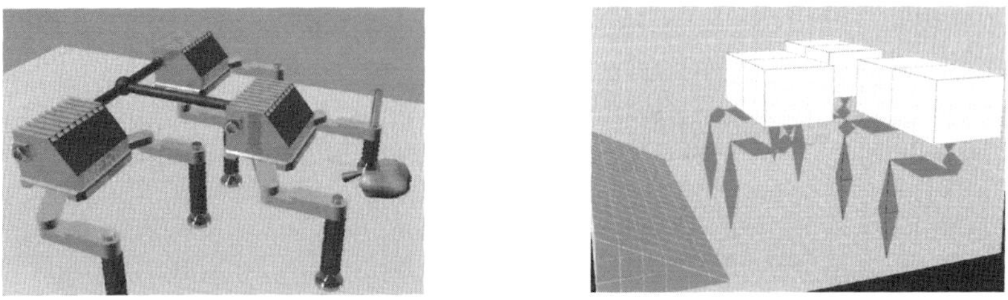

(a) (b)

Figure. 2 MERO2-The modular walking robot, moves over obstacle and along a slope (Computer simulation)

The structure of the foot of a walking robot in general and even more so of modular walking robot and especially the foot structure RRP (MERO 2- Fig. 2(a), (b)) must adapt to small land surface irregularities moving Fig. [8], [10], [11].

The active surface of the foot is relatively small and it is considered that the reaction force is applied in the gravity centre of this surface. The reaction force represents the resultant of the elementary forces, uniformly distributed on the foot sole surface. The gravity centre of foot active surface is called theoretical contact point.

Therefore, the pressure applied to the foot sole on terrain is transmitted through successively varying contact generative and the forces are transmitted smoothly from the terrain to the foot sole for a stable gait.

In order to determine the position of the robot, it is necessary to determine, at each of the support reaction force component values of the pitch and the leg, namely Fig. 2(a): [3], [9].

- normal component $\overline{N}$, perpendicular on the surface of terrain in the theoretical contact point;

- tangential component $\overline{T}$, frictional force, situated in the tangent plane at terrain surface in the theoretical contact point, it is necessary to determine the stable positions of walking robots.

For stability of robot movement, the magnitude of vector cannot be greater than the product of the magnitude of the normal component by the frictional coefficient between foot sole and terrain. If this magnitude is greater than the friction force, then the foot slips along the support surface down to the stable position.

In this situation will be activated sensor in the sole of the walking robot, shown in Fig. 3(a), (b).

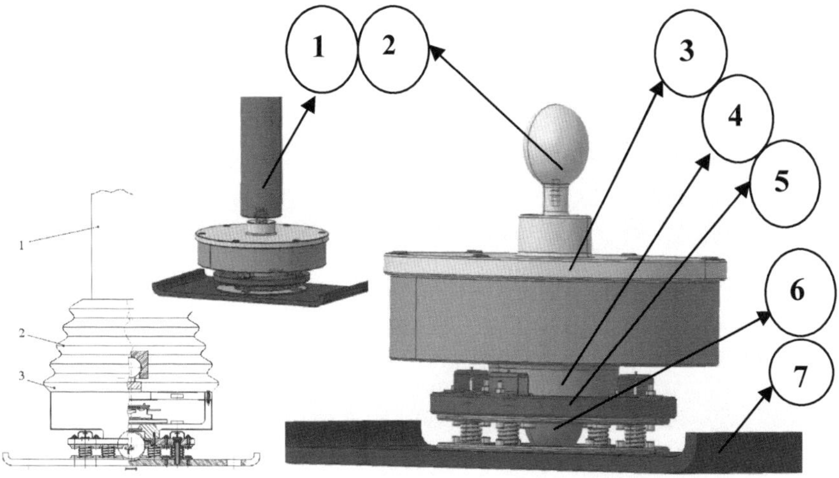

(a) (b)

Figure. 3 The leg of robot and intelligent sensor attached to the sole of the foot

2. Building of Experimental Intelligent Sensor for MERO2 Modular Walking Robot

When a modular walking robot foot lands on the ground, it suffers a shock more or less strong because the ground reaction force. This shock damping is required to transmit the necessary forces to land in order to maintain the desired position of the robot body and its verticality are controlled in accordance with stepping periodic leg movements.

If the ground surface is not horizontal, foot contact with the ground is unstable.

The structure sole of the foot of a walking robot in general, or a walking robot modular system structure displacement RRP (MERO2) [3], [9], [8], [10], in particular, must adapt to small land surface irregularities moving.

This structure allows the reaction forces are transmitted smoothly between the feet and the ground, so that the robot can be controlled in position and direction of travel.

Control system which controls the modular walking robot can be attached to a series of sensors linked to the interaction with the environment and elements of movement kinematics system and sensors to detect the reaction forces of the ground and feet.

The sole of foot of a walking robot, presented and still of interest among researchers. [1], [2], [4], [11], [12].

The shape of the foot sole was inspired by nature or requirements relating to the structure of the robot, stepping mode, both in form and sensor system that was equipped. [8].

For the new model of walking robot leg MERO2 whose structure is RRP (Fig. 1, Fig. 2(a)) , we present an intelligent paper sensor, attach the foot robot.

The sensing and motion control allow identification and measurement of contact forces with the ground through a transducer element whose elastic deformation is caused by a ball that sends robot step foot contact with the ground under the weight of the burden robot foot and additional load for transported technological installations. Contact forces are determined using strain gauges of the measurement module whose deformation generates an analog signal processed with the power regulator.

Measuring relative sliding is through an optical sensor without physical contact with the track that provides a measure displacement resolution 1000 dpi/inch extent at resolution of 0.024 mm.

The sensor is based on Laser Stream technology, which measures changes in position by acquiring consecutive images (frames interference) and mathematically determining the direction and magnitude of movement.

The sensing and motion control (force and slip) is attached to the sole (pos. **2**) of the foot walking robot. (pos. **1**). Fig. 3(a).

The sensing and motion control attached the foot's **1**, by spherical cup **2**, (Fig. 2(b)) is composed of sensor element elastic force / torque, **3** that were bonded strain gauges and optical sensor AVAGO type, **4**, **8** ball caught special plate between **5** and **7** seat of the foot is kept in contact with them through special bolts that keep the foot lifting phase tensioned springs **6**.

The screws secure the lower link 6de the sole seven springs.

Through bolt attached to the support plate is adjusted stroke limiting plate race spurs foot contact with the ground under the weight of the robot that lies foot in the down position.

Maintaining high spurs is done through spring tension due to stroke limitation by special plates fastened in the screw head.

To minimize the risk of slipping feet, the system is equipped with a spurs of the hydraulic system.

Depending on the signal from the force sensors / time and slip control system is given actuation signal spurs.

It refers to the complex phenomenon of slip (tangential component), which through the two different types of sensors (sliding on two axes and power) controlled through an integrated microcontroller system (power regulator) allows adapting them to displacement on unimproved terrain

To measure forces and torques during the move walking robot will be analyzed below sensor force / moment developed by the authors with an important role in the complex system control slipping when moving robot on unimproved terrain. [8].

Figure 4 shows diagrams efforts and the structure of sensing elements of the six-axis force/torque sensor for the leg of intelligent robot MERO component of the device sensing force/torque and slip.

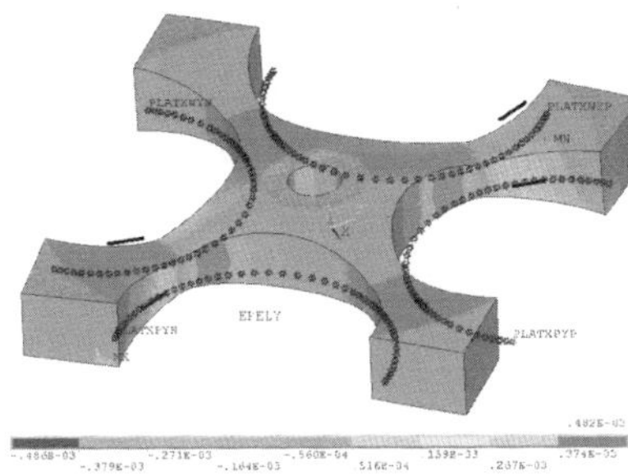
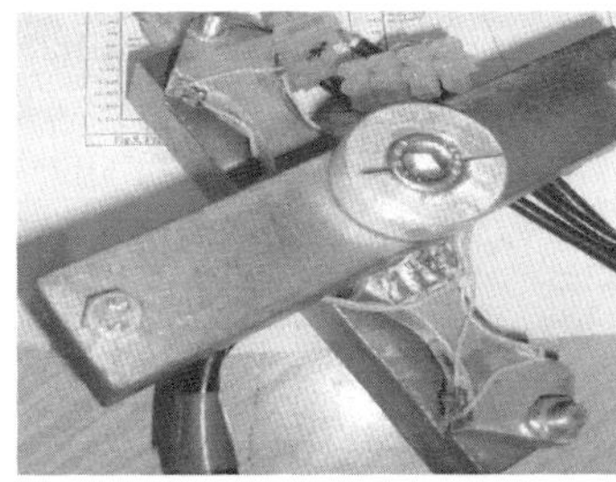

Figure 4 Force/ torque transducers located at the ends of the modular walking robot foot

Amplification analog signal generated by the strain gauges is a specialized differential instrumentation amplifier INA 125U providing an output voltage range of 0-2, 5Vcc which is applied to a 12bit analog to digital which results in a theoretical resolution measuring force 1/4096 = 0.02% of FSR (full scale range).

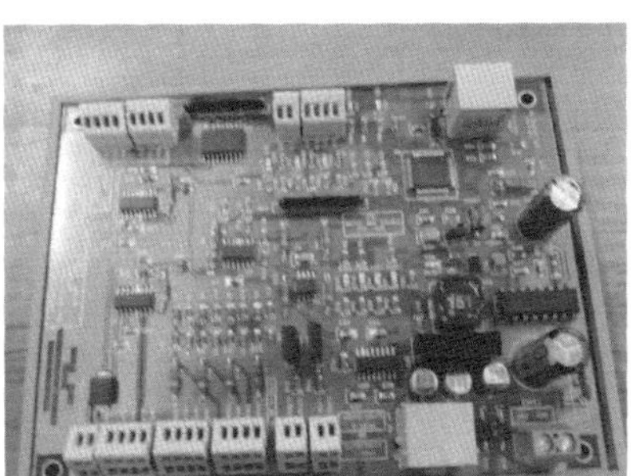
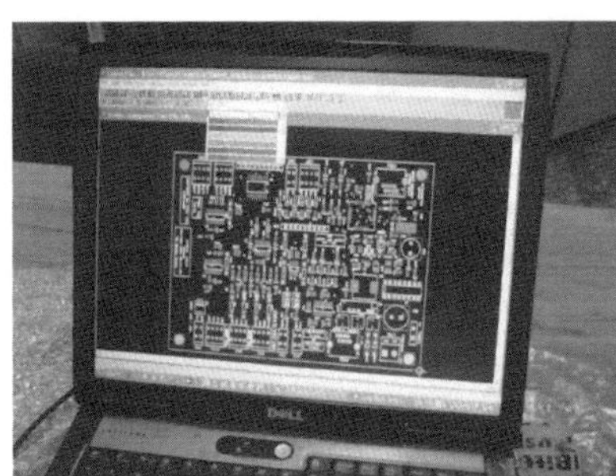

Figure 5 DX308 micro system

The sensing and motion control by force is a hardware and software system (how to up-grade) that has two different types of sensors (sliding on two axes and strength).The intelligence given by a micro-integrated controller provides a force control loop to keep it from slipping. (Fig. 5) It controls allowable values and is attached to the feet o robot has the composition and distribution mains pressurized fluid, proportional pressure regulator, electronic module setting reference point, power regulator, DX308 micro system (Fig. 5).

DX308 micro system is designed around a type ADUC832 controller manufactured by Analog Devices on a 80C51 core to which were added some additional ports.

Measuring relative sliding is through an optical sensor without physical contact with the track that provides a measure displacement resolution 1000 dpi/inch extent at resolution of 0.024 mm.

The sensor is based on Laser Stream technology, which measures changes in position by acquiring consecutive images (frames interference) and mathematically determining the direction and magnitude of movement.

The safety line is established in accordance with the prescribed value of the stability limit for a certain walking regime, or contingent of the reaction forces values, where the feet touch the support ground; these forces are not allowed to drop beneath the accepted values.

This system attached leg walking robot whose information is collected and processed by micro-system DX 308 and connect to the command and control system of the robot, increase safety and stability movement.

3. Conclusion

To avoid slipping walking robot driving on a slope or along the slope in paper presents smart sensor and main components.

It refers to the complex phenomenon of slip (tangential component), which through the two different types of sensors (sliding on two axes and power) controlled through an integrated microcontroller system (power regulator) allows adapting them to displacement on unimproved terrain.

The type transducers used in walking robots MERO offers both force control and robot protection.

Solutions found by the authors has enabled sensing and control device, which senses touch motion, force measurement

References

1. A. Schneider, U. Schmucker (2010) "Force Sensing for Multi-legged Walking Robots-Theory and Experiments – Part: *Mobile Robots, moving intelligence*".
2. A. Mahdi., N. Stephen. (2012) "Foot Force Criterion for Robot Stability" *Proceedings of the 13th International Conference on Climbing and Walking robots and the Support Technologies for Mobile Machines CLAWAR 2012,* Baltimore USA (2012).
3. I., Ion, I Simionescu, A., Curaj, A. Vasile (2011), "*MERO* Modular Walking Robots, Solution for Displacing Technological Equipments on Irregular Terrains" *Proceedings of the 13th World Congress in Mechanism and Machine Science,* Guanajuato, México, 19-25 June, 2011.

4. S. M Song and J. K. Waldron, (1987). "An analytical approach for gait study and its application on wave gaits", the *International Journal of Robotics Research,* 6, 60-71.
5. K. J. Waldron (1991) "Force and Motion Management in Legged Locomotion", *Proceedings of the 24th IEEE Conference on Decision and Control,* Fort Lauderdale, pp. 12-17.
6. V. Kumar, K. J Waldron, "Force Distribution in Walking Vehicle", Proc. 20th ASME Mechanism Conference, Orlando, Florida (1988).
7. I. Ion, I. Simionescu, A. Curaj. (2002) "Mobil Mechatronic System with Applications in Agriculture and Sylviculture" *The 2th IFAC International Conference on Mechatronic Systems,* December 8-12, - Berkeley University – USA (2002).
8. I. Ion, M. Gavan, C. Dinu, A. Vasile, O. Nicola. I.Dumitru, G.Stamatescu (2013) Device for sensing and motion control walking robot attached to the displacement 07/02/2013 nr.A/00132.
9. I. Ion, I. Simionescu, A. Curaj, L. Vasile (2011) "Distribution of reaction forces during movement walking robots" *FIELD ROBOTICS Proceedings of the 14th International Conference on Climbing and Walking Robots and the Support Technologies for Mobile Machines*, University Pierre et Marie Curie UPMC Paris France 6-8 September (2011).
10. I. Ion., A. Curaj, A. Vasile, I. Dumitru, G. Stamatescu Force Sensing and (2013) "Control during Movement and Object Manipulation in MERO walking robots", *The 15th International Conference on Climbing and Walking Robots,* Sydney, Australia 602-608.
11. R. B McGhee, G. I., Iswandhi, (1979) "Adaptive Locomotion of a Multilegged Robot Over Rough Terrain", IEEE *Trans. on Systems, Man, and Cybernteics,* Vol. SMC-9, No. 4, pp. 76-182.
12. I. Simionescu, I. Ion, L Ciupitu. "Mechanism of Industrial Robots", *Published by AGIR Press Bucharest*-In Romanian. (Tome1, 2008).

DESIGN AND IMPLEMENTATION OF A SCANSORIAL ROBOT

M. A. H. HASSAN[†] and M. O. TOKHI

Department of Automatic Control and System Engineering,
University Of Sheffield, Sheffield, UK
[†]cop11mah@sheffield.ac.uk

This paper presents a general framework for an autonomous climbing robot. The objective of the work is to design climbing robot by using robot arm which has the capabilities to grasp the surface, i.e. pole. The robot consists of two arms with wheels attached to each arm enabling the robot to climb or descend a steep surface such as poles or gap between walls. The arms are used for grasping to the surface while the wheels are used for movement upward or downward. The wheels offer friction contacts which allow to apply force on the surface for stable climbing and descending. Fundamental challenges to the development of real robotic systems include the hardware design, control system, grasping and manipulation technique and analysis of the scansorial robot.

Keywords – scansorial robot, grasping, manipulation, adhesion mechanism.

1. Introduction

A robot with climbing capabilities may be devised for a variety of applications; manipulation and handling, cleaning, maintenance, security and surveillance. Currently, climbing poles to access power transmission and telephone lines has become a routine task for technicians and municipal workers. This poses dangerous and hazardous circumstances which could result in serious injuries or loss of lives. The robotic pole climber proposed in [1] was inspired by human climbers to move up the poles for the purpose of cleaning, maintenance or inspection.

The design of a climbing robot needs to consider a proper adhesion mechanism to ensure the attachment of the robot on the surface is sustained. The most general adhesion mechanism is vacuum suction to produce a suction force toward the smooth or non porous surfaces [2]. Magnetic adhesion [3] is adopted for ferromagnetic surfaces where the robot remains attached to the surface. Another adhesion mechanism is creating vortex force (aerodynamic principle) by using a propeller, also known as "tornado in a cup" [4]. Recent studies have

This work is supported by Ministry of Education Malaysia

shown biologically inspired mechanisms for climbing robot using dry adhesion (e.g. gecko and cockroach) [5]. The climbing task for manmade structure or a natural environment requires a different type of adhesion mechanism. The structure can only be adhered by gripping on the surface as in [6], [7].

By considering the adhesion mechanism for a pole climbing robot, the appropriate mechanism is gripping on the surface. In order to grip the pole, a robot with multi-fingered grasping has to be developed. Multi-fingered manipulation has a significant relationship with legged locomotion. The robot hand grasps an object while a legged robot "grasps" its environment. There is a similarity between robot's fingers and legged robot on grasping object or environment by repositioning its fingers or legs. The relation between multi-fingered and legged robot also exists in climbing robots with multi-limbed mechanism.

The grasping process uses a force-closure to catch the object. The grasp achieves force-closure if forces and torques applied to the object are resisted [8]. In order to achieve the closure of an uninformed object, at least two fingers are required. Various models of contacts have been studied in [9], [10].

In designing the climbing robot, the center of gravity is a main aspect to be considered. The essential aspect that will affect the drive force and the stability of the robot is the location of the center of gravity. Moreover, for climbing, the climbing force can be improved by placing the center of gravity closer to the pole. Simple climbing locomotion that is stated with grasping is established to emulate the motion of the center of gravity of the arm from which values of the climbing locomotion are obtained.

2. System Description

Figure 1 shows the climbing robot developed in this work. A prototype robot has been designed and developed using seven dynamixel AX-12A servo motors, plastic frames and wheels. The climbing robot uses an Arduino Mega 2560 as the main microcontroller and MPU-6050 device (Gyro and Accelerometer) as the level position sensor. Each arm consists of two links with 2 DOF. The robot has 4 DOF in total. Table 1 shows the robot parameters and pole dimension.

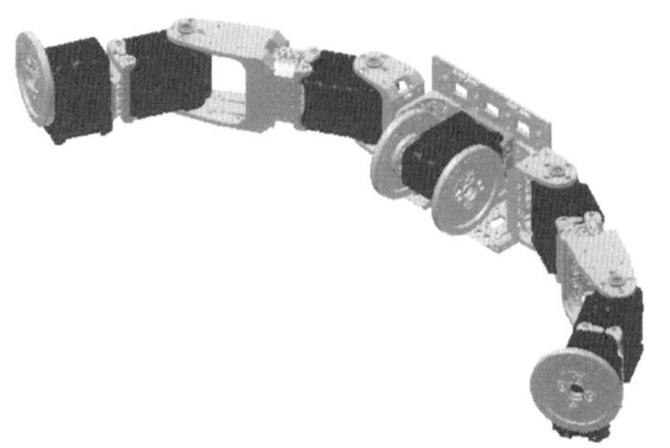

Figure 1. Scansorial Robot

Table 1. Scansorial Robot Parameter and Pole Dimension

Parameter	Value
Mass of the robot	0.70 kg
Total robot length	531 mm
Width	80 mm
Height	65 mm
Body length	141 mm
Link 1 (L/R)	105 mm
Link 2 (L/R)	90 mm
Pole diameter	110 mm

3. Kinematics and Dynamics of the Robot

Denavit and Hartenberg standardized the way to represent the modelling of robotic motions [20]. The following notations are used here: a_1 and a_2 are the lengths of link 1 and link 2, c_1 for $\cos\theta_1$, s_1 for $\sin\theta_1$, c_{12} for $\cos(\theta_1+\theta_2)$, and s_{12} for $\sin(\theta_1+\theta_2)$. Figure 2 represents the arm of the robot with two links.

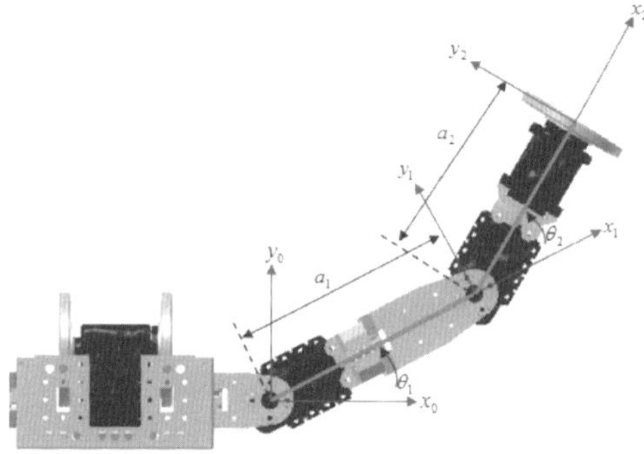

Figure 2. Two links coordinate frames attached to the arm

The forward kinematic of the robot arm can be represented as follows,

$$x = a_1c_1 + a_2c_{12} \tag{1}$$

$$y = a_1c_1 + a_2s_{12} \tag{2}$$

The angles θ_1 and θ_2 can be calculated using inverse kinematic as follows,

$$\theta_1 = atan2\left(\pm\sqrt{1-\left(\frac{x(a_1+a_2c_2)+ya_2s_2}{x^2+y^2}\right)^2}, \frac{x(a_1+a_2c_2)+ya_2s_2}{x^2+y^2} \right) \tag{3}$$

$$\theta_2 = atan2\left(\pm\sqrt{1-\left(\frac{x^2+y^2-a_1^2-a_2^2}{2a_1a_2}\right)^2}, \frac{x^2+y^2-a_1^2-a_2^2}{2a_1a_2} \right) \tag{4}$$

4. Locking Mechanism

Friction force is a main property for every climbing robot to perform the climbing task. The higher the friction force applied toward the surface the better the mobility of climbing locomotion. This property allows the robot to attach on the surface and is called self-locking condition. Self-locking can be obtained by friction and the sufficient amount of external forces to create a locking condition [11].

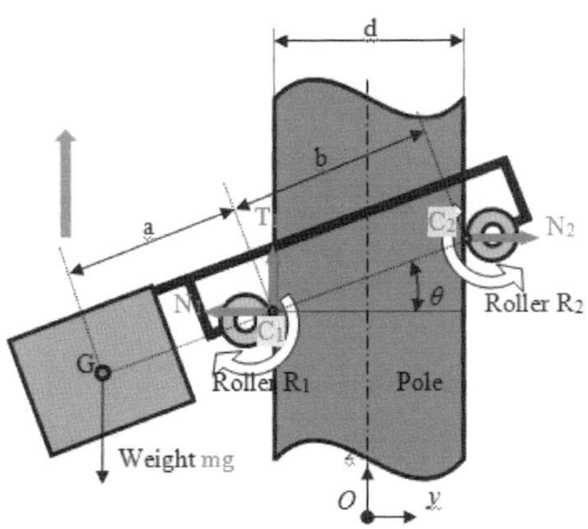

Figure 3. Rolling self-locking mechanism

Figure 3 shows the climbing robot with rolling self-locking. The Roller R_1 and roller R_2 are both designed to be actuated. Roller R_1 is in contact with the pole at C_1, which is transmitting a normal force N_1 and a tangential force T_1 from the pole to the robot. While roller R_2 is in contact with the pole at $C_{2,}$ which is transmitting a normal force, N_2. Roller R_2 is actuated to provide very low resisting torque. At point C_1, the addition of torques around x axis yields,

$$mga\cos(\theta) = b\sin(\theta)N_2 \text{ with } N_1 = N_2 \text{ and } T_1 = mg \tag{5}$$

The non-slipping condition at point C_1 can be derived based on Coulomb friction law with friction coefficient μ.

$$N_1 = \frac{amg}{b\tan(\theta)} \text{ with } T_1 \le \mu N_1 \tag{6}$$

Finally, the rolling locking mechanism can be represented as:

$$a \ge \frac{b\tan(\theta)}{\mu} \text{ with } \theta = \arccos(d/b) \tag{7}$$

The above equation 7shows that the geometry and friction properties are the factors to create the self-locking condition instead of the mass. If the distance a is sufficiently long, it will create the self-locking. From equation (7), it clear that the friction, μ is disproportional to the distance, a. The locking condition will be retained when the tilt angle θ decreases which is proportional to the distance a. In practice for a climbing robot, it is impossible to keep a perfect horizontal level while climbing due to structure deformation. Therefore, the distance b will increase and cause the angle, θ to be increased as well. As a result, it creates a locking condition that can be verified by equation (7).

5. Implementation of the Scansorial Robot

Accelerometer and gyroscope are used for the angle measurement in the system. It is known that accelerometers give slow response and corrupted with noise and gyroscopes have fast response but suffer from the time drift error due to the accumulated error by integration to obtain angle information. Figure 4 shows the angle measurements of both accelerometer and gyroscope. The noisy signal has been filtered using Kalman filter. The initial orientation of the accelerometer was set at 180 degrees (Figure 5).

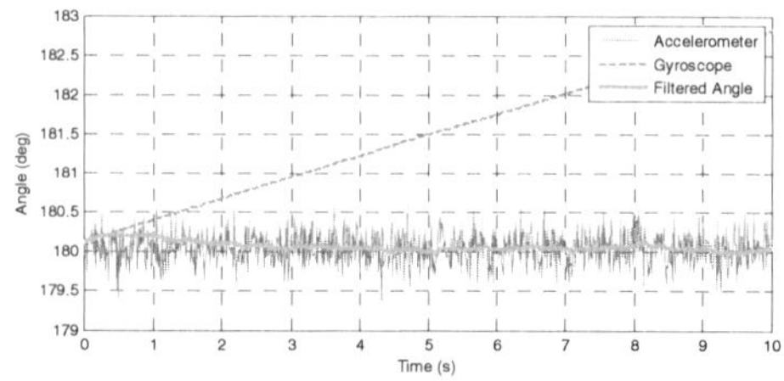

Figure 4. Kalman filter angle measurement

A PID controller was introduced to the robot system due to simplicity in terms of implementation and general adequacy of results at this stage. The PID controller comprises of three parameters (Kp, Ki and Kd) that denote the gains of proportional, integral and derivative of error. Error is represented by tilt angle in the system.

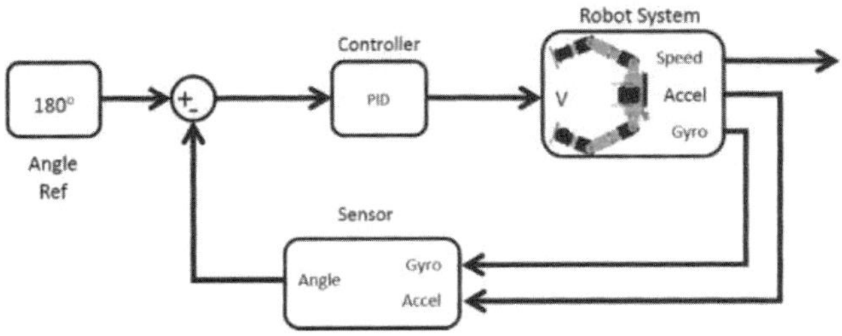

Figure 5. Scansorial robot with PID controller

The initial speed for the robot was set at 40 mm/s (15 rpm). As the robot started to climb, the arm ascended ahead of the robot body as in Figure 3. To compensate for the tilt angle error, PID controller parameters were set heuristically with Kp = 7.0, Ki = 1.0 and Kd= 1.0.

6. Result and Discussion

Figure 6 shows the implementation result from the experiment, which indicates that the middle wheel increased while left and right wheels decreased the speed to ensure the level position of the robot. Figure 7 shows the holding forces from both arms while the robot climbed. Figure 8 shows the tilt angle (pitch angle) about $\pm$ 5 degrees from the actual angle setpoint of 180 degrees. In practice, the climbing robot will not be able to keep a perfect horizontal level due to structure deformation. However, this position actually provides a good grip for the robot since it can be considered as self locking to pole.

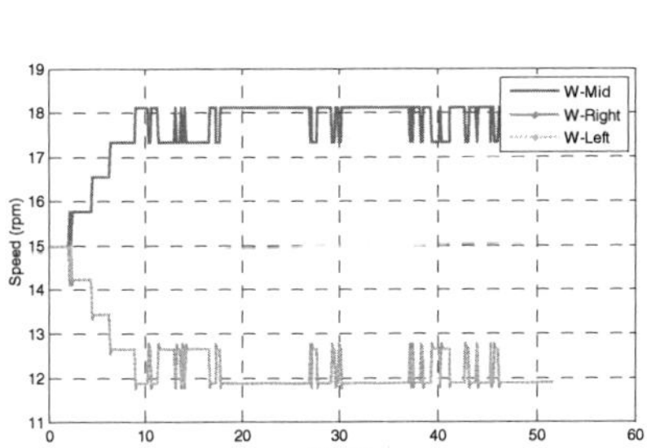

Figure 6. Wheels speed while climbing

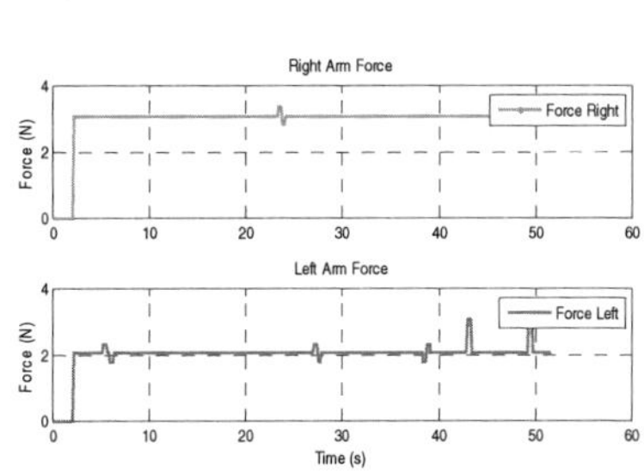

Figure 7. Holding Force while climbing

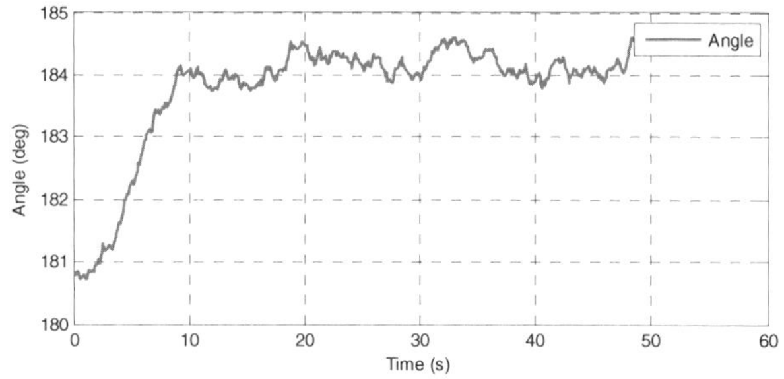

Figure 8. Pitch angle while climbing

Figure 9 shows the sequence of climbing during the experiment. From the initial position in Figure 9(a), the robot performed a grasp with both arms. As soon as the grasp was secured in Figure 9(c) the robot initiated the climbing task. The robot climbed at a steady speed in Figure 9(f)–9(h) while PID control system controlled the robot position level at minimum tilt angle deviation.

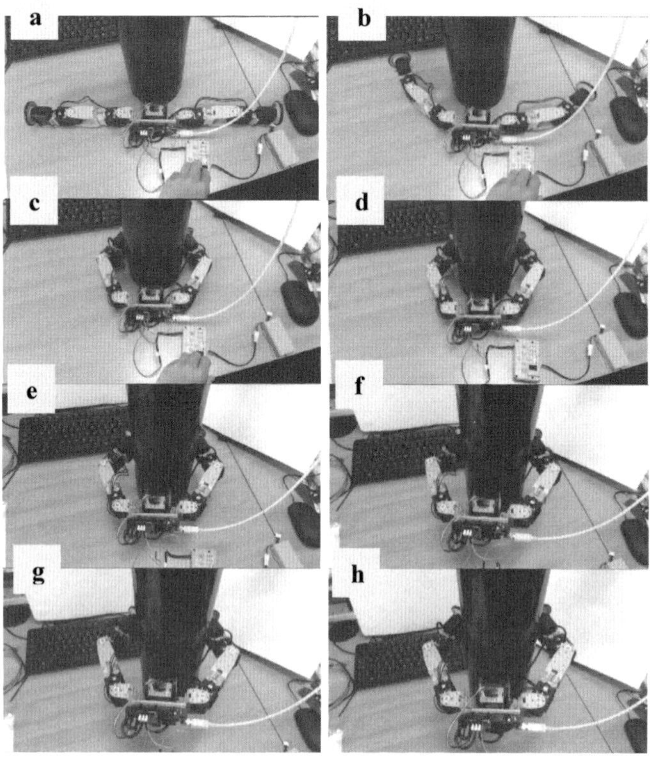

Figure 9: Experimental test of the climbing robot

7. Conclusion

The development of a climbing robot with grasping mechanism has been presented with experimental verification. A PID based control mechanism has been applied to the robot system by controlling the tilt angle with respect to the movement of the wheels. By controlling the angle, the robot can provide a good grip while climbing as well as maintaining the climbing stability. The experimental results have shown the ability of the robot to grasp/grip the pole and climb steadily. For future work, the second body with similar configuration will be attached with a flexible joint to increase contact points to the pole.

References

1. Sadeghi, A., Moradi, H., Ahmadabadi, M. N., Analysis, Simulation and Implementation of A Human Inspired Pole Climbing Robot, 2008.
2. Chen H., Sheng W., Xi N., and Tan J. Motion control of a micro biped robot for non-destructive structure inspection. *In Proceedings of the 2005 IEEE International Conference on Robotics and Automation*, pp. 480-485, Barcelona, Spain, April 2005.
3. Nishi, A. & Miyagi, H. Mechanism and control of propeller type wall-climbing robot, *Proceedings of the IEEE/RSJ International Conference on Intelligent Robots and Systems*, pp. 1724--1729, 1994.
4. Sitti, M. & Fearing, R. S. (2003). Synthetic Gecko Foot-Hair Micro/Nanostructures for Future Wall-Climbing Robots, *Proceedings of the IEEE Robotics and Automation Conference* , pp. 1164-1170, Sept. 2003.
5. C. Balaguer, A. Gimenez, and A. Jardon. Climbing robots mobility for inspection and maintenance of 3d complex environments. *Autonomous Robots*, 18(2):157-169, 2005.
6. Bell M., Balkcom D. A toy climbing robot. In Proc. of the 2006 IEEE Int. Conf. on Rob. and Aut., pages 4366-4368, Orlando, Florida, USA, May 2006.
7. Bicchi A., Kumar V. Robotic grasping and contact: A review. *In IEEE Int. Conf. on Robotics and Automation*, pages 348–353, San Francisco, CA, 2000.
8. Rimon R., Burdick J. On force and form closure for multiple finger grasps. In IEEE Int. Conf. On Robotics and Automation, 1996.
9. Mason R., Rimon E., Burdick J.. Stable poses of 3-dimensional objects. In IEEE Int. Conf. on Robotics and Automation, Albuquerque, NM, 1997.
10. Niku, Saeed B. "Robot Kinematics: Position Analysis." An Introduction to Robotics Analysis, Systems, Applications . Upper Saddle River, N.J.: Prentice Hall, 2001. 67-80. Print.
11. Lenormand, G., Migne´e, R. and Tinel, J. (1990), Ele´ments de technologie, Edition Foucher, Vanves, Chapitre 9 – Liaisons comple`tes de´montables obtenues directement par adherence. Section no. 9.1 "Arc-boutement", p.126.

A NOVEL MAP BUILDING METHOD AND EKF BASED LOCALIZATION FOR MOBILE ROBOT*

XIAOGANG WANG
College of Mechanical Engineering, BeiHang University, XueYuan Road No.37
Beijing, 100191, China

GUANGHUA ZONG
ARMS, XueYuan Road No.37
Beijing, 100191, China

Simultaneous Localization and Mapping (SLAM) algorithms with autonomous robots have received considerable attention in recent years. In general, those algorithms use odometry information and measurements from exteroceptive sensors of robots. The accuracy of the map building and the robot localization algorithms directly affect the overall success of the system. This paper proposed a novel method in map building and robot localization. Unlike traditional approach in building maps, which uses the recursive algorithm and is of low efficiency and occupies much memory, a gradient based algorithm is proposed which has the advantages of concise form, high efficiency and accuracy. Several experiments are taken to demonstrate the comparisons between the ordinary method and the novel one. In localization, Kalman Filter (KF) can evaluate the state of the system from the dynamic circumstances, but it must obey the rule that the noise is of Gaussian distribution, so the modified form (EKF) is adopted on many occasions. Compared to other algorithms such as Markov localization, Mont Carlo localization, EKF is more accurate. By taking the experiments, the applicability of the algorithm is verified. All these techniques have been implemented on mobile robot Pioneer P3-AT equipped with a 2D laser range scanner SICK LMS 200.

1. Introduction

Map building and localization play an important role in the field of robotics. Only with a precise map can the robot be capable of localizing itself in environment and performing navigation tasks credibly[1][2].

To achieve the above goals, we need an accurate map representing its surrounding environments[3]. For the moment, commonly used map are usually representated in three formats: grid-based map, topological map and geometric

* This work is supported by the National Natural Science Found of China (No. 51475018) and the Innovation Foundation of BUAA for PhD Graduates.

feature map. Of which, grid-based map is firstly proposed by Elfes and Moraves[4], it is convenient to build and to revise, but has low accuracy. Topological map is first proposed by Kuipers and Byun[5]. It is economize in computer resource but it is inconvenient to build and revise. Furthermore, when the error of the sensors is slightly big, it is difficult to recognize which is the wanted node. In this paper, the geometrical feature map is used which is well-knit and occupying low resource. Besides, it is intuitive[6].

As to robot localization, we use the data of inertial sensors and laser scanner. Because of the uncertainty in the process, many researchers use the probability methods[7][8]. Markov Localization can get the position by calculating the probability distribution of the robot position[9]. Mont Carlo Localization is the extended formation of Markov Localization. It just calculates the partial instead of all the probability distribution. Kalman Filter method is a common approach in localization and it is consist of two steps: making a preliminary estimation and rectification[10]. In this paper, EKF is adopted.

The paper is organized as follows. Section II presents the novel map building method based on geometric feature map. Section III presents the basic definition and algorithm of the KF and EKF. Section IV, the EKF localization method which is based on the information of laser scanner and odometer is presented. In section V, the result of the algorithm is verified by experiments. Finally, the paper is concluded in Section VI.

2. The Map Building Method

When building the map, we preliminarily cut data from the laser scanner into different parts according to a threshold. Considering the distances between two points is different because some points lie nearer to the robot while some are farther, so we should make a dynamic threshold to make data segmentation. The rule is as following: the distance of the point which is less than 4 meters has a threshold of d_{th} and out of 4 meters is $2.5 \cdot d_{th}$, of which, d_{th} is 80mm. The divided section is expressed as S_i $(i=1,2,3,\ldots,n)$, in which n is the number of the total sections. Figure 1 shows the preliminary segmentation of the original data.

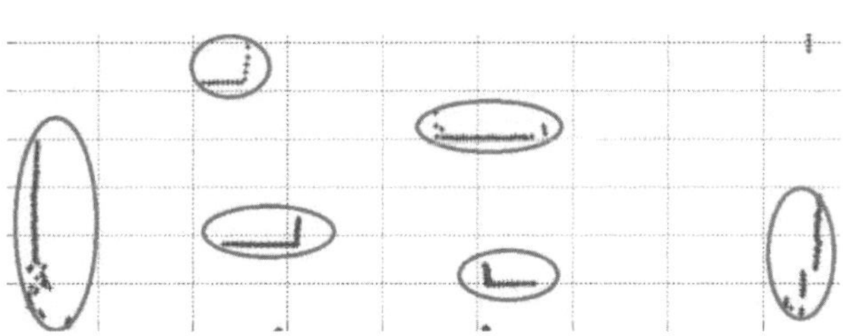

Fig. 1 Preliminary data segmentation

After the preliminary segmentation, we cannot make a line fitting directly because the data belong to the same section cannot fit to a straight line due to that they may belong to several ones with different gradients. Traditional method in further segmentation is the recursive method in which firstly draw a line with the first and the last points in a section and then calculate the farthest distance from remaining data. Secondly, segment the data into two sections and repeat the first step to find another maximum distance until the maximum meets some conditions. The recursive method not only costs time but has low accuracy.

In this paper, a gradient based method is proposed which calculates the gradient of the two adjacent points and make the gradient from the first two points as a criterion, the subsequent gradients from the next two points should be compared to the criterion gradient and if the difference exceeds 15°, we assume the correspondent point should belong to the next section. Up to now, S_i is divided into several sections which we mark them S_{ij} $(i=1,2,3,...,n; j=1,2,3,...,k)$. By taking the experiments, we demonstrate that the novel one is of higher efficiency and precision. Figure 2(a) shows abbreviated drawing of the principle. If α is larger than 15°, point 3 will belong to the next line. Figure 2(b) shows the result of the advanced segmentation.

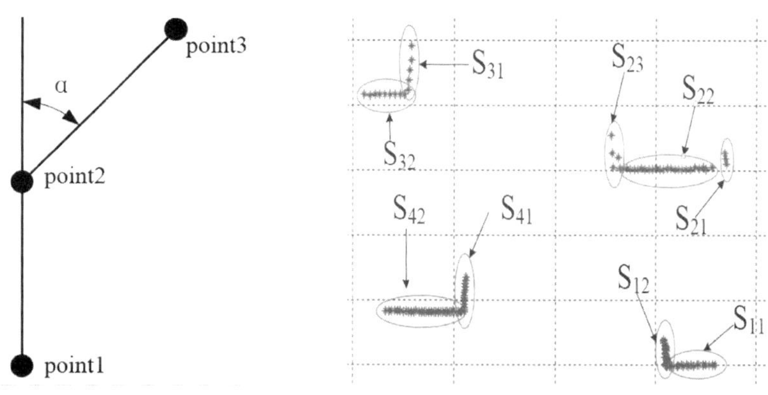

(a) Abbreviated drawing of the principle (b) The result of the advanced segmentation

Fig. 2 The illustration of the gradients based method

Once we accomplish the date segmentation completely, we can fit the lines. Here we use the least square method which is if of high efficiency and accuracy. In the least square method, we assume the function of a line is $y=kx+b$.

Figure 3 shows the comparison between the traditional method and modified method. From the two pictures we can see that the modified one is more accurate.

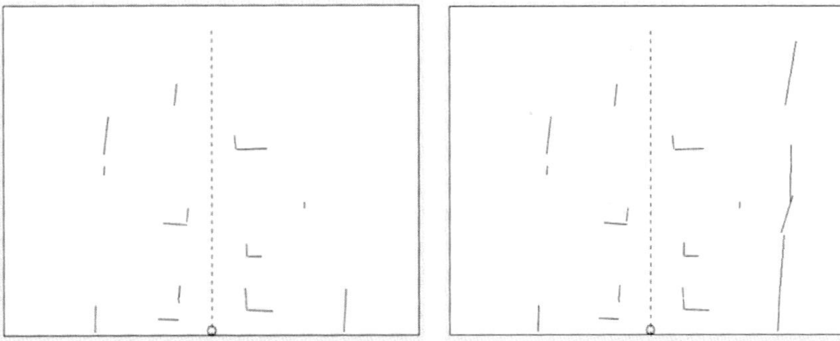

(a) The traditional method (b) The modified method

Fig. 3 The comparing result between the two methods

Table 1: The accuracy and efficiency of the two methods

No.	The traditional method	The novel method
Succeed times	6	10
Average time consuming	10.3 *ms*	6.4 *ms*

Table 1 shows the comparison of the two approaches. The experiment is carried out in ten times

3. The Introduction of the KF and EKF

3.1. *The Kalman filter*

The Kalman filter is an algorithm for fusing information in state-space. Essentially, it works by a two-step cycle, a prediction step and an update step, to recursively produce a state estimate with minimum mean squared error.

Given a system defined by the state x_k (at time k) and modelled by the Eq. (1):

$$\hat{x}_{k|k-1} = F_k \hat{x}_{k-1|k-1} + B_k u_k + q_k. \tag{1}$$

where:

F_k is the state transition model which is applied to the previous state x_{k-1};

B_k is the control-input model which is applied to the control vector u_k;

q_k is the process noise which is assumed to be drawn from a zero mean multivariate normal distribution with covariance Q_k and $q_k \sim N(0, Q_k)$.

At time k an observation (or measurement) z_k of the true state x_k is made according to Eq. (2).

$$z_k = H_k x_k + r_k. \tag{2}$$

where H_k is the observation model which maps the true state space into the observed space and v_k is the observation noise which is assumed to be zero mean Gaussian white noise with covariance R_k, of which, $r_k \sim N(0, R_k)$.

The Kalman filter determines the optimal Kalman gain as follow equations:

$$\tilde{y}_k = z_k - H_k \hat{x}_{k|k-1}. \tag{3}$$

$$S_k = H_k P_{k|k-1} H_k^T + R_k. \tag{4}$$

$$W_k = P_{k|k-1} H_k^T S_k^{-1}. \tag{5}$$

The vector $\tilde{y}_k$ is the error between the predicted and actual observation. The matrix S_k represents the innovation covariance. The updated state estimate is then determined as Eq. (6) and Eq. (7).

$$\hat{x}_{k|k} = \hat{x}_{k|k-1} + W_k \tilde{y}_k. \tag{6}$$

$$P_{k|k} = (I - K_k H_k) P_{k|k-1}. \tag{7}$$

3.2. *The extended Kalman filter*

In most real systems, they are not governed by linear equations. So the extended Kalman filter (EKF) was devised. For example, a state transition is modelled by a non-linear function *f* and an observation of the state modelled by a non-linear function as Eq. (8) and Eq. (9):

$$x_k = f(x_{k-1}, u_k) + q_k. \tag{8}$$

$$z_{k+1} = h(x_{k+1}) + r_k. \tag{9}$$

The function *f* can be used to compute the predicted state from the previous estimation and similarly the function *h* can be used to compute the predicted measurement from the predicted state. However, *f* and *h* cannot be applied to the covariance directly. Instead, a matrix of partial derivatives (the Jacobian) is computed.

In each timestep the Jacobian is evaluated with current predicted states. These matrices can be used in the Kalman filter equations. This process essentially linearizes the non-linear function around the current estimate.

4. The EKF Method Based on the Information of Laser and Odometer

EKF is a useful way in localization which firstly read the information of the odometer to be the robot's evaluate value of the position and angular orientation, then read the data of laser scanner and make a match between them to get the optimal position and angular orientation of the robot.

The specific steps are as follows:

1. Filter prediction: When the robot moves from the position x_k to the next position x_{k+1}, the difference in that period of time can be expressed as $u_k = [\Delta S_k, \Delta\theta_k]^T$, in which the ΔS_k is the length of the translation, while the $\Delta\theta_k$ is the difference of the angle. When define the motor system, we ignore the procedural noise. So the system defined by the state x_k is as Eq. (10):

$$x_{k+1} = f(x_k). \tag{10}$$

where $f(x_k)$ is the new pose according to the odometer and it is expressed as Eq. (11), Figure 4 is the illustration of the movement.

$$f(x_k) = \begin{bmatrix} x_{k+1} \\ y_{k+1} \\ \theta_{k+1} \end{bmatrix} = \begin{bmatrix} x_k + \Delta S_k \cos\theta_k \\ y_k + \Delta S_k \sin\theta_k \\ \theta_k + \Delta\theta \end{bmatrix}. \tag{11}$$

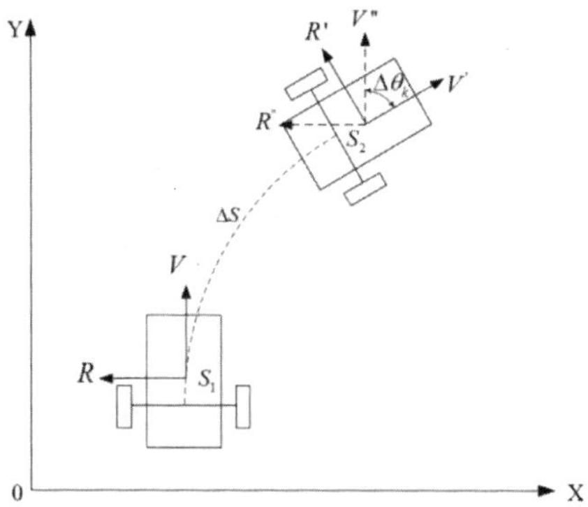

Fig. 4 The illustration of the movement

Linearize the coefficient of the $f(x_k)$ and filtered out the second order and more terms, we can get the Eq. (12):

$$\nabla f_{x_k} = \frac{\partial f}{\partial x_k} = \begin{bmatrix} \frac{\partial x_{k+1}}{\partial x_k} & \frac{\partial x_{k+1}}{\partial y_k} & \frac{\partial x_{k+1}}{\partial \theta_k} \\ \frac{\partial y_{k+1}}{\partial x_k} & \frac{\partial y_{k+1}}{\partial y_k} & \frac{\partial y_{k+1}}{\partial \theta_k} \\ \frac{\partial \theta_{k+1}}{\partial x_k} & \frac{\partial \theta_{k+1}}{\partial y_k} & \frac{\partial \theta_{k+1}}{\partial \theta_k} \end{bmatrix} = \begin{bmatrix} 1 & 0 & -\Delta S_k \sin\theta_k \\ 0 & 1 & \Delta S_k \cos\theta_k \\ 0 & 0 & 1 \end{bmatrix}. \tag{12}$$

2. Observation predication: We assume a line segment is L_i, parameters of the L_i in the robot coordinates are (d_i, θ_i), which is shown in Figure 5. In which, $\theta_i = \alpha_i + 90°$.

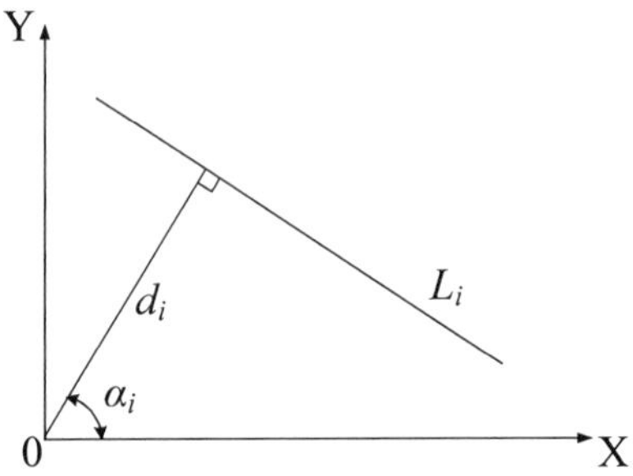

Fig. 5 The line segment L_i

So the observe equation is as Eq. (13):

$$z_{k+1,i} = \begin{bmatrix} d_k \\ \theta_k \end{bmatrix} = h(x_{k+1}, L_i) \tag{13}$$

in which:

$$h(x_{k+1}, L_i) = \begin{bmatrix} d_i - \hat{x}_{k+1}\cos\theta_i - \hat{z}_{k+1,i}\sin\theta_i \\ \theta_i - \hat{\theta}_{k+1} \end{bmatrix}. \tag{14}$$

Linearize the coefficient of the $h(x_{k+1}, L_i)$, we can get Eq. (15):

$$\nabla h_{x_{k+1}} = \frac{\partial h(x_{k+1}, L_i)}{\partial x_{k+1}} = \begin{bmatrix} \frac{\partial d_{k+1}}{\partial x_{k+1}} & \frac{\partial d_{k+1}}{\partial y_{k+1}} & \frac{\partial d_{k+1}}{\partial \theta_{k+1}} \\ \frac{\partial \theta_{k+1}}{\partial x_{k+1}} & \frac{\partial \theta_{k+1}}{\partial y_{k+1}} & \frac{\partial \theta_{k+1}}{\partial \theta_{k+1}} \end{bmatrix} = \begin{bmatrix} -\cos\theta_i & -\sin\theta_i & 0 \\ 0 & 0 & -1 \end{bmatrix}. \tag{15}$$

Equation (5) can be expressed as Eq. (16):

$$v_{k+1} = z_{k+1} - h(x_{k+1}, L_i). \tag{16}$$

3. Calculating the Kalman gain: The Kalman gain is expressed as Eq. (17):

$$W_{k+1} = P_{k+1} \nabla h_{x_{k+1}}^T S_{k+1}^{-1}. \tag{17}$$

4. Update the state: Repeat the step 1 to 3 and calculate all the Kalman gains, so we can get the new pose of the robot as following equation:

$$x_{k+1} = x_k + \frac{1}{n}\sum_{i=1}^{n} W_k [z_k - h(x_k, L_i)] = x_k + \frac{1}{n}\sum_{i=1}^{n} W_k v_k. \tag{18}$$

By executing the steps above we can get the calibrate position and orientation of the robot.

5. Experiment Results

In order to demonstrate the practicability of the algorithm, ten experiments are taken several times and Figure 6 shows the mobile robot: Pioneer P3-AT. During the experiments, each time is taken as following procedures: The robot firstly moves from the initial point with 10m and turns right 90 degrees, then repeating that step four times, the robot return to the start point. So the ideal coordinates of four corner points are (1000, 1000), (1000, 11000), (11000, 11000), (11000, 1000). During the movement, data of the sensors are recorded every 50ms.

Fig. 6 Pioneer P3-AT

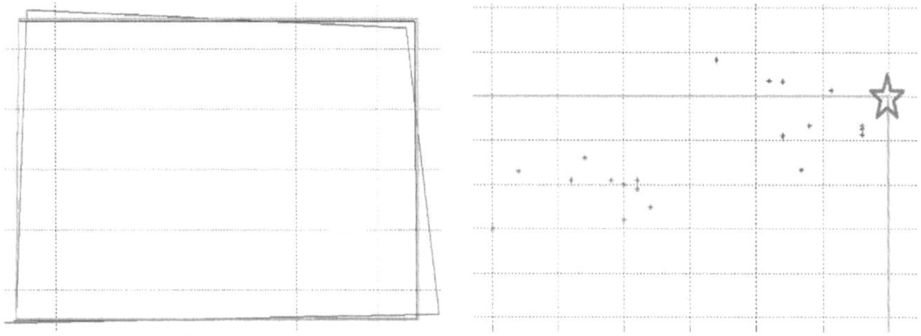

(a) Trajectories of the two methods (b) The comparison of two methods

Fig. 7 The experiment results

After the experiments, we calculate the arithmetic means of the four corners' coordinates. The trajectories of the two methods are shown in Figure 7(a), in which, the green line is the ideal trajectory, the red line is the non-calibrated trajectory and the blue line is calibrated trajectory. Figure 7(b) shows the points from one of the corners, pentagram is the precise position. From Figure 7(b) we can conclude that the calibrated positions keep close to the ideal value while the non-calibrated positions are less accurate.

6. Conclusion

This paper presents a novel algorithm in line extraction and an EKF-based algorithm which can be used in locating the robot. It shows that using the gradient based method is better than the previous one. Relying on the information of the laser scanner, the pose error accumulated by the odometer can be corrected and a precise pose of the robot can be gotten. Experiments are taken to verify the applicability of the algorithm and the results show the EKF based method has a satisfied accuracy.

References

1. Dissanayake G, Newman P, Clark S.A. Solution to the simultaneous localization and map building (SLAM) Problem [J]. IEEE Trans on Robotics and Automation, 2001:229- 241.
2. Guivant J E, Nebot E M. Optimization of the simultaneous localization and map-building algorithm for real-time implementation. IEEE Trans. Robotics and Automation [J]. 2001, 17(3):242-257.
3. Dedeoglu G Mataric M, Sukhatme G S. Incremental, online topological map building with a mobile robot. Proceedings of mobile robots, SPIE, 1999. 129-139 R. Tyrrell Rockafellar, Roger J-B Wets, Variational Analysis, Springer-Verlag, 2005, ISBN 3-540-62772-3, ISBN 978-3-540-62772-2, pg.117.
4. Elfes A, Moravec H. High resolution maps from wide angel sonar [C]. IEEE International Conference on Robotics and Automation, 1985:116-121.
5. KuiPers B, Byun Y T. A robot exploration and mapping strategy based on a semantic hierarchy of spatial representations [J]. Robotics and Autonomous Systems 1991:47-63.
6. Guivant J, Nebot E. Simultaneous localization and map building using natural features in outdoor environment [C]. Proceedings of Intelligent Autonomous System, 2000:581- 588.
7. N Vlassis, Y Motomura. Supervised linear feature extraction for mobile robot localization [C]. IEEE Int Conf on Robotics & Automation, 2000:2979-2984.
8. Arulampalam S, Maskell S. A tutorial on Particle filters for on-line non-linear/non-Gaussian Bayesian tracking [J]. IEEE Transactions on Signal processing, 2002, 50 (2):174-155.
9. D Schulz, W Burgard. Probabilistic state estimation of dynamic objects with a moving mobile robot [J]. Robotics and Autonomous Systems, 2001, 34:107-115
10. Rudy N. Robot Localization and Kalman Filters [M]. Utrecht University, 2003
11. Roumeliotis S I, Bekey G A. Bayesian estimation and Kalman filtering a unified framework for mobile robot localization [C]. Proceedings of the IEEE International Conference on Robotics and Automation, 2000: 2985-2992.

SECTION–4

BIOLOGICALLY-INSPIRED SYSTEMS AND SOLUTIONS

A BIO-INSPIRED BEHAVIOR BASED BIPEDAL LOCOMOTION CONTROL – B4LC METHOD FOR BIPEDAL UPSLOPE WALKING*

JIE ZHAO, QI LIU, STEFFEN SCHUETZ, and KARSTEN BERNS

Robotics Research Lab, University of Kaiserslautern, Kaiserslautern, 67655, Germany

Though over decades' development of bipedal robots, the terrains that bipedal robots can walk remains limited compared to what human can accomplish. To increase bipedal transversality on upslope terrain, this paper studies the biomechanical and biological aspects of human walking on inclined slopes with underlying the motor skills and reflexive systems. The control strategy can be divided into low and high gradient upslope walking. The strategy for the low gradient uphill walking is generated on the basis of an existing B4LC system. Furthermore, investigating the human walking on high gradient upslope terrain thoroughly unveils a new control strategy for bipedal walking on high gradient slope. Through validating the suggested method on a simulated biped upon different upslope terrains, the bipedal robot shows a naturally looking walking gait, achieving uphill walking up to 15° inclination which can compete with most advanced bipedal robots in the world.

Keywords: biped; B4LC; upslope walking; motor patterns; reflexes

1. Introduction

Compared to human walking on slope, the bipedal robots possess yet limited walking ability on those terrains. Following we will survey the scientific studies in this area. Chew *et al.*[1] introduced a *Virtual Model Control* approach to a planer bipedal robot for walking over rolling terrain, leading to a rhythmic gait during slope walking. However, the resulting walking looked unnatural and only the sagittal walking was taken into consideration. Huang *et al.*[2] introduced an approach for calculating the future ZMP trajectory for bipedal walking on uneven terrain. Braun *et al.*[3] proposed a state-dependent torque control for a biped in which the inherent dynam-

*This work was funded by the European Commission 7th Framework Program under the project H_2R (no.60069).

ics has been fully used. A spring-damper couple is constructed as strictly passive functions with fixed equilibrium points. Again, those ZMP-based control methods have many drawbacks as mentioned in.[4,5]

Looking over the mentioned studies on the upslope walking for bipedal robots, the achievement is yet limited. Current research focus only on the modification of ZMP based on the pre-defined trajectories, a more systematic solution for this situation is still missing.

2. Bio-inspired Control of Upslope Walking for Bipedal Robots

Rather than studying the mathematic model of system dynamics, the bio-inspired control approach absorbs all advantageous aspects of the human body, including its morphology, physiology, muscles, and receptors. The following content will introduce a new approach for walking on upslope based on the suggested B4LC method. Inspired by,[6] Sup *et al.*[7] developed a upslope walking controller for lower limb prosthesis. The knee and ankle kinematic data for healthy subjects walking on level and up 5°, 8°and 10°provide the biomechanical hint for the bipedal robots' upslope walking. Instead of changing the control parameters continuously, Sup *et al.*[7] suggested to insert a finite-state machine to decide the impedance parameters for the lower limb prosthesis in different inclinations. The suggested control approach utilizes the finite-state machine to switch the control parameters and behaviors to accommodate in various inclinations. According to the classification of the suggested controller, the control for upslope walking can be divided into *low gradient controller*, *5° controller*, and *10° controller*. Due to similarities of this *low gradient controller* with the existing B4LC system, we will not study it in this paper.

2.1. *Slope walking controllers*

A state machine to switch different controllers based on the detection of the ground inclinations is firstly developed, in which 5°, 10° 15° upslope controllers can be triggered separately.

In walking phase 1 (*Weight Acceptance*), the local reflex *Weight Acceptance Slope* is responsible for stretching knee joint through two strategies. The first one is to keep the current knee angle constantly at the end of the walking phase 5 (*Heel Strike*) until the ipsilateral ankle angle with a dorsiflexion over a threshold value. Then the second one keeps to flex the knee joint till the end of *Weight Acceptance*. Before approaching to a threshold, the distance from the CoM to the fore leg becomes relatively longer. Meanwhile, flexing the knee joint leads to rise of the CoM in altitude and requires more propelling power from the rear leg. Therefore, keeping CoM low before the fore ankle approaching to the threshold value θ_{sg} mitigates the forward movement of CoM. Additionally, a second controller stretches the leg by a gradual increase of stimulation. The target knee angle reduces along with a nonlinear speed on the basis of the knee angle at the end of the *Heel Strike*. The postural reflex *Forward Velocity Slope* generates torque at ankle joint in accordance with the CoM error. In walking on the high gradient slope, due to the gravitational potential and disturbances, *Forward Velocity Slope* is formulated as a PD-controller to stabilize the ankle joint. The target angle of the ankle $\alpha_{ankley,target}$ in the 10° indicates the desired dorsiflexion angle of the stance leg, as listed in Table 1. In the slope locomotion, the oscillation in the lateral direction becomes more disturbing. The development of posture reflexes for lateral stability, which mainly inherit the behaviors of original ones with slight modification, overwhelm this unexpected behavior.

Table 1. The target ankle angle of *Forward Velocity Slope* in different controllers.

Controller	The target ankle angle (deg)
10° uphill controller	$5 + \frac{1}{2}\theta_{sg}$
15° uphill controller	$10 + \frac{1}{2}\theta_{sg}$

In the walking phase 2 (*Propulsion*), two posture reflexes, *Upright Trunk Slope* and *Lateral Balance Ankle Slope* and two local reflexes, *Stabilize Pelvis* and *Forward Velocity Slope* keep active as in *Weight Acceptance*. Two new control units, motor pattern *Ankle Propel Slope* and the local reflex *Knee Propel Slope*, are developed to generate torques to propel the rear leg moving upwards and forwards.

Stretching the knee with an acceleration by the *Knee Propel Slope* results in facilitating the forward movement and lowering the CoM vertically at the same time. The stretching speed is proportional to current knee angle error, as described in Table 2. Stretching too fast at the beginning will result in an enormous upward kinetic energy acting on the body trunk, generating even a jump motion. Hence, the knee stretching starts with a relative low angular velocity to avoid this behavior. The motor pattern *Ankle Propel Slope* is only active after the contralateral leg touching on the ground. The *Ankle Propel Slope* inherits the algorithm mentioned in.[4] The distinction

Table 2. The knee angle error and corresponding speed in *Knee Propel Slope*

Difference between current and reference	Stretching speed
$difference > \frac{7}{8} reference\ angle$	$1 \cdot \dot{\alpha}_{knee\_stretching}$
$difference > \frac{5}{8} reference\ angle$	$4 \cdot \dot{\alpha}_{knee\_stretching}$
else	$8 \cdot \dot{\alpha}_{knee\_stretching}$

is, beside the desired velocity, the $xcom_{for.vel.,error}$ and the slope gradient, the $xcom_{for.vel.,error}$ of the opposite body side are introduced to generate a compensating torque working as an assistant propulsion during the dorsiflexion of the contralateral ankle.

Among all active control units in the walking phase 3 (*Stabilization*), *Upright Trunk Slope*, *Lateral Balance Ankle Slope*, *Stabilize Pelvis* and *Ankle Propel Slope* continue working as in previous phase. The slope gradient signal θ_{sg} is passed to the *Initialize Swing Slope* as a modulation signal to generate an additional torque in order to increase the swing velocity. The reasons for increasing the swing velocity can be summarized as following: First, in the slope locomotion mode, the knee angle at heel strike flexes more than that in walking on the level ground, leading to decrease the step length from the ipsilateral leg nearly 50 percent since the shank has the similar length as the thigh. To keep the step length long enough for both leg during this motion, the hip should stop swinging at the end of swinging behavior with a larger attack angle due to increased swinging speed. Second, in the swinging process, the perpendicular distance from the point of hip joint to the slope surface decreases with the ankle dorsiflexion of the supporting leg. Increasing the swinging velocity makes the foot of the swing leg pass the 'foot clearance' earlier and dwindles the possibility of the swinging leg hitting the ground.

In walking phase 4 (*Leg Swing*), the hip joint in the swinging process is controlled by the local reflex *Lock Hip Slope* until the leg comes to a swinging stop and touching down on the ground. This is a position control adjusting the hip angle through estimating the step length.

Compared to walking on level ground, the step length in the slope locomotion decreases with the dorsiflexion of the knee joint at the heel strike by reducing the required propelling torque. However, shorter step length could result in leaning forward seriously of the trunk. The calculation of target step length can be separated into two parts: the step length offered by the fore leg S_{fore} and by the rear leg S_{rear}. To implement this strategy, two prerequisite are needed: first, the rear leg is fully stretched; second, the shank of the fore leg is perpendicular to the level ground. The S_{rear} can be formulated by the parameters of the rear leg, as described in Eq. (1).

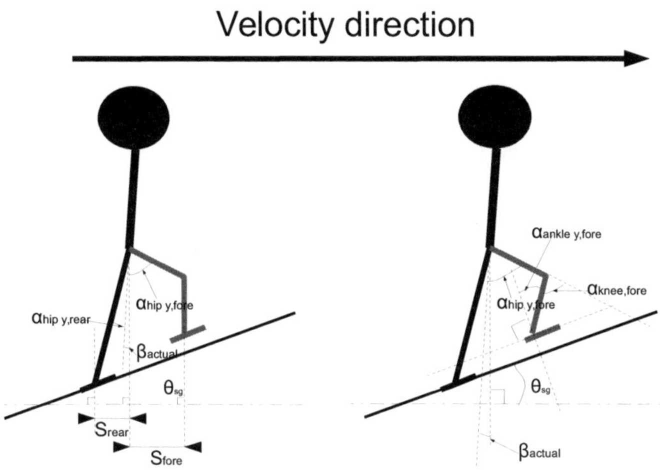

Fig. 1. The relevant parameters for the estimation of step length (left) and the geometry of target ankle angle in walking phase 4 (right).

$$S_{rear} = \sin(\alpha_{hip\ y,rear} + \beta_{actual}) \cdot (l_{thigh} + l_{shank}) \tag{1}$$

Where, $\alpha_{hip\ y,rear}$ indicates the current rear hip angle, l_{thigh} and l_{shank} denote the length of the thigh and shank, respectively. Then the S_{fore} can be estimated by the Eq. (2)

$$S_{fore,est} = \sin(\alpha_{hip\ y,fore} - \beta_{actual}) \cdot l_{thigh} \tag{2}$$

And the total step length $S_{estimation}$ equals the sum of the $S_{fore,est}$ and S_{rear}. When the difference between $S_{estimation}$ and S_{target} is larger than a threshold and is inversely proportional to the slope gradient, the position control of the *Lock Hip Slope* will take the geometrical hip angle as the target hip angle to avoid emergence of short step length. The geometrical hip angle is formulated by the Eq. (3).

$$\alpha_{hip\ y,fore} = \arcsin \frac{S_{fore,target}}{l_{thigh}} + \beta_{actual} \tag{3}$$

Where,

$$S_{fore,target} = S_{target} - S_{rear} \tag{4}$$

For the walking on the slope , on account of the rise of the ground, the local reflex *Lock Knee Slope* acts as an active position controller to reduce the forward step length of the swinging leg at the beginning. After the swing leg passes the 'foot clearance', *Lock Knee Slope* starts to stretch the swing leg to the target knee angle with increasing activation gradually. On the basis of the kinematics analysis on the inclined walkway, the target knee angle at the heel strike is set as shown in Table 3.

Table 3. The target knee angle at heel strike in different controllers

Controller	The target knee angle at heel Strike(deg)
10° uphill controller	$2.5 \cdot \theta_{sg}$
15° uphill controller	$5 + 2 \cdot \theta_{sg}$

In the slope locomotion mode, a cutaneous reflex *Ankle Swing Slope* is introduced. Through the position control, this reflex keeps the foot parallel with the ground surface during the whole walking phase 4 to reduce the forward step length. The target ankle angle $\alpha_{target\ ankle\ y,fore}$ is calculated as in the Eq. (5).

$$\alpha_{target\ ankle\ y,fore} = 90^\circ + \alpha_{hip\ y,fore} - \alpha_{knee,fore} - \beta_{actual} - \theta_{sg} \quad (5)$$

In walking phase 5 (*Heel Strike*), the posture reflex *Lateral Foot Placement Slope* and the local reflex *Lock Hip Slope* continues functioning. The posture reflex *Weight Acceptance Slope* described in the *Weight Acceptance* takes the charge of the knee control from the *Lock Knee Slope*. Before touching the ground of the foot, the knee joint increases the stiffness in order to keep the angle during foot landing. The local reflex *Heel Strike* starts regulating ankle dorsiflexion immediately after touching ground.

3. Experimental Results

Fig. 2. A part of the stride of the simulated biped walking on the 5° upslope

To validate the suggested control methods and evaluate their performance, a simulated bipedal robot with 21 degrees of freedom and the height up to 1.8 m (see Fig. 2) has conducted specific walking tasks on different slopes.[4] The experiments are conducting the bipedal walking on upslope at 8°, 10° and 15° respectively.

As can be seen in the left figure of Fig. 3, the main difference of the knee trajectories is that the maximum knee flexion during the swing phase decreases from 62° on level ground to 60° at 5° inclination. Due to the component of the gravitational force parallel to the slope, the rear leg supports more weight during stabilization phase, incurring a slower hip and knee flexion after toe off. The plantarflexion of the ankle joint increases from 22° on level ground to the 30° at 5° inclination which reflects the leaning forward of the whole body before propulsion.

The angle trajectories of the simulation results for walking on the 8°, 10° and 15° upslope are illustrated in the right figure of Fig. 3. The knee flexion during the heel strike phase among various slopes show significant differences. It increases from 4.5° to 38° along with an increase in upslope's inclination from 0° to 15°. This is induced by the local reflex *Lock Knee Slope* and shortened step length. Meanwhile, the shorter step length requires a smaller plantarflexion of the ankle joint before toe off.

As illustrated in Fig. 3, hip trajectories show a similar behavior as that of knee joint. The main difference is that walking on steeper upslope requires more hip flexion during heel strike and weight acceptance.

Considering the kinematics of ankle joint, one can find that ankle joints rotates for both plantarflexion and dorsiflexion in a smaller range when the upslope inclination increases. This is because in upslope walking, it requires faster propulsion and shorter duration in weight acceptance, resulting in avoidance of leaning forwards or backwards of the COM of the biped.

4. Conclusion

Inspired by studying the kinematics of human walking on those slopes, we have developed specific motor patterns and reflexes to implement varying upslope walking up to 15° in this paper. For future work, including systematical descriptions for both the motor patterns and reflexes during changing upslope and velocities, this way, can facilitate the introduction of optimization and learning schemes into the B4LC system, thus it leads to emergence of more human-like and efficient walking behaviors.

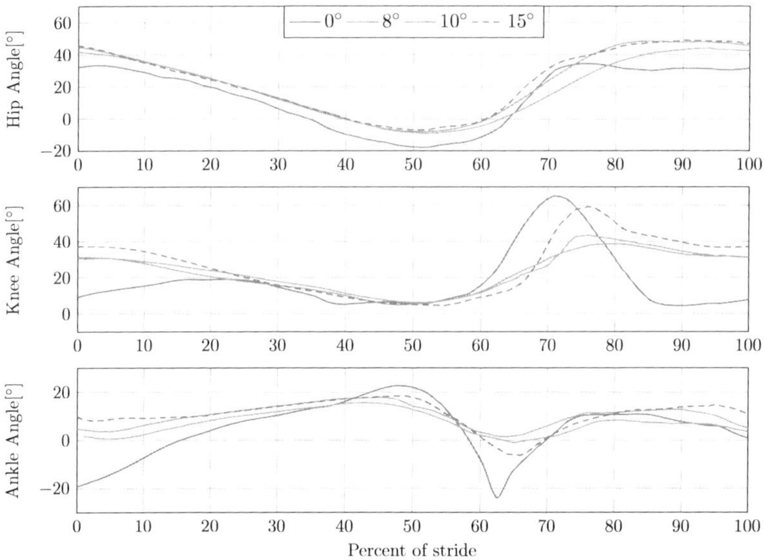

Fig. 3. The trajectories of the hip, knee and ankle joints among bipedal walking on level ground, 8°, 10° and 15° upslope. HS, CTO, CHS and TO indicate heel strike, contrateral toe off, contralateral heel strike, and toe off individually.

References

1. C.-M. Chew, J. Pratt, and G. Pratt, "Blind walking of a planar bipedal robot on sloped terrain," in *Robotics and Automation, 1999. Proceedings. 1999 IEEE International Conference on*, vol. 1, pp. 381–386, IEEE, 1999.
2. W. Huang, C.-M. Chew, Y. Zheng, and G.-S. Hong, "Pattern generation for bipedal walking on slopes and stairs," in *Humanoid Robots, 2008. Humanoids 2008. 8th IEEE-RAS International Conference on*, pp. 205–210, Dec 2008.
3. D. J. Braun and M. Goldfarb, "A control approach for actuated dynamic walking in biped robots," *Robotics, IEEE Transactions on*, vol. 25, no. 6, pp. 1292–1303, 2009.
4. T. Luksch, *Human-like Control of Dynamically Walking Bipedal Robots*. PhD thesis, University of Kaiserslautern, October 2009.
5. J. Zhao, S. Schütz, and K. Berns, "Biologically motivated push recovery strategies for a 3d bipedal robot walking in complex environments," in *IEEE - International Conference on Robotics and Biomimetics*, 2013.
6. A. S. McIntosh, K. T. Beatty, L. N. Dwan, and D. R. Vickers, "Gait dynamics on an inclined walkway," *Journal of biomechanics*, vol. 39, no. 13, pp. 2491–2502, 2006.
7. F. Sup, H. A. Varol, and M. Goldfarb, "Upslope walking with a powered knee and ankle prosthesis: initial results with an amputee subject," *Neural Systems and Rehabilitation Engineering, IEEE Transactions on*, vol. 19, no. 1, pp. 71–78, 2011.

ADAPTIVE WALKING ON UPHILL SANDY SURFACE BASED ON ORGANIC COMPUTING AND SOMATOSENSORY FEEDBACK

AHMAD AL-HOMSY*, JAN FROST and ERIK MAEHLE

Institute of Computer Engineering
Ratzeburger Alle 160, 23564, Luebeck, Germany
**E-mail: alhomsy@iti-luebeck.de*
www.iti-luebeck.de

Walking on uphill sandy surface is a challenging task and is considered one of the most difficult terrains that a walking robot can face. This paper will shed light on our applied decentralized approach to enable a hexapod robot to walk on uphill sandy surface effectively. In this paper we will show that the combination between our previously introduced approaches and our new strategy enables the hexapod robot to cope with this type of terrain. The suggested strategy is based on the synchronization between the moved legs that are touching the ground. This synchronization provides the robot's legs with the sufficient driving force during uphill walking. The novelty of our approach is the only evaluation of the local current consumption and angular position of each leg's joint as somatosensory feedback. It is based on an organic computing architecture and was tested on a low-cost version of the OSCAR walking robot.

Keywords: Adaptive Walking; Uphill Walking; Leg Synchronization; Sandy Surface; Somatosensory Feedback; Organic Computing.

1. Introduction

Walking on a sandy surface is not easy, especially when the terrain is inclined. Several studies have been conducted to enable legged robots to walk on sandy surface. A common method to detect sandy or soft ground is adding a force sensor mounted to the bottom of the robot's feet. This method is used by the TITAN-VIII robot,[1] the property of the walking ground (soft ground) is determined based on the sensed force, the robot reverses its forward walking into backward walking as soon as the force drops below the predefined threshold. In work closer to what we report here, Giguere et al.[2] presented how their six-legged amphibious robot identifies the environment. Their robot learns to identify the environment based

on the internal sensor information (vertical acceleration, motor current and angle of a particular leg) by using a probabilistic Bayesian classifier. A related work has been accomplished on the hexapod robot SandBot.[3] SandBot achieves adaptive walking on granular media based on changing the limb angular frequency that leads to change the type of the gait pattern.
Our introduced approach for walking on uphill sandy surface is based on a combination between our previous presented approaches, which are a sandy ground detection approach that is presented in[4] and a control of the leg position approach that is presented in.[5] These approaches have been tested previously just on flat ground and separately. In addition to the mentioned combination, a new suggested strategy has been added to enable the hexapod robot to walk effectively on flat sandy surface and additionally on uphill sandy surface. This approach depends only on the evaluation of the consumed current and angular position in the legs joints as proprioception sensing. This work has been implemented in the Organic Robot Architecture (ORCA)[6] and tested on the latest version of the hexapod robot OSCAR (Organic Self Configuring and Adaptive Robot), based on the Bioloid robot kit[a] which uses low-cost off-the-shelf servos.

2. Test Framework

The Organic Robot Control Architecture (ORCA) is designed to comply with Organic Computing principles. The main objective of Organic Computing (OC) is enabling the developed technical systems to exhibit self-x properties such as self-adapting, self-configuring, self-protecting, and self-healing.[7] Therefore, OC systems use controlled self-organization. ORCA is a modular and hierarchically organized software framework, which enables the design of organically inspired systems.[6] The architecture consists of two main units, the BCU (Basic Control Unit) and the OCU (Organic Control Unit). BCUs form the backbone of a system and provide sufficient functionality for the system under normal conditions, e.g. motion control or perception. OCUs are supplemented units to the BCUs. They monitor and observe the BCUs and other OCUs signals and have the authority to change the BCUs parameters as soon as the BCUs signal shows anomalies, ensuring e.g. the systems function in case of faults. For more information about ORCA architecture see the work.[6,8]
The ORCA architecture is tested on our demonstrator OSCAR robot that is based on the Bioloid kit as shown in Figure 1 (left). It has six roughly

[a]Bioloid Robot Kit http://www.robotis.com/xe/bioloid_en.

symmetrically distributed legs, which are divided into front, middle and hind legs. Each leg has three joints, which will be called α , β and γ in the remainder of this paper, with α being closest to, γ furthest away from the body, and β the middle joint. Each joint is actuated by a Dynamixel AX-12 servo motor, which makes it possible to read, among other parameters, consumed current, the actual servo position and speed. The consumed current is provided as the value at the time of reading. All robot components use two batteries (NiMH) as power supply. For testing purposes, the robot is controlled from an external PC via USB. The robot weight with batteries is about 2,5 kilogram. The foot is located at the tip of the leg with a hemispherical shape made of rubber. It is about 1.5 cm in diameter. Each robot leg achieves two phases during its cyclic movement, the swing and the stance phase. The leg moves between two extreme positions: the Anterior Extreme Position (AEP) and the Posterior Extreme Position (PEP). In the swing phase, the leg begins at the PEP. The β- and γ-joints lift the leg up. Then α-joint moves the leg forward to the AEP without touching the ground. At the AEP, β- and γ-joints move the leg down until it finds supporting foothold. The swing phase finishes and the stance phase begins. In the stance phase, the α-joint moves the leg back from AEP to PEP. During this movement, the leg keeps in touch with the ground and supports a part of the robot weight, which leads to propelling the robot forward. The stance phase finishes as soon as the leg reaches its PEP (see Figure 1 right for more clarification). This cyclic movement for each leg is not sufficient to achieve stable walking. A stable walking requires coordination between these six legs to prevent the robot from falling down. OSCAR coordinates its locomotion based on Cruse's rule,[9] where each leg independently switches from stance into swing phase when its two neighbouring legs are touching the ground.

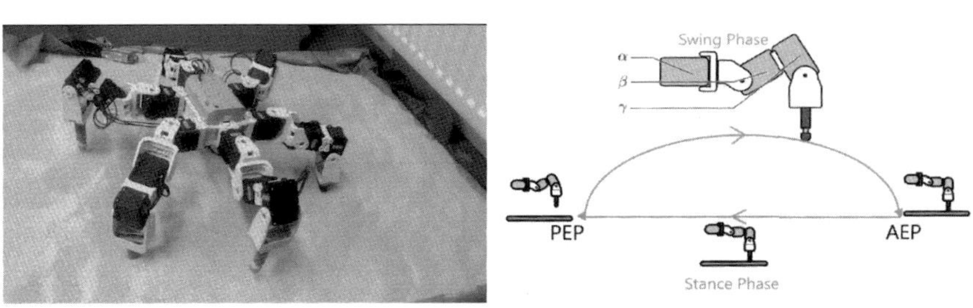

Fig. 1. Left: The test platform OSCAR based on Bioloid kit while walking on inclined sandy surface. Right: Schematic representation of the stance and the swing trajectory of the foot point between PEP and AEP.

3. Sandy Ground Detection

Sandy ground detection has been presented in our previous work.[4] For sandy ground detection, the β-current consumption is monitored at the end of the swing phase while the leg moves down. OSCAR's leg decides that the walking surface is hard as soon as the β-current consumption exceeds the maximum predefined threshold (T_{max}) without significant changes in the angular position of the β-servo motor at the moment of contact with the ground. As soon as the robot leg touches soft ground such as a sandy surface, the β-current consumption rises gradually, exceeding the minimum predefined threshold (T_{min}). In addition, there are significant changes in the β-angular position after the contact with the ground and before the current consumption exceeds the maximum predefined threshold (T_{max}). The leg stops moving down as soon as it reaches its maximum position or the current consumption exceeds T_{max}. The leg (L) that detects sandy surface is denoted with D^L_{sand}. D^L_{sand} is a Boolean value. It is defined at each leg's cycle at the end of swing phase. The ground type recognizing test reads:

$$D^L_{sand} = \begin{cases} true & \text{if } T_{min} < I^L_\beta \leq T_{max} \text{ and} \\ & \Delta P^L_\beta \geq D \\ false & \text{else} \end{cases} \tag{1}$$

Here, ΔP^L_β is the change in angular position after the consumed current exceeds T_{min}. T_{min} represents the maximum β-current consumption (I_β) while the leg moves down without contact to any object. D represents the amount of change in the angular position of the servo, after which the consumed current is assumed to rise above T_{max} on hard ground. T_{min}, T_{max} and D are defined by experiments. The robot control system decides that the walking terrain is a sandy surface when at least two cooperating legs detect sandy ground. The cooperating legs are the front and middle legs. This procedure ensures that the robot really faces soft or sandy ground. The robot detects the sandy surface based on the following condition:

$$D_{sand} = \begin{cases} true & \text{if } (D^{FL}_{sand} \text{ and } D^{FR}_{sand}) \text{ or } (D^{FL}_{sand} \text{ and} \\ & D^{ML}_{sand}) \text{ or } (D^{FL}_{sand} \text{ and } D^{MR}_{sand}) \text{ or} \\ & (D^{FR}_{sand} \text{ and } D^{ML}_{sand}) \text{ or } (D^{FR}_{sand} \text{ and} \\ & D^{MR}_{sand}) \text{ or } (D^{ML}_{sand} \text{ and } D^{MR}_{sand}) \\ false & \text{else} \end{cases} \tag{2}$$

As soon as the robot detects the sandy surface, OSCAR executes some new reactions that have been added in this work and are presented in the following section. Due to these reactions, the hexapod robot is now able to walk on all types of sandy surfaces such as flat, uphill and downhill sandy ground.

4. Reaction to Sandy Ground

Walking on compliant surfaces such as sand leads to the activation of the control of the leg position regardless of the sandy ground detection. Each leg controls the distance between the robots body and the surface independently from the other legs. This has been presented in detail in our previous work[5] and is active either if the robot walks on flat or inclined ground based on the following table:

$I_\beta \leq T_{min}$	The leg moves down without touching any object and continues its movement down until it reaches its maximal span.
$T_{min} < I_\beta \leq T_{max}$	The leg touches the ground and the OCU records the β-angular position. The leg is still moving down till the β-current consumption exceeds T_{max} or the leg reaches its maximum span. Then the leg has to adapt its position.
$I_\beta > T_{max}$	The leg stops moving down, the leg has to move back to the point of contact with the ground.

Each leg corrects its position to the point of contact with the ground and at the same time decides whether the walking substrate is soft or hard ground. In case of sandy ground detection, the robot system changes some parameters to achieve adaptive and effective walking, in this case each leg's OCU changes its BCU's parameters that leads to change the global behavior of the robot such as:

Decreasing the height of the robot body: In this case, the leg's OCU decreases the height of the leg step. The height of the robot body thus will be decreased. This procedure is triggered as soon as the robot detects a sandy surface independently of whether it walks on flat or on inclined surface. This increases the efficiency of the robot legs during the stance phase. The reason is: the produced torque in the α-joint is greater than the applied torque on it. This leads to pushing the robot forward effectively on sandy surface. Additionally, it guaranties that the robot does not overturn while walking on uphill/downhill surface.

Increasing the robot speed: Each leg's OCU increases the speed of the servo motors of its leg. This procedure reduces the time of leg synchronization.

Synchronization between the legs: This is considered the most important procedure. Walking on sandy surface in general and on inclined sandy surface in particular needs a special strategy. In the normal case, as soon as the leg reaches its AEP and touches the ground it executes its stance phase without waiting for the other stance legs that reach their aimed position (AEP) later. This leads to a wear driving force while walking on sandy surface due to the low friction coefficient between the leg and the surface. The low friction coefficient especially shows when walking on an uphill sandy surface. As a result, there is not a united driving force that pushes the robot forward effectively. Additionally, the α-joint is loaded and the robot sometimes stops moving forward while walking on uphill sandy surface.

To solve this problem, the robot has to synchronize its stance legs, in case of facing such a terrain, to produce a unified driving force that pushes the robot forwards. Our suggested approach for leg synchronization is activated as soon as the robot detects sandy surface. This approach work as following: Each leg's BCU communicates with other legs' BCUs. When one leg reaches its AEP, its BCU sends signal to other legs' BCUs informing them that it is at AEP and is ready to achieve the stance phase, but it has to wait for the other two legs at this point. As soon as the leg's BCU receives two signals from others legs's BCUs informing that they are at their AEP, the three legs retract backward together producing a unified driving force of the stance legs, which leads to push the robot forward on sandy flat and uphill/downhill surface.

5. Experimental Results

For experimental evaluation, our approach has been tested on an uphill sandy surface that presents the most difficult case. The test ground consists of an indoor area with 180x70 cm sandy basin with 15° inclination. The used sand is dry and fine, similar to the sand that is found in deserts. Figure 2 shows the used test ground.

To clarify how the robot detects the sandy ground and how each leg controls its position while walking on sandy inclined surface, Figure 3 shows the obtained results from the front right and left legs (FR and FL), which detect sandy surface in this experiment. The β-current consumption and the β-angular position are depicted while the leg moves down at the end of the swing phase, the stance phases are not depicted in this figure, because they are not relevant to the sandy ground detection and to the control of

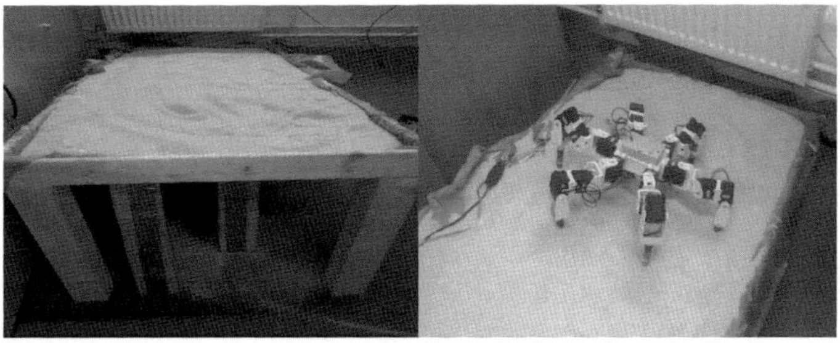

Fig. 2. The test ground, it is sandy surface with inclination 15°.

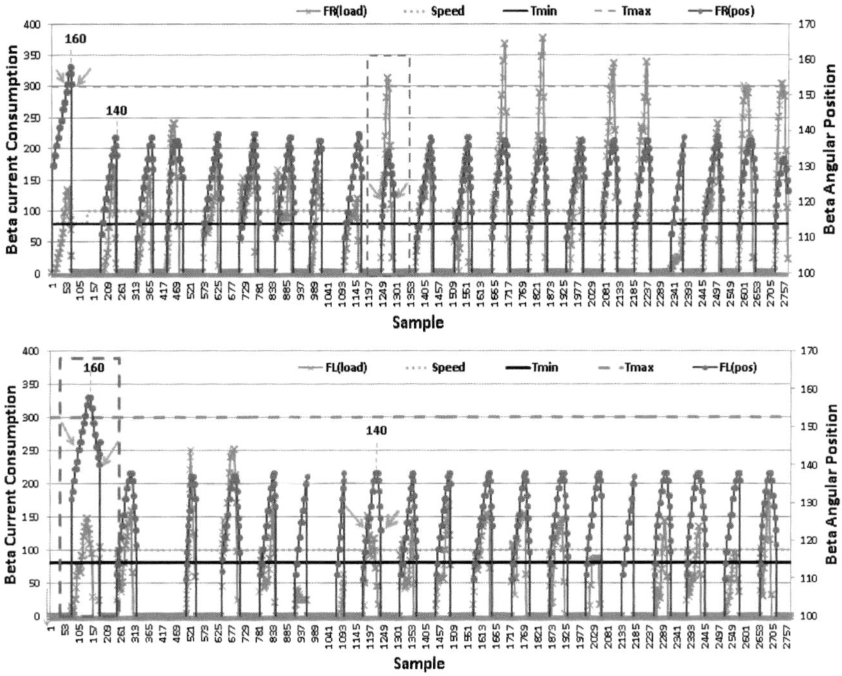

Fig. 3. The β-current consumption and the β-angular position while walking on sandy inclined surface.

the leg position. In this experiment, the β-joint moves the leg down from position 130° to 160°. In case the leg does not yet detect sandy surface. For more clarification, we will analyse the first step of the FR leg at the top of Figure 3. The FR leg touches the surface at data point 42 and β-angular position 153° (at the first arrow). The leg continues its movement until it reaches its maximal position, unless the β-current consumption rises above T_{max}. When the β-joint reaches its maximal position of 160°, the leg corrects its position to the point of contact with the ground at 153° (second arrow). The leg finishes its swing phase. Here, the leg has detected the sandy surface due to the obvious angular position changes in addition to the rising of the current consumption above the predefined threshold T_{min}. The robot detects the sandy surface, when at least two legs detect sandy surface. As shown in the plot at the bottom of Figure 3, the FL leg touches the surface at data point 113 and β-angular position 146° (at the first arrow in the first step). The leg continues moving down until it reaches its maximal position of 160°. The leg detects the sandy surface and tries to correct its position. It corrects its position to 146° (at the second arrow). Here the robot detects the sandy surface and activates several procedures that improve the walking on this terrain. It increases its speed by 20% (the speed is changed from 80 to 100 as shown in the dotted curve). Additionally, it decreases its body height. The β-joint now moves from 110° to 140°, due to decreasing the robot body height as shown in the following depicted steps in the FR and FL legs. After the two mentioned adaptations, the robot adapts its gait pattern based on using the synchronization between the stance legs.

Figure 4 shows the recorded gait pattern while OSCAR walking on uphill sandy surface. The bottom bars represent the times of contact with the ground. It begins at AEP and finishes when the leg switches into swing phase. The spaces between them are swing phases. The times of stance phases are depicted in the middle above the bars of (the leg on the ground) as shown in Figure 4. It begins at the AEP when the leg touches the ground and ends at the PEP. The leg switches into swing phase when the Cruse's rule is fulfilled. Sometimes, the leg waits at PEP until Cruse's rule is fulfilled and then it switches into swing phase. Therefore, the time of bottom bars (on ground) is bigger than the time of middle bars (stance phases). The top bars represent the time of the active movement of the leg during the stance phase. Before sandy ground detection, it begins and finishes with the stance phase as shown in Figure 4. After sandy ground detection, it depends on leg synchronization. In our case the robot detects a sandy surface at data point 667 (see the first solid vertical line). After this point, the robot synchronizes its legs. For more clarification, we will analyse the cases that are marked with three circles. The vertical dashed line represents the moment of movement of all three legs (MR, HL and FL), when all legs are at the AEP. The FL leg reaches its AEP first (circle 1), its stance phase is activated, but it waits for the other two legs. Then the MR leg reaches its AEP (circle 2), its stance phase is activated, but it waits until another leg reaches its AEP. Finally, the HL leg reaches its AEP (circle 3). At this moment, the three legs move together executing the stance phase. The same scenario is repeated for all legs. Similar cases are shown at the two dotted vertical lines. Due to the leg synchronization, the robot can now walk on sandy uphill surface more efficiently and quickly. It has been found through experiments that without this measure, our robot cannot walk on sandy uphill surface at all.

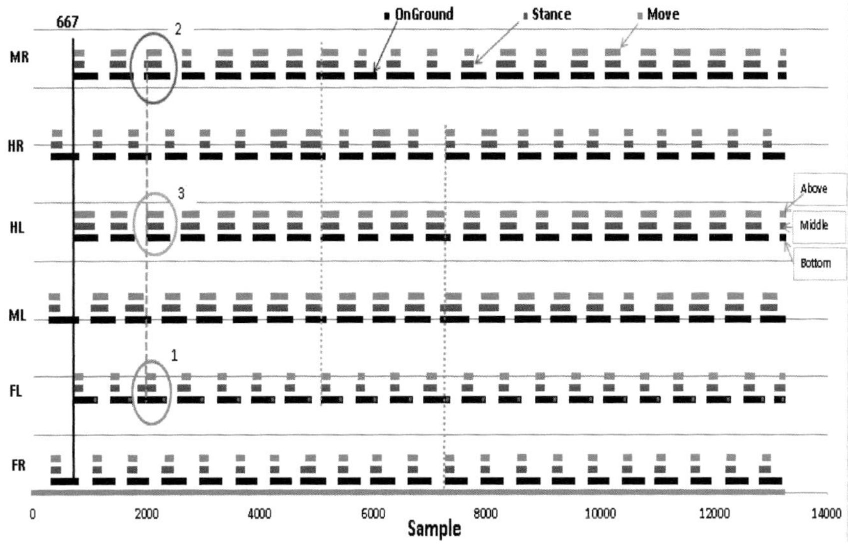

Fig. 4. Gait pattern while walking on sandy uphill surface.

6. Conclusion

This work aims to improve the robot walking on inclined sandy surfaces. It is based on our approaches for sandy ground detection and control of the robot leg position on compliant surfaces as well as on our new suggested strategy for walking on a sandy inclined surface. When the robot detects a sandy surface, it changes several parameters and achieves legs synchronization that helps the robot to walk on sandy inclined surface effectively due to unifying the driving forces of the stance legs. Moreover, our approaches are based on analyzing the consumed current and angular position without using any additional sensors, which would increase the cost of the robot and the complexity of the control system. Furthermore, this approach can be used as alternative sensing in the event of a malfunction of the used hardware sensors.

Acknowledgments

This work was funded in part by the German Research Foundation (DFG) within priority programme 1183 under grant reference MA 1412/8-2 and the ALBAATH University, Syria.

References

1. M. Ikeda, K. Izumi and K. Watanabe, Navigation strategy for a quadruped robot on soft flat ground, in *Control Automation and Systems (ICCAS), 2010 International Conference on*, 2010.
2. P. Giguere, G. Dudek, S. Saunderson and C. Prahacs, Environment identification for a running robot using inertial and actuator cues, in *Robotics: Science and Systems*, 2006.
3. C. Li, P. B. Umbanhowar, H. Komsuoglu, D. E. Koditschek and D. I. Goldman, Sensitive dependence of the motion of a legged robot on granular media, in *Proceedings of the National Academy of Sciences*, (Vol.106, no.9) 2009
4. A. Al-Homsy, J. Hartmann and E. Maehle, Slippery and sandy ground detection for hexapod robots based on organic computing principles and somatosensory feedback, in *Robotic and Sensors Environments (ROSE), IEEE International Symposium on*, 2012.
5. A. Al-Homsy, J. Hartmann and E. Maehle, Adaptive control of leg position for hexapod robot based on somatosensory feedback and organic computing principles, in *Proceedings of the 16th International Conference on Climbing and Walking Robots, Sydney, Australia - Nature-Inspired Mobile Robotics*, 2013
6. W. Brockmann, E. Maehle, K.-E. Grosspietsch, N. Rosemann and B. Jakimovski, ORCA: An organic robot control architecture, in *Organic Computing: A Paradigm Shift for Complex Systems*, (Springer, 2011)
7. C. Mueller-Schloer, Organic computing - on the feasibility of controlled emergence, in *Hardware/Software Codesign and System Synthesis, 2004. CODES+ ISSS 2004. International Conference on*, 2004.
8. E. Maehle, W. Brockmann, K.-E. Grosspietsch, A. E. S. Auf, B. Jakimovski, S. Krannich, M. Litza, R. Maas and A. Al-Homsy, Application of the organic robot control architecture ORCA to the six-legged walking robot OSCAR, in *Organic Computing: A Paradigm Shift for Complex Systems*, (Springer, 2011)
9. H. Cruse, What mechanisms coordinate leg movement in walking arthropods?, in *Trends in Neurosciences*, (Vol.13, no.1) (Elsevier, 1990)

FERROMAGNETIC SOFT ROBOT BIO-MIMETICALLY INSPIRED

HERNANDO LEON-RODRIGUEZ *

Industrial Engineering Department, Nueva Granada Military University, Karrera 11# 101-80 Bogota, Colombia.

SUKHO PARK, JONG-OH PARK

**Robot Research Initiative, Chonnam National University, 300, Yongbong-dong, Buk-Gu, Gwangju, Korea*

Nature has adapted to allow the survival of its species through the development of several forms, materials, techniques and processes. These features, such as flying, defensive and offensive mechanisms and many others have inspired researchers, creative and enthusiasts into developing artificial solutions that mimic their natural counterparts. Early developments of biomimetic involved large structures, however, with current advances in mechanics, electronics, and computer science the development of micro and nano structures is allowing the creation of more complex, flexible and portable structures resembling biological features found in nature. These advances in robotics may help solve or improve tasks otherwise difficult with traditional robotics. This paper focuses on the design and development of soft-robotic manta ray prototype whose motion is controlled using magnetic fields over a body composed of a ferro-fluid. The goal is to explore magnetics actuation as an alternate mean to control and reduce soft-robotics size as an alternative to current traditional mechanical or electromechanical biomimetic developments and even shape memory alloys

1. Introduction

The evolution found in nature has provided sufficient material for analysis and research in the development of solutions in areas such as aerodynamics, self-cleaning, smart materials, self-assembly and energy conversion among many others, resulting in biologically inspired designs mimicking nature's functionality [1]. In the field of robotics several mechanisms have been inspired by the biomechanics of living animals, many of them with remarkable characteristics that human creations cannot meet yet.

Due to advances in micro mechanics, electronics and smart materials smaller components can be currently designed and fabricated, thus resulting in the miniaturization of robotics [2]. As the development of Micro Electro-Mechanical Material technologies increased many smart materials started being developed with applications as active, passive or wireless actuators. These smart materials used as actuators can be seen in a Shape Memory Alloy (SMA), Piezoelectric Actuators (PZT), Ionic Exchange Polymer Metal Composites (IPMC), and even ElectroMagnetic Actuators (EMA), among many others [3].

Soft-robotics is an emerging field of technology that is taking advantage of the properties from the advances in smart materials. Among the applications, smart materials can be used for developing miniature actuators that can resemble movements found in nature. [2]

Some of these smart materials are being used in several biomimetic inspired robots such as tadpoles, jellyfish, walking robots, fish and propelling fins as alternate locomotive actuators. Within these developments, advances has also been made in the development of manta ray-like micro robots actuated with SMA wires that can flexibly bend the fins [4]. One approach is a 3d swimming tadpole mini-robot that can move freely in water arranged with actuation mechanism of one permanent magnet powered wirelessly using an EMA system. [2] Other approaches in smart materials is the conduct polymer material developed as actuator, structural component and power deliver mechanism. [5] Emerged field of autonomous soft robotic fish powered by fluidic elastomer actuation system is also considered. [6] Or fish-like aquatic robotics using flexible bimorphs made of macro-fiber composite (MFC) piezoelectric laminates for carangiform locomotion. [7]

This paper explores the design and the development of a biomimetic inspired ferro-fluid controlled aquatic manta raysoft-robot. The manta ray is one of the most efficient fish swimmers (11.2 Km/h) considering its size, weight (up to 1300 kg) and morphology. Its unique motion characteristics allow saving more efficiently energy than any underwater vehicles created by human [8]. The main subject of study is their swimming modes based on propulsion or oscillation mode (fins lift up /down), which its uses in a cruise or long distance and high speed; and a second mode denominated undulation mode where energy is saved for swimming at low speeds, through an amplitude of motion at least the half of oscillation mode as presented in Figure 1. This navigational feature is the main topic of interest and research in robotics.

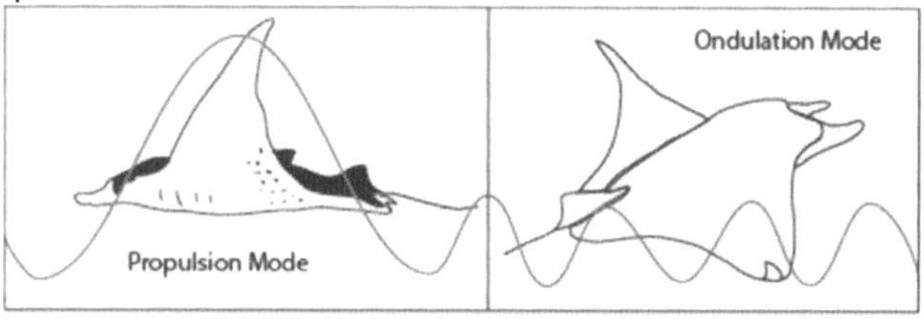

Figure 1. Manta ray's swimming modes.

2. Materials and Methods

2.1. *Biomechanical and mimetic analysis*

The manta ray is characterized for its inaudibly swimming modes, low flapping frequency in slow speed and high flapping amplitude in long distance. It also has a highly manoeuvrability in vertical direction, horizontal direction and can instantly change course while swimming upwards and downwards. In addition, its unique stability and locomotion is caused by the position of its pectoral fins close to its centre of gravity [9].

The pectoral fins of the manta ray are characterized for their adaptation, efficiency and several numbers of achievable degrees of freedom due to its skeletal structure that allows moving in both directions and make 360 degrees turns [8]. Its remarkable adaptation, control and power allow the manta ray to leap out of water and even glide as bird for a few milliseconds.

During locomotion displacements the manta ray performs different movements within a single flap, these are important because the proximal and distal parts of the pectoral fin do different motions in any given moment. Also, the flexibility of the fins can be classified into two sections: the relatively rigid proximal part and the relatively flexible distal part. The proximal part is the first third part nearest to the body while the remaining part is distal part as presented in Figure 2 [9] [10].

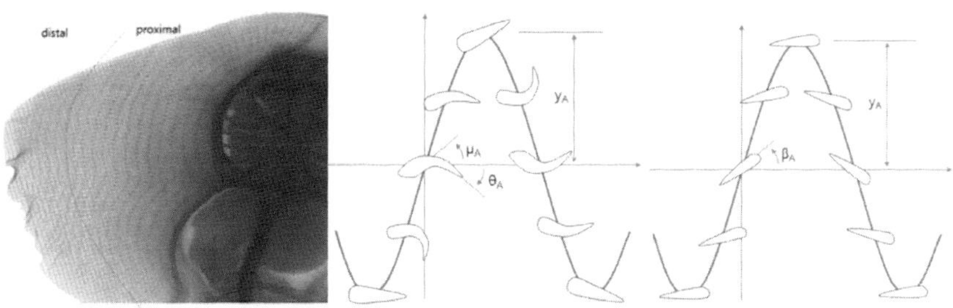

Figure 2. left: Manta ray's distal and proximal pectoral fin location; and simplified model of fin motions: center: front view, right: side view.[9].

2.2. *Design*

The manta swimming trajectory requires the movement of the pectoral fin in three dimensions for achieving both propulsion and undulation modes. The diagram model of its motion could be simplify in two ways, as presented from the frontal and side view as presented in Figure 2. A mechanical representation of the pectoral fin for exactly representing its movements would require a complex mathematical system composed of several linkages that would mimic the navigation modes. The fin deformation of Manta Ray requires mathematical model simplification with minimum degrees of freedom. [11]. This simplification is possible by locking the angular motion for simplicity, the model of proximal and distal parts will not consider; as a result, the angles μ_A, θ_A are not included, so the fin will act as a forehand flat surface described as follow:

$$y_A(t) = y.\sin(\omega t)$$
$$\beta_A(t) = \beta.\sin(\omega t + \varphi)$$

Where: β_A is the rotational angle, y_A is the maximum amplitude of translation displacement, ω is angular frequency and φ is phase difference.

2.3. *Passive actuator: ferrofluid*

Considering that the goals is to develop a soft-robot, the ferro-fluid is chosen as it allows a stable colloidal suspension of sub-domain magnetic particles within a liquid carrier. The particles average size of about 100Å (10 nm) allows miniaturizing the robot's actuator. These particles are coated with a stabilizing dispersing agent (surfactant) which prevents their agglomeration even when a strong magnetic field gradient is applied [12]. Typically ferro-fluid contains in fraction of volume 5% magnetic solid, 10% surfactant and 85% carrier. The surfactant must overcome the attractive van der Waals and magnetic forces between the particles. In addition, surfactants have a polar head and non-polar tail (or vice versa). However, ferrofluids lose their magnetic properties at sufficiently high temperatures, known as the Curie temperature.

2.4. *Movement control*

Soft robotics devices can mimic natural movements involving oscillations or undulations when the passive actuator is triggered through a magnetic field. The variations of amplitude and angle of deformations could be considered as a key factor resulting from the material's properties, such as the flexibility and the thickness that could be affected in the robot performance.

One common EMA system is a combination of 2D, 3D electromagnetic coils arranged at 90 degrees with one pair of Helmholtz, Maxwell, Uniform or Gradient coils which generates uniform magnetic field in x, y, z axes. As an example, these pair of coils generally consists of two identical circular magnetic coils, where the radius (r) of the coils is equal to the distance (d) between them for Helmholtz configuration. In addition, the applied currents in these coils flow in the same direction and have the same intensity in order to produce continuous magnetization. There are some other especial electromagnetic actuators designed to obtain similar results of magnetic field in all three axes with the advantage of increasing the region of interest (ROI).

In terms of the ferro-fluid material acting as an active actuator powered wirelessly by the magnetic field, can be expressed by the following equation:

$$F= \mu o V(M.\Delta)H$$

Where; μo is the magnetic permeability of free space, V is the volume of ferro-fluid particles, M is the magnetization of a ferrofluid, and the H is the magnetic field intensity applied [2]. The ferro-fluid responses when a magnetic field is applied, the magnetic particles placed in the soft-robot orient along the field lines almost instantly. The magnetization of the ferro-fluid responds immediately to the changes in the applied magnetic field and when the applied field is removed, the actuated areas back to original position.

In a gradient field the whole fluid responds as a homogeneous magnetic liquid which moves to the region of highest flux. This means that the ferro-fluid can be precisely positioned and controlled by an external magnetic field producing linear displacement acted in the soft robot. The forces holding the magnetic fluid in place are proportional to the gradient of the external field and the magnetization value of the fluid. This means that the retention force of a ferro-fluid can be adjusted by changing either the magnetization of the fluid or the magnetic field in the region or interest.

3. Experimental Results

Magnetic forces are been applied in all three axes as a preliminary test in order to produce horizontal motion and orientation on the manta ray with positive buoyance powered by the EMA system. The EMA system is composed of 1 Helmholtz coil and 1 Maxwell coil in x axis, 1 uniform coil in y axis, and 1 uniform coil and 1 gradient coil in z axis.

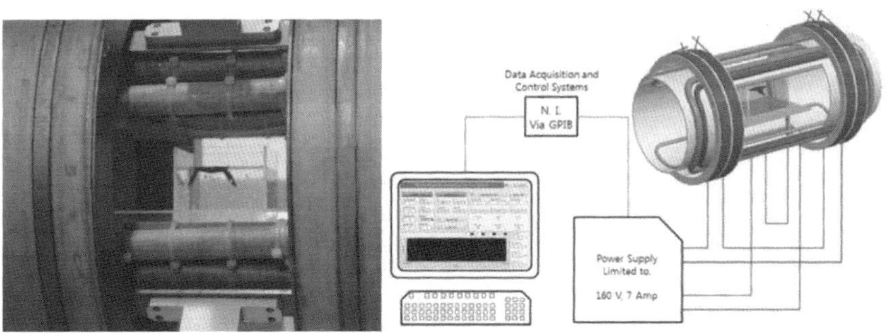

Figure 3. Left: Electro-magnetic actuation system; Right: Schematic diagram of EMA system power micro-robots.

Figure 3 (left) presents the electromagnetic actuator system used to execute the experiments with different robots shapes, however in this case the focus was on the manta ray. In the first trial the 5 cm robots were power only with z axes coils. The combinations of the two magnetic coils in the z-axis were controlled by a computer using Lab-view GUI and program via GPIB network protocol. The two coils were powered at 160 V dc with limited current of 7 Amp. The schematic diagram is showing in Figure 3 (right).

The process of constructing a soft-robot takes advantages from 3D printing of elastic polymers or even uses hands skills to assembly. The manta ray was constructed using flexible materials of thin film polymers. The biomimetic shape and scale is down into several times to achieve the model description. The pectoral fin and tail were made with 100 μm polyurethane film and 20 μm of polyethylene film sealed by a heat device which creating the contour shape of biomimetic robot, as presented in Figure 4.

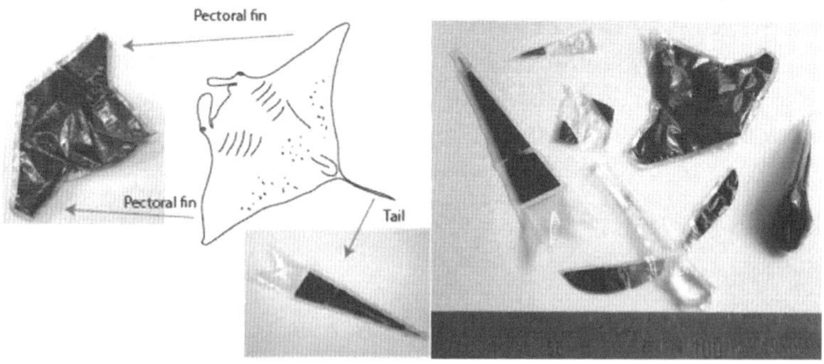

Figure 4. Left: soft robotic of manta ray like robot; right: different sizes of soft robotic prototypes.

The film materials are very flexible, thus allowing generating passively phase differences under the hydrodynamic force; furthermore they have a solid mechanical strength to recover back into its original shape. The motion is generated as a result of the induced magnetic field that allows changing the soft-robot form through the movement within the ferro-fluid.

Figure 5 presents the preliminary tests with 5 cm length prototypes exposed to the EMA system actuator. The picture allows visualizing a sequence of low frequency (2 Hz) magnetic field applied in the Z axis direction powered by the gradient coils. The sequences are showing a red line limit of maximum amplitudes reached by the robots when maximum positive voltages are applied. The voltages are sinusoidal oscillations in other to generate the undulations movements of the robot. In addition, similar performance is seen when negative values are applied. The reason of use half of sinusoidal voltages is due to the ferro-fluid does not have a positive of negative pole as a common magnet. So the returned forces to the original positions are done by the flexibility of the film. This result in high frequency voltage cannot be achieved in other to obtain better performances.

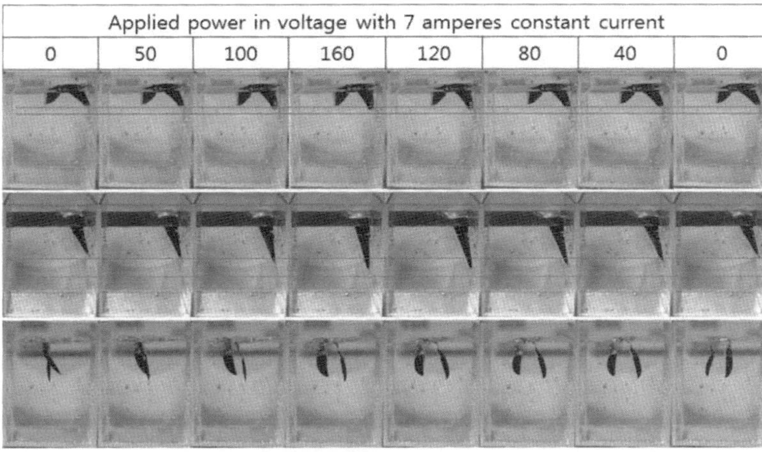

Figure 5. Biomimetic soft robot fin movement.

4. Discussion and Future Works

Several tests were performed with minimum success due to low effects of the actuator within the EMA system, as the ferro-fluid composition proved to required stronger magnetic fields to bend the fins; a possible solution is to use permanent magnets actuated by manual control or with high control theories involving for precise orientation and position.

After testing the manta ray it was observed that the performance and elasticity changed dramatically due to the evaporation of the ferro-fluid. This differences can be visually see between polyethylene film and polyurethane film, were last one keeps the ferrofluid for much longer as a liquid.

Ferro-fluids presented a remarkable actuation response in soft robots for future applications and improvements. However, to obtain a precise control it requires further analysis as nano particles for molecular motion. Also, the ferro-fluids could be considered as neutral monopole or magnetized dense liquid that follow the magnetic field with capillarity restrictions forces.

It is worth noting that during this exploration, ferro-fluids provided an interesting approach for the development of passive miniature actuators in the field of soft robotics, so the results of experimentation and applications could be consider as effort for further research.

Additional challenges arose from the tests, such as the difficulty to trace concentration material of ferro-particles in specific areas in order to improve the performance and also create a micro buoyancy space to execute in a flotation motions its horizontal displacements. So micro channels as alternative of powered the ferro-fluid is also possible in other to used its hydrodynamics properties.

References

1. Yoseph Bar-Cohen, Biomimetics—using nature to inspire human innovation, Institute of physics publishing Ltd, 2006.
2. Hyunchul Choi, Semi Jeong, Cheong Lee, Bang Ju Park, Seong Young Ko, Jong-Oh Park*, and Sukho Park*, Three-Dimensional Swimming Tadpole Mini-Robot using Three-Axis Helmholtz Coils, International Journal of Control, Automation, and Systems 2014.
3. Won-Shik Chu, Kyung-Tae Lee, Sung-Hyuk Song, Min-Woo Han, Jang-Yeob Lee, Hyung-Soo Kim, Min-Soo Kim, Yong-Jai Park, Kyu-Jin Cho, and Sung-Hoon Ahn; Review of Biomimetic Underwater Robots Using Smart Actuators; International journal of precision engineering and manufacturing vol. 13, no. 7, 2012, pp. 1281-1292.
4. Zhenlong Wang, Yangwei Wang, Jian Li, Guanrong Hang, A micro biomimetic manta ray robot fish actuated by SMA, IEEE International Conference on Robotics and Biomimetics, 2009.
5. James Louis Tangorra, Member, IEEE, S. Naomi Davidson, Ian W. Hunter, Peter G. A. Madden, George V. Lauder, Haibo Dong, Meliha Bozkurttas, and Rajat Mittal, The Development of a Biologically Inspired Propulsor for Unmanned Underwater Vehicles, IEEE journal of oceanic engineering, vol. 32, no. 3, July 2007.
6. Andrew D. Marchese, Cagdas D. Onal, and Daniela Rus; Autonomous Soft Robotic Fish Capable of Escape Maneuvers Using Fluidic Elastomer Actuators, SOFT ROBOTICS Volume 1, Number 1, 2014.
7. L Cen and A Erturk, Bio-inspired aquatic robotics by untethered piezohydroelastic actuation, BIOINSPIRATION & BIOMIMETICS, 2013 IOP Publishing Ltd.
8. Wanchao Chi and Kin Huat Low, Review and Fin Structure Design for Robotic Manta Ray (RoMan IV), Journal of Robotics andMechatronicsVol. 24 No. 4, 2012.
9. Z. Wang, Y. Wang, Jian Li, G. Hang, A micro biomimetic manta ray robot fish actuated by SMA, International Conference on Robotics and Biomimetics, pp. 1809-1813, 2009.
10. Justin T. Schaefer and Adam P. Summers, Batoid Wing Skeletal Structure: Novel Morphologies Mechanical Implications, and Phylogenetic Patterns, journal of morphology 264, pp. 298–313 (2005).
11. J. Gao, S. Bi, Y. Xu and C. Liu, Development and Design of a Robotic Manta Ray Featuring Flexible Pectoral Fins, International Conference on Robotics and Biomimetic; pp. 519-523, 2007.
12. Jian Li, Yan Huang, Xiaodong Liu, Yueqing Lin, Lang Bai, Qiang Li, Effect of aggregates on the magnetization property of ferrofluids: A model of gaslike compression, Science and Technology of Advanced Materials 8, pp. 448–454, (2007).

THE MODELING OF THE STANDING-UP PROCESS OF THE ANTHROPOMORPHIC MECHANISM*

S. F. JATSUN, L. YU. VOROCHAEVA, A. S. YATSUN, S. I. SAVIN

Department of mechanics, mechatronics and robotics, South-West State University, 305040, 50 let Oktyabry, 94, Kursk, Russia

The article presents the mathematical model of the anthropomorphic robot, which rises from an immovable surface, the stage-by-stage algorithm of the mechanism operation, the results of the numerical modeling, the dependence of the change torques generated by the actuators from the laws of the rotation angles of the mechanism links.

1. Introduction

The developing of the gaits, constructions, methods of the realization of the different types of the motion, the investigation of the dynamics features of the two-legged anthropomorphic walking robots and mechanisms are very popular among different countries and universities. The first works about this topic appeared in the end of the 20th century [1-5]. The modern authors [6-10] describe various techniques walk anthropomorphic robots with different number of degrees of freedom, rotational and translational kinematic pairs of connecting links of mechanisms, algorithms are proposed for the control of these devices increase the stability of their movement when walking.

At the same time, the articles do not really describe the problem non-directly associated with anthropomorphic mechanism walking and with the implementation of such movements as the getting up of the robot from a chair, squat, standing up from a prone position, tilt left/right or forward. Therefore, considered in this paper, the problem of standing-up of the anthropomorphic mechanism is topical.

2. The Mathematical Model of the Anthropomorphic Mechanism

For the mathematical description of standing-up anthropomorphic robot with a chair, consider the calculation scheme (Fig. 1). Let the legs of the robot move synchronously, therefore in the vertical plane *Oxy* the system is four-link mechanism. This mechanism links are connected with the cylindrical hinges.

* Work is performed with RNF, Project № 14-39-00008.

In order to simplify the process of the object motion modeling let's make the assumption that all robot links i=1-4 are absolutely solid bodies. Each of them is a rod of length l_i and mass m_i, which concentrates in the centre of the link symmetry C_i. The angle of inclination of the link to the positive direction of the axis Ox is φ_i. In the points O_2-O_4 the rotational motion drives 5-7 are set. They generate the torques M_{12} (M_{21}), M_{23} (M_{32}), M_{34} (M_{43}).

The generalized coordinates are coordinates x_{C1}, y_{C1} of the mass centre of the link 1, the angles φ_1, φ_2, φ_3, φ_4 of the links rotation i=1-4.

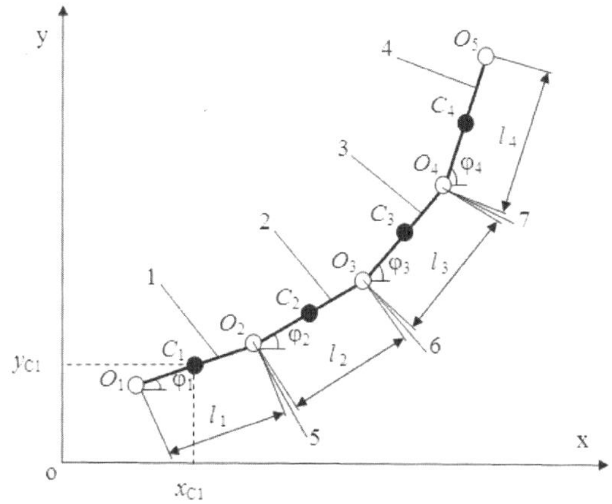

Figure 1. Calculation scheme: 1 - foot, 2 - leg, 3 - hip, 4 - body, 5-7 drives.

The system of the differential equation of the mechanism motion is formed using Lagrange's equation. To define the generalized forces the principle of possible movements are used.

3. The Stages of the Anthropomorphic Mechanism Motion of the Standing up

The work presents the standing up of the anthropomorphic robot from a chair. The chair surface is rough and locates at an angle α to the horizon.

The initial position

At the initial moment the mechanism is immovable, its position short of the Fig. 2. The robot link 1 interacts with the horizontal rough surface. The link 3 interacts with the chair surface. At that x_{C1}=const, y_{C1}=const, $\varphi_i=\varphi_i^0$=const, i=1-4, where φ_i^0 is some initial values of the angles.

When the robot links interact with the surfaces, the friction forces and normal reactions appear. In this model has the assumption that there are only two contact points, the first point O_2 is the point of the interaction with the surface, where the forces N_{O2}, F_{frO2} are concentrated. The second contact point O_4 is the point, where the forces N_{O4}, F_{frO4} are concentrated. In the work there is one more assumption: the friction forces F_{frO2} and F_{frO4} don't let the links to slide on the surface.

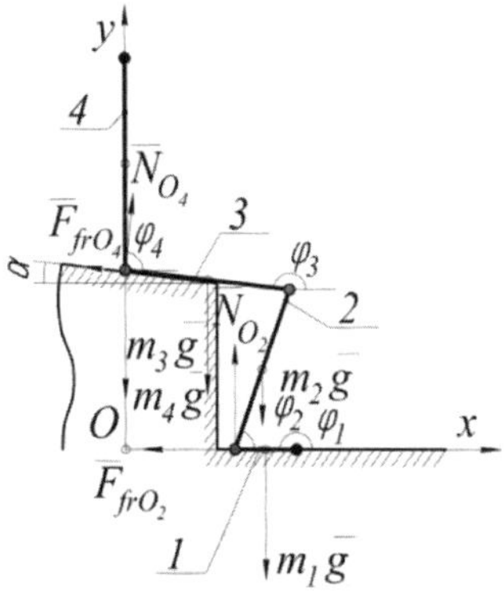

Figure 2. Calculation scheme of the robot in the initial position.

The first stage: the body inclination

On the first stage: x_{C1}=const, y_{C1}=const, $\varphi_i=\varphi_i^0$=const, i =1-3. In the mechanism points O_2 and O_4 the friction forces and normal reaction are applied. The link 4 under the action of the torque M_{34} rotates about the point O_4 till the following condition will be complied $\varphi_4=\varphi_4^1$, where φ_4^1 the value of the angle reached on this stage (Fig. 3).

The differential equation is:

$$\ddot{\varphi}_4 \cdot (J_4 + l_4^2 m_4 / 4) = M_{34} - m_4 g l_4 \cos\varphi_4 / 2 \tag{1}$$

The second stage of the motion: standing

During the second stage of the motion: x_{C1}=const, y_{C1}=const, $\varphi_1=\varphi_1^0$=const. Under the actions of the torques M_{12}, M_{23}, M_{34} the link 3carries out the break-off from the chair (with zeroing of the normal reaction N_{O4}) and the "straightening" of the apparatus till the links i=2-4 take on the vertical position: $\varphi_i=\varphi_i^2=\pi/2$ (Fig. 3).

On this stage the robot contacts with the rough surface in the points O_1 and O_2 of the link 1. It is supposed that in this points there are normal reactions N_{O1}, N_{O2}, the friction force is concentrated only in the point O_2 - F_{frO2}. The sliding on the surface of the link 1 doesn't occur.

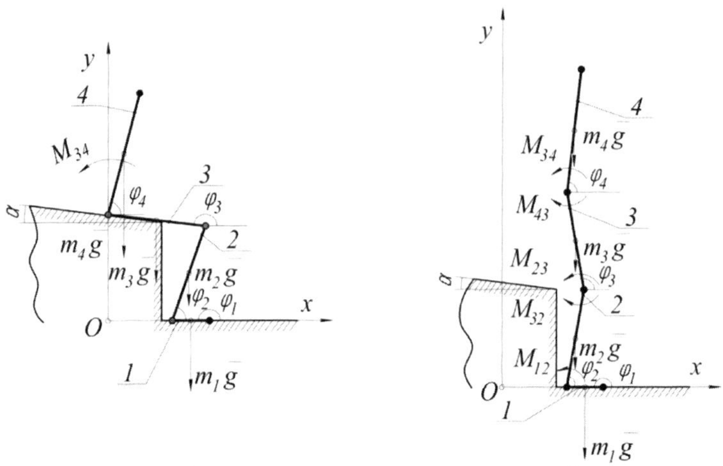

(a) (b)

Figure 3. The calculation scheme of the robot: (a) - first stage, (b) - second stage.

The robot motion on the second stage is described by the differential equation according to generalized coordinates: φ_2, φ_3, φ_4.

$$\begin{gathered}\ddot{\varphi}_2 \cdot (J_2 + m_2 l_2^{\,2}/4 + m_3 l_2^{\,2} + m_4 l_2^{\,2}) + \ddot{\varphi}_4 m_4 l_2 l_4 \cos(\varphi_2 - \varphi_4)/2 + \\ + \ddot{\varphi}_3 (m_3/2 + m_4) l_2 l_3 \cos(\varphi_2 - \varphi_3) + \dot{\varphi}_3^2 (m_3/2 + m_4) l_2 l_3 \sin(\varphi_2 - \varphi_3) + \\ + \dot{\varphi}_4^2 m_4 l_2 l_4 \sin(\varphi_2 - \varphi_4)/2 = M_{12} - M_{32} - m_2 g l_2 \cos\varphi_2/2 - \\ - m_3 g l_2 \cos\varphi_2 - m_4 g l_2 \cos\varphi_2\end{gathered} \quad . \quad (2)$$

$$\begin{gathered}\ddot{\varphi}_3 \cdot (J_3 + m_3 l_3^{\,2}/4 + m_4 l_3^{\,2}) + \ddot{\varphi}_4 m_4 l_3 l_4 \cos(\varphi_3 - \varphi_4)/2 + \\ + \ddot{\varphi}_2 (m_3 + 2m_4) l_2 l_3 \cos(\varphi_2 - \varphi_3)/2 - \dot{\varphi}_2^2 (m_3/2 + m_4) l_2 l_3 \sin(\varphi_2 - \varphi_3) + \\ + \dot{\varphi}_4^2 m_4 l_3 l_4 \sin(\varphi_3 - \varphi_4)/2 = M_{23} - M_{43} - m_3 g l_3 \cos\varphi_3/2 - m_4 g l_3 \cos\varphi_3\end{gathered} \quad . \quad (3)$$

$$\begin{gathered}\ddot{\varphi}_4 \cdot (J_4 + m_4 l_4^{\,2}/4) + \ddot{\varphi}_3 m_4 l_3 l_4 \cos(\varphi_3 - \varphi_4)/2 + \ddot{\varphi}_2 m_4 l_2 l_4 \cos(\varphi_2 - \varphi_4)/2 - \\ - \dot{\varphi}_2^2 m_4 l_2 l_4 \sin(\varphi_2 - \varphi_4)/2 - \dot{\varphi}_3^2 m_4 l_3 l_4 \sin(\varphi_3 - \varphi_4)/2 = M_{34} - m_4 g l_4 \cos\varphi_4/2\end{gathered} \quad . \quad (4)$$

Using the two-stepped algorithm of the robot standing up from the chair we are finding the torques generated by the drives, when the apparatus links move according given laws.

4. The Definition of the Laws of Change of the Rotation Angles of the Links

To determine the laws according which the links angles of the inclination change to the horizontal line, we use natural experiment. In this experiment the camera shoots the process when a person stands up from a chair (Fig. 4 shows the photo of a person with the scheme of the four-link mechanism on it). We got the experimental table of the values of the rotational angles of the links at equal intervals of the time. We use splines approximation to get the analytic functions.

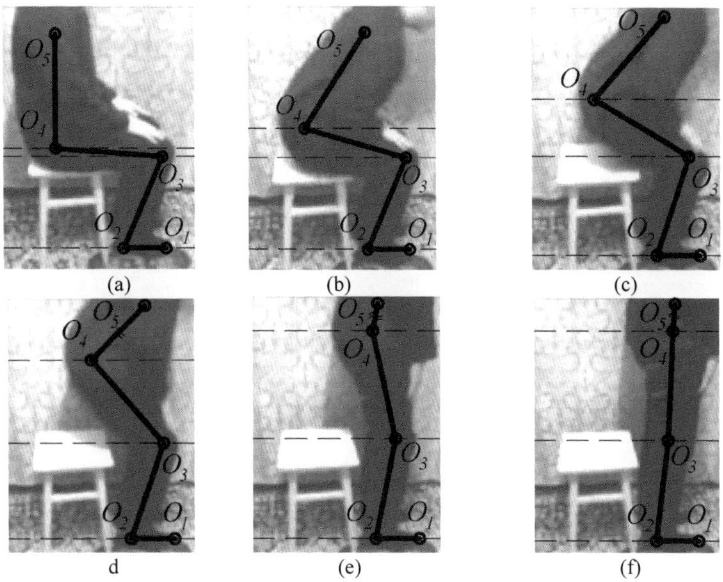

Figure 4. The intermediate position of the person when conducting field experiment: (a) - t =0 s, (b) - t =2,52 s, (c) - t =2,86 s, (d) - t =3,14 s, (e) - t =3,45 s, (f) - t =4,07 s.

The analytic functions of the change of the rotational angles of the links are:

$$\varphi_1(t) = \pi, \quad \varphi_2(t) = \begin{cases} 7\pi/18, if \ (t \geq 0) \wedge (t < 3.36) \\ f_2(t), if \ (t \geq 3.36) \wedge (t < 5) \\ \pi/2, if \ (t \geq 5) \end{cases}, \tag{5}$$

$$\varphi_3(t) = \begin{cases} 173\pi/180, if \ (t \geq 0) \wedge (t < 2.68) \\ f_3(t), if \ (t \geq 2.68) \wedge (t < 5) \\ \pi/2, if \ (t \geq 5) \end{cases}, \tag{6}$$

$$\varphi_4(t) = \begin{cases} \pi/2, if \ (t \geq 0) \wedge (t < 1.3) \\ f_4(t), if \ (t \geq 1.3) \wedge (t < 2.68) \\ f_5(t), if \ (t \geq 2.68) \wedge (t < 5) \\ \pi/2, if \ (t \geq 5) \end{cases} \tag{7}$$

where

$$f_i(t) = \sum_{k=1}^{7} A_{ik} t^k \tag{8}$$

- polynomial functions, i=2-5.

In the Fig. 5 there are the graphs of the derived analytic functions.

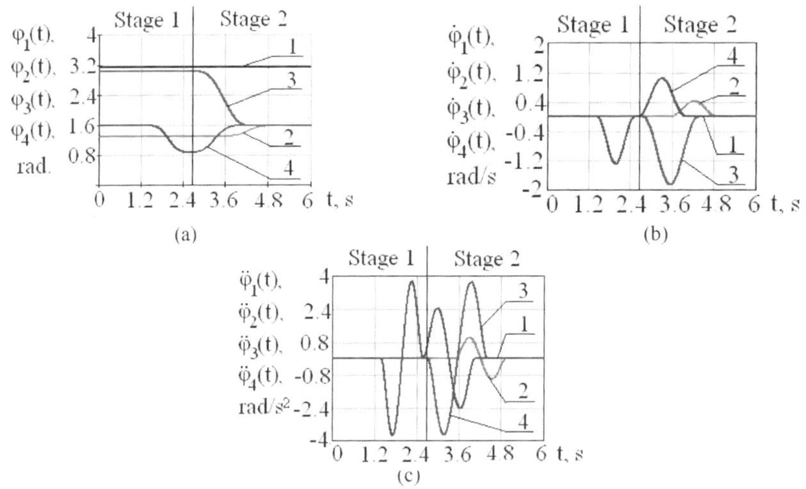

Figure 5. The graphs of dependencies: (a) - angles, (b) - angular velocities, (c) - angular accelerations; 1 - link 1, 2 - link 2, 3 - link 3, 4 - link 4.

5. The Results of the Numerical Modeling

The numerical modeling of the apparatus motion was done in the software MathCAD according special algorithm. Its feature is in the sequential switching of the motion stages at the moment when the rotation angles of the links become equal to the desired value.

The lengths of the simulated robot links are l_1=0.15 m, l_2=0.4 m, l_3=0.4 m, l_4=0.5 m. The modeling was carried out under the following initial conditions $\varphi_1^0 = \pi$ rad, $\varphi_2^0 = 7\pi/18$ rad, $\varphi_3^0 = 173\pi/180$ rad, $\varphi_4^0 = \pi/2$ rad, $\dot{\varphi}_i^0 = 0$ rad/s, i=1-4, $x_{C1}^0 = 0.335$ m, $y_{C1}^0 = 0$ m, $\dot{x}_{C1}^0 = 0$ m/s, $\dot{y}_{C1}^0 = 0$ m/s.

In the Fig. 6 there are the graphs of the derived drives torques when the links have different mass.

On the graphs it is clear that on the stage 1 one torque M_{34} is formed. This torque ensures the rotation of the link 4. In the beginning during a short period it is negative. It is caused by transient process. Then the torque changes the sign on the opposite. During the transition from the first stage to the second is the change of the direction of the torque M_{34} associated with changing the direction of rotation of the corresponding link. On the second stage all three torques are generated. The torque M_{34} first positive and then negative. The law of the change of the torque M_{12}, which makes the link 2 rotation, has the oscillatory nature, there are two peaks of the torque, the first negative, the second positive. The change of direction of the torque action is also caused by changing the direction of rotation of link 2.

The torque M_{23} also has the oscillatory nature. The amplitude of the oscillation is twice the largest values of the torques M_{12} and M_{34} independently of the values of the masses of the links.

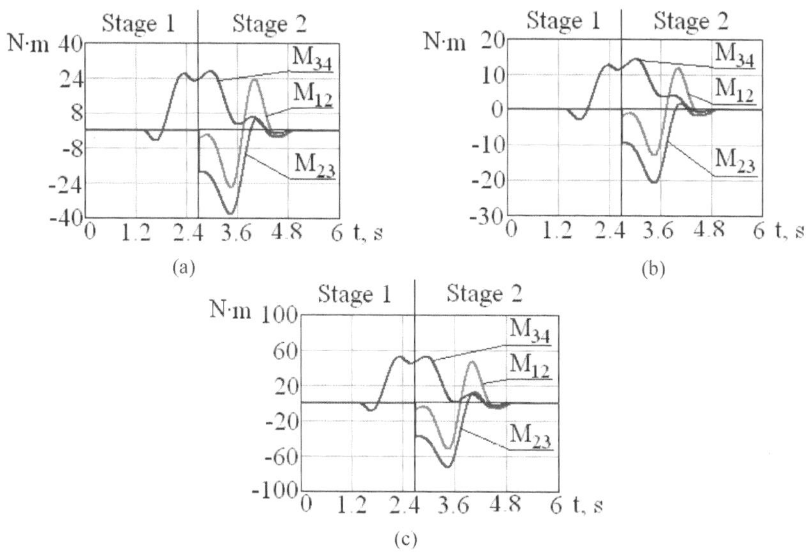

Figure 6. Graphs of torque drives: 1 - $M_{12}(t)$; 2 - $M_{23}(t)$; 3 - $M_{34}(t)$ if: (a) - m_1=1 kg, m_2=2.5 kg, m_3=5 kg, m_4=12 kg; (b) - m_1=0.5 kg, m_2=1,25 kg, m_3=2.5 kg, m_4=6 kg; (c) - m_1=2 kg, m_2=5 kg, m_3=10 kg, m_4=24 kg.

According this investigation we can conclude that in order to get as possible little values of the torques are generated by the drives, it is necessary to minimize the masses of the robot links.

The robot positions during the standing up from a chair, which are derived using numerical method shown in Fig. 7.

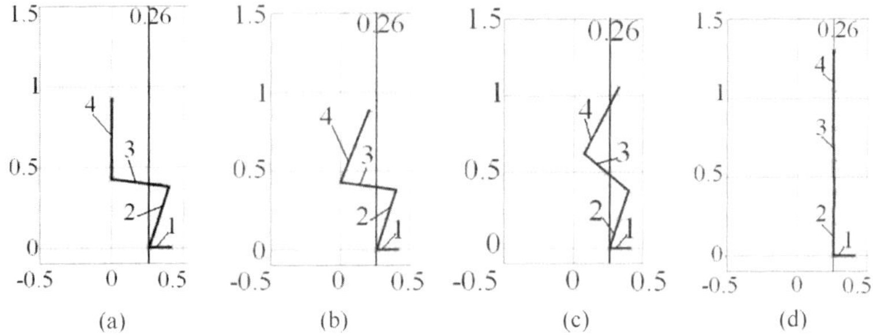

Figure 7. The position of the robot during standing up, obtained by the numerical method: (a) - initial position, (b) - the first stage, (c) - second stage, (d) - end position.

6. Conclusion

This article presents the mathematical model which describes the standing up process of the anthropomorphic mechanism from an immovable surface (a chair). There is also the algorithm of the stage-by-stage realization of the movement, the conditions of the switching between the stages, differential equations and imposed constraints. There are the results of numerical modeling as the graphs of the torques, which are formed by the drives, in dependence on the laws of the change of the rotational angles of the links.

References

1. V. V. Beletskiy, *Nauka*, 288 (1984).
2. M. Vukobratovich, *Mir*, 541 (1976).
3. A. M. Formalskiy, *Nauka*, 368 (1982).
4. L. A. Alekseeva, Yu. F. Golubev, *Izvestia AN SSSR, Tehnicheskaya kibernetika*. **3**, 72-80 (1975).
5. D. E. Ohocimskiy, Yu. F. Golubev, *Nauka*. 312 (1984).
6. I. V.Golubyatnikov, Yu. I. Fantalov, *Vestnik MGUPI*. **3**, 205-218 (2008).
7. I. V. Golubyatnikov, S. V. Shanygin, *Vestnik MGUPI*, 156-163 (2008).
8. Yu. F. Golubev, D. Yu. Pogorelov, *Fundamentalnaya i prikladnaya matematika*. **4(2)**, 525-534 (1998).
9. V. E. Pavlovsky, *Preprinty IPM im. M.V.Keldysha*. **101**, 32 (2013).
10. A. K. Platonov, V. E. Pavlovsky *Proc. of the XIII CISMIFTOMM Int. Symp. RoManSy'2000*. 131-138 (2000).

A BABY-MIMIC CRAWLING ROBOT*

XIULI ZHANG†, YAN LIANG, HAIBO WU, MINGCHENG E
School of Mechanical and Electrical Control Engineering, Beijing Jiaotong University, Beijing,100044,China

A baby-mimic quadruped robot with global flexible spine of variable cross-section stiffness mechanism scheme is proposed aiming at solving the problem of unstable body posture when the quadruped robot is walking. The trot gait of 'BabyBot' is realized by adopting center pattern generator method based on biological neural control mechanism. The dynamic simulation analysis is carried out and the experimental platform is set up. In addition, the walking experiments and performance evaluations of the baby-mimic quadruped robot are conducted. Experiments verify that adding the baby-mimic quadruped robot with a global flexible spine of variable cross-section stiffness has a positive effect on reducing the swing of the body posture (head and hind torso). Thereby, it is effective for improving walking coordination and posture stability for legged robots.

1. Introduction

With the advance of science and technology, the performance of robot develops towards high speed and adaptability. Baby-mimic robot is one of quadruped robot. So far, the structure design of most baby-mimic robots is rigid, its speed and bearing capacity are limited. In addition, the joints of robot have friction. So the robot has vibration and impact[1].The robot 'iCub' designed by Italian researchers is the prime example, it is inferior to real baby and its speed is low[2]. The Communication Research Laboratory in Japan develops an infant-like humanoid robot 'Infanoid'. It can imitate baby identification and communication behavior by means of establishing a learning mechanism of social intelligence[3]. Professor Asada Minoru from Osaka University develops two new Baby-mimic robots M3-neony and M3-synchy, separately they can combine speech and object recognition and is programmed to communicate both verbally and non-verbally with multiple subjects[4]. Furthermore, Kenichi Narioka and Koh Hosoda investigate how the body structure, especially the spinal posture influences the way of development of locomotion through a experiment[5].

* This work is supported by the Fundamental Research Funds for the Central Universities under Grant No. M15JB00250.

† zhangxl@bjtu.edu.cn

Therefore, for most existing baby-mimic crawling robot[6-9], main researches focus in non-verbally communication and interaction between the robot and people. However, there have been few studies to investigate the spine for stability and coordination and the unique posture of hind-leg.

Compliant mechanism has two typical types: local compliant mechanism and global compliant mechanism[10,11]. Boston Dynamics is developing a new quadruped robot "Cheetah". "Cheetah" realizes fast running with the aid of flexible spine on the pitch direction of bending deformation[12]. Hiroyuki Ishii built a bionic quadruped robot 'WR-2', with two passive degrees of freedom in the waist, pitch and yaw, which can realize trunk bending and rotation like rat[13]. Se-Hoon Park develops a quadruped robot 'ELIRO'. Experiments show that the rotation gait of the robot with the waist joint is more stable than traditional gait[14-16].

By analyzing the current quadruped robot with compliant spine, we can find that most quadruped robot employ adding a passive freedom or local spring in spine to realize the flexibility[17-21]. However, compared with local compliance, global compliance is closer to the spine when animals are crawling. In addition, global compliance makes robot more stable through the coordination of spine and four legs. Based on this situation, this paper designs a spine with variable cross-section stiffness of global compliance and the baby-mimic crawling robot 'BabyBot'. It is concluded that 'BabyBot' is more stable and adaptable than robot with global compliance through simulation comparison analysis and experiments of physical prototype.

2. Mechanical Design

We find that when baby is crawling, the shoulders, elbows, and hips of baby emerge periodic contraction and extension. Meanwhile spine has periodically oscillating motion along with the limbs, and that the lumbar vertebra has the largest swing but the knee has slight rotation. Based on the background of bionics as above, we put forward the compliant structure of 'BabyBot'. As shown in Fig. 1, the quadruped robot consists of trunk with I-shaped structure, fore-arm, hind-leg, 'BabyBot' has eleven freedoms, thereinto, the shoulders, elbows, and hips have six active freedoms. And it has five passive freedoms, roll, pitch and yaw of the pine, and a freedom achieved by an arc spring leg structure in hind hip. Figure 2 is the prototype of 'BabyBot'.

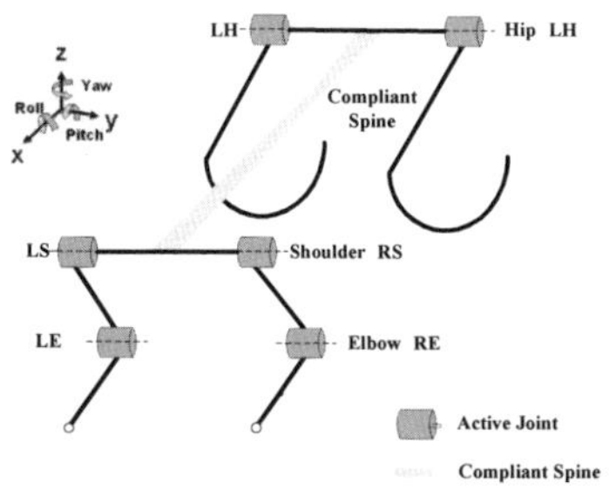

Fig. 1 Schematic diagram of 'BabyBot'

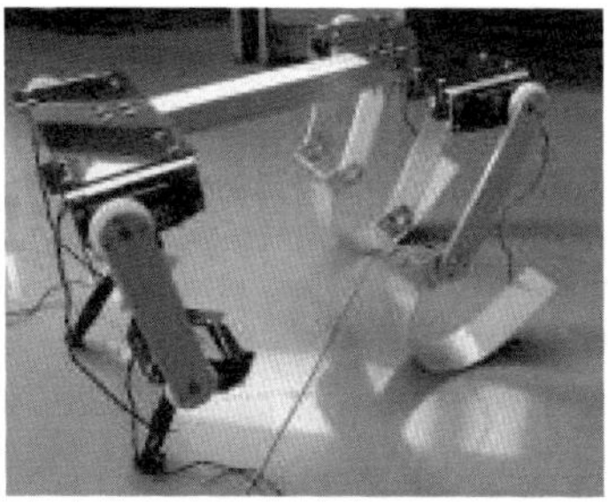

Fig. 2 'BabyBot'

3. Control Model

The quadruped robot 'BabyBot' with elbow joint for front arms and knee joint for hind legs walk with diagonal gait. A biologically-inspired central pattern generator (CPG) model is employed as the controller to generate angular positions for the six active joints. Thereinto, four neural oscillators based on left shoulder joint act on two shoulder joints and two hip joints. As is shown in Fig. 3, phase difference between other joints and base joints is ϕ_i, arrow points stand for joints lag. Elbow joints are controlled by two independent neural oscillators and only have a fixed phase interlocking relation with the corresponding shoulder joint.

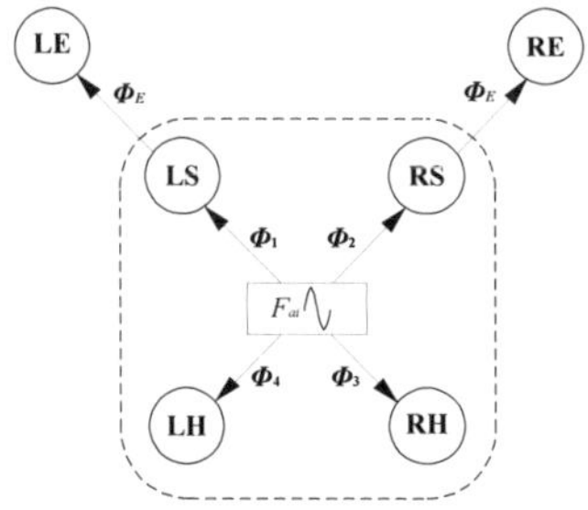

Fig. 3 CPG Model
(L: left, R: right, H: hip, S: shoulder, E: elbow)

$$\begin{cases} \theta_{i,j}(t) = A_{i,j}\sin(2\pi t/T + \phi_{i,j}) \\ \theta_{i,E}(t) = A_E \sin(2\pi t/T + \phi_{i,S} + \phi_E) \end{cases}$$
$$i = L, R \qquad j = S, H \tag{1}$$

4. Design of Compliant Spine

As the axial skeleton of the body, spine has functions of shock absorption, protection and movement etc. Each segment of spine has different degree, including lumbar-sacral region completely motionless, little chest movement, the neck and the waist is more flexible. We find that t a fixed center as the axis of rotation doesn't exist in the movement of spine, and is only average center of rotation when baby is crawling. Tournier investigates that the rotation center of spine is located in the third place of back spine[22]. So we the minimum radius of the spine is located in the third place of back spine, too. The spine structure is shown in Fig. 4.

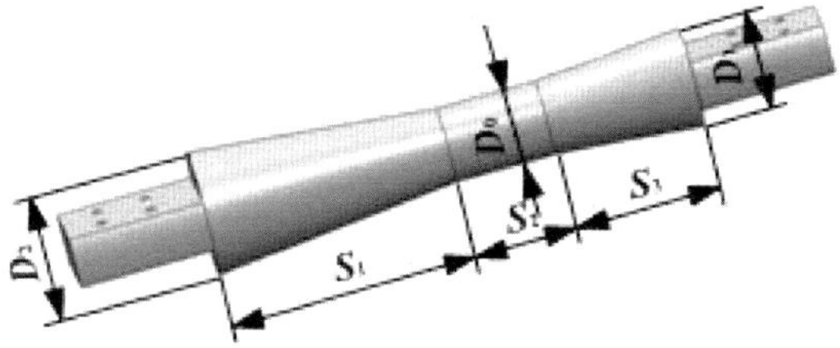

Fig. 4 Radial compliant spine

When baby is crawling, spine present cyclical swing, J.Gray investigates that the relation between vertical force of robot foot and V_4 is $F_{friction}$, $F_{riction} = V_4 \tan\alpha$ is angle between foot end and the shoulder rotation center and vertical direction[23]. After measuring, the friction coefficient μ is 0.219. At last we can calculate the minimum radius.

According to the theory of pseudo rigid body model, we build the pseudo rigid body model of spine. It is shown in Fig. 5.

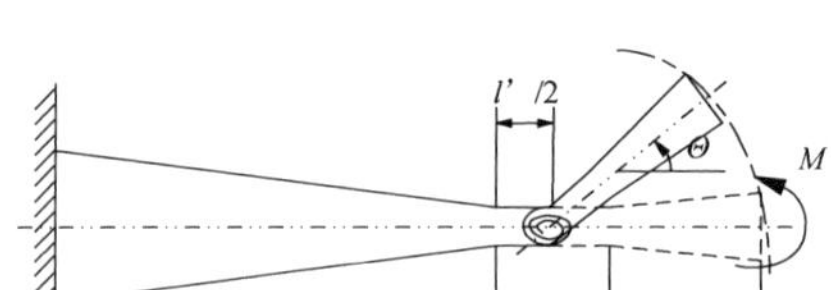

Fig. 5 PSGB model of spine

When the free end of spine is in the function of moment, the Bernoulli-Euler equation is

$$\frac{\mathrm{d}\theta}{\mathrm{d}s} = \frac{M_0}{EI} \tag{2}$$

After the integration of the separation of variables, the end angle of beam can be obtained.

$$\theta_0 = \frac{M_0 L^{'}}{EI} \tag{3}$$

$$\int_0^{\theta_0} \mathrm{d}\theta = \int_0^{L^{'}} \frac{M_0}{EI} \mathrm{d}s \tag{4}$$

$$M_0 = M = 2F_\mathrm{h} L \qquad F_h = F_z \tan\theta' \qquad I = \pi D^4 / 32$$

So

$$D > \sqrt[4]{\frac{32 M_0 L'}{\pi E \theta_0}} \tag{5}$$

To sum up, we calculate the range of the minimum radius of spine. Based on the background of bionics, we calculate the length of the variable cross-section compliant spine. Furthermore, the spine is made of organic material, we choose thermo-set polyurethane elastomer (PU). The diameter range of spine is 16.8-20.4mm. To ensure the strength of materials, we determine the minimum diameter is 30mm. Diameter of each segment of spine is shown in Table 1.

Table 1 The size of radial compliant spine

Variable	Value（mm）
D_0	30
D_1	40
D_2	50
S_1	100
S_2	40
S_3	60

5. Experiment Results

In order to evaluate the stability of compliant spine for the robot, we set up contrast experiment of rigid and flexible spine. The quadruped robot ‘BabyBot’ with elbow joint for front arms and knee joint for hind legs walk with diagonal gait. We choose the angle oscillation amplitude as the evaluation criteria of stability and adaptability of the robot.

We record the posture data of the yaw direction of trunk and the roll direction of front shoulder. They are shown in Fig. 6, Fig. 7. Statistical results are shown in Table 2 and Table 3. Compared with posture data of quadruped robot with rigid and flexible spine, deviation angle in yaw of front shoulder trunk of 'BabyBot' with flexible spine decreases 27.62%, and deviation decreases 75.68%. Deviation angle in roll of front shoulder trunk of 'BabyBot' with flexible spine decreases 2.12%, and deviation decreases 5.21%.

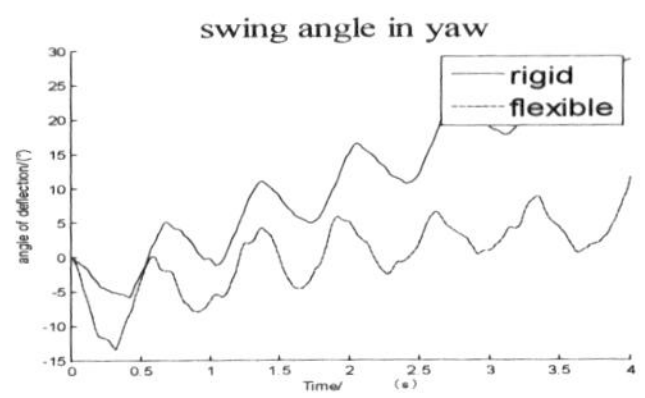

Fig. 6 Swing angle in yaw of front trunk

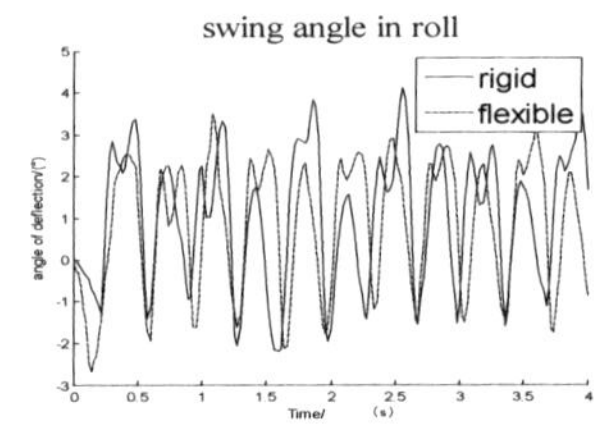

Fig. 7 Swing angle in roll of front trunk

Table 2 Data analysis of front and hind trunk

		front trunk		hind trunk	
		rigid	flexible	rigid	flexible
maximum deviation angle in roll (°)		6.320	6.186	4.604	4.184
variance		2.482	2.352	2.152	1.173
result	deviation angle	decrease 2.12%		decrease 9.12%	
	variance	decrease 5.21%		decrease 45.49%	

Table 3 Data analysis of front shoulder trunk

	front shoulder trunk		result
	rigid	flexible	
maximum deviation angle in roll (°)	34.444	24.930	decrease 27.62%
variance	106.125	25.811	decrease 75.68%

Deviation angle in roll of hind trunk of 'BabyBot' with flexible spine decreases 9.12%, and deviation decreases 45.49%. Experiments show that baby-mimic quadruped robot employed radial compliant spine can decrease the deviation angle in roll and yaw. Because of this, the swing in roll and yaw are much smoother and the robot walks more adaptable and stable.

6. Conclusion

This paper investigates the problem of baby-mimic quadruped robot being less stable and not adaptable. We design a baby-mimic quadruped robot 'BabyBot' to improve motion coordination and posture stability. The shape and dimensions of the radial flexible spine are designed based on the theory of mechanics and pseudo rigid body model. The experiment of 'BabyBot' shows that quadruped robot with global compliant spine is more stable and adaptable.

References

[1] Chen Zhitai. The dynamics of compliant mechanism research[D]. Beijing: Press Beijing Univ. Tech, 2005.

[2] Righetti, L. and A. J. Ijspeert. Design methodologies for central pattern generators: an application to crawling humanoids[J], in Robotic Science and Systems. 2006, pp. 191-198.

[3] Kozima, H. and H. Yano. A robot that learns to communicate with human caregivers[C]. in Proceedings of International Workshop on Epigenetic Robotics, 2001, pp.47-52.

[4] http://www.ndtv.com/article/sci-tech/m3-neony-japan-s-latest-offering-in-robot-baby-series-37776

[5] Kenichi Narioka and Koh Hosoda(2011). Motor development of an pneumatic musculoskeletal infant robot[C]. IEEE ICRA, pp. 963-968, 2011 Shanghai.

[6] K. Tsujita, K. Miki. "A study on trunk stiffness and gait stability in quadrupedal locomotion using musculoskeletal robot"[C]. The 15th International Conference on Advanced robotics Tallinn University of Technology Tallinn, Estonia, 2011.

[7] S. Rasakatla, K. M. Krishna, B. Indurkhya. "Design, construction of a compliant gait of "ModPod": A modular hexapod robot"[C]. IEEE International Conference on Robotics and Biomimetics (ROBIO). Tianjin, China. Dec 14-18, 2011.1341-1345.

[8] D. Kuehn, F. Grimminger, F. Beinersdorf, et al. "Additional DOFs and sensors for bio-inspired locomotion: towards active spine, ankle joints, and feet for a quadruped robot"[C]. International Conference on Robotics and Biomimetics. 2011: 2780-2786.

[9] L. L. Howell. Compliant Mechanisms (Chinese Edition) [M]. Beijing: High Education Press, 2007.

[10] G. K. Ananthasuresh, S. Kota. Designing Compliant Mechanisms Mechanical[J] Engineering, 1995, 117(11):93-96.

[11] L. L. Howell. Compliant Mechanism[M]. New York, McGraw-Hill, 2001.

[12] T. Matsuki, I. Mizuuchi, S. Kagami, et al. "Quadruped robot with spine structure and simulation environment"[C]. Proceedings of the 16th Annual Conference of the Robotics Society of Japan.1998: 85-88.

[13] Hiroyuki ISHII et al..Design and development of biomimetic quadruped robot for behavior of rats and mice[C]. 31st Annual International Conference of the IEEE EMBS Minneapolis, Minnesota, USA, September 2-6, 2009.

[14] S.H. Park and Y.J. Lee. Turning gait planning of a quadruped walking robot with an articulated spine[C]. Procs. of the ICCAS2004, pp. 1926-1930, 2004.

[15] Tsujita K, Miki K. Stability analysis on quadrupedal gaits according to body's flexibility using musculoskeletal robot[C]//Robotics and Biomimetics (ROBIO), 2011 IEEE International Conference on. IEEE, 2011: 1609-1614.

[16] Kani M H H, Derafshian M, Bidgoly H J, et al. Effect of flexible spine on stability of a passive quadruped robot: Experimental results[C]//Robotics and Biomimetics (ROBIO), 2011 IEEE International Conference on. IEEE, 2011:963-968.

[17] Niemitz C. Kinematics and ontogeny of locomotion in monkeys and human babies[J]. Zeitschrift für Morphologie und Anthropologie, 2002:383-400.

[18] Righetti L, Anna Nylén K R, Ijspeert A J. Is the locomotion of crawling human infants different from other quadruped mammals?[J]. Under review, 2008.

[19] Jarakorn A, Kunlabud A, Siwaraphornsakul T, et al. Impact reduction mobile robot(IRR): Design of compliant legs[C]. Advanced Robotics (ICAR), 2011:671-676.

[20] Tournier C, Aunoble S, Le Huec JC, et al. Total disc ar-throplasty: consequences for sagittal balance and lumbar spine movement[J]. Eur Spine, 2007, 16(3):411-421.

[21] Narioka K, Hosoda K. Motor development of an pneumatic musculoskeletal infant robot[C]. IEEE ICRA, Shanghai, 2011:963-968.

[22] Lee D V, Bertram J E, Todhunter R J. Acceleration and balance in trotting dogs[J]. Journal of Experimental Biology, 1999, 202(24): 3565-3573.

[23] Gray J. Studies in the Mechanics of the Tetrapod Skeleton[J]. Journal of experimental biology, 1944, 20(2): 88-116.

ANALYSIS, DESIGN AND EXPERIMENTS OF A ROUGH WALL CLIMBING ROBOT BASED ON GRABBING CLAWS*

FENGYU XU, JINGJIN SHEN, BEI WANG
College of Automation, Nanjing University of Posts and Telecommunications, Nanjing, 210003, P. R. China.

In this study, we analyzed the bionic structure of cockroach legs and observed the morphological characteristics of the hooks on cockroach legs in reference to structural bionic principles. We investigated the interaction mechanism of the bionic hook with the bulges on rough surfaces and deduced the mechanical conditions of the hook stably grabbing the wall bulges. Then, we presented an initial design of a wall-climbing robot based on a grabbing claw and proposed a climbing model. Furthermore, we performed several experiments to demonstrate the grasping stability. The grasping claw of the climbing robot can offer a more reliable way of adsorption for complicated rough surfaces.

1. Introduction

The climbing on high-altitude rough construction built using rock, concrete, and irregular water brush stone is difficult to achieve because of severe ashing condition on the superficial area. Construction accidents have occurred in recent years, causing serious damage such as collapse of bridges, some examples of which are the accidents that have occurred at St. Nazaire Bridge in France [1], Kohlbrand Estuary Bridge in Germany, Luling Bridge in the United States [2, 3], and Phoenix Bridge in Hunan, China. Therefore, periodic inspection is needed to guarantee the safety of special structures.

Scientists proposed the bionic grabbing "adsorption" method and developed a variety of bionic robots by investigating the excellent characteristics of the claws of bumblebees and beetles. Spenko et al. [4] created the RiSE series of robots, one of which was a climbing RiSE robot. This mechanism was hung on a wall using some tiny steel hooks. Goran et al. [5] designed a bionic climbing robot called DynaClimber based on the biological climbing model, which

* This project is supported by the National Natural Science Foundation of China (51005046), Natural scientific research fund of Nanjing University of posts and telecommunications (NY214071), and Jiangsu Province Natural Science Fund Project of colleges and Universities (13KJB460013).

achieved dynamic climbing on vertical walls. Sebastian et al. [6] designed a 6-foot climbing robot called SpaceClimber using the bionics principle. This robot was able to walk freely in an extreme environment and on a slope. Sintov et al. [7] proposed a hook-type wall-climbing robot. This robot could maintain the same pose on a rough wall surface for a long time, and therefore was applicable to surveillance, rescue, observation, and recreational activities. Lam et al. [8–10] designed Treebot, a tree-climbing robot that could cross treelike branches and had a claw-type and new flexible motion structure, and introduced the stability control algorithm by investigating the peristaltic motions of insects. Researchers at Stanford University also developed a bionic wall-climbing robot called SpinyBot [11]. The tiny hooks on the feet of this robot could "hook" the bulges on the rough surfaces to help the robot complete its climb. This robot was suitable for rough concrete and brick wall surfaces. SpinyBot was significantly different from conventional wall-climbing robots in terms of adsorption principle, motional shape, and had strong adaptability to wall surfaces.

In this study, we proposed an attachment scheme through the small claw to overcome the affection of ash and vibration of the wall. We designed a flexible bionic hook and grabbing claw to grab the rough bulges on the wall surface. The claw consisted of a large number of micro-biohooks, which provided adsorption to the wall surface. The influences of wall vibrations on the robot stability were thereby avoided.

2. Interaction Mechanism of Bionic Hook with Rough Bulges on Wall Surface

Traditional magnetic adsorption and vacuum adsorption robots are difficult to use on high-altitude building surfaces, such as cliffs, irregular water brush stone, rugged concrete, and rocks with large cracks (Fig. 1). By contrast, animals with sharp claws, such as cockroaches and ants, can freely move on these surfaces. The tiny spine structure on the soles of these animals can hook tiny bulges on the wall.

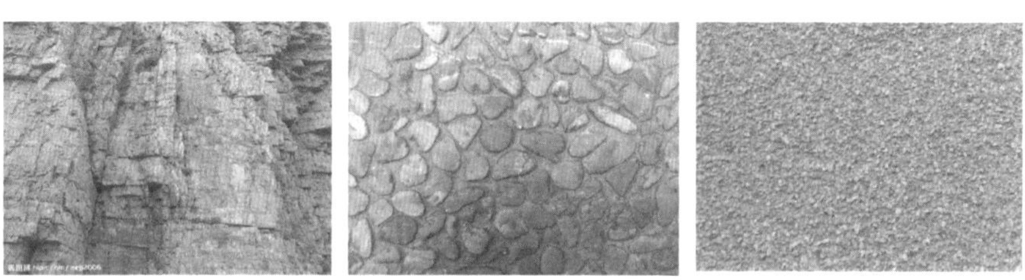

(a) Cliff surface (b) Irregular water brush stone surface (c) Rough wall surface

Fig. 1. Rough surfaces

We designed a micro bionic structure to obtain a flexible rough wall-oriented climbing mechanism. Using cockroaches as the bionic prototype, we analyzed the spine structure of cockroach legs and the interaction mechanism of the hook with the rough wall surface. Thus, we deduced the judging conditions for the hooks to stably grab the bulges on wall surfaces.

2.1. *Bionic research on spine structure of cockroach leg*

Cockroach feet and legs have many claws and hooks for attachment on wall surfaces. Fig. 2 shows that these hooks grasp the concave and convex points of the wall surface rather than piercing the wall. The cockroach leg is roughly constituted by trochanter, femur, setae, tibia, spines, and tarsal. The spines are the tiny hooks at the end of the legs. Using these spines, cockroaches can firmly seize the bulges on rough surfaces.

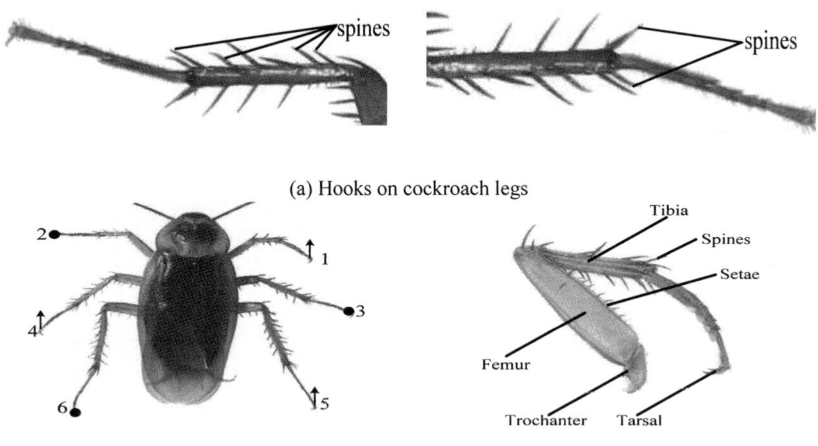

(a) Hooks on cockroach legs

(b) Body structure of a cockroach

(c) Leg structure of a cockroach

Fig. 2. Leg structure of a cockroach

A cockroach walks fast even on an uneven surface. It does not reduce its climbing velocity even on a rugged wall surface. It walks with three feet on the floor, that is, the propodeum and metapedes on one side, and the propodeum on the other side, which combine into a triangular support structure. As the three legs touch the ground and kick backward, the other three legs are raised for forward creeping. A cockroach uses the propodeum to grab an object and pull its body forward as well as the metapedes to support and raise its body. The metapedes push the body forward and rotate it. The body of a cockroach is positioned forward and slightly turns outward while walking, thereby enabling rapid movement. The movement forms a serrated curve instead of a straight line. Using such an "alternating triangle" gait, the triangular support is achieved as three feet leave the ground. Thus, a cockroach can walk on various types of ground in a balanced state through the gravity center on the base of the supporting feet.

2.2. *Judgment condition for bionic hook stably grabbing bulges on rough wall surface*

We selected sampling points on an arbitrary rough wall. Then, we measured the distance from the wall surface to the base surface and obtained the original data to describe rough bulges. We obtained the curve to describe the wall bulges by fitting the data collected using a polynomial. Three types of curves (A, B, and C) were deduced (Fig. 3(a)–(c)) according to the characteristics of curves and the cycle division on the curves (the period between the two peaks is a cycle).

The type B curve has a concave groove; that is, a y value corresponds to multiple x values. Thus, the claw can directly hook the concave groove. For the type C curve, the lower hook instead of the upper hook is incapable of grabbing the concave groove. The analysis of the type C curve can refer to that on the type A curve in addition to the concave groove. The hook can grab the types B and C curves stably. Therefore, only the type A wall surface should be analyzed specifically.

For the type A curve and the stress analysis in Fig. 3(d), the equations $F\sin\beta < N\mu$ and $F\cos\beta = N$ should be satisfied for the hook to attach to the bulge. Thus, $\beta < \arctan\mu$; that is,

$$\alpha < \theta + \arctan(\mu) \tag{1}$$

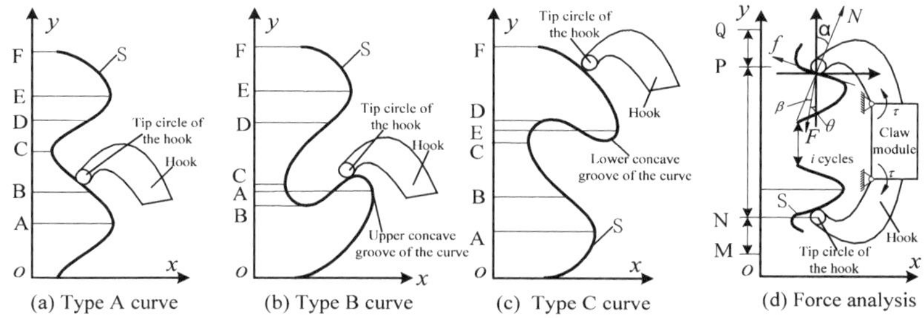

Fig. 3. Mechanism of hook grabbing bulges on wall surface

where α is the normal angle, namely, the angle between the normal directions of the hook–bulge contact point with vertical plane; μ is the frictional coefficient; and θ is the grabbing angle, namely, the angle between the grabbing force with the negative direction of the y axis. The condition of Eq. (1) should satisfy the principle of friction self-locking. AE is a cycle of the type A curve. The upper hook is incapable of completing the grab in section CE. The grabbing condition should be judged using Eq. (1). The grab is successful if Eq. (1) is met as the hooks slide along the wall surface from top to bottom.

3. Grasping Claw Based on Tensional Spring

We built a wall-climbing robot model based on a grab claw modeled on wall-climbing animals (i.e., cockroaches). The robot was mainly used to detect cracks on the concrete surface of cable-stayed bridge towers, viaduct bridge piers in remote mountainous areas, or rough cliff surfaces.

Figure 4(a) illustrates the driven model claw. The principle of this model can be described as follows: an eccentric wheel mechanism is driven by a steering engine, a small roller is installed between the pressure plate and the hooked claw, and a torsional spring is placed on each hooked claw. When the pressure plate is pressed downward, every hooked claw stretches. The claw tip is separated from the microprotuberances on the rough surface, deforming the torsional spring. When the pressure plate is drawn back, every hooked claw grasps the rough surface independently through the respective torsional springs. The pressure plate continues to draw back when a certain claw comes in contact with the microprotuberances on the wall surface while the tips of other sharp claws continue to look for graspable microprotuberances.

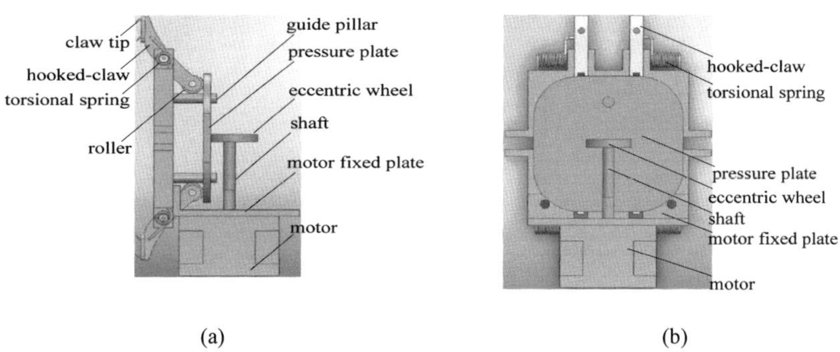

Fig. 4. Driven model of torsional spring

Figure 4(b) shows the distribution of the hooked claws. Two pairs of claws are arranged on the upper and lower parts of the module. Another pair of hooked claws can be arranged on the left and right sides of the module to improve the stability of the robot. The torsional spring drives the independent movement of each hooked claw, thereby increasing the reliability and probability of the tight grasp of microprotuberances on the wall surface.

In the climbing model, the microprotuberances on the rough wall surface provide the robot with an attachment force pointing toward the wall surface when the hooked-claw grasps the wall. The constraint imposed by the wall surface on the robot is increased, which helps the robot to resist the wind load in high altitudes and the adverse effect of wall surface vibration on the climbing performance of the robot.

4. Bionic Robot and Experiments

We design a simple robot prototype by analyzing the climbing gait of the cockroach. On this basis, the stressing condition of the microhook is experimentally investigated.

4.1. *Experiments of the climbing robot*

To verify the feasibility of the microhook grabbing the wall surface, we designed a simple climbing prototype.

(a) Experiments on water brush stone surface

(b) Experiments on vertical rough wall surface

(c) Experiments on concrete wall

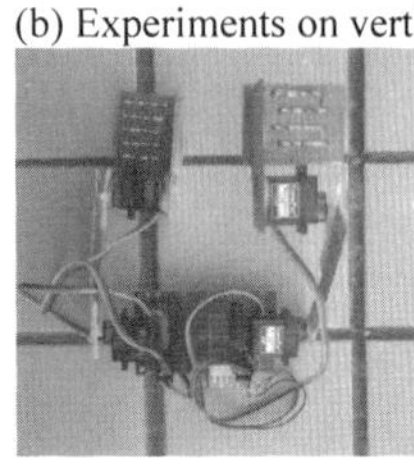

(d) Experiments on smooth wall

(e) Experiments on the ground

Fig. 5. First generation of claw and initial experiments

In a laboratory, an initial climbing robot was designed using a plastic plate, and climbing experiments were conducted on rough, inclined, and vertical rough or smooth concrete walls (Fig. 5). The results indicated that the robot could climb over small obstacles, and the climbing ability of the robot satisfied the requirements for various wall surfaces, such as cliff surface, rough brick, and water brush sandstone (Fig. 5(a)). As there are all kinds of gravels on the surface of water brush sandstone, no cases of failure claw grasping are appeared (Fig. 5(a) and (b)). For some smooth bricks, the claws can only grab the joined seams (Fig. 5(d)). However, for the smooth concrete walls, some cases shows that the grasping claw often failed to gasp the wall. When the surface is smooth, even for the horizontal ground, the claw can't grasp the surface steadily.

4.2. *Experiments on loading ability of a single claw*

Loading experiments were performed in the laboratory to verify the feasibility of a single claw. The stress condition on the hook was measured using a spring dynamometer as the hook grabbed rough objects. The results revealed that in case of self-locking, the hook did not slide regardless of tension. For the hook without quenching, the stress increase induced the bending deformation of the hook (Fig. 6(b)). If the tension was large enough, the quenched hook revealed cracks and was even ruptured. However, the bearing stress was improved significantly; in the absence of self-locking, a sliding phenomenon was observed. For the same rough object, a smaller hook top size indicated a lower occurrence probability of sliding. Therefore, the dimension of the claw must be within the scope of design, and the dimension of the tip point should follow the basic principle of "the smaller, the better."

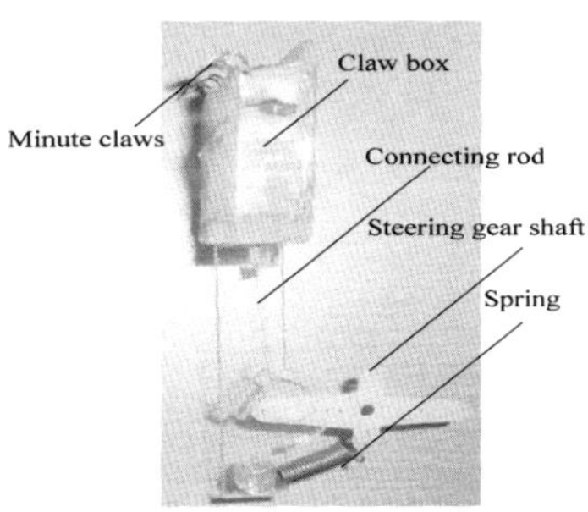

(a) Normal shape and morphology of claws

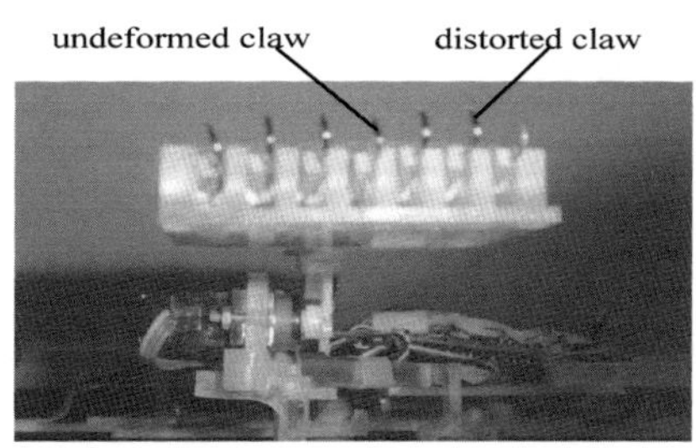

(b) Distorted claws

Fig. 6. Deformation of bionic claws

5. Conclusion

We studied the interaction theory between the claw tip and the microprotuberance on the rough wall surface after examining the spine structure of cockroach legs. We developed an innovative grasping bionic hooked claw and a claw-based climbing model of a wall-climbing robot. The proposed sub claw contained multiple bionic hooks that could move freely. Moreover, the results of the experiments initially demonstrated that the proposed grasping claw could steadily grasp the microprotuberances on the rough wall surface.

References

1. Elachachi S M, Breysse D, Yotte S, Cremona C. A Probabilistic multi-scale time dependent model for corroded structural suspension cables. Probabilistic Engineering Mechanics, 2006, 21, 235-245.
2. Mehrabi A B, Asce M, Ligozio C A, Ciolko A T, Wyatt S T. Evaluation, Rehabilitation Planning, and Stay-Cable Replacement Design for the Hale Boggs Bridge in Luling, Louisiana. Journal of Bridge Engineering, 2010, 15, 364-372.
3. Mehrabi A B In-Service Evaluation of Cable-Stayed Bridges, Overview of Available Methods and Findings. Journal of Bridge Engineering, 2006, 11, 716-724.
4. Spenko M J, Haynes G C, Saunders J A. Biologically Inspired Climbing with a Hexapedal Robot. Journal of Field Robotics, 2008, 25, 223-242.
5. Lynch G A, Clark J E, Lin P C, Koditschek D E. A bioinspired dynamical vertical climbing robot [J]. The International Journal of Robotics Research. 2012, 31, 974-996.
6. Bartsch S, Birnschein T, Rommermann M, Hilljegerdes J, Kuhn D. Development of the Six-Legged Walking and Climbing Robot SpaceClimber. Journal of Field Robotics, 2012, 29, 506-532.
7. Sintov A, Avramovich T, Shapiro A. Design and motion planning of an autonomous climbing robot with claws. Robotics and Autonomous Systems, 2011, 59, 1008-1019.
8. Lam T L, Xu Y S. Motion planning for tree climbing with inchworm-like robots. Journal of Field Robotics, 2013, 30, 87-101.
9. Lam T L, Xu Y S. Biologically inspired tree-climbing robot with continuum maneuvering mechanism. Journal of Field Robotics, 2012, 29, 843-860.
10. Lam T L, Xu Y S. Climbing Strategy for a Flexible Tree Climbing Robot-Treebot. IEEE Transactions on Robotics, 2011, 27, 1107-1117.
11. Asbeck A T, Kim S, Cutkosky M R, Provancher W R, Lanzetta M. Scaling hard vertical surfaces with compliant microspine array. International Journal of Robotics Research, 2006, 25, 1165-1179.

DESIGN AND IMPLEMENTATION OF A SMART ROBOTIC SHARK WITH MULTI-SENSORS

SHIFENG CHEN†

State Key Laboratory of Management and Control for Complex Systems, Institute of Automation, Chinese Academy of Sciences, Beijing 100190, P. R. China

JUNZHI YU, XIANGBIN LI, JUN YUAN

State Key Laboratory of Management and Control for Complex Systems, Institute of Automation, Chinese Academy of Sciences, Beijing 100190, P. R. China

Smart intelligent robotic fish has shown promising advantage in underwater searching. This paper addresses the smart robotic shark design and control issues with multi-sensors. In particular, we propose a new design of a two-link mechanism robotic shark equipped with gyroscope, pressure sensor, infrared sensor, and light sensor. Then three-dimensional motion control, depth control, autonomous obstacle avoidance, and light navigation are developed. In particular, a bio-inspired Central Pattern Generator (CPG) based control method is adopted to smoothly control the robotic shark's locomotion in all the above realization. All motion control methods are implemented in real time with a hybrid control system based on embedded microprocessor (STMicroelectronics STM32F407). Latest aquatic experiments demonstrate a fairly good result in improving the robotic shark's intelligence. The developed scheme affords an alternative to smart robotic fish design in complex underwater environments.

Keywords: Robotic shark, CPG, Multi-Sensors, Motion control.

1. Introduction

Underwater vehicles are well recognized in various applications, such as exploration of marine resources, underwater target search, as well as military purposes. Among a variety of underwater vehicles [1], bio-inspired swimming robot [2] has shown superior performance in efficient propulsion and high maneuverability compared with conventional underwater vehicles propelled by rotary propellers [3].

† This work was partly supported by the National Natural Science Foundation of China (61375102, 61165016, and 61333016), by the Beijing Natural Science Foundation (3141002), and by 2013 Project Based Personnel Exchange Programme under Grants 201314.

Smart robotic fish plays an important role in underwater exploration, especially in cave search for its limited space. The existing studies almost have been focused on theoretical aspect and development of large robotic fish whose body length is up to 50 cm or longer. Among those with the propulsive mechanism of multi-linkage [4], the number of joints is almost larger than three. There have been few or limited studies on the robotic fish with two joints or only one. With the propulsive configuration of two links, how to achieve similar fishlike motions like forward swimming, turning, diving, and surfacing remains challenging.

This paper, on the basis of our previous research on fishlike swimming, aims at designing and implementing a two-link smart robotic shark with multi-sensors [5]. Specifically, a robotic shark with two tail joints and a pair of pectoral fins is designed to validate the functionality of the two-link mechanism. In the meantime, multiple sensors are used to improve its intelligence. A bio-inspired central pattern generator (CPG) controller [6-9] is then adopted to achieve stable 3D underwater movements. Finally, aquatic tests on the real robot verify the effectiveness of the proposed mechanism and control methods.

The rest of this paper is organized as follows. The overall mechatronic design of the smart robotic shark with multi-sensors is overviewed in Section 2. Section 3 gives a brief introduction of motion controller combining sensor information and the CPG controller for stable fishlike swimming. Experimental results are provided in Section 4. Finally, Section 5 concludes the paper with an outline of future work.

2. Prototype of the Robotic Fish

2.1. *Mechatronic design*

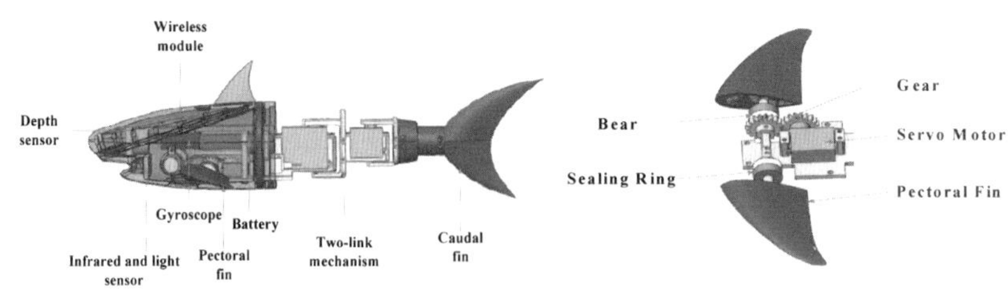

(a) (b)

Figure 1: (a) Mechanical configuration of the robotic shark. (b) Mechanical design of pectoral fin propulsion.

Figure 1(a) shows a schematic diagram of the developed robotic fish prototype. As can be seen, the robotic fish has multiple fins including a caudal fin, a pair of pectoral fins. Mechanically, the robotic shark consists of a rigid head housing multi-sensor, control circuits, and lithium battery pack, a flexible body with two joints, and a caudal fin. Each joint is actuated by one servo motor, whose control signal is produced by the fine-tuned CPG controller. Coordinated multiple joints allow the robotic fish to swim forwards/backwards, submerge, surface, and turn. The detailed technical parameters of the robotic shark are shown in Table 1.

Table 1: Technical parameters of the robotic shark.

Parameter	Description
Size (L×W×H)	350 × 61 × 83 mm3
Weight	970 g
Joint drive	HS-5565MH and HS-82MG
Sensor type	Infrared sensor, pressure sensor light sensor, gyroscope
Controller	Stm32f407
Operating conditions	7.4 V

Due to the limited size of the robotic shark, there is not enough space to mount two servo motors to drive each pectoral fin. Thus a servo motor along with a pair gears is adopted. As shown in Figure 1(b), the servo motor drives the two pectoral fins through the pair of gears. They will be rotated to the same angular position simultaneously.

2.2. *Hardware development*

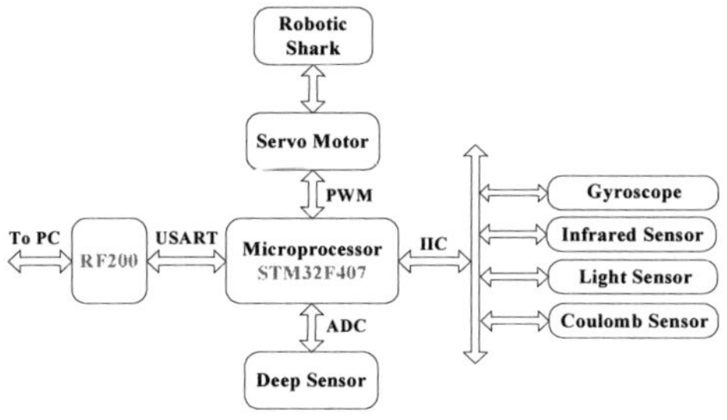

Figure 2: Hardware design of the online swimming control.

At the level of hardware design, as shown in Figure 2, STMicroelectronics STM32F407 microcontroller is selected as the main control chip. A bidirectional RF module RF200 that can transfer message between the embedded system and the PC is chosen. Multi-sensors are selected to improve the intelligence of the robotic shark including gyroscope, depth sensor, infrared sensor, light sensor, and coulomb sensor. Particularly, A MPU9150 is chosen as the gyroscope part. In addition, the BH1750FVI is adopted as the light sensor for its large measurement range. A micro-pressure sensor is used as the depth sensor with a range of 0 to 2 m. Three infrared sensors are installed on each side of the robotic shark to detect block area. Moreover, a MAX17044 is designed to monitor the capacity of the Li-Po battery. The main control circuit board and infrared sensor board are shown in Figure 3.

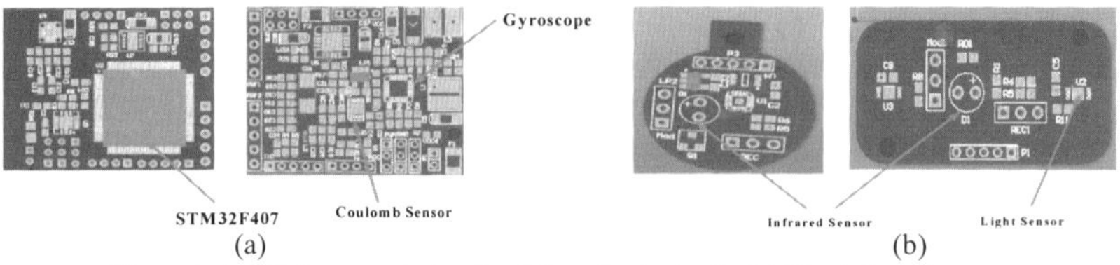

Figure 3: (a) Main control circuit of the robotic shark. (b) The infrared sensor.

3. Motion Controller Combining Sensory Information and the CPG Controller

3.1. *Three-dimensional motion control based on gyroscope compensation*

In general, to achieve three-dimensional (3-D) locomotion aided by the pectoral fin, the robotic fish should maintain a certain forward speed. Particularly, this speed is related to the length of the multi-link mechanism. In order to explore propulsive mechanism and performance of the two-link robotic fish, we propose a 3-D motion control method based on gyroscope compensation.

Particularly to achieve up and down swimming, a pair of pectoral fins is used to realize this. When the robotic shark wants to swim downside, we should set the angle of the pectoral fins as shown Figure 4(a). In that case when the robotic shark swim forward, the water will generate a downward force to the pectoral fins that will make the robotic shark swim downside. Otherwise if it is set as shown in Figure 4(b), the robotic shark will get an upward force that drive it to swim upside. Moreover, if we want to swim forward as fast as we can, we should make the pectoral fins be parallel to the water surface as parallel as we can, for it will get little drag force from the water.

Given that there are tough issues such as the interference of water, output signal error of CPG [10-14], servo motor rotate error, the robotic shark can hardly swim in strictly straight direction in water. Thus an output compensation based on gyroscope feedback is proposed in this paper to reduce the swimming error. Particularly upon the start of forward swimming, we obtain the direction angle it will swim to by sampling the output of the gyroscope. Then we start forward swimming and calculate the current direction angle in real time. Due to its locomotion property, the output direction angle will oscillate around a certain value. If the robotic shark swim as strictly straight as it is mean to, the value should be the start direction angle. Assume that the compensate period is T, in the sample period, we average the sampled direction angle and compare it with the destination direction angle, if it is on the left of the destination, then the robotic fish will turn right and vice versa. Then keep turning until the current direction angle is equal to the destination and then continue to swim forward. With the method above, the final direction error is reduced in real time swimming. As a result, the robotic shark will finally achieve 3-D locomotion.

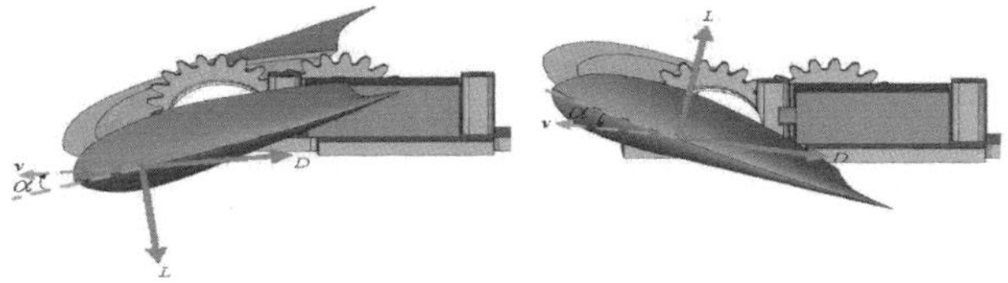

(a) (b)

Figure 4: (a) is the pectoral angle of swimming downside. (b) is the pectoral angle of swimming upside.

3.2. *Depth control of robotic shark based on pressure sensor feedback*

For precise three-dimensional locomotion, depth control is fundamental and significant to the robotic shark. With the pressure sensor equipped in the designed robotic shark, the depth of the robotic shark can be acquired and used as a feedback signal for depth control [15, 16].

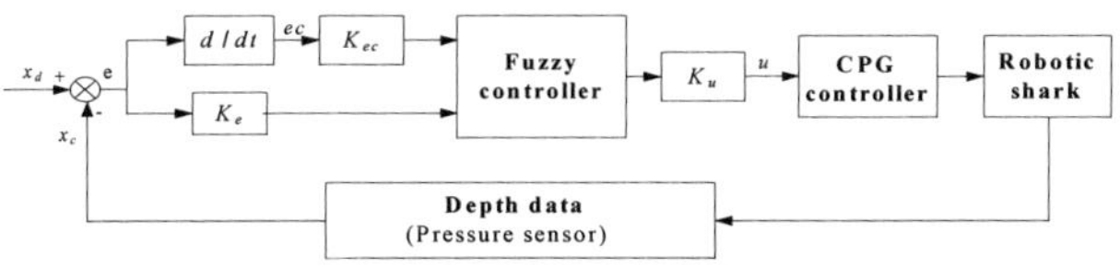

Figure 5: Depth control framework combing the fuzzy logic controller and the CPG controller.

In this paper, a fuzzy logic controller is developed to obtain the input of CPG controller. The control framework that combines the fuzzy logic controller and the CPG model is described in Figure 5. $e(k)$ denotes the input error, while $ec(k)$ represents the derivative of the input error. d_d is a predetermined depth position that the robotic shark finally will reach. $d_c(k)$ is the real-time depth of the robotic shark. With the fuzzy controller above, the robotic shark can finally reach a predetermined depth after a series of up-and-down movements.

3.3. *Autonomous obstacle avoidance motion control based on infrared sensor feedback*

Due to uncertainty of water environment, it is necessary for the robotic shark to have ability to detect the information of the environment, especially obstacle avoidance in autonomous navigation. In this paper, three infrared sensors are installed on the right, front and left side of the robotic shark to detect obstacles underwater. If there are obstacles on the left of the robotic fish, it will turn right, vice versa. If there are obstacles in the front of the robotic fish, it will turn back. If there is no obstacle, the robotic fish will swim forward. An output table is designed based on the status of the infrared sensors. Let 1 be obstacle detected and 0 be none. The status of the three infrared sensors is from 000 to 111 while first bit represents the status of the left infrared sensor, middle bit is front one and third bit is the right one. The detailed description is tabulated in Table 2.

Table 2: Output offset value based on status of infrared sensors.

Status	Direction
000	Forward
001	Turn Left
010	Turn Back
011	Turn Left and Forward
100	Turn Right
101	Forward
110	Turn Right and Forward
111	Backward

3.4. *Autonomous navigation based on light intensity sensor detection*

Light navigation is very useful in real-world applications, especially when we want to swim out a dark area, such as a pipe or cave.

In this paper, we installed two light detect sensor in the front of the robotic shark, and by sampling the light intensity data of the two sensor, the robotic fish can swimming ahead automatically. If the light intensity of the left is strong than that of the right, the robotic shark should turn left and swimming forward, otherwise it should turn right and swim forward. If the two is almost the same, the robotic shark should swim forward directly. By the rule stated above, the robotic shark will finally be autonomously navigated with the help of light sensors.

4. Experimental Results

To verify the feasibility of the designed robotic shark and proposed control method, we have carried out some aquatic experiments in the swim pool. In a pool with the size $500 \times 400 \times 150$ cm^3, we conduct some experiments including forward swimming, up-and-down swimming, automatic obstacle avoidance, light navigation and depth control. A sequence of autonomous navigation based on all the sensors is demonstrated in Figure 6, where the robotic shark can successfully perform free-swimming in the pool without remote control. In the experiments, the speed of the robotic shark has 9 levels from 0.25 BL/s to 2.5 BL/s and with the battery capacity of 3400 mAh, the robotic shark can keep about 40 minutes for continuous free-swimming.

Particularly the two-link mechanism is really enough for three dimensional free-swimming.

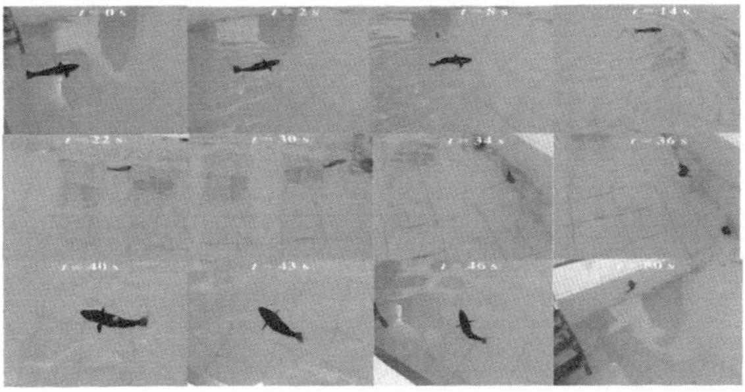

Figure 6: Snapshot sequence of the robotic shark navigation

5. Conclusion and Future Work

In this paper, we have designed a smart robotic shark with multi-sensors using a two-link mechanism. Specifically, a hybrid control system framework combing one microprocessor and multi-sensor is firstly built. Then a three-dimensional motion control method based on CPG mode and gyroscope compensation is adopted. In addition, depth control with pressure sensor feedback, autonomous obstacle avoidance with infrared sensor, light navigation is performed to improve the intelligence of the robotic shark. Finally, autonomous navigation swimming control experiments have been carried out to demonstrate the effectiveness of the proposed control framework and methods.

The ongoing and future work will focus on the intelligence improving and energy optimization of the robotic shark, including adding embedded vision and improving CPG model with energy optimization so that the robotic fish can adapt to more complicated underwater environments.

References

1. Paull L, Saeedi S, Seto M, AUV navigation and localization: A review, *IEEE Journal of Oceanic Engineering,* 39(1), 131–149, 2014.
2. Bogue R, Underwater robots: a review of technologies and applications. *Industrial Robot: An International Journal*, 42(3), 2015.
3. Asare V A, Shende R S, Mechatronics and Motionability of an Underwater Robot: A Review, *Mechatronics*, 2(3), 2014.
4. Faudzi A A M, Razif M R M, Nordin N A M, A Review on Development of Robotic Fish, *Journal of Transport System Engineering*, 1: 12-22, 2014.
5. M. Wang. Locomotion Modeling and Control of Biomimetic Robotic Fish Based on Central Pattern Generators [PhD Thesis]: Institute of Automation, Chinese Academy of Sciences, 2010.

6. J. Yu, M. Tan, S. Wang, and E. Chen, Development of a biomimetic robotic fish and its control algorithm, *IEEE Transactions on Systems, Man, and Cybernetics*, Part B: Cybernetics, vol. 34, no. 4, pp. 1798–1810, 2004.
7. J. Yu, M. Tan, and J. Zhang, Fish-inspired swimming simulation and robotic implementation, *in Proc. 41st International Symposium on Robotics (ISR) and 2010 6th German Conference on Robotics* (ROBOTIK), Munich, Germany, June 2010, pp. 1158–1163.
8. J. Yu, M. Wang, Z. Su, M. Tan, and J. Zhang, Dynamic modeling and its application for a CPG-coupled robotic fish, *in Proc. IEEE Int. Conf. Robot. Autom.*, Shanghai, China, May 2011, pp. 159–164.
9. Z. Wu, J. Yu, and M. Tan, CPG parameter search for a biomimetic robotic fish based on particle swarm optimization, *in Proc. IEEE Int. Conf. Robot. Biomim.*, Guangzhou, China, Dec. 2012, pp. 563–568.
10. J. Duysens and H. W. A. A. Van De Crommert, "Neural control of locomotion; Part 1: The central pattern generator from cats to humans," *Gait and Posture*, 1998, 7(2): 131–141.
11. G. Wang, D. Zhang D, L. Lin, H. Xie, T. Hu, and L. Shen, CPGs control method using a new oscillator in robotic fish, *Science China Technological Sciences,* 2010, 53(11): 2914–2919.
12. C. Zhou and K. H. Low, Kinematic modeling framework for Biomimetic undulatory fin motion based on coupled nonlinear oscillators, *in Proc. IEEE/RSJ Int. Conf. Intell. Robot. Syst.*, Taipei, Taiwan, Oct. 2010, pp. 934–939.
13. Yu J, Tan M, Chen J, and Zhang J, A survey on CPG-inspired control models and system implementation, *Neural Networks and Learning Systems, IEEE Transactions on,* 25(3), 441–456, 2014.
14. Wang M, Yu J, Tan M. CPG-based Sensory Feedback Control for Bio-inspired Multimodal Swimming, *International Journal of Advanced Robotic Systems*, 11, 2014.
15. K. H. Low, "Locomotion consideration and implementation of robotic fish with modular undulating fins: analysis and experimental study," *in Proc. Int. Conf. Intelligent Robots and Systems*, 2006, pp. 2424–2429.
16. P. J. Lee, C. H. Yen, C. L. Chan, M. S. Lee, and R. C. Wang, Implementation of a fuzzy control based intelligent robot fish, *International Journal of Fuzzy Systems*, vol. 11, 2009, pp. 287–297.

CONTROL OF QUADRUPED WALKING ROBOTS USING MOTOR PRIMITIVES*

BOXING WANG
Department of Control Science and Engineering, Zhejiang University
Hangzhou, 310027, China

CHUNLIN ZHOU
Department of Control Science and Engineering, Zhejiang University
Hangzhou, 310027, China

Dynamic movement primitives provides an approach to generate and modulate periodic trajectories real time. Based on DMP, We proposed an open loop position control quadruped controller which is able to make the robot trot on a flat plane. In this modular controller, each Single leg module is responsible for one leg motion generation, and this motion generation is interpreted in Cartesian space. Such design makes it easy to modulate the four legs phase relation and independent of mechanical structure. Speed control is achieved through frequency control which even allows moving backwards without a stop smoothly. Turning control is achieved through amplitude control. Our controller is easy to apply and expand but quite effective of phase coordination and online trajectory modulation. Results of simulated experiments on moving, turning and speed control are presented.

1. Introduction

Quadruped robots are useful in many tasks due to their ability to walk smoothly on rough terrain[1]. Although the stability problem isn't that prominent in quadruped robots, but there still needs lots of work to do. Among different discussions about how to keep quadruped move quick and stable, biologically inspired strategy has become a promising approach.

Biologically inspired strategy is actually a feedback control strategy with a dynamical system as its controller. Most popular biologically inspired strategy is central pattern generator[2] (CPG). A typical approach to express CPG is to design an autonomous dynamical system of coupled nonlinear oscillators. This autonomous dynamical system can be a network[3] or just two coupled oscillators[4] , and its working space can also be different, either joint space[4, 5] or

* This work is supported by National Natural Science Foundation of China (No. 51305396) and Zhejiang Provincial Natural Science Foundation of China (No. LQ13F030001)

leg coordination Cartesian space[3]. Most significant advantage of CPG is that it can be easily extended and keep the ability of disturbance robustness[6-8]. Meanwhile, as the CPG provides coordination strategy naturally, the controller don't need to consider the leg coordination a lot when it performs as a relative high-speed continuous dynamic gait.

CPG is not easy to design as it's a nonlinear dynamic system. And the curve of output signal of most CPGs cannot be arbitrarily reshaped. Dynamical movement primitives (DMP)[9, 10] provides a good solution for this problem. The concept of primitives, motor primitives or movement primitives comes from biological research, which means basic elements that can generate complicate movements[11].

In this paper, we use DMP as our core control strategy to control a quadruped robot. The controller is established to achieve middle speed moving and the output signal is explained in leg coordination Cartesian space.

2. Dynamic Movement Primitives

Dynamical movement primitive (DMP) is a line of research for modeling attractor behaviors of autonomous nonlinear dynamical system with the help of statistical learning techniques. Both point and limit cycle attractor are presented but only the last one is introduced here, as we only need periodic motions. More details can be found in the research of Ijspeert's group[9].

2.1. *Basic structure of DMP*

A DMP can be divided into three parts: Canonical system, forcing term, and Transformation system. The former's output will be the latter's input, making the whole DMP like a cascade system. So we can conclude the three parts' equations as follows:

$$\phi = p_1(t) \tag{2.1}$$

$$f = p_2(\phi) \tag{2.2}$$

$$y = p_3(f, \dot{z}, \dot{y}) \tag{2.3}$$

In these equations, (2.1) stands for Canonical system, (2.2) stands for forcing term, and (2.3) stands for Transformation system, where y and z are the state variables.

Actually, (2.1) is a proportional function with output regulated into [0,2π], which makes the Canonical system a periodic phase oscillator.

Forcing term will drive the Transformation system to reach the trajectory we want. Its equations (2.2) is a set of algebraic equations, which is the only nonlinear part of DMP.

$$f(\phi) = p_2(\phi) = \frac{\sum_{i=1}^{N} \psi_i \omega_i}{\sum_{i=1}^{N} \psi_i} r \tag{2.4}$$

$$\psi_i = \exp(h_i(\cos(\phi - c_i) - 1)) \tag{2.5}$$

Where the exponential basis functions in equation (2.5) are fixed von Mises basis functions, essentially Gaussian-like functions. And ω are adjustable weights, r is a scale factor. Representing arbitrary nonlinear functions as such a normalized linear combination of basis functions has been a well-established methodology in machine learning or statistical learning techniques. So we can arbitrarily change the output of it.

Transformation system can be seen as a low-pass filter for forcing term, and here it is a linear time-invariant system modeled from spring-damping system.

$$\begin{aligned} \tau \dot{z} &= \alpha_z(\beta_z(g - y) - z) + f \\ \tau \dot{y} &= z \end{aligned} \tag{2.6}$$

α_z and β_z are positive constants, g is the attract point of the linear system.

3. Proposed Controller

This section introduces the controller we designed for quadruped walking by using DMP as its core. The actuators we used are servo motors, so the output of our controller is position information.

3.1. *Pre-design trajectory*

In this work, we only use two main coordinate to observe the motion of the quadruped: the body coordinate frame Σ_b attached to the center of the robot, and the global coordinate frame Σ_g attached to the ground. The pre-design trajectory of the free end is considered in Σ_b. The whole image of the robot can be seen in Figure 1.

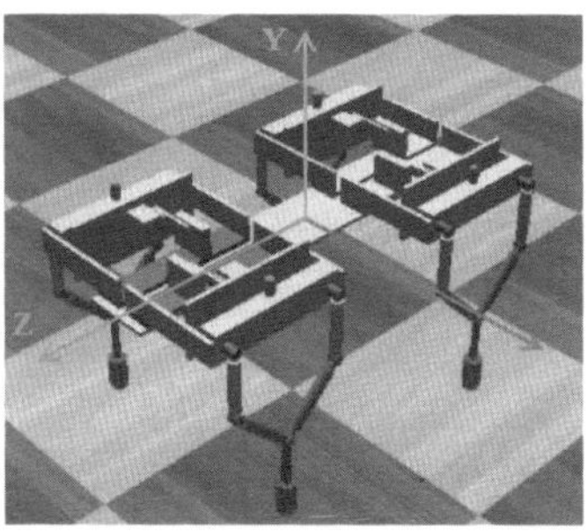

Figure 1. The body coordinate frame Σ_b and the robot Horizon in simulation.

The trajectory is designed in YZ plane in Σ_b or so called Sagittal Plane. Its shape is shown in Figure 2. Since we want the robot to keep its height unchanged when this foot touches the ground and considering our gait is actually a kind of static period gait, making bottom a line is straightforward.

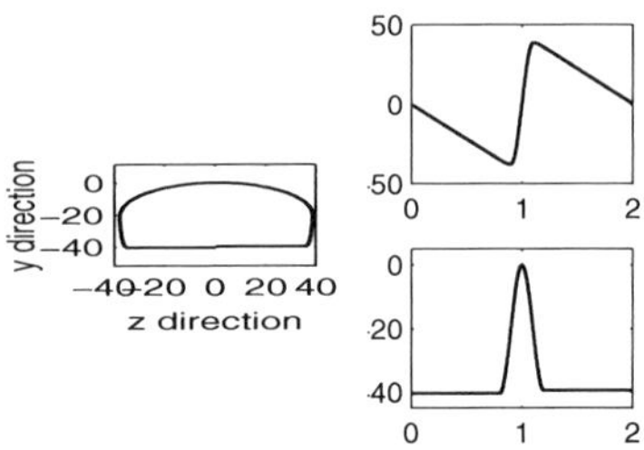

Figure 2. Left graph is the predesign-trajectory in Sagittal plane. Right graph is decomposition of y and z direction. Remind that the origin of this graph is not the origin of YZ or Sagittal Plane. We define a new origin here just to make it easier to generate such a pre-design trajectory.

3.2. *Single leg module and the overall controller*

In order to make DMP suitable for quadruped control, we use two DMPs to construct a new elementary module called Single leg module to control a single leg. Its general structure is shown in Figure 3.

Coordination through different Single leg modules is various. Like the fundamental topological structures of multiples coupled oscillators, there can be chain type, radial type, ring type and fully connected type. In this work, we use unidirectional radial type (Figure 4).

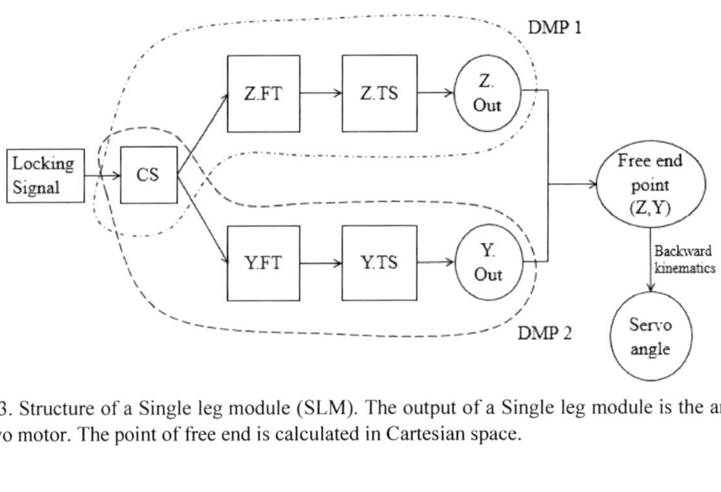

Figure 3. Structure of a Single leg module (SLM). The output of a Single leg module is the angle of the servo motor. The point of free end is calculated in Cartesian space.

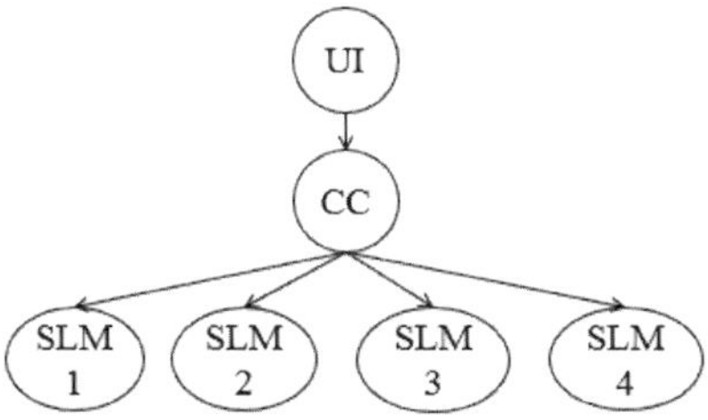

Figure 4. Unidirectional radial type used in our work. SLM is short for Single leg module; CC is short for Central controller; UI is short for User interface.

Central controller (CC) in Figure 4 is a high level controller which is responsible for external base signal generating and User interface (UI) input explanation. So the coordination between four Single leg modules (SLM) is achieved by synchronizing them to the same external base signal separately.

Although four-leg animals have different gaits, we choose trot gait, because it's the most stable one for our quadruped robot. Since our control strategy is open loop control actually, the parameters can only be decided by trial and error and obviously a little different from the primary gaits [6].

4. Simulation Results and Discussion

Simulations were done in Webots[12] and NI labview Robotics module using a quadruped robot on a flat plane. Prototype is constructed by using NI myRIO controller while more hardware experiments will be done in the future.

4.1. *Walking in trot gait through open loop control*

We choose trot gait as our prime gait, but we do a little modification based on the phase relation in primary gaits. Here in our experiments, we use:

$$p_{LF} = 0 \quad p_{RF} = \pi \quad p_{LH} = 0.95\pi \quad p_{RH} = -0.05\pi$$

So we minus 0.05π to each hind leg's phase. The adjustment we made here can be seen as a kind of linear interpolation between walk, trot and pace.

We measure the height of the center of robot as the main evaluation of the stability of the robot. The record is showed in Figure 5. The initial height of robot is a little higher, because the mechanical zero point of this parallel structure is at the edge of its work space and we need to move the free end into its work space first. Right graph of Figure 5 shows the phase evolution of four legs. The output result is a typical result of a PID controller. Although there is no overshoot, the phase lock is quite rapid.

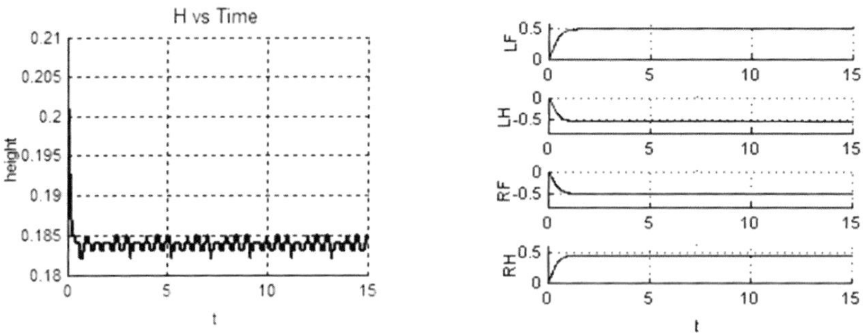

Figure 5. Left: Height record of the center of the robot in moving test. The unit of the height is meter, and the unit of time is second. Right: Phase evolution of four legs in moving test.

4.2. *Speed control*

Speed control can be achieved through Center controller, which actually changes the frequency of the external oscillator. In simulation environment, we double the frequency at 5s. The displacement along forward direction is shown in Figure 6.

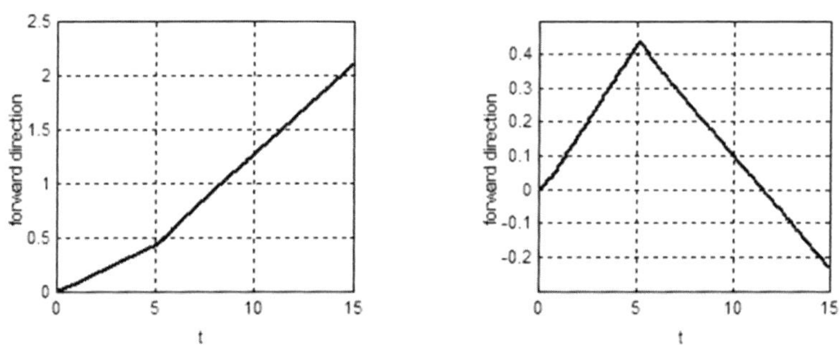

Figure 6. Record of displacement along forward direction in speed up test. The unit of the displacement is meter, and the unit of time is second. Left: At 5s the moving velocity increased as we double the frequency of the reference oscillator. Right: At 5s opposite the frequency value.

Another extreme example of speed controlling is to make the robot move backwards from forwards just by turning the external reference signal into a decreasing slope signal. See Figure 6 right graph.

4.3. *Turning control*

Reminding that we decomposed the trajectory into two directions in SLM, here we just need to modify the z direction amplitude of one side legs to achieve turning control. So that we can make the average speed of two side different, then the robot can turn without any effects to phase relations. Figure 7 shows the robot moving trajectory projected on the walking plane.

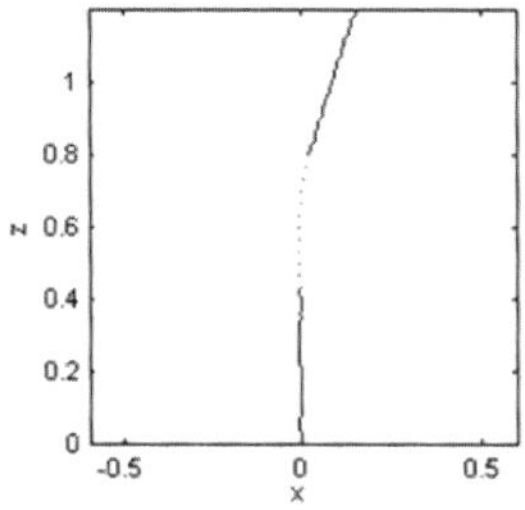

Figure 7. 2D trajectory recorded in turning test. This trajectory is the projection of robot 3D trajectory on the walking plane in Σ_g . Dotted line indicates the robot was doing a turning motion.

5. Conclusion

In this paper, we proposed a quadruped controller based on DMP. This controller can learn a given trajectory and regenerate it real time. Moreover, this controller is able to adjust the output frequency and two independent degrees' amplitude. And due to the DMP's natural coordination ability and unidirectional radial coordination type we used, we can control the robot to move in a certain trot gait on a flat plane The output of the controller is explained in Cartesian space, which makes it suitable for different kinds of leg design and easy to change the shape of output trajectory.

Although our controller is an open loop controller, the robot is able to speed up or even move backwards without any stop. Turning is achieved by taking the same techniques used in differential wheels robot. All the tests about the above tasks are finished in simulation environment, showing that it's a simple but effective way to control a quadruped robots.

Loss of sensors feedback makes our controller cannot walk on uneven ground stably. More research will be done about how to adjust the phase relation and ratio of stance and flight phase. More hardware experiments will be done.

Acknowledgments

Thanks to Mr. Jing Pan of National Instruments corporation for the help of setting up the hardware, and the assistance of simulation in NI labview robotics module.

References

1. L. Bruzzone and G. Quaglia, *Mechanical Sciences* **3** (2), 49-62 (2012).
2. A. J. Ijspeert, *Neural Networks* **21** (4), 642-653 (2008).

3. D. T. Tran, I. M. Koo, Y. H. Lee, H. Moon, S. Park, J. C. Koo and H. R. Choi, *Robot Auton. Syst.* **62** (10), 1497-1516 (2014).
4. H. Kimura, Y. Fukuoka and A. H. Cohen, *Philosophical Transactions of the Royal Society A: Mathematical, Physical and Engineering Sciences* **365** (1850), 153-170 (2007).
5. H. Kimura, Y. Fukuoka and A. H. Cohen, *Int. J Robot. Res.* **26** (5), 475-490 (2007).
6. C. L. Zhou and K. H. Low, *IEEE-ASME T Mech.* **17** (1), 25-35 (2012).
7. A. J. Ijspeert, A. Crespi, D. Ryczko and J. M. Cabelguen, *Science* **315** (5817), 1416-1420 (2007).
8. A. Sprowitz, A. Tuleu, M. Vespignani, M. Ajallooeian, E. Badri and A. J. Ijspeert, *Int. J Robot. Res.* **32** (8), 932-950 (2013).
9. A. J. Ijspeert, J. Nakanishi, H. Hoffmann, P. Pastor and S. Schaal, *Neural Comput.* **25** (2), 328-373 (2013).
10. S. Schaal, P. Mohajerian and A. Ijspeert, *Progress in Brain Research*, **165**, pp. 425-445.(2007)
11. S. Degallier and A. Ijspeert, *Biol. Cybern.* **103** (4), 319-338 (2010).
12. O. Michel, *Int. J Adv. Robot Syst.* **Num. 1** (Vol. 1), 39-42 (2004).

DESIGN AND GAIT ANALYSIS OF A TORTOISE-LIKE ROBOT WITH SOFT LIMBS*

SHIXIN MAO†

Robotics Research Centre, School of Mechanical and Aerospace Engineering, Nanyang Technological University, 639798, Singapore.

ERBAO DONG*, LIPING ZHOU, HU JIN, YI TAN, GUO LIU, MIN XU

Department of Precision Machinery and Precision Instrumentation, University of Science and Technology of China, Hefei, Anhui Province, 230026, China

KIN HUAT LOW

Robotics Research Centre, School of Mechanical and Aerospace Engineering, Nanyang Technological University, 639798, Singapore.

This paper presented a tortoise-like robot with various gaits in different terrestrial environment. Firstly, the locomotion mode of tortoise was particularly investigated, including triangular gait, diagonal gait and synchronous gait. The trials for acquiring the trajectory of tortoise's limb prepared principal information for the fabrication of robot's limb, structural design and corresponding control. Secondly, the exploration of soft limbs presented an efficient method to develop new bionic robot .The optimization of structure and the driving system help to reduce the number of motors for the robotic system. Finally, the robot prototype is able to realize multi-gaits driven by crank shaft-rope actuation system via sequence control method. The results of terrestrial locomotion experiments verify the adaptability of the robot in different environmental terrains.

1. Introduction

Reptiles played a role in the evolutionary changes from aquatic to terrestrial locomotion [1, 2], such as crocodilians [3], salamanders [4], turtles [5], lizards [6]. They survived in the earth for millions of years, forming high environmental adaptability [7, 8]. They are able to switch between two locomotion modes: crawling and swimming, which is related with their environmental adaptability [2, 7].

---

* This work is supported by National Natural Science Foundation of China (No. 51105349, 61375095 and 51275501).

* Erbao Dong :corresponding author to provide phone: +86-551-63601482, Email: ebdong@ustc.edu.cn

Williams investigated the voluntary locomotion of tortoise over arrange of experimental conditions, displayed the correlation between observed frequency of locomotion and body weight, and walking surface in a manner as well as the step cycles of individual limbs and the inter limb intervals [9]. Alexander presented the gaits of quadrupebal animals, including turtles (the gaits of turtles appear to reduce unwanted displacements to the minimum possible for animals with such slow muscles) [10]. Peter summarized two basic motion gaits: symmetrical gait (diagonally opposite limbs move together) and asymmetrical gait (contralateral limbs move together) [11].

Scientists presented various reptile-like robots benefitting from the bionic inspiration. In the 1940's and 1950's, Walter developed lots of robots which demonstrated great sophistication [12]. Walter's tortoise robot ("CORA") based on animal-like neural systems and behaviors. Kim displayed a turtle-like robot using a smart soft composite structure for swimming gait under aquatic environment [13]. Yoshimasa designed an amphibious robot inspired by the locomotion of both turtles and tortoises, i.e. swimming gait and walking gait, but the robot didn't possess the movement ability [14]. Kazuya Yoshida tried to design a robot (i.e. jumping tortoise) for locomotion on micro gravity surface [15].

2. Locomotion Principle of Tortoise

2.1. *Bionic investigations for terrestrial locomotion*

In order to acquire the locomotion principle of tortoise for the gait planning of the robot, we built an simple experimental system with camera, water tank, coordinate paper and temptation (small shrimp), The locomotor gaits of live tortoise can be obtained via many trials, such as triangular gait displayed in Figure 1(a). To easily describe the locomotion of tortoise, we named its each limb as *LF, LH*, *RF*, and *RH*, which respect represent left front limb, left hind limb, right front limb, and right hind limb as shown in Figure 1(b), and arrows indicate the movement of feet off of the ground; the center of gravity lying on a line or within a stable area made up of limbs upon ground.

A stable pose is one principal factor for achieving the locomotion of reptiles. Grey Walter once discussed the stability principle of tortoise locomotion in 1944 [13], i.e. the triangle of support. there should be a stable area (quadrangle or triangle) or line formed by the limbs (feet) upon the ground surface, with the center of gravity lying within a tripod (quadrupod) or balancing on a line, as described in Figure 1(c, d and e). On the other hand, the conditions to maintain the stability depend on the locomotion of the animals.

From the bionic investigations on the locomotion gaits of the employed tortoise, its locomotor modes primarily comprise: symmetrical gait (e.g. diagonal gait) and asymmetrical gait (e.g. triangular gait, and synchronous gait).

The difference among these locomotor modes and the step sequence relationship among limbs can be defined as:

1. Triangular gait: 1→1', 2→2', 3→3', 4→4', i.e. *LF* → *RH* → *RF* → *LH*, as shown in Figure 1(c). The center of gravity lying within a tripod of three limbs upon the ground.
2. Diagonal gait: 11→22→33, i.e. *LH*, *RF* → *LF*, *RH* → *LH*, *RF*, as shown in Figure 1(d). The center of gravity balancing on a line of two limbs upon the ground.
3. Synchronous gait: 11→22→33, i.e. *LF*, *RF* → *LH*, *RF* → *LF*, *RF*, as shown in Figure 1(e). The center of gravity lying within a tripod of two limbs upon the ground and body.

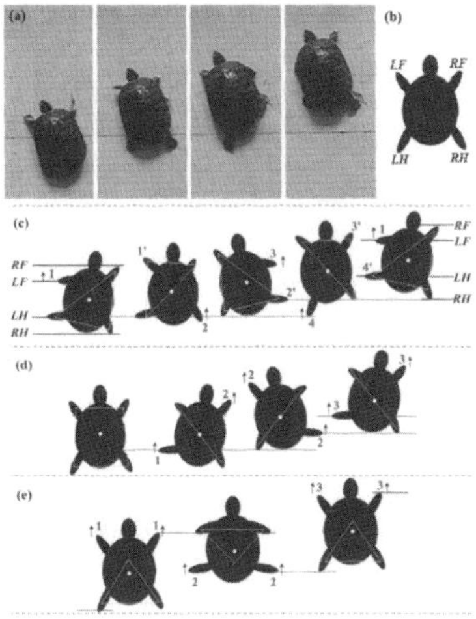

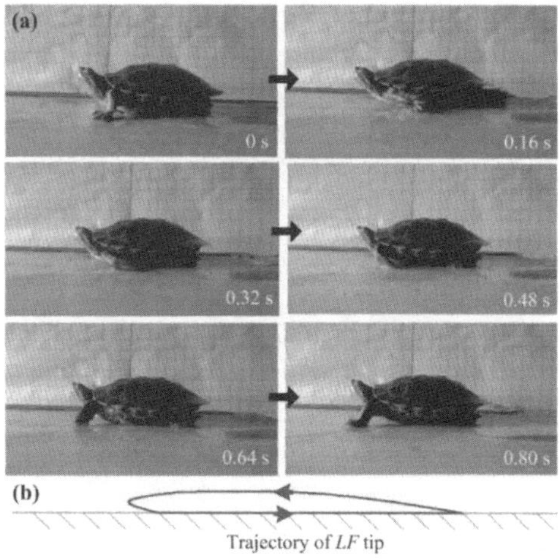

Figure 1 (Left). Bionic gaits analysis: (a) live tortoise crawling with triangular gait, (b) tortoise limbs abbreviations,(c) triangular gait, (d) diagonal gait, (e) synchronous gait.

Figure 2 (Right). Obtaining the motion trajectory of tortoise's limb tip: (a) a circular motion of limb tip, (b) the trajectory of *LF* tip during a circular motion; and arrows show the moving direction.

2.2. *Trajectory of tortoise's limb*

Utilizing the bionic experiment system used above, we can obtain a circular motion of live tortoise's limb tip, as shown in Figure 2(a). And then the trajectory of *LF* tip during a circular motion can be extracted, it's the anticlockwise closed curve, as displayed in Figure 2(b), arrows indicate the moving direction. Meanwhile, the raising height of the limb tip is not high enough for a robot acquiring good environmental adaptability, so the trajectory of the limb tip for the bionic should be optimized, which will be discussed in the following section.

3. Design and Analysis for the Robotic System

3.1. *Structural design for the robotic system*

To construct an entire robotic system, many key issues should be considered, such as bionic design for optimization, deformable structure for environmental capability, actuation system for locomotion and energy saving and so on. According to the investigation of a live tortoise (e.g. its locomotion gaits, the trajectory of limb tip), there are many challenges to develop a bionic robot that can behave with multi-gait in different environments, such as the design, actuated system, control methods etc. Thus, we focus on the exploration of the soft limbs before the design for entire robotic system.

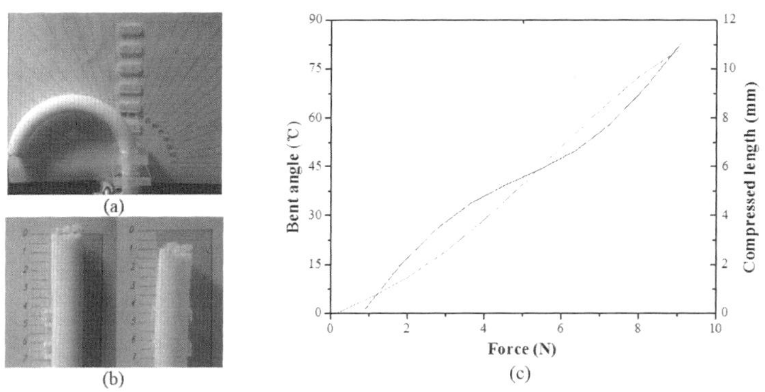

Figure 3. (a) the bending ability testing trial, (b) the compressed soft limb, (c) the bent angle (line with red color) and compressed length (line with blue color) of soft limb under different pull force.

We made soft limb with 160mm in length and 15mm in diameter of cross section using special silicone, and tested its capability for bending and being compressed, as displayed in Figure 3(a, b). And Figure 3(c) presented the relationship between the bent angle and pull force (line in red color) as well as the relationship between the compressed length and continuous pull force (line with blue line), the soft limb can bend more than 80°C at maximum and the compressed length is able to reach 11.5mm when pulled by 10N force. Furthermore, in order to develop appropriate limbs for realizing locomotion gaits of robotic system, we made different soft limbs with multiple sizes. The length-diameter (cross section circle) of two kinds of soft limb (hollow and solid) comprised 13:1 and 21:1. The bending ability of different design limb was tested, as shown in Figure 4. In consideration with the size and capability of the soft limbs, we choose the limb with 210mm in length and 10mm in diameter as the robotic limb.

3.2. *Actuation system*

The cam-rope actuation system determines the output of the robot limb, and the length of rope L changes along with the rotation angle ω. In addition, the trajectory of robot limb always depends on the cam-rope actuation system; the actuated cam should be able to meet the characteristics of circular motion for bionic robot application, because the shape of common cam is complex. In order to simplify the mechanical structure for high efficiency, we choose the optimized actuation structure, *i.e.* crank shaft structure. So the actuation system can be simplified to as the physical model of the crank shaft-rope driving system, as shown in Figure 5.

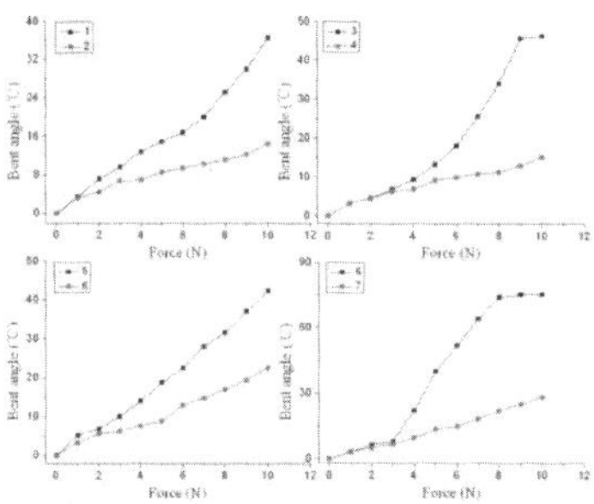

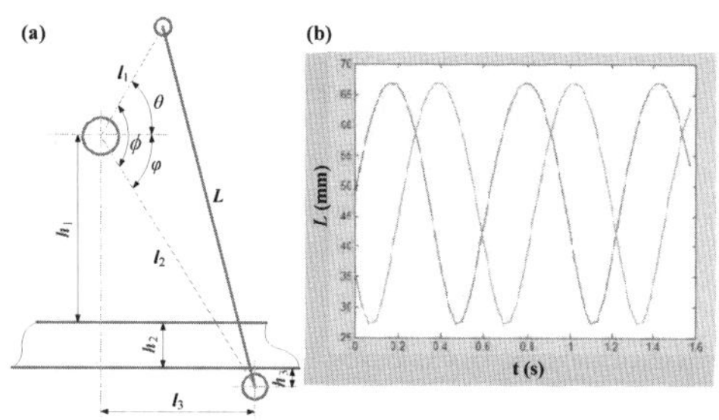

Figure 4 (Left). The bending capability of soft limbs with different sizes and design.

Figure 5 (Right). (a) diagram of L-t and 180° phase L-t (b)

As displayed in Figure 5(a), it is easy to obtain the variable relationship as follow:

$$\begin{aligned} \tan\varphi &= \frac{h_1 + h_2 + h_3}{l_3} \\ \theta &= \omega t + \theta_0 \\ \phi &= \theta + \varphi = \omega t + \theta_0 + \varphi \end{aligned} \tag{1}$$

where θ_0 and φ are constant.

According to the cosine law, the trajectory of the rope L can be expressed:

$$\begin{aligned} L &= \sqrt{l_1^2 + l_2^2 - 2l_1l_2\cos\phi} \\ &= \sqrt{l_1^2 + l_2^2 - 2l_1l_2\cos(\omega t + \theta_0 + \varphi)} \end{aligned} \tag{2}$$

Meanwhile, the trajectory of the 180° phase rope L is:

$$L = \sqrt{l_1^2 + l_2^2 - 2l_1l_2\cos(\omega t + \theta_0 + \varphi + 180^0)} \tag{3}$$

Here, we assume θ_0=0, h_1=30, h_2=7, h_3=3, l_1=20, l_2=47, l_3=25, on the basis of Equation (1-3), the trajectories of L-t and 180° phase L-t can be obtained, as presented in Figure 5(b). This model simplification will be useful for the design of soft limbs and the complete structure.

3.3. *Trajectory of single limb tip and its optimization*

As described in Section 2.2, the trajectories of the tortoise's limb tip played an import role in fulfilling its locomotion gaits and environmental adaptability. With help of bionic trajectory investigation, the actuation system, robotic design and soft limb construction, we fabricated a single soft limb system to test its motion trajectory for further motion optimization and terrestrial locomotion. Figure 6(a) displayed a sequence of single limb movement during a circular motion, and the attached video presented the whole process. The trajectory of this soft limb tip can be extracted through video processing techniques, it's the clockwise closed curve, as displayed in Figure 6(b), and arrows indicate the moving direction.

4. Experiments and Results

As discussed above, based on the bionic investigations of tortoise for terrestrial locomotion and design and analysis for the robotic system, we will discuss the experiments of locomotor gaits in different environmental terrains and related results.

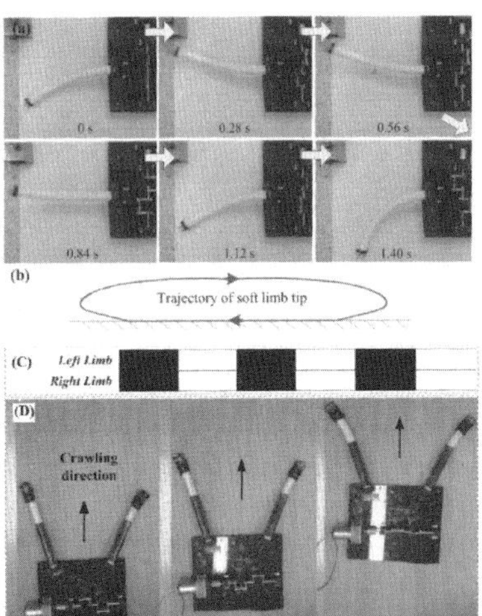

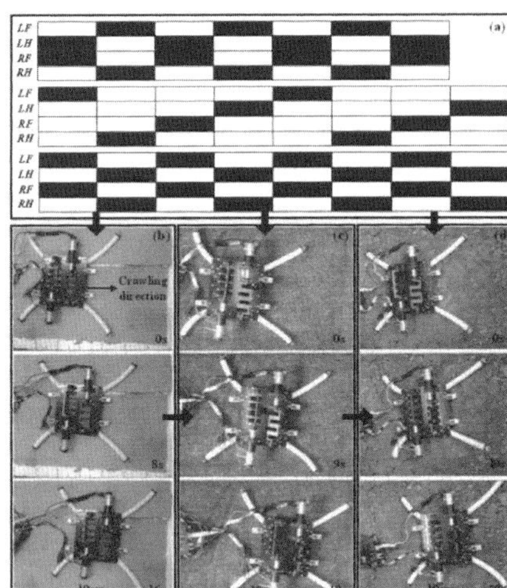

Figure 6 (Left). (a) a circular motion of soft limb, (b) the optimized trajectory of soft limb tip, control pattern for the synchronous gait (c) and related experiment verification (d);

Figure 7 (Right). Control strategy (a) and verification experiments

4.1. *Verification of synchronous gait*

Synchronous gait is basic locomotor modes for the natural tortoise, thought which the tortoise can take advantage of low center of gravity to improve the movement efficiency. Benefitting from such bionic investigation, we conducted the testing experiment of a robotic system with double soft limbs (*Left limb* and *Right limb*), as displayed in Figure 6. Using the control strategy and step sequence method, the control pattern (see Figure 1(e)) of the double soft limbs can lead to rhythmical synchronous movement. This simplified robotic prototype completed the displacement of 199 mm within 80s, as presented in Figure 6(d). The experiment verifies the efficiency of gait planning and actuation mechanism and build foundation for terrestrial locomotion exploration.

4.2. *Terrestrial Locomotion Experiments*

Figure 7(a) presented the gait sequence for fulfilling synchronous gait, triangular gait, and diagonal gait respect; the frame with black color stands for actuation phase of the limb, the frame with white color means dormant phase of the limb. The locomotion modes (i.e. synchronous gait, triangular gait, and diagonal gait) can be realized in different environmental terrains (flat ground, grassland, and mudflat) through corresponding control sequence, as shown in Figure 7(b, c, and d). The attached video vividly displayed the movement of robot prototype. The results also indicate that this tortoise-like robot possess terrestrial adaptability.

4.3. *Results*

These experiments discussed above verified that the tortoise-like robot possess different locomotion gaits and terrestrial adaptability in three kinds of environmental terrains. Furthermore, in order to compare the moving capability of the robot under different environments, we acquired the displacement and corresponding information of the robot prototype through video processing.

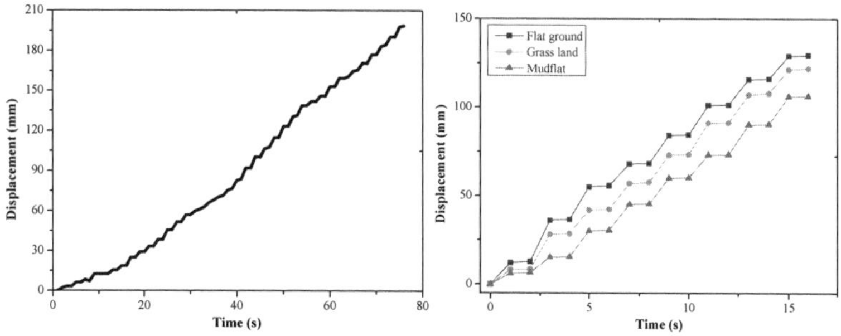

Figure 8. Displacement of the prototype with double limbs on flat ground (upper diagram); and the displacement (under diagram) of tortoise-like prototype in different environmental terrains (flat ground, grassland and mudflat).

Figure 8 displayed the displacement of the prototype with double limbs on flat ground (upper diagram), and the displacement (under diagram) of tortoise-like prototype in different environmental terrains (flat ground, grassland and mudflat). The robot is able to complete the displacement of 129.5mm, 121.9mm, 106.2mm on the flat ground, grassland and mudflat respect within 16s, the corresponding average velocity are 8.09mm/s, 7.62mm/s, 6.64mm/s.

5. Conclusion

We developed a tortoise-like robot with terrestrial adaptability actuated by soft limbs. The locomotion principle of tortoises has been addressed through exhaustive investigation. Three basic kinds of locomotor gaits (i.e. triangular gait, diagonal gait, and synchronous gait) were presented clearly for the control and locomotion planning of the bionic robot. The exploration and discussion about soft limbs presented an efficient way to develop soft or deformable robots. The development of the driving system improved the efficiency of actuation, optimized the structure and reduced the number of motors for the robotic system. The trials for acquiring the trajectory of tortoise's limb prepared principal information for the fabrication of robot's limb and related control. The results of terrestrial locomotion experiments indicate the adaptability of the robot in different environmental terrains.

References

1. Garland Jr, Theodore, and Jonathan B. Losos. "Ecological morphology of locomotor performance in squamate reptiles." *Ecological morphology: integrative organismal biology* (1994): 240-302.
2. Fischer, Martin S., Cornelia Krause, and Karin E. Lilje. "Evolution of chameleon locomotion, or how to become arboreal as a reptile." *Zoology* 113, no. 2 (2010): 67-74.
3. Reilly, Stephen M., and Jason A. Elias. "Locomotion in Alligator mississippiensis: kinematic effects of speed and posture and their relevance to the sprawling-to-erect paradigm." *The Journal of experimental biology* 201.18 (1998): 2559-2574.
4. Ijspeert, Auke Jan, et al. "From swimming to walking with a salamander robot driven by a spinal cord model." *Science* 315.5817 (2007): 1416-1420.
5. Wyneken, Jeanette. "Sea turtle locomotion: mechanisms, behavior, and energetics." *The biology of sea turtles* 1 (1997): 165-198.
6. Elphick, Melanie J., and Richard Shine. "Longterm effects of incubation temperatures on the morphology and locomotor performance of hatchling lizards (Bassiana duperreyi, Scincidae)." *Biological Journal of the Linnean Society* 63, no. 3 (1998): 429-447.
7. Walker, Brian, Crawford S. Holling, Stephen R. Carpenter, and Ann Kinzig. "Resilience, adaptability and transformability in social--ecological systems." *Ecology and society* 9, no. 2 (2004): 5.
8. Frankenberg, Eliezer, and Yehudah L. Werner. "Adaptability of the daily activity pattern to changes in longitude, in a colonizing lizard, Hemidactylus frenatus." *Journal of Herpetology* (1981): 373-376.
9. Williams, T. L. "Experimental analysis of the gait and frequency of locomotion in the tortoise, with a simple mathematical description." *The Journal of physiology* 310.1 (1981): 307-320.
10. Alexander, R. McN. "The gaits of bipedal and quadrupedal animals." *The International Journal of Robotics Research* 3.2 (1984): 49-59.

11. Lutz, Peter L., John A. Musick, and Jeanette Wyneken, eds. *The biology of sea turtles*. Vol. 2. CRC press, 2002.
12. Langton, Christopher G., and Katsunori Shimohara, eds. *Artificial Life V: Proceedings of the Fifth International Workshop on the Synthesis and Simulation of Living Systems*. Vol. 5. MIT Press, 1997.
13. Kim, Hyung-Jung, Sung-Hyuk Song, and Sung-Hoon Ahn. "A turtle-like swimming robot using a smart soft composite (SSC) structure." *Smart Materials and Structures* 22.1 (2013): 014007.
14. Kawamura, Yoshimasa, et al. "Design and development of amphibious robot with fin actuators." *International Journal of Offshore and Polar Engineering* 20.3 (2010): 175-180.
15. Yoshida, Kazuya. "The jumping tortoise: a robot design for locomotion on micro-gravity surface." *EUROPEAN SPACE AGENCY-PUBLICATIONS-ESA SP* 440 (1999): 699-704.

CONTRIBUTIONS OF STANCE AND SWING LEG MOVEMENTS TO HUMAN WALKING DYNAMICS

GUOPING ZHAO* and ANDRE SEYFARTH

Institute of Sport Science, Technical University of Darmstadt, Darmstadt, 64283, Germany
** E-mail: zhao@sport.tu-darmstadt.de*
www.lauflabor.de

Minimalistic gait models assuming massless legs (e.g. models based on spring loaded inverted pendulum (SLIP) or inverted pendulum (IP) template) have been widely used to interpret human gaits. Although these models can describe basic gait features like center of mass (CoM) trajectories or ground reaction force (GRF) patterns, it is rather challenging to investigate swing leg control strategies with these massless leg models. In this paper, based on experiment data, we analyse how much swing leg, stance leg and upper body movement contribute to total GRF during single-support phase of walking. The results show that, in vertical direction, swing leg and upper body create in-phase M-shape force patterns, but stance leg does not contribute to the M-shape force pattern. In walking direction, the inertia forces created by swing and stance leg cancel each other out, while stance leg and upper body create similar inertia forces in both shape and magnitude. The results suggest there is a phase locking mechanism for swing leg, stance leg and upper body movement. It can help to refine current conceptual models which better describe human walking.

Keywords: GRF contribution; swing leg; stance leg.

1. Introduction

Currently, two different minimalistic template models–spring loaded inverted pendulum (SLIP)[1,2] and inverted pendulum (IP)[3]–are widely used for describing human gait pattern. A series of models (e.g. SLIP with trunk[4] or with curved foot,[5] IP with ankle[6]) have been developed based on these two templates.

These massless leg models can describe several basic gait characteristics,[7–9] but they have some inherent drawbacks. First of all, in massless leg models, contributions from swing and stance leg inertia forces to GRF pattern are all attributed to stance leg function. Secondly, massless leg conceptual models cannot describe human locomotion with perturbations which require to accelerate limbs (consequently create a torque on the trunk) to keep balance. That is, accelerating limb masses can potentially help human to stabilize the trunk. In addition, it is quite challenging to investigate swing leg control strategies with massless leg models. For example, in order to stabilize a walking model, we can calculate the capture region based on the capture point concept,[10] but it is unclear how to control the swing leg properly so that the foot can reach the capture region in time without affecting trunk pitch stability considerably.

Thus, in order to find out the influence of swing and stance leg movements on human walking dynamics, here we investigate how much swing leg, stance leg, and upper body movement contribute to total GRF pattern based on experiment data of human walking at five different speeds.

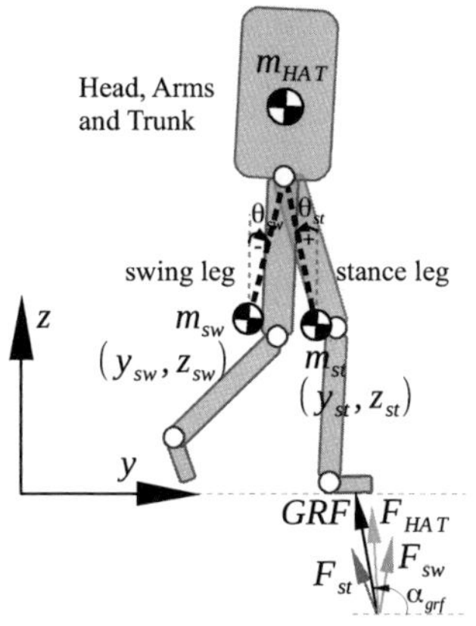

Fig. 1. Seven-link model to investigate GRF components.

2. Method

2.1. *Experiment*

Nine young healthy subjects (4 males, 5 females, age 24.8±2.3 years, body mass 71.7±12.7 kg, height 1.75±0.09 meters, preferred transition speed (PTS) from walking to running 2.06±0.13m/s) participated in the experiments. Subjects were instructed to walk on an instrumented treadmill (HEF Tecmachine, Andrezieux Boutheon, France). The belt speed was controlled at five different speeds (25%, 50%, 75%, 100% and 125% of each subject's

PTS, that is, ~0.5m/s, ~1.0m/s, ~1.5m/s, ~2.1m/s and ~2.6m/s). Three dimensional GRF data were collected by force sensors embedded in the treadmill (Kistler, Winterthur, Switzerland) at a frequency of 1kHz. Three dimensional kinematic data of lower limbs were collected by 8 high-speed cameras (Qualisys, Gothenburg, Sweden) at a frequency of 240Hz. Fifteen markers were used for each subject to capture the motion. Force and kinematic data were synchronized by a trigger signal provided by the treadmill. The time lag and time drift between the two systems are 2.5ms and 0.02ms/s. They were corrected after the measurements.[11]

In order to get individual PTS, each subject was asked to walk on the treadmill based on a predefined speed curve (increase from 1.0m/s by 0.2m/s every 2 seconds until 3.0m/s, and then decrease in a same manner). After four trials, each subject chose a speed at which he/she felt most comfortable to switch the gait between walking and running. All subjects wore athletic shoes and had enough time for warming up and getting familiar with the measurements.

2.2. *Data pre-processing*

All data were processed using MATLAB (2014a, The MathWorks Inc., Natick, USA). GRF raw data were down sampled to 240Hz after filtered by a 49-51Hz bandstop and a 70Hz low-pass filter (zero-lag 4th order Butterworth). The position data of all markers were filtered by a 40Hz zero-lag 4th order Butterworth low-pass filter. The velocity and acceleration of all markers were filtered by a 15Hz zero-lag 4th order Butterworth low-pass filter. Take-off (TO) moment was defined as the moment when the decreasing vertical GRF crosses 10N. Touch-down (TD) moment was defined, in the same way, as the moment when the increasing vertical GRF crosses 10N. In total, 1284 gait cycles were separated and analysed in this paper.

2.3. *GRF component*

As shown in Fig. 1, GRF can be divided into three parts:

$$\begin{aligned}\mathbf{GRF} =& \mathbf{F}_{sw} + \mathbf{F}_{st} + \mathbf{F}_{HAT} \\ =& m_{leg}\begin{bmatrix}\ddot{y}_{sw} \\ \ddot{z}_{sw} + g\end{bmatrix} + m_{leg}\begin{bmatrix}\ddot{y}_{st} \\ \ddot{z}_{st} + g\end{bmatrix} + m_{HAT}\begin{bmatrix}\ddot{y}_{HAT} \\ \ddot{z}_{HAT} + g\end{bmatrix}\end{aligned} \tag{1}$$

where the subscripts sw, st and HAT indicate swing leg, stance leg and upper body (head-arms-trunk), respectively; m_{leg} is leg mass; g is the gravity acceleration; $(y, z)^T$ denote the position of the segment center of mass

(CoM), y-axis and z-axis denote walking direction and vertical direction, respectively. In this paper, based on cadaver studies,[12] CoM accelerations of two legs and HAT were calculated by the movement of markers attached on limbs and GRF. See Appendix for more details.

3. Results

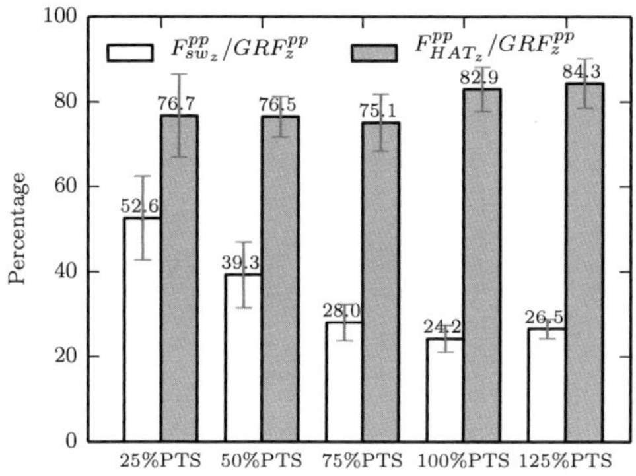

Fig. 2. Means and standard deviations of nine subjects' peak-to-peak (pp) value of F_{sw_z} and F_{HAT_z} to GRF_z during single-support phase at five different walking speeds. The means are indicated by bars and the standard deviations by error bars. The subscript sw, HAT and z denote swing leg, head-arms-trunk, and vertical direction, respectively.

For preferred walking speed (75% PTS, around 1.5m/s), Fig. 2 shows that the ratio of $F_{HAT_z}^{pp}$ to GRF_z^{pp} is about 75.1%, which means that the swing leg dynamics explain about 24.9% of the typical M-shaped GRF variation (because F_{st_z} keeps almost constant, Fig. 3). For slow walking speeds (25% and 50% PTS), swing leg can explain about 23%. For fast walking speeds (100% and 125%PTS), the ratios of $F_{HAT_z}^{pp}$ to GRF_z^{pp} are higher than preferred walking speed (around 83~84%). $F_{sw_z}^{pp}/GRF_z^{pp}$ is higher at low walking speeds (52.6% at 25%PTS, 39.3% at 50%PTS) than at preferred walking speed (28%).

Figure 3 shows that in vertical direction, the force contribution created by upper body is dominant for the GRF at all speeds. Surprisingly, swing leg also contributes to the characteristic M-shape force pattern in walking in contrast to stance leg. For all speeds, F_{sw_z} has a similar M-shape pattern to GRF_z: one peak at the early and one peak at the late of single-support phase, but F_{st_z} keeps almost constant (about leg weight, except for the

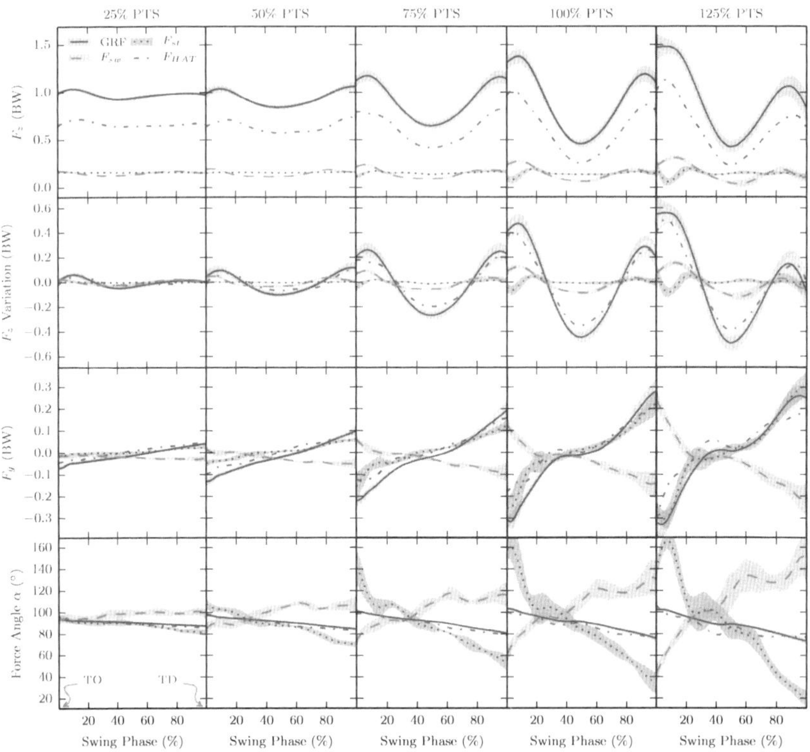

Fig. 3. GRF components. F_z and F_y denote the force in vertical and walking direction. F_{sw}, F_{st} and F_{HAT} denote the force caused by swing leg, stance leg and HAT (head, arms and trunk) dynamics, respectively. All forces are normalized by body weight (BW). TO and TD denote take-off and touch-down, the beginning and end of single-support phase. Bands on each waveform indicate the standard derivation (±SD).

early single-support phase of 100%, and 125% PTS). In walking direction, as shown in Fig. 3, the forces created by the movement of stance leg (F_{st_y}) and HAT (F_{HAT_y}) are similar to each other in both magnitude and shape. In contrast, the GRF contribution of swing leg (F_{sw_y}) has a similar shape and magnitude as F_{st_y} but is opposite in direction. The direction of force contribution created by HAT is almost the same as the GRF direction (Fig. 3). The forces caused by accelerations of both legs do not have a strong influence on total GRF direction.

4. Discussion

The motivations of this study was to understand the effect of leg movement on the dynamic behavior of human walking during single-support phase. In order to achieve this, we analysed the contributions of stance leg, swing leg and upper body movements to total GRF at five different walking speeds based on experimental data. The results suggest the phase locking mechanism among stance leg, swing leg and HAT. For example, in vertical direction, the swing leg inertia force has an in-phase M-shape pattern with respect to HAT inertia force and GRF, but stance leg inertia force does not have a pronounced similar M-shape pattern.

In vertical direction, although the force created by HAT is dominant at all walking speeds, which agrees with the assumption of point mass models, swing leg also contribute to M-shape force patterns (Fig. 3). In SLIP model, the M-shape of GRF is just attributed to the axial compliance of stance leg. Here we see that the swing leg contributes to this stance leg function. For example, $F_{sw_z}^{pp}$ is about 39.3% of GRF_z^{pp} at 50%PTS (around 1.0m/s, Fig. 2). In addition, the virtual leg length of swing leg also has a similar M-shape comparing to F_{sw_z}. This suggests to use a spring loaded pendulum model to mimic the swing leg dynamics.

At low speed walking (25% and 50%PTS), F_{st_z} is almost constant and equal to leg weight during whole single-support phase. This agrees with the IP model which assumes a rigid stance leg. But for preferred and high speed walking (75% to 125%PTS), F_{st_z} has a pronounced peak at the beginning of single-support phase. In this case, the stance leg is more like a mass spring damper system, which could absorb landing impacts.

From the force patterns in walking direction (Fig. 3) we can find that the swing leg and stance leg are coupled. The forces generated by two legs cancel each other out during the single-support phase. This effect was previously found in the M-SLIP model.[13] From the trunk stabilization point of view, the coupling effect of two legs could simplify the upper body control. It has been widely observed that there is a high peak power output from the ankle joint during push-off in order to generate enough speed to swing the leg forward.[14] Thus, stance leg dynamics have a strong impact for the initiation of swing phase at push-off. On the other hand, swing leg dynamics also have a strong impact at touch-down. With the coupling mechanism of two legs, impact on one leg could mainly just affect the movement of the other leg rather than the trunk.

5. Future Work

In this paper we found that swing and stance leg have significantly different behaviors in both vertical and walking direction. A next step could be to analyse the double stance phase to understand the transition between swing leg and stance leg. And based on the results of this paper, it becomes possible to build more realistic conceptual walking models which could better describe the swing leg and upper body balance control strategies.

Acknowledgment

This work was supported by the EU project BALANCE under Grant Agreement No. 601003. The authors would like to thank Susanne W. Lipfert for conducting the experiments. We are very thankful to Maziar A. Sharbafi for helpful advice and discussions.

Appendix A.

According to the anthropometric data given by Ref. 12, leg CoM position was calculated by the positions of markers which are attached on hip/greater trochanter, knee, ankle/lateral malleolus, and 2nd metatarsal, respectively. Leg CoM velocity and acceleration were calculated by differentiating the position data. There are two ways to calculate $\mathbf{F}_{HAT}$:

(1) Calculate HAT CoM acceleration based on the markers attached on trunk (greater trochanter and glenohumeral joint) and arms (elbow and wrist joint). This method assumes the trunk is rigid, which neglects e.g. spine compliance and wobbling mass effects.
(2) Calculate $\mathbf{F}_{HAT}$ by subtracting $\mathbf{F}_{sw}$ and $\mathbf{F}_{st}$ from **GRF** (measured by force plates). This method avoids the error of estimating trunk CoM acceleration by trunk markers' positions. But it introduces the error of $\mathbf{F}_{sw}$ and $\mathbf{F}_{st}$ from **GRF** to $\mathbf{F}_{HAT}$.

In order to check whether the calculation method of $\mathbf{F}_{HAT}$ affects the results a lot, we calculated F_{HAT_z} at normal walking speed (75%PTS) in both methods (Fig. A1). Fig. A1 shows that although the error at the beginning and end of swing phase is a little bit high, the whole shape of two curves do not change a lot. In over 95% of swing phase the error is lower than 5% of body weight. Because the shape of $\mathbf{F}_{HAT}$ is not sensitive to the method, in this paper, $\mathbf{F}_{HAT}$ is calculated by the second method (subtracting force leg forces).

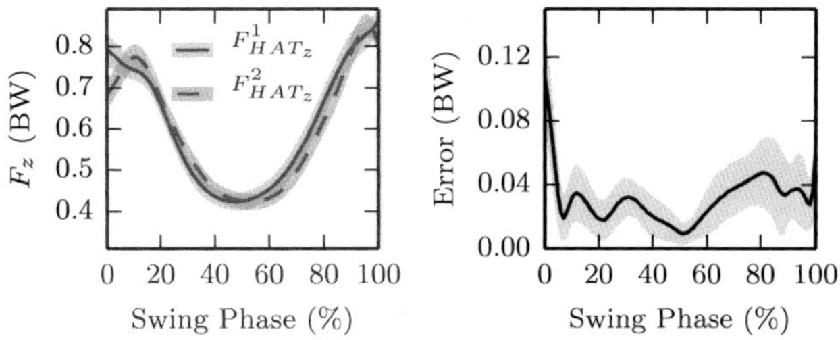

Fig. A1. Comparison of two methods. $F^1_{HAT_z}$ and $F^2_{HAT_z}$ denote the force calculated by first (marker positions) and second (subtracting leg forces) method, respectively. The error shows the difference between $F^1_{HAT_z}$ and $F^2_{HAT_z}$.

References

1. R. Blickhan, *Journal of Biomechanics* **22**, 1217 (1989).
2. T. McMahon and G. Cheng, *Journal of Biomechanics* **23**, 65 (1990).
3. T. McGeer, *The International Journal of Robotics Research* **9**, 62 (1990).
4. H.-M. Maus, S. W. Lipfert, M. Gross, J. Rummel and A. Seyfarth, *Nature communications* **1**, p. 70(January 2010).
5. H. Hong, S. Kim, C. Kim, S. Lee and S. Park, *Journal of Biomechanics* **46**, 77 (2013).
6. J. Ahn and N. Hogan, *PLoS ONE* **7**, p. e47963(11 2012).
7. M. G. Pandy, *Philosophical Transactions of the Royal Society of London. Series B: Biological Sciences* **358**, 1501 (2003).
8. H. Geyer, A. Seyfarth and R. Blickhan, *Proceedings. Biological sciences / The Royal Society* **273**, 2861(November 2006).
9. H.-M. Maus, S. Revzen, J. Guckenheimer, C. Ludwig, J. Reger and A. Seyfarth, *Journal of The Royal Society Interface* **12** (2014).
10. J. Pratt, J. Carff, S. Drakunov and A. Goswami, Capture point: A step toward humanoid push recovery, in *Humanoid Robots, 2006 6th IEEE-RAS International Conference on*, Dec 2006.
11. S. W. Lipfert, M. Gnther and A. Seyfarth, *Journal of Biomechanics* **42**, 786 (2009).
12. D. A. Winter, *Biomechanics and Motor Control of Human Movement*, fourth edn. (John Wiley & Sons, Hoboken, NewJersey USA, 2009).
13. F. Peuker, A. Seyfarth and S. Grimmer, Inheritance of slip running stability to a single-legged and bipedal model with leg mass and damping, in *Biomedical Robotics and Biomechatronics (BioRob), 2012 4th IEEE RAS EMBS International Conference on*, June 2012.
14. S. Lipfert, M. Günther, D. Renjewski and A. Seyfarth, *The Journal of Experimental Biology* (2013).

BIO-ROBOTIC MODEL AS A SCIENTIFIC TOOL FOR EXPERIMENTALLY INVESTIGATING HYDRODYNAMIC FUNCTIONS OF FISH CAUDAL FIN

ZIYU REN

School of Mechanical Engineering and Automation, Beihang Univerisity, 37, Xueyuan Street, Beijing, 100191, People's Republic of China

QICHAO ZHU

School of Mechanical Engineering and Automation, Beihang Univerisity, 37, Xueyuan Street, 100191, Beijing, People's Republic of China

TIANMIAO WANG

School of Mechanical Engineering and Automation, Beihang Univerisity, 37, Xueyuan Street, 100191, Beijing, People's Republic of China

WEN LI*

School of Mechanical Engineering and Automation, Beihang Univerisity, 37, Xueyuan Street, 100191, Beijing, People's Republic of China

Bio-robotic models become increasingly important for understanding biological system in field such as biomechanics. Fish caudal fin is a prominent example of biological propulsion, in which the caudal peduncle, fin ray and fin membrane together form a dynamic locomotory system. In this paper, we developed a bio-robotic model to mimic the fin ray structure and kinematics of Bluegill Sunfish (*Lepomis macrochirus*). We coupled controlled oscillations in both heave and pitch directions to the robot to model the caudal peduncle motion of swimming fishes. Synchronized multi-axis force transducer and particle image velocimetry were then used to quantify the hydrodynamic forces and wake flow. We found that the addition of three-dimensional fin kinematics significantly enhanced the lift force without deceasing thrust force compared with the no fin motion. The vortex wake directs water both axially and vertically and forms jet like structure with notable wake velocity. According to the bio-robotic model experimental data, we hypothesized that fish may actively control the caudal fin rays to achieve considerable lift force when swimming at low speed, however, negative at high speed.

* Corresponding author (email: liwen@buaa.edu.cn.)

1. Introduction

The analysis of non-traditional biomimetic propulsion swimming under controlled conditions have attracted mathematicians [1], fluid engineers [2], roboticists [3-8], material engineers [9-11] and biologists [12-16] interests in studying the principals underlying unsteady locomotion in aquatic animals. Recent advances in understanding fish caudal fin propulsion, in which oscillatory motion of a foil-like structure generate hydrodynamic forces, have included the use of robotic devices which exhibit a rich variety of dynamic behaviors similar to the fin motions of live swimming fishes [17-19]. Most previous biomimetic robotic propulsors focused on the thrust performance in the horizontal plane. The motion of robotic caudal fin has been modeled as a two-dimensional flapping (heave & pitch) as a simple extension of the undulatory body wave in most previous relevant studies [2]. The significant impact of heave & pitch motions on the hydrodynamic propulsion of foil have been investigated [20].

However, fins of most bony fishes move in three-dimensions, both caudal peduncle and fin rays move during fish swimming [21-23]. The peduncle motion is de-rived from the undulatory body wave generated by myotomal body musculature [23, 24]. While the fin rays are moved by intrinsic caudal musculature that are distinct from the body muscles. Bony fish were able to actively control both caudal peduncle motion and deform the fin surface through individually controlling the fin ray root muscle. Previous work of fish biomechanics investigated the effect of stiffness and motion program of fin rays only [25-27]. Nevertheless, no studies yet considered the peduncle motion while studying the fin locomotor functions [7, 18, 28, 29], the integrated locomotor functions by caudal fin rays and peduncle remain unexplored.

As summarized by a recent review article [30], "bio-robotics is becoming important scientific tools and can be used to investigate locomotion and to test hypothesis". How do the kinematics of the peduncle and the fin rays together determine the locomotor performance of a caudal fin? How does the flow speed affect the locomotor forces and the wake flow generated by the caudal fin? Can we use a robotic experimental device to mimic both the caudal peduncle and fin ray motions and is there an optimal phase relationship between these two? To our knowledge, no experimental studies have yet addressed above fish biomechanics issues, nor does any existing robotic system could allow such questions to be investigated.

In this paper, we described a scientific tool for investigating hydrodynamics function of the caudal fin that mimic key relevant features of fish kinematic functions, therefore to investigate the hydrodynamics of caudal fin locomotion. We first designed and fabricated a robot to mimic the fin rays of Bluegill Sunfish (Lepomis macrochirus), and programmed it with motions that match the biological counterpart. A heave and pitch robotic system was then implemented on the towing system, which allow coupling the fish caudal peduncle motion to the fin motions and move the robot in axial direction under controlled speeds. We also simultaneously measured forces, torques and kinematics of the robot at varied flow speeds. Wake flow analysis were conducted in both the horizontal (XY) and vertical (YZ) planes to provided two separate measurements of caudal fin from orthogonal laser light-sheet orientations. Finally, we discussed the use of bio-robotics as a scientific tool for investigating hydrodynamics function of fishes, and addressed the biological relevance of current experimental results as well as formulating several predictions and hypothesis in fish biomechanics.

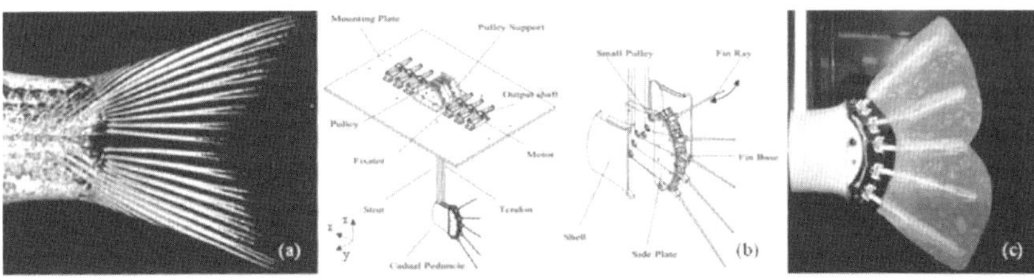

Figure 1. (a) Computer Tomography of Bluegill Sunfish caudal peduncle and fin rays [25], (b) Robotic fin demonstration in Solidworks; (c) Snapshot of the robotic caudal fin.

2. Materials and Methods

2.1. Bio-robotic caudal fin

The mechanical design of the robotic caudal fin was based on caudal fin anatomical study of the Bluegill Sunfish [18, 23, 25] (Lepomis macrochirus), as can be seen from Figure 1a. For simplification, five articulated fin rays were de-signed to actuate the robotic caudal fin. In order to generate controlled movement patterns of fin rays, each ray was individually actuated by a servo motor (MG995, HuiSheng Inc., China) via nylon tendons with diameter of 0.6mm that was attached to each side of the fin base. Two fixators were mounted on each output shaft, which can be used for adjusting the preload force of the tendons. Tendons from the output shaft of the servo motors went along the vertical strut which is connected with caudal peduncle through group of idle pulleys, and eventually actuate the fin ray to move side-to-side. A peduncle shaped shell covered both tendons and pulleys, which also makes the flow around the caudal peduncle similar to that of a live fish (Figure 1(b)). Fin rays were fabricated with polylactic acid by using a 3D printer (Makerbot replicator2, MakerBot Inc., NY, USA). Robotic fin membrane is made of silicone material with thickness of 0.3mm. Silicone adhesive mixed properly with painting thinner was applied to paste silicone membranes on both sides of fin rays. A snapshot of the real robotic caudal fin is provide in Figure 1(c). The quantitative data of material properties of both biological and robotic fin rays and fin membrane were provided in Table 1. It should be noted that the Young's modulus of biological and robotic fin rays are in the same order, therefore the mechanical property of the robotic fin is biologically relevant to that of a live fish.

The fin ray base was designed to actively move side to side, and have passive compliance along the folding direction. The servo motors were controlled by PWM (pulse-width-manipulated) signal individually from a microcontroller STM32F103ZET6 (STMicroelectronics Inc., EU) for controlling angular position of each fin ray movements. For each experimental trial, motion data were programmed into the controller. The controller and a DAQ card (PCI-6284, National Instrument Inc., USA) that was used for data acquisition were synchronized by a trigger signal.

2.2. Kinetic model of caudal fin and peduncle

To our knowledge, no previous study considered peduncle motion while studying the fin motion. There are several caudal fin movement patterns been observed, such as flat, cupping, W, rolling and undulation [25]. The biological fin motion data were digitalized in three dimensions according to high-speed videos of live fish steady swimming [26]. Among all observed caudal fin motions, rolling and undulation are particularly interested as they were hypothesized to generate superior propulsive force than other patterns according to estimation of wake flow from DPIV. To model these two motions (rolling and undulation), a reference frame was settled on vertical plane, with the origin on the tip of the lowest ventral fin ray (as shown in Figure 1(b)). Referring to the method used in previous study [33, 34], the mathematical control model of rolling and undulation of the caudal fin trailing edges can be expressed as follows:

$$y_r(z,t) = \frac{a_r z}{l} \sin(2\pi f t) \tag{1}$$

$$y_u(z,t) = a_u \sin(2\pi f t + z / \lambda f) \tag{2}$$

where y_r and y_u are the horizontal displacement of the fin ray tip for rolling and undulation patterns, respectively. z denotes the vertical coordinate of the fin ray tip, t denotes the time. a_r and au indicate the amplitudes of the fin ray tip in rolling and undulation pattern, respectively. l is the chord length of the caudal fin, f is the motion frequency and λ is the undulation wavelength. The rotation angle of the fin base (see Figure 1(b)) can be derived from:

$$\theta_i(t) = arcsin(y_i(t) / r_i) \tag{3}$$

where θ_i is the rotation angel of the ith fin base, and r_i is the length of the ith fin ray. y_i denotes the horizontal displacement of the ith fin ray which is derived from (1) or (2), depending on the motion patterns. According to the design of the mechanism of the robotic fin, the rotation angle of the server motor is proportion to that of the fin base. The caudal peduncle motion of live fish can be resolved into heave and pitch motions and can be expressed follows:

$$y_h(t) = h\sin(2\pi ft) \quad (4)$$

$$y_p(t) = p\sin(2\pi ft + 90°) \quad (5)$$

where y_h denotes the caudal peduncle displacement on horizontal plane and y_p denotes the rotated angle around the vertical axis. h indicates the heave amplitude, and p indicates the pitch amplitude. In this paper, the phase difference between heave and pitch motions was set to 90° for all experimental trials.

Table 1. Shape and mechanical characterization parameters of the robotic caudal fin and the biological fin [25].

Variable	Value
Span length (mm)	183
Chord length (mm)	160
Young's modulus of robotic fin ray (Nm^{-2})	4.9×10^9
Young's modulus of biological fin ray (Nm^{-2})	10^8~10^9
Fin membrane thickness (mm)	0.6
Elongation rate of robotic fin membrane	300%

2.3. Hydrodynamic characterization

Figure 2(a) shows the schematic view of the experimental apparatus for the hydrodynamic characterization. The water tank has a dimension of 7.8m in length, 1.2m in width and 1.1m in height. A guide rail that was actuated by a 4000 watt AC motor with a travel distance of 7.5m, a position accuracy of 0.1mm and a maximum speed of 3m/s set vertically above the water tank. A servo towing system was used to generate precisely controlled towing speed. A carriage assembled with the guide rail integrated capabilities of both translational and rotational movements. The translational and rotational motions were actuated by servo motors and were used for generating heave and pitch motions for bio-robotic caudal fin model. The robotic caudal fin was designed to move at mid-depth of the water tank to avoid the interference effect of the free surface and the bottom of the tank. More details of the towing system and water tank can refer to our previous work [3, 5, 31].

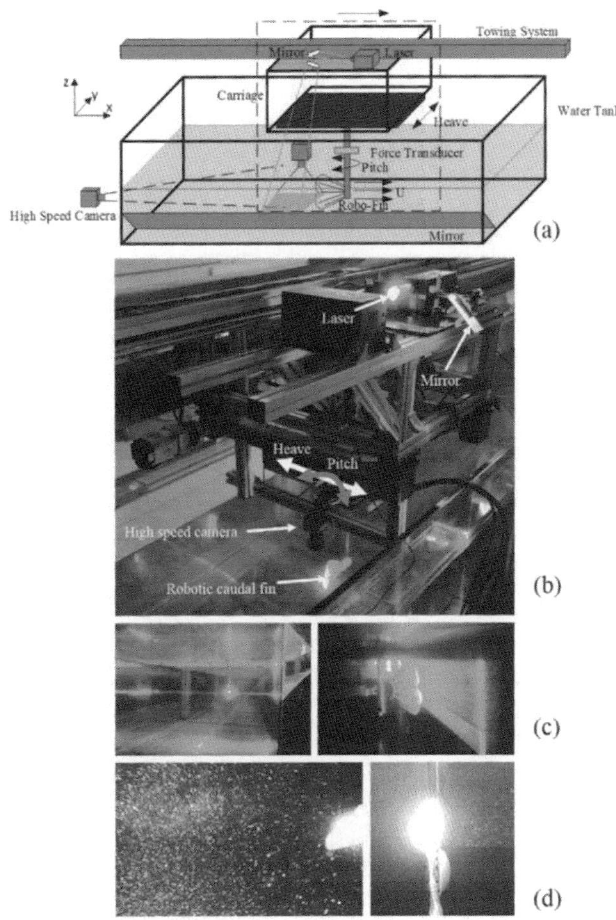

Figure 2. (a) Schematic view of the experimental apparatus; (b) Snapshot of experimental apparatus with laser system in operation; (c) Snapshots of caudal fin with laser planes across mid-plane and vertical plane; (d) High speed camera image of the illuminated particles in horizontal and vertical views.

To measure the hydrodynamic forces generated by the robotic caudal fin, a multi-axis force transducer (mini-40, ATI Industrial Inc., Canada) assembled with the heave &pitch robotic carriage connected the bio-robotic caudal fin model. The force transducer allows three forces and three torques to be measured simultaneously. The force data were then collected by the DAQ card (PCI-6284, National Instrument Inc., USA). As shown in Figure 2(b), the high speed camera shown was used to record the images of particles movement in water. Flow was visualized by seeding the water with near-neutrally buoyant glass beads 10μm in diameter which reflected light from a 4W laser with wavelength of 532nm. Laser sheets were projected horizontally in water by several mirrors. The horizontal laser sheet was around 1mm thick and 150mm wide and was positioned at the mid-sagittal plane of the caudal fin. The vertical laser sheet was approximately 1mm thick and 100mm wide. Figure 2(c) shows snapshots of particles in both horizontal and vertical planes. Particle images of both vertical and horizontal were recorded with high speed cameras at frequency of 125Hz. Particle images by high speed camera in horizontal and vertical plane can be seen in Figure 2(d). The whole experimental system was controlled by a Labview program (National Instrument Inc., USA). The motion of the bio-robotic caudal fin, the force data and the high speed particle images can be simultaneously recorded by a trigger from the Labview program.

3. Results

3.1. Kinematics

Systematic tests were performed for the robotic caudal fin hydrodynamics at fixed heave and pitch motion amplitude and frequency. The heave motion was conducted at f=1Hz, h=2.5Hz, while the pitch motion was set to f=*1*Hz, θ=10°. By controlling the towing system, caudal fin hydrodynamic force measurements were conducted at three different towing speeds: U=0cm/s, 5cm/s and 10cm/s. For caudal fin ray movements, the frequency was set to *1Hz* which is consistent with the frequency of the peduncle motion. All these motion parameters were selected as they are biological relevant to real fish swimming kinematics. Figure 3a shows the snapshot of the undulation motion (coupling with heave &pitch) of the robotic fin model at one instant. Figures 3(b)~3(d) show the trailing edge trajectories of three fin movement patterns during one flapping cycle. The trailing edge trajectories under different movement patterns were extracted from the high speed videos using a self-developed Matlab program. Three movement patterns were investigated in this paper: 1) no fin motion (Figure 3(b)), in which all fin rays were held still without angular movements through a tail beat cycle; 2) undulation pattern (Figure 3(c)) with wavelength of 160mm, amplitude of 20mm; 3) rolling pat-tern (Figure 3(d)) with maximum amplitude of 20mm and minimum amplitude of 10mm. To investigate the effect of phase angle between movements of the caudal peduncle and the fin rays on the hydrodynamic force of the bio-robotic caudal fin, the phase angle was systematically investigated from 0° to 315° with increment of 45° while keeping the rest motion parameters under undulation mode constant under towing speed of 0cm/s.

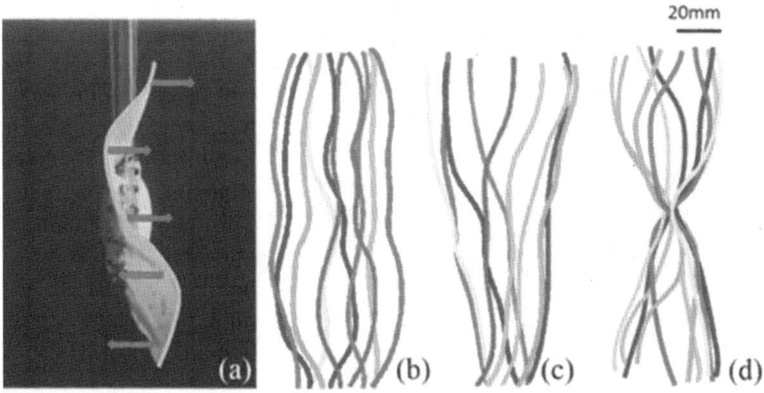

Figure 3. (a) Snapshot of dynamic moving caudal fin from the high speed camera, the yellow curve indicates the trailing edge of the caudal fin; and the red arrows indicate direction of individual fin ray movement. At *U*=0cm/s, trailing edge trajectories of caudal fin under (b) no fin motion, (c) rolling, and (d) undulation are demonstrated.

3.2. Hydrodynamic force

We first investigate the mean thrust force and mean lift force (force averaged from one flapping cycle) generated by three different movement patterns under different towing speeds. All these tests were performed with phase angle of 0°. Data of mean thrust force and mean lift force are shown in Figures 4(a) and 4(b). It can be observed that mean thrust force and lift force of bio-robotic caudal fin varied with towing speed *U*. For all three movement patterns, *i.e.*, no fin motion, rolling and undulation, mean thrust force decrease as *U* increases. This can be considered as a directly consequence as the drag force increased due to an increasing towing speed. At *U*=0cm/s, no fin motion pattern generated the maximum mean thrust force, with a force magnitude of 0.063±0.001N, 5% greater than the undulation pat-tern, and significantly larger than the rolling pattern by 26%. As the towing speed increased to *U*=5cm/s, however, the mean thrust force produced by undulation motion with a magnitude of 0.008±0.003N exceeded the other two motions, while the thrust magnitude of no fin motion and rolling were almost zero. As *U*=10cm/s, the mean thrust force became negative, which in turn indicate that the thrust force generated by the caudal fin was insufficient to over-come the drag force. The mean thrust force of three motion under this condition were -0.026±0.002N, -0.027±0.004N and -0.010±0.001N. The undulation again produced maximum mean thrust force that were around 60% larger than those of no fin motion and rolling. It is noteworthy that, the mean lift force of rolling and undulation decreased significantly as the towing speed increased. The rolling motion produced mean lift force of 0.050±0.006N under *U*=0cm/s, which was significantly greater than those produced under 5cm/s and 10cm/s. The undulation motion generated mean lift force of 0.055±0.009N under U=0cm/s, which was 17% and 400% larger than those under 5cm/s and 10cm/s. From aspect of average force, the mean lift forces generated by undulation were larger than those of the rolling under 0cm/s and 5cm/s. However, when flow speed increased to 10cm/s, the magnitude of the lift force generated by rolling movement surpassed that of undulation. In particular, the no fin motion generates very small mean lift force at all flow speeds which was consistent with the experimental data by Esposito *et al.* [25] and the theoretical work conducted by Affleck *et al.* [32]

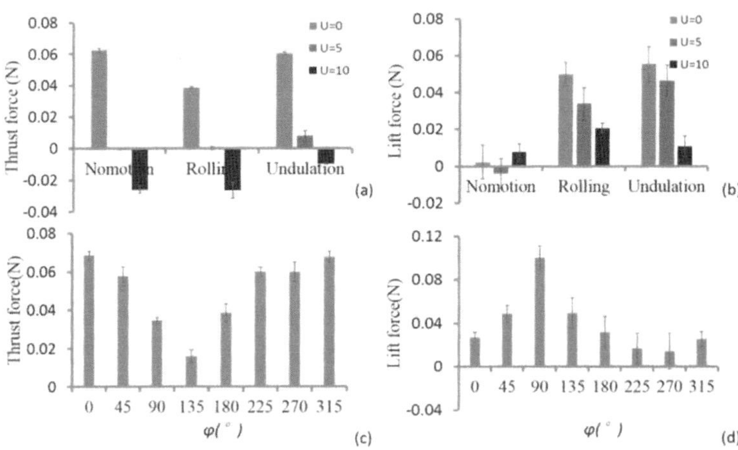

Figure 4. Mean thrust (a) and lift force (b) under U=0, 5 and 10cm/s for three fin movement patterns: no fin motion, rolling and undulation. Thrust force (c) and lift force (d) as function of phase difference (φ =0°~315°) between caudal peduncle and fin ray motion were tested.

The force data from experimental groups with different phase angle were summarized in Figures 4(c) and 4(d). Pronounced periodical trend can be observed. The mean thrust force reached its maximum value at φ=0° and decreased as the phase angle increased until 135°, at which the thrust force reached a minimal peak with magnitude of 0.016±0.003N. The peak-to-valley ratio was about 4.5. While mean lift force increased with phase difference until 90°, at which the mean lift force reached a maximum peak with magnitude of 0.100±0.01N, and then decreased until the phase angle reached 270°. The peak-to-valley ratio was about 8.0.

Instantaneous force during two flapping cycles at φ=90° and φ=225° were shown in Figure 5, under which maximum and minimal mean lift force were generated among all phase tests. The forces are significantly different in terms of both profile and magnitude. Pitch motion seemed to play a leading role in affecting instantaneous force. When phase angle was 90°, thrust force gradually decrease as pitch motion reached its maximum velocity, which was when it passed through the mid-position (Figure 5(a)). As the pitch motion reversed direction, the thrust force gradually increased from a minimum negative value. When pitch motion again reached its maximum velocity, the thrust force almost became zero. With the pitch motion from the mid-point to the other side of the extreme lateral position, the thrust force continued increased. But before the maximum peak thrust force, the profile always became flattened. Similar experimental results could be found in previous literature. The trend of lift force profile was almost the same with the thrust force besides this phenomenon. As for φ=225°, the trend of force profile was nearly reversed. The maximum force at φ=225° always occurred around the time when minimum force reached at φ=90°, and minimum force at φ=225°. In addition, no apparent "flattened" phenomenon was observed.

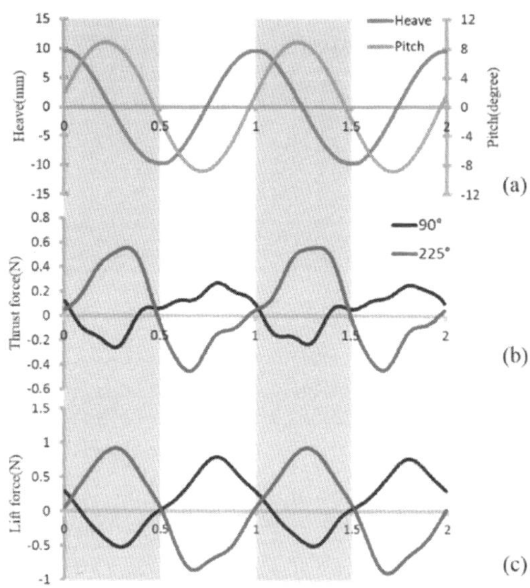

Figure 5. Instantaneous force for two caudal fin motion cycles. (a) Heave and pitch motions for two motion cycles; (b) Thrust force at φ =90° and 225°; (c) Lift force comparison at φ =90° and 225°.

3.3. *Wake flow*

We conducted particle image velocimetry (PIV) measurements on the caudal fin with the no fin motion and undulation pattern from both horizontal and vertical planes under zero towing speed (U=0). The average wake jet velocities on horizontal and vertical planes are provided in Figures 6 and 7. In Figure 6, the average wake flows on horizontal plane were totally different between two patterns. For no fin motion, caudal fin "push" the flow downstream (Figure 6(a)) with a relatively large speed, in contrast the magnitude of wake flow speed of undulation pattern was relatively small. The result even indicates that the flow was "suck" forward towards upstream (Figure 6(b)). For a further comparison, we chose to compare average wake velocity on the horizontal in U direction which we marked in Figures 6(a) and 6(b) with a white dotted line. The average wake velocity profiles of the two motions are shown in Figure 6(c). In the horizontal plane, jet speed of undulation along the white line is close to zero, while the control motion generates significant jet flow speed with a peak at around middle position of an entire stroke (Figure 6(c)).

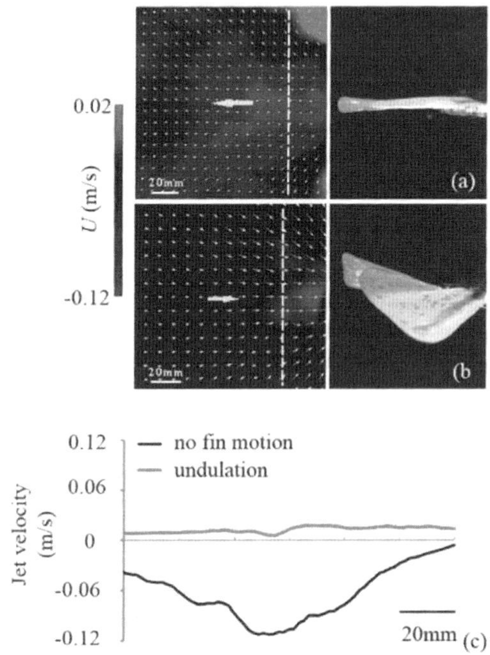

Figure 6. Average speed in the horizontal plane based on the particle image velocimetry analysis; (a) no fin motion Horizontal; (b) fin undulation at φ=90°; (c) average horizontal jet velocity in U direction.

On vertical plane, we recorded the wake flow of the upper lobe of the fin only (as the laser plane covers the upper part of the fin only). The jet flow direction of no fin motion pattern was downward (Figure 7(a)), which has opposite direction compared with the flow generated by the undulation pattern (Figure 7(b)). This phenomenon is also clearly reflected by jet velocity, as can be seen from Figure 7(c).

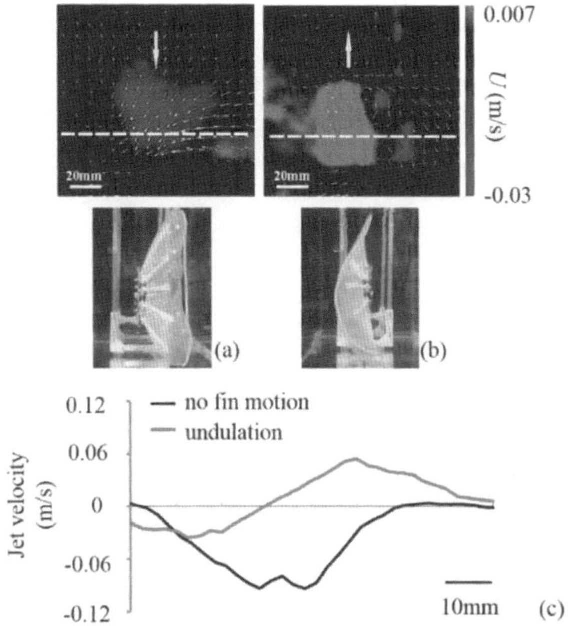

Figure 7. Average speed in the vertical plane based on the particle image velocimetry analysis; (a) no fin motion Horizontal; (b) fin undulation at φ=90°;(c) average vertical jet velocity in U direction.

4. Discussion and Conclusion

Bio-robotics could become part of the scientific process of making progress and new hypothesis while studying the fish biomechanics [30]. Measuring forces and flow from freely swimming live fishes is always challenging [13]. Although there were some previous functional locomotor studies of biological caudal fin, the inability to manipulate and isolate key biomechanical parameters has hindered our understanding of caudal fin propulsion, for example, the hydrodynamic effect of caudal peduncle, the fin ray movement patterns and the swimming speed. The bio-robotic model described in this paper was proven to have capabilities of generating kinematics matching that of freely-swimming fishes. The current experimental apparatus allows for collecting useful scientific data of fish biomechanics, such as XYZ forces and wake flow in both horizontal and vertical planes, under controlled flow speeds. We summed up current experimental results and formulated several new predictions:

(i) The addition of heave and pitch motions of caudal peduncle were not taken into account by predecessors [23, 25, 26], which have significant impact on the hydrodynamic performance of overall caudal fin propulsion.

(ii) The lift force decrease dramatically as flow speed increase. Therefore, we hypotheses that fish may actively change their fin locomotor shapes to achieve considerable lift force at low swimming speed. At higher swimming speed, however, three-dimensional fin motions were not able to generate significant lift force.

(iii) Using the no fin motion as a control, robotic fins with undulation motion generate significantly larger lift forces. In addition, the time-averaged wake jets in both horizontal and vertical planes are notably different from that of no fin motion. Altering the fin ray motion, caudal peduncle motion, or the phase difference between these two would result in significant different thrust and lift forces.

The addition of a mechanism that generates controlled caudal peduncle-like flapping motion to a robotic fin ray system provides a new experimental avenue for studying the hydrodynamics of fish caudal fin, as well as other bio-robotic model with three-dimensional locomotion. Besides understanding fundamental questions in biomechanics of fish swimming, the design and experimental results may also be leverage to future non-traditional propellers [20]. The experiments described here focused on swimming of caudal fin in steady swimming state, but of equal interest is hydrodynamics of caudal fin during non-steady-state, for example, acceleration, deceleration and turning *etc*. Future work will include more systematic parametric studies of the caudal fin kinematics and conduct experiments under non-steady-state conditions.

4.1. Effect of flow speed

Testing hydrodynamic performance of caudal fin movement patterns under different flow speeds would be helpful for answering two important questions: 1) how the flow speed affect the forces produced by the caudal fin movement and 2) which is the optimal movement pattern of caudal fin under a given flow speed. According to current experimental results, increasing towing speed remarkably decrease the lift force both in undulation motion and rolling motions. One possible explanation may locates on the fact that, both rolling and undulation motion patterns produce jet flow in the vertical plane. The oncoming water flow speed to the caudal fin increased as fish swimming speed, while accelerated water flow speed may decrease the angle between jet flow and horizontal plane, therefore result in decrease of lift force. We also found that, when fish swims under high speed, the three-dimensional fin motions has little hydrodynamic functions on the maneuverable force in vertical plane. In contrast, when fish hovers or swims under a relatively low speed, the caudal fin movement may play a critical role in generating lift force, and therefore provide considerable pitch maneuvering. In fact, for many high swimming speed fishes like tuna, the caudal fin is relatively stiff and is not able to generate complex three-dimensional motion pattern. Three-dimensional caudal fin ray locomotion have been observed in many bony fishes such as bluegill under low swimming speed, however, rarely been reported at high steady swimming speed [26].

4.2. Effect of adding peduncle motion

The motion of the fish caudal fin rays is always coupled with the motion of peduncle. Although the caudal fin peduncle motion have be commonly treated as a "trailing edge" of the fish body undulation, the overall hydrodynamic performance of the caudal fin rays and the caudal peduncle remain unexplored. Additional heave and pitch motion can significantly change the kinematics of the fin rays, such as the amplitude, trailing edge trajectories, *etc*. Besides, different phase angle between caudal peduncle motion and fin ray motion have significant impact on lift and thrust force. We found that the lift force reached the maximum peak value at φ=90°, however, very small thrust force was generated; but for other case (*i.e.*, φ=270°), the thrust force was large, while rather small amount of lift force was generated. When fish hovering under still flow condition, considerable lift force could be produced while generating little thrust force (these additional thrust can be balanced by other fins such as pectoral fins). When the fish chases a prey in the vertical plane, both significant thrust and lift force are required, the addition of phase angle at φ=45° between the caudal peduncle and fin rays would be quite beneficial. Based on current experimental data, we predict that real fish may actively modulate the phase angle between the caudal peduncle motion and the fin ray motion for different behaviors, such as prey, escape and migration.

Acknowledgements

Many thanks to Cai Yingjie, Wang Yueping and Wang Zaijun for their help in implementing the experimental apparatus and programming motion of the robotic fin. This work was supported by the National Science Foundation support projects, China under contract number 61403012 (to Li Wen), Beijing Science Foundation support projects under contract number 4154077 (to Li Wen) and National Science Foundation support projects, China under contract number 61333016 (to Tan Min).

References

1. S. Alben, C. Witt, T.V. Baker, E. Anderson, and G.V. Lauder, *Phys Fluids*, 24 (5), 051901, (2012).
2. J. Anderson, K. Streitlien, D. Barrett, and M. Triantafyllou, *Journal of Fluid Mechanics*, 360, 41-72, (1998).
3. L. Wen, T.M. Wang, G.H. Wu, and J.H. Liang, *Bioinspir Biomim*, 7 (3), 036012, (2012).
4. L. Wen, J.C. Weaver, and G.V. Lauder, *J Exp Biol*, 217 (Pt 10), 1656-1666, (2014).
5. L. Wen, T.M. Wang, G.H. Wu, and J.H. Liang, *Ieee-asme T Mech*, 18 (3), 1027-1038, (2013).

6. Y. D, K. KS, K. S, K. J, and L. S, *Review of Scientific Instruments*, 82 (7), 075114 - 075114-075117, (2011).
7. K.H. Low, and C.W. Chong, *Bioinspir Biomim*, 5 (4), 046002, (2010).
8. C. Zhou, Z.Q. Cao, S. Wang, and M. Tan, *Sci China Technol Sc*, 53 (11), 2896-2904, (2010).
9. Z.L. Wang, G.R. Hang, Y.W. Wang, J. Li, and W. Du, *Smart Materials and Structures*, 17 (2), 2900-2912, (2008).
10. T. Wang, *Smart Materials and Structures*, 22 (7), 2896-2912, (2008).
11. T.M. Wang, Q. Shen, L. Wen, and J.H. Liang, *Sci China Technol Sc*, 55 (12), 3359-3369, (2012).
12. D. Barrett, M. Triantafyllou, D. Yue, M. Grosenbaugh, and M. Wolfgang, *Journal of Fluid Mechanics*, 392, 183-212, (1999).
13. G.V. Lauder, 'Swimming hydrodynamics: ten questions and the technical approaches needed to resolve them': 'Animal Locomotion', *Springer*, pp. 3-15, (2010).
14. G. Wu, Y. Yang, and L. Zeng, *J Exp Biol*, 210 (Pt 12), 2181-2191, (2007).
15. G.H. Wu, Y. Yang, and L.J. Zeng, *Review of Scientific Instruments*, 77 (11), 114302, (2006).
16. A. Taheri, and S. Orangi, *Sci China Technol Sc*, 53 (11), 2883-2895, (2010).
17. J. Oeffner, and G.V. Lauder, *J Exp Biol*, 215 (Pt 5), 785-795, (2012).
18. G.V. Lauder, B. Flammang, and S. Alben, *Integr Comp Biol*, 52 (5), 576-587, (2012).
19. L. Wen, and G. Lauder, *Bioinspir Biomim*, 8 (4), 046013, (2013).
20. M.S. Triantafyllou, and G.S. Triantafyllou, *Scientific american*, 272 (3), 64-71, (1995).
21. G.V. Lauder, E.G. Drucker, J.C. Nauen, and C.D. Wilga, *Vertebrate Biomechanics and Evolution*, 117-135, (2003).
22. J.C. Nauen, and G.V. Lauder, *J Exp Biol*, 205 (Pt 12), 1709-1724, (2002).
23. B.E. Flammang, and G.V. Lauder, *J Exp Biol*, 211 (Pt 4), 587-598, (2008).
24. C.D. Wilga, and G.V. Lauder, *J Exp Biol*, 205 (Pt 16), 2365-2374, (2002).
25. C.J. Esposito, J.L. Tangorra, B.E. Flammang, and G.V. Lauder, *J Exp Biol*, 215 (Pt 1), 56-67, (2012).
26. B.E. Flammang, and G.V. Lauder, *J Exp Biol*, 212 (Pt 2), 277-286, (2009).
27. G.V. Lauder, J. Lim, R. Shelton, C. Witt, E. Anderson, and J.L. Tangorra, *Marine Technology Society Journal*, 45 (4), 41-55, (2011).
28. O.M. Curet, N.A. Patankar, G.V. Lauder, and M.A. MacIver, *Bioinspir Biomim*, 6 (2), 026004, (2011).
29. O.M. Curet, N.A. Patankar, G.V. Lauder, and M.A. Maciver, *Journal of the Royal Society, Interface / the Royal Society*, 8 (60), 1041-1050, (2011).
30. A.J. Ijspeert, *Science*, 346 (6206), 196-203, (2014).
31. T.M.W. Li Wen, G H Wu and J H Liang, *Sci China Technol Sc*, 55(10), 2941-2952, (2012).

32. R.J. Affleck, 'Book Some points in the function, development and evolution of the tail in fishes*' *Wiley Online Library*, pp. 349-368, (1950).
33. H. Zhou, T.J. Hu, H.B. Xie, D.B. Zhang, and L.C. Shen, *Sci China Technol Sc*, 53 (11), 2966-2971, (2010).
34. G. Wang, D.B. Zhang, L.X. Lin, H.B. Xie, T.J. Hu, and L.C. Shen, *Sci China Technol Sc s*, 53 (11), 2914-2919, (2010).

APPLYING ITERATIVE LEARNING CONTROL INTO PREISACH-TYPE HYSTERESIS COMPENSATION FOR BIO-INSPIRED LOCOMOTION *

TIANJIANG HU†, GUANGMING WANG, LINCHENG SHEN

College of Mechatronics and Automation, National University of Defense Technology, Changsha 410073, P. R. China

This paper concentrates on hysteresis nonlinearity that comprehensively exists in bio-inspired locomotion of clawing, walking, running, flying and/or swimming robots. A memory clearing iterative learning control scheme is employed into hysteresis-compensating. A second-order state-space model with Preisach-type hysteresis is addressed to describe the dynamic features of biomimetic joints and components. The memory clearing operator enables feasibility of iterative learning control (ILC) methods, whether the state initial errors are accurately set or not. The simulation examples validate that combining a pre-operator with an iterative controller can effectively eliminate effects caused by the Preisach-type hysteresis in bio-inspired robotic systems.

1. Introduction

In the past decade, biologists and robotists made effective collaborations and meaningful achievements in the field of bio-inspired robotics [1]. Researches show that hysteresis has a closer relationship with biomimetic underwater robots, regardless of oscillating or undulating joints and components [2]. Actually, the joints and components can be widely used for understanding and imitating animal-inspired locomotion of clawing, walking, running and/or flying robotic systems, besides swimming robots. In particular, piezoceramic, piezoelectric and electrostrictve actuators that typically possess hysteresis nonlinearity are testified into bio-inspired prototype design and implementation [3].

To our best knowledge, presence of hysteresis introduces uncertainties and nonlinearity in dynamic modeling and thus difficulties in achieving a good control design. For instance, hysteresis behavior may cause substantial tracking error — as much as 15% of the total displacement range — when piezoceramic

---

* This work is was supported in part by the National Science Foundation of China under Grant 61005077, 51105365 and 61273347.

† Corresponding author: Tianjiang Hu, E-mail: t.j.hu@nudt.edu.cn, actjhu@gmail.com

actuators are used to achieve precision positioning with some typical parameters [4-5]. Hysteresis nonlinearities and memory characteristics impose as one of the most difficult challenges in control designs. Under such circumstances, several control approaches based on hysteresis inversion have been developed [6]. However, model fitting or identification of hysteresis is very difficult for its local memory. Especially, it is hardly possible to achieve an appropriate inverse model in practice. As a result, advanced control methods are to be proposed and developed. In practically all dynamic systems with hysteresis, there is a repetitive cycling of some input or control variable that results in a non-unique but strictly repetitive cycling of the response. Therefore, iterative learning control offer great potentials and are promising to overcome hysteresis nonlinearity and its induced inaccuracies. Li *et al.* [7] integrated ILC into a piezoactuator system by using a self-tuning regulator to improve the estimation of system parameters. The pity is that this method did not explicitly address the hysteresis behavior. Hu *et al.* [2] exploited how to determine the ILC learning gain to control an electrostrictive actuator with some assumptions on the hysteresis effect. Hinnen *et al.* [8] proposed an improved ILC aided by an initial state correction procedure for the piezoelectric tube scanner with hysteresis. Leang [4-5] has successfully applied the P-type ILC algorithm into compensation on hysteresis-caused positioning error in piezo-based systems, such as atomic force microscopes.

This paper further studies the applicability of ILC into dynamic systems with Preisach hysteresis based on the previous proof on both continuity and repeatability for hysteretic systems [6]. A memory clearing ILC scheme is employed into the second-order dynamic systems that possess principle features of bio-inspired 2DOF joints or components within robotic prototypes. Simulations are carried out to validate the effectiveness of the hysteresis-compensating ILC approach.

2. Dynamic Systems with Preisach Hysteresis

In this study, the two degree of freedom (2DOF) bio-inspired joints or components are concentrated on due to its basic foundation and generality for clawing, walking, running, flying and swimming robotic prototypes.

2.1. *System model*

Considering the 2DOF joints of bio-inspired locomotion, we consider the second-order mechanical systems described by

$$m\ddot{x}_k(t) + b\dot{x}_k(t) + cx_k(t) + h_k(t) = u_k(t) \tag{1}$$

$$y_k(t) = x_k(t) \tag{2}$$

where the subscript k indicates the operation cycle, $x(t)$ the state vector, $y(t)$ the system output, $u(t)$ the input vector, $h(t)$ denotes the hysteresis nonlinearity.

2.2. *Preisach-type hysteresis*

Hysteresis is a form of nonlinearity with memory. The mentioned term $h(t)$ can be expressed as

$$h(t) \triangleq H\left[x \mid h_{0^-}\right](t) \tag{3}$$

where h_0^- is the memory of the previous hysteresis output history. In (3), the hysteresis output is subjected to the input $x(t)$ as well as the initial state h_0^-. This hysteresis term is highly nonlinear with memory and can induce inaccuracy in tracking control performance. In published hysteresis models [9-12], the Preisach model straddles the boundary between physics and mathematics; in recent decades, it has been well studied, widely accepted and successfully applied. For instance, Hu et al employed the Preisach-type hysteresis model to address the phase lagging effect within the fish-like swimming prototypes [2].

In view of Preisach model, hysteresis is a weighted parallel connection of a finite number of elementary hysteresis nonlinearities (namely hysterons or non-ideal relays), as shown in Figure 1. The input $v(t)$ is individually filtered by each non-ideal relay (see Figure 1(a)), and then, all the hysteron outputs are integrated into the hysteresis with some specified coefficients (see Figure 1(b)).

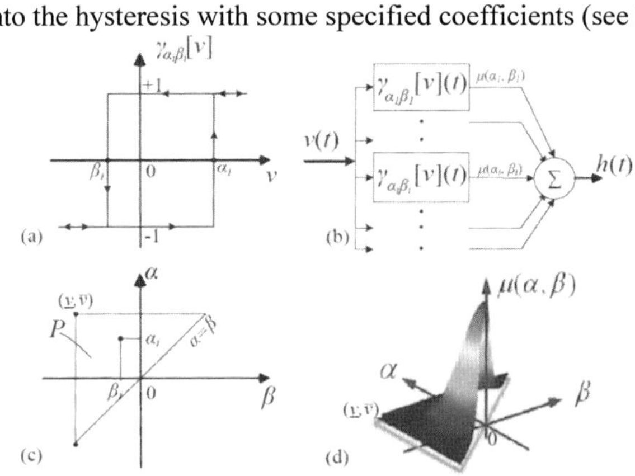

Fig. 1 The schematic of the Preisach model (adapted from [5]) (a) the elementary relay (Preisach hysteron) with switching values$\beta_i \leq \alpha_i$, (b) the model of hysteresis where the output is the collection of finite weighted hysterons, (c) the Preisach plane with the point (α_i, β_i) and (d) the Preisach distribution function.

From Figure 1(a), the non-ideal relays are regarded as elementary shifted rectangular hysteresis operators with different coercive fields. The switching fields α and β, at which the hysteretic loop jumps up from -1 to +1 and jumps down from +1 to -1 respectively, characterize each operator. Each relay can assume a value of +1 or -1 depending on not only the current input $v(t)$, but also the output of the last memory, as follows,

$$w(t)=\gamma_{\alpha\beta}[v](t)=\begin{cases}+1 & v(t)>\alpha \\ \gamma_{\alpha\beta}[v](t^{-}) & \alpha\le v(t)\le\beta \\ -1 & v(t)<\beta\end{cases} \tag{4}$$

where $t\in[0,T]$ and t^{-} means the time just before the current time t. Based on this definition of the elementary hysteron, the Preisach model is expressed as

$$h(t)=H[v](t)=\iint_{P}\mu(\alpha,\beta)\gamma_{\alpha\beta}[v](t)\,d\alpha\, d\beta \tag{5}$$

where $P\triangleq\{(\alpha,\beta)\mid \underline{v}\le\beta\le\alpha\le\overline{v}\}$ is the Preisach plane with $[\underline{v},\overline{v}]$ as its boundary (see Figure 1(c)), and $\mu(\alpha,\beta)$ is the Preisach distribution function (see Figure 1(d)).

3. Design of Hysteresis-Compensating Iterative Learning Control

3.1. *Memory clearing operator*

The Preisach-type hysteresis is proved to be non-Lipschitz continuous by the counterexample of [2]. Even though the system input $v_k(t)$ are guaranteed to be identical during iterations, the hysteresis component may possess great difference. In a word, the Lipschitz condition between the hysteresis $h(t)$ and the system input $v(t)$ cannot be satisfied in dynamic systems with Preisach hysteresis. Since the Lipschitz condition is necessary for their applications, traditional ILC methods are not feasible in the aforementioned systems.

Under such circumstances, a memory clearing iterative learning control scheme is considered in this study. From the viewpoint of practical applications in robotic joints with hysteresis, the conventional ILC methods should be preceded by a memory clearing operator during each iteration, to ensure the initial state as $h_{0,0^-}$. The memory clearing operator O_{mc} was defined in [2], as follows,

$$O_{mc}:\ v_d(t),t\in[0,T]\rightarrow v_d^{*}(t),t\in[0,T+t_r] \tag{6}$$

with

$$v_d^{*}(t)=\begin{cases}a\left(t-\sqrt{\left(v_d(0^{-})-\underline{v}\right)/a}\right)^{2}+\underline{v} & t\in[0,t_r) \\ v_d(t-t_r) & t\in[t_r,T+t_r]\end{cases} \tag{7}$$

in which, $t_r>0$ and

$$a = \left(\frac{\sqrt{v_d(0^-) - \underline{v}} + \sqrt{v_d(0) - \underline{v}}}{t_r} \right)^2 \tag{8}$$

3.2. *Hysteresis compensating control law*

The hysteresis-compensating iterative learning controller for the specified hysteretic system (1)-(3) is given in the form of

$$u(t) = u^{fb}(t) + u^{ilc}(t) \tag{9}$$

where $u^{fb}(t)$ is the feedback PD controller to ensure stability, and $u^{ilc}(t)$ is the feed-forward input which is updated using some ILC law. In light of [2], the A-type ILC law [13] is used as follows

$$u_{k+1}^{ilc}(t) = u_k^{ilc}(t) + L(\bullet)\left(x_d(t+\Delta) - x_k(t+\Delta)\right) \tag{10}$$

4. Simulation and Discussion

In this study, the dynamic parameters are set as m=1, b=c=0, and the Preisach density is given as

$$\mu(\alpha,\beta) = e^{-2(\alpha+\beta)^2}, \quad -1 \le \alpha \le \beta \le 1$$

The operation cycle is T=6s and the desired trajectory is given as

$$y_d(t) = e^{-2(T-t)/T}\sin(2\pi t), \quad t \in [0,T]$$

Hysteresis nonlinearity is characterized by the hysteresis loop that reflects the relationship between the system state $x(t)$ and the hysteresis output $h(t)$. Figure 2 exhibits this hysteresis loop during the first iteration.

In order to ensure the Lipschitz condition of the dynamic system with Preisach hysteresis, the memory clearing operator is applied with the parameter t_r=2 in this simulation. The rectified desired trajectory $y_d^*(t)$, $t\in[0, t_r+T]$ can be achieved with (7) with several initial states of {-1.0, -0.7, -0.3, 0.2, 0.4, 1.0}. Note that the rectifying section can smoothly transit different history states into the desired starting state. Furthermore, the rectified desired trajectory passes through the left-bottom corner of the Preisach plane, to ensure each iteration with a Minimum Hysteresis Initial Memory State. Therefore, this rectifying section is feasible to clear the effects caused by the hysteresis memory.

In this simulation, the position and velocity gains for the feedback controller are chosen as k_p=60 and k_v=10, while the learning gains are chosen as l_p=40 and l_v=10. Meanwhile, the sampling interval is set as Δ=20ms.

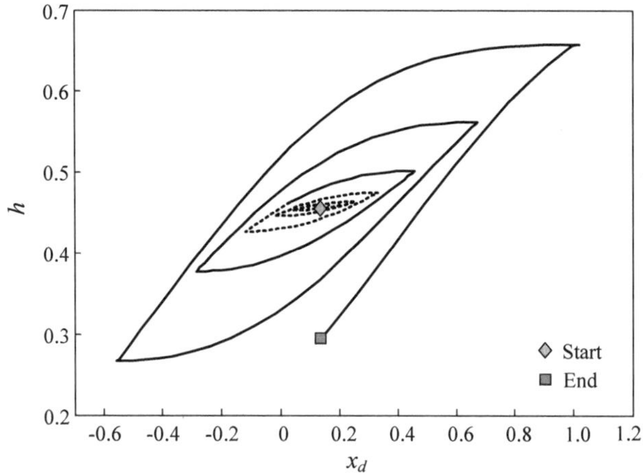

Fig. 2 Hysteresis loop between $x_d(t)$ and $h(t)$ during the first iteration.

Two iterative learning procedures are carried out to validate the memory clearing operator's performance on eliminating the hysteresis effect. At the first procedure, the conventional ILC (10) is applied into the dynamic system (1)-(3). To validate the effectiveness of the conventional A-type ILC in hysteretic systems, the initial memory state is set as a random real number within a specified bound [14]. Figure 3 exhibits the corresponding RMS errors against iterations. Obviously, the tracking accuracy is enhanced with iterative learning. After only a few iterations, RMS error decreases fast into a limited region, compared with the original tracking error achieved by the independent feedback control. Certainly, the tracking accuracy is related with the extent, to which the initial states vary, and it is consistent with properties of physical hysteresis.

It clearly illustrates the convergence of the tracking errors in the hysteretic system, and thus, ILC can effectively compensate the hysteresis nonlinearity for dynamic systems. But a specified steady RMS error cannot be furthermore reduced after several iterations, as shown in Fig. 3. This steady RMS error is related with the hysteresis initial state, which will be uncovered in the coming procedure.

As for this example, the memory clearing operator, O_{mc}, is introduced into the conventional A-type ILC (10). In viewpoint of practical applications, the Minimum Hysteresis Initial Memory States can hardly be identical during iterations, so the starting point of hysteresis loops is iteratively reset with some errors in a specified boundary. Fig. 4 shows the corresponding RMS errors against iterations in the second procedure.

Compared with the simulation result in Fig. 3, the steady RMS error is located in a more limited region as shown in Fig. 4. In detail, this steady error is controlled into the noise level with the A-type ILC preceded by the memory clearing operator. It is sure that the tracking accuracy is furthermore improved by the memory operator, rather than the independent ILC method. Simulations validate the correctness of Theorem 1 and the convectional ILC methods' applicability in dynamic systems with Preisach hysteresis nonlinearity as well.

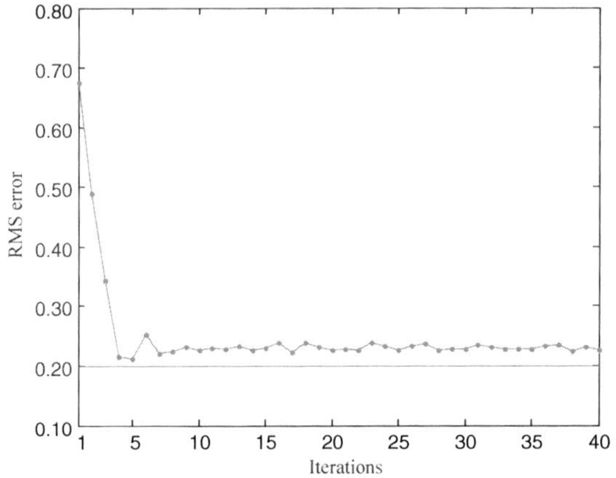

Fig. 3 RMS errors against the iteration times for the conventional ILC without the memory clearing operator, O_{mc}.

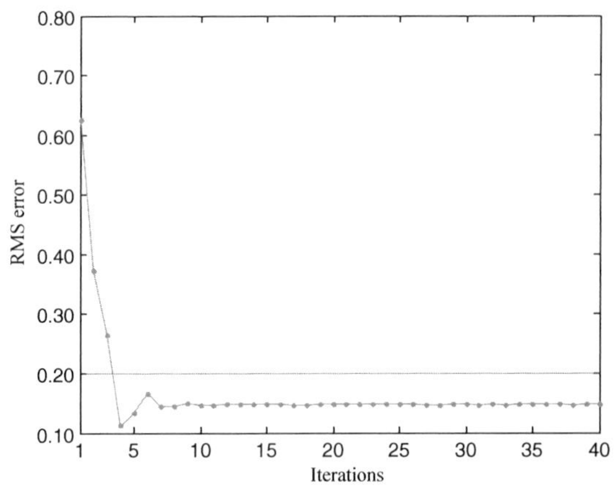

Fig. 4 RMS errors against the iteration times for the conventional ILC with the memory clearing operator, O_{mc}.

5. Concluding Remarks

In this paper, a hysteresis-compensating control scheme is initially considered for bio-inspired locomotion of various robots, in terms of walking, running, flying, swimming and clawing robotic systems. The second-order dynamic system with Preisach-type hysteresis is used to characterize the basic joints and components for bio-inspired prototypes. The simulation results validate that the memory clearing ILC scheme can effectively compensate the Preisach hysteresis within the second-order dynamic nonlinear systems. To be honest, it is definitely a theoretical case study and rather abstract with the simple example. In the near future, we would more closely connect the theoretical studies with practical applications. For example, the effects of the iterative method on the behavior of the actual walking of a legged robot could be more enlightening.

Acknowledgments

The first author would like to thank Prof. Danwei Wang and Mingxun Sun for their sincere guidance and constructive collaboration during Dr. Hu's academic visit to Nanyang Technological University, Singapore.

References

[1] A. J. Ijspeert. Biorobotics: Using robots to emulate and investigate agile locomotion. Science, 2014. 346(6206): pp. 196-203.

[2] Tianjiang Hu, K. H Low, Lincheng Shen, and Xin Xu. Effective Phase Tracking for Bioinspired Undulations of Robotic Fish Models: A Learning Control Approach. IEEE/ASME Transactions on Mechatronics, 2014, Vol. 19(1): pp. 191-200.

[3] Y. Zhang, J. He, S. Zhang, E. Dong, J. Yang, Research on biomimetic fish fin driven by NiTi shape memory alloy, Robot, 2007. Vol. 39(3):207-213.

[4] K. K. Leang and S. Devasia, Feedback-Linearized Inverse Feedforward for Creep, Hysteresis, and Vibration Compensation in Piezoactuators. IEEE Transactions on Control Systems Technology: Special Issue on Dynamics and Control of Micro- and Nano-Scale Systems. Vol. 15(5): pp. 927-935, 2007.

[5] K. K. Leang and S. Devasia, Design of hysteresis-compensating iterative learning control for piezo-positioners: Application to atomic force microscopes. Mechatronics. Vol. 16(3-4): pp. 141-158, 2006.

[6] T. J. Hu, et al. Iterative Learning Control for a Class of Systems with Hysteresis. International Conference on Control, Automation, Robotics and Vision (ICARCV), Hanoi Vietnam, December 2008.

[7] C. J. Li, H. S. M. Beigi, S. Li, and J. Liang, "Nonlinear piezo-actuator control by learning self-tuning regulator," ASME J. Dyn. Syst. Meas. and Cont., vol. 115, pp. 720–723, 1993.

[8] K. J. G. Hinnen, R. Fraanje, and M. Verhaegen, The application of initial state correction in iterative learning control and the experimental validation on a piezoelectric tube scanner. Proceedings of the Institution of Mechanical Engineers Part I-Journal of Systems and Control Engineering. Vol. 218(I6): pp. 503-511, 2004.

[9] M. Brokate and J. Sprekels, Hysteresis and phase transitions. New York, USA: Springer, 2003.

[10] M. P. Mortell, R. E. O'Malley, A. Pokrovskii, and V. Sobolev, Singular perturbations and hysteresis. Society for Industrial and Applied Mathematics: Philadelphia, USA, 2005.

[11] J. W. Macki, P. Nistri, and P. Zecca, Mathematical models for hysteresis. SIAM Review. Vol. 35(1): pp. 94-123, 1993.

[12] I. D. Mayergoyz, Mathematical models of hysteresis and their applications. Amsterdam, Holland: Elsevier, 2003.

[13] D. W. Wang, On D-type and P-type ILC designs and anticipatory approach. International Journal of Control. Vol. 73(10): pp. 890-901, 2000.

[14] G. Heinzinger, D. Fenwick, B. Paden, and F. Miyazaki, Stability of learning control with disturbances and uncertain initial conditions. IEEE transactions on automatic control. Vol. 37(1): pp. 110-114, 1992.

ON DESIGNING AN ACTIVE TAIL FOR BODY-PITCH CONTROL IN LEGGED ROBOTS VIA DECOUPLING OF CONTROL OBJECTIVES

STEVE W. HEIM, MOSTAFA AJALLOOEIAN, PETER ECKERT, MASSIMO VESPIGNANI, AUKE IJSPEERT

Biorobotics Laboratory (BIOROB), Institute of Bioengineering, École Polytechnique Fédérale de Lausanne (EPFL), Lausanne, Switzerland

This work explores the use of active tails for steady-state legged-locomotion. Simple models are proposed which capture the dynamics of an idealized running system with an active tail. Analysis suggests that the control objectives of injecting energy into the system and stabilizing body-pitch can be effectively decoupled via proper tail design: a long, light tail. Thus the overall control problem can be simplified, using the tail exclusively to stabilize body-pitch. We show in simulation that models with long-light tails are better able to reject perturbations to body-pitching than short-heavy tails with the same moment of inertia. We also present the results of an active tail mounted on the quadruped robot Cheetah-Cub. The results show greatly improved forward velocity and reduced body-pitching and validate the long-light tail design: shorter, heavier tails are much more sensitive to control parameter changes.

Keywords: tail, legged locomotion, bio-inspiration, design principle

1. Introduction

In many cases animals achieve locomotion through a combination of periodic feed-forward control signals, generated via central pattern generators (CPGs),[1] decentralized feedback in the form of reflexes,[2,3] and mechanical design with natural dynamics displaying good stability properties.[4] Understanding how animals move often leads to great simplifications in the design process[5] of machines. With this in mind, we turn our attention to tails. Though most terrestrial vertebrates possesses a tail, it's primary purpose is usually something other than locomotion.[6] Nonetheless there are some notable exceptions: geckos and other lizards have also been shown to use their tails extensively for aerial posture-control[7] and the concept has been demonstrated in robotics as well.[8,9] Among larger animals, cheetahs appear to use their large tail to aid in cornering at high speeds, another concept explored in robotics.[10] Kangaroos also appear to use their tails during hopping for controlling posture.[11,12] As Libby et al.[8] we limit our study to the sagittal plane; we focus however on steady-state locomotion, taking into consideration the dynamics of both flight and stance phases. We approach the problem through model-analysis and simulation of a simple idealized system by which we extract a basic design principle; we then validate this principle by applying it to a more complex system in hardware, using the Cheetah-Cub robot,[13] shown in Fig. 1.

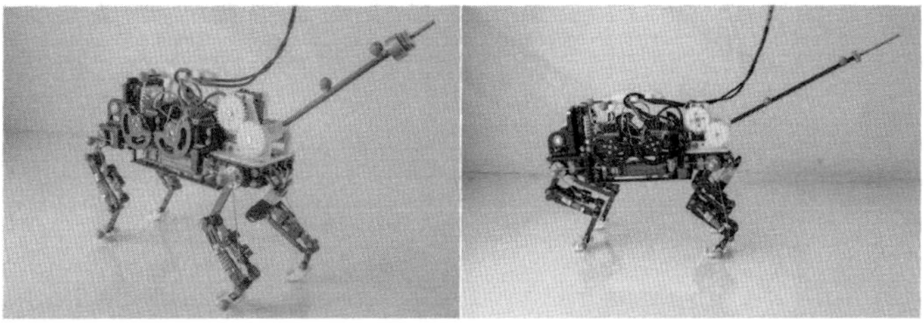

Fig. 1. Experiments were conducted using a newly built version of the Cheetah-Cub robot,[13] with an added 1 degree of freedom tail module. Shown are two different tails, a short-heavy (left) and long-light (right) tail.

2. Hypothesis

We define a tail as *an appendage which alters the dynamics of the system without directly applying contact forces*, i.e. without touching the ground. In this sense, when a kangaroo (or robot) uses its tail for support while standing or shuffling,[14] we consider it as an additional leg. Conversely, any limb capable of inertial manipulation could occupy the same role, as discussed in.[9] We hypothesize that a proper tail design allows a much simpler control strategy by decoupling the control problem. This is similar in concept to Raibert's original control scheme[15] in which hopping-height, forward-velocity and body-pitch were each controlled independently by means of leg-thrust, landing angle-of-attack and hip-torque during the stance phase, respectively. This decoupled control scheme is not only simple but also very effective. However, body-pitch stabilization is partly simplified by the body having high moment of inertia (MoI) and the hip-joint being located near the body center of gravity (CoG). More importantly, hip-torque is made available exclusively for body-pitch stabilization during the stance phase.

Most legged-robots however use rotary motors driving articulated legs, in which case hip-torque is required for generating leg-thrust. In biology also, several studies indicate that hip-torque plays a key role in generating leg-thrust.[16] For the sake of simplicity, we group the two control objectives of hopping-height and forward-velocity into a single objective, that of energy-input. We hypothesize that a properly designed active tail can allow for an alternative path to decoupling the body-pitch and energy-input control objectives: if the tail actuator has negligible impact on the energy-input, it can be used exclusively for controlling body-pitch. Similarly, if the coupling between leg actuators and body-pitching is negligible, they can be used exclusively for energy-input. We examine the requirements for decoupling these control objectives through simple, idealized models.

3. Model

We examine a simple idealized model which still captures the salient dynamics of the system,[17] basing our model on the Spring-Loaded Inverted Pendulum (SLIP) model.[18] We extend it as shown in Fig. 2 to include three rigid bodies (body, leg and tail).We derive equations of motion (EoM) using Lagrange's second method and obtain

$$M_{(q)}\ddot{q} = H_{(q,\dot{q})} + B_{(q)}u_{(t)} + J^{\mathsf{T}}_{c(q)}\lambda_{GRF} \tag{1}$$

where q is the vector of generalized coordinates, $\dot{q}$ and $\ddot{q}$ are the velocities and accelerations respectively, $M_{(q)}$ is the mass matrix, $H_{(q,\dot{q})}$ contains differentiable forces, $B_{(q)}$ is the control matrix, $u_{(t)}$ is the vector of control inputs, λ_{GRF} is the ground-reaction force (GRF) in cartesian-coordinates, and $J_{c(q)}$ is the contact Jacobian of the foot, used to project λ_{GRF} into generalized coordinates. During stance we assume no slipping of the foot, meaning the vector of *minimal coordinates* is reduced from 6 DoFs to 4. Instead of switching between two separate minimal vectors q, we add two constraint equations during the stance-phase in the form of the ground-reaction force λ_{GRF}, which enforce the no-slip assumption.[19]

For the analysis, we then solve Eq. (1) explicitly, yielding

$$\ddot{q} = M^{-1}_{(q)}(H_{(q,\dot{q})} + B_{(q)}u_{(t)} + J^{\mathsf{T}}_{c(q)}\lambda_{GRF}) \tag{2}$$

$$\ddot{q} = \bar{H}_{(q,\dot{q})} + \bar{B}_{(q)}u_{(t)} \tag{3}$$

In the simplest case, we use a flywheel instead of a tail (see *b*) in Fig. 2), i.e. a tail with MoI but no length or mass, and with both tail and hip joints located directly on the body CoG. The matrices $\bar{H}_{(q,\dot{q})}$ and $\bar{B}_{(q)}$ as well as the control vector $u_{(t)}$ are shown in Eq. (4) and Eq. (5).

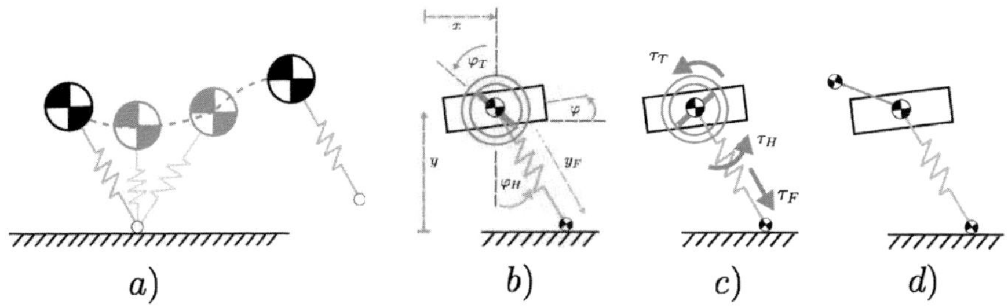

Fig. 2. Our models build on the SLIP model, shown in *a*). Coordinates and control inputs are shown in *b*) and *c*), respectively, for the flywheel model, which is an idealized tail with MoI but no mass. A full tailed model is shown in *d*).

$$\bar{H} = \begin{pmatrix} \frac{J_F \cos(\varphi_H)(g\sin(\varphi_H) - 2\dot{\varphi}_H \dot{y}_F)}{J_F + m_B y_F^2} \\ \frac{g(J_F \cos 2\varphi_H + J_F + 2m_B y_F^2) + 4J_F \dot{\varphi}_H \sin(\varphi_H)\dot{y}_F}{2(J_F + m_B y_F^2)} \\ 0 \\ y_F \dot{\varphi}_H^2 - g\cos(\varphi_H) \\ \frac{m_B y_F (g\sin(\varphi_H) - 2\dot{\varphi}_H \dot{y}_F)}{J_F + m_B y_F^2} \\ 0 \end{pmatrix} \tag{4}$$

$$\bar{B}_{(q)} u_{(t)} = \begin{pmatrix} \frac{\sin(\varphi_H)}{m_B} & \frac{y_F \cos(\varphi_H)}{J_F + m_B y_F^2} & 0 \\ -\frac{\cos(\varphi_H)}{m_B} & \frac{y_F \sin(\varphi_H)}{J_F + m_B y_F^2} & 0 \\ 0 & -\frac{1}{J_B} & -\frac{1}{J_B} \\ \frac{1}{m_B} & 0 & 0 \\ 0 & \frac{1}{J_F + m_B y_F^2} & 0 \\ 0 & 0 & \frac{1}{J_T} \end{pmatrix} \begin{pmatrix} \tau_F \\ \tau_H \\ \tau_T \end{pmatrix} \tag{5}$$

Here, m and J refer to mass and moment of inertia, g is gravity and the sub-indices $_B$,$_F$,$_H$,$_T$ indicate the body, foot, hip and tail respectively. If we examine the explicit EoM for this case, we see that the flywheel DoF has no influence on the rest of the system dynamics. More importantly, in the control matrix $\bar{B}_{(q)}$ we see that the flywheel torque τ_T only affects the tail-rotation φ_T and the body-pitch φ but not the other body DoFs x and y. This changes drastically if the joint is displaced from the body CoG or a full tail is used. As long as the coupling between body-pitch and translation is not too strong, the coupling can effectively be ignored and treated as a disturbance, and a simpler control loop can be applied to body-pitch, with the tail-torque τ_T as the sole control input. In order to do this, the tail design should approximate the dynamics of a flywheel.

4. Simulation Results

For the simulation we use the MATLAB framework developed by C. Remy,[19] with minor modifications. The framework disposes of gradient-based optimization tools to synthesize *locally-optimal, open-loop control trajectories* for a periodic gait, which require a relatively good initial guess. Since the models presented in section 3 are very sensitive to changes in parameters, finding initial guesses becomes a tedious manual operation. For this reason we proceed in two steps: quasi-periodic initial guesses are first generated using a particle-swarm optimization (PSO) framework, and gradient-based optimization is then applied to these results. To generate the initial-guesses we simply maximize distance traveled before falling, while limiting torques to avoid giant leaps. This method reliably produces solutions which are close enough to being both periodic as well as locally optimal that they can be used as an initial guess. We then perform Floquet analysis on the return maps and generate a closed-loop controller which stabilizes a first-order linearization of the return-map.[19] We compare the response of systems with different tail-designs to body-pitch velocity perturbations, with both open-loop and closed-loop controllers. The MoI of the tail is the same in each model, while length and mass are adjusted accordingly. The results are shown in Fig. 3. As expected due to the increased coupling,

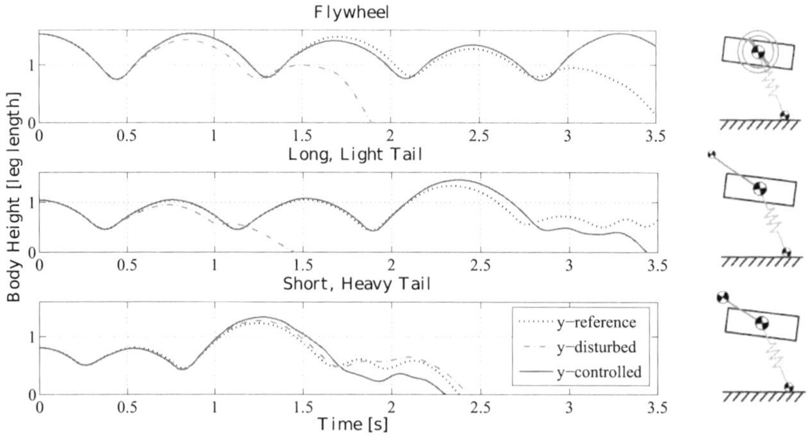

Fig. 3. We compare the responses of several models to a small perturbation to the body-pitch velocity: plotted are the open-loop reference with no perturbation (black, dashed), open-loop response to perturbation (red, dot-dashed) and closed-loop response (blue, solid). In the three cases, the moment of inertia of the tails remains the same.

the closed-loop controller is progressively less effective as off-center of the tail increases, despite maintaining the same MoI and thus the same inertial effectiveness.[9] These results match the predictions we make in section 3.

5. Hardware Results

A 1-DoF rigid tail was built and mounted on the Cheetah-Cub robot.[13] Cheetah-Cub is a small quadruped with two actuated DoFs per leg, hip and knee-flexion, roughly the size of a house-cat. The Cheetah-Cub features compliant pantographic legs which make it robust to impacts, and more importantly allow for stable and fast *open-loop* locomotion: feedforward inputs are fed directly to the servo-motors, generated from a simple CPG network. Stability is provided by the mechanical morphology of the robot. This allows the CPG to be tuned for performance, i.e. forward speed, without worrying about stability issues. Though previous research with the Cheetah-Cub has focused on a trotting gait, we chose a bounding gait, i.e. with the fore pair and hind pair of legs moving in unison. We chose bounding since body-pitching (and therefore the effect of the tail) is more pronounced and relevant in this gait. The CPG nodes of the hips are tuned to be in-phase left-right, and anti-phase fore-hind. The tail CPG node is connected directly to the hind-hip nodes in anti-phase. For more details, see.[20] Our tails are made with a hollow carbon-fiber tube, weighted at the end with aluminum and brass weights. This design not only makes it easy to adjust the tail parameters but is also well represented as a point-mass on a massless rod. The parameters are reported in TABLE 1. For reference, the entire robot without the tail weighs 1.25 kg and measures 21 cm from hip to hip. In our experiments, we hand-tune the CPG-parameters to find a gait which represents a local optimum in terms of forward-velocity for both the short-heavy tail as well as the long-light tail. We then change the values of individual CPG-parameters and observe the effect. In general, we observe two situations: either the effect is the same using either tail, as is the case for changing the phase-lag between the tail and hind-hip CPG nodes (forward velocity drops in the same way for both tails), or the parameter change has a much greater effect on the performance of the robot with short-heavy tail than it does with a long-light tail. This is the case, for example, when changing the tail-offset or tail amplitude. We show in Fig. 4 and TABLE 1 the experimental results for changing the tail-offset.

For comparison, we also show the performance of the robot with a rigid, passive tail, i.e. where the tail is kept at a fixed angle of 30 degrees with respect to the body. We use this as a comparison instead of with no tail to

Table 1. We tested the effect of the active, open-loop tail on the performance of Cheetah-Cub using a bounding gait. A sample of the data-set is shown here, and visualized in Fig. 4. The performance metrics we use are forward velocity and body-pitching; the sample size and standard deviation are also shown for each metric.

Tail Parameters	**Plot**	**Tail Position** [degrees]	**Forward Velocity** [$\frac{m}{s}$]			**Body Pitching per Stride** [degrees]		
			mean	standard deviation	runs	mean	standard deviation	number of strides
Short-heavy tail length: 12.8 [cm] mass: 53 [g] MoI: 8.68 [kg cm^2]	a)	offset: 35	0.558	0.012	4	8.7	1.6	32
	b)	offset: 65	0.517	0.036	8	6.6	2.1	59
Long-light tail length: 16.8 [cm] mass: 31 [g] MoI: 8.75 [kg cm^2]	c)	offset: 35	0.554	0.023	8	5.7	1.0	58
	d)	offset: 65	0.556	0.011	8	4.6	1.0	63
Passive Tail length: 12.8 [cm] mass: 53 [g] MoI: 8.68 [kg cm^2]	e)	offset: 35	0.479	0.034	6	8.5	3.5	53

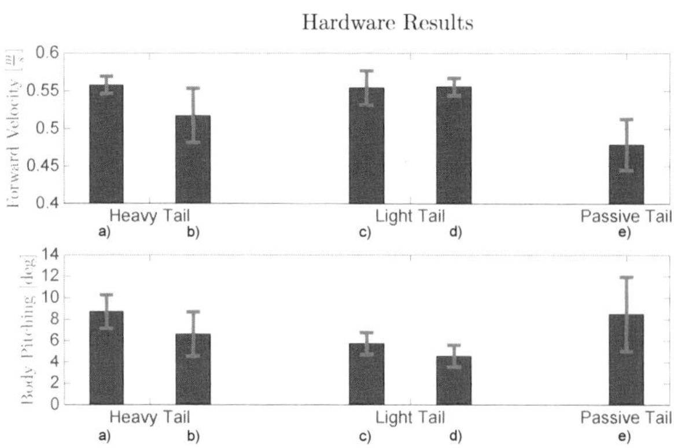

Fig. 4. The results shown in TABLE 1 are visualized here. Of greatest interest is the comparison of the two examples shown with active-tails: they have almost the same moment of inertia, but the shorter tail is 71.0% heavier and thus more sensitive to control parameter changes. We show here one of the more critical parameters: the offset of the tail with respect to the body.

keep the total MoI as similar as possible across all three configurations. In the trials shown, all CPG node parameters are kept constant across each configuration, except for those specified: in the tail CPG node the tail-offset, i.e. the midpoint of the tail oscillations, is changed between trials a) and c) compared to trials b) and d). We find that performance is very similar between the two tails in the perfectly tuned cases a) and c). However, with the short-heavy tail, changing the tail-offset negatively affects velocity, decreasing by 7.35%; whereas it remains practically the same with the long-light tail. Considering that the entire robot weighs 1.25 kg and even the heavy tail represent only 4.4% of the total body-mass, such a large effect on performance is impressive. These results match our predictions: since both tails have almost the same MoI they are equally effective at stabilizing body-pitch, however the long-light tail is less sensitive to configuration changes, making the robot potentially more versatile in a larger portion of state-space.

6. Conclusions and Outlook

Using an idealized model we have extracted a design principle for tails: for a specific MoI, a long-light tail results in much lower degree of coupling between energy-input and body-pitch control objectives. We verified this principle in simulation and demonstrated its applicability to more complex systems through hardware experiments with the Cheetah-Cub quadruped robot. From our analysis, a flywheel rather than a tail would most effectively decouple the control objectives, however a tail can be more suitable due to a number of design constraints.[8,10] Most notably, a tail can have a higher MoI-to-mass ratio and is generally easier to integrate into a design or existing platform due to the dimensional requirements of a flywheel.

As a next step, we are working on stability analysis to identify the effect of an active tail on regions of attraction, in terms of both control and design parameters. Comparisons with closed-loop control in hardware would also be of great interest, particularly across different platforms.[9] Although we focused on the advantage of a simplified control problem by decoupling the control objectives, the advantages of exploiting the coupling is also interesting: it has already been shown that a tail can be used specifically for energy-input,[21] and indeed we suspect some species of kangaroos and wallabies may make this trade-off between versatility and efficiency.[22]

Acknowledgments

We would like to thank all the members of the BioRob lab for constructive conversations and feedback. In particular we thank François Longchamp and Daniel Chapuis for their contribution in the hardware development, and Daniel Renjewski for advice on stability analysis. Funding was received from the European Community's Seventh Framework Program (FP7/2007-2013 -Challenge 2-Cognitive Systems, Interaction, Robotics; grant agreement number 248311 AMARSi) and from the Swiss National Science Foundation through the National Centre of Competence in Research Robotics.

References

1. A. J. Ijspeert, *Neural Networks* **21**, 642 (2008).
2. H. Geyer and H. Herr, *Neural Systems and Rehabilitation Engineering, IEEE Transactions on* **18** (2010).
3. D. Owaki, T. Kano, K. Nagasawa, A. Tero and A. Ishiguro, *Journal of The Royal Society Interface* **10**, p. 20120669 (2013).
4. J. Rummel and A. Seyfarth, *The International Journal of Robotics Research* **27**, 919 (2008).
5. P. Holmes, R. Full, D. Koditschek and J. Guckenheimer, *Siam Review* .
6. G. C. Hickman, *Mammal review* **9**, 143 (1979).
7. A. Jusufi, D. I. Goldman, S. Revzen and R. J. Full, *Proceedings of the National Academy of Sciences* **105**, 4215 (2008).
8. T. Libby, T. Y. Moore, E. Chang-Siu, D. Li, D. J. Cohen, A. Jusufi and R. J. Full, *Nature* **481**, 181 (2012).
9. A. M. Johnson, T. Libby, E. Chang-Siu, M. Tomizuka, R. J. Full and D. E. Koditschek, Tail assisted dynamic self righting, in *Proceedings of the Fifteenth International Conference on Climbing and Walking Robots*, July 2012.
10. R. Briggs, J. Lee, M. Haberland and S. Kim, Tails in biomimetic design: Analysis, simulation, and experiment, in *Intelligent Robots and Systems (IROS), 2012 IEEE/RSJ International Conference on*, 2012.
11. R. M. Alexander and A. Vernon, *Journal of Zoology* **177**, 265 (1975).
12. G.-H. Liu, H.-Y. Lin, H.-Y. Lin, S.-T. Chen and P.-C. Lin, *Journal of Bionic Engineering* **11**, 541 (2014).
13. A. Spröwitz, A. Tuleu, M. Vespignani, M. Ajallooeian, E. Badri and A. J. Ijspeert, *The International Journal of Robotics Research* **32**, 932 (2013).
14. S. M. O'Connor, T. J. Dawson, R. Kram and J. M. Donelan, *Biology letters* **10** (2014).
15. M. H. Raibert, *Legged robots that balance* (MIT press, 1986).
16. H. Tan and A. M. Wilson, *Proceedings of the Royal Society B: Biological Sciences* **278**, 2105 (2011).
17. R. Full and D. Koditschek, *Journal of Experimental Biology* **202**, 3325 (1999).
18. R. Blickhan, *Journal of biomechanics* **22**, 1217 (1989).
19. C. D. Remy, K. Buffinton and R. Siegwart, A matlab framework for efficient gait creation, in *Intelligent Robots and Systems (IROS), 2011 IEEE/RSJ International Conference on*, 2011.
20. S. Heim, Design and control of an active tail for robot locomotion (2014), MSc thesis, http://biorob.epfl.ch/students/heim.
21. A. De and D. E. Koditschek, Parallel composition of templates for tail-energized planar hopping, in *Robotics and Automation (ICRA), 2013 IEEE International Conference on*, 2015.
22. C. P. McGowan, R. V. Baudinette and A. A. Biewener, *Comparative Biochemistry and Physiology Part A: Molecular & Integrative Physiology* **150**, 151 (2008).

CONTROL ALGORITHM FOR WALKING ROBOT WITH MOSAIC BODY

ALEKSEY V. PANCHENKO

Theoretical Mechanics And Mechatronics, Lomonosov Moscow State University, Moscow, Russia
E-mail: PanchekoAV@vap.ru

IGOR A. ORLOV* and VLADIMIR E. PAVLOVSKY**

Keldysh Institute of Applied Mathematics, Russian Academy of Sciences, Moscow, Russia
**E-mail: orlovbel@gmail.com*
***E-mail: vlpavl@mail.ru*

This article is devoted to the kinematics of the multilink body of a multilegged walking robot. The number of degrees of freedom for multilink body is obtained and the reverse kinematic problem is solved. The control algorithm for mosaic robot body is provided. All algorithms have been verified via multibody computer simulation.

Keywords: Hexapod; Articulated Body; Kinematics.

1. Introduction

Due to new microelectronic elements and modern powerful actuators and batteries the hexapod robots and other multilegged machines became widespread during past decades. Originally many multilegged robot models were inspired by insects - beetles, cockroaches, spiders, water striders and etc. However, mechanical models can imitate real insects only to some limited extent. The biomechanics of a real insects still remains more complicated. There is a complex ramification of further evolution of robots similar to insects. The way of increasing kinematic complexity is among them.

On Fig. 1 there is an example of a robot with a so called mosaic body. The robot's body consists of nine rigid segments connected to each other with twelve three dimensional rotational joints. The robot has twelve insect legs with 3 d.o.f for each one. The corner segments have two legs connected, and the central segment doesn't have any legs at all.

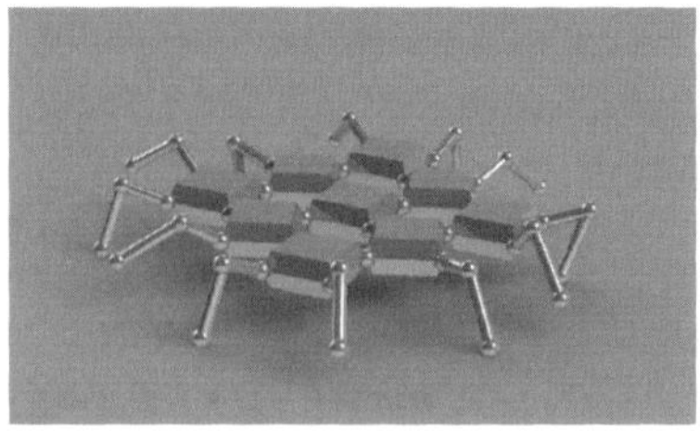

Fig. 1. Multilegged robot with 3 by 3 segments mosaic body.

2. Mosaic Body Kinematics

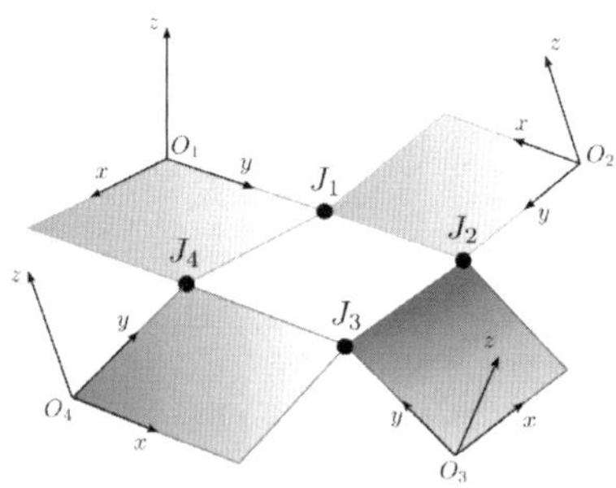

Fig. 2. Structural element of mosaic body.

The primitive structural units of a mosaic body are:

- two by two segment cells;
- pair of connected segments;
- single segment (leaf) connected via one or two joints.

Let's consider the rotation of a single leaf of the 2 by 2 body cell on the Fig. 3. The rotation matrix for angles $(\psi_i, \theta_i, \phi_i)$ of joint J_i can be written in the following form:

$$R(\psi_i, \theta_i, \phi_i) = R_z(\psi_i)\, R_x(\theta_i)\, R_y(\phi_i), \tag{1}$$

where R_z, R_x, R_y orthogonal rotation matrices with the reference to denoted axis indexes.

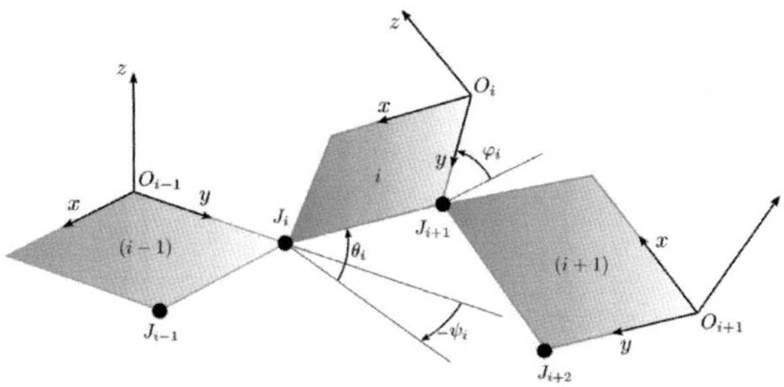

Fig. 3. Rotation of a corner segment.

The orientations of the segments with numbers $i-1$ and $i+1$ after rotation around J_iJ_{i+1} of the ith segment actually haven't changed. This relation can be written in a matrix form. For the segments coordinate systems depicted on the Fig. 3 the following equation is true:

$$R(\psi_i,\theta_i,\phi_i)\cdot R_z\left(-\frac{\pi}{2}\right)\cdot R(\psi_{i+1},\theta_{i+1},\phi_{i+1})\cdot R_z\left(-\frac{\pi}{2}\right)= \tag{2}$$

$$=R(\psi_i,\theta_i,\tilde{\phi}_i)\cdot R_z(-\frac{\pi}{2})\cdot R(\tilde{\psi}_{i+1},\tilde{\theta}_{i+1},\tilde{\phi}_{i+1})\cdot R_z\left(-\frac{\pi}{2}\right),$$

where $\tilde{\psi}_{i+1},\tilde{\theta}_{i+1},\tilde{\phi}_{i+1}$ are unknown angles, and angle $\tilde{\phi}_i:=\phi_i+\delta_i$ is known and δ_i is rotation angle around J_iJ_{i+1} axis. The matrix $R(\tilde{\psi}_{i+1},\tilde{\theta}_{i+1},\tilde{\phi}_{i+1})$ is calculated from Eq. (1) and its elements can be labelled as follows:

$$R(\tilde{\psi}_{i+1},\tilde{\theta}_{i+1},\tilde{\phi}_{i+1})=\begin{pmatrix} a_{11} & a_{12} & a_{13}\\ a_{21} & a_{22} & a_{23}\\ a_{31} & a_{32} & a_{33}\end{pmatrix} \tag{3}$$

The new unknown angles $\tilde{\psi}_{i+1},\tilde{\theta}_{i+1},\tilde{\phi}_{i+1}$ can be found in explicit form:

$$\begin{aligned}\tilde{\psi}_{i+1} &= arctan\left(-\frac{a_{12}}{a_{22}}\right)\\ \tilde{\phi}_{i+1} &= arctan\left(-\frac{a_{31}}{a_{33}}\right)\\ \tilde{\theta}_{i+1} &= arctan\left(cos(\tilde{\psi}_{i+1})\cdot\frac{a_{32}}{a_{22}}\right)\end{aligned} \tag{4}$$

Every segment in $2x2$ cell can rotate independently from three other segments in the cell. Four leaf rotation angles give four independent internal degrees of freedom. The other two independent degrees of freedom are rotation angles around two diagonals: J_1J_3 and J_2J_4.These two angles allow to rotate the pairs of neighbouring segments of the $2x2$ cell. To find new joint angles in case of rotating several segments like a rigid body the similar considerations are used. The six rotation angles - four angles for every leaf and two angles for rotation around two diagonals, define the configuration of the $2x2$ mosaic cell uniquely.

By adding to the $2x2$ cell the pairs of coupled segments the number of internal degrees of freedom is increased by three with each pair of segments. By adding single segment in the corner of the mosaic body the number of internal degrees of freedom stays the same.

For the arbitrary mosaic body with the size of NxM elements the number of internal d.o.f. J_{NxM} can be calculated by the following formula:

$$J_{NxM} = 3 \cdot (N + M) - 6 \tag{5}$$

The Eq. (5) can be derived by induction for the number of body segments.

However, complicated kinematic constraints appear for mosaic bodies with both dimensions greater than two. In this paper we consider examples with only one dimension greater than two.

3. Mosaic Body Control

As it was shown in previous section the arbitrary NxM mosaic body is controlled by $3 \cdot (N + M) - 6$ angles. To be able to control the mosaic body the rotation angles must be calculated somehow. The natural way is to get rotation angles with the help of stowage algorithm which aligns all joints of the robots body on some smooth surface. In other words, the mosaic body control algorithm tries to locate all body joints and selected control points on some smooth surface. The stowage algorithm uses iteration procedure at each time step to minimize the deviation of effector points from stowage surface. The kinematics of mosaic body allows to apply the same control algorithm for arbitrary NxM bodies without sufficient procedure complications.

4. Modeling

The verification of the kinematic formulas and algorithms was provided by the multi body system software package Universal Mechanism.[1] The model of a robots $3x2$ segments mosaic body was implemented with the system of a rigid bodies. Mosaic body consists of six square segments with size a. Main geometrical and dynamical parameters of the model are provided in Table 1.

Table 1. Mosaic robot model parameters

Parameter	Value
Body segment size	a = 0.2m
Body segment mass	m = 1kg
Body segment main moments of inertia	$(I_{xx} = I_{yy} = I_{zz} = 1) kg \cdot m^2$
Leg segments lengths	$(p_1 = 0.05m,\ p_2 = 0.1m,\ p_3 = 0.2m)$
Leg segments masses	$(m_1 = 0.1kg,\ m_2 = 0.1kg,\ m_3 = 0.2kg)$
One longitudinal folding angle	$\alpha_{Long} = \frac{\pi}{5} sin(t)$
Two lateral folding angles	$\left(\alpha_{Lat_1} = \frac{\pi}{4} sin(t), \alpha_{Lat_2} = \frac{\pi}{4} sin(t)\right)$
Four corner segments rotation angles	$(\alpha_{Leaf_1} = \alpha_{Leaf_2} = \alpha_{Leaf_3} = \alpha_{Leaf_4} \equiv 0)$

The implemented body stowage algorithm can be described with following pseudocode:

for each time moment t **do**
  resetBodyAngles(); \\Set all body joint angles to zero
  $(\alpha_{Long}, \alpha_{Lat_1}, \alpha_{Lat_2}, \alpha_{Leaf_1}, \ldots, \alpha_{Leaf_4})$:= getFoldingAngles(Map);
  rotateLongitudina(α_{Long});
  rotateLateral($\alpha_{Lat_1}, \alpha_{Lat_2}$);
  rotateLeafs($\alpha_{Leaf_1}, \alpha_{Leaf_2}, \alpha_{Leaf_3}, \alpha_{Leaf_4}$);
  updateLegAngles();
  executeJointAngles();\\Send calculated angles to servo controller
next t

The visualization of the mosaic robot model with sample rotation angles is depicted on Fig. 4. The Universal Mechanism software package allows to verify dynamics for mosaic body model as well. All reaction forces and control momentums can be obtained during the simulation.

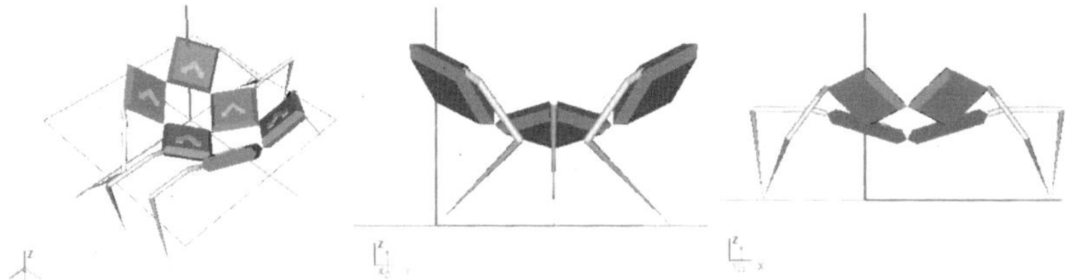

Different views on robot configuration at time $t = \frac{\pi}{2}$.

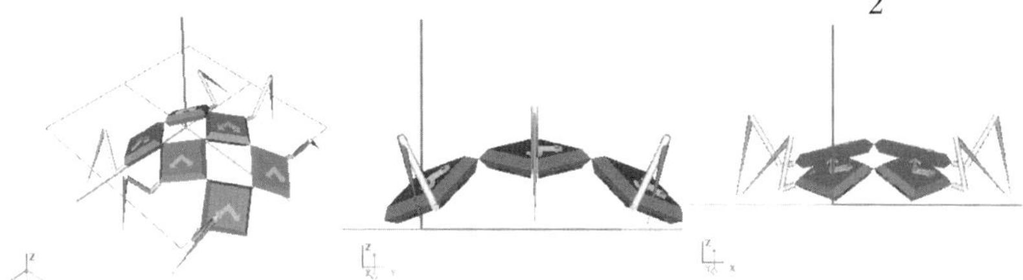

Different views on robot configuration at time $t = \frac{3\pi}{2}$.

Fig. 4.

5. Conclusion

Multilink body can be implemented as an array of unified mechanical modules able to link to each other with the help of controllable rotational joints and capable to be quickly attached to the legs with different kinematics. Mosaic body kinematics provides simple additive procedure for incrementing body's segments number and further body configuration control. Modular robot body could be easily rebuild to perform any specific task. Many types of multilegged walking robots can be easily imitated by one mosaic body robot.

Figure 5, Fig. 6 and Fig. 7 demonstrate possible real-world application scenarios.

Fig. 5. Longitudinal mosaic adaptation to the stair obstacle.

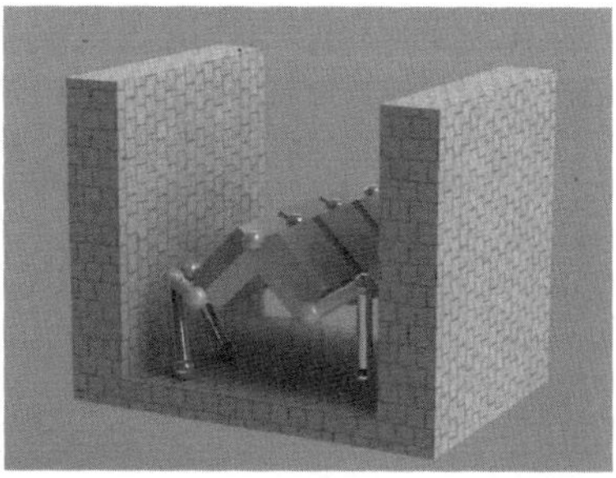

Fig. 6. Lateral mosaic body adaptation to narrow passage.

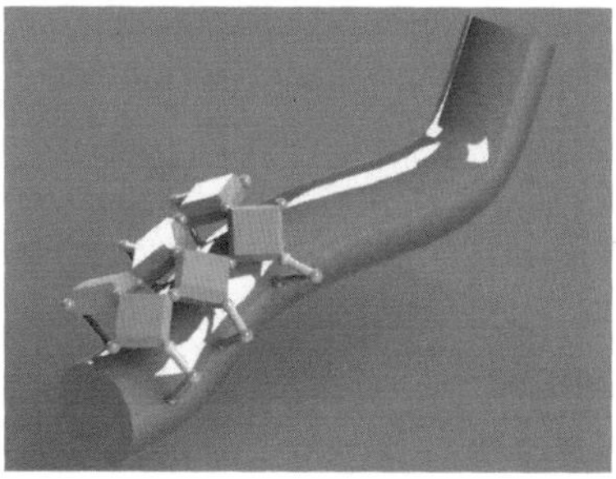

Fig. 7. Lateral and longitudinal mosaic body adaptation to support surface.

Acknowledgments

The author receives great pleasure in acknowledging the support and the help of Robotics Laboratory of the Keldysh Institute of Applied Mathematics, Russian Academy of Sciences, Moscow, Russia.

References

1. Program package Universal Mechanism. http://www.universalmechanism.com
2. P. Brazevic, A. Iles, D.E. Okhotsimsky, A.K. Platonov, V.E. Pavlovsky, A.V. Lensky: Development of multi-legged walking robot with articulated body. CLAWAR99, Portsmouth, UK, 1999, 205-212
3. U.F. Golubev and V.V. Korianov: Motion design for six-legged robot overcoming the vertical column by means of friction forces. CLAWAR'03, Catania, Italy, 2003, 609-616.

SECTION–5

INNOVATIVE DESIGN OF CLAWAR

A NOVEL INSPECTION ROBOT MOVING ON HIGH-VOLTAGE POWER TRANSMISSION LINE*

TAO GUANGHONG

School of Mechanical Engineering and Automation, Northeastern University, Shenyang, China,110819

FANG LIJIN

School of Mechanical Engineering and Automation, Northeastern University, Shenyang, China,110819

Regarding the working task requirement, a novel robot mechanism for inspecting power transmission lines with multi-unit mechanisms in series is designed. Several key mechanical design features are briefly described. The dynamics and stiffness characteristics of the parallelogram structure (PS) driven by two flexible cables are researched. The analysis results show that the stiffness of the robot in vertical plane can be controlled by actively changing elevation angle and cable tensile force of PSs.

1. Introduction

In order to guarantee the security and stability in the process of operation, the power transmission line should be inspected and maintained regularly. Currently, this work is mostly done by a worker with a telescope on the ground or a helicopter with a video camera. To improve the inspection quality, the inspection worker sometimes has to climb the high-voltage power transmission line to complete the inspection and maintain task. The manual inspection method has many disadvantages, such as long inspection cycle, high working intensity and risk. The helicopter method is very expensive and the images provided from the helicopter are usually unclear. As new inspection equipment, robot has a great market application prospect and gradually becomes research hotspot.

Since 1980s, a number of relative achievements have been achieved. In 1988, Sawada et al. presented a mobile robot that can navigate power transmission lines automatically. This robot has ability to maneuver over the

* This work was supported by The University Innovation Team of Liaoning Province Research Program (LT2014006)

obstructions by a type of subsidiary equipment on the ground wire. It is equipped with a cambered auxiliary arm that acts as a guide and allows it to negotiate transmission towers [1]. In 1989, Aoshima S I et al. presented a concept of inspection robot with multi-unit structure for telephone line or power transmission wires. Every unit has a vertically lifting arm. It is neither able to change the distance between the wheel on the line nor able to work on the line with larger slope [2]. Although never applied in the field, these early researches greatly influence the later work. Recent years, many research teams have presented robotic technologies for inspecting transmission lines, and some prototypes have been developed, such as the "LineScout" from Canada Hydro-Quebec research institute [3], the "Expliner" from Japan Kansai Electric Power Company (KEPCO) and the Japanese Electric Power Systems, Inc. (JPS) [4]–[5], the dual-arm or three-arm inspection robot prototypes from the state key laboratory of robotics, Shenyang institute of automation, Chinese academy of sciences (SIACAS) [6], Chinese Wuhan University [7], Chinese Northeastern University [8], Chinese Hunan University [9], etc.

This paper describes ongoing work on the research and design of a novel robot. Parallel flexible drive mode is applied to the unit mechanism for its lightweight and adjustable stiffness characteristic. The dynamics and stiffness characteristics of the unit mechanism are researched and presented for subsequent entity prototype and control system design.

2. Design of the Robot

2.1. *Robot architecture*

A novel inspection robot mechanism with PS in series is designed which combines the very best of serial robot, parallel mechanism and wire driven mode.

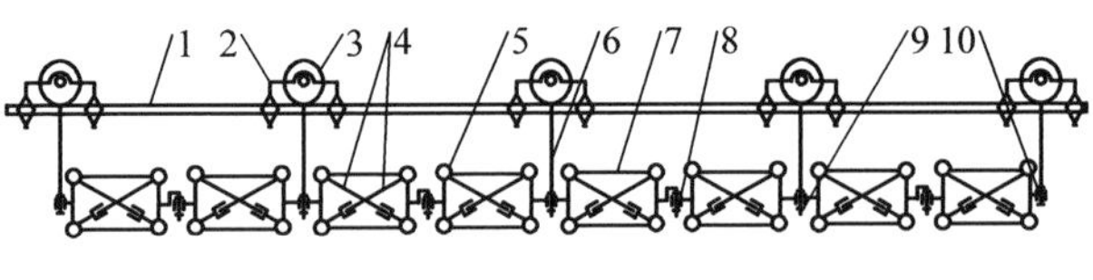

1—Transmission line, 2—Claw mechanism, 3—Walking wheel mechanism, 4—Cable, 5—Hinge joint, 6—Robot arm 7—Parallelogram structure, 8—Revolution joint in a unit 9—Composite revolution joint between the units, 10—Revolution joint at the end of the robot.
Fig. 1. Robot structure

In Fig. 1, Two PSs (7) between the two adjacent robot arms (6) is deemed to be a unit structure. The robot is composed of a set of 4 identical modules. The robot arms and the adjacent PSs can rotate independently for the composite joint (9). A composite mechanism is also installed on every arm, which is composed of walking wheel mechanism (3) and claw mechanism (2). The two PSs in one unit structure are connected by the revolution joint (8); the arm at the either end of the robot is mounted on the parallelogram structures by the revolution joint (10); the rod in the parallelogram structures is connected by the hinge joint (5). Parallelogram vertices are respectively connected by two cables (4).

The parallelogram design enables the robot arms consistently to be vertical when moving on the transmission line with slope angle. The robot parameters are shown in Fig. 2 and Tab. 1. After a few iterations, the first generation virtual conceptual prototype model is assembled and shown in Fig. 3. The following sections present a variety of detailed innovation aspects of the robot mechanical design.

Table.1. Mechanism parameters

Parameters	Length or range of motion	Parameters	Length or range of motion
l_1	0.2m	l_2	0.25m
l_3	0.1m~0.44m	l_4	0.1m~0.44m
θ_1	-90°~90°	θ_2	-120°~120°
a	0.075m	b	0.075m
h	0.4m	d	0.07m

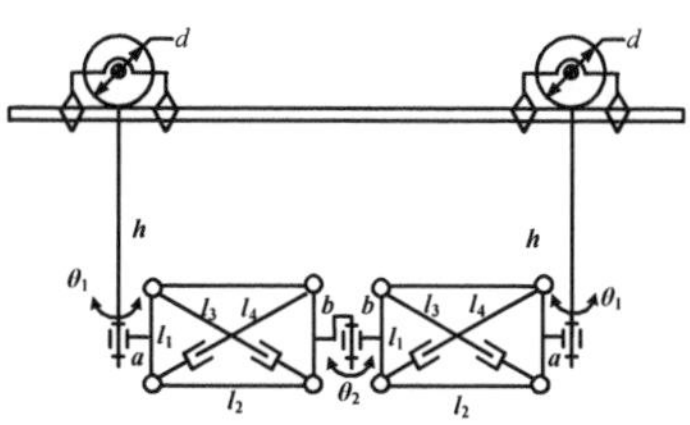

Fig. 2. Unit structure parameters of the robot

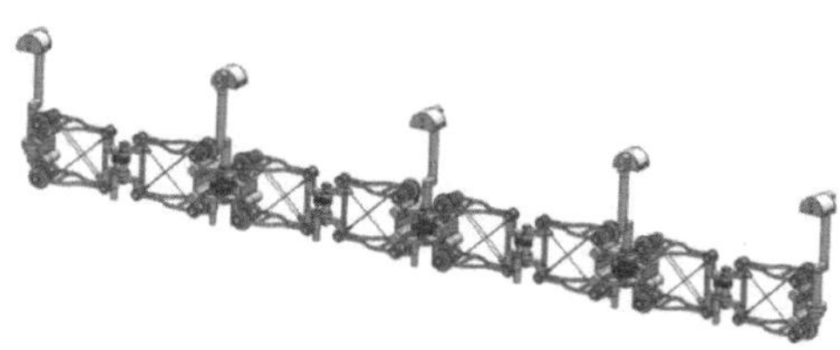

Fig. 3. Virtual conceptual prototype mode

2.2. *Revolution joint*

Figure 4(a) is the schematic diagram of mechanism of the revolution joint between the units. Because the unit mechanisms in the joint's both sides and the arm shaft (5) need to have the capacity of rotating respectively, a servo motor (1), a gear shaft (4) and a gear (2) are added on the revolution joint in a unit as shown in Fig. 4(b). Two servo motors work cooperatively to meet the motion requirements.

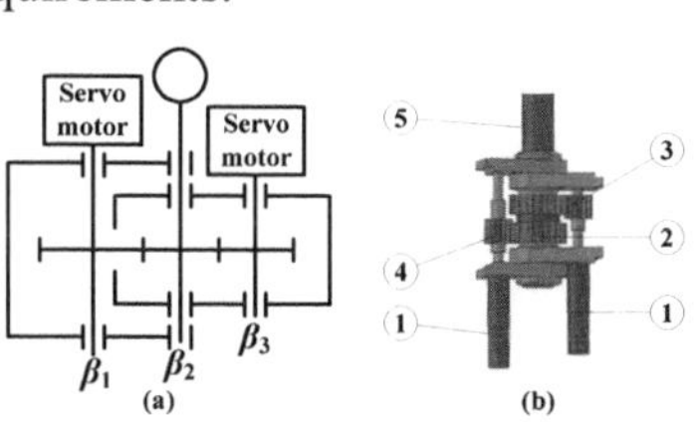

Fig. 4. Revolution joint between the units design

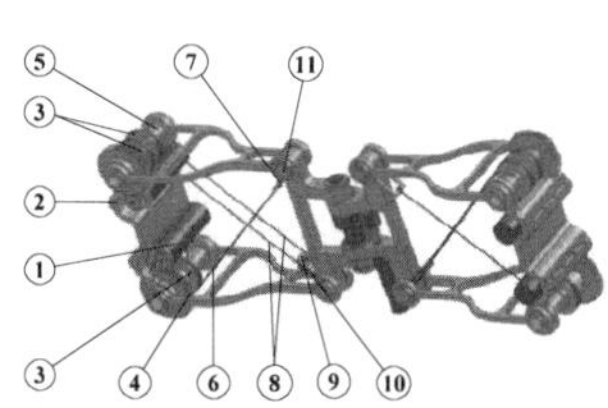

Fig. 5. Unit mechanism

2.3. *Unit mechanism*

As shown in Fig. 5, two identical PSs are combined to be a unit mechanism and are symmetrically disposed. Two cable driven servo motors (1) are all mounted on the wide stand (2) in one side of PS. The roller (3) and big gear (4) are all mounted on the shaft (5) with flat key. The roller (3) is actuated by big gear (4). Two driven cables are interactively installed. One end of cable (6) is coupled with roller and another side is coupled with the ring (7). But two ends of the cable (8) threading the small roller (9) are all coupled with the two rollers (3). This cable driven design can prevent the two cable from interference, at the same time the cable (8) has equal tensile force. The small roller (9) is hinged at the clevis (10). The ring (7) and the clevis (10) are hinged with short shaft (11).

2.4. *Crossing modes*

Obstacle-crossing ability is a critical and essential technology to the robot walking on the line. Three representative crossing modes are provided to enable the robot to adapt the various situations. When crossing or avoiding the obstacles, the robot with multi-unit in series always gasps the line at two or more points to obtain better stability. When coming across damper, the robot could cross it with the straightforward principle shown in Fig. 6. The robot could cross single clamper or double clampers with the under principle shown in Fig. 7.

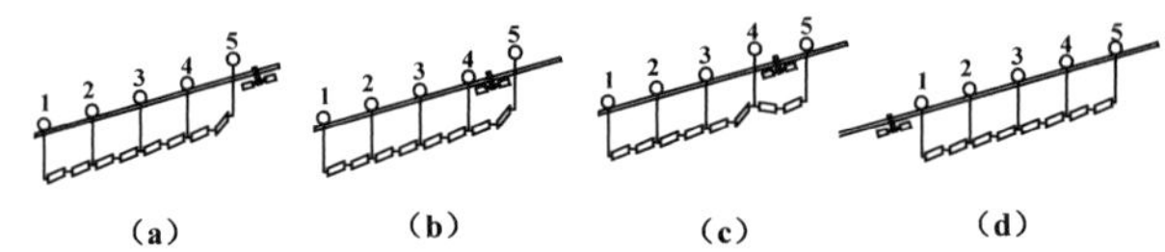

Fig. 6. Crossing mode above the line – front view

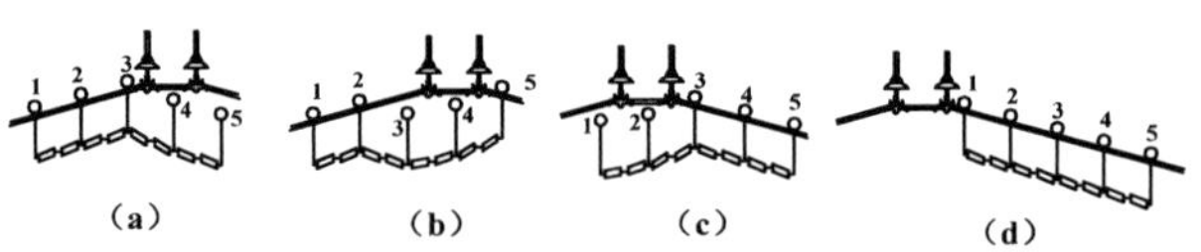

Fig. 7. Crossing mode under the line – front view

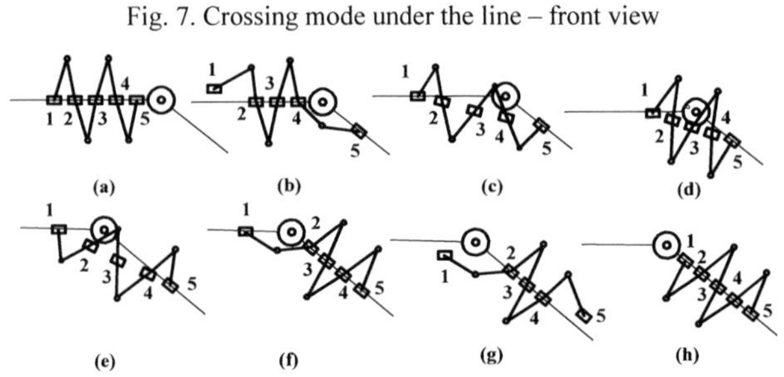

Fig. 8. Crossing mode in stretch mode – top view

On the account of geographical conditions, the transmission line may have horizontal angle suspending on the tower. In this case, the robot will have to move on a power line with horizontal angle. The stretch mode was planned for this condition, as in Fig. 8. The presented robot mechanism with multi-unit in series has ability to cross typical obstacles on the 500kV transmission line and can meet the requirements of crossing obstacles on the transmission line with certain slope or level angle have been proved in [10].

3. Dynamics and Stiffness Analysis of the Parallelogram Structure

3.1. *Position analysis and dynamics analysis*

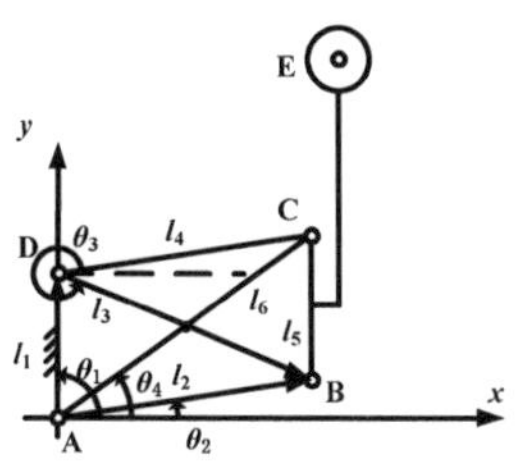

Fig. 9. Parallelogram structure

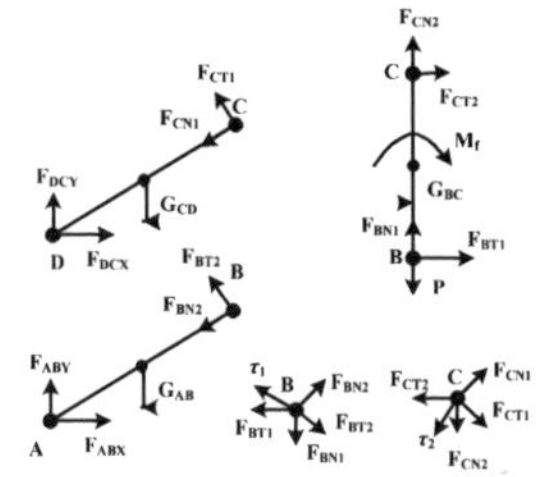

Fig. 10. Force analysis

There must be a fixed rod when the PSs do pitching motion. Take for example the rod-AD is the fixed rod, as in Fig. 9, l_3 is the active driven cable, l_6 is the passive driven cable. $\dot{l}_3$ and $\ddot{l}_3$ are stretching velocity and acceleration of the active driven cable. $\dot{l}_6$ and $\ddot{l}_6$ are stretching velocity and acceleration of the passive driven cable.

In order to reduce the number of unknown forces in the moment equation, the forces are set in tangential and normal direction, as shown in Fig. 10. G_{AB}, G_{BC}, G_{CD} is separately the gravity of every rod, M_f and P are the moment of couple and gravity applied on the rod transforming from external load, τ_1 and τ_2 are tensile forces of l_3 and l_6. Dimension of all the unknown forces in the figure is arbitrary and the actual values are determined by calculated values. Rod-AB, rod-BC, rod-CD, point-B, point-C are separately seen as the research subjects. Assumptions: the cables have a negligible mass, the link points have negligible friction, the mass center of every rod is in the mid-position.

Table. 2. Load of parallelogram mechanism

Conditions	Maximum load moment (Nm)	Load gravity(N)
one arm leaves the line	45.3125	125
two arms leave the line	205.875	270

Parameter set: the mass of rod m_{ad}, m_{bc} is 1.5kg, m_{ab}, m_{cd} is 0.5kg, revolution joint between the units and arm are 2kg, revolution joint in a unit is 0.5kg, revolution joint at the end of the robot and arm is 1.5kg, control unit and power supply of every parallelogram mechanism is 2kg. The maximum load of PS can be solved in different crossing modes, as shown in Table 2. The value range of τ_2 is set to be 0~50N, $\dot{l}_3$ is set to be 0.01 m/s, $\ddot{l}_3$ is set to be 0 m/s^2. With Newton-Euler method, the relationships of τ_1, l_3 and τ_2 are solved out and shown in Fig. 11.

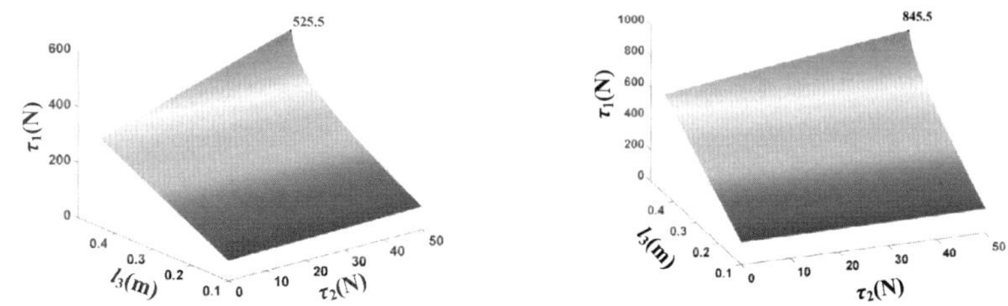

(a) One arm leaves the line (b) Two arms leave the line

Fig. 11. Tensile force of the drive cable

Figure 11 illustrates that the variation trends of the active driven cables are basically the same in both cases. When l_3 and τ_2 is max value, τ_1 is maximal and separately is 525.5N, 845.5N. Active driven force τ_1 increases as passive driven force τ_2 and l_3 become large. It means that tensile force of active driven cable gets bigger along with the mechanism elevation angle decreasing. When the length of l_3 is lesser, τ_2 has little impact on τ_1. But τ_1 increases quickly along with τ_2 increasing when l_3 is closed to maximum.

3.2. *Stiffness analysis*

As PS has only one rotating DOF, considering the angular displacement only is enough for the stiffness analysis. The relationship between tensile force of cables and applied moment has to be confirmed before calculating joint stiffness. $\dot{\boldsymbol{l}}$ (2×1) is velocity of cable and $\dot{q}$ is angular velocity. According to principle of virtual work, tensile force $\boldsymbol{\tau}$ (2×1) and applied moment M do the same work:

$$\dot{\boldsymbol{l}}^{\mathrm{T}}\boldsymbol{\tau}=\dot{q}^{\mathrm{T}}M \tag{1}$$

The tensile force of cable is in direct proportion to its deformation, we can obtain: $\delta\tau_i=K_l\delta l_i$, where K_l is the line stiffness of cable. The stiffness matrix of the PS joint can be defined as K_S, then $\delta M=K_S\delta_q$. Then we can obtain:

$$K_S\delta q=\delta\boldsymbol{J}^{\mathrm{T}}\boldsymbol{\tau}+K_l\boldsymbol{J}^{\mathrm{T}}\boldsymbol{J}\delta q \tag{2}$$

where $\boldsymbol{J}$ is a matrix(2×1):

$$\boldsymbol{J}^{\mathrm{T}}=\left[\frac{\mathrm{d}l_3}{\mathrm{d}\theta_2},\frac{\mathrm{d}l_6}{\mathrm{d}\theta_2}\right] \tag{3}$$

So K_S can be expressed as:

$$K_S = \delta \boldsymbol{J}^T \boldsymbol{\tau} / \delta q + K_l \boldsymbol{J}^T \boldsymbol{J} \tag{4}$$

Here, $\boldsymbol{J}$ can be written into $\boldsymbol{J}=[\nabla l_1, \nabla l_2]$, where $\nabla$is gradient operator. The length of cable is mapped to its gradient vector ∇l_i. Now we can obtain the stiffness matrix of the joint as follow:

$$K_S = \tau_1 H(l_1) + \tau_2 H(l_2) + K_l \boldsymbol{J}^T \boldsymbol{J} \tag{5}$$

where $H(l_i)$ is the Hessian Matrix of cable length.

Equation (5) illustrates that the stiffness of the PS joint varies as the elevation angle of the joint, tensile force of the two cables changing. Based on mechanism size and the primary prototype design, the value range of q is set to be -π/3~π/3, K_l is 5kN/m~50kN/m, applied moment is the maximum load when two arms leave the line, tensile force of passive cable τ_2 is 10N. The stiffness of PS joint is shown in Fig. 12.

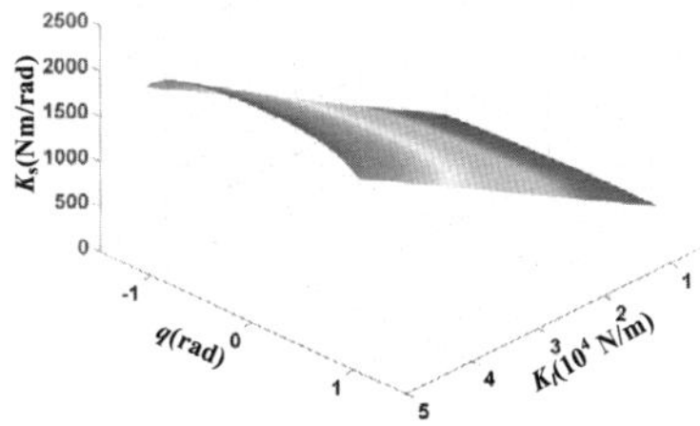

Fig. 12. Relationship between joint stiffness and cable stiffness

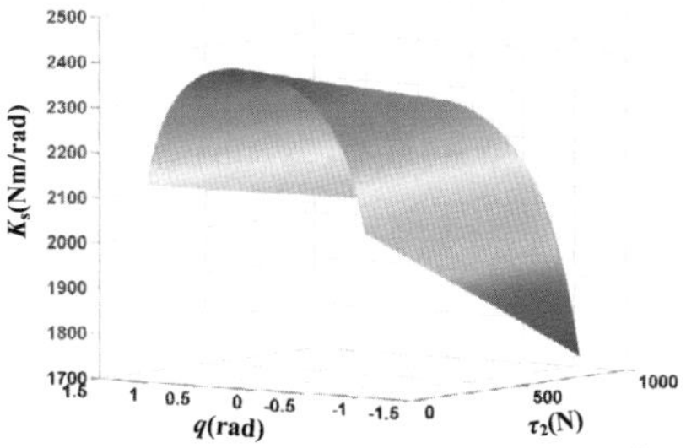

Fig. 13. Relationship between joint stiffness and passive driven force

when K_l is set to be 50kN/m, value range of q is set to be -π/3~π/3, applied moment is the maximum load when two arms leave the line, value range of τ_2 is 0~1000N. The stiffness of PS joint is shown in Fig. 13.

Figures 12 and 13 illustrate that joint stiffness is maximum when the PS is in a horizontal position and it is decreasing in both sides. The joint stiffness K_S increases when the cable stiffness K_l becomes large, the joint stiffness decreases when τ_2 becomes large. Moreover, the joint stiffness decreases rapidly when the joint elevation angle is close to -π/3. Through the above analysis, we can get the conclusion that the joint stiffness of PS can be easily controlled by changing its elevation angle and the tensile force of passive cable. Moreover, the stiffness of the robot in vertical plane can be controlled by actively changing elevation angle and the tensile force of passive cable to preferably adapt to different working environment.

4. Conclusion

A novel inspection robot mechanism for high-voltage transmission line and its crossing principle have been represented. The simulation results show that the stiffness of the joint is not only affected by the line stiffness of cables but also affected by the elevation angle and the tensile force of cables. Further, the result shows the robot has the advantage that its stiffness in vertical plane can be controlled by changing the robot pose and passive driven cable.

References

1. Sawada J, Kusumoto K, Maikawa Y, et al. *A mobile robot for inspection of power transmission lines, IEEE Transactions on Power Delivery*, **6**, 309 (1991).
2. Aoshima S I, Tsujimura T, Yabuta T. *A wire mobile robot multi-unit structure, IEEE/RSJ International Workshop on Intelligent Robots and Systems. Tsukuba*, 414(1989).
3. Pouliot N, Montambault S. *Field-oriented developments for LineScout technology and its deployment on large water crossing transmission lines, Journal of Field Robotics*, **29**, 25(2012).
4. Debenest P, Guarnieri M, Takita K, et al. *Expliner-robot for inspection of transmission lines, in: IEEE International Conference on Robotics and Automation. Pasadena*, 3978(2008).
5. Debenest P, Guarnieri M. *Expliner-From Prototype Towards a Practical Robot for Inspection of High-Voltage Lines, in: 1st International Conference on Applied Robotics for the Power Industry. Montreal*, (2010).
6. Guo Weibin, Wang Hongguang, Jiang Yong, et al. *Obstacle navigation planning for a power transmission line inspection robot, Robot*, **34**, 505(2012).
7. Wu Gong-ping, Cao Heng, Pi Yuan. *An autonomous inspection robot for high voltage bundled transmission line and its application, Engineering Journal of Wuhan University*, **45**, 96(2012).
8. Fang Lijin, Wei Yongle, Tao Guanghong. *Design of a novel dual-arm inspection robot with flexible cable, Robot*, **35**, 319(2013).
9. Qiaokang Liang, Yaonan Wang, Dan Zhang, et al. *Multifunctional Robotic System for Live Power Transmission Lines, International Journal of Robot and Automation*, **29**, 175(2014).
10. Fang Lijin, Tao Guanghong. *Research and design of a novel inspection robot mechanism with multi-unit in series, Journal of Northeastern University*, **35**, 1173(2014).

DESIGN OF A HYDRAULICALLY ACTUATED ARM FOR A QUADRUPED ROBOT

BILAL UR REHMAN*, MICHELE FOCCHI, MARCO FRIGERIO, JAKE GOLDSMITH, DARWIN G. CALDWELL and CLAUDIO SEMINI

Department of Advanced Robotics, Istituto Italiano di Tecnologia, via Morego, 30, 16163 Genova, Italy
**E-mail: bilal.rehman@iit.it*
www.iit.it

A common disadvantage of quadruped robots is that they are often limited to load carrying or observation tasks, due to their lack of manipulation capability. To remove this limitation, arms can be added to the body of the robot, enabling manipulation and providing assistance to the robot during body stabilization. However, a suitable arm for a quadruped platform requires specific features which might not *all* be available in off-the-shelf manipulators (e.g. speed, torque-controlled, light-weight, compact, without external control unit). In this paper, we present a systematic approach to design a robotic arm tailored for an 80kg quadruped robot. A full robot with arms and legs (aiming for a *centaur*-style robot) was simulated performing a range of "representative" tasks to estimate joint torques and velocities. This data was then extensively used to select the design parameters, such as the joint actuators to develop a novel, compact (0.743m fully extended), light-weight (12.5kg), and fast (maximum 4m/s no-load speed at end-effector) hydraulically actuated robotic arm with 6 torque-controlled degrees of freedom. The enclosed video presents preliminary experimental results.

Keywords: Manipulator Design, Hydraulically Actuation, Torque Controlled, Hydraulic Quadruped, Multi-legged Robot.

1. Introduction

How does a robot transverse highly uneven terrain? And what does it do when it gets to its destination? This is an area that is expected to be covered by legged robots. On the whole, quadrupeds have the advantage (over bipeds) of improved stability, whilst not becoming overly complex (like hexapods). Traditionally quadrupeds have been limited to load carrying or observation tasks, as they have no manipulation ability. This paper presents a "best-of-both-worlds" approach, by a bespoke arm system which can be mounted on the hydraulic quadruped robot HyQ,[1] in a single or bimanual configuration, creating a centaur robot. In its normal standing posture, the body of HyQ is around 1m from the ground, weighs roughly 80kg and features 12 torque-controlled joints powered by hydraulic actuators. The arm system will add a new dexterous manipulation capability to the already wide range of abilities of the HyQ such as trotting, running, jumping, stabilization, external disturbance rejection,[2] quick reflex action,[3] and careful foot hold planning during navigation through unstructured terrain.[4] This enables new tasks to be performed, including: removal of obstacles, grasping and manipulating objects, opening doors, or robot balance assistance (Fig. 1). Today, a huge variety of robotic arms are commercially available. Most

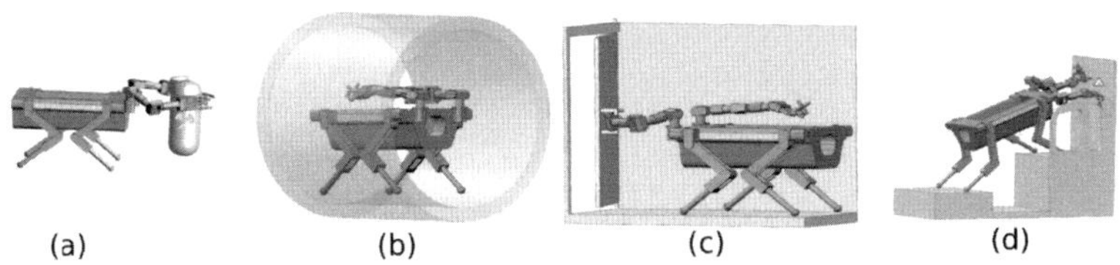

Figure 1. Conceptual tasks for the centaur robot (a) Carrying an object (b) Passing through a narrow passage (c) Opening a door (d) Body stabilization of the robot

of them suffer from limitations: some are very heavy and only allow position control (e.g the hydraulically actuated arms: $Hydra - MP$ by KNR and $ATLAS - 7R$, $CONAN - 7P$ by Schilling Robotics). Others, are torque controlled but need a bulky external control unit ($LWR3$ by DLR[5]), are limited in joint-speed ($MK1$ by HDT[6]) or have low payload lifting capacity (Schunk's $LWA3$) or, like WAM sold by Barrett Technology,[7,8] have a heavy base ($27kg$). The limitations, of these existing manipulators reduce the number of potential application scenarios and the possibility to be integrated onto an agile mobile robot such as HyQ.

The main contribution of this work, is the design of a novel **Hy**draulically actuated **Arm** (HyArm) which addresses all the above-mentioned limitations present in off-the-shelf manipulators (see Fig. 2) (e.g. fast, torque-controlled, light-weight, compact, without external control unit). In this paper, we present a systematic approach to design a robotic arm tailored for an 80kg quadruped robot. We developed a simulation of a centaur robot to acquire design parameters for the arm, such as joint torque and speed by simulating a range of dynamic "representative" tasks on the centaur robot. The arm can also be easily mirrored to provide a dual arm system (see section 5). Throughout this document, we will mainly focus on the design of the arm with only a short discussion regarding the

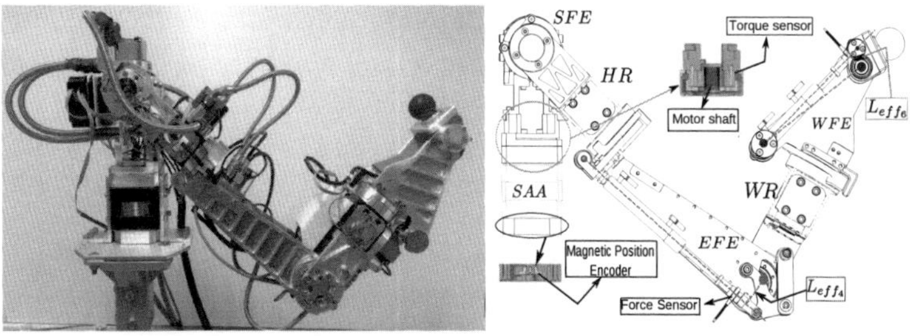

Figure 2. The HyArm: (left) Real robot side view (right) CAD model showing actuator arrangement (see section 4 for further details)

control. The paper is structured as follows: Section 2 describes the design specifications, Section 3 presents simulations of "representative" tasks. The mechanical design of *HyArm* is presented in Section 4. Finally Section 5 draws the conclusions and presents future works directions. Section 6, provide youtube link of simulation and real robot experiments (under torque controlled)

2. Design Specification

The first crucial step in the design process of the arm is to define the specifications that make the arm suitable for our desired applications (e.g. perform fast manipulations, lift objects, and help the robot to balance).

The rationale followed in the design is to be compact, without any external control unit, modular, light-weight, and being able to move a heavy payload. We chose to have the joints *hydraulically actuated* and *torque-controlled*. Indeed hydraulic actuators are well known for their robustness when coping with high force peaks during impacts. This feature becomes essential when performing highly dynamic motions whilst interacting with the environment. On the same line, is preferable to have a torque controlled actuator, because due to the physical nature of the impacts, they are better dealt within the torque domain.[9] Therefore, implementing torque control is a priority for our design. Another advantage of using hydraulics is that the HyQ already has an hydraulic supply. The arm will be equipped with position and force/torque sensors at each joint, in order to implement impedance/torque control schemes.[9] We choose to have six *DoF* in the arm, which is the minimum requirement to define position/orientation in 3D space. From a purely kinematic point of view, the arm is designed as a chain of revolute joints. Inspired by human arm anatomy, each actuated joint is labelled as follows: Shoulder Adduction/Abduction (*SAA*),

Shoulder Flexion/Extension (SFE), Humerus Rotation (HR), Elbow Flexion/Extension (EFE), Wrist Rotation (WR) and Wrist Flexion/Extension (WFE) as shown in Fig. 3.

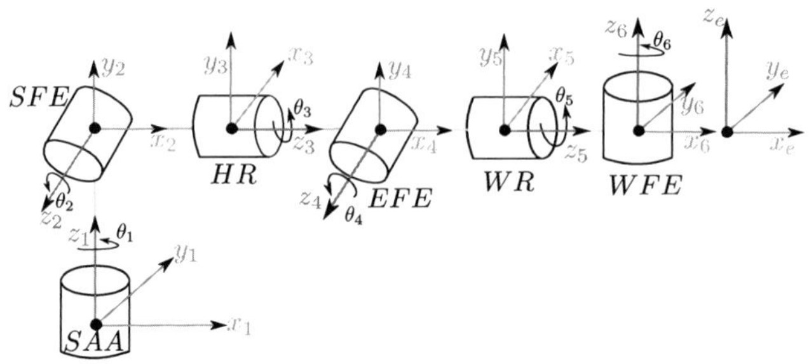

Figure 3. HyArm kinematics: Shoulder Adduction/Abduction (SAA), Shoulder Flexion/Extension (SFE), Humerus Rotation (HR), Elbow Flexion/Extension (EFE), Wrist Rotation (WR), Wrist Flexion/Extension (WFE) All the joints are shown at zero configuration

3. Simulation

This section presents the simulation results for a set of "representative" tasks allowing the estimation of torque and velocity profiles for actuator selection (see subsection. 3.1). The robot model used in the simulations is a floating base quadruped robot with two arms (see Fig. 4 of Centaur). We

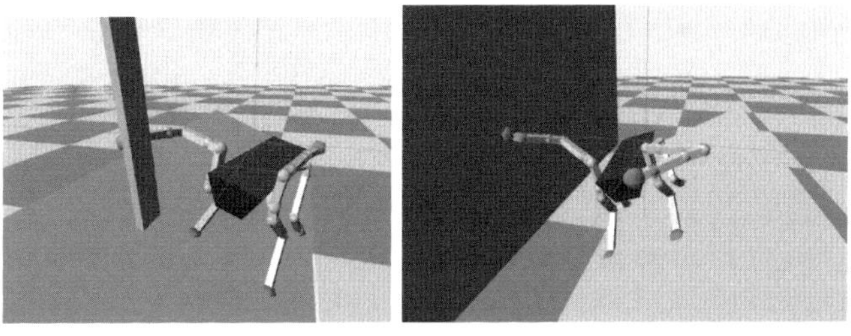

Figure 4. Picture of the Centaur simulation during the pull-up (left) and push-up (right) tasks.

used SL[10] that is a real-time simulation environment for rigid bodies, where we implemented the controller as well. To build the dynamics for the simulated robot, we used *RobCoGen*,[11] a model based code generator to provide kinematics and dynamics of articulated robots.[12] Given the kinematic tree of the robot and its inertia properties, RobCoGen automatically generates forward/inverse dynamics and kinematic transforms targeting different software platforms. The parameter for the HyQ robot are taken from the CAD model. To calculate inertia properties for the robotic arm simulation (considering the design specifications given in section 2), we selected each link as represented by an aluminum cylindrical link of mass $2kg$, length $0.175[m]$, diameter $0.075[m]$ with material density of $2700[kg/m^3]$.

3.1. *Representative tasks*

We designed the "representative" task trajectories to be demanding in terms of torque or velocity, for a single or a combination of joints. We developed minimum jerk trajectories for the end-effector of the arm in the Cartesian space (unless otherwise specified). These then resulted in the motion of the arm joints. The remaining joints are kept in a default configuration. An impedance control law defined both for position and orientation sets the virtual forces/torques (F,T) at the end-effector. We also simulated the robot falling to estimate fall time. Fig. 5 summarizes required torque and velocity plots for each joint for all simulated tasks. The explanation of each simulated task is presented as follows:

(a) **Lifting an object**: This task simulates the centaur robot lifting an object located at the end-effector when the arm is fully extended. It demands high torque for shoulder and elbow joints. We set three different kind of trajectories for shoulder joints: *horizontal* (SAA), *vertical* (SFE), *humerus rotation*. Each scenario has been simulated *with* and *without* payload which are two different estimations for maximum joint *torque* and *velocity*, respectively. We set a conservative payload of $5kg$ at end-effector moving at speed of about $1.5m/s$ for the shoulder joints. These trajectories were also simulated without payload at a speed of $4m/s$ (three times faster than the robot falling time).

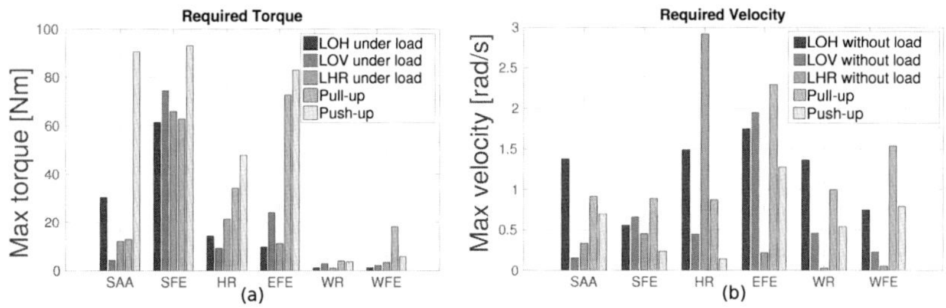

Figure 5. (a)Required torque and (b) velocity for lifting an object, biceps curl, pull-up and wall push-up tasks for each arm joint. LO:**L**ifting an **O**bject, V:**V**ertical, H:**H**orizontal, HR:**H**umerus **R**otation

(b) **Pull up**: This task demands a high torque output for the shoulder and elbow joints. In this task an arm is holding a vertical beam and pulling the robot torso (up to $0.3m$) towards the beam while standing on a slope of 0.5 rad inclination (Fig. 4 (left)). This task resembles opening a door or pulling an object.

(c) **Wall push-up**: The wall push-up task (Fig. 4(right)) demands high torque for shoulder and elbow joints. For this task the robot is standing on a 0.5 rad inclined slope and performing a push-up motion against a wall which resembles to provide assistance to robot while climbing stairs or to balance.

4. HyArm Design

The simulated tasks of Section 3, provided the required peak torques, and velocities for each DoF. According to simulated data, we intended to select commercially available hydraulic actuators which are as light weight and compact as possible.

The HyArm joints are actuated with a combination of rotary and linear hydraulic actuators Fig. 2. The benefit of using this combination is to achieve large joint ranges while still ensuring both a compact and lightweight design. The HyArm shoulder joints ($3DoF$) are equipped with rotary motors to improve compactness (also keeping the shoulder CoM closer to the base, which reduces the arm inertia) with a constant torque output. The elbow joint is actuated by a hydraulic cylinder. A four bar linkage (inspired by the excavator bucket joint) has been designed for this joint to achieve a good trade-off between joint range-of-motion and output torque. This choice has the advantage that the whole elbow assembly is part of the upper arm. The wrist joints play an important rule in determining end-effector position and orientation. For the WR joint we selected a rotary actuator to achieve wider range-of-motion, while the WFE joint is actuated by a cylinder. A standard lever mechanism provides the required range-of-motion and torque for the WFE joint.

Referring to Fig. 2(right) (elbow and wrist assembly) using known geometric calculations (law of cosines) we developed the kinematic relationship between the angles θ_i and the effective lever arm L_{eff_i}. The effective lever arm is the quantity which allows to map joint angular velocity $\dot{\theta}_i$ and torque τ_i into cylinder linear velocity $\dot{x}_{cyl_i} = L_{eff_i}(\theta_i)\dot{\theta}_i$ and force $F_{cyl_i} = L^{-1}_{eff_i}(\theta_i)\tau_i$, where $i = 4, 6$ represents the joint numbers (EFE and WFE respectively). The HyArm is equipped with position encoder, torque and force sensors (see Fig. 2(right)) to achieve torque control. Table. 1 presents an overview of the arm specification. The enclosed video (see section 6) presents preliminary experimental results. The HyArm is demonstrating torque controlled capability to change joint impedance and user and robot interaction while performing a continuous motion with different speeds.

Table 1. System Overview of HyArm Robot

Specification	Value	Joint	Total range-of-motion	Torque/force
Arm extension	$0.743m$	SAA	-3.14/0.52 [rad]	$126[Nm]$
Weight	$12.5kg$	SFE	-0.74/0.83 [rad]	$120[Nm]$
Maximum payload	$10kg$	HR	-1.57/0.068 [rad]	$120[Nm]$
Actuation system	Hydraulics	EFE	0/2.21 [rad]	$225[Nm_{peak}]$
Degree of freedom	$6Dof$ (3+1+2)	WR	-2.01/1.57 [rad]	$60[Nm]$
Control mode	Position/torque	WFE	-1.57/0.52 [rad]	$100[Nm_{peak}]$

5. Conclusion and Future work

We presented the design process of a compact and light-weight hydraulic robotic arm that is fully torque controlled, compact, light-weight and without an external control unit (Table. 1). Hence, perfectly suitable to be mounted on quadruped robots which are meant to perform sophisticated dynamic tasks. The design evolution of the HyArm is presented step-by-step starting from task simulations to estimate the joint torques and velocities necessary to select the actuators and for the mechanical development. Each joint is equipped with position encoder and torque/force sensors to be torque controlled with powerful hydraulic actuators.

Future works will mainly focus on robustness testing, implementation of more robust controllers, and experimental studies of dynamic tasks and flow estimation for hydraulic supply with the quadruped robot (see Fig. 6).

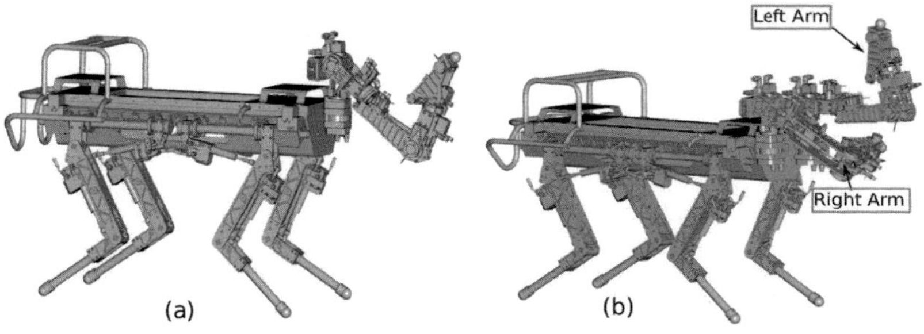

Figure 6. Future work: (a) The HyQ robot with the HyArm (b) The HyQ with dual arm system (Centaur robot)

Hardware changes will deal with improving hydraulic hose routing, weight reduction and the design of a light-weight hydraulic gripper. An intensive study will be performed to find a suitable attachment position for one or dual arm system with the HyQ robot. This will involve walking simulation of the robot with arm(s) while performing manipulation task and testing with real robot.

6. Appendix

The youtube link of simulation and real robot experiments (under torque controlled): `http://youtu.be/JhbHPZc-NGU`

Bibliography

1. C. Semini, N. G. Tsagarakis, E. Guglielmino, M. Focchi, F. Cannella and D. G. Caldwell, Design of HyQ - a hydraulically and electrically actuated quadruped robot, *IMechE Part I: Journal of Systems and Control Engineering* 225, 2011.
2. V. Barasuol, J. Buchli, V. J. D. Negri, E. R. D. Pieri, D. G. Caldwell and C. Semini, On trajectory generation and active impedance for running trotting *Dynamic Walking 2014*
3. M. Focchi, V. Barasuol, I. Havoutis, J. Buchli, C. Semini and D. G. Caldwell, Local reflex generation for obstacle negotiation in quadrupedal locomotion, in *Int. Conf. on Climbing and Walking Robots (CLAWAR)*, 2013.
4. A. Winkler, I. Havoutis, S. Bazeille, J. Ortiz, M. Focchi, D. Caldwell and C. Semini, Path planning with force-based foothold adaptation and virtual model control for torque controlled quadruped robots, in *IEEE International Conference on Robotics and Automation (ICRA)*, 2014.
5. B. Rainer, J. Kurth, G. Schreiber, R. Koeppe, A. Albu-Schffer, A. Beyer, O. Eiberger, S. Haddadin, A. Stemmer, G. Grunwald and G. Hirzinger, The KUKA-DLR lightweight robot arm - a new reference platform for robotics research and manufacturing in robotics (isr), in *41st int. symposium on and 2010 6th german conference on robotics (robotik)*, 2010.
6. http://www.hdtglobal.com/services/robotics/.
7. Barrett technology: http://www.barrett.com/robot/products-arm.htm.
8. W. T. Townsend and J. K. Salisbury, *Robots and Biological Systems: Towards a New Bionics?* (1993).
9. T. Boaventura, C. Semini, J. Buchli, M. Frigerio, M. Focchi and D. G. Caldwell, Dynamic torque control of a hydraulic quadruped robot, in *IEEE International Conference in Robotics and Automation*, 2012.
10. S. Schaal, The SL simulation and real-time control software package Technical Report, (Online) Accessed February 2011 at *http://www-clmc.usc.edu/publications/S/schaal-TRSL.pdf*, (2006).
11. M. Frigerio, J. Buchli and D. G. Caldwell, Code generation of algebraic quantities for robot controllers, in *IEEE/RSJ International Conference on Intelligent Robots and Systems (IROS)*, 2012.
12. R. Featherstone, *Rigid Body Dynamics Algorithms* (Springer US, 2008).

SOFTWARE MODULARITY FOR ROBOTIC BY USING LABVIEW PROGRAMMING

L. CANTELLI, G. MUSCATO

Dipartimento di Ingegneria Elettrica Elettronica e Informatica, Università degli Studi di Catania, Viale A. Doria 6, CATANIA, 95125 ITALY

Many robotic systems use similar sensors and actuators and navigation and localization algorithms. The adoption of a modular approach can greatly improve reusability of sub-systems and helps team work and system maintenance. Graphical programming environments can furthermore support this process, avoiding writing textual code, by using simple graphical objects. In this paper we show three different experiences we had by using a modular approach in the software programming of three different robotic systems. LabVIEW software environment was used and its main advantages will be briefly described.

1. Introduction

Modularity is a fundamental issue to allow a widespread adoption of robots into our society and to increase dissemination of robots in a wider number of applications. Nowadays the Internet of Things concept is becoming increasingly popular and most of hardware sensors, actuators and general devices can be easily connected each other in a standard way. As a consequence it becomes increasingly important to use easy and standard software tools to allow the different modules to be interfaced together.

ISO has just set up a new working group (ISO TC184/SC2/WG10 Modularity for service robots) in order to establish new standards concerning modularity for robotics.

Many projects have already considered software modularity, as OROCOS (Open Robot Control Software) [1], CLARAty [2] and ROS, as one of the widely adopted. For a review of these environments the reader is referred to [3] and [4]. However, even if ROS is widely adopted and open-source, it usually needs a long training time for those roboticists or students that have not a strong computer science background.

Other environments have arisen in the last years, allowing a more easy approach, by using graphic programming, as Microsoft Robotics Development Studio (MRDS) and the National Instrument LabVIEW Robotic module [5].

However MRDS has not been updated since 2012 and in 2014 the Robotics division of Microsoft Research was suspended.

In our laboratory in the last years several different robotic architectures have been implemented by using LabVIEW programming environment, sometime in conjunction also with NI Hardware boards. Being a graphical programming environment LabVIEW allows easily building programs in a modular way, by using blocks that can be connected together. In this way it is also really easy to write programs simple to be understood and to be reused for similar applications. Moreover the great readability of the modules allows them to be maintained and upgraded without difficulty and moreover new researchers can be involved rapidly within a project, with a very short training time.

For the most time-critical operations, as the low level motion control modules, the LabVIEW Real-Time Module was also adopted. It is an add-on used for the creation and debugging of critical and deterministic applications, running on embedded hardware targets and allows LabVIEW to execute applications that require real-time performances.

The paper is structured as follows: section II concerns the description of the Robovolc, a multipurpose six-wheeled mobile robot mainly adopted for volcanic exploration; section III describes the U-Go tracked platform; in section IV another architecture concerning an image processing system for the control of a manipulator is presented; finally, in section V some considerations conclude the paper.

2. ROBOVOLC

ROBOVOLC is a project aimed to the development of a robotic platform to perform explorations and inspections in volcanic environments [6]. The project has been initially funded by the European Commission, but since then has evolved renewing during the years the sensors and control architecture of the different modules.

The main platform of the ROBOVOLC system is the rover (Fig. 1). This is a six wheeled UGV with an articulated chassis, actuated by means of six independent DC motors. Recently the control architecture of the rover has been totally renewed, by using an highly modular architecture.

In particular a NI sbRIO-9636 embedded control and acquisition board has been adopted as low level controller, aimed to the RT control of the motor and to the management of most of the sensors. This embedded control and acquisition device integrates a real-time processor, a user-reconfigurable FPGA, and I/O on a single printed circuit board. It features a 400 MHz industrial processor, a Xilinx Spartan-6 LX45 FPGA, 16 single-ended/8 differential 16-bit analog input channels at 200 kS/s, four 16-bit analog output channels, and 28 digital I/O lines. The sbRIO-9636 includes a built-in 10/100 Mbit/s Ethernet port and host built-in web (HTTP) and file (FTP) servers. It also offers integrated

USB, CAN, SDHC, RS232 serial, and RS485 serial ports, for controlling peripherals devices.

The main algorithms that have been implemented are skid steering, speed control of single wheels and the Integral Redistribution-Traction Control Architecture [7].

Figure 1. The ROBOVOLC with the manipulator arm and the prodder.

A KUKA KR 5 sixx R850 robotic arm was also mounted on the robotic platform for experimenting coordinated motion of the robot and of a prodder for material detection. Specific software has been developed to remotely guide the robotic manipulator. The software is based on the KUKA RSI tool and Ethernet connection between the robot controller and the sbRIO on the Robovolc. Figure 3 shows the mobile platform with the intelligent prodder on the arm [8].

3. U-Go

U-Go is a tracked robot, in differential drive configuration. Each track is actuated by a DC brushed motor, and thanks to the differential configuration, each of them can run with different velocities, in order to control the robot direction [9].

The robot has a central compartment, with dimensions 99.1 cm × 48.5cm × 28.5cm. Each track has a width of 18cm and the distance between the two tracks measures 70cm. This robot has been mainly designed for application in agriculture, as Vineyard and greenhouse spraying and as a vehicle to transport apposite sensors for humanitarian demining operations.

The robot comprises inside:

- A robot control board National Instrument: NI sbRIO 9626.

- Two batteries of 12 V and 130 Ah in series.
- Two DC 800W brushed motors and gear reduction with a ratio of 1/32.
- Two DRUM digital drives by Elmo Motion Control.
- Two encoders Kubler 1024ppr, placed on rotors axis of the motors.
- Four voltage regulators, to give the correct energy to all devices on board including the government units and sensors.
- A battery charger.

Moreover as other sensors, U-Go includes a "Leica gs10" GPS system, a SICK LMS200 laser scanner and an MTI Xsense inertial platform. To allow tele-control and telemetry, a powerful WI-FI connection is present and a digital joystick is used for sending linear and angular velocity references.

All these modules are connected to the NI sbRIO 9626 board and programmed in LabVIEW (Fig. 2). Many software interfaces have been reused from the ROBOVOLC system. Moreover other localization and navigation algorithms were implemented, as a self-calibrating Extended Kalman Filter and a potential field obstacle avoidance algorithm [10].

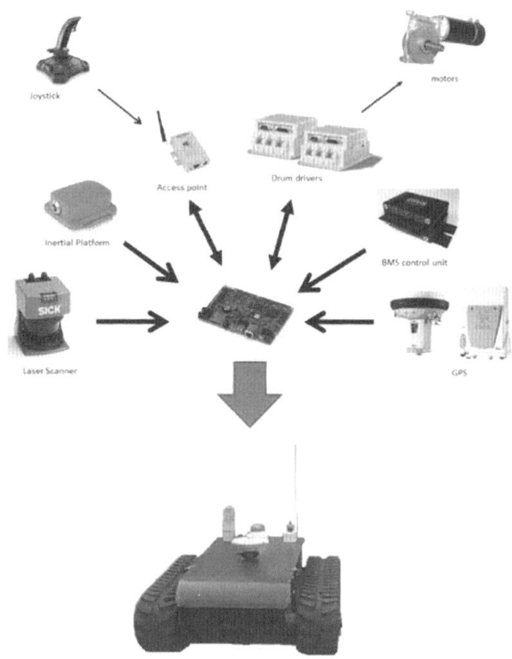

Figure 2. The different modules interfaced to the U-go tracked platform.

A simulation environment was also realized, using LabVIEW Robotics and the Robot Simulator model Builder. In this way it was possible to simulate most of the algorithms and of the modules of the system, before implementing them on the real robot (Fig. 3).

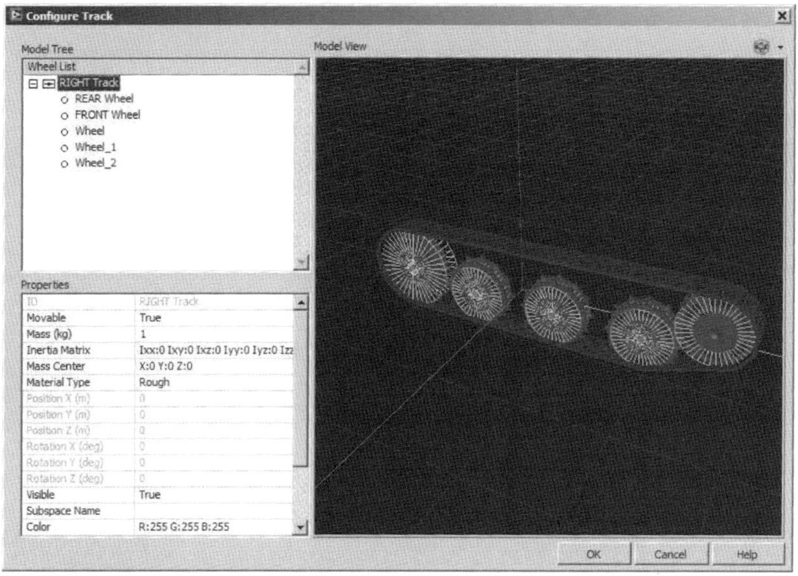

Figure 3. A screenshot of the simulation environment with a track.

4. The RAPOLAC System

Shaped Metal Deposition (SMD) is an innovative technique for rapid manufacturing of mechanical parts. It uses a robotic arm and a Tungsten Inert Gas (TIG) arc welding system, in order to obtain the desired component final shape, adding progressively the needed material. Within the EC project RAPOLAC our group developed an automatic control system for the robotic arm, integrated with all the sensors and the welding system [11].

Several different systems had to be interfaced: A KUKA KR 5 sixx R850 six axis manipulator, a Fronius TIG welding system, a CCD camera, an IR camera, a pyrometer and a microphone.

Figure 4 reports a block diagram of the whole system, showing also the user interface, implemented in LabVIEW that allow controlling all the processes and monitoring all the parameters.

Also in this case, the adoption of LabVIEW allowed fast programming and the reuse of several sub-modules from other projects.

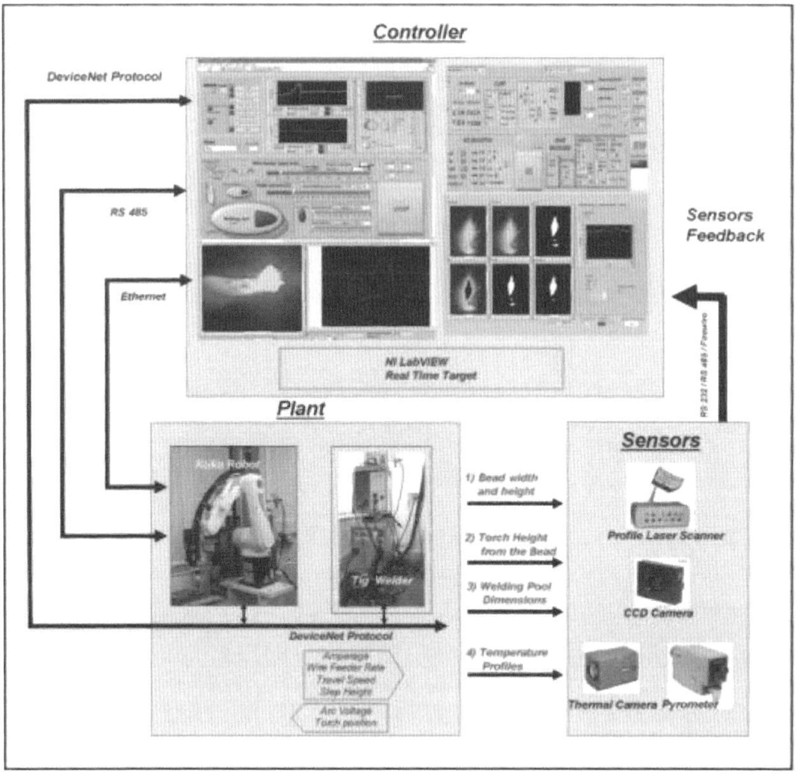

Figure 4. Experimental setup developed for the RAPOLAC system.

5. Conclusions

Software modularity sometimes needs more time to be spent in the initial phases of a project, but this is rewarded by the time saved later for maintenance and updating. In a robotic system, once that a modular approach is adopted, it is possible to easily reuse sensors, actuators and navigation, localization, path planning algorithms for a different project. It is also much easier to add or change sensors or actuators within a robotic architecture.

The three projects share several common modules as it is summarized in Figure 5. Moreover, from an academic perspective, the involvement within a project of students without a solid computer science background is much more simplified.

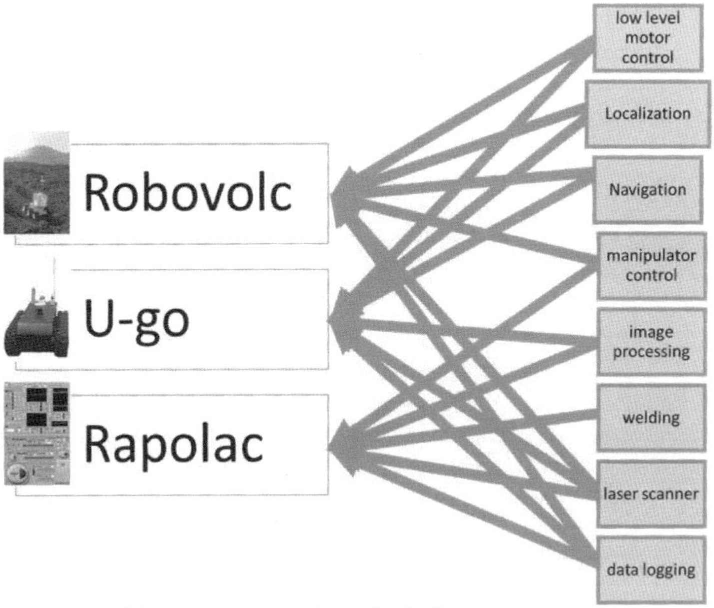

Figure 5. Common modules adopted for the three projects.

References

1. H. Bruyninckx, "Open robot control software: the OROCOS project." Proceedings of the IEEE International Conference on Robotics and Automation ICRA2001, Vol. 3, 2001.
2. I. A. Nesnas, R. Simmons, D. Gaines, C. Kunz, A. Diaz-Calderon, T. Estlin, D. Apfelbaum, "CLARAty: Challenges and steps toward reusable robotic software", International Journal of Advanced Robotic Systems, Vol. 3, n.1, pp. 23-30, 2006.
3. P. Iñigo-Blasco, F. Diaz-del-Rio, M. C. Romero-Ternero, D. Cagigas-Muñiz, S. Vicente-Diaz, "Robotics software frameworks for multi-agent robotic systems development", Robotics and Autonomous Systems, Vol. 60, n.6, pp. 803-821, 2012.
4. H. Bruyninckx, M. Klotzbücher, N. Hochgeschwender, G. Kraetzschmar, L. Gherardi, D. Brugali, "The brics component model: A model-based development paradigm for complex robotics software systems", Proceedings of the 28th Annual ACM Symposium on Applied Computing, pp. 1758-1764, March 2013.
5. K. Muecke, B. Powell, "Feasibility of LabVIEW as a scalable robot programming language", ISRCS 2011, 4th IEEE International Symposium on Resilient Control Systems, pp. 161-165, August 2011.

6. G. Muscato, F. Bonaccorso, L. Cantelli, D. Longo and C.D. Melita, “Volcanic environments: Robots for Exploration and Measurement”, IEEE Robotics & Automation Magazine, Vol.19, No.1, March 2012, pp. 40-49, DOI: 10.1109/MRA.2011.2181683.
7. D. Caltabiano, D. Longo and G. Muscato, “A new traction control architecture for planetary exploration robot”, Proceedings of the Eight International Conference on Climbing and Walking Robots and the support Technologies for Mobile Machines CLAWAR 2005, London U.K., 13-15 September 2005.
8. S. Baglio, L. Cantelli, C. Costantino, F. Giusa, G. Muscato, “A survey on instrumented prodding techniques for landmine recognition”, 10th International Symposium on Humanitarian Demining coupled with the 11th IARP WS HUDEM’2013, 23 April 2013, Šibenik, Croatia.
9. D. Longo, A.Pennisi, R.Bonsignore, G. Schillaci, and G. Muscato,“A small autonomous electrical vehicle as partner for heroic viticulture”, Acta Horticulturae (ISHS) 978:391-398 Ed. S. Poni; ISBN 9789066052093; ISSN 0567-7572, 2013.
10. L. Cantelli, M. Mangiameli, C.D. Melita, G. Muscato, “UAV/UGV cooperation for surveying operations in humanitarian demining”, 11th IEEE International Symposium on Safety Security and Rescue Robotics, Oct 21-26, 2013 Linkoping, Sweden.
11. F. Bonaccorso, L. Cantelli, G. Muscato, “An Arc Welding Robot Control for a Shaped Metal Deposition Plant: Modular Software Interface and Sensors”, IEEE Transaction on Industrial Electronics, vol.58, N.8, Aug. 2011, pp. 3126-3132, http://dx.doi.org/10.1109/TIE.2011.2114311.

RISE-ROVER: A WALL-CLIMBING ROBOT WITH HIGH RELIABILITY AND LOAD-CARRYING CAPACITY*

JIZHONG XIAO, BING LI

Electrical Engineering Department
The City College, City University of New York
160 Convent Avenue, New York, NY 10031, United States
{jxiao, bli}@ccny.cuny.edu

KENSHIN USHIRODA, QIANG SONG

InnovBot LLC, Zahn Center for Entrepreneurship at Grove School of Engineering
The City College, City University of New York
160 Convent Avenue, New York, NY 10031, United States
{kenshinushiroda, songqiangty}@gmail.com

This paper presents Rise-Rover, a new generation wall-climbing robot with high reliability and load-carrying capacity on vertical surfaces for Non-Destructive Testing (NDT) of concrete and steel infrastructure. One Rise-Rover drivetrain module can operate on both smooth and rough vertical/inclined surfaces independently. The on-board electronics and PID controller monitor the pressure reading and adjust impeller speed to provide stable suction force for the wall-climbing robot. The use of duct fan and tether further increase the operation reliability. Rise-Rover is remotely controlled by Android smartphone via Wifi, and the User Interface (UI) provides good usability and convenience. The experimental test verified the good performance of the Rise-Rover prototype.

1. Introduction

The civil infrastructure (e.g., buildings, bridges, tunnels, dams, concrete towers) that was built in the mid-20th century or earlier in the United States and across the world is reaching its life expectancy, leaving questions about their structural integrity and deterioration levels, thus have strong needs for routine inspection and maintenance to ensure sustainability. In addition to visual inspection of surface flaws, the inspectors need to detect subsurface defects (i.e., cracks, delamination, and voids) using NDT instruments such as ground penetration radar (GPR) or impact echo device to inspect the structural integrity of concrete

* This work is supported by U.S National Science Foundation I-Corps program and STTR Phase-1 grant: "Wall-climbing robots for nondestructive inspection to ensure sustainable infrastructure".

and steel structures. It is a very challenging job to inspect vertical surfaces and other difficult to access places. The current practice of manual inspections is time consuming, expensive and often requires the use of extensive scaffolding, leading to human safety concerns. Climbing robots with the ability to maneuver on vertical surfaces are needed to automate the inspection process, which will provide vertical mobility and allow inspections to be performed significantly faster, safer and more thorough, at a lower cost, by eliminating the costly scaffolding.

2. Related Work

To successfully accomplish NDT inspection task, the climbing robots should have reliable adhesion, high maneuverability, large payload/weight ratio, and adaptability on a variety of wall surfaces. The most challenging work in climbing robot design is the adhesion method to ensure that the robots grip the target surface firmly and without sacrifice of maneuverability. So far, researchers around the world have developed adhesion mechanisms that mainly fall in six categories: 1) magnetic adhesion devices for climbing ferrous surfaces [3]; 2) vacuum suction techniques for operation on smooth and nonporous surfaces [4]; 3) attraction force generators based on aerodynamic principles [1, 2, 5]; 4) grippers; 5) bio-mimetic approaches inspired by climbing animals [6]; and 6) other emerging techniques such as compliant electroadhesion. The book chapter [7] by Dr. Jizhong Xiao provides a comprehensive review on the climbing robot technologies.

The immediate competitor comes from International Climbing Machines (ICM) which produces a large robot platform using an AC vacuum pump to generate strong suction and using rolling tread with thick foam to create perfect seal perimeter as shown in Figure 1. It is so far the most mature robot prototype that has the potential to be used in NDT inspection of civil infrastructure [8]. The ICM robot can generate very strong adhesion on smooth and rough surfaces and overcome small surface irregularities thanks to the conformable thick form. However, the major weakness of this robot is that it has one vacuum chamber that requires perfect sealing enclosed by the thick foam tread. The robot falls when the vacuum breaks, and it happened when the robot crosses a ditch or deep groove/crack — which are very common on brick/concrete walls.

Figure 1. ICM robot uses rolling seal and AC suction pump for strong adhesion.

3. Mechanical System of Rise-Rover

3.1. *Mechanism*

Figure 2 shows the innovative design of Rise-Rover climbing robot prototype. The robot uses two drivetrain modules on each side, and two ducted fans in the middle of the chassis. Each drivetrain module can operate on wall surface independently and two modules can carry payload in the middle payload area.

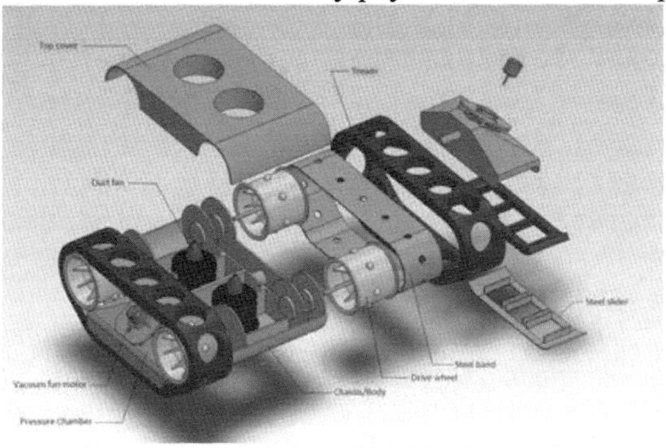

Figure 2. Exploded view of the Rise-Rover robot.

The design of a drivetrain module is shown in Figure 3. The drivetrain consists of two driving wheels, a foam tread, a rotor package with air chamber and impeller, and a perforated spring steel band (belt). The backside of the foam tread is lined with the belt which acts to provide structure and prevent deformation to the tread. The holes in the belt allow the driving wheels to pull the belt with matching teeth, and also serve as ports for the air to be evacuated from the circular openings in the foam. The ultra-high-density foam is adaptable to surface irregularity and provide a rolling seal to the chamber. The chamber is evacuated by a 2.75-inch diameter vacuum impeller powered by an ultra-fast and highly efficient brushless motor. The chamber is divided into three sections and contains two pressure valves which close upon the break of a seal, virtually increasing the number of independent chamber seals to three.

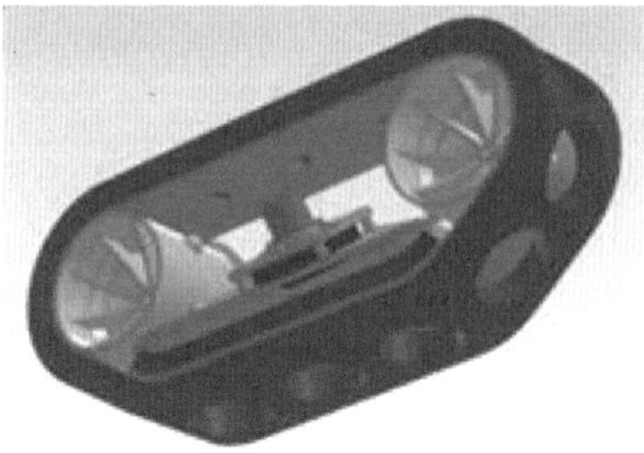

Figure 3. A drivetrain with foam tread and vacuum rotor package.

The rotor package is an improved design variance of the one used at City-Climber robot (our old generation wall-climbing robot). The rotor package produces adhesion for the drivetrain to scale vertical walls while the high-density foam is conformable to surface irregularity and helps to muffle the impeller noise. The design has been undergo several iterations of improvement with multiple validation tests. We evaluated the relationship between impeller speed and adhesion force, and the improvement in noise/disturbance reduction, etc. The most critical test is to determine if the track can rotate when it is pushed against the wall with maximum suction force. The experiment comes out that the driving motor, when powerful enough, is able to overcome the friction on the track and provide adequate mobility. We also tested the valves function well in dividing the chamber into multiple sections for independent sealing effects.

3.2. *Design for reliability*

For any wall-climbing robots working on the vertical surfaces, reliability is a very critical factor. The climbing robots must be able to supply necessary adhesion to operate on various wall surfaces, attach to a wiring harness for safety, and implement fault tolerance features to deal with deep cracks/ditches, ledges/overhangs on wall surfaces.

Since our Rise-Rover robot has three individual chamber seals on each tread, any straight-line gaps cannot cross all six chamber seals at any given time. Thus, the robot is able to cross over ditches, which our previous version City-Climber failed to cross over.

Another novel design for reliability is duct fans to push the robot in contact with the wall when suspended in midair as shown in Figure 4. The thick foam tread is deformable to cope with surface irregularities that may exist. The propellers of two duct fans generate push force to allow re-attachment of the robot to vertical walls when it loses contact with the surface.

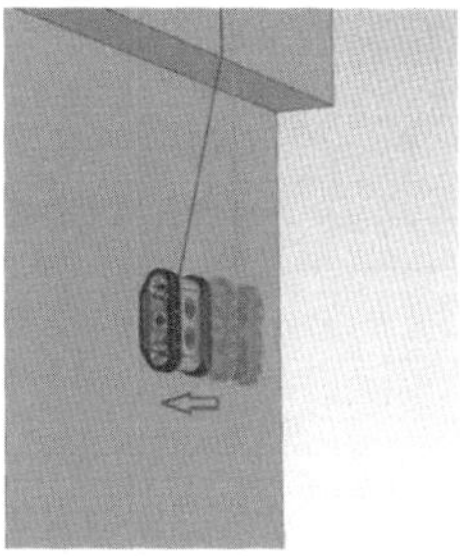

Figure 4. Duct fans push the robot in contact with wall surface.

4. Electronics and Software

4.1. *Electronics design*

The Rise-Rover platform was designed to initially use minimal control and sensing to reduce the weight of the overall system and simplify the electronics configuration. The requirements of the system called for a programmable controller capable of driving two brushed dc motor, and four DC brushed wheel motors, as well as the onboard Live-Camera or NDT devices to be installed in the middle payload compartment. The integrated controllers onboard are the ATmega32u4 running Arduino Leonardo, and Atheros AR9331 running Linux with the OpenWrt wireless stack. The board has built-in Ethernet and Wifi support, a USB-A port, micro-SD card slot, 20 digital input/output pins (of which 7 can be used as PWM outputs and 12 as analog inputs), a 16 MHz crystal oscillator, a micro USB connection, an ICSP header, and a 3 reset buttons. This microcontroller incorporates all of these specifications into a board that is 73 mm × 53 mm and weighs 32 g. The suction motors are powered by 6 cells of 3.7 V battery, and wheel motors are driven by 3 cells of 3.7 V battery.

The electrical system diagram of Rise-Rover is shown in Figure 5. The user level control is running on Android platform, and it provides the user interface (UI) and data transmission from the climbing robot, also further transfer to the data server. The middle level control is based on the embedded Linux system which handles on-board peripheral devices (surveillance camera, extended NDT device, etc.), and the data transmission. The lower level control is based on the Arduino, which process the real-time motor control, and odometry calculation for left/right wheel motors, left/right vacuum motors and monitor pressure sensor reading.

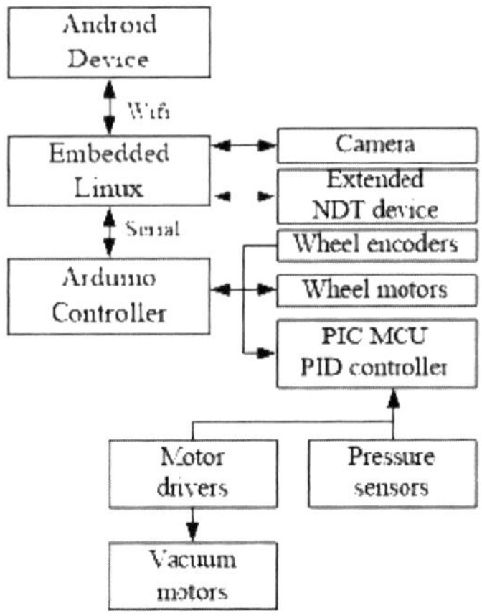

Figure 5. Electrical System Diagram.

The high-speed blushless DC motor, along with the PID control using pressure measurement as feedback, is designed for the vacuum suction system. To support more than 60K RPM high speed and make it compactable, we design and integrate the controller and driver in one PCB board (50 mm x 30 mm).

4.2. *Software design*

The software development strategy for Rise-Rover has been to build robust software, incorporating modular code wherever possible. A variety of control schemes were investigated, and the PID control with the pressure feedback is used to adjust the vacuum motor speed to generate reliable suction on wall surfaces.

The Android device is connected directly to the Wifi router on climbing robot, to control the locomotion of the robot, and display the onboard live camera scream wirelessly. The resolution and fps (frames per second) is set smaller as the distance between climbing robot and Android device is bigger. The settings menu for the App includes the settings for locomotion speed, acceleration, camera, Wifi, etc.

5. Experiments

After several design/test iterations, we have successfully developed our Rise-Rover prototype as shown in Figure 6. The innovations are the fault tolerant design using two drivetrain modules with soft/conformable treads to carry a much larger payload of NDT instruments. Each module has multiple suction/chamber seals that allow the robot to cross over gaps/grooves/cracks without loss of adhesion, and the introduction of duct fans allows for re-attachment to the wall surface for increased reliability.

The performance of the Rise-Rover is evaluated and the specs are summarized as follows:

- Dimension of each drivetrain module: 8 inches x 21 inches x 5.5 inches
- The maximum normal suction force generated by each module: 30 lbs
- Whole unit self-weight of Rise-Rover (two drivetrain modules + payload chamber): 24 lbs
- Pull-up force of Rise-Rover (i.e., payload carrying on vertical wall): 16 lbs
- Locomotion speed of Rise-Rover: 30 meters/minute
- Power consumption: Peak 4KW

  The Rise-Rover test videos can be seen at:
- http://tinyurl.com/Rise-Rover-1
- http://tinyurl.com/Rise-Rover-2
- http://tinyurl.com/Rise-Rover-Android-app

Figure 6 shows Rise-Rover that can operate on both smooth and rough surfaces. The black tread is made from a commercially available high-density thick foam. The custom-designed tread using light-weight foam encapsulated by silicon rubber is soft and conformable to surface irregularities while still allowing the robot to move and maintain strong suction against the surface.

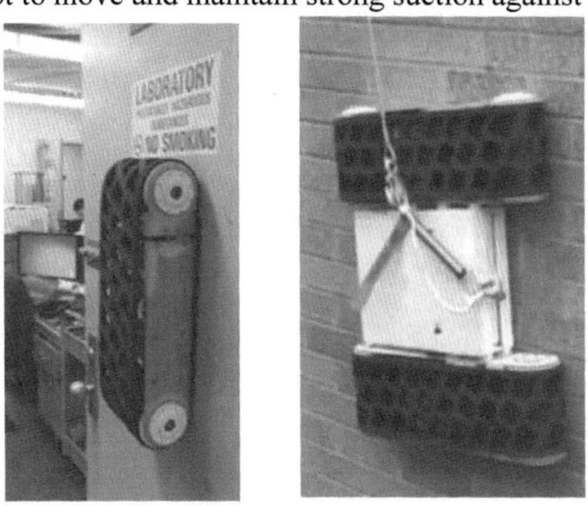

Figure 6. Rise-Rover climbing robot test.

6. Conclusions

This paper presents a novel wall-climbing robot, Rise-Rover, which adopts many fault tolerant features to increase the reliability and has the capability to carry NDT instruments for subsurface flaw detection. We have conducted several design/test iterations and developed a drivetrain module using a rotor package and soft tread. The module not only produces strong adhesion enabling the independent operation on vertical walls but also helps to overcome surface irregularities and reduce the acoustic/vibration interference. The experimental tests show that the Rise-Rover works effectively on the vertical surface with small gaps or ditches. With the self-weight of 24 lbs, the Rise-Rover can carry 16 lbs on vertical wall at the locomotion speed of 30 meters/minute.

Acknowledgments

This work is supported by U.S National Science Foundation I-Corps program grant and STTR Phase-1 program grant: "Wall-climbing robots for nondestructive inspection to ensure sustainable infrastructure".

References

1. J. Z. Xiao, A. Sadegh, M. Elliot, A. Calle, A. Persad, H. M. Chiu. *Design of Mobile Robots with Wall Climbing Capability. Proceedings of the 2005 IEEE/ASME International Conference on Advanced Intelligent Mechatronics*, Montery, CA, USA. 2005: 438~443.
2. J. Z. Xiao, A. Sadegh. *City-Climber: a new generation of wall-climbing robots, Chapter 18. In: Climbing and Walking Robots---Towards New Applications*. Vienna, Austria: I-Tech Education and Publishing, 2007. 383~402.
3. S. Hirose and H. Tsutsumitake, *Disk Rover: A Wall-Climbing Robot Using Permanent Magnet Disks, IEEE/RSJ Int. Conf. on Intelligent Robots and Systems*, USA, 1992, pp. 2074-2079.
4. J. Savall, A. Avello, and L. Briones, *Two Compact Robots for Remote Inspection of Hazardous Areas in Nuclear Power Plants, IEEE International Conference on Robotics and Automation*, pp.1993-1998, 1999.
5. A. Nishi, *Development of Wall-Climbing Robots, Computer and Electrical Engineering*, 22(2), pp. 123-149, 1996.
6. M.P. Murphy and M. Sitti, *Waalbot: An Agile Small-Scale Wall-Climbing Robot Utilizing Dry Elastomer Adhesives, IEEE/ASME Transactions on Mechatronics*, 2007, vol. 12(3), pp. 330-338.
7. J. Z. Xiao and H. Wang, *Advances in Climbing Robots, in Chapter 22, Contemporary Issues in Systems Science and Engineering, IEEE Press/Wiley*, Edited by M.C. Zhou, H. X. Li, and M. Weijnen, in press.
8. E. Strickland, *Crawler Bot Could Inspect Nuclear Power Stations, IEEE Spectrum*, September 5, 2013.

SECTION–6
INSPECTION AND INNOVATIVE SENSING

AN INNOVATIVE TORQUE SENSOR DESIGN FOR THE LIGHTEST HYDRAULIC QUADRUPED ROBOT

H. KHAN*, F. CANNELLA, D G. CALDWELL
and C. SEMINI

Department of Advanced Robotics, Istituto Italiano di Tecnologia
Via Morego 30, 16163 Genova, Italy
**E-mail: hamza.khan@iit.it*
http://www.iit.it/en/advanced-robotics.html

High-performance legged robots that are required to navigate on unstructured and challenging terrain benefit from torque-controlled joints. High-fidelity torque measurements are crucial for proper joint torque control. Commercially available torque sensors are expensive and often hard to integrate into compact and light-weight robot leg designs. Custom-made sensors on the other hand often suffer from asymmetric behaviour with respect to direction of rotation or poor linearity, especially for small and compact applications. This work is motivated by the need to achieve reliable torque measurements for the newly developed, small-size hydraulically actuated quadruped robot *MiniHyQ*. The main contribution of this work is the development of a new innovative design of a strain gauge based torque sensor with a high degree of linearity, symmetry, and scalability (both in dimension and measuring range). Furthermore, the glueing and wiring of the strain gauges are easy thanks to the geometry of the sensor that allows direct access to the mounting surfaces, even in compact dimensions. We show the design's symmetric (clockwise and counterclockwise rotation) and linear behaviour through virtual prototyping and experimental tests. Furthermore, we show how a small-scale instance of the sensor design is successfully installed on the MiniHyQ robot.

Keywords: Quadruped Robot Design; Torque Sensor; Strain Gauge based Sensor.

1. Introduction

In order to advance research faster, legged robots must become more manageable. So that dynamic experiments can be performed faster and more easily. The simplest way to achieve this goal is by reducing the size and weight of the robot. Currently, it is still area of great interest for designers to build a portable highly dynamic and versatile quadruped robot to run fast in all terrains. One of the key points of a successful of this kind of locomotion is to improve the joint compliance: passive or active. The first one is based on decoupling actuator and the driven link from the using an elastic element,[4] while the second one depends on the feedback of force and torques. Even if the active one requires a more complex control, it for sure reduces the mechanical complexity and permits greater performance. To implement the compliance controllers[5] it is necessary the measurement of the torque at each robots leg joints. The development of torque sensor for robot has been of great interest for decades, as the commercially available torque sensors are expensive, need extra space in joints and difficult to customize. Thus, a variety of sensors were designed and developed for robotics systems with acceptable performance by different researchers.[9] It gets more complicated when it is needed to be customized for smaller and compact applications. Nevertheless a torque sensor with all the requirements together (as the linearity, the symmetrical behavior in both the clockwise (CWR) and counterclockwise (CCWR) rotations, high noise-to-signal ratio and temperature compensation) rarely were designer.[10,13] Last, but not least, the strain gauges installation is one of the issues that affects the shape of the sensors and their performances. In this paper a new design of torque sensor that collects all the aforementioned characteristics is presented. The virtual prototyping design (in particular the finite element analysis) was used to reach the final shape and the experimental set up was built to test the performances.

2. Related Work

Considering the advantages of flexible joints (the compliance permits, for instance, to make safer the human robot interaction, the energy harvesting and to move in unstructured environment, in particular the uneven terrain), in the last decades the robotic research began to deal with the compliant joints. Consequently, the control moved from the rigid joint to the fully sensorized joints based on the applications of torque sensors as Visher et al. and Aghili et al.[1,12] ; as well as more recently Wolf et al.[14] have shown. In particular, Tsetserukou et al. classify the torque sensors in three categories: electrical, optical and based on electromagnetic phenomena.[11] Considering that in this paper we will apply the aforementioned torque sensor to a hydraulically quadruped, the high stresses and shocks push to take in account only the first ones and essentially them that are associated with strain gauges. Further developments were targeted towards the human-robot interaction,[7] the motion stabilization[6] or miniaturization.[3] Till now, despite to the different applications, the basic design was not improved too much, because the non linearity and the non symmetrical behavior (in CWR and CCWR) was not guaranteed.[2]

3. Why Compact Torque Sensor is Needed for MiniHyQ Robot?

To the authors best knowledge, MiniHyQ is the lightest and smallest hydraulic quadruped robot that has been built so far. It is shown in comparison of existing hydraulic quadruped robots (Table 1). MiniHyQ is around 3 times lighter than most of the existing hydraulic quadruped robots. It has almost 30% higher joint torque density (robot mass to joint torque ratio) and 40% wider joint range of motion in leg-sagittal plane comparing to HyQ.[8] MiniHyQ is fully torque controlled, measured directly at

Table 1. A comparison of Hydraulic Quadruped Robots

Name	**Mass** (offboard, onboard pump)	**Dimensions** (LxWxH)	**DoF** (per leg), **Joint Torque Controlled**
SCalf	78kg, 123kg	1.1m x 0.49m x 1m	3, Yes
HyQ	75kg, 98kg	1m x 0.5m x 1m	3, Yes
Baby Elephant	90kg, 130kg	1.2m x 0.6m x 1m	3, No
BigDog	N.A, 110kg	1.1m x 0.4m x 1m	4, Yes
JINPOONG	80kg, 120kg	1.1m x 0.4m x 1.2m	4, No
RLA-1	60.2kg, N.A	1.1m x 0.67m x 1m	3, No
LS3	N.A	bigger than BigDog	3, N.A
Wildcat	N.A	N.A	3, N.A
Spot	N.A, 74kg	smaller than BigDog	3, N.A
MiniHyQ	24kg, 35kg	0.85m x 0.35m x 0.77m	3, Yes

the joint. MiniHyQ's each leg has three active joints. Two of its joints use linear cylinder with in series load cell to measure joint torque. Its Hip Flexion/Extension(HFE) joint (third joint) uses rotary motor which provides constant joint torque of 60 Nm at 20MPa. It is needed to find a compact and high performance torque sensor which fits in the desired limit space shown in Fig. 2(b).

4. Torque Sensor design

4.1. *Design evolution*

In order to find compact solution due to limited space, we started with scaling down and investigating the traditional strain gauge based designs.

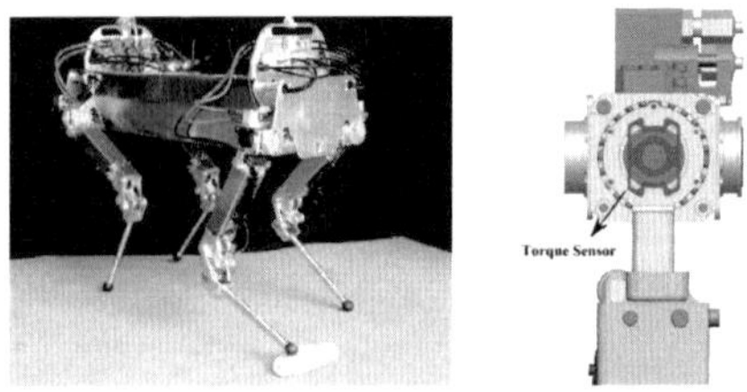

(a) (b)

Fig. 1. MiniHyQ **(a)** picture of robot and **(b)** CAD model of HFE joint, it shows where custom designed strain gauges based torque sensor is fitted.

Four spokes based Torque Sensor It is classic example of strain gauge based torque sensor. The main drawback of this design, due to four spokes it is not easy or possible to glue and wire strain gauges on it. It provides reasonable output but it is not feasible for compact applications.

Two spokes based Torque Sensor Design with two spokes, seems promising for gluing strain gauges and symmetric output. But due to size constrains, it was not possible to bring peak strain or stress in middle of spoke to act like beam. Even though it is brought in middle by changing spoke thickness and its round size, but in trade off it reduced horizontal flat surface less than 5mm (in our case recommended horizontal length for mounting strain gauge is minimum 5mm). So, both of the previous torque sensors already in use in the HyQ[8] were not suitable to be scaled. A new design was necessary.

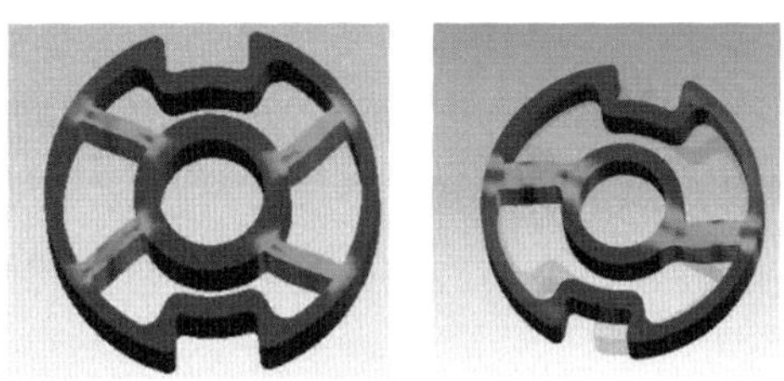

(a) (b)

Fig. 2. Traditional starin gauge based torque sensors **(a)** Four spokes and **(b) Two spokes.**

4.2. *Final design*

The design of this torque sensor is the synthesis about three needs: the easy access to the maximum strain rate point to glue the strain gauges, the symmetry and linearity of the sensor response. That was reached re-designing the former torque sensors with four wings equal in geometry, as shown in the Fig. 3; moreover, the half bridge strain gauge was mounted in two opposite wings of the sensor to sum two opposite signals; that guarantees the symmetry during both CWR and CCWR load application, because the possible difference between the tension/compression are canceled, as shown in the Fig. 4.

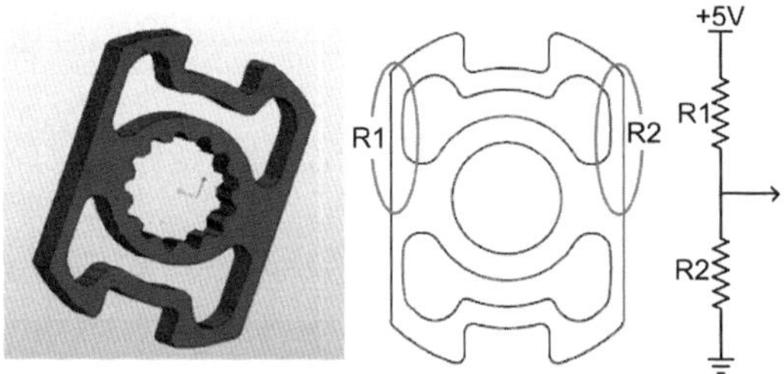

Fig. 3. *(left)* The 3D design of the torque sensor; *(right)* the half bridge strain gauge positions.

5. Results

The symmetric behavior in both the rotations (CWR and CCWR) and its linearity are designed by virtual prototyping design (in particular Finite Element Analysis) and tested performing the experimental tests.

5.1. *Virtual Prototyping Design Results*

The model (built with 50000 elements, with 39NiCrMo3 material) permitted to simulate the torque sensor design performance; so the numerical model was built and the simulation were launched. The loads and the constraints were the input data. It is shown in the Fig. 4(a), (b) and the outputs were the strains of surfaces where the strain gauges should be glued, as shown in the Fig. 4(c), (d). The two results are similar in distribution and in values; moreover there is a wide area where to glue the strain gauge with low gradient. Also the stress was checked: 200MPa, that is lower of than 450MPA the yield point.

5.2. *Experimental Results*

The torque sensor was machined and the full bridge strain gauges was easily glued on the both sides, to maximize the signal and to compensate the

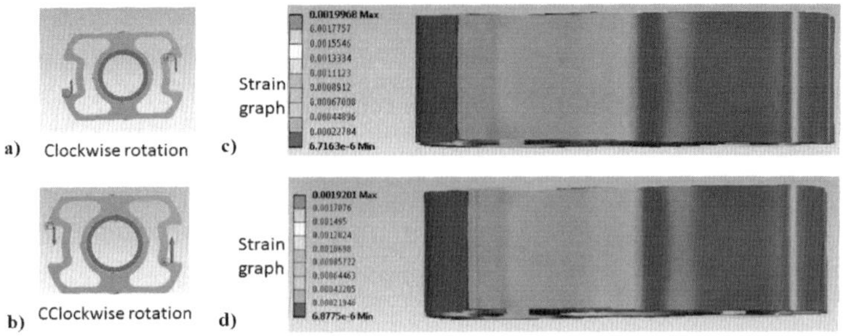

Fig. 4. **(a)** and **(b)** loads a couple of forces, in red) and constrain (the blocked shaft, in blue) of the Finite Element Model. **(c)** and **(d)** the strain contours of the upper surface.

temperature, as shown in the Fig. 5(a). To verify the design performances, an experimental test was carried out loading the torque sensor with a force in order to reproduce the working conditions. The torque was applied with a beam and the masses hanged on its ends, as shown in the Fig. 5(b). The applied torque is defined as $\tau = m * g * b$ where g is the gravity acceleration. The obtained relationship between the torque input and strain gauges output is linear, symmetry and with good sensitivity, as shown in the Fig. 6.

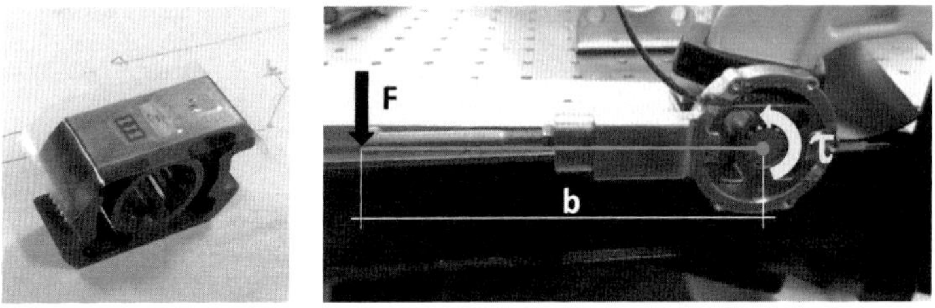

(a) (b)

Fig. 5. (a) physical model of the torque sensor equipped with strain gauge. (b) the test rig with schema of applied forces (in black) and arm length (in white) that created the applied torque (in yellow).

The obtained relationship between the torque input and strain gauges output is linear, symmetry and good sensitivity, as shown in the Fig. 6.

6. Discussion and Conclusion

The Fig. 5(a) shows that the strain gauges installation is easy to do, because the surfaces have not any obstacle to be reached, despite the small size of

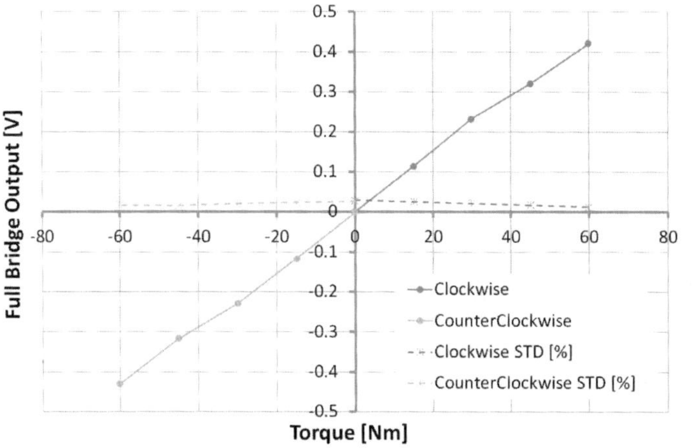

Fig. 6. Experimental results of torque sensor while varying torque CWR and CCWR.

this sensor. The experimental tests results permit to highlight the torque sensor performance: there are linearity and the symmetrical behavior, as shown in the Fig. 6. Thus the final design satisfies the requirements for the robotic application. In fact, this torque sensor will be easily used in the same way in CWR and CCWR, because it does not needs any adjustment in the control. Moreover even the temperature it is compensated thanks to the full bridge strain gauges installation. So a new design of the torque sensor is presented and all the characteristics are elucidated. The requirements for being used in the robotics (arms and/or leg) are satisfied. To do that, a Finite Element Model was built and the experimental tests were carried out in order to measure the performance: in particular the symmetric and linear behavior. Moreover this design is compact and it permits the easy access to the surfaces for installing the strain gauges.

Future Works The future work envisages the optimization of the shape and the material in order to increase the sensitivity and to reduce the size. This will be generalize and a tool will be provided for designers to help easily customize this an innovative according their desired requirements.

References

1. F. Aghili, M. Buehler, and J. M. Hollerbach. A joint torque sensor for robots. In *ASME International Mechanical Engineering Congress & Exposition*, 1997.
2. M. DImperio, F. Cannella, C. Semini, DG. Caldwell, D. Catelani, and R. Ber-

netti. Finite element analysis within component design process of hydraulic quadruped robot. In *ASME 2014 International Design Engineering Technical Conferences and Computers and Information in Engineering Conference*, pages V007T05A005–V007T05A005. American Society of Mechanical Engineers, 2014.

3. I. Fujii, T. Inoue, Dzung Viet Dao, S. Sugiyama, and S. Hirai. Tactile perception using micro force/moment sensor embedded in soft fingertip. In *Sensors, 2006. 5th IEEE Conference on*, pages 558–562, Oct 2006.
4. R. Van Ham, T. Sugar, B. Vanderborght, K. Hollander, and D. Lefeber. Review of actuators with passive adjustable compliance / controllable stiffness for robotic applications. In *Robotics & Automation Magazine, IEEE*, 2009.
5. G. Hirzinger, N. Sporer, A. Albu-Schaffer, M. Hahnle, R. Krenn, A. Pascucci, and M. Schedl. Dlr's torque-controlled light weight robot iii-are we reaching the technological limits now? In *Robotics and Automation, 2002. Proceedings. ICRA '02. IEEE International Conference on*, volume 2, pages 1710–1716 vol.2, 2002.
6. Zhibin Li, N.G. Tsagarakis, and D.G. Caldwell. A passivity based admittance control for stabilizing the compliant humanoid coman. In *Humanoid Robots (Humanoids), 2012 12th IEEE-RAS International Conference on*, pages 43–49, Nov 2012.
7. A. Parmiggiani, M. Randazzo, L. Natale, G. Metta, and G. Sandini. Joint torque sensing for the upper-body of the icub humanoid robot. In *Humanoid Robots, 2009. Humanoids 2009. 9th IEEE-RAS International Conference on*, pages 15–20, Dec 2009.
8. C. Semini, N G. Tsagarakis, E. Guglielmino, M. Focchi, F. Cannella, and D G. Caldwell. Design of HyQ - a hydraulically and electrically actuated quadruped robot. *IMechE Part I: J. of Systems and Control Engineering*, 225(6):831–849, 2011.
9. S. Shams, Dongik Shin, Jungsoo Han, Ji Yeong Lee, Kyoosik Shin, and Chang-Soo Han. Compact design of a torque sensor using optical technique and its fabrication for wearable and quadruped robots. *2011 IEEE/RSJ International Conference on Intelligent Robots and Systems*, Sep 2011.
10. Y. Sun, Y. Liu, and H. Liu. Temperature compensation for six-dimension force/torque sensor based on radial basis function neural network. *2014 IEEE International Conference on Information and Automation (ICIA)*, Jul 2014.
11. D. Tsetserukou, R. Tadakuma, H. Kajimoto, N. Kawakami, and S. Tachi. Development of a whole-sensitive teleoperated robot arm using torque sensing technique. pages 476–481, 2007. cited By 3.
12. D. Vischer and O. Khatib. Design and development of high-performance torque-controlled joints. *IEEE T. Robotics and Automation*, 11(4):537–544, 1995.
13. N. Wei, H. Sun, Q. Jia, X. Ji, and H. Shi. Analysis and design optimization of a compact and lightweight joint torque sensor for space manipulators. *Advances in Mechanical Engineering*, 2013, 2013.
14. S. Wolf and G. Hirzinger. A new variable stiffness design: Matching requirements of the next robot generation. In *ICRA*, pages 1741–1746. IEEE, 2008.

SCANNING MANIPULATOR WITH TERRAIN SURFACE MAPPING FOR DEMINING TASKS*

ROEMI FERNÁNDEZ[1], JAVIER GAVILANES[1], HÉCTOR MONTES[1,2], CARLOTA SALINAS[1], PABLO GONZÁLEZ DE SANTOS[1], MANUEL ARMADA[1]

[1]Centre for Automation and Robotics CAR CSIC-UPM
Agencia Estatal Consejo Superior de Investigaciones Científicas
Ctra. Campo Real Km. 0,200, La Poveda, Arganda del Rey, 28500
Madrid, Spain
[2]Faculty of Electrical Engineering, Technological University of Panama, Panama

Landmines, cluster munitions and explosive remnants of war still pose a threat to the physical, psychological, economic and social welfare of many peoples. With the aim of contributing to alleviate this terrible problem, this paper introduces a robotic manipulator capable of handling a metal detector while scans the surface of interest searching for mines during humanitarian demining tasks. A set of mini-cameras installed on the robotic arm acquires the data required for adapting the scanning to the irregularities of the terrain in such a way that the metal detector remains at a constant height from the ground during all the operating procedures. Results of preliminary tests show the potential of the proposed approach.

1. Introduction

Every day some 10 people around the world lose their life or limb to a landmine or another explosive remnant of war [1]. Approximately 80% of these victims are civilians [2]. About 110 million landmines are still lodged in the ground of 60 countries – waiting. The process of detection and deactivation of these devices represents a complex problem that includes several operational stages. From these activities, the close-in detection is one of the most dangerous and tedious tasks, since it is performed by human operators who inspect every centimetre of terrain equipped with prodders, metal detectors, and/or dual sensors consisting of a combination of ground penetrating radars and metal detectors. A small mistake during these activities may cost deminers their life.

Wheeled, tracked or legged robots endowed with proper array of sensors for mines detection constitute a good solution for automating this risky and time-consuming task. Mobile robots are capable of operating continuously during

---

* This work is supported by the European Community's Seventh Framework Programme (FP7/2007-2013 TIRAMISU) under Grant Agreement No. 284747.

long periods of time without losing repeatability of movements and avoiding, in case of accidents, the injury of operators or the loss of human lives. In order to assist human deminers in close-in detection tasks, several prototypes have been proposed in the last years such as: MR-2, a teleoperated EOD vehicle with a detector arm of 4 DoF and a LFR arm used to carry all the auxiliary sensors [3-4]; GRYPHON-IV, a teleoperated quad with a 3 DoF pantograph arm equipped with a metal detector, a GPR and an artificial vision system [5-6]; HUNTER, a caterpillar EOD robot provided with a Cartesian arm with 3 DoF for scanning tasks [7]; AMRU5, an hexapod walking robot for humanitarian demining [8]; TITAN-IX, a quadruped walking robot that is capable of reconfiguring its legs for manipulation tasks and its gait in case that one of its leg is broken [9-11]; SILO 6, an autonomous hexapod walking robot designed for humanitarian demining tasks that carries onboard a 5 DOF robotic arm with IR sensors [12]; RMA teODor, a tracked mobile robotic platform equipped with a multi-channel metal detector system [13]; and FSR Husky, a lightweight mobile robot that carries a 2 DOF mine-clearance arm with metal detector and artificial nose [14].

Although most of the aforementioned prototypes are equipped with some kind of robotic arm in order to handle the sensors for mine detection, just a few of these researches address the strategy for inspecting the field. This paper is focused in the implementation of an instrumented robotic arm for scanning the polluted terrain. The manipulator, which is an updated version of the SILO 6 robotic arm [15], has 5 DOF and it is endowed with a metal detector for landmines detection and a set of mini-cameras for mapping the terrain that has to be scanned. The paper is organized as follows. Section 2 describes the design of the robotic arm, including the sensors utilised for scanning the terrain. Some preliminary experimental tests are gathered in Section 3. Finally, Section 4 summarizes the major conclusions.

2. Instrumented Scanning Manipulator

The robotic manipulator has 5 DOF in elbow-up configuration, three of them for positioning the metal detector in the Cartesian space, and the other two, located at the wrist of the manipulator, for adapting the metal detector attitude in case of terrain inclinations. This robotic arm is installed on-board of a six-legged robot that acts as mobile platform (see Figure 1(a)). By means of a joint coordination of both systems it is possible to achieve an accurate positioning of the metal detector on the terrain. The proposed robotic arm configuration provides sufficiently mobility for positioning and orienting the metal detector with respect to the ground and reduces possible undesirable contacts with other parts of the robot or objects in its environment. The robotic arms structure is made of aluminum and all the rotational joints are driven by DC motors. Figure 1(b) shows the instrumented scanning manipulator.

One of the most important factors for detecting buried landmines with metal detectors in an effective way is to keep the sensor head as closer as possible to the ground. The closer the detector head is to the ground, the deeper the electrical field is projected and the greater is the possibility to detect the mines. Therefore, the robotic arm should move the sensor head parallel to and at a constant height from the ground at all times, following the terrain irregularities, and bearing in mind that generally, metal detectors' manufacturers recommend keeping the sensor head at a distance less than 5cms above the ground during scanning tasks [16].

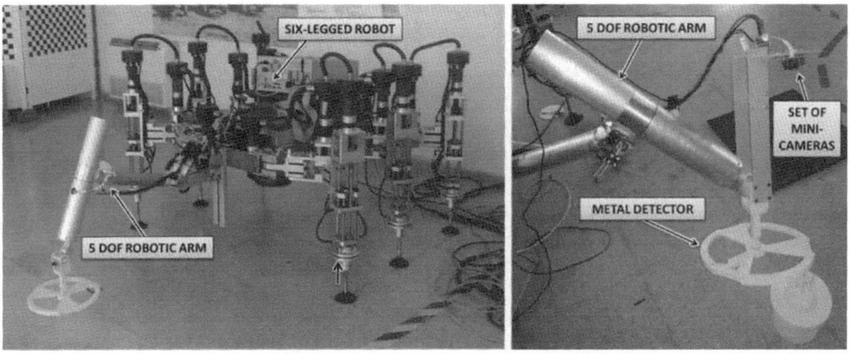

(a) (b)

Figure 1. (a) Six-legged robot carrying on-board the scanning manipulator. (b) Robotic arm endowed with metal detector and a set of mini-cameras.

In order to provide to the control system of the robotic arm the required information for planning its movements during the close-in detection activities, the sensory system approach proposed is based on a mini-TOF camera, the pmd[vision] CamBoard nano and a VRMagic multi-sensor camera. The mini-TOF camera is able to acquire a depth map and an amplitude image at the resolution of 120 x 160 pixels, as well as *x*, *y* and *z* coordinates to each pixel in the depth map. The VRMagic multisensor camera is equipped with two pixel-synchronous CMOS sensors. This multi-sensor camera with size 42mm x 38mm x 21mm is able to acquire RGB at the resolution of 754 x 480 pixels and at a maximum frame rate of 69Hz. The image data is coordinated on a dedicated FPGA module. Therefore, the mini-TOF camera provides a point cloud of the terrain that will allow keeping the metal detector at a constant height above the ground and performing an efficient scanning of the contaminated terrain, whereas the multi-sensor camera provides colour information that can be useful for detection and discrimination of objects on the terrain. Table 1 summarizes the main technical specifications of the mini-cameras. A close-up view of the set of the mini-cameras is shown in Figure 2.

Táble 1. Main technical specifications of the mini-cameras.

Mini-TOF camera		Multi-sensor camera	
Frame rate	≤ 90 fps (typical)	Sensor	Aptina MT9V024
Field of view	90° x 68°	Sensor size	1/3" wide
Dimensions	37 x 30 x 25 mm	Frame rate	60 Hz
Mass	35 g	Pixel size	6 μm x 6 μm
Power supply	16-30 VDC	Dimensions	42 x 38 x 21 mm
Illumination wavelength	850 nm	Interface	USB 2.0

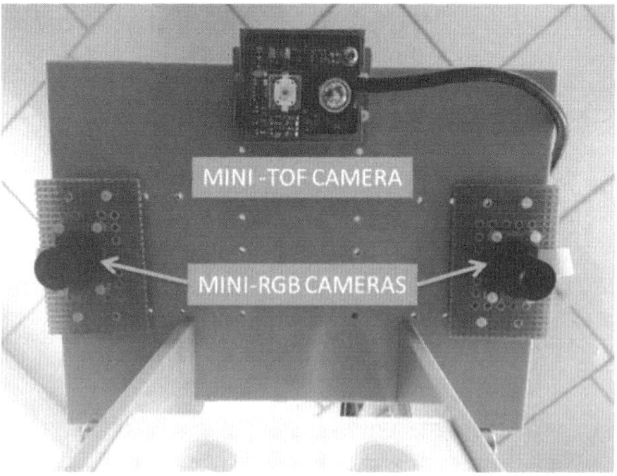

Figure 2. Close-up view of the mini-cameras installed on the robotic arm.

Although mini-TOF data is fundamental for keeping the metal detector at a constant height from the ground during all the demining tasks, it is still necessary to automatically match this information with RGB images acquired with the multi-sensor camera in a common reference frame. For accomplishing this procedure it should be taken into account that mini-TOF data and the RGB images come from sensors that exhibit different field of view and different pixel array size. Thus, data will only depict the same content partially, and the pixel correspondence will not be direct. To overcome this problem, the random sample consensus (RANSAC) algorithm is adopted for the multisensory registration [17].

3. Preliminary Experimental Tests

Several preliminary experimental tests have been carried out in order to validate the proposed set of mini-cameras for mapping the terrain surface. This information will be crucial not only for adapting the scanning to the irregularities of the terrain, but also for detecting and discriminating features of interest that are present on the field during the demining tasks.

Figure 3 shows the RGB images acquired with the multi-sensor camera, while Figure 4 displays the amplitude image and the z-axis range data acquired with the mini-TOF camera. Figure 5 presents the resulting RGB image after mosaicking the two images acquired by the multi-sensor camera. This resulting image is then utilised together with the amplitude image from the mini-TOF camera for conducting the registration procedure. Figure 6 illustrates the result obtained from this registration procedure. In this way, direct correspondence between the RGB information and the range data is achieved. Finally, Figure 7 shows resulting surface terrain mapping.

(a) (b)

Figure 3. RGB images acquired with the multi-sensor camera. (a) Left-hand side image. (b) Right-hand side image.

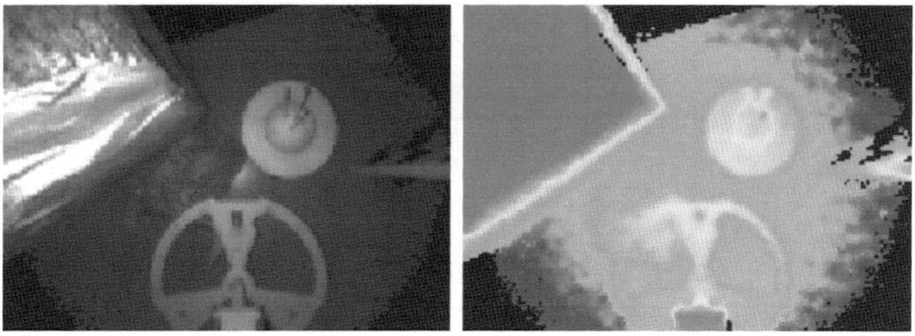

(a) (b)

Figure 4. (a) Mini-TOF amplitude data. (b) Z-axis range data acquired with the mini-TOF camera.

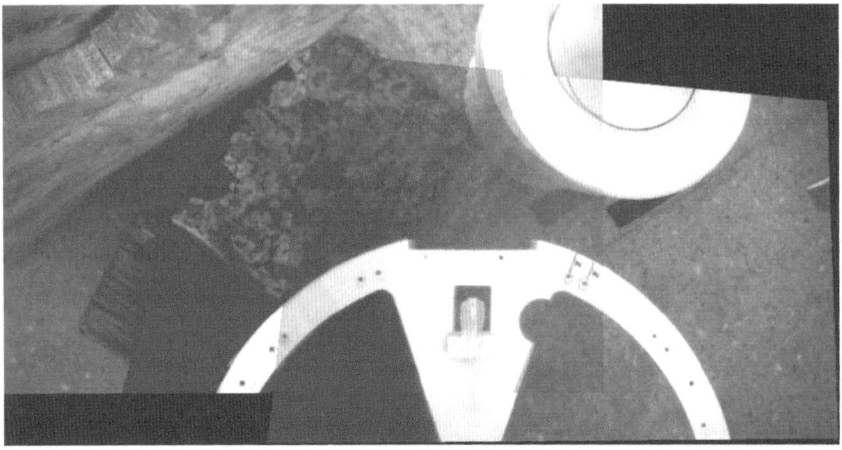

Figure 5. Resulting RGB image after the mosaicking.

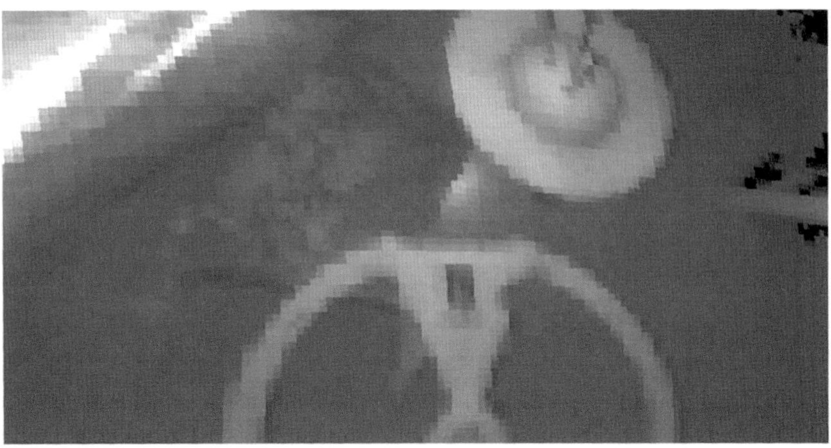

Figure 6. Mini-TOF amplitude image resulting after the registration procedure.

4. Conclusions

This paper presents an instrumented robotic arm for close-in detection tasks carried out during humanitarian demining activities. The manipulator has 5 DOF and it is equipped with a metal detector and a set of mini-cameras for acquiring RGB and 3D information of the polluted terrain. With this information the robotic arm will be controlled to keep the metal detector as closer as possible to the ground, improving the rate of landmines detection.

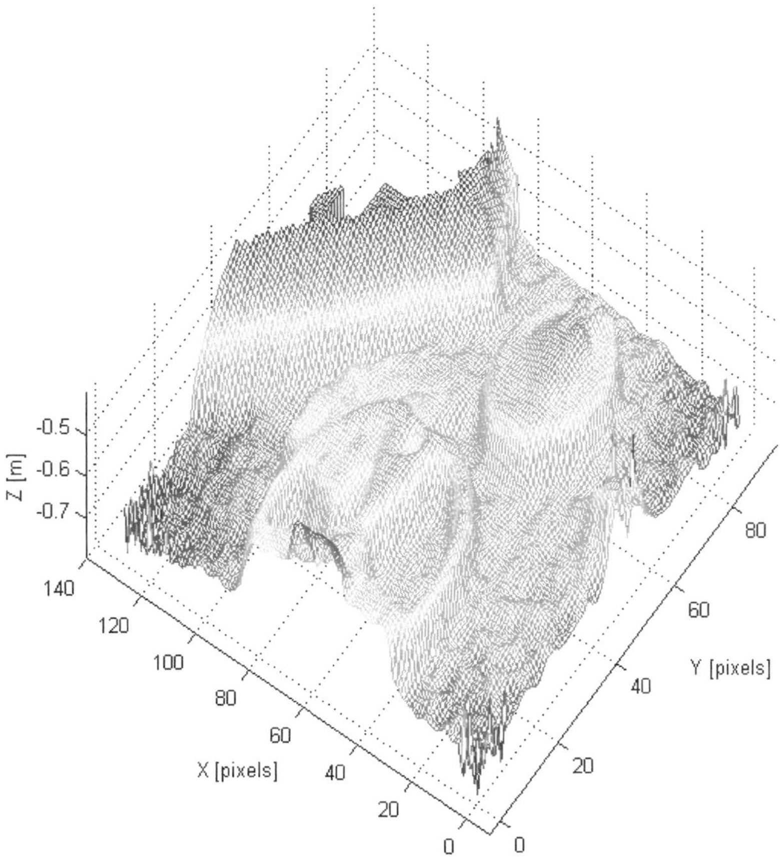

Figure 7. Terrain surface mapping.

Acknowledgments

The authors acknowledge funding from the European Community's Seventh Framework Programme (FP7/2007-2013 TIRAMISU) under Grant Agreement No. 284747 and partial funding from the RoboCity2030-III-CM project (Robótica aplicada a la mejora de la calidad de vida de los ciudadanos. Fase III; S2013/MIT-2748), funded by Programasa de Actividades I+D en la Comunidad de Madrid and cofunded by Structural Funds of the EU. Dr. Héctor Montes also acknowledges support from Universidad Tecnológica de Panamá.

References

1. Landmine & Cluster Munition Monitor. Why landmines are still a problem. International campaign to Ban Landmines. http://www.icbl.org/en-

gb/problem/why-landmines-are-still-a-problem.aspx (2015).
2. Landmine & Cluster Munition Monitor. Landmine Monitor Report (2008).
3. Y. Das, K. Russell, N. Kircanski, A.A. Goldenberg, "An articulated Robotic scanner for mine detection – a novel approach to vehicle mounted systems," SPIE, 1999 Conference (Aerosense), Orlando, Florida (1999).
4. ESI, "Mr-2 terrain mapping and in-ground mine detection," http://www.esit.com/index.php/product/mobile-robots/190-mr-2, 2008.
5. E.F. Fukushima, M. Freese, T. Matsuzawa, T. Aibara and S. Hirose, *International Journal on Smart Sensing and Intelligent Systems* **1**, 735 (2008).
6. M. Freese, S.P.N. Singh, W. Singhose, E.F. Fukushima, S. Hirose, *Field and Service Robotics, Springer Tracts in Advanced Robotics* **62**, 3 (2010).
7. E. Colon, G. De Cubber, H. Ping, J-C Habumuremyi, H. Sahli and Y. Baudoin, *International Journal of Advanced Robotic Systems* (2007).
8. I. Doroftei and Y. Baudoin, *International Journal Industrial Robot* **39**, 441 (2012).
9. S. Hirose and O. Kunieda, *The International Journal of Robotics Research* **10**, 3 (1991).
10. S. Hirose and K. Kato, "Quadruped walking robot to perform mine detection and removal task", Proceedings of the 1st International Symposium CLAWAR'98, Brussels, Belgium, (1998).
11. S. Hirose, K. Takita, K. Kato, A. Torri, M. Ogata and S. Sugamuna, "Quadruped walking robot centered demining system – development of TITAN-IX and its operation," Proceedings of the 2005 IEEE International Conference on Robotics and Automation ICRA'2005, Barcelona Spain (2005).
12. P. González de Santos, J. Cobano, E. García, J. Estremera and M. Armada, *Mechatronics* **17**, 417 (2007).
13. H. Balta, H. Wolfmayr, J. Braunstein, and Y. Baudoin, "Integrated mobile robot system for landmine detection", Proceedings of the 12th IARP Workshop HUDEM 2014, Croatia (2014).
14. J. Prado and L. Marques, "Energy efficient area coverage for an autonomous demining robot," ROBOT 2013: First Iberian Robotics Conference, Springer International Publishing **253**, 459 (2014).
15. R. Ponticelli, E. Garcia, P. Gonzalez de Santos and M. Armada, *Industrial Robot: An International Journal* **35**, 133 (2008).
16. R. Fernández, H. Montes, C. Salinas, P. Gonzalez de Santos and M. Armada, *Industrial Robot: An International Journal* **39**, 450 (2012).
17. Fischler, M.A.; Bolles, R.C. *Commun. ACM* **24**, 381 (1981).

MAPPING REPETITIVE STRUCTURAL TUNNEL ENVIRONMENTS FOR A BIOLOGICALLY-INSPIRED CLIMBING ROBOT

GAVIN PAUL[1], SHUYUAN MAO[2], LIYANG LIU[3] and RONG XIONG[4]

Authors[1,3] are with the CAS, University of Technology Sydney and Authors[2,4] are with CSC at Zhejiang University{[1]gavin.paul, [3]liyang.liu} at uts.edu.au, {[1]maoshuyuan, [4]rxiong} at iipc.zju.edu.cn

This paper presents an approach to using noisy and incomplete depth-camera datasets to detect reliable surface features for use in map construction for a caterpillar-inspired climbing robot. The approach uses a combination of plane extraction, clustering and template matching techniques to infer from the restricted dataset a usable map. This approach has been tested in both laboratory and real-world steel bridge tunnel datasets generated by a climbing robot, with the results showing that the generated maps are accurate enough for use in localisation and step trajectory planning.

Keywords: 3D Mapping; Depth cameras; Tunnel Environments; Biologically Inspired; Climbing Robot

1. Introduction

Recent advances in robotics and sensor technology increasingly enable evermore complicated tasks to be automated. When a climbing robot[1] must perform tasks in a three dimensional(3D) environment, for example inspecting the health of structures[1–3] a geometric map is crucial for online localisation, motion planning and accurately determining foot placements that avoid certain areas and collisions. Building an accurate map in repetitive structured environment such as a tunnel is challenging due to noisy and incomplete sensor data and the lack of visual features.

Many mobile robot systems that perform 3D geometric mapping and SLAM[4] have been developed using laser range finders and/or depth cameras, combined with scan-matching algorithms to construct 3D volumetric maps. To overcome occlusions, a 3D map can be generated from multiple locations using manipulator-mounted sensor exploration approaches.[5–7] In these cases, data is registered and overlapping parts of several views are fused together. However, a robot manipulator with the flexibility to traverse complicated surfaces, and carry its own adhesion mechanism, has been found to be subject to sag in its links,[1] making sensor data taken from multiple poses more complicated to fuse than for a rigid industrial manipulator. Noisy and incomplete sensor data, combined with inaccuracies in the manipulator model, or flexible joints/links, often mean that even after exploring an environment the geometry cannot be determined with sufficiently high accuracy.

Prior knowledge about structural elements in a workspace has previously been used for an eye-in-hand sensor configuration mounted on a static base to identify and localise these elements.[8] However, our prior work made several assumptions that simplified the problem: a) the exact shape of the templates was known *a priori* so fused point to template plane ICP could be performed, and b) a static base meant template locations could be confined in a limited bounding box. For a mobile climbing robot in a tunnel, where scans can only happen at certain locations, planes must be detected from the noisy data necessarily gathered by looking at surfaces at near-parallel angles (i.e. into the tunnel), and templates need to be created online since tunnel sizes vary at all locations.

This paper presents an approach that combines plane extraction, plane clustering, and template matching techniques so as to infer a smooth tunnel map given incomplete noisy depth datasets. There are two main contributions of this paper. First, we exploit organised point clouds and local surface normal features to implement a region growing algorithm for plane extraction which enhances the speed significantly. Then we combine prior knowledge encoded in a set of variable shaped templates to achieve geometrically accurate mapping of the environment which can be used to localise the robot in the environment. The reliable maps generated are currently being used to map and localise a caterpillar-inspired climbing robot in real-world repetitive structural tunnel environments. The remainder of this paper is organized as follows. Section 2 describes the proposed approach to eye-in-hand map generation, region growing, then template matching. Section 3 presents experimental results using data collected for both simulated and real-world tunnels and discusses the limitations and possible drawbacks to the current approach. Finally, Section 4 provides conclusions.

2. Methodology

Given a depth sensor mounted on a robot's end effector in an "eye-in-hand" configuration, and the nDOF manipulator pose, $\mathbf{q} = [q_1, \dots q_n]^T$, it is possible to compute the position and orientation of the sensor, ${}^0T_s(\mathbf{q})$ using forward kinematics. When the sensor is a Kinect-like depth camera it gives images comprised of a matrix of $M \times N$ depths values, $D = d_{m,n} \forall m \in M, n \in N$. A 3D point is computed using ${}^0T_s(\mathbf{q})$ for each image pixel with an index in m, and n, such that in the cameras coordinate frame a $\mathbb{R}^3$ point, $\mathbf{p_{m,n}}$ corresponds to the image's mth collum and nth row.

2.1. *Plane Extraction*

In a repetitive structural tunnel, we know *a priori* that the environment is bounded with four sets of main coplanar plane patches as well as many smaller plates with rivets connecting these. Extracting planes largely eliminates sensor noise, and since the templates required are a set of bounded planes, plane extraction is fundamental to template matching. The plane extraction problem is, given N 3D points, $\{p_i\}_{i=1,\ldots N}$, extract a set of planes, $\Pi_i = \{\mathbf{p_i}, \mathbf{n_i}\}$ for $i = 1, \ldots M$. Each plane detected should consist of a set of data including a point that is within that i'th plane's region, $\mathbf{p_i}$ and a normal, $\mathbf{n_i}$ of i'th plane. The plane model for Π_i is

$$\mathbf{n_i} \cdot (\mathbf{r} - \mathbf{p_i}) = 0. \tag{1}$$

Region growing based methods[9–11] are the most popular approach for plane segmentation since they extract bounded plane patches from point clouds. Our plane growing algorithm is based on[10] and consists of several steps: first, pick three or four adjacent points as a seed plane model and test their neighbor points, if a neighbor point combines then the plane model is still a plane, so grow the plane group and update the plane model. Iterate the test/add step until no point can be added, then a new plane is grown.

The popular coplanar criterion is mean square error (MSE) of plane fitting and can feed into the plane model update. This operation is time consuming since a plane fit is required for each test. We present a novel coplanar judgement and model update which utilises each point's local surface normal. The coplanar criterion used are, the normal angle between local surfaces at point and plane, as well as the point-to-plane distance.

$$\begin{aligned} &\|n^* \cdot (p_{new} - r^*)\| \leq \epsilon \\ &arccos(n_{new}, n^*) \leq \lambda \end{aligned} \tag{2}$$

where ϵ and λ are maximum distance and angle thresholds. The plane model is updated by averaging all points and their normals, thus significantly simplifying region growing computations.

Calculating surface normals at each point is generally time consuming. However, in organised point clouds, the adjacent points in the image are the adjacent points in space, and the order is maintained from the original depth image. We use a fast normal estimate algorithm[12] to compute surface normals using the cross product between two tangential vectors to the local surface and using integral images to rapidly compute tangential vectors.

A plane model update consists of updating $(\mathbf{p}^*, \mathbf{n}^*)$. It is straightforward to prove that the plane centroid is on the optimal plane under MSE. Additionally, optimal plane normals, n^* are usually computed with principal component analysis; however, an approximate plane normal can be computed by averaging all points' normal as

$$(\mathbf{p}^*, \mathbf{n}^*) = \left(\frac{1}{N} \sum_{i=1}^{N} p_i, \frac{1}{N} \sum_{i=1}^{N} n_i \right) \tag{3}$$

then adopting a iterative model update,

$$\begin{aligned} \mathbf{p}^*_{\mathbf{N+1}} &= \tfrac{1}{N+1}(Np^*_N + p_{N+1}) \\ \mathbf{n}^*_{\mathbf{N+1}} &= \tfrac{1}{N+1}(Nn^*_N + n_{p_{N+1}}) \end{aligned} \tag{4}$$

so only N,r^*_N,n^*_N need to be stored for incremental plane model updates.

2.2. *Robust Tunnel Detection*

After plane extraction a map of the tunnel (Fig. 1) needs to be generated. In order to ensure that maps are constrained within expected bounds and to enable efficiency gains during later trajectory and step planning procedures, tunnel detection uses a combination of prior knowledge and template matching. Tunnel detection requires that at least four planes are detected. In the case of looking down a tunnel almost parallel to the surfaces (i.e. the image plane is not perpendicular to any surfaces), and due to the nature of a depth camera that projects light and requires a reflection from the surface, there are often several spurious points detected and the points on the surfaces are noisy and patchy. However, it turns out that the largest detected planes are the walls, roof and floor planes. Generally the tunnel can be assumed to follow a rectangular frame as shown in Fig. 1, even though the interior walls consist of jagged zone planes of varying thickness, and there are several challenging zones. Tunnel detection must overcome these issues in order to robustly generate a tunnel map that incorporates prior knowledge and is idealised for step and path planning.

In order to determine the tunnel width and height, the angle between the normals of two candidate planes, Π_i and Π_j can be determined, $\mathbf{n_i} \cdot \mathbf{n_j}$.

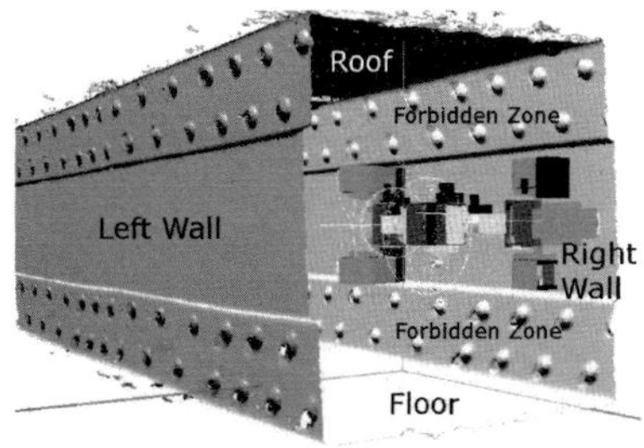

Fig. 1. Simulated robot attached to the right wall by its foot (right footpad) while the sensor-mounted hand (left footpad) looks into the tunnel. Note the simulated tunnel dataset shown here has been manually reconstructed from raw field scan data. Zones adjacent to tunnel edges contain rivets, and are therefore referred to as forbidden zones since the robot must avoid stepping on them. The middle zone plates are the robot's moving base. Multi-thickness zones can cause inconsistent tunnel dimensions to be measured. Rivets on forbidden zones and surface rust can also skew plane normals detected.

Then, if the normals are approximately in opposite directions, the distance between the two planes at the points detected as the center of the planes can be found by the average of the two point to plane distances as,

$$D \approx \frac{1}{2}\left(\frac{\mathbf{n_i}\cdot\mathbf{p_j} - \mathbf{n_i}\cdot\mathbf{p_i}}{|\mathbf{n_i}|} + \frac{\mathbf{n_j}\cdot\mathbf{p_i} - \mathbf{n_j}\cdot\mathbf{p_j}}{|\mathbf{n_j}|}\right), \tag{5}$$

and simplified to, $\frac{1}{2}(\mathbf{n_i}+\mathbf{n_j})\cdot(\mathbf{p_i}-\mathbf{p_j})$, since the normals are of unit length.

In order to robustly detect a rectangular tunnel shape in the presence of uneven wall surfaces, classification and clustering methods are utilised. Tunnels have geometric structure properties that are known *a priori* and exploitable, such as that multiple zones associated with each wall are a plane shifted by small distances, and all zones are symmetric about the tunnel axis that extends into the tunnel. The zone plane normals and positions can thus be dealt with independently since they are uncorrelated. To prepare candidate zone planes, stringent plane parameters are chosen so as to detect planes that are small and accurate. This results in a large set of small planes from the same zone. Initially, each small plane is classified and assigned to one of the four wall classes: left, right, top and bottom, based on the plane's normal orientation. The distribution of the $\mathbb{R}^3$ axial components of the normals in each class can be assumed to follow a Gaussian distribution. The sample mean becomes the wall normal, and is made more accurate after outlier rejection based on histogram analysis. The zone's position on a wall is then determined by forcing each zone plane normal to be equal to the sample mean, and performing a statistical analysis on newly formed planes' positions. Clustering of the positions is used to categorise zones and the cluster's mean becomes the zone's position. For each pair of parallel walls in a tunnel (top-bottom, left-right), the clusters that are the furthest apart are assigned to be the distance between that set of parallel walls.

If there are too few planes detected that are sufficiently large, then the robot moves to a different pose, and takes another scan until the tunnel is detected. Once four walls are detected and dimensions are found to be within the template threshold for tunnel width and height (given the context from previous scans), then a map is generated and combined with prior knowledge about forbidden zone locations.

3. Results

Two experiments where conducted in steel bridge tunnel environments using a 7DOF climbing robot with two cameras mounted to the end effector: a Structure Sensor depth camera, and a Logitech C930e RGB camera. Fig. 2a shows the robot attached to the roof while scanning into the lab tunnel. Experiment 1 was conducted in the laboratory, with the robot attached to the roof or walls, depth data was collected (Fig. 2b) and fused with RGB and triangulated to generate a mesh (Fig. 2c), then analysed to detect the tunnel. Experiment 2 was conducted in the field in a steel bridge tunnel and depth images were collected at 3 different distances (i.e. 1.8m, 3m and 5m) from the manhole as the robot walked along the tunnel. Fig. 2d and Fig. 2e show field data: a depth image and the fused mesh and tunnel.

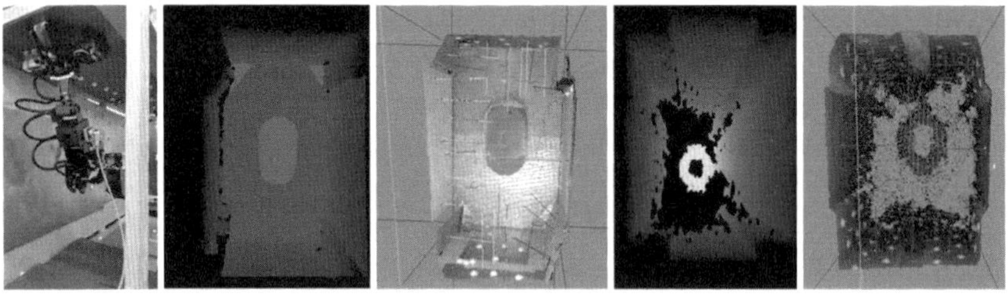

(a) (b) (c) (d) (e)

Fig. 2. *a*) The robot scanning the tunnel from the roof; *b*) Lab: depth image of tunnel (aspect ratio altered); *c*)Lab: fused RGB-D and detected tunnel boundary (blue) and traversable planes (red); *d*) Field: depth image (aspect ratio altered); *e*) Field: fused RGB-D and resulting tunnel boundary (blue) and main traversable planes (red)

Planes are extracted from VGA (i.e. 640 × 480) frames and all plane areas and contours are computed in under 100ms, which is real-time enough for the intended application. In experiments, the values for parameters in our Plane Growing and Clustering (PG&C) algorithm are as follows: the region growing's point normal to plane normal angle is 4°, and point to plane distance is 15mm; the point to candidate point distance is 40mm; the min. points that constitute plane is 40; the min. allowable plane area is $0.004m^2$; the max. usable range from the sensing plane is $5m$; the candidate search of n^2-1 neighboring points has $n=5$; the max. allowable differences are 40mm for parallel wall normals, and 10mm for similar walls.

The results are presented in Table 1. Our PG&C approach is compared to two alternative algorithms: a RAndom SAmple Consensus & Least Squared Error (R&L) fitting approach[13](p.59) that iteratively perform RANSAC and returns the result with the LSE; and Qhull algorithm[14] (i.e. MATLAB's convhulln function to determine the bounding minimum convex set containing the points, then an axial alignment). Processed was done on an Intel i7-2620M with 8GB RAM. Although speed is not directly comparable since PG&C was written in C++ while the alternative algorithms are in MATLAB, both PG&C and the Qhull approach can detect the tunnel from a 640x480 depth image and create a tunnel map at a rate of approximately 1FPS, whereas the R&L, which is set to iterate 100 times and output the LSE result, runs at 0.07FPS.

Table 1. Experimental Results.

Approach	PG & C		Qhull		RANSAC & LSE	
Exp. 1: Lab	$\bar{w}(\sigma)$	$\bar{h}(\sigma)$	$\bar{w}(\sigma)$	$\bar{h}(\sigma)$	$\bar{w}(\sigma)$	$\bar{h}(\sigma)$
Pose 1	824(2.9)	1310(1.8)	849(94.1)	1260(7.8)	752(0.4)	1199(0.7)
Pose 2	827(3.8)	1307(0.8)	803(6.1)	1260(16.1)	751(0.9)	1198(0.8)
Pose 3	824(0.6)	1312(4.2)	827(32.3)	1279(44.8)	751(0.5)	1198(1)
Exp.2: Field	$\bar{w}(\sigma)$	$\bar{h}(\sigma)$	$\bar{w}(\sigma)$	$\bar{h}(\sigma)$	$\bar{w}(\sigma)$	$\bar{h}(\sigma)$
5m to plate	841(1.1)	1234(2.4)	825(10.4)	1174(2.4)	761(2.6)	1099(3.3)
3m to plate	841(3.7)	1225(15.9)	824(14)	1169(11.1)	756(4.6)	1093(9.2)
1.8m to plate	844(1.7)	1213(15.8)	814(5.1)	1155(7.7)	762(1.6)	1083(59.7)

The dimensions of the lab test rig in Experiment 1 are width=0.82m x height=1.3m. The robot was moved to 3 different poses with 10 depth frames captured at each. Our PG&C approach's average width and height measurements (i.e. $\bar{w}$ and $\bar{h}$) were within 12mm of the correct dimensions, and only had a maximum standard deviation, σ of 4mm. The Qhull method detected $\bar{w}$ and $\bar{h}$ within 49mm of the correct result, however values varied significantly with σ up to 45mm. The R&L method was the slowest but most invariant due to the 100 iterations and internal LSE calculation. However,

R&L constantly underestimated the dimensions ($\bar{w}$ by 70mm and $\bar{h}$ by 100mm) since parts of another inner surface were often taken to be part of an incorrect inner surface. In Experiment 2 data was collected in the field at 3 locations as the robot walked in a steel bridge with dimensions, [0.84m x 1.2m]. The locations are indicated by their distance to the manhole shown in Fig. 2e. A larger distance from the manhole plate means more tunnel depth data is available for detecting the tunnel, which improves detection accuracy. Conversely, close proximity to a manhole results in less data being available to detect the tunnel, and thus poorer results. In the field, PG&C detected the width correctly and was within 34mm of the correct height, although the variance was larger than the lab (16mm) due to the rough, rusted surface condition. The Qhull method once again slightly underestimated the dimensions with the largest σ=15mm. R&L once again underestimated the dimensions but with a low variance, except for at 1.8m where insufficient data resulted in several incorrect detections.

The presented approach has been shown to work well in both lab, and in the real field environment, where the sensor data was often noisy and sparse for metal surfaces whose condition and reflectivity varied significantly. The approach has been shown to have a similarly low variability to a RANSAC method run over 100 LSE iterations, similar speed as a QHull method, and the suitably high accuracy in both lab and field environments.

4. Conclusions

This paper presents an approach that uses a combination of plane extraction, and template matching techniques to infer from the restricted dataset a usable map. Surface normal features are used which enhance the plane growing speed significantly. A set of variable shaped templates are built using prior knowledge, plane clustering and classification. A sufficiently accurate tunnel map is thus generated from noisy and spurious sensor data, and used to localise a climbing robot in a real-world steel bridge environment. Future work will combine the approach with an exploration approach for automatically detecting the tunnel, and extend this method to become a localisation solution for a climbing robot which is subject to sag.

Acknowledgments

This work is supported by the National Nature Science Foundation of China under Grant 61473258, by the Joint Center for Robotics Research between Zhejiang University, Hangzhou, China and the University of Technology,

Sydney, NSW, Australia, by the Australian Research Council Linkage Grant (LP100200750), and by the NSW Roads and Maritime Services.

References

1. P. Ward, G. Paul, P. Quin, D. Pagano, C. Yang, D. Liu, K. Waldron, G. Dissanayake, P. Brooks, P. Mann, W. Kaluarachchi, M. P. and L. Matkovic, Climbing robot for steel bridge inspection: Design challenges, in *9th Austroads Bridge Conference*, (Sydney, 2014).
2. F. Rochat, P. Schoeneich, B. Luthi, F. Mondada, H. Bleuler and R. Moser, Cy-mag3d: A simple and miniature climbing robot with advance mobility in ferromagnetic environment., in *Emerg Trends Mob Robot - Proc. 13th Int. Conf. Climbing Walk Robot Support Technol Mobile Machines*, 2010.
3. M. Eich and T. Vogele, Design and control of a lightweight magnetic climbing robot for vessel inspection, in *19th Mediterranean Conference on. MED. Control Automation*, (Corfu, 2011).
4. D. Ferguson, A. Morris, D. Haehnel, C. Baker, Z. Omohundro, C. Reverte, S. Thayer, W. Whittaker, W. Burgard and S. Thrun, An autonomous robotic system for mapping abandoned mines, in *Proc. Conf. on Advances in Neural Information Processing Systems*, (Vancouver, 2003).
5. P. Quin, G. Paul, A. Alempijevic, D. Liu and G. Dissanayake, Efficient neighbourhood-based information gain approach for exploration of complex 3d environments, in *Int. Conf. on Robotics and Automation (ICRA)*, 2013.
6. P. Quin, G. Paul, A. Alempijevic and D. Liu, Nearest neighbour exploration with backtracking for robotic exploration of complex 3d environments, in *Proc. Australasian Conference on Robotics and Automation*, 2013.
7. G. Paul, S. Webb, D. K. Liu and G. Dissanayake, *Robotics and Autonomous Systems* **59**, 543 (2011).
8. S. Sehestedt, G. Paul, D. Rushton-Smith and D. Liu, Prior-knowledge assisted fast 3d map building of structured environments for steel bridge maintenance, in *Int. Conf. Automation Science and Engineering (CASE)*, 2013.
9. D. Hähnel, W. Burgard and S. Thrun, *Robotics and Autonomous Systems* **44**, 15 (2003).
10. J. Xiao, J. Zhang, B. Adler, H. Zhang and J. Zhang, *Robotics and Autonomous Systems* **61**, 1641 (2013).
11. C. Feng, Y. Taguchi and V. Kamat, Fast plane extraction in organized point clouds using agglomerative hierarchical clustering, in *IEEE Int. Conf. on Robotics and Automation (ICRA)*, 2014.
12. S. Holzer, R. B. Rusu, M. Dixon, S. Gedikli and N. Navab, Adaptive neighborhood selection for real-time surface normal estimation from organized point cloud data using integral images, in *IEEE/RSJ Int. Conf. on Intelligent Robots and Systems (IROS)*, 2012.
13. M. YÖRÜK, Performance comparison of point and plane features for slam, PhD thesis, Middle East Technical University,2014.
14. C. B. Barber, D. P. Dobkin and H. Huhdanpaa, *ACM Transactions on Mathematical Software (TOMS)* **22**, 469 (1996).

SECTION–7

LOCOMOTION

APPLICATION OF LOCAL SLOPES IN THE STUDY OF METASTABLE WALKING

A. T. SAFA and M. NARAGHI*

School of Mechanical Engineering, Amirkabir University of Technology, Tehran, Iran
*E-mail: atehranisafa@gmail.com, *E-mail: naraghi@aut.ac.ir*

A. ALASTY

School of Mechanical Engineering, Sharif University of Technology, Tehran, Iran
E-mail: aalasti@sharif.edu

Metastable walking is a relatively new concept in the field of walking robots. According to the theory, a walking machine is inherently subjected to stochastic disturbances due to the uncertainties about the terrain. As a result, it cannot demonstrate a limit cycle walking; but, it is still able to exhibit remarkably long-living periods of continuous walking.[1] In this paper, metastable walking over irregular terrains has been revisited. By introducing the concept of 'local slopes' and highlighting their role in limit cycle walking stability, we have shown how this concept contributes to the study of metastable walking over an uneven terrain. As an instance, a simulation study is performed to demonstrate that metastable walking of a simple passive dynamic walker is achievable on irregular surfaces with the average slopes beyond the maximum slope of a simple ramp which provides the biped with a stable periodic walking. This would help to reduce the energy inputs applied to control bipedal robots on uneven terrains.

Keywords: Metastable walking; Limit cycle walking; Local slopes.

1. Introduction

Stable walking is still one of the major challenges that bipedal robots face. This is in serious contrast with their ultimate utility which is frequently defined as walking through rough, uneven terrains. In this regard, in the past decade, one of the main research streams in the field of legged robots has been the stabilization of walking robots on unsmooth environments.[2–5]

It is a bitter truth that a walking machine is naturally subjected to disturbances due to randomized variations in the robot's terrain. These accidental variations continuously compel the walking robot to stand a disturbance at each step; since the moment of the swing leg impact changes frequently. This makes the robot unable to demonstrate a perfect limit cycle walking in the real world. As a consecutive result, the classical meaning of limit cycle walking stability is vanished, owing to the nonexistence of any equilibrium points. As a replacement, the concept of 'metastable walking' is proposed to analyze the stability of bipedal walking on uneven terrains.[1] 'Metastable walking' belongs to that style of motion in which falling is the ultimate fate but this happens after a long period of continuous walking.

To study 'metastable walking', researchers usually use a random distribution for the robot's terrain. This distribution is performed around a nominal ground, i.e. an even terrain on which the biped represents a stable limit cycle walking. But, recently, it has been shown and proven that a stable limit cycle walking could be performed on various types of repetitive terrains.[6,7] These terrains consist of a series of parallel 'local slopes' with equal vertical distances and also describe a general form of a ramp-stair surface.

The present paper attempts to bridge between the theory of 'metastable walking' and its predecessor, 'stable limit cycle walking', by using the concept of 'local slopes'. In this regard, we suggest the following procedure in the study of metastable walking on uneven terrains:

(1) Finding the walking system's fixed point.
(2) Finding the interval of the local slopes that stabilize the already found fixed point.
(3) Finding metastable walking on those uneven terrains that consist of the already found local slopes.

The rest of the paper is devoted to give the details of the above procedure. In this regard, the simplest passive dynamic walker is employed.[8]

2. Stable Walking on Local Slopes

Recently, it has been shown and proven that a wide range of the unstable long-period limit cycles of the simplest passive walker, introduced by Garcia et.al.,[8] would be stabilized by choosing a proper ramp-stair surface instead of an ordinary ramp.[6,7] This helps the biped to experince higher stable velocities.

A common type of a ramp-stair surface consists of a series of parallel local slopes with equal vertical distances. Figure 1 describes a schematic of the walking surface and the walker. In this two-dimensional walking model, the angle between each local slope and the horizontal line is shown by ψ

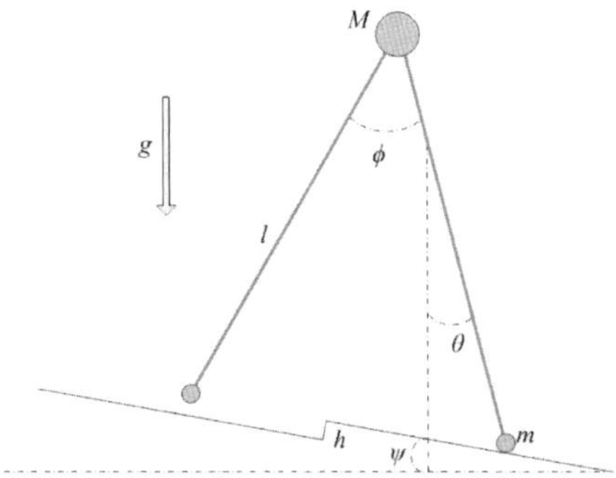

Fig. 1. The walking model which consists of the robot and the walking surface.[7]

and the vertical distance between the two local slopes is labeled h. For $h = 0$, we have an ordinary ramp and $\psi = 0$ indicates a series of regular stairs. The walker consists of mass-less legs, infinitesimal point-masses as the robot's feet and a huge point-mass located at the hip. Here, the merely source of energy is gravity; so the robot is passive. Also, all assumptions about the passive walkers are respected in this study.[8]

The mathematics of the motion is described by a set of impulsive differential equations. The differential part, describing the swing and the stance legs dynamics, is as follows:

$$\ddot{\theta} = \sin\theta \tag{1}$$

$$\ddot{\phi} = \sin\theta + (\dot{\theta}^2 - \cos\theta)\sin\phi \tag{2}$$

Note that these equations are written in a dimensionless form by means of scaling time as $(t' = t\sqrt{g/l})$. From a given initial state, numerical integration of the equations of motion continues until the heel-strike is detected. The following algebraic equation determines the end of integration:

$$\cos(\theta + \psi) - \cos(\phi - \theta - \psi) - \eta = 0 \tag{3}$$

Where $\eta = h/l$. Note that the positive status of $\{\psi, \eta\}$ is in accordance with Fig. 1. After the heel-strike, the state variables of the system jump through the phase space according to the following switching rule:

$$\dot{\theta}^+ = (\cos\phi^-)\dot{\theta}^- \tag{4}$$

$$\dot{\phi}^+ = (1 - \cos\phi^-)\dot{\theta}^+ \tag{5}$$

$$\theta^+ = \theta^- - \phi^- \tag{6}$$

$$\phi^+ = -\phi^- \tag{7}$$

The solving procedure could be summarized by using Poincaré map, which maps the states from just after one heel strike to the next one. This map will be two dimensional because of the dependence of ϕ and $\dot{\phi}$ on θ and $\dot{\theta}$ at Poincaré section. We have:

$$\left\{\theta_{i+1}^{+}, \dot{\theta}_{i+1}^{+}\right\} = P(\theta_i^{+}, \dot{\theta}_i^{+}) \tag{8}$$

The fixed points of Equation (8) correspond to the robot's hybrid limit cycles. The preferred solution is usually a long period-one limit cycle which has a strong chance for stability.[6,8] To find this desired solution, firstly we uncover a fixed point by putting $\eta = 0$ (i.e. considering the walking surface as a simple ramp). Afterwards, the stability of the found fixed point would be guaranteed if a proper pair of $\{\psi, \eta\}$ is chosen.[6,7] In this regard, the following relation has to be regarded:[7]

$$\eta = 2\sin\frac{\phi^*}{2}\sin(\psi - \gamma) \tag{9}$$

Where ϕ^* indicates the absolute value of ϕ at the system's fixed point and γ is the slope angle of the considered ramp. For the understudied walker, for each γ, it has been shown that we are able to find an interval of local slopes that guarantees stability.[6] These local slopes are called 'Ground Attractors' (GAs).[6] GAs exist upto $\gamma = 0.26\,rad.$, beyond which the flight phase begins.[6]

3. Metastable Walking on Local Slopes

During walking on an uneven terrain, it can be supposed that the walker meets a new local slope at each foot-strike. Here, we show that if the met local slopes are GAs then metastable walking is feasible. To prove the claim, we consider rough terrains with the average slopes of $\gamma_{mean} = 0.1$ and $\gamma_{mean} = 0.2\,rad.$ The interval of GAs for these uneven terrains are $0.386 \leq \psi \leq 0.433\,rad.$ and $0.813 \leq \psi \leq 0.843\,rad.$ respectively. Note that equation (9) can be used to define the values of η for these GAs.

We study two kinds of uneven terrain: Rectangular and triangular. In the rectangular form, the value of ψ is fixed at each foot-strike and η arbitrarily varies. On the other hand, in the triangular form, ψ randomly changes. In our simulation study, the random parameter is chosen from a truncated normal (Gaussian) distribution; so we do not let a local slope be out of GAs. Also, the mean of the normal distribution is set at the center of GAs.

Quantification of metastability is performed by means of stochastic measurement tools, like 'mean first passage time' (MFPT) which describes the time expected for our robot to walk before falling down.[1] To estimate MFPT, we have repeated each simulation 10 times and calculated the average number of successful steps taken by the robot. The initial condition is set to be the best, i.e. the system's fixed point which is obtained by finding a long-period walking solution on a smooth surface with the slope angle of γ_{mean}.

Figure 2 illustrates the simulation output. The red icons belong to the triangular surfaces and the blue markers describe the results for the rectangular uneven terrains. Note that in this study, when MFPT passes the $100,000^{th}$ step, the simulation is halted. According to the figure, by increasing the overall slope angle, γ_{mean}, the sensitivity to irregularity rises. Also, it seems that a rectangular terrain is more sensitive than a triangular surface. Furthermore, as the value of standard deviation, σ_s, decreases the number of successful steps before falling down increases. This is in complete agreement with our expectation about the behavior of the robot which would infinitely take steps for $\sigma_s = 0$.

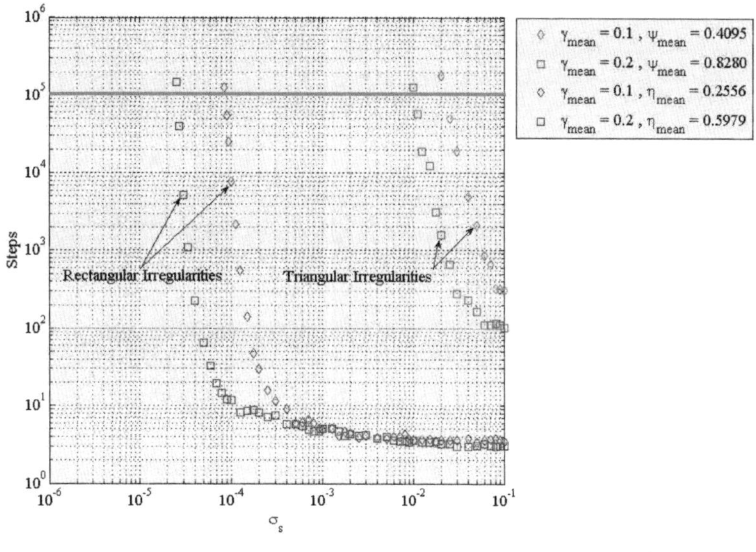

Fig. 2. The MFPT diagram for triangular and rectangular rough terrains with the parameters shown in the box. The red icons belong to the triangular surfaces and the blue markers describe the results for the rectangular uneven terrains.

We use our MFPT diagram to guess two values for σ_s assuring one million successful strides. By selecting $\gamma_{mean} = 0.2\,rad.$ and doing a simple linear extrapolation, the values of 7×10^{-3} for a triangular surface and 2×10^{-5} for a rectangular terrain are chosen. Figure 3 indicates the distribution of $\{\theta, \dot{\theta}\}$ for these two kinds of uneven terrain in the phase space. The initial condition, which is set as the system's fixed point, is shown by a black star. Also, to confirm the average slope angle travelled by the robot, we have calculated γ_{mean} through the following relation:

$$\gamma_{mean} = \frac{\sum_{i=1}^{n} (\frac{\phi_i^+}{2} - \theta_i^+)}{n} \tag{10}$$

Where n is the number of steps. The figure clarifies that metastable walking for the simplest walker on uneven terrains with the average slope angle of 0.2 rad. is feasible.

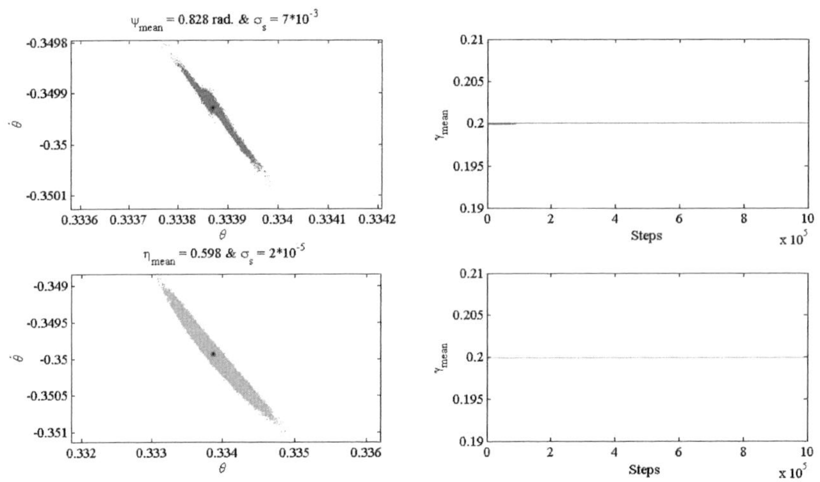

Fig. 3. The phase space for metastable walking on the triangular (up) and the rectangular (down) irregular grounds with the average slopes of 0.2 rad. The number of successful steps would be more than 1,000,000.

We are interested in comparing the metastable walking on the triangular surface with the rectangular ground. Figure 4 demonstrates the contrast between these two kinds of irregularity. Here, again, the magenta and cyan colors exhibit the phase space generated by walking through uneven triangular and rectangular grounds respectively. The black star shows the system's fixed point for $\gamma = 0.2\,rad.$ and the yellow area illustrates the 'Basin of Attraction' (BoA) of the fixed point for $\psi = 0.828\,rad.$ and $\eta = 0.598$.

Note that the BoA explains the set of all initial conditions leading to the system's fixed point. The figure suggests two interesting conjectures. First, during a metastable walking behavior, the walker will not fall until exiting from the BoA of the system's fixed point. Note that the fixed point and its BoA are defined if the values of $\gamma_{mean}, \psi_{mean}, \eta_{mean}$ are known. Second, because of a longer distance from the borders of the BoA, it seems that walking on the triangular terrain is more likely to be far from falling if other kinds of disturbances are included.

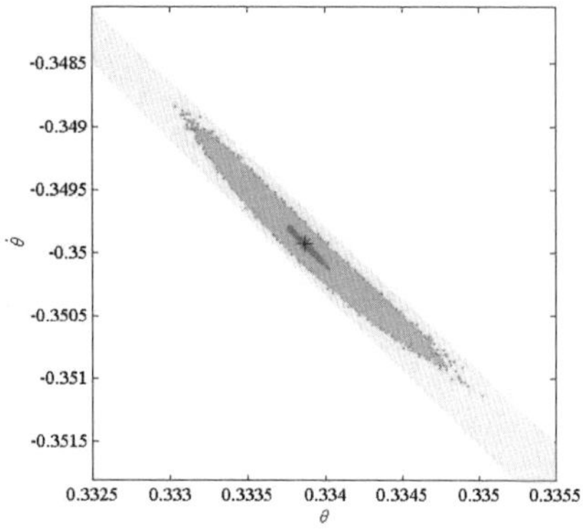

Fig. 4. Yellow area: BoA for $\gamma = 0.2\,rad.$, $\psi = 0.828\,rad.$ The black star illustrates the fixed point. Cyan area: Phase space for 1,000,000 steps on an uneven rectangular terrain with $\gamma_{mean} = 0.2\,rad.$, $\eta_{mean} = 0.598$, $\sigma_s = 2 \times 10^{-5}$. Magenta area: Phase space for 1,000,000 steps on an uneven triangular terrain with $\gamma_{mean} = 0.2\,rad.$, $\psi_{mean} = 0.828\,rad.$, $\sigma_s = 7 \times 10^{-3}$.

Figure 5 is presented to demonstrate the first speculation. In this case, we have only changed the standard deviation of the triangular surface to $\sigma_s = 8 \times 10^{-3}$. Beginning with the same initial condition defined as the system's fixed point, the robot could merely take 315,150 steps. The steps are indicated via cross markers in the figure. According to the figure, The final steps are out of the BoA.

4. Conclusion

It has been shown that metastabe walking could be studied more systematically by employing the concept of local slopes. In this regard, the following procedure is proposed: First define your system's fixed point, then define the GAs and finally find the metastable walking on those uneven terrains

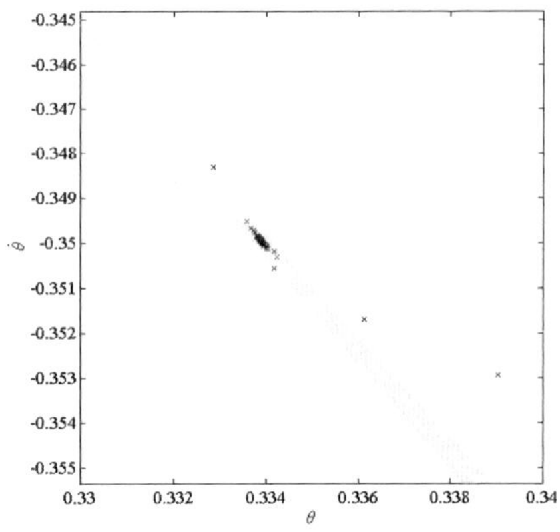

Fig. 5. Yellow area shows the BoA for $\gamma = 0.2\,rad.$, $\psi = 0.828\,rad.$ while dark crosses indicate the initial conditions just after the heel-strike for walking on an uneven triangular terrain determined by $\psi_{mean} = 0.828\,rad.$ and $\sigma_s = 8 \times 10^{-3}$. The number of successful steps is 315,150.

that consist of the found GAs. As an example, we have demonstrated the metastable walking for the simplest passive biped on some steep irregular grounds with the mean slope of $\gamma_{mean} = 0.2\,rad.$ Based on this example, we believe that neglecting GAs would end in the consumption of more energy input for a bipedal locomotion on an unsmooth environment.

References

1. K. Byl and R. Tedrake, Metastable walking machines, *The International Journal of Robotics Research* **28** (2009).
2. F. Iida and R. Tedrake, Minimalistic control of biped walking in rough terrain, *Autonomous Robots* **28** (2010).
3. I. R. Manchester, U. Mettin, F. Iida and R. Tedrake, Stable dynamic walking over uneven terrain, *The International Journal of Robotics Research* **30** (2011).
4. M.-Y. Chen and K. Byl, Analysis and control techniques for the compass gait with a torso walking on stochastically rough terrain, in *ACC*, (2012).
5. J. Solomon, M. Locascio and M. Hartmann, Linear reactive control for efficient 2d and 3d bipedal walking over rough terrain, *Adaptive Behavior* **21** (2013).
6. A. T. Safa and M. Naraghi, The role of walking surface in enhancing the stability of the simplest passive dynamic biped, *Robotica* **33** (2015).
7. A. T. Safa, M. Naraghi and A. Alasty, A different switching surface stabilizing an existing unstable periodic gait: an analysis based on perturbation theory, *Nonlinear Dynamics, Accepted.*
8. M. Garcia, A. Chatterjee, A. Ruina and M. Coleman, The simplest walking model: stability, complexity, and scaling, *Journal of biomechanical engineering* **120** (1998).

ENVIRONMENTAL PERCEPTION FOR HUMANOIDS USING GAUSSIAN PROCESS REGRESSION

DINGSHENG LUO, YAOXIANG DING, ZHAN LIU and XIHONG WU*

Key Lab of Machine Perception (Ministry of Education)
Speech and Hearing Research Center, School of EECS, Peking University
Beijing, 100871, China.
E-mails: {dsluo,yxding,Liuz,wxh}@cis.pku.edu.cn
**Corresponding Author, IEEE Senior Member*

As humanoids are expected acting in the real world to humanly and intelligently complete some high-level tasks, precisely perceiving the changes of environments is thus an essential premise. Due to the complexity of real-world environments, how to exactly perceive the environmental changes with limited sensors equipped on the robot becomes a challenging problem. Though this problem can be solved by establishing direct sensory mappings or employing probabilistic filtering methods, the nonlinearity and uncertainty caused by the high degree-of-freedom of the robots and the complexity of the environment result in tough modeling difficulties. In this paper, an alternative learning approach for addressing such modeling problems under the Gaussian process regression framework is proposed. The evaluations under perceptual tasks for two representative environmental changes, i.e. unknown pushing and sloped terrains, reveal the potential of the proposed approach in coping with the nonlinearity and uncertainty of environmental perception for humanoids.

Keywords: Environmental Perception; Intelligent Behavior; Humanoid Robots; Gaussian Process Regression.

1. Introduction

Environmental perception is one of the elementary functionality for intelligent robots. Like human or animals, an intelligent robot acting in a real world necessarily follows from the feedback information based online control mechanism. The robot gathers information about the environment all the time with its equipped sensors. Based on the primitive sensory information, how the environment is changing is then estimated. As a consequence, the robot could either evaluate its past behaviors and decide the next action in the task performing at hand, or design new strategies to cope with events caused by outside emergency.

Though various kinds of sensory facilities have been settled on robot systems, environmental perception still remains a challenging task. Mechanisms that describes the relationship between primitive sensory signals and environmental state variables are necessary. One possible approach for establishing such mechanisms is to build the mapping from sensory signals to state variables, by which the environmental changes could be directly revealed [1]. Another approach is to introduce the probabilistic filtering techniques, such as Kalman filter, by which the transition function from state variables to sensory signals as well as the system dynamic model are built [2-3]. Both approaches can serve as the appropriate solutions for efficient online environment perception task. However, for robot systems with large number Degrees of Freedom (DOFs) such as humanoid robots, the actual dynamics are heavily nonlinear and high-dimensional. As a result, designing suitable sensory mappings or system dynamics may be quite difficult since the models can hardly be described by linear transformations or simple differential equations. It turns out that a suitable way for dealing with such nonlinearity and uncertainty is to employ probabilistic learning methods, through which the above models are learned from data, as several past works have been done on this topic [4-6].

Gaussian Process (GP) regression is a general probabilistic approach for representing high dimensional nonlinear functions, and has been successfully employed in many robotic research areas [6-8]. Considering the representational power of strongly nonlinear functions and the ability for quantifying uncertainties in estimations, the GP regression is thus an appropriate choice in this research. The environmental perception task studied in this paper are the detection of unknown pushes from outside and the detection of sloped terrain while walking. In past works, these tasks can be solved by designing dynamical models or utilizing specific sensory facilities [2][9]. Nevertheless, GP regression can serve as an alternative approach that avoids introducing unnecessary simplifications and constraints in perceptual models. In [10], GP regression is introduced for sloped terrain estimation of quadrupedal walking with laser sensing, which justifies the usefulness of GP regression for learning such complicated sensory mappings. In our work the idea will be further extended on humanoid robots. Based on the idea of integrating multiple sensory information, the GP regression approach is verified to be a powerful potential candidate for coping with nonlinearities and uncertainties in humanoid robot environmental perception.

2. Task Description

2.1. *Involved sensors*

Sensors involved in this research including a 3-axes gyroscopes and a 3-axes accelerometers that integrated into the Inertial Measurement Unit (IMU) settled near the Center-Of-Mass (COM) of the robot. With these sensors, the body tilt angles can be calculated by the integration of angular velocities based on the chain multiplication rule of the rotational matrix [11]. In this paper, for simplicity it is assumed that the environmental changes are only taken place in the forward direction of the sagittal plane. In each of time slides, the sensory information received by the robot is a 3-element vector

$$z = [q\ \dot{q}\ a], \tag{1}$$

where q denotes the body tilt angle, $\dot{q}$ denotes the angular velocity and a denotes the acceleration.

2.2. *The task of detecting pushing forces*

Pushing force detection is the subject of push recovery that is one of the fundamental topics in humanoids researches [12-13]. Obviously, in order to efficiently react to the unpredicted forces, it is essential to give accurate estimations of them as soon as they inflict on the robot. Since individual sensor usually has limited performance due to strong sensory noise and complicated disturbance, the pushing forces estimated with multiple sensory information will be investigated in this research, through some mapping or transformation function which usually might be nonlinear, i.e.

$$f = \phi(z), \tag{2}$$

where f denotes the estimation of pushing forces with a mapping function ϕ. Here, we only focus on the case that the pushing forces are instantly applied on the COM from the back of a standing robot. The objective is to detect the strength of forces as earlier as possible.

2.3. *The task of detecting sloped terrains while walking*

Sloped terrains are representative environmental changes for humanoid walking. Ideally, the sloped terrain angles can be calculated from both of body tilt angles changes and proprioceptive measurements by employing simplified modeling and assumptions [9]. But difficulties remain for giving elaborate estimations. More than the problems stated for the pushing force detection, another important problem to be considered, even ignoring the effects of sensory noises. Without visual feedbacks, the robot can only passively "run into" a sloped terrain. As the moments of touching are different, the sensory information changing process after the first touch should also be different. This effect reveals the fact that detecting the terrain slope changes is not only just the matter of sensory information modeling, but also involves in the walking control of robot. That is to say, the mapping

$$\theta = \psi(z) \tag{3}$$

is not enough since the sensory measurements z are not sufficient to give identical estimations of terrain slope angles. The estimations also rely on the walking control states of the robot.

2.4. *The walking controller*

In this paper, a pendulum based model is adopted for normal walking control as [13], and the relative displacement of COM could be calculated by:

$$p = x - b\ddot{x}, \tag{4}$$

where p is the planned Zero Moment Point (ZMP) trajectories, x is the displacement of COM, and b is the parameter. Since no force sensors are assumed for detecting ground contact forces, the ZMP trajectories and the foot positions are pre-scheduled rather than controlled. The joint angles are calculated through inversed kinematics. A discrete phase variable t_{ph} is introduced for synchronizing the evolution of the walking process. The walking control state of robot is identically decided from t_{ph}. As a result, in the sloped terrain detection task, the sensory mapping becomes well-defined when t_{ph} is considered, i.e.

$$\theta = \psi(z, t_{ph}). \tag{5}$$

3. GP Regression Based Environmental Perception

3.1. *The basic setup*

In each of environmental perception tasks, the objective is to give accurate and prompt estimation of environmental changes. In general, there are two candidate approaches. One is the probabilistic filtering algorithms based approach, such as the Kalman filter, which can give efficient and accurate performances in practice. To deal with the modeling problem, the GP regression approach can be used for learning the sensory and dynamical model needed for filtering [6]. The other approach is to directly obtain the mappings from sensory observations to environmental change variables based on supervised learning approaches. The online estimation of environmental changes can be performed based on these learned mappings. As the second approach avoids the problem of environmental analysing and modeling, in this paper, the direct sensory mappings are the targets to be learned. The sensory mapping as (2) and (5) are firstly obtained with GP regression, then the learned model is used to online estimate the environmental changes.

3.2. *The general description of GP*

GP learning is a general kernel based supervised learning framework. A GP is an infinite set of random variables, in which arbitrary finite subset are jointly Gaussian. As a nonparametric model, GP only introduces minor assumptions of the function to learn, that is, the Gaussian form of outputs. As a kernel based method, GP can be used to model highly nonlinear functions, which is therefore desirable for learning sensory mappings.

For GP regression, given observation set $\{x, y\}_n$ which denotes n-element input-output pairs, observation noise σ_n, kernel function k and new test input x_*. Denote

$$k(X,X) \triangleq K; \qquad k(x_*,X) \triangleq K_*; \qquad k(x_*,x_*) \triangleq k_*. \tag{6}$$

where $K(X,X)$ is n by n Gram matrix with X being the l by n input data matrix with l being the dimension of one input, $k(x_*,X)$ is n by 1 column vector and $k(x_*,x_*)$ is a scalar. Then the joint distribution

$$p(y, y_*) \sim \mathcal{N}\left(0, \begin{bmatrix} k(X,X) & k(X,x_*) \\ k(x_*,X) & k(x_*,x_*) \end{bmatrix}\right) \tag{7}$$

holds, where y_* is the test output. Thus the conditional distribution is as

$$p(y_*|X,k,x_*) \sim \mathcal{N}(\mu, \sigma^2), \tag{8}$$

$$\mu = K_*^T K^{-1} y, \qquad \sigma^2 = k_* - K_*^T K^{-1} K_*. \tag{9}$$

The mean function gives the estimated output value, and the variance function gives the uncertainty about this estimation. The kernel function is commonly the squared-exponential

$$k(x_p, x_q) = \sigma_f^2 \exp(-\frac{1}{2l^2}(x_p - x_q)^2) + \sigma_n^2 \delta_{pq}, \tag{10}$$

where δ_{pq} is the Kronecker delta. The hyper-parameters σ_f, l and σ_n can usually be learned by maximum likelihood estimation (MLE) from the training data. Therefore the training dataset for GP regression is used for both of learning and prediction.

The explicit expression of estimation uncertainty given by variances is one of the major advantages for GP learning. The trade-off between estimation accuracy and computational efficiency may be the major obstacle for using GP regression to do online estimation, especially for robot systems whose computational power is limited. It is often desirable to employ local or sparse modification of the standard GP regression algorithms for doing such online estimations [14].

3.3. *GP model learning*

As described above, the GP model learning involves the MLE procedure for optimizing the hyper-parameters in the kernel function. For the pushing force detection task, this involves first to collect n training sample pairs

$$X = \{z, f\}_n, \tag{11}$$

where z denotes the sensory inputs and f denotes the labeled forces as outputs. The whole training set is then used for doing GP model learning, then the estimation of function given by equation 2 is obtained. However, for the slope terrain detection task, the mapping not only depends on sensory inputs, but also depends on the phase variable t_{ph} as shown in equation (5). This fact is a benefit for both of model learning and online estimation, since the overall training data can be segmented into bins according to t_{ph}:

$$X = \{\{z, \theta\}_{n_i}\}_m, \tag{12}$$

where m is the possible value of t_{ph} and $i \in \{1, 2, \ldots, m\}$, and different models can be learned using different bins of data. As a result, the number of models to be learned is increased, but the number of data points needed for learning a single model is decreased. This helps to reduce the learning complexity of each model and to increase the efficiency for online estimation. In the learning process, a empirical threshold is given to restrict the variance of Gaussian model not too high so that to keep a more efficient estimator.

3.4. *GP regression-based environmental perception*

Based on the learned GP models, the online estimations can be given by using equation (9). The mean value gives the estimated output for a new test input, and the variance value gives its uncertainty.

For sloped terrain detection task, the estimation for a new test input is given by using the model with the same phase variable t_{ph}. Therefore before

Algorithm 3.1 Online Estimation Based on GP Regression

Input: the learned GP models, the sample data set
the new test input x_t, t_{ph}(if available)
Output: the estimated f_t or θ_t
1. If t_{ph} is available, identify the GP model and the sample data set;
2. Calculated the mean μ and variance σ based on (9);
3. If $\sigma^2 <$ threshold $f_t = \mu$ or $\theta_t = \mu$; Else $f_t = f_{t-1}$ or $\theta_t = \theta_{t-1}$;
4. Return f_t or θ_t.

calculating equation (9), t_{ph} of the new test input is firstly identified. The overall procedure for online estimation is described in Algorithm 1.

4. Experiments

4.1. *Experimental setup*

The experiments are performed in the Webots simulation environment using the DARwIn-OP humanoid robot model. And the sensory readings from gyroscopes and accelerometers are acquired from the build-in functionality of Webots. Both inputs for training and testing accord with equation (1).

In the task of detecting pushing forces, the forces ranging from $\{20N, 25N, 30N, 35N\}$ are generated using the physics plugin of Webots and are applied forward and horizontally at the COM of robot. For a single pushing process, the robot initially keeps standing still when no push is applied. Then the robot is pushed by a instant pushing force, after which the data is labeled as the strength of the push and collected as training data. The processes end at the fixed time after the forces are applied. Overall 98 training samples are collected. 17 samples are collected for each label category of pushing forces, and 30 samples for standing without a push. 44 test samples from the whole pushing process for each of $\{27N, 32N\}$ are collected via the similar pushing processes as for training data, but keep the whole processes for testing. Note that the robot falls down when suffer from the pushing forces of $\{30N, 32N, 35N\}$ and does not fall otherwise. No variance threshold is employed in this task.

In the task of detecting sloped terrains, the terrains with angles ranging from the degrees of $\{5, 10, 15\}$ are set in front of the robot. The robot keeps walking straightly towards the sloped terrains. The data are labeled as 0 when the robot does not touch the sloped terrain, and are labeled as the terrain angle after the touches are triggered. The number of walking phase variable t_{ph} is set to 75. Overall 6210 training samples are collected. The numbers of samples for different terrain angles are $\{1762, 1606, 1041\}$. 1801 samples are collected for non-sloped terrain walking. For collecting data for each of terrain angle, the robot is set to start walking from ten different distances to the sloped terrains in order to cover different moment of touching. The differences among distances are $0.001m, 0.002m, \ldots, 0.01m$. 291 and 251 test samples are respectively collected from the angles of $\{7, 12\}$ degrees, which also contains the processes from normal walking to meeting with the sloped terrain. Note that the robot can walk without falling on the slope terrain with the angles of $\{5, 7, 10\}$ degrees, and falls with the angles of $\{12, 15\}$ degrees. The variance threshold is set to be 3.

For GP learning and estimation, the GPML Matlab toolbox is used [15]. The Gaussian kernel function given by equation (10) is employed for both tasks. For the force detection task, the whole training dataset is used for learning a single GP model. While in the sloped terrain detection task, 75 GP models are learned based on t_{ph}. The estimations for test data are based on algorithm 1. The angular velocities values are discounted by 50 to tune the relative weights for input features.

4.2. *Results and discussions*

The experimental results are illustrated in Fig. 1. As shown from the estimated results, the GP based estimation algorithm gives prompt estimations for both of pushing forces and terrain angles with rather satisfying accuracy. Comparing the sensory input data with the estimated results, the promptness for force estimations is not surprising since the sensory data also show significant variation as soon as the pushes are applied. On the contrary, for sloped terrain detection task, even though the sensory inputs change rather gradually, the promptness of GP based estimations are reserved. It can be seen that GP regression can utilize multiple sensory information to give accurate and prompt estimation results.

It can be observed for force estimations, the variances do not change very much for all of the test points and keep rather high values. While for terrain angle estimations, the variances fluctuate more significantly, and turn to rather small value for a few test points (the points where the mean estimations change). This phenomenon is interesting to be observed since intuitively sloped terrain estimation is a more difficult task. As known, it is generally easier to train a local model instead of a global one. In the sloped terrain detection task, as the GP models are learned locally according to the phase variable t_{ph}, the number of data required for learning and estimation with a single model is significantly reduced. It shows that

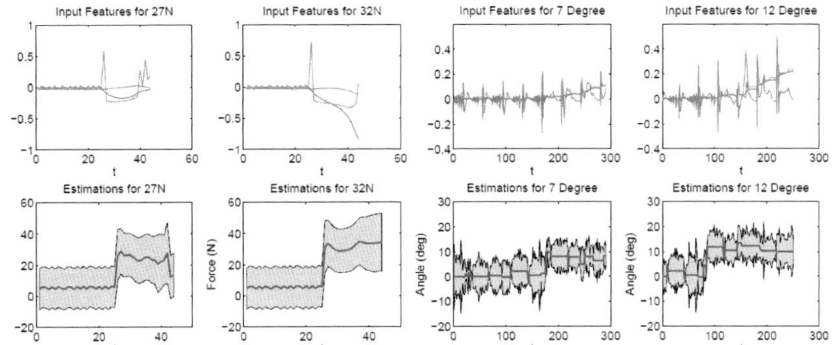

Fig. 1. The experimental result for two environmental perception tasks. The upper four figures are sensory inputs of test data, while the four figures below illustrate the estimated results.

the idea of employing the phase variable t_{ph} into learning and estimation is necessary other than the consideration of efficiency.

It can also be observed from the experimental results that the biases and errors in estimations are not ignorable. One major source for these errors lies in the ambiguity from training data. This problem is not only caused by the insignificance of input features, but also from the difficulty in labeling training data.

5. Conclusions

In this paper, an environmental perception approach with GP regression for humanoid robots is proposed, aiming to solve the modeling problem of nonlinear and noisy indirect mapping from sensory information to environmental changes. The applications on two typical environmental perception tasks, i.e. pushing force and sloped terrain detection with multiple inertial sensory information, are analyzed. The experimental results reveal the effectiveness and potential of our approach in environmental perception.

Acknowledgement

The work is supported in part by the National Natural Science Foundation of China (No. 90920302, No.61121002), the National Science & Technology Support Program of China (No. 2012BAI12B01), the Seeding Grant for Med and Info Sciences of Peking University (No. 2014-MI-10) and the Key Program of National Social Science Foundation of China (No. 12 & ZD119).

References

1. O. Höhn and W. Gerth, Probabilistic balance monitoring for bipedal robots, The International Journal of Robotics Research, vol. 28, no. 2, pp. 245-256, (2009).
2. B. J. Stephens, State estimation for force-controlled humanoid balance using simple models in the presence of modeling error,, 2011 IEEE International Conference on Robotics and Automation (ICRA), pp. 3994-3999, (2011).
3. C. G. Atkeson, State estimation of a walking humanoid robot, 2012 IEEE/RSJ International Conference on Intelligent Robots and Systems (IROS), pp. 3693-3699, (2012).
4. P. Abbeel, A. Coates, M. Montemerlo, A. Y. Ng, and S. Thrun, Discriminative training of kalman filters, in Robotics: Science and Systems, pp. 289-296. (2005).
5. J. A. Ting, E. Theodorou, and S. Schaal, A kalman filter for robust outlier detection, 2007 IEEE/RSJ International Conference on Intelligent Robots and Systems (IROS), pp. 1514-1519. (2007).
6. J. Ko and D. Fox, GP-BayesFilters: bayesian filtering using gaussian process prediction and observation models, Autonomous Robots, vol. 27, no. 1, pp. 75-90, (2009).
7. D. J. Lizotte, T. Wang, M. H. Bowling, and D. Schuurmans, Automatic gait optimization with gaussian process regression, in IJCAI, pp. 944-949, (2007).
8. D. Nguyen-Tuong and J. Peters, Local gaussian process regression for real-time model-based robot control, 2008 IEEE/RSJ International Conference on Intelligent Robots and Systems (IROS), pp. 380-385, (2008).
9. S.-J. Yi, B.-T. Zhang, and D. D. Lee, Online learning of uneven terrain for humanoid bipedal walking, in Proceedings of AAAI, pp. 1639-1644, (2010).
10. C. Plagemann, S. Mischke, S. Prentice, K. Kersting, N. Roy, and W. Burgard, Learning predictive terrain models for legged robot locomotion, 2008 IEEE/RSJ International Conference on Intelligent Robots and Systems (IROS), pp. 3545-3552. (2008).
11. J. Pratt, J. Carff, S. Drakunov, and A. Goswami, Capture point: A step toward humanoid push recovery, 2006 IEEE-RAS International Conference on Humanoid Robots, pp. 200-207, (2006).
12. B. Stephens, Humanoid push recovery, 2007 IEEE-RAS International Conference on Humanoid Robots, pp. 589-595, (2007).
13. S.-J. Yi, B.-T. Zhang, D. Hong, and D. D. Lee, Online learning of a full body push recovery controller for omnidirectional walking, 2011 IEEE-RAS International Conference on Humanoid Robots, pp. 1-6, (2011).
14. J. Quinonero-Candela and C. E. Rasmussen, A unifying view of sparse approximate gaussian process regression, The Journal of Machine Learning Research, vol. 6, pp. 1939-1959, (2005).
15. C. E. Rasmussen and H. Nickisch, Gaussian processes for machine learning (GPML) toolbox, The Journal of Machine Learning Research, vol. 11, pp.3011-3015, (2010).

A MECHANISM OF PARTICLE SWARM OPTIMIZATION ON MOTOR PATTERNS IN THE B4LC SYSTEM

QI LIU, JIE ZHAO, STEFFEN SCHUETZ, KARSTEN BERNS

Robotics Research Lab, University of Kaiserslautern, Kaiserslautern, 67655, Germany

This paper presents an optimization mechanism for feed-forward control units termed as motor patterns in the Bio-inspired Behavior-Based Bipedal Locomotion Control (B4LC) system. In specific locomotion phases, motor patterns *ActiveHipSwing* and *LegPropel* produce torques at the hip and ankle joints respectively. Biped is activated to swing leg ahead and push body forward by corresponding motor patterns, in which the parameters determine the profile of generated torques according to the sigmoid function. To obtain efficient and stable locomotion control, we employ Particle Swarm Optimization (PSO) method to tune motor patterns' parameters by formulating locomotion stability, energy consumption and walking speed as fitness functions. The optimization procedure takes place on a 3-dimensional simulated bipedal robot. Simulation results prove that suggested approach reinforces the walking behavior of biped with respect to stability, velocity control as well as enhances the energy efficiency significantly.

Keywords: Particle Swarm Optimization; Motor pattern; Locomotion control

1. Introduction and Related Work

As humans have possessed advanced skills in locomotion, the optimization of bipedal robots becomes an intensive issue that they are expected to achieve human-like behaviors with high efficiency and robust stability.

According to the employed locomotion control methods, the bipedal optimization approaches can be divided into two categories, which includes the model-based and model-free optimization. The model-based optimization is a computational approach that generates optimal motions or joint force profiles based on complex robot dynamic models.[1] In those issues, the classical control methods are normally used to plan the trajectory and mo-

*This work was funded by the European Commission 7th Framework Program under the project H_2R (no.60069).

tion. The optimization problems regarded as nonlinear programming problems are solved by utilizing efficient algorithms, *e. g.* sequential quadratic programming methods and augmented Lagrangian methods.[2]

Nevertheless, the challenge for model-based approaches is to establish complex dynamical models, and the bipeds developed correspondingly can hardly imitate the human behaviors in terms of stability and energy efficiency. Therefore, model-free optimization approaches have been proposed. The robot controlled by biologically-inspired methods can easily adapt to searching the parameters of these controller with stochastic optimization approaches. Shafii *et al.* has introduced a system that Particle Swarm Optimization (PSO) is used to optimize the angular trajectories of a stable biped walking.[3] Yang *et al.* has presented a method that stable gait generation in bipedal locomotion by using Genetic Algorithm (GA) optimization based Fourier Series Formulation.[4] This approach has the advantage over the classical control method on achieving human-like motions.

Inspired by humans locomotion, Luksch has addressed a behavior-based control framework, in which the bipedal control structure is organized in hierarchical levels by combining feed-forward and feedback control units, namely motor patterns and reflexes respectively.[5] However, due to the interaction of behaviors, the derivations of parameters incur intensive variations on the whole system. Because of this complex interconnectivity, the performance of bipedal robot in terms of stability, energy consumption are hardly optimized compared to humans behaviors. Since motor patterns play significant roles in stabilizing locomotion and generating propulsive torques for joints, such as swinging the leg forward, propelling the body and accepting the body weight, we will optimize the parameters of motor patterns based on PSO method in this work.

2. The Control Structure and Motor Pattern

As described in Fig. 1, the control structure defined by the flow of stimulation, inhibition and modulation among six classes of control units is organized in hierarchical levels.[5] The local control behaviors motor patterns generate torque commands to the corresponding joints in feed-forward manner. A sigmoid function is applied as basis for motor patterns, as depicted in Eq. 1:

$$\hat{\tau} = A \cdot \begin{cases} \frac{1}{2} + \frac{1}{2}\, sin(\ \pi(\frac{t}{T_1} - \frac{1}{2})\ ) & 0 \leq t < T_1 \\ 1 & \mathrm{T}_1 \leq t < T_2 \\ \frac{1}{2} - \frac{1}{2}\, sin(\ \pi(\frac{t-T_2}{T_3-T_2} - \frac{1}{2})\ ) & \mathrm{T}_2 \leq t < T_3 \end{cases} \tag{1}$$

where A, T_1, T_2 and T_3 are the parameters of motor patterns representing the maximum torque of torque command, the starting time of maximum torque, the ending of maximum torque and the total time of torque command respectively. The values of parameters are defined experimentally by comparing the kinematic data of bipedal robot and humans. As shown in Fig. 1, the modification of the parameters results in a variety of torque trajectories.

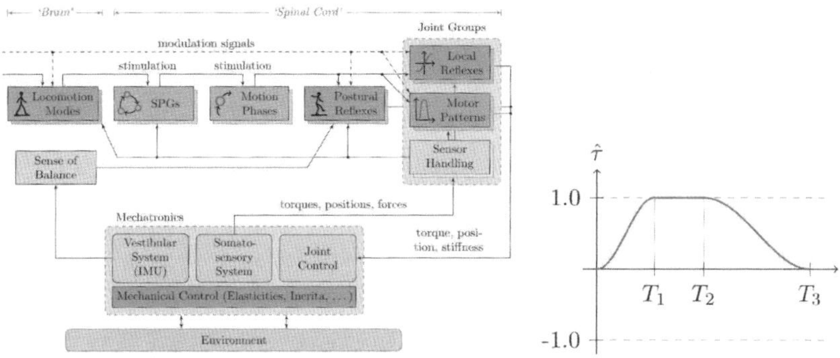

Fig. 1. The left side of figure represents the hierarchical organization of control units, while the right side shows that torque trajectories produced by motor pattern.[5]

3. Particle Swarm Optimization Algorithm

Inspired by the studies of fish and bird flocks, Kennedy and Eberhart first introduced the algorithm of PSO.[6] The movements of a particle is guided in the direction of its own best known position and the best know position of the entire swarm. The velocity of a particle in one iteration can be derived, as described in Eq. 2.

$$v_p = \omega v_p + c_1 \cdot Rand_1(\ B_p - x_p\ ) + c_2 \cdot Rand_2(\ B_g - x_p\ ) \tag{2}$$

where v_p and x_p represent the velocity and the position of a particle respectively. B_p is the known best position of a particle in the past iterations, while B_g the known best position of entire swarm in the past iterations. c_1 and c_2 mean the parameters of the acceleration constants respectively. $Rand_1$ and $Rand_2$ are the random values ranged within $[0, 1]$. The inertia weight ω controls the impact of the previous velocities on the current velocity. The larger inertia weight tends to favor global searching, while the smaller one leads to the local searching strategy.

The fitness functions summarizing the distance of a given solution approaching to the objectives are calculated during an iteration. The best position of particles B_p and the best position of the entire swarm B_g are updated by comparing the fitness values in each iteration. The position of a particle x_p is updated according to Eq. 3.

$$x_p = x_p + v_p \tag{3}$$

4. Implementation of Optimization in B4LC System

The optimization unit is built in a way that it can be integrated into an existing behavior-based control system. A behavior control unit termed as Optimization Module is inserted into the system by connecting the stimulated Motion Phase and the corresponding motor patterns, as shown in Fig. 2.

Activated by the stimulation signals from Motion Phase, the Optimization Module feeds generated outputs into each motor pattern as parameters T_1, T_2 and A. The cycling time of motor pattern T_3 is kept constant as introduced in.[5]

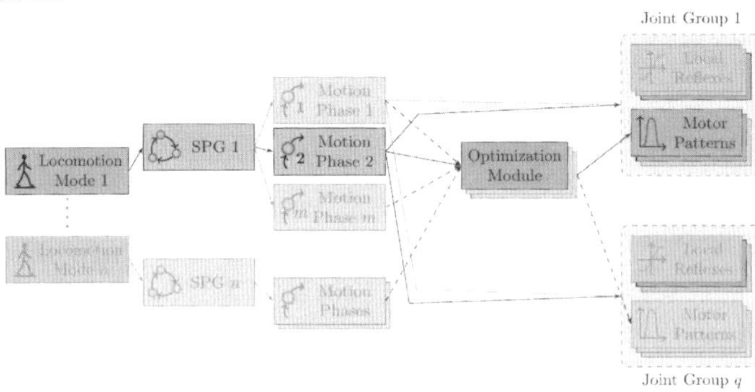

Fig. 2. Optimization modules in the B4LC control system.

The parameters of motor pattern *Leg Propel* at the ankle joints and the motor pattern at the hip joints *Active Hip Swing* are optimized. A 6-dimensional vector $\vec{P}_{i,t}$ that consists of the parameters of *Leg Propel* $[T_{a1}, T_{a2}, A_a]$ and *Active Hip Swing* $[T_{h1}, T_{h2}, A_h]$ stands for the position of a particle, where i is the number of particles and t represents iteration times. The particle position are initialized randomly in the searching space. To increase the learning efficiency, the searching space are constrained in a range based on the experimental studies of bipedal locomotion, as depicted in Eq. 4.

$$\begin{matrix} T_{a1} \in [0.0\ ,\ 0.6] & T_{a2} \in [0.6\ ,\ 0.9] & A_a \in [0.0\ ,\ 1.0] \\ T_{h1} \in [0.0\ ,\ 0.4] & T_{h2} \in [0.4\ ,\ 0.7] & A_h \in [0.0\ ,\ 1.0] \end{matrix} \tag{4}$$

To achieve a natural walking while maintaining stability, the fitness functions are formulated by considering optimization goals including robustness, stability, and speed and energy consumption in form of:

$$F = k_1 f_1 + k_2 f_2 + k_3 f_3 + k_4 f_4 \tag{5}$$

where k_1, k_2 , k_3 and k_4 are the weightings of the fitness functions, and

$$f_1 = s_w / s_{max} \tag{6}$$

$$f_2 = \sum_{i=0}^{T} (\Delta Xcom_r + \Delta Xcom_l) / s_w \tag{7}$$

$$f_3 = \sum_{i=0}^{T} |(V_{ref} - V_i)|^2 / T \tag{8}$$

$$f_4 = \sum_{i=0}^{T} \tau_i^h / T \tag{9}$$

f_1 is defined as the function that is linear to the walking steps s_w with respect to maximum walking steps s_{max} the robot can achieve. f_2 evaluates the stability of the robot by taking the error value of extrapolated Center of Mass ($Xcom$) into consideration, where $\Delta Xcom_r$ and $\Delta Xcom_l$ are the right and left error value at each time step respectively, and T is the maximum time step in one iteration. Furthermore, f_3 represents the deviation of the walking speed respecting reference, with V_i meaning the walking speed at time step i, and V_{ref} representing the reference equals 1.2m/s. In addition, f_4 contains the accumulated resistant torques generated by the reflex *Lock Hip*, which is utilized to stop the flexion of hip joints when swinging legs moved by *Active Hip Swing* to a desired angle. τ_i^h means the braking torques generated by *Lock Hip* at time step i.

5. Optimization Process and Setup

The optimization scenario is set up on the simulated robot with 21 degrees of freedom and 1.8m height.[7] The robot is designed to start the experiments with a stable walking on a flat ground. The parameters generated

Table 1. The parameters before and after optimization.

parameters	before optimization	after optimization
Leg Propel T_1	0.5	0.3
Leg Propel T_2	0.7	0.75
Leg Propel A	0.65	0.47
Active Hip Swing T_1	0.2	0.09
Active Hip Swing T_2	0.56	0.65
Active Hip Swing A	0.7	0.41

by Optimization Module known as the position of a particle are fed into

Leg Propel and *Active Hip Swing*. When the pitch and roll angle of the robot exceed thresholds or a robust walking behavior emerges after successful walking in long time, a test of a particle is finished. Until all the particles in one iteration are tested, the best position of the entire swarm and the best position of each particle are updated. Consequently, the velocities and the positions of particles start a new iteration in accordance with updates.

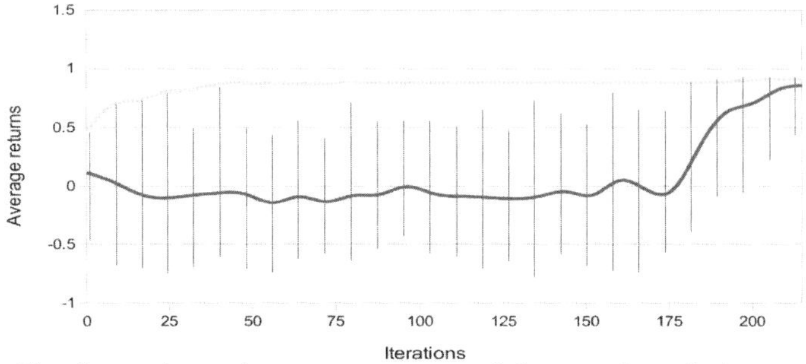

Fig. 3. The figure shows the average returns of fitness values during optimization progress. The yellow line presents the average fitness value of the known best particles in the past iterations. The blue line indicates the average fitness value of particles in each iteration. The vertical lines stand for the maximum and minimum value of the fitness value in each iteration.

The average fitness values for each iteration are shown in Fig. 3. At the beginning, we use the high inertia weight, which is equal to 1, to perform global searching. After about $50th$ iteration, the fitness values of the best known particles are constant, and the average fitness value of the particles has no more significant improvement. To explicitly search the optimal parameters, we decrease the inertia weight after 160 iterations to perform local searching. The average fitness value in each iteration is sharply increased and navigates closer to the best known fitness value, and the robot is able to perform stable and robust walking. The parameters before and after optimization are shown in Table 1.

6. Simulation Results

To compare the bipedal performance after and before optimization approach, we have implemented 30 times of walking experiments with optimized and non-optimized parameters respectively. The motor torques generated at the ankle and hip joint are illustrated in Fig. 4. The data is normalized to one gait cycle and averaged over 30 successive steps of walking on level ground. The torque profiles present smaller peek during activation of *Leg Propel* and *Active Hip Swing* after optimization. This leads to a longer swing phase and shorter stance phase compared to the robot with non-optimized parameters.

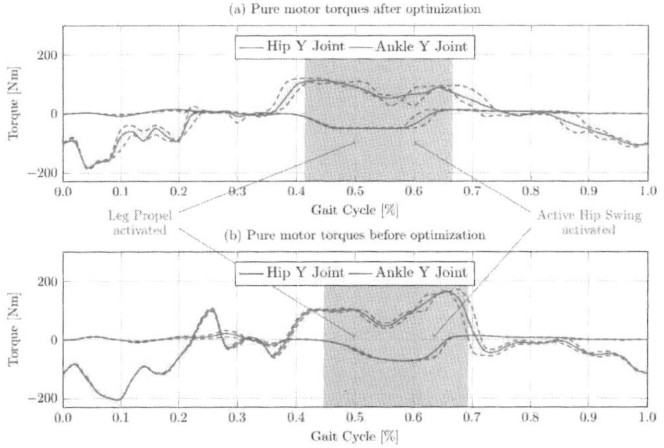

Fig. 4. The motor torques generated from hip joints around y-axis and ankle joints around y-axis using optimized and non-optimized parameters over the course of one gait cycle. The solid line presents the mean value over 30 successive steps. The dashed lines stands for the minimum and maximum values. The blue area and the red area indicate the activation of the Leg Propel and Active Hip Swing respectively.

Because of the improper torques generated, the average walking speed is not able to follow the reference. It can be seen from Fig. 5 that the robot is initialized to walk at the speed of 1.2m/s. However, as walking distance increases, the speed enhances to around 1.3m/s. This result in major increase of velocity deviation calculated in Eq. 8.

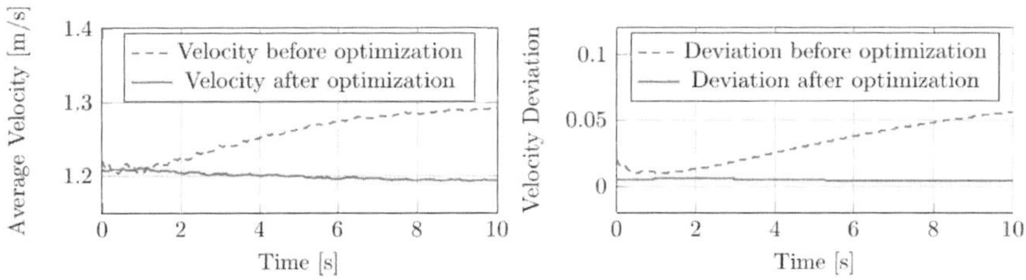

Fig. 5. The left side of figure shows the pitch angle before and after optimization, while the right side presents the average walking speed and the speed deviations with optimized and non-optimized parameters.

In term of energy consumption, the result of locking-hip torques both after and before optimization progress are shown in Fig. 6. By using non-optimized parameters, the robot boosts the leg in a shorter duration to the desired position. The robot with optimized parameters generates less torques to prevent over-swinging of the leg to achieve a suitable angle of attack at heel strike. One can observe that the angle of attack at hip joint dwindles since less energy is generated from Leg Swing and therefore the over-swing at hip joints appears to be insignificant compared to previous.

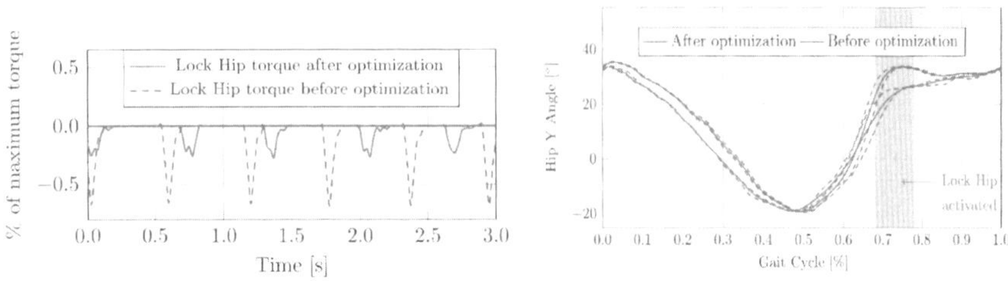

Fig. 6. The left side of figure indicates the percentages of the torque commands generated from Lock Hip reflex over the maximum hip torque; the right side of figure shows the hip angular trajectories around y-axis over the course of one gait cycle.

7. Conclusion

In this paper we described the development of a PSO-based optimization approach for motor patterns in the B4LC system. The simulated biped is optimized in terms of stability, energy consumption and walking velocity. The optimized parameters have been used to conduct the normal walking experiments. The locomotion performance with optimized parameters have been compared to that with non-optimized parameters.

References

1. M. Srinivasan, "Computer optimization of a minimal biped model discovers walking and running", Nature 439.7072 (2006): 72-75.
2. Y. Xiang, "Physics-based modeling and simulation of human walking: a review of optimization-based and other approaches", Structural and Multidisciplinary Optimization 42.1 (2010): 1-23.
3. N. Shafii, "Biped walking using coronal and sagittal movements based on truncated Fourier series", RoboCup 2010: Robot Soccer World Cup XIV. Springer Berlin Heidelberg, 2011. 324-335..
4. L. Yang, "A uniform biped gait generator with offline optimization and online adjustable parameters", Robotica 25.05 (2007): 549-565.
5. T. Luksch, "Human-like control of dynamically walking bipedal robots", Verlag Dr. Hut, 2010.
6. R. Eberhart, "A new optimizer using particle swarm theory", Proceedings of the sixth international symposium on micro machine and human science. Vol. 1. 1995.
7. J. Zhao, "Biologically motivated push recovery strategies for a 3D bipedal robot walking in complex environments", Robotics and Biomimetics (ROBIO), 2013 IEEE International Conference on. IEEE, 2013.

THE KEY FACTORS FOR A ONE-LEGGED ROBOT TO KEEP BALANCE WHEN HOPPING IN PLACE*

XUAN LIU

Computer Science institute, Beijing University of Posts and Telecommunications, 10 Xi Tu Cheng Road. Haidian District, Beijing, 100876, China

CHENGLONG FU

Beijing Key Lab of Precision/Ultra-precision Manufacturing Equipments and Control, Tsinghua University, Beijing, 100084, China

YU ZHOU

Beijing Key Lab of Precision/Ultra-precision Manufacturing Equipments and Control, Tsinghua University, Beijing, 100084, China

In order to study the key factors that influence the balance of a hopping one-legged robot, we decompose the hopping balance model into two parts: leg swing control in flight phase and stability control in stance. During flight phase, since the horizontal velocity is not easy to measure by sensors and might cause cumulative error and time delay, we use body attitude and angular velocity as approximate input parameters to replace horizontal velocity in leg swing algorithm. In the experiments of stance phase, we discover the close relation between duty factor and spring rate as well as state transition threshold of pneumatic cylinder on the leg. We combine the two experiments and preliminary implemented the balance of one-legged robot hopping in place. The experiments demonstrated that body attitude and angle velocity in flight phase and duty factor in stance phase are the key factors for balance. Comparing to works made by Raibert[2] in balance controlling under high translation speed, our work provides references to study balance of robot while hopping under low horizontal velocity.

1. Introduction

As we know, adding series elastic damping to provide compliance and energy storage, and reduce energy consumption has become a widely accepted method in robot foot structure research ([5], [6]). It is why most vertebrate animals rely on the flexing of their elastic muscles and tendons rather than bones or joints to

* This work is supported by the National Natural Science Foundation of China under Grant 51175288 and 61375099, and Beijing National Natural Science Foundation under Grant 3152014.

provide energy for locomotion. However, as mentioned in MIT research of cheetah robot[6], there is still vast optimization design space in parameters, control methods and hardware configurations. One-legged hopping robot is a type of classic springy leg robot. Its 3D hopping in balance is one of the most unstable motion modes for legged robot.

Most previous work by Raibert[2] presented a successful designation and a series of controlling methods for one-legged robot to hop in balance. However, the work still left us space to make studies on the influence of factors such as state transition threshold and performance of pneumatic cylinder. Moreover, though a large number of research results on 2D models emerged over the past 2 decades, only few attempts were made to extend to 3-D locomotion[7], and even fewer on real platforms.

In this paper, we studied the key factors that influence the balance of robot hopping in place by doing control experiments on Tsinghua one-legged robot.

Following the classic approach by Raibert[2], we divided the model into two parts: leg sweep control in flight phase (section 2) and stability control in stance phase (section 3). In the experiment in flight phase, the goal of leg control is to avoid body tumbling and to recover to a stable status by neutralizing the falling trend influence caused by center-of-mass (COM) shift. Two solutions of leg sweep control in flight phase have been presented in 'Legged Robot that balance'[2] and 'Capture Point'[3]. Both of them took horizontal velocity and acceleration as the main factors causing COM shift. On the other hand, the paper 'Control of One-Legged Robot Hopping in Place'[4] has presented a hypothesis that 'body attitude and angular velocity are the main reasons of tumbling'. Being inspired by this hypothesis, we simplify the model of robot during stance into the movement of a two linked single axle joint. Using inverse kinematics, we demonstrate that body attitude and angular velocity can be used to replace horizontal velocity as the input of controlling algorithm, and discovered that the 'body attitude-angular velocity' algorithm has higher agility when hopping in place from the control experiments.

In the experiment of stance phase, we focused on the significance of a low duty factor [5]. By controlling variables of the experiment, we analyzed following items:

- The regularity of hopping motion and value of stance duty factor.
- The hopping height and value of stance duty factor.

We combined experiments of two phases, and realized the stable hopping in place without physical support. The test results show that the parameters in PD control algorithm in flight phase and duty factor of stance phase are key factors for robot to keep balance while hopping in place. Our experimental platform is presented in Figure 1(A). All factors discussed are presented with our Factors Relation Graph in Figure 1(B).

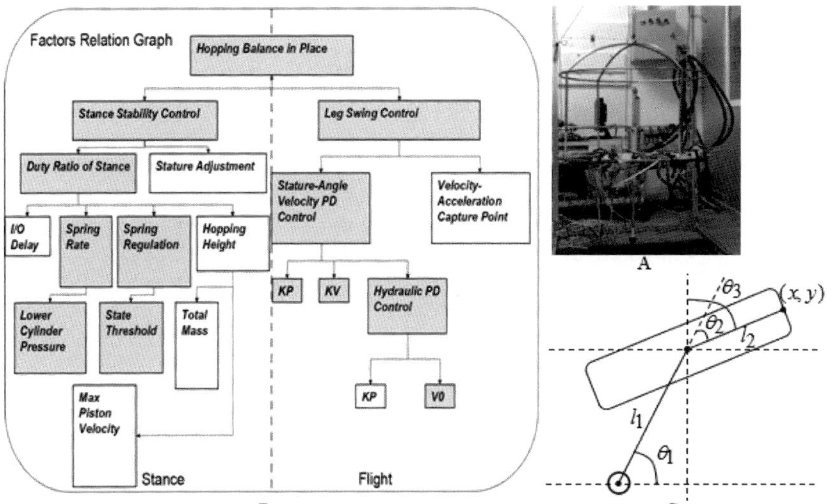

Figure 1. (A) the one-legged hopping robot in Tsinghua Robot and Automation Lab. (B) the relation among the two key factors and related factors in flight phase and stance phase respectively. Shadowed areas are considered more significant than the others. (C) a simplified structure model of robot leg with two linked single axle joint.

2. Key Factors in the Leg Sweep Control During Flight Phase

2.1. *The designation of control model*

The mathematical model of leg control in flight phase is summarized herein on the basis of capture point method in calculating foothold$(x_{capture}, y_{capture})$ [3] and the body attitude-angular velocity PD control[4].

2.1.1. *The calculation of capture point*

The capture point algorithm focuses more on the response to horizontal disturbances[3]. Given equation of capture distance[3], we expand it into 3D space as below:

$$x_{capture} = \dot{x}\sqrt{\frac{z_0}{g}},\ y_{capture} = \dot{y}\sqrt{\frac{z_0}{g}}, a^2{}_{capture} = x^2{}_{capture} + y^2{}_{capture}. \quad (1)$$

Where $\dot{x}$ stands for the horizontal velocity component along the direction of x-axis; z_0 is the vertical dimension between COM and foot. The most direct method to count the value of $\dot{x}$ is making discrete-time integration of acceleration as the equation presented below:

$$\dot{x} = \sum_{t_0}^{t_n} \ddot{x} dt, \dot{y} = \sum_{t_0}^{t_n} \ddot{y} dt. \tag{2}$$

Where t_0 is the starting point of the time slot and t_n is the last time slot for integration. However, there exists accumulative error in the progress of integration. As a result, we replace the instantaneous velocity with average velocity at the beginning of each hopping period to compensate for the error. The average velocity can be presented as below:

$$\bar{\dot{x}} = \frac{d\cos\theta_\alpha}{T}, \bar{\dot{y}} = \frac{d\sin\theta_\alpha}{T} \tag{3}$$

Where θ_α is the direction of the leg in hydraulic control. $d = distance$ is the axial motion of robot body, and $d = l_1|\cos\alpha| + l_2|\cos\beta|$ (l_1 and l_2 are the distance between the joint and center of body, α and β are the angle between leg and body at the begin and end of stance). Vector (x, y, z) is the coordinate of joint of hip actuator and springy leg.

According to the geometrical properties of the robot model, we will get:

$$a_{capture} = \sqrt{\frac{-K^2 + \sqrt{K^4 + 4K^2 l^2}}{2}} \tag{4}$$

Where $K = \frac{(\dot{x}^2 + \dot{y}^2)}{g}$ and l is the distance between the joint and center of body.

By using inverse kinematics we can get $\theta_\beta = \sin^{-1}\frac{a_{capture}}{l}$ and $\theta_\alpha = \tan^{-1}\frac{y_{capture}}{x_{capture}}$. Where θ_β is the sweep angle of the robot leg in hydraulic control.

2.1.2. *The control based on body attitude and angular velocity*

Using horizontal velocity $\dot{x}$ as the input parameter of leg controlling algorithm in flight phase has been proved reasonable in several previous works [2][3]. In this paper, section 2.1.1 has shown two ways to acquire an approximate $\dot{x}$. However, accumulative error may occur while doing integral acceleration; and the calculation of average velocity will lead to frequent I/O operations and cause time delays. These problems should not be tolerated under our experimental condition. However, if certain physical quantities directly acquired from robot sensors can approximately equal horizontal velocity, then we'll be able to improve the performance of the controlling algorithm in real experiment. Thus we make further steps in designing body attitude and angular velocity PD control method as below:

$$\theta_{\alpha x} = Kp \cdot Pitch + Kd \cdot gyr_y, \theta_{\alpha y} = -(Kp \cdot Roll + Kd \cdot gyr_x), \theta_\alpha = \sqrt{\theta_{\alpha x}^2 + \theta_{\alpha y}^2}. \quad (5)$$

Where θ_α is swing angle, θ_β is direction which can be determined by the ratio of $\theta_{\alpha x}$ and $\theta_{\alpha y}$. $\theta_{\alpha x}$ and $\theta_{\alpha y}$ are component of θ_α.

The relationship between body attitude, angular velocity and horizontal velocity are demonstrated herein:

According to Figure 1(C), the springy leg is considered as rigid body at every instantaneous time point. Using inverse kinematics and geometric relation in Figure 1(C) we get equation as follows:

$$\dot{x} = (l_1 \cos(\theta_3 - \theta_2) + l_2 \cos(\theta_3 - \theta_2))\dot{\theta}_3 - 2l_2 \cos(\theta_3 - \theta_2)\dot{\theta}_2 \quad (6)$$

According to Eq. (6), when robot body can be regarded as particle, then $l_2 = 0$, and $\dot{x} = l_1(\dot{\theta}_3 - \dot{\theta}_2)\cos(\theta_3 - \theta_2)$ is same to one of the reasons of Eq. (4) and Eq. (5) in the 'capture point' [3].

When θ_2 and $\dot{\theta}_2$ can be approximately regarded as constant, according to Eq. (6), $\dot{x}$ can be linear represented approximately by θ_3 and $\dot{\theta}_3$.

2.2. *Experiment designation*

2.2.1. *Control experiments for capture point algorithm and body attitude-angular velocity PD control algorithm*

The performance indicators of experiment are listed as below:

- Counting time duration from the start point T1 in which a disturbance is detected to the end point T2 when a response of leg is performed. |T1-T2| is set as the response delay. A lower delay means higher agility.
- The accuracy of the direction of sweep leg locomotion. In theory both methods should be 100% accurate. However, the measuring error of gyroscope inevitably reduces the rate of accuracy.

Initially setting KP = 0.5, Table 1 shows the difference of response delay and accuracy between the two algorithms when applying disturbance to the robot:

Table 1. The data below was taken from 10 experiments respectively.

Method	Average|T1-T2| (ms)	Accuracy (%)
body attitude-angular velocity	50	90
capture point	90	70

Table 1 indicates a clear advantage of body attitude-angular velocity algorithm in both time delay and accuracy of sweep leg motion.

2.2.2. *Determine initial voltage of hydraulic valves*

Set KP=0.5 in hydraulic PD control and change value of V0 before every new experiment. By applying disturbance from different directions, we get data of desired hydraulic elongation and real time hydraulic elongation. The results are depicted below.

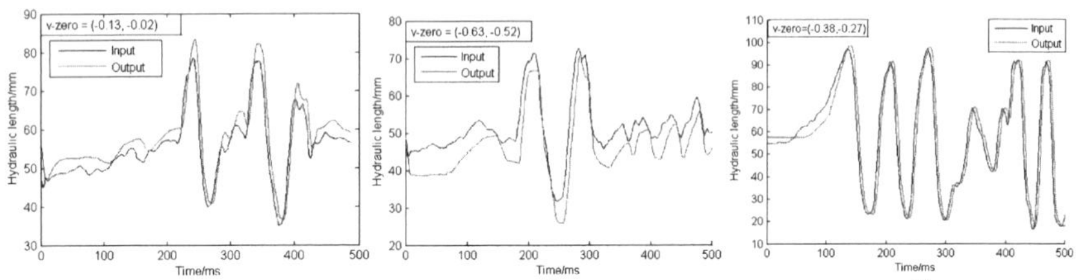

A B C

Figure 2. time-hydraulic length, initial voltages are A(-0.13,-0.02), B(-0.63,-0.52), C(-0.38,0.27).

According to results shown in Figure 2, it is discovered that the curves reach an approximate optimal following performance when V0=(-0.38,-0.27).

2.3. *Conclusion of experiments*

It was confirmed that horizontal velocity can be replaced with body attitude and angular velocity as the key input of leg sweep control during the hopping in place movement. Moreover, we have determined several conditions for the experiment, which will be applied to preset initial conditions in section 3.

3. Key Factors in the Stability Control During Stance Phase

3.1. *Control model designation*

3.1.1. *Pneumatic control*

Figure 3 is our control flow graph on the basis of Raibert's[1] method:

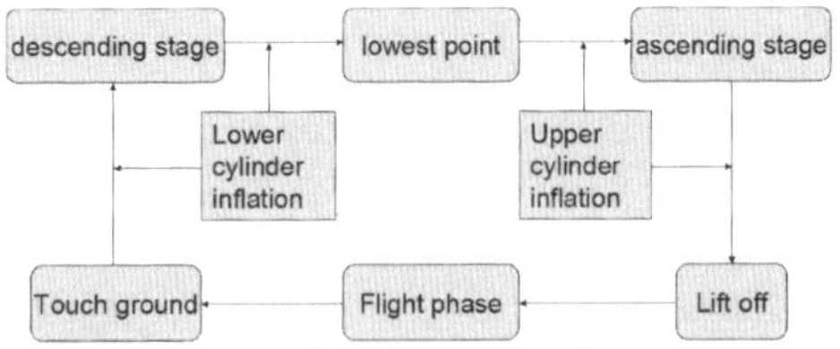

Figure 3. the operation of pneumatic cylinder for each transition between two states.

3.1.2. *Hydraulic control to maintain stance stability*

Since angular momentum of legged systems is conserved during flight, the body attitude of the robot cannot be changed during flight. Thus it can only be controlled during stance[2]. As a result we set θ_d, the angle between body platter and level ground as a constant.

The goal of coordinate transformation is to make body attitude remain level when the COM of body shifts. The way to calculate (u,v,w) is same as that in flight phase. Vector $(\theta_R, \theta_P, \theta_Y)$ is the current body attitude. We use rotation matrix R to transform vector (u,v,w) into (x,y,z):

Substitute R to $(x, y, z)^T = R(u, v, w)^T$, and translate the coordinate result into direction and sweep angle to control two hydraulic actuators.

$$\theta_d = 0, \Delta\theta = \theta_{\beta,xyz} - \theta_{\beta,uvw} = (\frac{\pi}{2} - \arcsin\frac{z}{l}) - (\frac{\pi}{2} - \arcsin\frac{w}{l}). \quad (7)$$

In this case, our method is similar to PD control because foothold can be considered as ground fixed during stance. Therefore, the error of body attitude can be adjusted in the next period.

3.2. *The designation and data of experiments*

Table 2. (A) duty factors, time duration of stance phase(Ts) and flight phase(Tf) under different thresholds. Preset spring length as 30mm and stiffness as 5/8 (air pressure 0.5MPa for upper cylinder and 0.8MPa for lower cylinder). (B) duty factors, Ts and Tf under different air pressures in upper cylinder. Preset spring length as 30mm and state transition threshold is 476mm on the basis of result in 3.2.1. (C). a brief statistical result of number of stable cycles under different duty factors.

Table A				Table C	
Threshold D(mm)	Ts(s)	Tf(s)	Duty	Ts:Tf	Stable cycles
376	0.21	0.24	7:8	1:1	8~12
476	0.15	0.33	1:2	4:5	8~15
576	0.22	0.33	2:3	3:5	>12
676	0.23	0.27	8:9	1:2	>15

Table B			
Upper cylinder pressure(MPa)	Ts(s)	Tf(s)	Ts:Tf
0.30	0.21	0.27	1:1
0.40	0.18	0.30	3:5
0.50	0.15	0.33	1:2
0.60	0.18	0.26	3:4

The experiment in stance is consisted of 3 sub experiments below:

- State transition threshold optimization experiment: Focus on the influence of transition threshold in the falling stage between flight phase and stance phase. From Table 2(A), the performance leads to a relative optimal result when we set threshold as 476mm.
- Stiffness adjustment experiment: The duty factors of stance phase are compared by applying the air pressure in upper cylinder when doing whole experiment repeatedly. As can be seen from Table 2(B), the performance leads to a relative optimal result when we set air pressure as 0.50MPa.

- Influence of stance duty factor to the stability of hopping motion: Preset parameters of key factors in flight phase fixed and then conduct complete hopping experiment to observe the performance of stability control. As can be seen form Table 2(C), the number of stable cycles increases as the duty factor decreases, which demonstrates the significance of this factor in keeping stability of one-legged robot hopping in place.

3.3. *Analysis and conclusion*

From the experiment results above, we discover following characters of duty factor in stance phase and make conclusions on them as below:

- The state transition threshold is the primary indicator for the duty factor--a reasonable range of value of duty factor can not only maintain the stability of pneumatic spring but also keep the flexibility of robot to different conditions.
- A lower duty factor can lead to longer time or even a convergence of stable hopping.

Figure 4. 7 frames in a whole period of the hopping movement in the practical experiment.

4. Overall Analysis and Conclusion

This paper initially realizes the balance of one-legged robot hopping in place by joining leg sweep control experiment in flight phase and stability control experiment in stance phase. Video links and pictures of practical experiment are given at the end of this page.[a]

The results of our experiments indicate that PD control algorithm based on body attitude and angular velocity and duty factor of stance phase are the two main factors for robot to keep balance while hopping in place. The main indicator of PD control algorithm is value of KP, KV and initial voltage. And the main indicator for duty factor is state transition threshold. Due to the limitation of hardware configuration and mechanical structure, we still need to collect more data in future experiments to increase the objectivity and reliability of the conclusions.

[a] You can obtain this file from our WWW pages at: http://learn.tsinghua.edu.cn:8080/2007990009/3D_HoppingRobot.mp4

Although we've basically achieved the anticipated goal, the performance gap between our robot and the one in MIT Leg Lab still makes it difficult for us to increase hopping height. As a result our robot has not enough time to increase range of leg sweep motion to move forward or backward. We are planning to improve the disadvantages of mechanical structure if budget permits.

5. Future Work

During the experiment of leg sweep control in flight phase, we've found it easy to realize hopping motion in place with external physical support, which is similar to a child toddling with the help from his/her parents. As a result we propose a hypothesis that robot can learn to walk like a child by feeling external physical support. The robot will gradually get rid of the support by repeatedly adjusting its moving strategy. If such kind of mode can be proved efficient, the motion control of robot will become closer to human motion system. Since conditional reflex arc requires less calculation for motions control, under this method the robot will be more sensitive to external disturbance and respond more rapidly.

References

1. M. H. Raibert, H. Benjamin Brown, E. Hastings, J. Koechling, K. N. Murphy and S. S. Murthy, "Dynamically stable legged locmotion", Mit Artificial Intelligence Laboratory. (1989).
2. M. H. Raibert, "Legged Robots That Balance", *99 Gnral and Mllano // mahma* 4(1986).
3. J. Pratt, J. Carff and S. Drakunov, 'Capture point: A step toward humanoid push recovery", *IEEE-RAS International Conference on Humanoid Robots*. (2006).
4. Y. Zhou, C. Fu, C. Yun, Z. Yu and K. Chen, "Control of one-legged robot hopping in place", *IEEE International Conference on Robotics & Biomimetics* (2013).
5. A. Ananthanarayanan, A. M and K. S., "Towards a bio-inspired leg design for high-speed running", *Bioinspiration & Biomimetics* 7.4 (2012).
6. S. Seok, A. Wang, M. Y. Chuah, D. Otten, J. Lang and S. Kim, "Design principles for highly efficient quadrupeds and implementation on the MIT Cheetah robot", *IEEE International Conference on Robotics and Automation*. (2013).
7. A. Wu and H. Geyer, "The 3-D spring–mass model reveals a timebased deadbeat control for highly robust running and steering in uncertain environments", *IEEE Transactions on Robotics*, vol. 29, no. 5, pp. 1114–1124, (2013).

MOVING CoM FORWARD INCREASES SPEED IN BIPED WALKING USING COUPLED ELASTIC ACTUATION

XINGANG PAN

Department of Automotive Engineering, Tsinghua University, Beijing, 100084, China
E-mail: panxg1994@163.com

MINGGUO ZHAO

Tsinghua National Laboratory for Information Science and Technology (TNList), and Department of Automation, Tsinghua University
E-mail: mgzhao@mail.tsinghua.edu.cn

Coupled elastic actuation has been proven to be a simple and effective method to realize natural biped walking. However, former researches found it hard to achieve high speed merely by adjusting control parameters. For the purpose of realizing fast walking in this method, we put forward a simple approach: moving CoM(center of mass) forward. Through numerical simulation, we found that a positive lateral offset of CoM could effectively increase walking speed by enlarging step length and shortening step period, both of which result from the reduction of swing leg retraction. To testify the simulation results, experiments based on a prototype were conducted. The dimensionless speed of the robot changes from 0.264 to 0.424 as CoM moves forward, which confirms our speed-increasing strategy.

Keywords: Center of Mass; Walking speed; Biped robot; Coupled elastic actuation.

1. Introduction

In the field of biped locomotion, achieving high speed walking has long been an important and challenging issue. Few biped robots could walk at a speed comparable to human's fast walking. In the past dozen years, a series of planar biped robots were built to realize high-speed walking. Pratt increased the speed of Spring Flamingo by more quickly swinging the swing-leg using trajectory tracking method.[1] Chevallereau et al. designed a controller based on the concept of virtual constraints and hybrid zero dynamics and implemented it on a biped robot Rabbit, which achieved a dimensionless speed of 0.43.[2] Geng et al. enhanced the record to 0.53 on RunBot using a neuronal sensor-driven controller and an online reinforcement learning algorithm.[3] This record was kept until 2010, when our lab designed the biped robot with the fastest dimensionless speed in the literature: Stepper-2D, which was controlled using virtual slope walking method.[4] However, even its dimensionless speed, 0.716, is only half of human's Olympic Records. More speed-increasing methods still remain to be found.

Another biped walking method proposed by our lab was coupled elastic actuation, which realizes natural gait through simple control.[5] Although its speed could be adjusted by changing control parameters, the highest speed is slow. It was found that the low speed results from excessive swing leg retraction, which is caused by this method's inherent feature that the stance legs move much slower than the swing legs. To overcome this disharmony, we believe that the advisable choice is adjusting physical parameters instead of control parameters.

The effects of some physical parameters on speed have been investigated. Asano[6] and Liu[7] found that enlarging the foot radius can increase walking speed. However, considering the limitation of foot size, foot radius cannot be too large. As Mcgeer pointed out, a proper foot radius is 1/3 leg length.[8] Hass[9] and Liu[7] found that the robot walks faster when the leg mass centralizes to the hip. Both researches are based on passive models and have little to do with swing leg retraction. Recently, Geng proposed a strategy to realize fast walking in passive ankle robots: torso inclination, which reduces the fluctuation of the instantaneous speed.[10] It was also found that a positive horizontal offset of CoM could increase speed in passive dynamic walking.[9] Its capability to adjust the available potential energy of two legs shows potential to modulate the swing leg retraction.

In this paper, we achieve fast walking in biped robot using coupled elastic actuation by means of increasing the lateral offset of CoM. The simulation results show how and to which extent the lateral position of CoM could affect the walking speed. The mechanism of this speed-increasing strategy is analysed. We also carried out experiments on a prototype, which validate our simulation results.

2. Model

The biped walking model in this paper is the same with the model in Wu's work,[5] which is shown in Fig. 1. The details of the dimensionless parameters of the model studied here is shown in Table 1.

Wu has clearly described the coupled elastic actuation and has given

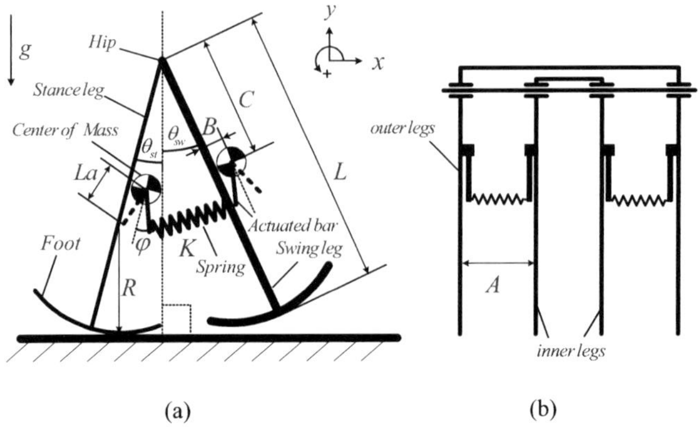

Fig. 1. Sketch of the quasi-2D model. (a) is the lateral view and (b) is the front view

Table 1. Symbols and values of parameters.

Parameter	Symbol	Dimensionless value
foot radius	R	0.25
spacing between the swing and stance leg	A	0.25
lateral position of CoM	B	-
vertical position of CoM	C	-0.35
natural length of the spring	l_0	0.25
inertia of one leg	I	0.06
mass of one leg	m	1
length of the leg	L	1
length of the actuated bar	La	0.13
spring coefficient	K	-
amplitude of the actuated bar	Φ	15

the dynamics of one step.[5] In this paper we assume that the bars' motion phase happens instantaneously. So the equation (4) in reference 5 becomes $\phi = \Phi$. Other equations are all the same.

3. Numerical Simulation Results

3.1. *Effects of B on walking speed*

To study the influence of the lateral offset of CoM on walking speed, we change B continuously and fix other parameters as shown in Table 1. The spring coefficient K is set to be 10, 15, 20, 25, 30. Considering that gait bifurcation and chaos are not the normal walking behavior, we only analyze period-one gait.

The effects of B on the walking speed, the step length and the step period are shown in Fig. 2. As B increases, the step length grows larger while the step period is shorten. Both changes are monotonous and contribute to the increase of walking speed. The spring coefficient K shows little capacity to adjust walking speed, but with a growing K, the maximum available value of B is enlarged, which makes higher speed possible. Although a too large B is hard to achieve in practice, a moderate B could significantly enlarge the speed. Typically, for K=10, the walking speed doubles when B increases from 0 to 0.1.

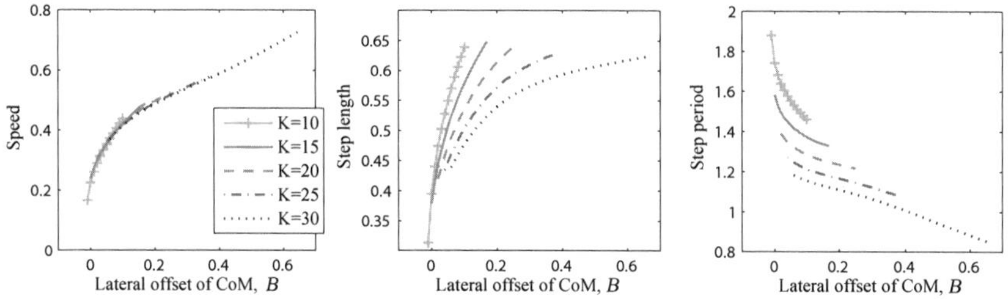

Fig. 2. Walking indexes with respect to B, when $K = 10, 15, 20, 25, 30$.

3.2. *Analysis*

To reveal how B affects walking speed, we first analyze the reason for the low speed when B=0. The limit cycle of the stable gait with B=0 is shown in Fig. 3. The maximum value of θ_{sw} reaches 0.54 while at heel strike θ_{sw} is only 0.18, indicating that the swing legs loss 2/3 angle in retraction. The excessive swing leg retraction has two disadvantages: (1) the step length is short, (2) because of the time spent in the retraction, the step period is relatively long. Thus, the excessive swing leg retraction should account for the low walking speed.

To understand why the excessive retraction happens, we observe the states of both legs. As shown on the limit cycle with B=0, when the swing legs are lifted to the highest position, the stance legs have just reached the vertical position at a low angular speed, in which case a big retraction is needed to realize heel strike. The fall behind of the stance legs can be explained by the feature of coupled elastic actuation. After the rotation of bars, the swing legs are dragged by the stretched springs and their angular speed rises quickly. However, for the stance legs, the influence of the springs

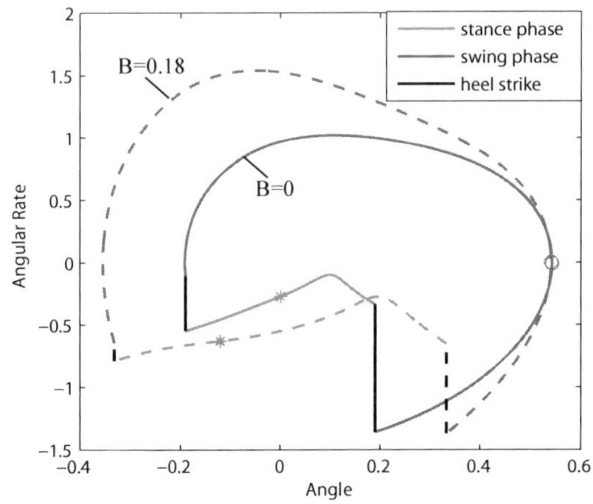

Fig. 3. Limit cycles of the stable gaits with K=15, B=0 (solid line) and 0.18 (dashed line). For each limit cycle, the circle represents the state of the swing legs at the moment when θ_{sw} reaches the maximum value and the asterisk represents the state of the stance legs at the same moment.

is negative. To overcome this discrepancy, the angular speed of the stance legs should be enlarged.

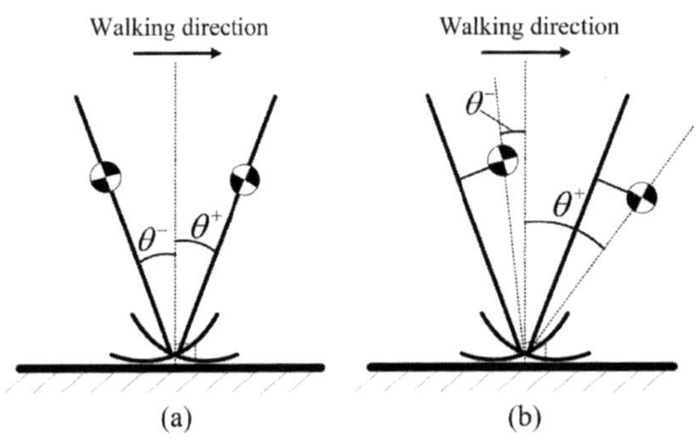

Fig. 4. The limiting positions of the stance legs with (a)B=0 and (b)B>0.

Fig. 4 (a)(b) shows the two limiting positions of the stance legs in the stance phase with B=0 and B>0. During θ^-, gravity does negative work to the stance legs while in the next θ^+ it does positive work. When B=0, θ^-=θ^+ . As B increases, θ^- is shortened and θ^+ becomes larger, indicating

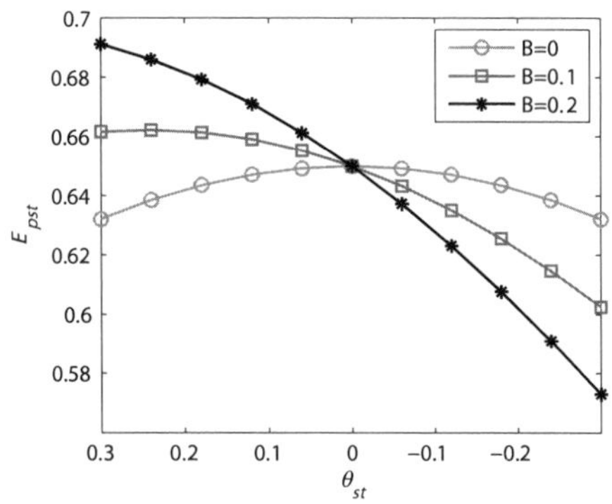

Fig. 5. The gravitational potential energy of the stance legs E_{pst} in the stance phase (note that θ_{st} decreases as the walker moves forward), with B=0, B=0.1 and B=0.2.

that more gravitational potential energy of the stance legs is transformed into kinetic energy. This can be clearly seen in Fig. 5, which shows the gravitational potential energy of the stance legs E_{pst} in the stance phase with different B. The bigger B is, the more E_{pst} decreases, imputing more kinetic energy to the stance legs. From Fig. 3 we can see that when B=0.18, the stance legs move faster and the big retraction has been effectively reduced, enabling a larger step length and a shorter step period, thus increasing the walking speed.

4. Prototype Experiment

In order to validate the effects of B on walking speed, we conducted experiments using a prototype. The prototype is shown in Fig. 6. There are springs connecting the ends of the bars, which are driven by the motors fixed on legs. Trigger switches are installed at the button of the semicircular feet to detect heel strike. Each leg is designed to have two parts so that the leg length can be adjusted by motors. To avoid foot scuffing, the swing legs would be shortened by a few millimeters during swing phase and recover at next heel strike.

The way we used to adjust B is adding additional weight on the prototype and shifting its lateral position. As shown in Fig. 6, four 30cm rulers are fixed on the legs to provide attachment points for the additional weight. We let the prototype walk 10 meters with different B and record the time used and the number of steps. Fig. 7 shows series of snapshots of the walking robot. Readers can find the walking video

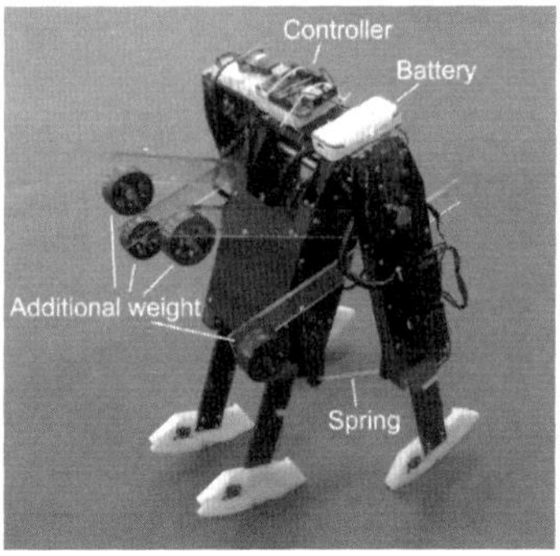

Fig. 6. Prototype. The robot weighs 5.80kg and the additional weight weighs 0.955kg in total. The leg length of the prototype is 0.433m.

Fig. 7. Series of snapshots of the walking robot with additional weight.

on `http://v.youku.com/v_show/id_XODkyMzAwMDA0.html`. The calculated dimensionless indexes including speed, step length and step period are shown in Fig. 8.

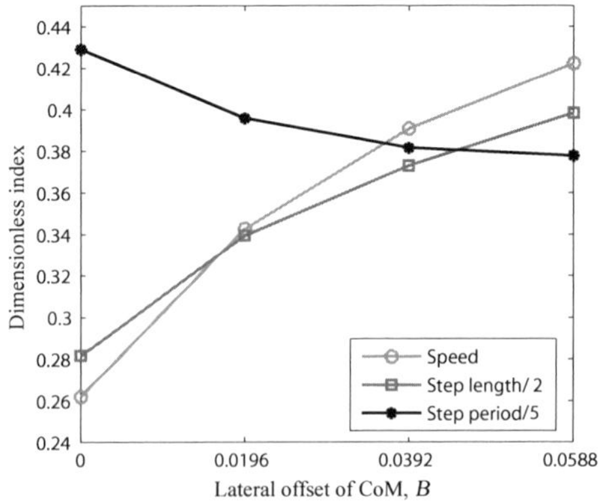

Fig. 8. Dimensionless experiment results.

From Fig. 8 we can see that all the three indexes show the same trends as the simulation results: as B becomes larger, the speed and the step length increase monotonically and the step period gets shorter. With only a small increase of B (from 0 to 0.0588), the walking speed is enlarged by 61%. The highest dimensionless speed, 0.424, is comparable to Spring Flamingo and Rabbit,[1,2] while our control method is much simpler.

5. Conclusions

In this paper, we proposed a simple and effective method, moving CoM forward, to achieve faster walking in biped robots using coupled elastic actuation. Through simulation, we revealed how the lateral offset of CoM affects walking speed: by reducing the excessive swing legs retraction. Both the simulation and the prototype experiments indicate that shifting CoM forward could obviously enlarge the step length and shorten the step period, thus greatly increasing the speed of biped walking driven by coupled elastic actuation.

Acknowledgment

This work is supported by National Natural Science Foundation of China under Grant 61273357.

References

1. J. Pratt, Exploiting inherent robustness and natural dynamics in the control of bipedal walking robots, PhD thesis, Massachusetts Institute of Technology, 2000.
2. C. Chevallereau, G. Abba, Y. Aoustin, F. Plestan, E. Westervelt, C.Canudas-de-Wit, and J. Grizzle, Rabbit: A testbed for advanced control theory, *IEEE Control Syst.*, 23(5):57-79, Oct. 2003.
3. T. Geng, B. Porr, F. Wörgötter, Fast biped walking with a sensor-driven neuronal controller and real-time online learning, *The International Journal of Robotics Research*, 25(3):243-259, Mar. 2006.
4. H. Dong, M. Zhao, N. Zhang, High-speed and energy-efficient biped locomotion based on Virtual Slope Walking, *Autonomous Robots*, 30(2):199-216, 2011.
5. B. Wu, M. Zhao, Coupled elastic actuation for biped walking, in *Proceedings of the 17th International Conference on Climbing and Walking Robots (CLAWAR)*, 2014.
6. F. Asano, Z. W. Luo, On energy-efficient and high-speed dynamic biped locomotion with semicircular feet, *IEEE International Conference on Intelligent Robots and Systems*, pp.5901-5906, 2006.

7. N. Liu, J. Li, T. Wang, The effects of parameter variation on the gaits of passive walking models: simulations and experiments, *Robotica*, 27:511-528, 2008.
8. T. McGeer, Passive dynamic walking, *International Journal of Robotics Research*, 9(2):62-82, Apr. 1990.
9. J. Hass, J. M. Herrmann, T. Geisel, Optimal mass distribution for passivity-based bipedal robots, *International Journal of Robotics Research*, 25(11):1087-1098, Nov. 2006.
10. T. Geng, Torso inclination enables faster walking in a plannar biped robot with passive ankles, *IEEE Transactions on Robotics*, 30(3):753-758, 2014.

EXPERIMENTAL EXAMINATION OF WALKING DIRECTIONAL CONTROL FOR A SIX-LEGGED ROBOT

HIROAKI UCHIDA, HIROKAZU ITHO, NAOKI NUMATA

Department of Mechanical Engineering, Kisarazu National College of Technology, 2-11-1 Kiyomidai-higashi, Kisarazu, Chiba 292-0041, Japan

The present paper discusses a control method for the location (x,y) and yaw angle of the body of a six-legged robot using a laser range finder. During walking, the location and yaw angle of the body are calculated by referring to these two fixed points obtained by the laser range finder. In the experiments, the validity of the proposed walking directional control method was examined for straight walking.

1. Introduction

Multi-legged robots are expected to be used in extreme environments that are dangerous for humans, such as minefields, nuclear power plants, and disaster areas [1][2]. We have examined six-degree-of-freedom control methods for the body of a six-legged robot and proposed a three-degree-of-freedom posture control method for the height of the body, the pitch angle, and the roll angle using virtual impedance. The effectiveness was verified through 3D simulations [3] and walking experiments. Moreover, an walking directional control method for the location of the body (x, y) and the yaw angle was verified through 3D simulations [4][5]. However, in applying the proposed walking directional control method, we need to correctly detect the location of the body and the yaw angle in global coordinates. Reference [6] reports the estimating method for the location (x, y) and the yaw angle of the body. In this study, we examine the control method for walking in both directions (x, y) and the yaw angle of the body using the data obtained by a laser range finder (LRF). In this method, it is not necessary to design the desired orbit of the rotating link at the supporting leg beforehand. A six-legged robot is able to realize flexible walking and adapt to the walking environment. Moreover, it becomes possible to control the six axes of the body by applying the proposed control method with our proposed posture control method. The six-legged robot can realize walking that is suitable for its specific environment. In this study, we experimentally examine the validity of the proposed walking directional control method using the data obtained by the LRF.

2. Six-legged Robot

Figure 1 shows the six-legged robot used in this study. Each leg of the six-legged robot consists of three joints, a rotating part, a thigh part, and a shank part. The advantages of the leg mechanism of the six-legged robot are that the pitching angle, the rolling angle, and the height of the body are controlled by the thigh links, and the walking direction (x, y) and yaw angle of the body are controlled by the rotating links. Table 1 shows the dimensions of the six-legged robot.

Fig. 1 Six-legged robot.

Table 1 Robot dimensions

Height	500 mm	Length of Thigh	168 mm
Length	666 mm	Length of Shank	312 mm
Width	710 mm	Moment of Inertia I_z	1.966 Kgm^2
Weight	24.9 kg		

3. Location Estimation Method

3.1. *Setting the outside fixed points*

The location of the body (x,y) and the yaw angle are estimated using the LRF for the case of a moving six-legged robot. The location and yaw angle in global coordinates are calculated by setting two fixed points in the walking area. Figure 2 shows the model for measuring the external environment using the LRF on the body of a six-legged robot. Here, y is the walking direction, x is the vertical direction to y, and in the LRF, the angle from the x-axis is $\theta_i(i=1,\cdots,127)$. The distance between the LRF and the measurement object is $r_i(i=1,\cdots,127)$. The fixed points are set so that they are easy to detect because they are used as reference points to calculate the position and the yaw angle of the body. Then, we show the calculation method as follows. We suppose that the rate of change $diff(\Delta_{ri})$ becomes maximum around point θ_i, and θ_i is set as a fixed point. The rate of change $diff(\Delta_{ri})$ is calculated as follows:

$$diff(\Delta_{ri})=(\Delta_{ri}-\Delta_{r(i-1)})^2 \tag{1}$$

where

$$\Delta_{ri}=r_i-r_{i-1}$$

In this study, one fixed point is detected in each range $0 < \theta_i < \pi/2$ and $\pi/2 < \theta_i < \pi$ of the LRF. The fixed point is the maximum value of the rate of change $diff(\Delta_{\theta i})$ calculated by Eq. (1).

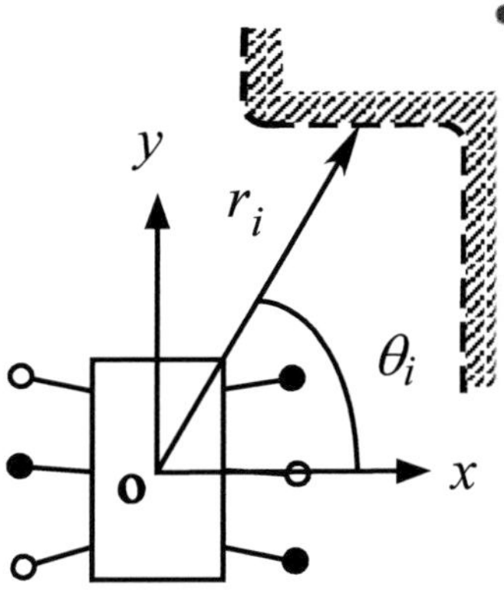

Fig. 2 Coordinate and measurement model using the LRF.

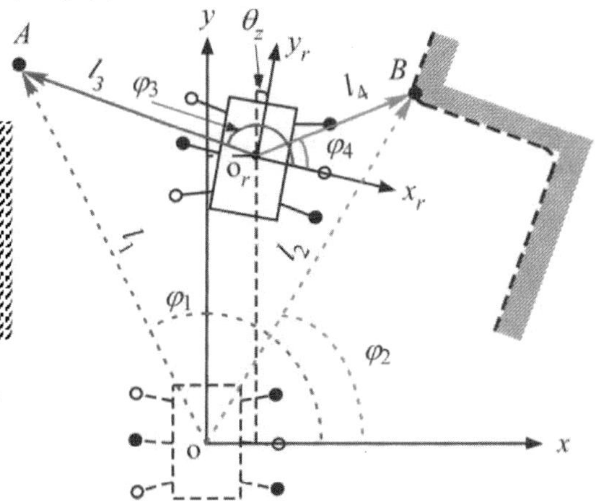

Fig. 3 Measurement model using outside fixed points.

3.2. *Calculation of location and yaw angle using two fixed points*

This section describes the method used to calculate the location and yaw angle of the body in global coordinates using the fixed points described in Section 3.1. Using the two fixed points, the location and yaw angle of the body are calculated from the variation of the position of the LRF resulting from the robot motion. Figure 3 shows the measurement model, where φ_1 and φ_2 are the angles between A and the *x*-coordinate and between B and the *x*-coordinate, respectively, at the start position. Moreover, φ_3 and φ_4 are these respective angles at the measurement position. l_1 and l_2 show the distance between the LRF at the start point and fixed points A and B, and l_3 and l_4 show the distance between the LRF point during walking and at fixed points A and B, respectively. As a result, the yaw angle θ_z, *x*-position, and *y*-position are calculated by the following equations:

$$\theta_z = -\{(\varphi_3 - \varphi_1) - (\varphi_4 - \varphi_2)\}/2 \tag{2}$$

$$x = \{l_1 \cos\varphi_1 - l_3 \cos(\varphi_3 + \theta_z) + l_2 \cos\varphi_2 - l_4 \cos(\varphi_4 + \theta_z)\}/2 \tag{3}$$

$$y = \{l_1 \sin\varphi_1 - l_3 \sin(\varphi_3 + \theta_z) + l_2 \sin\varphi_2 - l_4 \sin(\varphi_4 + \theta_z)\}/2 \tag{4}$$

For the measurement of the LRF, an error signal exists. Then, in this study, the error signal is filtered such that the external environment measured at the start point agrees with the external environment at the time of walking. The external environment is measured several times before walking, and fixed points A and B are set to a state without error. At the time of walking, fixed points A and B are obtained by two methods; one method is the measurement by the LRF, and the other method estimates the values using the location (x, y) and the yaw angle calculated by Eqs. (2)–(4). The measurement values of the fixed points are compared with the estimated values at each time. If the measurement value is different from the estimated value by more than the threshold, the location and the yaw angle are set by the estimated values.

4. Walking Directional Control Method

4.1. *Mathematical model of the body in the walking direction*

Figures 4 and 5 show the relationship between the forces working at the rotational parts of the three support legs. Here, the support legs are Legs II, III, and VI, and the swing legs are Legs I, IV, and V. Moreover, the support legs are assumed to support the body firmly and to not slip. Here, y is the walking direction, x is the vertical direction to y, and θ_z is the angle about the z axis. Moreover, the force $F_i\left(i=1,\cdots,6\right)$ from the support leg acts on the rotational axis of the support leg. M is the weight of the body, I_z is the inertia around the z axis, $\theta_i\left(i=1,\cdots,6\right)$ is the angle of the rotating link, $r_i\left(i=1,\cdots,6\right)$ is the distance between the center of gravity of the body and the center of rotation of the rotating link. The motion equations for the x-direction, y- direction, and about θ_z are shown as follows:

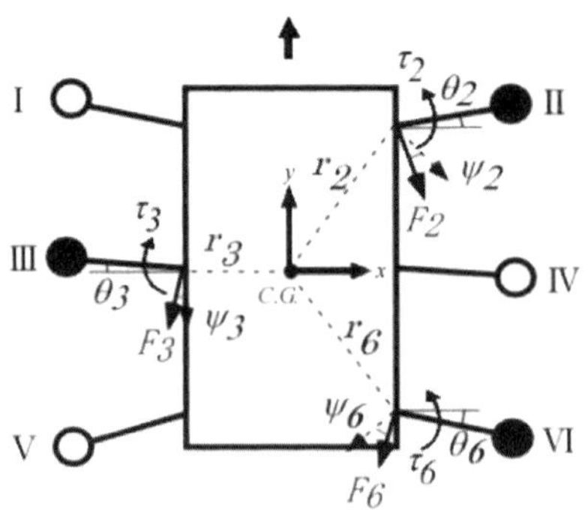

Fig. 4 Mathematical model of walking direction in the three support legs.

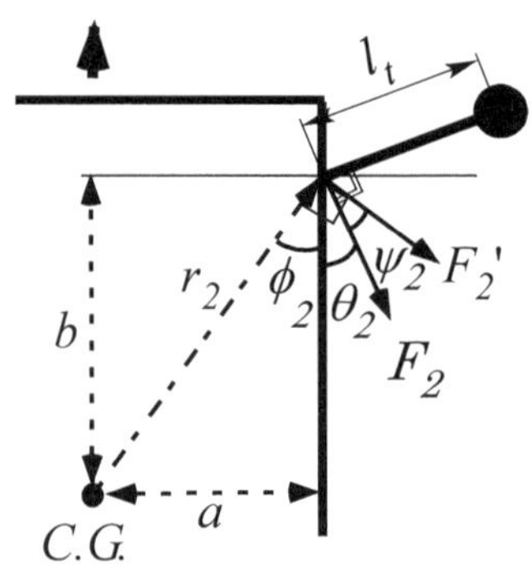

Fig. 5 The relationship between ϕ, θ, and ψ

$$\begin{cases} M\ddot{x} = F_2 \sin(-\theta_2) + F_3 \sin\theta_3 + F_6 \sin(-\theta_6) \\ M\ddot{y} = F_2 \cos(-\theta_2) + F_3 \cos\theta_3 + F_6 \cos(-\theta_6) \\ I_z\ddot{\theta}_z = F_2 r_2 \cos\psi_2 - F_3 r_3 \cos\psi_3 + F_6 r_6 \cos\psi_6 \end{cases} \tag{5}$$

By defining the state vector $x = \left[x, y, \theta_z, \dot{x}, \dot{y}, \dot{\theta}_z\right]^T$ and input vector $u = \left[\tau_2, \tau_3, \tau_6\right]^T$, the following state equation and output equation are obtained:

$$\dot{x}(t) = Ax(t) + B(t)u(t) \tag{6-a}$$

$$y(t) = Cx(t) \tag{6-b}$$

where

$$C = \begin{bmatrix} 1 & 0 & 0 & 0 & 0 & 0 \\ 0 & 1 & 0 & 0 & 0 & 0 \\ 0 & 0 & 1 & 0 & 0 & 0 \end{bmatrix}, \quad \tau_i = l_t F_i \left(i = 1, \cdots, 6\right),$$

$$A = \begin{bmatrix} 0 & 0 & 0 & 1 & 0 & 0 \\ 0 & 0 & 0 & 0 & 1 & 0 \\ 0 & 0 & 0 & 0 & 0 & 1 \\ 0 & 0 & 0 & 0 & 0 & 0 \\ 0 & 0 & 0 & 0 & 0 & 0 \\ 0 & 0 & 0 & 0 & 0 & 0 \end{bmatrix}, \quad B(t) = \begin{bmatrix} 0 & 0 & 0 \\ 0 & 0 & 0 \\ 0 & 0 & 0 \\ \dfrac{\sin(-\theta_2)}{l_t M} & \dfrac{\sin\theta_3}{l_t M} & \dfrac{\sin(-\theta_6)}{l_t M} \\ \dfrac{\cos(-\theta_2)}{l_t M} & \dfrac{\cos\theta_3}{l_t M} & \dfrac{\cos(-\theta_6)}{l_t M} \\ \dfrac{r_2 \cos\psi_2}{l_t I_z} & -\dfrac{r_3 \cos\psi_3}{l_t I_z} & \dfrac{r_6 \cos\psi_6}{l_t I_z} \end{bmatrix}$$

Here, the input matrix $B(t)$ becomes a time-variant matrix.

4.2. *Optimal servo system*

The servo system that the system shown in Eq. (6) follows to the desired value is designed.

$$\begin{cases} \dot{z}(t) = r(t) - Cx(t) \\ \dot{x}(t) = Ax(t) + B(t)u(t) \end{cases} \tag{7}$$

Equation (3) is described in equation form as follows:

$$\dot{x}_g(t) = A_g x_g(t) + B_g(t)u + f_g r \tag{8}$$

The control input to the actuator driving the rotating link is obtained to minimize the following equation:

$$J=\int_0^\infty \left[x_g(t)^T Q x_g(t)+u(t)^T R u(t)\right]dt \tag{9}$$

Here, $Q(n\times n)$ and $R(m\times m)$ are the weighting matrices given by the design specification, and $Q\geq 0, R>0$. The control input to minimize Eq. (9) is shown as follows:

$$u_b^o(t)=-R^{-1}B_g^T(t)P(t)x_g \tag{10}$$

Here, $P(t)(n\times n)$ is the solution to the following Ricatti equation:

$$P(t)A_g+A_g^T P(t)-P(t)B_g(t)R^{-1}B_g^T(t)P(t)+Q=0 \tag{11}$$

5. Walking Experiments

5.1. *Straight walking by three support legs and control method*

A six-legged robot walks using three support legs and three swinging legs. The orbit of the six-legged robot is straight walking at 8 cm per step, where one cycle of walking is 10 s in duration, and the total number of steps is five. In this case, the y-direction is the walking direction. The six-legged robot used in the present study has three rotating joints, a rotating part, a thigh part, and a shank part in each leg. The reference signals for each joint angle are calculated based on inverse kinematics from the toe orbit. Each leg joint is controlled by PD control to follow the reference signal. The proportional and differential gains are determined by trial and error. In the control method, the control input of the rotating part is taken as the sum of the input of the PD control and the input of the proposed control method for walking directional control in order to reduce the error. The feedback (FB) control input obtained by Eq. (10) is expected to improve the control performance in cases where the reference signal is a step function because the control system shown in Eq. (8) is type I. However, in the case of walking, the reference signal in the walking direction is a ramp function, and a steady error occurs. Therefore, we though of a control method where each rotational link is controlled by control inputs based on the sum of the PD control (which is a control method that follows the reference signal obtained by inverse kinematics) and the walking directional control. In this control method, the errors for the position and yaw angle of the body are compensated by the walking directional control.

5.2. *Experimental results and discussion*

Figure 6 shows the experimental results. Figure 6(a) shows the experimental results in the y-direction, which is the robot's moving direction. Figure 6(b) shows the experimental results for the yaw angle. In the experiments, the six-legged robot did not move until 10 sec, and after 10 sec, the six-legged robot performed a straight-line walk. The red solid line shows the walking directional control method, the blue dashed-dotted line shows only PD control, and the green dotted line shows the reference signal. In the walking directional control, a Kalman Filter was designed for the system, as shown in Eq. (6), and the velocity components of the body were estimated. Also, the weighting matrix Q in Eq. (9) was $Q = diag(\ 1{,}1500{,}250{,}0{,}\cdots{,}0)$ in order to minimize the errors between the reference signals and the control values. The control system shown in Eq. (10) is time-variant system, however, in experiments, we used the fixed FB gain obtained by fixed θ_i of B matrix in Eq. (6). As the walk advances, it is seen that the error accumulates, and the six-legged robot turns to the left with the PD control method controlling each leg link of the robot to follow each reference signal. However, in the case where FB control inputs of the walking directional control method are added, the control performance of the yaw angle improves. From this, we verified the validity of the proposed control method controlling the positions and yaw angle of the body in the experiments.

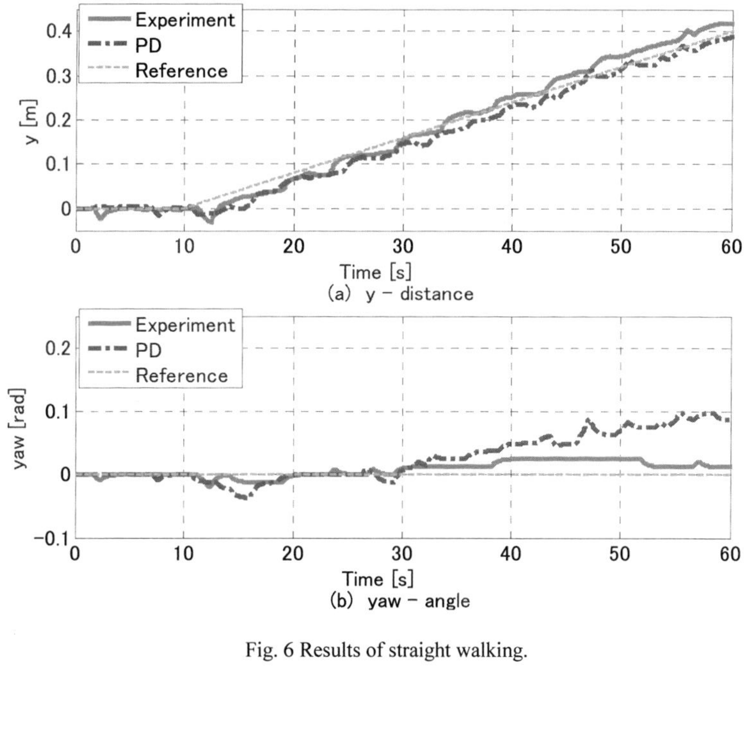

Fig. 6 Results of straight walking.

6. Conclusion

The present study describes a method for measuring the location (x,y) and yaw angle of the body of a six-legged robot using an LRF. We also show the walking directional control method using the data obtained by the LRF. As a walking direction control method for a six-legged robot, a linear-quadratic-integral control method was applied to the mathematical model of the walking directional control. Using the six-legged robot, the effectiveness of the proposed control method was examined in the case where the walking velocity in the y-direction was constant. As a result, the validity of the proposed control method was confirmed by walking experiments.

For future work, the estimation method for the location (x, y) of the robot and the yaw angle will be examined using a Kalman filter. Moreover, the trajectory-following control method for the location and yaw angle will be verified by 3D simulations and walking experiments using a six-legged robot.

References

1. E. Celaya and J.M. Porta,, "A Control Structure for the Locomotion of a Legged Robot on Difficult Terrain", *IEEE Robotics & Automation Magazine*, pp. 43–51(1998).
2. Y. Go, X. Yin and A. Bowling, Navigability of Multi-Legged Robots, *IEEE/ASME Transactions on Mechanics*, 11-1, pp. 1–8(2006).
3. H. Uchida and K. Nonami, Attitude Control of a Six-legged Robot in Consideration of Actuator Dynamics by Optimal Servo Control System, *Climbing & Walking Robots Towards New Application, I-Tech,* pp. 299–312 (2007).
4. H. Uchida and N. Shiina, Walking Directional Control Six-legged Robot by Time-Varying Feedback System, *Autonomous Control Systems and Vehicles, Springer,* pp. 251–263 (2013).
5. H. Uchida, N. Shiina and H. Kondoh Trajectory Following Control for Omni-Directional Walking of a Six-legged Robot, *Proc. 15th Int. Conf. Climbing and Walking Robots and the Support Tech. for Mobile Machines,* pp. 697–704 (2012).
6. H. Uchida, K. Takahashi and M. Suzuki, Method for Estimating Location and Yaw Angle of a Six-legged Robot for Omni-directional Walking Control, *Proc. 16th Int. Conf. Climbing and Walking Robots and the Support Tech. for Mobile Machines,* pp. 773–780 (2013).

MODULATION OF DYNAMIC MOVEMENT PRIMITIVES FOR BIPED LOCOMOTION

JOSÉ ROSADO

Department of Computer Science and Systems Engineering, Coimbra Institute of Engineering, IPC, 3030-199 Coimbra, Portugal

FILIPE SILVA

Department of Electronics, Telecommunications and Informatics, Institute of Electronics and Telematics Engineering of Aveiro, University of Aveiro, 3810-193 Aveiro, Portugal

VÍTOR SANTOS

Department of Mechanical Engineering, Institute of Electronics and Telematics Engineering of Aveiro, University of Aveiro, 3810-193 Aveiro, Portugal

ZHENLI LU

School of Electrical Engineering and Automation, Changshu Institute of Technology, Changshu, China

In recent years, several studies have suggested that improved performance of modern robots can arise from encoding commands in terms of motor primitives. In this context, dynamic movement primitives (DMP) is a powerful tool for motion planning based on demonstrations, well-suited for robot learning. In this work, we study on-line adaptation of biped locomotion patterns when employing DMP. The goal is to demonstrate and evaluate how new movements can be generated by simply modifying the parameters of rhythmic DMP learned in task space. This formulation allows recreating new movements such that the DMP's parameters directly relate to task variables, such as step length, hip height, foot clearance and forward velocity. Several experiments are conducted using the V-REP simulator, including the adaptation of the robot's gait pattern to irregularities on the ground surface and stepping over obstacles.

1. Introduction

In order to properly function in a real world the gait of humanoids robot must be able to constantly and quickly adapt to new situations. In this context, there is an increasing need to move away from robots that are pre-programmed explicitly towards those endowed with learning and adaptation abilities. A considerable research effort has been dedicated to the use of human demonstrations as input for teaching robots to perform from simple movements to complex skills. In particular, robot learning from demonstration is a powerful approach promoting movements that look natural and predictable [1].

Learning biped walking behaviors from human demonstrations have been addressed using different frameworks, number of examples and tasks. The literature provides a few works considering imitation learning in humanoid self-balancing robots [2-4] and others biped walking [5,6], both in simulation and on real robots. Existing learning algorithms, such as imitation learning and reinforcement learning, are difficult to efficiently scale up when directly applied to humanoid robots without any *a priori* knowledge. This is due to the many degrees-of-freedom, the high-dimensional state space, the lack of rigorous models and the dependence on external factors. A possible solution is to identify effective policy representations well suited for robot learning. In this context, nonlinear dynamical system and non-parametric regression techniques are promising candidates to achieve continuous learning and on-line adaptation.

This paper focuses on the problem of adaptation of biped locomotion patterns using rhythmic DMP learned, in the task space, from a single demonstration. For that purpose, the DMP parameters are first related with task variables, such as step length, hip height, foot clearance and forward velocity. Later, during the reproduction phase, the modulation of these parameters is driven by sensory feedback, *e.g.*, foot-contact events and visual information. The paper is organized as follows: Section 2 reviews the mathematical formulation of rhythmic movement primitives and their modulation. Section 3 describes the propose approach for the generation of biped gait patterns. Section 4 describes the computer simulations performed to evaluate the robot's adaptive behavior. Section 5 concludes the paper and discusses future work.

2. Methods

2.1. *Rhythmic movement primitives*

Dynamical system motor primitives have become a robust policy representation, for both discrete and periodic movements, that facilitates the process of learning and improving the desired behavior [7]. The basic idea behind DMP is to use an analytically well-understood dynamical system with convenient stability properties and modulate it with nonlinear terms such that it achieves a desired point or limit cycle attractor. The approach was originally proposed by Ijspeert *et al.* [8] and, since then, other mathematical variants have been proposed [9].

In the case of rhythmic movement, the dynamical system is defined in the form of a linear second order differential equation that defines the convergence to the goal g (baseline, offset or center of oscillation) with an added nonlinear forcing term f that defines the actual shape of the encoded trajectory. This model can be written in first-order notation as follows:

$$\tau\dot{z} = \alpha_z\left[\beta_z(g-y)-z\right]+f$$
$$\tau\dot{y} = z \qquad (1)$$

where τ is a time constant the parameters $\alpha_z, \beta_z > 0$ are selected and kept fixed, such as the system converge to the oscillations given by f around the goal g in a critically damped manner, y is the position and z is the speed. The forcing function f (nonlinear term) can be defined as a normalized combination of fixed basis functions:

$$f(\phi) = \frac{\sum_{i=1}^{N} \psi_i \omega_i}{\sum_{i=1}^{N} \psi_i} r$$
$$\psi_i(\phi) = \exp\left(-h_i(\cos(\phi - c_i) - 1\right) \qquad (2)$$

where ω_i are adjustable weights, r characterizes the amplitude of the oscillator, ψ_i are von Mises basis functions, N is the number of periodic kernel functions, $h_i > 0$ are the widths of the kernels and c_i equally spaced values from 0 to 2π in N steps (N, h_i and c_i are chosen a priori and kept fixed). The phase variable ϕ bypasses explicit dependency on time by introducing periodicity in a rhythmic canonical system. This is a simple dynamical system that, in our case, is defined by a phase oscillator:

$$\tau\dot{\phi} = \Omega \qquad (3)$$

where Ω is the frequency of the canonical system. In short, there are two main components in this approach: one providing the shape of the trajectory patterns (the transformation system) and the other providing the synchronized timing signals (the canonical system).

In order to encode a desired demonstration trajectory y_{demo} as a DMP, the weight vector has to be learned with, for example, statistical learning techniques such as locally weighted regression (LWR). Instead of using a single global model to fit all of the training data, local models attempt to fit the data only in a region around the location of the query using a distance weighted regression. The motivation for exploring locally weighted techniques came from their suitability for online robot learning, because of their fast incremental learning.

The input to the learning algorithm is the demonstration trajectory defined by the triplets of position, velocity and acceleration: $y_{demo}(t)$, $\dot{y}_{demo}(t)$, and $\ddot{y}_{demo}(t)$. Re-arranging equation (1) and equating y to y_{demo}, z to $\tau\dot{y}_{demo}$ and $\dot{z}$ to $\tau\ddot{y}_{demo}$, we obtain:

$$f_{target} = \tau^2 \ddot{y}_{demo} - \alpha_z\left[\beta_z\left(g - y_{demo}\right) - \tau\dot{y}_{demo}\right]$$
$$= \frac{\sum_{i=1}^{N} \psi_i \omega_i}{\sum_{i=1}^{N} \psi_i} r \tag{4}$$

LWR finds, for each kernel function ψ_i, the weight vector ω_i that minimizes the following quadratic error criterion:

$$J_i = \sum_{k=1}^{M} \psi_i^k \left[f_{target}^k - \omega_i r\right] \tag{5}$$

where k represents an index associated to the discrete time steps. This is a weighted linear regression problem that can be solved using a batch or an incremental regression. The incremental solution can be derived using recursive least squares with a forgetting factor λ [10]. Given the target data f_{target}, the update of the weights ω_i is performed for each time-step k as follows:

$$w_i^{k+1} = w_i^k + \psi_i P_i^{k+1} r e^k$$
$$P_i^{k+1} = \frac{1}{\lambda}\left(P_i^k - \frac{{P_i^k}^2 r^2}{\frac{\lambda}{\psi_i} + P_i^k r^2} \right)$$
$$e^k = f_{target}^k - w_i^k r \tag{6}$$

On the one hand, when the forgetting factor λ is one, batch and incremental learning regressions provide identical weights ω_i. On the other hand, when the forgetting factor is less than one, the differences appear since the incremental regression tends to forget older data and give more weight to recent ones.

2.2. *Modulation of the DMP parameters*

DMP exhibit a desirable property in the context of robot learning from demonstration: the system does not depend on an explicit time variable, giving them the ability to handle spatial or temporal perturbations. This property makes them attractive in order to create smooth kinematics control policies that can robustly replicate and adapt demonstrations. Additionally, its formulation includes a few parameters which allow changing the learned desired behavior.

Based on this formulation, the main goal is to demonstrate how new movements can be generated by simply modifying the parameters of a rhythmic DMP. These parameters can potentially be used to adapt the learned movement to new situations in order, for example, to adapt the final goal position, the movement amplitude or the duration of the movement. Therefore, the amplitude, frequency and offset parameters will be modified by scaling the corresponding parameters r, ω and g, respectively.

Figure 1 shows the time evolution of the rhythmic dynamical system associated to a canonical system with frequency $\Omega = 2Hz$ and whose weights ω_i are learned by incremental regression to fit two different trajectories: the first is the sum of the first two harmonics of a rectangular wave defined in the interval between $t = 0$ and $t = 8s$; the second trajectory is the sum of the first three harmonics of a triangular wave defined in the interval between $t = 8s$ and $t = 16s$. For comparison purposes, the reference input signal is superimposed and vertical dashed lines marks events where the parameters are changed. The dashed-lines in the plot indicate, from left to right, the instant in which the learned signal doubles the amplitude r (at t = 4 s), change the reference signal (at t = 8 s) and modulates the baseline g (at t = 12.5 s). The most important observation is that the changes of parameters result in smooth variations of the trajectory $y(t)$ to be reproduced by the robot.

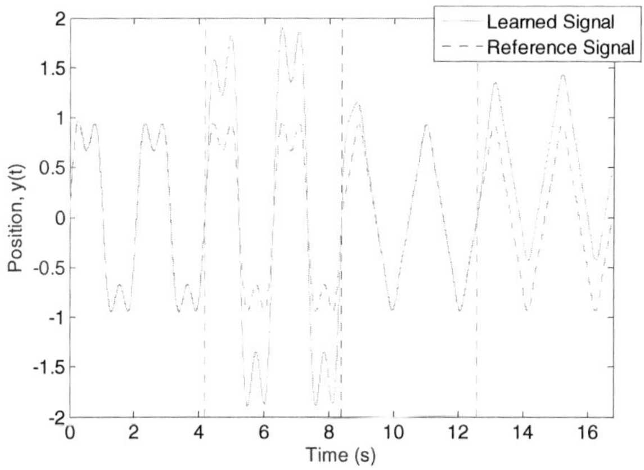

Fig. 1. Time courses of the rhythmic dynamical system with continuous learning (solid line) and the input reference signal (dashed-line). The vertical dashed-lines in the plot indicate, from left to right, the instant in which the learned signal doubles the amplitude r (at t = 4 s), change the reference signal (at t = 8 s) and modulates the baseline g (at t = 12.5 s).

3. Trajectory Formation of Locomotion Gaits

In this study, the proposed solution is to learn the DMP in task space and relate their parameters to task variables. To concretely formulate the dynamical model, a task coordinate system is fixed to the hip section that serves as a reference frame: the y-axis is aligned with the direction of movement and the z-axis oriented downwards). Then, one canonical system per leg and multiple transformation systems associated with the x-, y- and z-coordinates of the robot's end-effectors are adopted. Intra-limb coordination results from planning trajectories in the Cartesian space, constraining the leg to act as one unit. Phase coordination between legs is provided by two separate canonical oscillators coupled such that left and the right limbs move 180 degrees out of phase.

In line with this, a total of six DMP are learned to match the Cartesian trajectories of the lower extremities of both feet. Figure 2 shows the close match between the reference signals and the learned ones as defined in the reference frame. The gray shaded regions show the phases of double-support. Once the complete desired movement $\{y, \dot{y}, \ddot{y}\}$ is learned, new trajectories with similar characteristics can be easily generated. The DMP parameters are directly relate to task variables, such as step length, hip height, foot clearance and forward velocity. For example, the frequency is used for speed up or slow down the motion, the amplitudes of the DMP associated with the y- and z-coordinates are used to modify the step length and the hip height (or foot clearance) of the support leg (or swing leg), respectively. At the end, the outputs of these DMP are converted, through inverse kinematics, to the desired joint trajectories used as reference input to a low-level feedback controller.

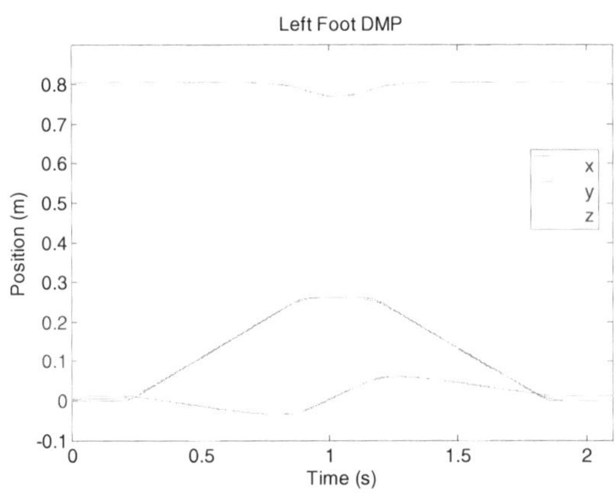

Fig. 2. Result of learning the single-demonstration: the task is specified by the x, y and z-coordinates of the robot's foot in the reference frame. Gray shaded regions show double-support phases.

4. Numerical Simulations

A series of experiments were performed using the Asti model provided on the V-REP software [11]. In the first experiment, the humanoid robot walks over a level surface when it finds a small step of 2 cm high used to approximate irregularities of the environments. The learned DMP is modulated online to properly incorporate the sensory information from the force sensors mounted on the robot feet. More concretely, the dynamic event corresponds to foot-contact information at the instant of impact of the swing foot with the ground. Figure 3 (left side) illustrates snapshots of the robot walking response with these adjustments of the DMP parameters. It was found that using the locomotion pattern as defined by the learned DMP, without any modulation, the robot tolerates step irregularities up to 0.5 cm height. Here, the proposed strategy is to change the baseline and the amplitude of the DMP corresponding to the *z*-coordinate of the swing foot immediately after the foot-contact event occurs.

In the second experiment, it is examined how the adaptation of the DMP parameters could provide the changes in the gait pattern so as to clear an obstacle placed in its path. In this context, the robot system is supposed to receive visual information regarding the obstacle location in order to modify the basic gait pattern. Several computer simulations of varying heights and proximity were conducted. Figure 3 (right side) shows a sequence of captured images of the robot stepping over a 5 cm obstacle height. The obstacle, placed at an arbitrary location, is cleared by the sequential modulation of the following parameters: the step length and the hip height when approaching the obstacle, and the trajectory of the swing foot (foot clearance) while stepping over the obstacle.

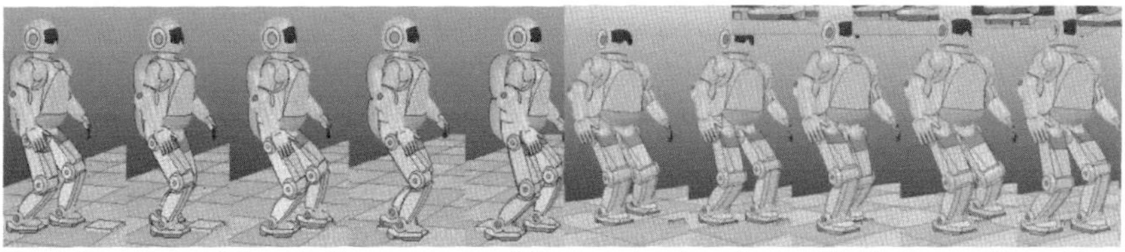

Fig. 3. Snapshots of the robot's response: walking on a level surface and it finds a small step 2 cm high that disturbs its balance (left); and avoiding an obstacle 5 cm high (right).

5. Conclusions

This paper presents a study in which online modulation of the DMP parameters are used for providing adaptation of biped locomotion. The proposed approach is based on the assumption that DMP are learned in the task space from a single demonstration. Therefore, the rhythmic DMP directly relate to locomotion parameters, such as step length, hip height, foot clearance and forward velocity. Numerical simulations show that the proposed formulation is well-suited for biped locomotion, namely to achieve robust steady-state walking in terrains with irregularities and to provide anticipatory adaptation for stepping over obstacles of different heights. Ongoing research aims to integrate sensory feedback into the DMP formulation by adapting the reference trajectory.

Acknowledgements

This work is partially funded by FEDER through the Operational Program Competitiveness Factors - COMPETE and by National Funds through FCT - Foundation for Science and Technology in the context of the project FCOMP-01-0124-FEDER-022682 (FCT reference Pest-C/EEI/UI0127/ 2011).

References

[1] Argall B, Chernova S, Veloso M, Browning B., "A survey of robot learning from demonstration", Robotics and Autonomous Systems 57(5): pp. 469-483, 2009.

[2] Kulic, D.; Ott, C.; Lee, D.; Ishikawa, J. and Nakamura, Y., "Incremental learning of full body motion primitives and their sequencing through human motion observation", The International Journal of Robotics Research. 31(3): 330-345, 2012.

[3] Kormushev, P, Nenchev, D, Calinon, S, and D. Caldwell, "Upper-body kinesthetic teaching of a free standing humanoid robot", Proceedings of IEEE International Conference on Robotics and Automation, 2011.

[4] Lee, D., C. Ott, and Y. Nakamura, "Mimetic communication model with compliant physical contact in Human–Humanoid interaction," The International Journal of Robotics Research, 29: 1684-1704, 2010.

[5] Chalodhorn, R, D. Grimes, K. Grochow, R. Rao, "Learning to Walk by Imitation in Low-Dimensional Subspaces". Advanced Robotics 24(1-2): 207-232, 2010.

[6] Nakanishi, J, Morimoto J, Endo G, Cheng G, Schaal S, Kawato M., "A framework for learning biped locomotion with dynamical movement primitives", International Journal of Humanoid Robots, 2004.

[7] Kober J, Peters J., "Policy search for motor primitives in robotics", Machine Learning, 2010.

[8] Ijspeert A, Nakanishi J, Schaal S., "Movement imitation with nonlinear dynamical systems in humanoid robots", In Proceedings of the 2002 IEEE International Conference on Robotics and Automation; 1398-1403, 2002.

[9] Ijspeert A, Nakanishi J, Hoffmann H, Pastor P, Schaal S., "Dynamical movement primitives: learning atttractor models for motor behaviors. Neural Computation", 25: 328-373, 2013.

[10] Schaal S and Atkeson C, "Constructive incremental learning from only local information. Neural Computing", 10, 2047-2084, 1998.

[11] Rohmer, E., Singh, S., & Freese, M., "V-REP: A versatile and scalable robot simulation framework", IEEE/RSJ International Conference on Intelligent Robots and Systems, pp. 1321-1326, 2013.

LOCOMOTION COST IN TROTTING GAIT OF FOUR-LEGGED ROBOT

K. SHIBUYA*, T. YAMADA, and R. URAMATSU

Dept. of Mechanical and Systems Engineering, Ryukoku University,
1-5 Yokotani, Seta-Oe, Otsu, Shiga, 520-2194, Japan
**E-mail: koji@ryukoku.ac.jp*
http://mec3342.mecsys.ryukoku.ac.jp/sibuya/index.htm

This paper deals with locomotion cost (energy consumption per unit distance) in the trotting gait of our four-legged robot compared to that of horses. First, we designed a new toe with a spring to absorb the impacts at touchdown, which we expect to increase the stability of the robot. We attached the toe to the robot leg and achieved its trotting gait without any support. We measured the locomotion cost in the trotting gait of the robot at several speeds. Based on the results, we found that the locomotion cost decreases when the locomotion speed increases. We concluded that this phenomena is caused by a decrease in the number of steps.

Keywords: locomotion cost; four-legged robot; trot; horse.

1. Introduction

The goal of this paper is to measure the locomotion cost (energy consumption per unit distance) for our four-legged robot and compare the result to that of real horses. To date, many four-legged robots have been developed. For example, Hirose et al. developed a series of quadruped robots and proposed useful control methods and the stability criterion[1,2]. Gonzalez et al. studied the control problem of their quadruped robot in detail[3]. Those studies mainly focused on control problems such as the creation of gait motion and robot stability. In contrast, many researchers have recently focused on energy efficiency of legged robots because, in general, legged robots consume much more energy than wheeled vehicles. For example, Ackerman et al. found that elastically suspending a load can increase the energy efficiency of legged robots[4]. Zarrouk and Fearing estimated the cost of transport of a hexapedal robot to find out the optimum locomotion paths for battery performance[5].

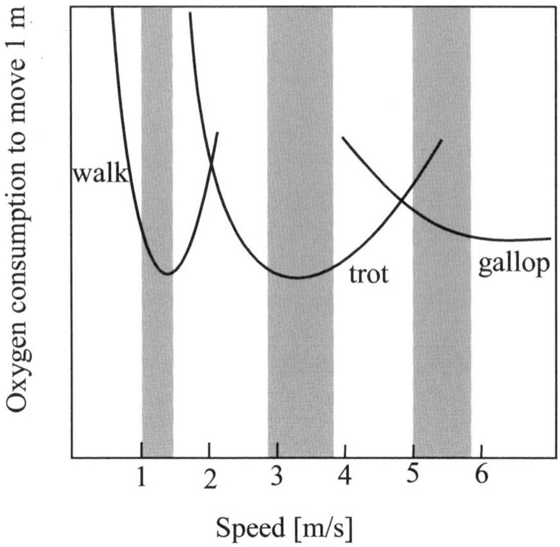

Fig. 1. Schematic drawing of relationship between oxygen cost and speed (redrawn from Ref. 1 by the authors)

In the research field of biology, locomotion costs of various animals have been studied. In 1981, Hoyt and Taylor reported the experimental results of the oxygen consumption of real horses in moving 1 m (we call it "oxygen cost" in this paper)[6]. Figure 1 shows a schematic drawing of their result. The solid curved lines are the approximated lines from the experimental results, and the gray areas indicate the region of speed in which horses usually walk, trot and gallop. They found that the oxygen cost for all is minimal. Particularly, the values for walking and trotting are almost the same. The horse usually moves at a speed for which the oxygen cost is minimal. Taylor et al. recently reported experimental results on the energy cost per unit mass, in which they found that the minimum energy cost per unit mass of animals decreases as their mass increases[7].

We are focusing on the above phenomena related to a horse's oxygen consumption and question whether it can be applied to real four-legged robots, although the oxygen cost of horses is not identical to the locomotion cost of a robot. If the relationship between the speed and the oxygen cost of hoses is applicable to four-legged robots, they should choose the most energy-efficient speed to reduce the locomotion cost that depends on its gait. In this paper, we report and briefly discuss an experimental result of energy cost of our four-legged robot, which was calculated from the electronic current and angular velocity all of its joints in the trotting gait.

2. Outline of Four-legged Robot

The left picture in Fig. 2 is a photograph of the robot used in the experiment described below. We are developing a four-legged robot that has almost identical link-length ratios as those of horses[8]. The foreleg and hind leg have two (shoulder and knee) and three joints (hip, knee, and ankle), respectively. Thus, the robot has ten joints in total. The dimensions of length, depth and height are approx. 330 mm, 540 mm, and 600 mm, respectively. The total mass is approximately 25 kg.

The robot originally had a head and a neck joint to study the effects of the neck-swing motion on the body posture during locomotion[8]. However, we removed them in this paper because the head is too heavy to walk smoothly, which prevents smooth locomotion.

A spring was installed in each toe (middle and right legs of the right picture in Fig. 2) because the landing impact seemed to prevent smooth locomotion. As a result, the robot trotted more smoothly than the previous version (left leg in the right picture in Fig. 2).

Fig. 2. Left: four-legged robot used in this research, Right: previous leg (left), new hind leg (middle), and new foreleg (right)

Each joint is driven by a D.C. motor, which is connected to a PC via a servo amplifier. A rotary encoder is installed in each motor to measure the joint angle and angular velocity.

The toe trajectory is set based on real horses, and by calculating inverse kinematics, trajectories of each joint can be calculated. To calculate the joint angle of the hind legs, we set the direction of the toe of the hind leg. The joint angles are controlled according to the calculated joint trajectories. However, we do not use any control method (ex. Z.M.P.) to achieve stabilized locomotion.

3. Calculation Method of Locomotion Cost

To calculate the locomotion cost, we have to measure the energy consumption during the locomotion, which can be achieved by measuring electric current I_{ij} supplied to the D.C. motors, where i and j are the number of motors and the number of sampling data. Although there are several methods to measure the electric current, we decided to use the information obtained from the servo amplifiers because the output voltage of the amplifier is PWM, and it is difficult to measure the current directly. Of course, some errors between the real current and the information of the amplifier exist. Thus, we conducted some preliminary experiments and obtained an equation to calculate the real current value from the information of the amplifier.

Then, we calculated the instantaneous power. From the information obtained from the encoder, we can calculate the angular velocity of the motor ω_{ij}. By multiplying I_{ij}, ω_{ij}, and the torque constant of the motor K_T, we can obtain the instantaneous power W_{ij} as follows:

$$W_{ij} = K_T \cdot I_i \cdot \omega_{ij}. \tag{1}$$

By assuming that W_{ij} is constant during the sampling period Δt (10 ms in this paper), we can calculate the energy during the sampling period ΔE_{ij} as follows:

$$\Delta E_{ij} = W_i \cdot \Delta t = K_T \cdot I_{ij} \cdot \omega_{ij} \cdot \Delta t. \tag{2}$$

By summing all ΔE_{ij} of all the motors during the locomotion, we can obtain the total energy E:

$$E = \sum_{j=0}^{m} \sum_{i=0}^{n} \Delta E_{ij}, \tag{3}$$

where m is the number of motors and n is the number of sampling data. Finally, by dividing E by the locomotion distance L, we can calculate the locomotion cost C:

$$C = \frac{E}{L}. \tag{4}$$

In addition to mechanical energy used for locomotion, the calculated locomotion cost includes such wasted electrical energy as Joule heat in motors and their amplifiers.

4. Experiment

4.1. *Method*

We conducted an experiment using the four-legged robot introduced in Section 2. As described, we determined toe trajectory before the experiment. Figure 3 shows an example of the trajectories used in this study. D in this figure is a step distance, and we altered it to change the average speed of the robot. The number of D is twelve, which varies from 0.055 m to 0.080 m every 0.005 m.

The walking cycle and the maximum toe height are 1 s and 0.01 m, respectively, which were fixed values. At the moment the robot reached 0.5 m, we stopped the control software. The moment was judged by visual inspection. The locomotion time was measured by the operating time recorded in the PC, and the average speed was calculated by dividing 0.5 m by the locomotion time.

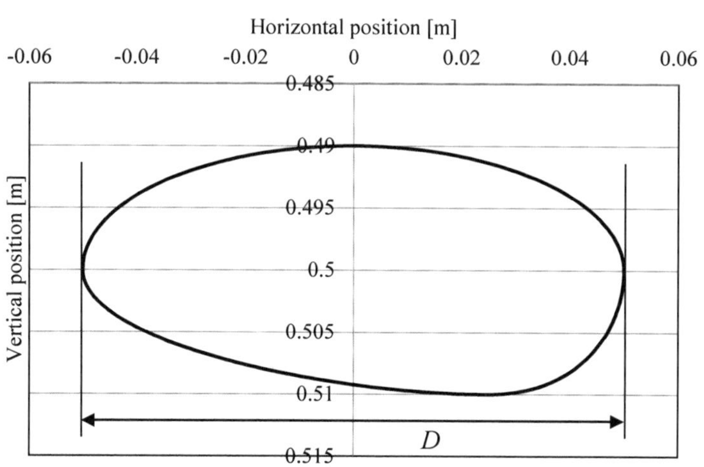

Fig. 3. Example of toe trajectory

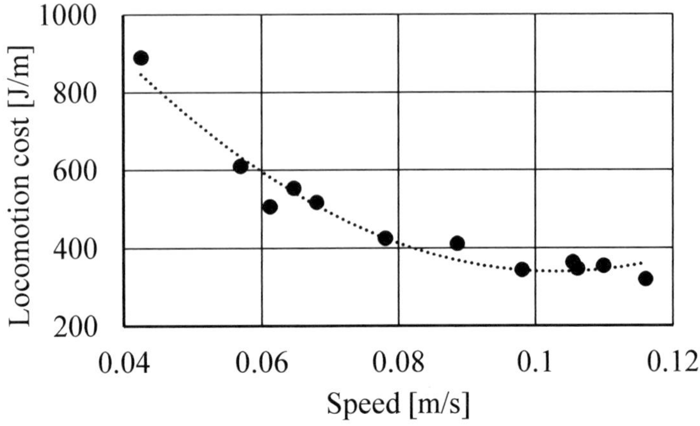

Fig. 4. Relationship between energy cost and speed

4.2. *Experimental result*

Figure. 4 shows the relationship between the average speed and the locomotion cost. The dotted curved line is an estimated line. From this figure, we can observe that the locomotion cost decreases when the average speed increases, although we cannot find any clear evidence as to whether the minimal point exists. However, the shape of the graph resembles the left half of the solid lines in Fig. 1. We considered that the experimental result partially represents the same tendency of the horse's oxygen cost.

5. Discussion

5.1. *Reason for decrease of locomotion cost*

One of the reasons of the decrease of the locomotion cost is the decrease of the number of steps. By increasing the value of D in Fig. 3, the number of steps decreases. This can be confirmed by the graphs of the energy consumption shown in Fig. 5, which show the results of $D = 0.06$ m and 0.08 m. The number of peaks apparently different from each other (14 in (a) and 8 in (b), except the first peaks). We considered that at the peaks the robot touched down because the motor needed the largest torque that was proportional to the supplied electrical current in order to endure its self-weight and impact force. That well explains the reason for the decrease of locomotion cost with increase of locomotion speed. Although the oxygen cost is not completely identical to the locomotion cost of a robot, this is an

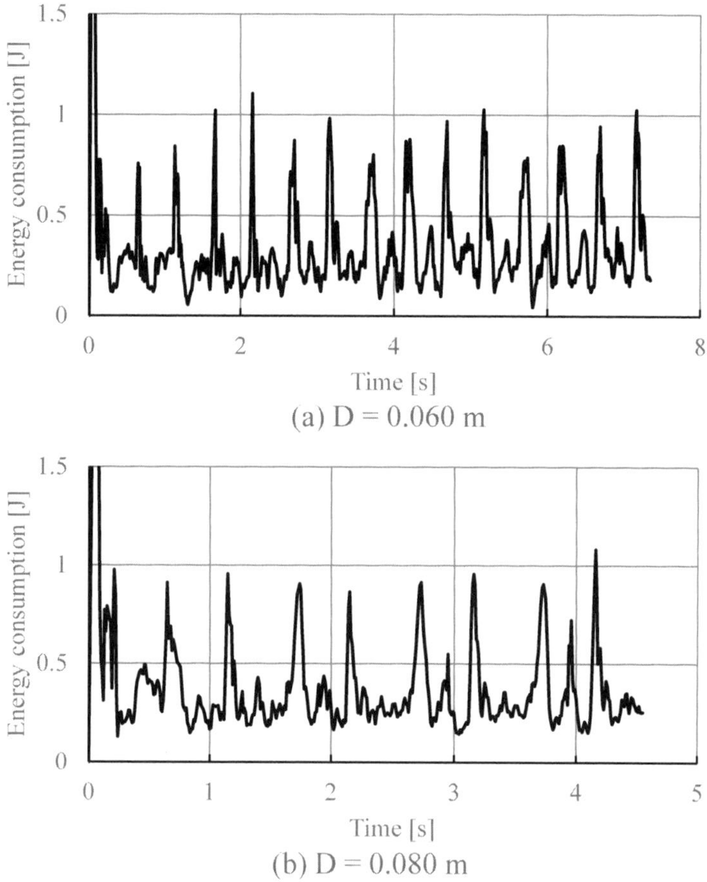

Fig. 5. Energy consumption during trotting

interesting result.

5.2. *Locomotion cost of higher speed*

Then, we have a question: if the speed exceeds 0.12 m/s, will the locomotion cost increase? At present, we do not have any experimental results to answer this question. However, we consider that it might be true because to increase the average speed the robot has to increase the walking cycle (fixed to 1 s in this paper), which will demand more electrical current.

The reason the robot cannot increase the average speed to more than

0.12 m/s is hardware limitations. In addition to improving the robot hardware, reducing the dimensions of the robot (building a smaller robot) may be one of the solutions for this problem. Also, the locomotion distance was too small to discuss the locomotion cost. Because the robot does not adopt any control method to stabilize its gait, it often falls down soon after the beginning of the locomotion. Thus, we must develop a stabilization method to achieve long-distance locomotion.

6. Conclusion

This paper reported the locomotion cost of a four-legged robot in the trotting gait and compared it to a horse's oxygen cost. The result seems to support half of the phenomena shown in Fig. 1. Building a mathematical model of the robot and calculating theoretical value by simulation is one of the remained works. In addition, obtaining locomotion cost in other gaits (walking and galloping), improving the hardware and building a control method to achieve long-distance and the high-speed locomotion remain as future works.

References

1. Hirose, S., Tsukagoshi, H., and Yoneda, K. Normalized energy stability margin: Generalized stability criterion for walking vehicles, *Proc. of the International Conference on Climbing and Walking Robots*, 71-76, Brussels, Belgium, (1998).
2. Kato, K. and Hirose, S. Development of the quadruped walking robot, TITAN-IX -mechanical design concept and application for the humanitarial de-mining robot, *Advanced Robotics*, **15**(2), 191-204, (2001).
3. Pablo Gonzalez de Santos, Garcia, E., and Estremera, J. Quadrupedal Locomotion and introduction to the control of four-legged robots, *Springer*, (2006).
4. Ackerman, J. and Seipel, J. Energy Efficiency of Legged Robot Locomotion With Elastically Suspended Loads, *IEEE Transactions on Robotics*, **29**(2), 321-330, (2013).
5. Zarrouk, D., and Fearing, R. S. Cost of locomotion of a dynamic hexapedal robot. *IEEE Int. Conf. Robotics and Automation (ICRA)*, 2548-2553, Karlsruhe, Germany, (2013), DOI: 10.1109/ICRA.2013.6630925
6. Hoyt, D. F., Taylor, C. R. Gait and the energetics of locomotion in horses, *Nature*, **292**(16), 239-240, (1981).
7. Langman, V. A., Rowe, M. F., Roberts, T. J., Langman N. V., and Taylor, C. R. Minimum cost of transport in Asian elephants: do we really need a bigger elephant?, The Journal of Experimental Biology, 215, 1509-1514, (2012).
8. Shibuya, K., and Matsuhira, K. Effects of Neck Swing Motion on the Body Posture of a Four-legged Robot, *Proc. of 13th International Conference on Climbing and Walking Robots and the Support Technologies for Mobile Machines (CLAWAR2013)*, 503-510, Sydney, Australia, (2013).

A NEW PROCEEDING OF CONTROL STRATEGY FOR A PARALLEL STRUCTURAL BIPED ROBOT

SHUCEN DU and JOSEF SCHLATTMANN

Workgroup on System Technologies and Engineering Design Methodology, Hamburg University of Technology, Denickestrasse 17(L) Hamburg, 21073, Germany

In the applications of simulator, accurate machining and soldering, a parallel hexapod can usually be found. However, it is still an innovative research object to implement this structure in a personal biped walking application. This article represents a realization of a walking mechanism implying parallel Hexapods as moving parts. Working principle and control structure are described in this work. To realize a stable and riding comfortable walking mechanism, the planning and control of stability is fatal. This aspect is inevitable tightly related to the pose recognition algorithm. To solve a forward kinematic problem (FKP) is always a difficulty in the recognition of pose. In this article, a method integrating MEMS inertial measurement units and lengths measurements is designed. With the force sensors, the center of pressure can also be confirmed. This fusion of sensory system is further to realize a steady and agile control of this biped walking robot.

1. Introduction

The appearance of biped walking robots enables a possibility to realize a locomotion that overcomes uneven surface or obstacles. The legs of biped robots can be comprised of serial arranged links or parallel arranged links. Many comparisons of two mechanisms are made, e. g. Y. D. Patel and P. M. George in [1], Y. D. Patel and P. M. George in [2] and so on.

The advantages of serial structure are widely thought as: larger work space, higher work space/ robot size ratio and easier solving dynamic and so on. Contrarily, a parallel mechanism shows its advantages in: stiffness, payload/ weight ratio, repeatability, small inertia and so on. For individual transportation, the demand of velocity and large step length are minor, but the safety and quick response to force interference (e.g. movements of passenger) become more important. Also, along the development of digital processors and sensors, the difficulties of modeling and solving dynamics can be compensated in a rank. For personal transportation, a biped robot formed by parallel mechanisms could be therefore a competitive alternative.

In this article, we represent CENTAUROB [3], a prototype of biped walking robot, developed in the Workgroup on System Technologies and Engineering Design Methodology of Hamburg University of Technology. The position control, pose recognition and balance control strategy will also be given.

2. The Biped Walking Robot-CENTAUROB

CENTAUROB is serially comprised of two six-DOF Gough-Stewart platforms, which are also known as parallel hexapod [4]. Two parallel hexapods form the two legs of the robot and support the whole structure in turn. One platform as the hip carries motor driving cards, battery package and master computer. The feet are distinctive C-shape. The construction and 3D model are given as Figure 1.

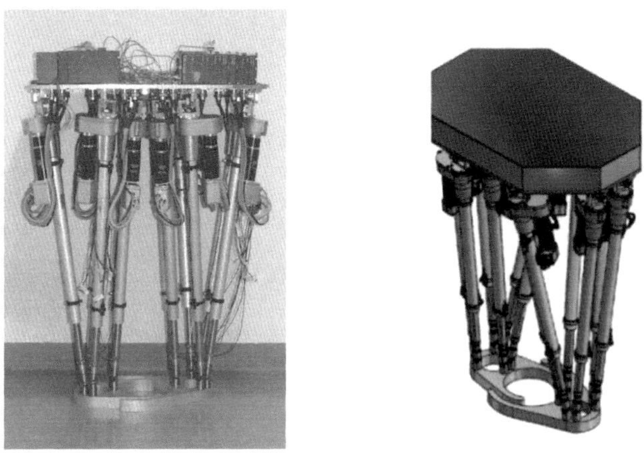

Figure 1. Photo and 3D model of CENTAUROB.

Between platforms, 12 linear actuators driven by motor units are connected to the platforms with 24 joints: 12 3-DOF-joints at hip's side, 12 2 universal joints at feet's side. As illustrated in Figure 1. To confirm the DOF of a motion mechanism, Eq. (1) is used.

$$F = 6(n - e - 1) + \sum_{g=1}^{e} f_g - \sum f_{id} \tag{1}$$

In which, take one leg as example, n = 14 mean the links' number, e = 18 is the number of joints. For 3-DOF-joints, the f = 3, and for universal joints, f = 2. A linear unit possesses f = 1. The sum of identical DOF Σf_{id} =0. We can therewith confirm a 6DOF mechanism from this calculation.

3. Inverse Kinematic of Hexapod

The coordinate systems of the robot are defined as below: the inertial frame {W}, body-fixed frames {A}, {B}, {C} and {D}. As illustrated in Figure 2.

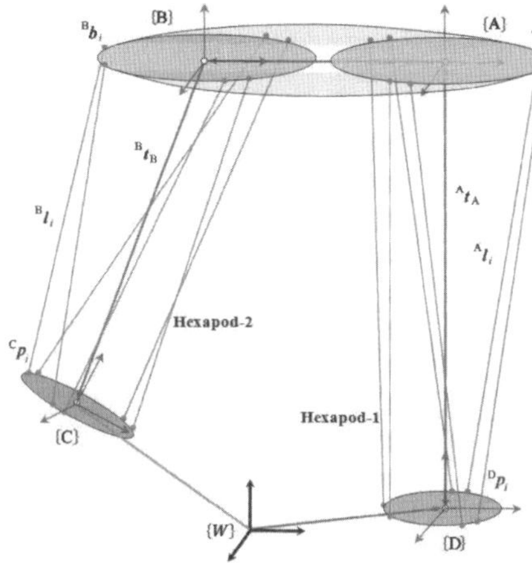

Figure 2. Coordinate system of CENTAUROB.

In Figure 2, vectors ${}^{A}\boldsymbol{t}_{A}$ and ${}^{B}\boldsymbol{t}_{B}$ mean the displacement from corresponding coordinate centers on the hip to both feet' s centers, ${}^{A}\boldsymbol{b}_{i}$ and ${}^{B}\boldsymbol{b}_{i}$ mean the distribution of joints on the hip, ${}^{C}\boldsymbol{p}_{i}$ and ${}^{D}\boldsymbol{p}_{i}$ mean the distribution of the joints on feet, ${}^{A}\boldsymbol{l}_{i}$ and ${}^{B}\boldsymbol{l}_{i}$ indicate the vector of the linear units. The left superscripts show the reference coordinate. ${}^{A}\boldsymbol{l}_{i}$ and ${}^{B}\boldsymbol{l}_{i}$ are from the joints on the hip and pointing to the joints on the feet.

To convert a vector from one frame to another, a rotation matrix is needed. We define a complex angular displacement takes place in the turn of x, y and z. The symbol ${}^{Y}{}_{X}\mathbf{R}$ implies the conversion from coordinate system X to Y. Taking the leg connected to frame {A} as an example, an expression of the position relationships among the vectors is derived in Eq. (2).

$$ {}^{A}\mathbf{l}_{i} = {}^{A}\mathbf{t}_{A} + {}^{A}{}_{D}\mathbf{R}\, {}^{D}\mathbf{p}_{i} - {}^{A}\mathbf{b}_{i} \tag{2} $$

And the length of each linear unit can be expressed as Eq. (3),

$$ l_{i} = \|{}^{A}\boldsymbol{l}_{i}\| = \|{}^{A}\boldsymbol{t}_{A} + {}^{A}{}_{D}\mathbf{R}\, {}^{D}\boldsymbol{p}_{i} - {}^{A}\boldsymbol{b}_{i}\| \tag{3} $$

Or expressed as

$$ l_{i} = \sqrt{{}^{A}\boldsymbol{t}_{A}^{T}\cdot{}^{A}\boldsymbol{t}_{A} + {}^{A}\boldsymbol{p}_{i}^{T}\cdot{}^{A}\boldsymbol{p}_{i} + {}^{A}\boldsymbol{b}_{i}^{T}\cdot{}^{A}\boldsymbol{b}_{i} + 2\,{}^{A}\boldsymbol{t}_{A}^{T}\cdot\left({}^{A}\boldsymbol{p}_{i} - {}^{A}\boldsymbol{b}_{i}\right) - 2\,{}^{A}\boldsymbol{p}_{i}^{T}\cdot{}^{A}\boldsymbol{b}_{i}} \tag{4} $$

In which, ${}^{A}\boldsymbol{p}_{i} = {}^{A}{}_{D}\mathbf{R}\, {}^{D}\boldsymbol{p}_{i}$.

To drive the motors correctly, the length vary velocities of linear units are needed. These velocities are considered referred to the hip frame {A}. And a minimal coordinate system is chosen as

$$ {}^{A}y = \left\{ {}^{A}\boldsymbol{t}_{A}^{T}, \theta_{A}, \psi_{A}, \varphi_{A} \right\}^{T}. \tag{5}$$

The time derivative of Eq. (2) in frame {A} is

$$ {}^{A}\dot{\boldsymbol{l}}_{i} = {}^{A}\dot{\boldsymbol{t}}_{A} + {}_{D}^{A}\mathbf{R} \cdot {}^{A}\tilde{\omega}_{D} \cdot {}^{D}p_{i} \tag{6}$$

Project the vector in Eq. (6) to the unit vector of Eq. (2), length vary velocities of the linear units are obtained, as followed equation.

$$ \dot{l}_{i} = {}^{A}\boldsymbol{l}_{i} \cdot {}^{A}\dot{\boldsymbol{l}}_{i} / l_{i} \tag{7}$$

In Eq. (4) and Eq. (7), "·" means the inner product computation. Now substitute the transmission ratio of the gearbox and lead screw, γ, to equation Eq. (4) and Eq. (7), a mapping from platforms' motion in Cartesian space to the motor motion space is built in Eq. (8).

$$ \begin{cases} qc_{i} = 2000 \cdot \gamma \cdot (l_{i} - l_{i0}) \\ n = 60 \cdot \gamma \cdot \dot{l}_{i} \end{cases} \tag{8}$$

Here, qc_i is the incremental encoder counting number; l_{i0} is the initial length of lead screws; n is the rotational velocity of motor, whose unit is revolutions per minute.

4. Generation of the Trajectory

A statically walking process for the CENTAUROB could be seen as a combination of three phases: starting step (Figure 3 a—b), periodic steps (Figure 3 b—d) and ending step (Figure 3 d—e). Periodic steps are formed of one leg supporting sub-phase and two legs supporting sub-phase, which happen in turns on two legs. A whole steps process is illustrated below.

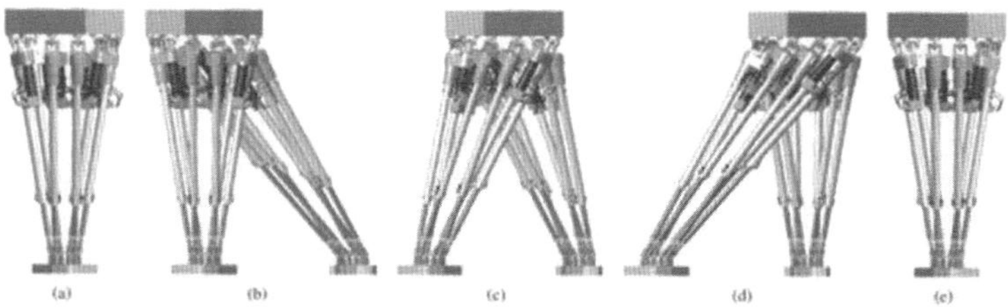

Figure 3. A whole step of the CENTAUROB.

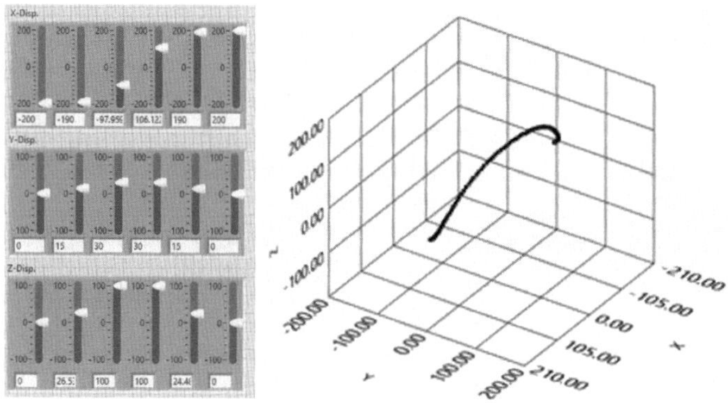

Figure 4. Use Spline interpolation to generate trajectory of swing foot in Cartesian space.

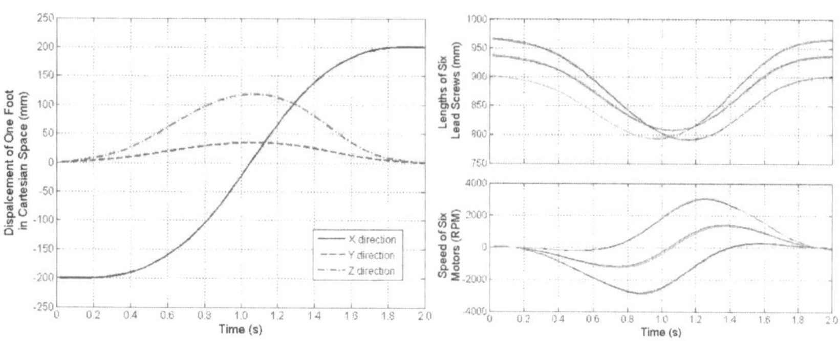

Figure 5. Trajectories in Cartesian space and in the motor motion space.

For steps of the robot, the position trajectory, velocity trajectory and acceleration trajectory should be continuous. In some cases, such as obstacles to avoid, the trajectory should be on-line manually adjustable through setting some characteristic points (left part of Figure 4). A Spline interpolation method is used to generate the Cartesian-spatial trajectory to avoid sudden change of acceleration. The corresponding mapping in the motor motion space is given in Figure 5. As soon as the trajectory is generated, the balance check module will be operated. Here the foot rotation indicator is used as criteria. This choice is excessive for a static walking, but makes it accessible for the future dynamic walking pattern. The calculation of FRI is represented in the work of [5]. Once the calculated FRI fell out of the safe supporting area, the pattern would be modified. The definition of dangerous and safe area of supporting area can be found in Figure 6.

Some regular patterns fitting different occasions will be generated off-line, such as, normal step, stair climbing, turning and so on. These regular patterns will be online divided into segments and converted to motor motional space according to the mapping in Eq. (8). Eventually, the converted segments in motor motion space will be sent to the decentralized motor servo drivers one after another. In an accidental situation, such as unbalance is warned according to the center of pressure (CoP) sensor, a fixing value would be added to the unconsumed regular patterns. These fixed pattern segments will be converted as new trajectory. This stability control strategy will be further described in the following section.

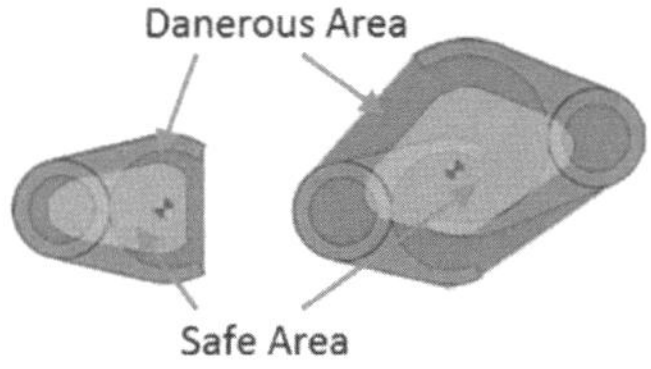

Figure 6. Foot/ feet supporting area and FRI point location.

5. Control Strategy and Sensory System

The control system could be divided into two levels: online stability control in the PC; the three-closed-loop PID control in the motor servo drivers.

The stability control level is formed by pose recognition sensors and CoP location sensors. The raw data will be acquired and filtered by an ATmega2560 based data acquisition board. Then, processed data will be packed and uploaded to PC for a stability control algorithm. The real time CoP location will be continuously checked. So long as the CoP is not on the boundary of the supporting area, it coincides with FRI [5]. This fact allows us to affect CoP. The strategy of stability control can be found in Figure 7.

12 strain gouge force sensors, 3 length sensors and 3 inertial measurement units (IMU) are used in this level. Force sensors are installed between the lead screws and universal joints. Length sensors are superseded by incremental encoders in present phase of project, for the sake of installation space. IMUs are placed on hip and feet. One force sensor provides the absolute force value along each lead screw. The pose information acquired from the length sensors and IMUs provide the forces' directions. These two kinds information provide the resultant force and torque applied on the ankle. For a static foot which fully

contacts the ground, the counter force and torque will be known. Thus a CoP location will be obtained.

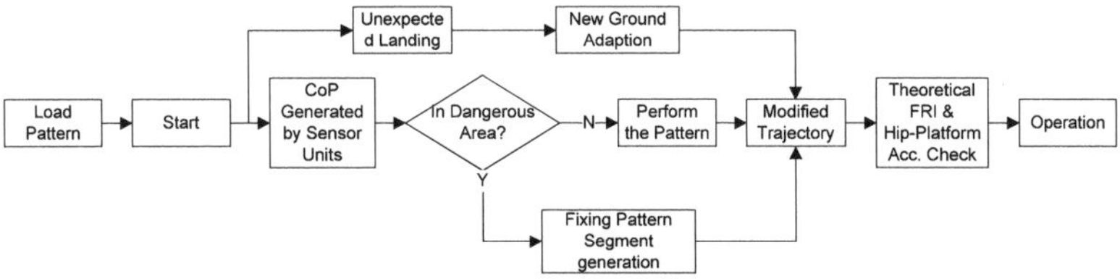

Figure 7. The online stability control strategy.

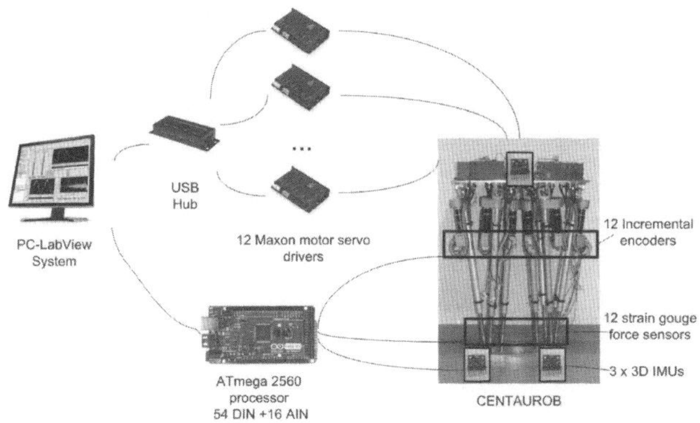

Figure 8. Structure of control system and sensory system.

A barrier of sensing the real position of the end effector of a parallel structure, like Stewart-Platform, is to solve the forward kinematic problem (FKP). In this work, to avoid the complexity of FKP iterations, the real position is acquired through the combination of a three dimensional IMU and length measurement of linear units. IMU provide the orientation, combined with the feedback of three length measurements, the spatial relation between the feet and hip will be deterministic. At this level of the project, the lengths are sensed through the incremental encoders integrated in the motor units. A zero position of the encoder will be set according to end switches installed on the lead screw hub.

6. Conclusions and Prospect

In this paper, a prototype of biped walking robot is represented. The uniqueness is that, two legs are formed of two hexapods, which provide high loaded and statically walking. The working principle, trajectory generation method and control strategy are provided. The sensory system is designed and constructed in a different way from the former design provided in [6].

For present phase, the simulation of motion, mechanical and mechatronical constructions and open looped tests are finished. The close looped walking test will be accomplished. For the trajectory planning, only continuity of the velocity and acceleration are concerned for now. Based on the criteria of time and energy consumption, an optimal trajectory planning strategy will be further developed.

In the motor servo driver level, the PID control is done and separated from the PC. This fact blocks the developer from realizing a direct control, such as pulse width modulation (PWM). Another control's bottle-neck is that, encoder information, as the length measurement, is uploaded through servo driver, which prolong the sampling period of pose recognition. In the next step, length measurement will be bypassing the servo drivers. This is foundation of an agile and precise stability control.

References

1. Z. Pandilov and V. Dukovski. "COMPARISON OF THE CHARACTERISTICS BETWEEN SERIAL AND PARALLEL ROBOTS." *Acta Technica Corvininesis-Bulletin of Engineering* 7.1 (2014).
2. Y. D. Patel and P. M. George. Parallel Manipulators Applications — A Survey. *Modern Mechanical Engineering*, *2*(03), 57, (2012).
3. J. Schlattmann. *Laufmaschine und Verfahren zur Steuerung einer Laufmaschine*. German Patent Nr. 19637501, (2000).
4. J-P. Merlet. "Parallel Robots." *Solid Mechanics and Its Applications*". *Second Edition.* (2006).
5. A. Goswami, "Postural stability of biped robots and the foot-rotation indicator (FRI) point." *The International Journal of Robotics Research* 18.6 (1999): 523-533
6. T. T. Mekonnen and J. Schlattmann, "MEMS based Position and Motion Sensors for Controlling Complex Parallel Mechanisms", *SENSOR+TEST Conference 2009 - SENSOR 2009 Proceedings II*. P.10 (2009): 271-276

CONTROLLER DESIGN FOR AN UNDERACTUATED BIPED WALKING ROBOT*

WENQI HOU, YANGZHEN CHEN, TAIHUI ZHANG, ZHIJUN WANG and HONGXU MA

College of Mechatronic Engineering and Automation, National University of Defense Technology, Deya Road 109# Kaifu district, Changsha, 0731, China

In this paper, we aim at realizing stilt-like biped walking. With the assumption that the Jacobian of task space is full row rank during the single support phase, the dynamics in task space is presented with the Moore-Penrose pseudo-inverse. Considering the inevitable parameter estimation errors, we design a task space controller without calculating the inversion of inertial matrix. We demonstrate the stability of the controller by Lyapunov method. With the designed the controller, we realized stilt-like biped walking on a virtual planar biped walker. The contact consistency is preserved during the biped walking.

1. Introduction

Stilt-like biped walking is considered as efficient and robust to rough terrain. The control of biped walking with point foot aroused researchers' much attention[1-8]. To realize stilt-like biped walking, there are three prime questions need to be answered: When a step should be taken? Where the foot should be placed? How to take the desired step? In our previous work[9], the former two questions has been answered on a variable spring-loaded inverted pendulum (VSLIP) model. In this work, we focus on the third question to realize stilt-like biped walking on a more practical model.

Different methods are utilized to address the position tracking problem in single support phase. Jessy W. Grizzle and Ian R Manchester *et al*[1, 4, 5, 8, 10]. implemented input-output linearizing method for the single support phase to realize position tracking. However, the performance of this method is heavily dependent on the accuracy of the robot's physical parameters. Since there is no actuation between the stance foot and ground, the dynamics of the stilt-like

* This work is supported by the Crossing Specialties Co-educating Program for PhD students in National University of Defense Technology (No. kxk140101), and the National High Technology Research and Development Program of China (No. 2011AA040801), and the National Natural Science Foundation of China (No. 61473304).

robot in single support phase is underactuated. Sliding Model Control (SMC) is full of robustness to the system uncertainties[11]. Sergey G. Nerservo *et al.* provided sufficient conditions for asymptotic stabilization of underactuated nonlinear systems using SMC[12]. Cenk Oguz Saglam *et al.* designed controller based on the SMC method for the single support phase, to have finite time convergence[6]. However, the controller designed in [6] and [12] need to calculate the inversion of the inertial matrix, which make it sensitive to parameter estimation errors.

In this paper, we design a task space controller without calculating the inversion of inertial matrix for the single support phase of biped walking. The rest of the paper is organized as follows. We first present the model of a planar 5-link point foot biped walker, and derive its dynamics in the task space. In section 3, we design a controller for the single support phase, and prove the stability of the controller. The designed controller is implemented on a virtual biped walker to test its performance in section 4. Section 5 concludes the paper and provides the directions for future work.

2. Model

In this work, we aim at realizing stilt-like biped walking and restrict the attention to planar motion. A planar 5-link biped walker with point feet is shown in Fig. 1. The robot comprises of 5 rigid links connected by 4 ideal actuated revolute joints. The motion of the robot's hip is modeled as VSLIP.

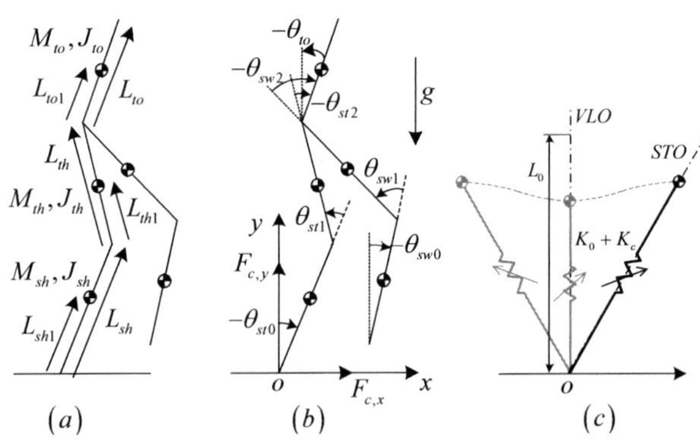

Figure 1. A planar 5-link biped walker standing on one of the point feet: (a) physical parameters of the robot; (b) generalized coordinates; (c) a 2D-VSLIP approximates the motion of the robot's hip.

2.1. *Single support phase*

During the single support phase, if there is no slipping between the stance foot and ground, a set of generalized coordinates can be selected as $q = [\theta_{st0}; q_c]$, where θ_{st0} is the vertical deflection of stance shank, and $q_c = [\theta_{st1};\theta_{st2};\theta_{sw2};\theta_{sw1}]$ is the joint configuration of the robot. The dynamics of the robot during single support phase is[1, 6, 13]:

$$M(q)\ddot{q} + C(\dot{q},q)\dot{q} + G(q) = B^T\tau \tag{1}$$

where $M(q) \in R^{5\times5}$ is the inertia matrix of the robot, $C(\dot{q},q) \in R^{5\times5}$ is the centripetal and Coriolis forces, $G(q) \in R^{5\times1}$ is the gravity forces, $B = \begin{bmatrix} 0_{4\times1} & I_{4\times4} \end{bmatrix}$ is the selecting matrix of the actuated joints, $\tau \in R^{4\times1}$ is the actuated joint torque vector.

Let $P_{tsk} = f(q)$ be the task space vector, the task velocities can be calculated:

$$\dot{P}_{tsk} = \frac{\partial f}{\partial q}\dot{q} = J_{tsk}\dot{q} \tag{2}$$

If $rank(J_{tsk}) = 4$,by using the Moore-Penrose pseudo-inverse of the task Jacobian[14-16]:

$$\dot{q} = J^{\#}_{tsk}\dot{P}_{tsk} \tag{3}$$

$$\ddot{q} = J^{\#}_{tsk}\ddot{P}_{tsk} + \frac{d(J^{\#}_{tsk})}{dt}\dot{P}_{tsk} \tag{4}$$

where $J^{\#}_{tsk} = J^T_{tsk}(J_{tsk}J^T_{tsk})^{-1}$. The dynamics in task space can be expressed as:

$$\tilde{M}\ddot{P}_{tsk} + \tilde{C}\dot{P}_{tsk} + \tilde{G} = F \tag{5}$$

where $\tilde{M} = (J^{\#}_{tsk})^T M J^{\#}_{tsk} \in R^{4\times4}$, $\tilde{C} = \left[(J^{\#}_{tsk})^T C J^{\#}_{tsk} + (J^{\#}_{tsk})^T M \frac{d(J^{\#}_{tsk})}{dt}\right] \in R^{4\times4}$,

$\tilde{G} = (J^{\#}_{tsk})^T G \in R^{4\times1}$, $F = (J^{\#}_{tsk})^T B^T\tau \in R^{4\times1}$. Note that $\dot{\tilde{M}} - 2\tilde{C}$ is skew symmetric. To see this, let $x \in R^{4\times1}, x \neq 0$, consider:

$$\begin{aligned}
& x^T(\dot{\tilde{M}} - 2\tilde{C})x \\
&= x^T\left[(J^{\#}_{tsk})^T \dot{M} J^{\#}_{tsk} + 2(J^{\#}_{tsk})^T M \frac{d(J^{\#}_{tsk})}{dt} - 2(J^{\#}_{tsk})^T C J^{\#}_{tsk} - 2(J^{\#}_{tsk})^T M \frac{d(J^{\#}_{tsk})}{dt}\right]x \\
&= x^T\left[(J^{\#}_{tsk})^T \dot{M} J^{\#}_{tsk} - 2(J^{\#}_{tsk})^T C J^{\#}_{tsk}\right]x \\
&= (J^{\#}_{tsk}x)^T(\dot{M} - 2C)(J^{\#}_{tsk}x)
\end{aligned} \tag{6}$$

For $\dot{M} - 2C$ is skew symmetric, then $x^T(\dot{\tilde{M}} - 2\tilde{C})x \equiv 0$ .Note that, J_{tsk} be full row rank is the prerequisite for Eq. (6). That means the dimension number of task space should be equal to the number of actuated joints.

2.2. *Double support phase*

Many researchers assume the double support phase is instantaneous[1, 5, 6], and model the interaction between the swing foot and the ground as inelastic rigid impact, which results a discontinuity in the velocity components of the robot's states. However, for a physical dynamic system, the position and velocities are continuous. By modeling the legs as massless spring, some researchers studied the control problem during the double support phase for biped walking [2, 3, 8, 10]. The interaction between the foot and the ground is modeled as linear spring model. However, as a stilt-like biped walking, the duration of double support phase is too short to realize control task. In this paper we assume that one leg is playing the role as the supporting leg and the other one is playing the role as the swing leg at any time, even when both the two legs are contacting with the ground. As a result, the roles exchanging is instantaneous, meanwhile, we model the foot-ground contact as nonlinear spring-damp model, that is:

$$\begin{cases} F_y = max\left(0, K_n \left|y\right|^{1.5} + D_n \left|y\right|^{0.5} \dot{y}\right) \\ F_x = clip\left(-K_t \left|y\right|^{0.5} x - D_t \left|y\right|^{0.5} \dot{x}, -\mu F_y, \mu F_y\right) \end{cases} \tag{7}$$

Further details for this contact model is explained in [13, 17]. Then the walking control of the robot only consists of a striding controller and a trajectory tracking controller.

3. Controller Design

In our previous work[9], a striding controller was designed for a SLIP biped walker. In this section, we will design a trajectory tracking controller for the planar 5-link point foot biped walker to realize biped walking.

There are four tasks for single support phase control: $P_{tsk} = [\theta_{tor}; y_{hip}; x_{swft}; y_{swft}]$, where θ_{tor} is the vertical deflection of the torso, y_{hip} is the hip's height, x_{swft} and y_{swft} are the horizontal and vertical distance from the stance foot to the swing foot, respectively. Apparently, these tasks are independently. So $rank(J_{tsk}) = 4$ will always be hold. Let

$$e = P_{tsk}^{des} - P_{tsk} \in R^{4\times 1} \tag{8}$$

where P_{tsk}^{des} is the desired task space trajectories vector. And define

$$S = \dot{e} + Ke, K = \text{diag}(k_1, k_2, k_3, k_4), k_i > 0 \tag{9}$$

Consider a Lyapunov function given by:

$$V = \frac{1}{2} S^T \tilde{M} S \tag{10}$$

As aforementioned $\dot{\tilde{M}} - 2\tilde{C}$ is skew symmetric, then the Lyapunov derivative can be derived as:

$$
\begin{aligned}
\dot{V} &= \frac{1}{2} S^T \dot{\tilde{M}} S + S^T \tilde{M} \dot{S} \\
&= \frac{1}{2} S^T \left(\dot{\tilde{M}} - 2\tilde{C} \right) S + S^T \tilde{C} S + S^T \tilde{M} \dot{S} \\
&= S^T \left[\tilde{C} \left(\dot{e} + Ke \right) + \tilde{M} \left(\ddot{e} + K\dot{e} \right) \right] \\
&= S^T \left[\tilde{C} \dot{P}_{tsk}^{des} - \tilde{C} \dot{P}_{tsk} + \tilde{C} Ke + \tilde{M} \ddot{P}_{tsk}^{des} - \tilde{M} \ddot{P}_{tsk} + \tilde{M} K \dot{e} \right] \\
&= S^T \left[\tilde{C} \left(\dot{P}_{tsk}^{des} + Ke \right) - \tilde{M} \ddot{P}_{tsk} - \tilde{C} \dot{P}_{tsk} + \tilde{M} \left(\ddot{P}_{tsk}^{des} + K\dot{e} \right) \right] \\
&= S^T \left[\tilde{C} \left(\dot{P}_{tsk}^{des} + Ke \right) + \tilde{G} - F + \tilde{M} \left(\ddot{P}_{tsk}^{des} + K\dot{e} \right) \right]
\end{aligned} \tag{11}
$$

If $BJ_{tsk}^{\#} \in R^{4\times4}$ is invertible, a controller can be

$$
\tau = (BJ_{tsk}^{\#})^{-T} \left[\tilde{M}_0 (\ddot{P}_{tsk}^{des} + K\dot{e}) + \tilde{C}_0 (\dot{P}_{tsk}^{des} + Ke) + \tilde{G}_0 + \Gamma \text{sign}(S) \right] \tag{12}
$$

where $\tilde{M}_0 = \left(J_{tsk}^{\#} \right)^T M_0 J_{tsk}^{\#} \in R^{4\times4}$, $\tilde{C}_0 = \left[\left(J_{tsk}^{\#} \right)^T C_0 J_{tsk}^{\#} + \left(J_{tsk}^{\#} \right)^T M_0 \dfrac{d\left(J_{tsk}^{\#} \right)}{dt} \right] \in R^{4\times4}$,

$\tilde{G}_0 = \left(J_{tsk}^{\#} \right)^T G_0 \in R^{4\times1}$, $\Gamma = \text{diag}(\gamma_1, \gamma_2, \gamma_3, \gamma_4), \gamma_i > 0$, and M_0 , C_0 , G_0 are the nominal values of M , C , G , respectively. Let: $\Delta\tilde{M} = \tilde{M} - \tilde{M}_0$, $\Delta\tilde{C} = \tilde{C} - \tilde{C}_0$, $\Delta\tilde{G} = \tilde{G} - \tilde{G}_0$, then

$$
\dot{V} = S^T \left[\Delta\tilde{M} \left(\ddot{P}_{tsk}^{des} + K\dot{e} \right) + \Delta\tilde{G} + \Delta\tilde{C} \left(\dot{P}_{tsk}^{des} + Ke \right) \right] - \Gamma \mid S \mid \tag{13}
$$

For a physical system, there always be a proper vector Γ, such that:

$$
\gamma_i \geq \left[\mid \Delta\tilde{M} \mid \left(\mid \ddot{P}_{tsk}^{des} + K\dot{e} \mid \right) + \mid \Delta\tilde{G} \mid + \mid \Delta\tilde{C} \mid \left(\mid \dot{P}_{tsk}^{des} + Ke \mid \right) \right]_i \tag{14}
$$

Then $\dot{V} \leq 0$, That means the designed controller(12) can guarantee the task space trajectories converge to the desired ones asymptotically. Additionally, this controller doesn't need to calculate the inversion of inertial matrix, so it's more robust to the parameter estimation errors than the controllers in[5, 6, 12].

4. Simulation

We built a 3D biped walker by SolidWorks 2012 as shown in Fig. 2(a). The physical parameters of the biped walker are shown in table 1 as the "model values". To validate the designed controller, we built a virtual 5-link planar biped walker by Matlab SimMechanics toolbox as shown in Fig. 2(b). The physical parameters of the virtual walker are set as the "model values". Considering the inevitable parameter estimation errors for a practical system, especially the mass distribution, the nominal values used in the controller is shown in Table 1 as the "nominal values". Simulation begins with the initial states shown in Table 2.

With the initial configuration in Table 2, the walker's COM is locating at the right side of the supporting foot, due to the gravity the walker will move forward, that is the primary actuation of walking. By taking steps, the horizontal velocity of the hip increases gradually and comes to a stable cycle eventually. Which can be seen in Figure 3.(a).

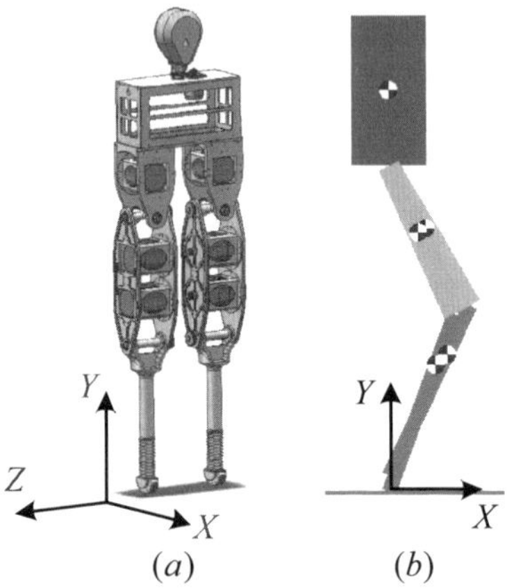

Figure. 2. 3D biped walker and 2D simulation model

Table 1. Physical parameters of the biped walker

		$M_{to}(kg)$	$J_{to}(kg\cdot m^2)$	$L_{to}(m)$	$L_{to1}(m)$
Torso	Model values	28.212	0.461344	0.439	0.208
	Nominal values	29	0.45	0.439	0.22
		$M_{th}(kg)$	$J_{th}(kg\cdot m^2)$	$L_{th}(m)$	$L_{th1}(m)$
Thigh	Model values	6.546	0.096527	0.42	0.231
	Nominal values	7	0.1	0.42	0.24
		$M_{sh}(kg)$	$J_{sh}(kg\cdot m^2)$	$L_{sh}(m)$	$L_{sh1}(m)$
Shank	Model values	2.792	0.068624	0.49	0.35
	Nominal values	3	0.07	0.49	0.36

Table 2. Initial states of the robot. Set the right leg as the stance leg initially

	θ_{st0}	θ_{st1}	θ_{st2}	θ_{sw2}	θ_{sw1}
$q(rad)$	-0.364	0.792	-0.428	0.997	-0.540
$\dot{q}(rad/s)$	0	0	0	0	0

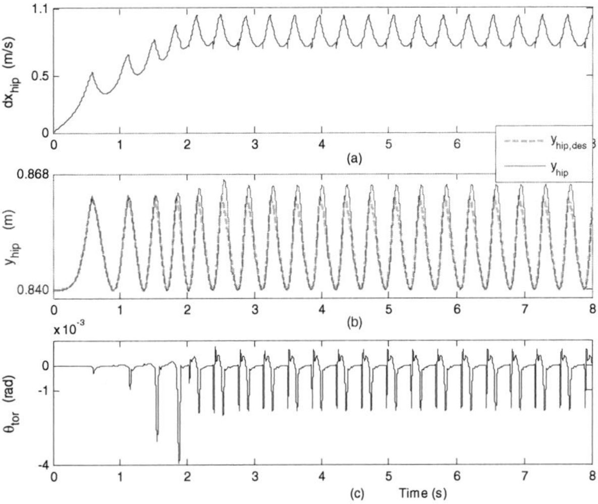

Figure. 3. Simulation results: (a) the horizontal velocity of the hip; (b) the vertical position of the hip: the desired trajectory(dashed line) and the simulation result(real line); (c) the torso's pitch angle.

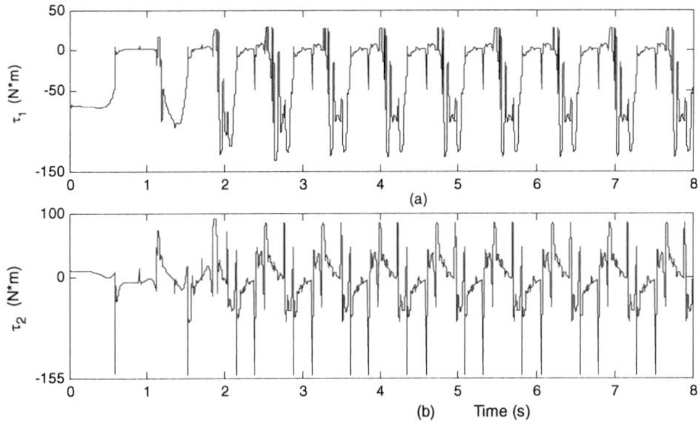

Figure. 4. Actuation torques of one leg (the right leg) during the walking: (a) the knee joint; (b) the hip joint

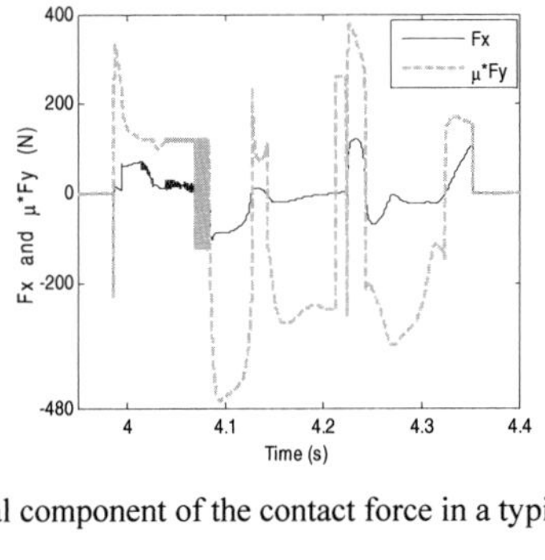

Figure. 5. The tangential component of the contact force in a typical single support phase

The vertical position of the hip is shown in Figure 3(b) by the real line, while the desired value is shown by the dashed line. It can be found that the tracking error of the hip's vertical position is larger at the beginning of single support phase than the rest. That's because when the swing foot comes in contact with the ground, the vertical velocity of the hip is not rigorously equal to zero, but a positive value, due to the inevitable tracking errors. Furthermore, to preserve the unilateral constraint between the support foot and the ground, the actuation torque of the knee joint cannot be too much large as shown in Figure 4, The torso is kept upright during the walking, which can be seen in Figure 3(c).

Figure 4 shows the actuation torques of one leg (the right leg) during the walking. Considering the ability of the actuators in practice, the actuation torques are restricted in the range of $-150N \cdot m$ to $150N \cdot m$.

Figure 5 shows the contact forces in a typical single support phase. It can be found that the contact consistency is preserved, i.e. there is no slipping and rebound between the support foot and ground. For the magnitude of the tangential component of the contact force is less than the friction force.

5. Conclusions

Our primary goal in this paper is to realize stilt-like biped walking by answering the question: How to take the desired step? With the assumption that the jacobian of task in single support phase is full row rank, we present the dynamics in task space form. Considering the parameter estimation errors, we design a sliding model controller for the position tracking problem. The stability of the controller is proved via Lyapunov function. Finally, we validate the controller on a virtual planar biped walker and realize stilt-like biped walking with contact consistency. The future work will focus on the physical biped walking experiment.

References

[1] E. R. Westervelt, J. W. Grizzle, *et al. Feedback Control of Dynamic Bipedal Robot Locomotion.* Boca Raton, London, New York: Taylor & Francis Group,(2007).

[2] J. Rummel, Y. Blum, *et al.* Stable and Robust Walking with Compliant Legs. *IEEE International Conference on Robotics and Automation*, pp:5250-5255, (2010).

[3] J. Rummel, Y. Blum, *et al.* Robust and efficient walking with spring-like legs. *Bioinspiration & Biomimetics*, 5, pp:1-13, (2010).

[4] I. Poulakakis,J. W. Grizzle. The Spring Loaded Inverted Pendulum as the Hybrid Zero Dynamics of an Asymmetric Hopper. *IEEE TRANSACTIONS ON AUTOMATIC CONTROL*, 54(8), pp:1779-1793, (2009).

[5] I. R. Manchester, U. Mettin, *et al.* Stable dynamic walking over uneven terrain. *The International Journal of Robotics Research*, 30(3), pp:265-279, (2011).

[6] C. O. Saglam,K. Byl. Stability and Gait Transition of the Five-Link Biped on Stochastically Rough Terrain Using a Discrete Set of Sliding Mode Controllers. *IEEE International Conference on Robotics and Automation*, pp:5655-5662, (2013).

[7] P. v. Zutven, D. Kostic, *et al.* Foot placement for planar bipeds with point feet. *IEEE International Conference on Robotics and Automation*, pp:983-988, (2012).

[8] L. C. Visser, S. Stramigioli, *et al.* Control Strategy for Energy-Efficient Bipedal Walking with Variable Leg Stiffness. *IEEE International Conference on Robotics and Automation*, pp:5624-5629, (2013).

[9] W. Hou, H. An, *et al.* Robust walking control of a planar spring mass bipedal robot. *ICMMC*, (2014).

[10] L. C. Visser, S. Stramigioli, *et al.* Robust Bipedal Walking with Variable Leg Stiffness. *IEEE RAS/EMBS International Conference on Biomedical Robotics and Biomechatronics*, pp:1626-1631, (2012).

[11] Y. Shtessel, C. Edwards, *et al. Sliding Mode Control and Observation*: Springer,(2014).

[12] S. G. Nersesov, H. Ashrafiuon, *et al.* On estimation of the domain of attraction for sliding mode control of underactuated nonlinear systems. *International Journal of Robust and Nonlinear Control*, 24, pp:811-824, (2014).

[13] M. Azad,R. Featherstone. Balancing and Hopping Motion of a Planar Hopper with One Actuator. *IEEE International Conference on Robotics and Automation*, pp:2019-2024, (2013).

[14] J. Nakanishi, R. Cory, *et al.* Operational Space Control: A Theoretical and Empirical Comparison. *The International Journal of Robotics Research*, 27, pp:737-757, (2008).

[15] L. Saab, O. E. Ramos, *et al.* Dynamic Whole-Body Motion Generation under Rigid Contacts and other Unilateral Constraints. *IEEE Transactions on Robotics*, 29(2), pp:346-362, (2013).

[16] S. Faraji, S. Pouya, *et al.* Compliant and Adaptive Control of a Planar Monopod Hopper in Rough Terrain. *IEEE International Conference on Robotics and Automation*, pp:4803-4810, (2013).

[17] M. Azad,R. Featherstone. Modeling the contact between a rolling sphere and a compliant ground plane. *Australasian Conference Robotics and Automation*, (2010).

DYNAMIC BIPEDAL WALKING WITH DOUBLE SUPPORT PHASE

A. SAFONOV, M.SHYSH

University of Magdeburg, Faculty of Engineering, Institute of Mobile Systems, P.B. 4120, 39106 Magdeburg, Germany

A. TELESH, U. SCHMUCKER

Fraunhofer IFF, Department Virtual Engineering, Sandtorstr. 22, 39106 Magdeburg, Germany

This paper presents the dynamic walking of the bipedal robot, in which the long double support phase is used for compensation of the external disturbances (including contact with support surface) for stabilization of the step cycle. Motions of the robot by the single support phase are based on free (ballistic) trajectories of the mass center (ZMP is located under the foot). During the double support phase robot first slows down due the damping features of the control system, then it stops and then accelerates again to the ballistic trajectories. Experiments were performed using a robot "ROTTO" [3].

1. Introduction

Implementation of the dynamic bipedal walking today is still one of the most important problems of robotics. In our previous works we relied on the use of only a single support phase (double support phase was instantaneous) [4,5,6]. This approach leads to the problem of synchronization of the robot movements in the frontal and sagittal planes under disturbances (especially by the contact with the support surface during the landing). Compensation of all disturbances was accounted for single support phase in which the mathematical model of the system is much less controllable (there are more uncontrolled degrees of freedom than in the double support state). At the same time a small support polygon (it is only one foot) greatly limits the controllability of the system.

Also the large legs mass that centered below the knees resulted to an increase of the negative influence of the cross-links in the control system of the robot walking.

The aim of this work is getting a greater controllability and stability reserves through the optimization of the mechanical construction of the robot and using the long double support phase in the step cycle.

2. Mechanics Optimization

The problem of heavy legs in this work is solved using a special design of the linear drive with a flexible hydraulic transmission [7]. This allowed replacing of the masses in the system, to shift the mass center of the legs up to the rotation axis and to reduce the moment of inertia (Fig. 1(b)). Reaction torque in the upper joint in the heavy and lightweight leg (Fig. 1(c)) shows the decrease of cross-links influence.

Walking is implemented with mechanism of the robot ROTTO. Kinematic structure is shown in Fig. 1(a). Joints [q3 ... q10; q13 ... q15] are position controlled. Drives in the joints [q1; q12] are force controlled. Joints [q2; q11] are not actuated and move freely.

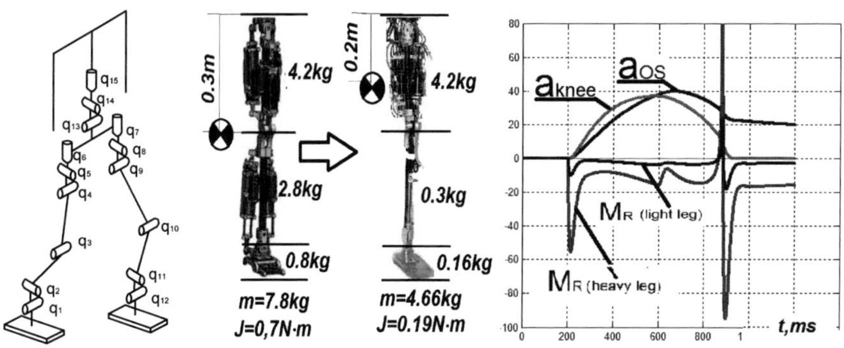

(a) (b) (c)

Fig. 1 – Mechanics optimization of robot ROTTO: (a) kinematic model; (b) parameters of the lightweight leg; (c) reaction torque in the upper joint of the leg

3. Walking Strategy

The mass arrangement, in which the masses of the legs are concentrated near the hip joint, allows us to consider the walking as a single-mass model in which the effects of the legs motions can be neglected. Consider the following single-mass model: let the movement of mass center of the robot correspond to the motion of an equal mass $\mathbf{m}_{CG}$ (Fig. 2). Mass $\mathbf{m}_{CG}$ mounted on a mass less rod which is attached to the joint in the point **A** on a ground. **A** is also the center of the support polygon. Location of the point **A** changes immediately (points $\mathbf{A}_1$ or $\mathbf{A}_2$) by the changing of the single and double support states.

For clarity, as the mechanism states while walking we use the values of the position and velocity projections of the mass $\mathbf{m}_{CG}$ on a surface plane **ZX** (Fig. 2(b)).

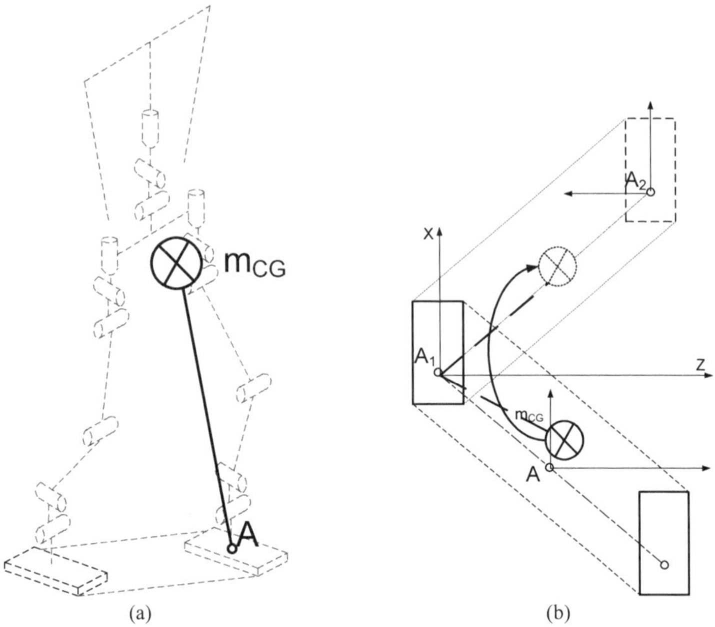

Fig. 2 – Equivalent robot model

Motion of the system is described with equations:

$$\begin{vmatrix} \ddot{q}_1 \\ \ddot{q}_2 \end{vmatrix} = \begin{vmatrix} \dfrac{M_1 + l \cdot m \cdot \cos(q_2) \cdot \left(g \cdot \sin(q_1) + 2 \cdot \dot{q}_1 \cdot \dot{q}_2 \cdot l \cdot \sin(q_2) \right)}{m \cdot l^2 \cdot \cos(q_2)} \\ \dfrac{M_2 + m \cdot g \cdot l \cdot \sin(q_2) \cdot \cos(q_1) - m \cdot l^2 \cdot \dot{q}_1^2 \cdot \sin(q_2) \cdot \cos(q_2)}{m \cdot l^2} \end{vmatrix} \tag{1}$$

$$\begin{vmatrix} X_{CG} \\ Z_{CG} \end{vmatrix} = \begin{vmatrix} -l \cdot \sin(q_2) \\ l \cdot \sin(q_1) \cdot \cos(q_2) \end{vmatrix} \tag{2}$$

Step is divided into 4 intervals: stay, acceleration, step and damping. Each stage of the step is characterized through the motion trajectories **Z_{CG}(t),X_{CG}(t), dZ_{CG}/dt(t), dX_{CG}/dt(t)** and the final values **[Z_{CG},X_{CG},dZ_{CG}dt,dX_{CG}dt]** (Fig. 3).

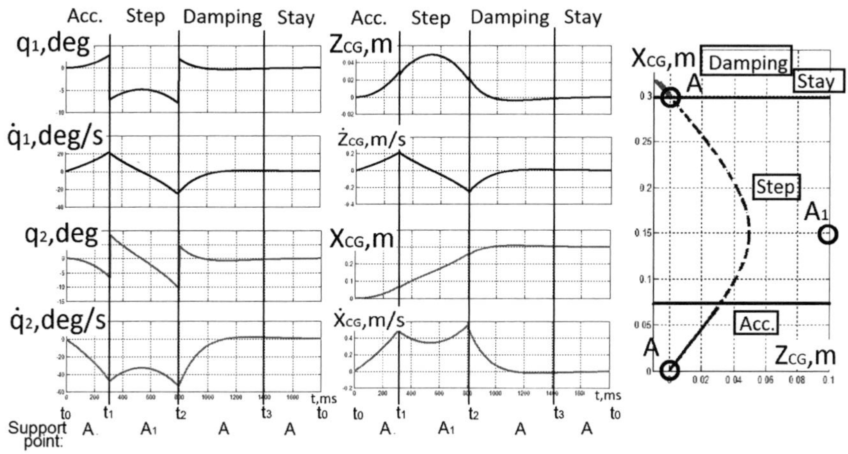

Fig. 3 – Motion trajectories of the mass center during the step

This gait feature that each step ends damping phase, after which the robot stops. Thus, the robot takes a rest state, such as before the step. Consider each stage of the step:

Stay. At this stage the robot stands motionless in double support state under the control of the stabilization system. The mass center of the robot is held in the center of the support polygon.

Acceleration. During acceleration $(\mathbf{t_0\text{-}t_1})$ the robot is still in double support state. In this stage the mass center of the robot accelerates along the trajectories (Fig. 3) under the action of the control system. Acceleration trajectories are synthesized offline considering following ballistic movements in single support phase. Under constant torques $\mathbf{M_X(t)=M_{XACC}}$, $\mathbf{M_Z(t)=M_{ZACC}}$ a single-mass system in double support state is brought to the states $\mathbf{Z_{CG}}(t_1)\mathbf{=0.3Z_{A1}}$ with speeds $\mathbf{dZ_{CG}/dt(t_1)}$, $\mathbf{dX_{CG}/dt(t_1)}$, which by the **Step** in the single support phase will lead the system to states $\mathbf{X_{CG}(t_2)=L}$ by $\mathbf{Z_{CG}(t_2)=0}$.

Step. At the end of the acceleration the robot's mass center has the states $\mathbf{[X_{CG}(t_1),dX_{CG}/dt(t_1),Z_{CG}(t_1),dZ_{CG}/dt(t_1)]}$, which cause the following motion of the system in the single support phase (Fig. 3). Swing leg at the same time moves along the trajectories (Fig. 4).

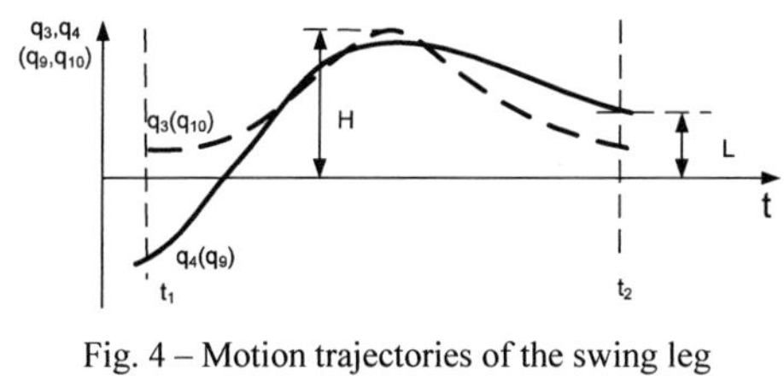

Fig. 4 – Motion trajectories of the swing leg

The motion trajectories for the joints of the swing leg are calculated offline using inverse kinematics. The parameter **L** corresponds to the step length and position of robot's mass center in displaced support polygon, **H** is the height of the feet by the swing.

Damping. After execution of the trajectories robot's swing foot will meet with the support surface. The robot's mass center will have a speed **[$dX_{CG}/dt(t_2)$,$dZ_{CG}/dt(t_2)$]**, acquired during the step. At this stage the stabilization system is striving to reduce these speeds to zero and keep the mass center of the system at the center of the support polygon. In this case torques in the feet joints are:

$$\begin{aligned} M_X(t) &= k_P \cdot Z_{CG}(t) + k_D \cdot Z_{CG}(t); \\ M_Z(t) &= k_P \cdot X_{CG}(t) + k_D \cdot X_{CG}(t) \end{aligned} \tag{3}$$

So the stabilization system damps the mechanism and brings it again to rest the state.

4. Robot Control System

To implement the earlier described robot movements was build the control system (Fig. 5). This control system is a position control loops for the projections of mass center coordinates on the axes **Z** and **X**. In the control loops are used PD - position controllers. Controller parameters are chosen experimentally.

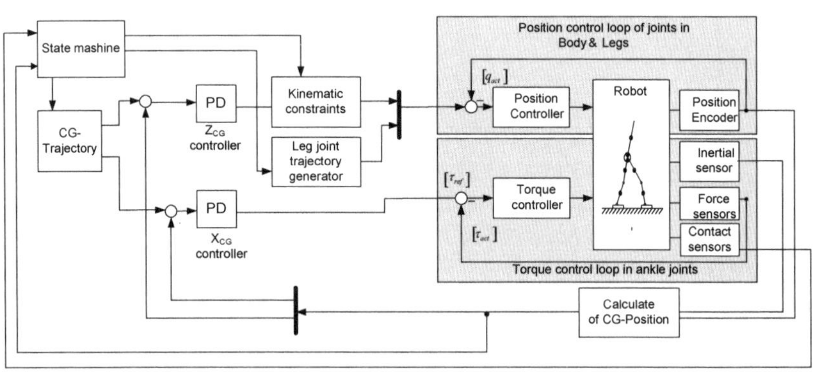

Fig. 5 – Robot control system

Because the robot joints are position controlled, the robot structure in the frontal plane is much rigid. For the damping it is necessary to provide the flexibility of the mechanical structure. Without damping by changing the support leg robot sharply rebounds from the support surface that does not allow him to take a double-support state. Overcoming this obstacle we use in the control system kinematic constraints for the robot leg joints (4). Thereby the behavior of rigid system (Fig. 6a) is changed to a "flexible" system (Fig. 6b). Such system is easier to stabilize and dampen it by the landing, and reliably hold in the double-support phase.

$$\begin{aligned}
L_L &= L_0 + dL_L + dL_{ML}, \\
L_R &= L_0 + dL_R + dL_{MR}; \\
dL_L &= k_L \cdot Z_{CG}, \\
dL_R &= -k_L \cdot Z_{CG}; \\
dL_{ML} &= k_M \cdot M_X, \\
dL_{MR} &= -k_M \cdot M_X.
\end{aligned} \tag{4}$$

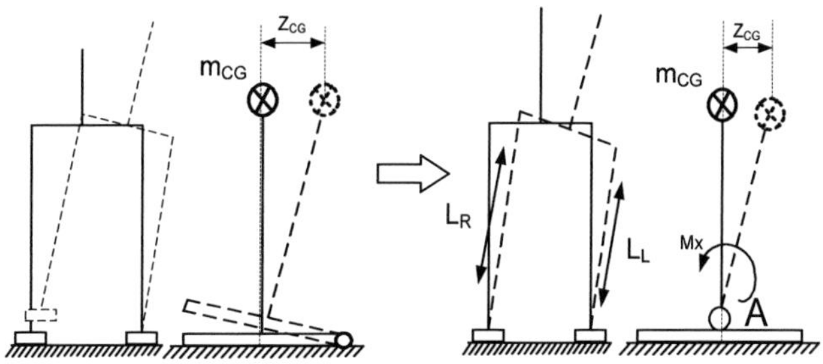

Fig. 6 - rigid system with straight legs (a) and damping due to the bending of the knees (b)

As a reference for the control loops are used trajectories (Fig. 3). Determination of current stage the step is determined by the state machine based on the values **[X_{CG},dX_{CG}/dt,Z_{CG},dZ_{CG}/dt]** and information about the contact of the robot with the support surface from sensors in the feet.

5. Experimental results

To evaluate the performance of the proposed walking concept was carried out an experiment using robot Rotto. The control system based on the real-time interface "Matlab /XPC-Target".

Movement trajectories of the robot's mass center (Fig. 7) correspond to the step length of approx. 22 cm. Trace of the mass center on the support surface (plane ZX) illustrates the movement of the robot in space while walking. The graphs show 6 steps.

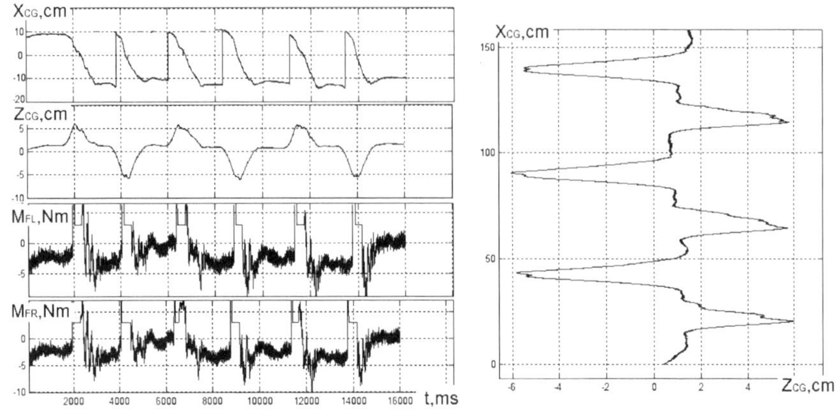

(a) (b)

Fig. 7 – Movement trajectories of the robot's mass center – (a) and Trace on the support surface

Repetitive curves on the phase planes for the frontal and sagittal planes show orbital stability gait of the robot. The shape of the phase portrait in the frontal plane in the closed form with center in the origin of coordinates says that the robot has the stops between the steps during the walking.

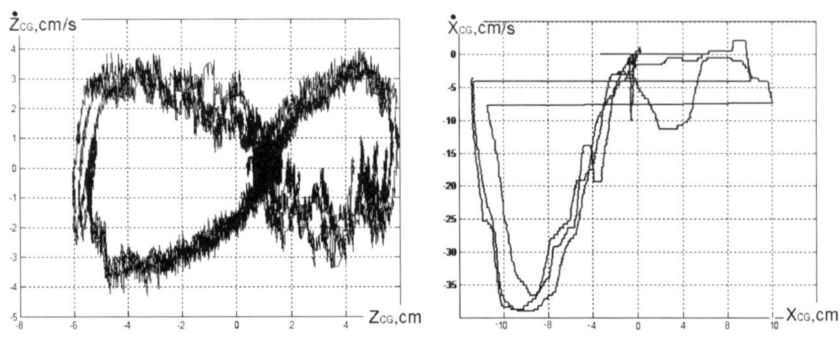

(a) (b)

Fig. 8 – Phase portraits of the robot mass center along Z-axis – (a) and X-axis – (b)

6. Conclusion

This approach allowed us to implement a more stable step cycle due to a more steady work of the control system. Here through the double-support phase the states of the system after landing always reduced to a given initial states for the single support phase.

On the one hand, such realization of the walking has allowed minimizing the disturbance in the robot control system caused by the contact process with the support surface during the changing of the support leg. On the other robot was brought to a full stop between the steps that causes a low walking speed (approx. 0.5 m / s at a step length of 25 cm) and inefficient use of energy.

Further it is planned the improvement of the gait such a way that the damping of the robot between the steps are not going to full stop, and to get the necessary initial conditions for the next step. This will allow excluding stage of "stay" from the step cycle, reducing the duration of the double support phase, and making the walking as a result faster and more efficient, and the proposed method to use only in case of large disturbances, or the end of walk.

References

1. www.uni-magdeburg.de/ieat/robotslab/
2. M. Vukobratovic. Shagayushie Roboti i antropomorfnie mehanismi. Mir, Moscow, 1976. In Russian
3. Y. Zavgorodniy, „Konstruktion und Steuerung von Schreitrobotern mit ballistischem Laufverhalten“, Ph.D. Thesis. Magdeburg, 2009
4. A.Telesh, „“,Ph.D. Thesis. Magdeburg, 2012
5. A.Melnykov, „“,Ph.D. Thesis. Magdeburg, 2012
6. Synthesis biped dynamic walking based on offline trajectory generation. A. Safonov, F. Palis, V. Bilous, A. Telesh, and U. Schmucker. The 15th International Conference on Climbing and Walking Robots and the Support Technologies for Mobile Machines (CLAWAR 2012). July 23-26, 2012, Johns Hopkins University, USA
7. Linear elastic actuator with a flexible hydraulic transmission A. Safonov, V. Bilous, M. Boiko, U. Schmucker and A. Telesh. 17th International Conference on Climbing and Walking Robots and the Support Technologies for Mobile Machines. Poznan, Poland, 21 – 23 July 2014

FOCUSED MODULARITY: RAPID ITERATION OF DESIGN AND FABRICATION OF A METER-SCALE HEXAPEDAL ROBOT

D. MILLER, I. FITZNER, S. B. FULLER, and S. REVZEN*

EECS Dept., U. Michigan, Ann Arbor, MI 48109, USA
**E-mail: shrevzen@umich.edu*

Robots are faced with increasingly complex, hard to simulate environments. Multilegged robots exhibit dynamics that are often difficult to model well. Their development thus depends on experimental verification. Here we propose *"focused modularity"* – an approach combining benefits of both modular robotics and more traditional rapid fabrication. We made robot mechanisms by laser cutting low-cost foam board, while the complexity of motor control and electronics is *focused* in a small number of servomotor *modules*. This dramatically reduced the cost and time of design iteration, and allowed incremental improvements to make up for deficits in material strength and manufacturing tolerance. We present meter-scale hexapods whose chassis and drive train were manufactured for under 20 USD and under four hours of skilled labor. To date we have produced over 40 revisions of the design, including 25 iterations of drive mechanism design and 7 full robot prototypes built. This represents a substantially more thorough exploration of the design space than could be possible using conventional development approaches. We suggest a library of suitable robot designs to be developed, permitting a robot to be built on the fly with functionality for a particular task as the situation demands.

Introduction

Classical engineering designs closed loop interactions of robots with their environment using accurate validated models, and sound application of mathematically provable control methods. For legged robots in uncontrolled environments, the requisite models of the environment and robot-environment interaction are rarely reliable. At the same time, these robots are coming into broader use where cost drives down manufacturing tolerances and reliability of robot components. Our challenge is to obtain reliable legged robot behaviors from cheaper, inaccurately characterized parts.

Here we explore one approach toward this goal, which we call *"focused modularity"*. While traditional *modular* robots aim to construct the entire robot from modular parts, we *focus* our modularity in few high-complexity reusable subsystems. We concentrated the electronics and actuation in a small number of general-purpose motor modules that attach to the chassis via a snap-on interface or large bolts, and daisy-chain for power and communications.

This approach differs from robotics kits[1] or classical modular robots,[2] in that we assume availability of rapid manufacturing tools, rather than relying entirely on pre-fabricated parts. We use modularity to localize complexity and speed up assembly, rather than as an overarching paradigm. Using our approach we built a meter-scale hexapedal robot — far larger than commonly available 3D printing allows. This robot can walk, steer, and turn in place on surfaces indoors and outdoors in the U. Michigan Ann Arbor campus, testifying to its viability as a legged platform.

The robot chassis and drive-train of our robots borrow from centimeter scale and millimeter scale robots made with the Smart Composite Microstructures (SCM) SCM uses a laser to cut planar materials. Thest are laminated together and folded to produce 3D linkages. SCM and related methods produced many robotic prototypes, such as a 3 cm six-legged hexapod walker made of fiberglass-reinforced composites,[3] a larger 10 cm hexapod built from fiberboard,[4] a 1 cm scale hovering robot fly,[5] centipede,[6] and other structures such as linked chains.[7] The speed of laser machining contributes to rapid design iteration. However, the maximum size of robots manufactured by SCM hitherto is approximately 10 cm. We contribute the first human-scaled mechanisms built using these techniques. Our primary

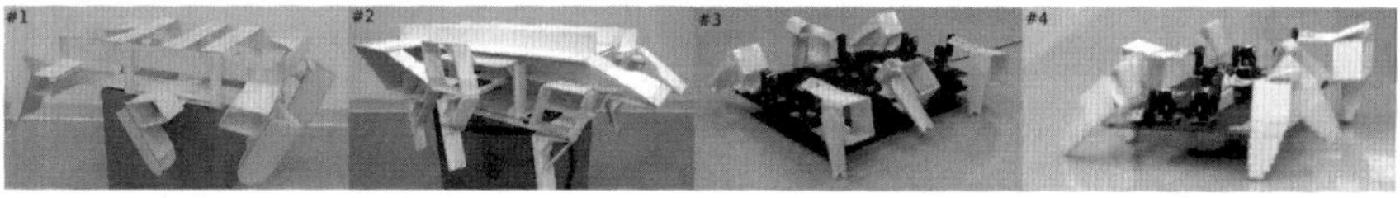

Fig. 1. Four generations of the robot chassis. #1 explored kinematics and had no actuators; #2 is scaled up Berkeley design; #3 first version able to walk; #4 improved leg and drive-train design. The structural material is foamboard.

contribution is showing how focusing modularity gives rapid iterations of mechanical design at low cost, opening the door to: **(1)** designing robots (legged or otherwise) based on experimentation rather than simulation; and **(2)** fabricating them on-the-fly when they are needed. A secondary contribution is an SCM-like process for rapid fabrication of low cost mechanisms at human scales. We primarily describe our #4 and #6 design iterations, which were 76 cm long ambulatory robotic hexapods. Each chassis, including legs and drive-train, cost less than 20 USD, and required less than four hours of skilled labor to build. Figure 2 lists design changes in terms of drive mechanism changes, complete chassis re-builds, and other leg and mechanism revisions. Over 40 revisions were made and tested at a cost (in time and money) normally associated with only a few iterations of robot design. Rapidly and cheaply exploring the design space and *experimentally* testing the results, we produced an effectively moving legged robot, whose operation was incrementally improved in the actual real world environment in which it operates.

Fig. 2. Design Revision History

Chassis version	1	2	3	4	5	6	7	total 7
Drive-Train revisions	5	6	2	4	5	1	1	total 24
Leg / Other revisions	5	15	2	5	10	3	2	total 42
Label in Fig.1	#1	#2	#3		#4			

Background on SCM

Our goal was to create a family of robot designs that can be fabricated quickly and inexpensively from few materials with few tools. Smart Composite Microstructures (SCM)[8] provides a rich collection of robot mechanisms designed for the centimeter and millimeter scale.

The process consists of: (1) Cutting a rigid material in a laser cutter leaving gaps for flexure joints. (2) Laminating the compliant flexure material on that lower rigid plate. (3) Placing and cutting an upper rigid plate. (4) Curing the materials to make them adhere. (5) Releasing the mechanism and locking it in shape.

Material Selection

To make an SCM-like process compatible with parts at the meter scale, we first selected an appropriate plate material. Our criteria were availability, low cost, and high rigidity. Additionally, material needed to be machinable in our Universal Laser Systems model PLS6.150D, 150 W CO_2 laser cutter.

Candidates for the plate material were foam-core board (Elmer's Products Inc. 50.8 cm × 76.2 cm × 0.72 cm foam-core board), corrugated plastic board (SABIC Polymershapes Coroplast COR-2436 91.4 cm × 61.0 cm × 0.40 cm), and corrugated cardboard (Home Depot 55.9 cm × 55.9 cm × 53.3 cm box). Price per square meter for foam-core was \$14.03 (OfficeMax), Corrugated Plastic \$22.21 (DisplayShops) and corrugated cardboard \$1.05 (Home Depot) .

We quantified each material's resistance to bending via its mass-specific flexural rigidity, the ratio $EI/M = (Pl^3)/(3M\delta)$, which assumes an Euler cantilever beam bending model[9] with load P and length l.

We bended single layers of each material by securing 5×20 cm samples to a table with a rectangular steel mass such that 15 cm of the sample extended off the table. We tested 3 samples of each material at 2 loads (143 ± 0.1 g, 257 ± 0.1 g), taping the entire length of the end of the sample to the entire length of the test mass. We measured the resulting deflections with a ruler (error ± 1 mm). For non-isotropic materials, we tested both strong and weak directions (ie. along the corrugations and across the corrugations).

	Plastic		Cardboard		Foam-Core
	(strong)	(weak)	(strong)	(weak)	
M (g)	7.2 ± 0.2	7.2 ± 0.2	0.24 ± 0.02	0.19 ± 0.02	6.4 ± 0.5
$\frac{EI}{M}$ ($\frac{\mathrm{m}^3}{\mathrm{s}^2}$)	$30. \pm 2.$	10.3 ± 0.4	38.3 ± 2.3	$31. \pm 2.5$	110 ± 20

The foam-core board is superior to the other materials in terms of mass-specific rigidity, and was therefore selected. It consists of a rigid foam sandwiched between sheets of fiberboard.

We selected composite fiber tape (3M, Scotch #8959; 26 k N/m strength; 6% yield strain) as our flexure material as the plastic is compliant to bending while the fibers provide support against tensile loads. Adhesive tape allowed us to skip the curing and lamination step of the SCM process.

Experience showed us that a common mode of failure is when foam-core fiberboard de-laminates from the foam filling. We protected against this by "frapping" – the wrapping of additional tape fully around the edge of the foam-core plate (see Fig. 3). For this we used a unidirectional fiber tape (3M, Scotch, #898, 66.5 k N/m strength; 5% yield strain) for frapping.

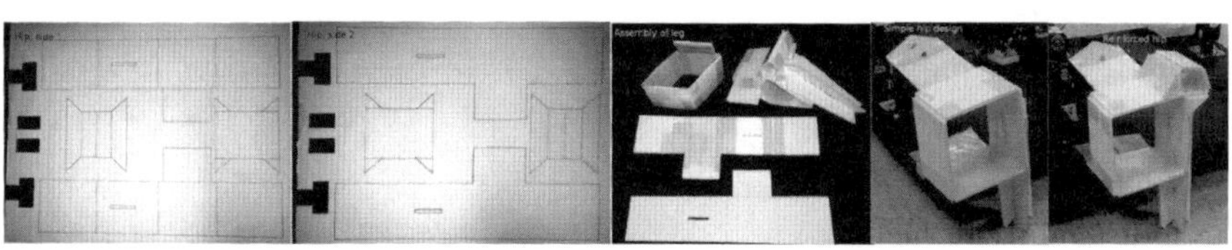

Fig. 3. Laser cuts of a hip (left two subfig). Assembly of a leg (middle) showing initial, frapping and folded stages. Two hip designs tested (right).

In chassis versions prior to #6 we used hot glue (Surebonder All purpose Glue Sticks) to adhere the foam-core plates together in locations where rigid bonds were needed. Later versions were without glue, requiring fewer materials and tools.

Fabrication

The fiberboard of the foam-core is flexible, allowing it to be used as a flexure material itself. Such flexures can easily be reinforced with fiber tape to give them higher durability. However, they are limited in their direction of folding – able to fold away from the cut side but not toward it. To allow folding in both directions, we cut alignment features into the boards, allowing them to be flipped over in the laser cutter and cut accurately from both sides ("T" shaped slots in Fig. 3).

Our procedure was: (1) Cut anchoring slots and anchor foam-core on alignment fixture; (2) Affix alignment fixture in laser cutter; (3) Cut first side; (4) Flip plate and lock on alignment fixture; (5) Cut second side; (6) Release part, tape and fold.

The pattern of cuts to make a hip is shown in Fig. 3. Cuts made on both sides create the outline of the hip piece, whereas cuts made on only one side create bending joints.

Design

Figure 2 lists the many design iterations we went through, starting with scaled versions of RoACH.[10] In all revisions the hip was driven by a crankshaft through hinged plates and a parallelogram linkage that converted the circular motor motion into a stepping motion. Our #1 chassis (Fig. 1) was designed by an undergraduate senior design team based on RoACH, and the body-long drive plates on each side of the robot twisted and bent compromising to robot's ability to move. Kinematic singularities and low foot clearance also interfered with smooth operation. Simply scaling up the Berkeley designs failed. We believe this is because structural properties like plate stiffness do not scale geometricaly. The increase in mass-specific compliance with a $\times 10$ increase in size made the drive-train design unusable.

For #2 we widened the flexures to resist torsional loads perpendicular to the hinge axis, and added stiffener ribs to many of the plates. This design (Fig. 1) was also unable to bear its own weight, but it had more favorable leg clearance in dry-dock.

In the #3 chassis (Fig. 1) we eliminated the robot-spanning drive plates, opting for six motors driving each hip independently. This choice represents a scaling effect: with increasing size, it becomes more favorable to add motors than to add robot-spanning drive transmissions.

The design of the legs for revisions 3 and 4 of the chassis (#3 in Fig. 1) is shown in Fig. 3, with mechanism illustrated in Fig. 4. While reliable at

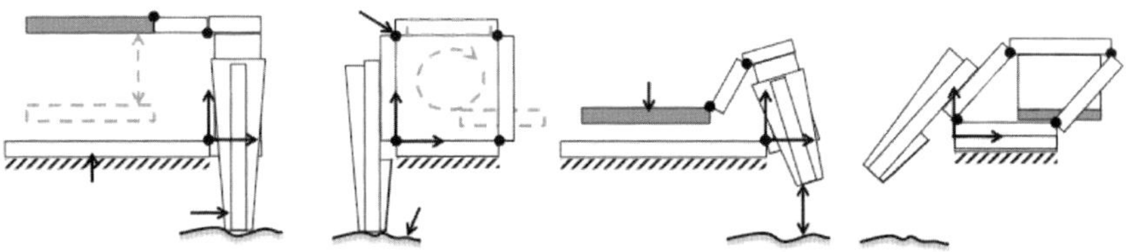

Fig. 4. The hip mechanism depicted in the stance (left pair) and swing (right pair) positions, as seen from the front and side. Motion of the crankshaft moves the link (gray filled box) in a circle (dashed arrows)

slow speeds on flat ground, one of the drive train parallelogram linkages occasionally flipped and jammed in an anti-parallelogram configuration when the leg encountered high loads. An additional prism on the top plate of the drive-train parallelogram linkage resolved this issue by preventing it from buckling, allowing chassis 5 to be reliable at over 40 RPM (#4 in Fig. 1).

In the alternating tripod gait we used middle legs bear larger loads. We reinforced the middle hips on each side of the robot against off-axis twisting by adding an additional four-bar linkage attached to the hinge above the leg (Fig. 3 rightmost). We validated this addition by measuring structural stiffness of each leg design along the vertical axis, as expressed by a Hooke's-law spring constant, by manually moving the motor to equally spaced angular "steps" in its cycle and using a force gauge on each leg to judge the amount of force required to displace the foot, with displacements measured by a ruler to ± 1 (mm) (see Fig. 5). During the stance phase (i.e. when the leg is vertical and supporting weight) the reinforced legs were significantly stiffer and able to support weight. Insights gained from this reinforcing structure may play a role in future designs, even though chasis 5 and up do not use it.

Kinematics

To move forward and back we produced an "alternating tripod gait" – front, middle and hind legs on alternating sides moving in sync. This gait is used by most hexapedal animals at moderate to high speeds within their speed range.[11] Motion of each foot is theoretically determined by the 1-DOF kinematic constraints of its linkage, which is similar to that of.[10] Under no load (dry-dock) conditions this is the trajectory graphed of Fig. 5 (with accuracy of ± 1 cm from possible errors in linkage dimension measurements). Leg moves vertically in the saggital plane around mid-stance, then rapidly swings outward at the end of stance to recycle forward. We designed the

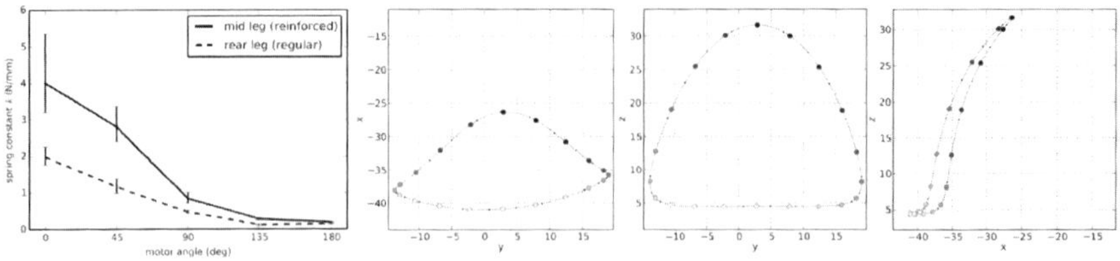

Fig. 5. Stiffness of legs to weight-bearing loads by motor phase for the two designs (n=1; left panel). Note higher stiffness in stance (0°) than swing (180°), as desired.The reinforced flexure has a higher stiffness and can support a larger load.Error bars based on displacement measurement uncertainty. Kinematic model of chassis #4 toe (orthographic views; corresponding points colored indentically in all projections; z vertical; y in direction of motion).

drive train linkage to give a $> 60°$ angle in mid-swing, for foot clearance of half a leg length.

Software & Control

The control software for our robot used the `pyckbot` library developed for the 2010 ICRA Planetary Challenge.[12] `pyckbot` comprises: (1) the `logical` library which identifies a variety of module architectures and busses, and represents the modules as a collection of python `Module` objects; and (2) the `joy` framework, which provides a `pygame` based event loop used to read user commands from keyboard, mouse, game controllers and mini slider boards, and a cooperative multithreading based collection of behavior building-blocks.

When configured in free rotation mode, the Dynamixel MX64 motors we used accept desired speed commands. Built in position commands did not allow full rotation of the motors. Instead, each motor was wrapped by a state-estimator and proportional feedback controller object, which proved sufficient to keep the motors in sync with centrally commanded reference angles and each other.

Cost

The creation of the robot's chassis 4 required two 76.2 cm × 101.6 cm pieces of 0.5 cm thick foam-core boards, one 76.2 cm × 50.8 cm piece of 1.3 cmthick foam-core board, 150 ± 25 cm of cross-fiber reinforced tape and 280 ± 50cm of length-wise reinforced fiber tape. At the current market prices (amazon.com), this is a total of 18.01 USD in foam-core and 1.26 ± 0.21 USD in tape – less than 20 USD altogether. The focused costs, i.e. the costs associated with the modules, were approximately 300 USD per leg. Thus, through focused modularity, over 98% of the materials cost is preserved from one design iteration to the next.

Each hip/leg mechanism took approximately 10 min to laser-cut and 10 min to assemble (times rounded to nearest 5 min). The two middle hips required an extra 10 min each to manufacture and attach the reinforcing four-bar linkages. Each hip/leg assembly required 5 min to attach to the foam base plate. The base plate needed mounting holes and slots cut to accommodate the frapping tape, which required 10 min to prepare. Each motor was attached to the base and linked to the hips in 5 min. This gave an overall build time of approximately 3.5 hr for the plate-and-flexure mechanisms, the designs of which we hope to iterate.

Speed

We measured walking speeds of chassis 4 as a function of crank-shaft rotation speed in RPM, by using a videocamera calibrated to an object of known size. Results at RPMs 15, 25 and 32 were 15, 21, and 30 ± 3(cm/s). At higher speeds, the motors lost coordination and the gait was less effective – presumably because of higher loads. Chassis 5 was able to walk at a nominal 42 RPM without losing gait coordination; this difference is being further investigated.

Conclusion

In this work we demonstrated a manufacturing process capable of fabricating a functioning ambulatory robot chassis and mechanisms at the meter scale, allowing them to easily interact with human-scaled environments.

The focus here has been to develop the underlying manufacturing process and show that it facilitates rapid iteration. Future work will provide experimental validation of robustness and performance characteristics of robots produced by this process. The aim will be to achieve design-test iterations of robot mechanisms, conducting actual experiments in difficult to model real-world environments, at rates comparable to those possible in software development. Such rapid design iteration could lead to dramatic performance improvements, while still maintaining low cost and short robot production time.

Our robot mechanisms come from a rich family of designs previously explored by SCM, suggesting many other robots can be produced with our paradigm. Future work will enrich the library of robot designs and improve fabrication rate and quality. Complexity and cost of the robot are focused in the actuation and computation modules, with over 98% of the cost being a one-time investment in the modules. Since the robot can be built rapidly, using only a laser cutter and a few inexpensive materials, we have shown that focused modularity indeed allows a robot to be fabricated from a library of designs quickly and at low cost. This suggests building robots on-the-fly when their functionality is required.

Acknowledgment

Supported in part by Army Research Office grant W911NF1210284 to S. Revzen. We thank J. Brown, S. Desousa, E. Eklov, D. Litz, A. Myers, D. Zhao, C. Schaffer and E. Revzen.

References

1. J. C. Larsen, Locomotion through morphosis: development of the modular robotic toolkit - locokit, Master's thesis, U. South Denmark (2011).
2. M. Yim, P. J. White, M. Park and J. Sastra, Modular self-reconfigurable robots, in *Encycl. of Complexity and Systems Science*, 2009 pp. 5618–5631.
3. A. M. Hoover, E. Steltz and R. S. Fearing, Roach: An autonomous 2.4g crawling hexapod robot, in *IEEE Int. Conf. on Intelligent Robots and Systems*, (Nice, France, 2008).
4. P. Birkmeyer, K. Peterson and R. S. Fearing, Dash: A dynamic 16g hexapedal robot, in *Intelligent Robots and Systems (IROS). IEEE/RSJ International Conference on*, 2009.
5. K. Y. Ma, P. Chirarattananon, S. B. Fuller and R. Wood, *Science* **340**, 603 (2013).
6. K. L. Hoffman and R. J. Wood, *Autonomous Robots* **31**, 103 (2011).
7. J. Whitney, P. Sreetharan, K. Ma and R. Wood, *Journal of Micromechanics and Microengineering* **21** (2011).
8. A. M. Hoover and R. S. Fearing, Fast scale prototyping for folded millirobots, in *IEEE Int. Conf. on Intelligent Robotics and Automation*, 2008.
9. R. C. Hibbeler, *Statics and mechanics of materials* (Prentice Hall, 2011).
10. D. W. Haldane, K. C. Peterson, F. L. G. Bermundez and R. S. Fearing, *IEEE Int. Conf. on Robotics and Automation* , 3279 (may 2013).
11. P. Holmes, R. J. Full, D. E. Koditschek and J. M. Guckenheimer, *SIAM Review* **48**, 207(June 2006).
12. S. Revzen, J. Sastra, N. Eckenstein and Yim, Ckbot platform for the icra 2010 planetary challenge, in *Workshop "Modular Robots: The State of the Art", Proceedings of IEEE ICRA conference*, 2010.

GAIT PLANNING AND STABILIZATION FOR BIPED ROBOT WITH ONE ACTUATOR

WENYUAN LIANG, LI MICHAEL LIU, JIANFEI LI
College of Engineering, Peking University, Beijing, China
E-mail: lwy123@pku.edu.cn

In this paper, we focus on the problems of gait planning and stabilization for a three-link planar biped robot with periodic gaits in level ground. The key in designing this under actuated bipedal robot is to construct a feedback controller to realize stable locomotion. To achieve the control objectives, a time independent periodic trajectory based on the dynamic equations is designed for gait planning, and a nonlinear control including adaptive control, force control and feedback sliding mode control is designed to realize gait stabilization. In this paper, the simulation results on gait generation and stabilization are presented.

1. Introduction

For under actuated robot [1-4], researchers [5] found that the energy efficient of it is between that of all passive walkers [5] and full actuated robot [6]. Recent re-searches apply the nonlinear control on the control problems of under actuated robots. Different than the ZMP principle [7] which is over-conservative leading to higher energy cost and slow motion [1], nonlinear control is well applied to realize a periodic, energetically efficient and dexterous dynamic walking, which is close resemblance to human walking with dynamically stable.

In this paper, it deals with the problem of under actuated robot which is a three-link biped robot with one actuator, and made improvements on gait planning and stabilization. First, different than the off-line gait planning in [1, 3, 4], we use on-line gait planning which is based on on-line differential iteration and on-line parameter estimate method. Second, in view of fact, we consider that the impact event is a continuous process in a short space of time rather than an instantaneous impulse model. It is also known that the biped walking will be not sustainable only with motion control [3]. Thereby, we design the controller for gait stabilization including force control, adaptive control and sliding mode control.

In this paper, simulation results on gait planning and stabilization are presented and discussed. The force control simulation is not included and will be presented in the following papers. Section 2 describes the dynamics, gait planning is presented in section 3, and the controller and simulation results are given in section 4.

2. Dynamics of Biped Robot

We consider the biped robot with two symmetric legs and upper torso shown in Figure 1, which is a three-link planar robot with one actuator.

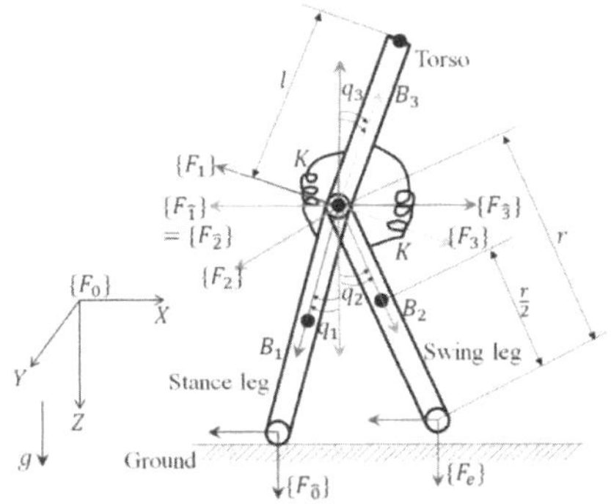

Figure 1. Structure of three-link biped robot

2.1. *Structure*

This robot is an under actuated robot with point feet, and the torque is applied between two legs. In Figure 1, swing leg is denoted with B_1 link, swing leg with B_2 link, and torso with B_3 link. The actuator is fixed at point O. The torsional spring between two legs and torso is used for keeping torso uprights. In this example, the gravity is $9.81 m/s^2$, and the spring stiffness coefficient is k=23.977*Nm/rad.* The following Table 1 shows the values of other parameters.

Table 1. Parameters definition for three-link biped robot

Parameters	Mass of link [kg]	Length of link [m]	Center of Mass [m]
Stance leg	$m_1=m=5$	$l_1=r=1$	$d_1=r/2$
Swing leg	$m_2=m$	$l_2=r$	$d_2=r/2$
Upper torso	$m_3=M_t=15$	$l_3=l=1$	$d_3=l$

2.2. *Swing phase model*

In Figure 1, the link angles are denoted with q_1, q_2, q_3 respectively. Under the assumption of non-sliding walking, the dynamics in swing phase is a continuous motion model that can be expressed with a second order Euler-Lagrange equation,

$$M \cdot \begin{bmatrix} \ddot{q}_1 \\ \ddot{q}_2 \\ \ddot{q}_3 \end{bmatrix} + C \cdot \begin{bmatrix} \dot{q}_1 \\ \dot{q}_2 \\ \dot{q}_3 \end{bmatrix} + G = \begin{bmatrix} u \\ -u \\ 0 \end{bmatrix} \tag{1}$$

where u is the torque.

2.3. *Impact phase model*

The impact event, happening in a short space of time in practice, is also a continuous procedure. Thereby, we use force control (see Figure 3) during the impact procedure to control the velocity of joints, ensuring that the velocity after impact is close to the velocity of the gait starting. Force (torque) command is calculated based on the required leg impedance control, which is a necessary for realizing stabilization during impact. Meanwhile, the leg position is

$$q_1^+ = q_2^-, \quad q_2^+ = q_1^- \tag{2}$$

where the subscripts $(.)^-$ and $(.)^+$ denote the values prior and immediately after the impact event respectively.

3. Gait Planning

In this section, our main objective is to find a set of time independent periodic gait, and use on-line differential iteration to plan the gait.

3.1. *Swing leg trajectory based on tracking states*

Through using a new set of variables to parameterize the biped robot, the virtual constraint in (3) represents the simplest time independent gait. In (3), the trajectories of swing leg and upper torso are tracking the state of stance leg.

The stance leg plays the role of active joint, and the trajectory of swing leg is defined with a one-dimensional Bezier polynomial of fifth degree function [9]. Then ϕ_2 will track the state of stance leg θ. Thereby, it is able to change the Bezier polynomial to determine different gaits by:

$$\begin{bmatrix} q_1 \\ q_2 \\ q_3 \end{bmatrix} = \begin{bmatrix} \theta \\ \phi_2(\theta) \\ \phi_3(\theta) \end{bmatrix}, \quad \begin{bmatrix} \dot{q}_1 \\ \dot{q}_2 \\ \dot{q}_3 \end{bmatrix} = \begin{bmatrix} \dot{\theta} \\ \phi_2' \cdot \dot{\theta} \\ \phi_3' \cdot \dot{\theta} \end{bmatrix}, \quad \begin{bmatrix} \ddot{q}_1 \\ \ddot{q}_2 \\ \ddot{q}_3 \end{bmatrix} = \begin{bmatrix} \ddot{\theta} \\ \phi_2'' \cdot \dot{\theta}^2 + \phi_2' \cdot \ddot{\theta} \\ \phi_3'' \cdot \dot{\theta}^2 + \phi_3' \cdot \ddot{\theta} \end{bmatrix} \tag{3}$$

where $\phi_i' = \partial\phi_i / \partial\theta$, $\phi_i'' = \partial\phi_i' / \partial\theta$, and $\dot{\theta} = d\theta / dt$, i=1, 2, 3.

3.2. *Torso trajectory based on on-line planning*

Combined with (1) and (3), the 3-order derivative of ϕ_3 can be yielded

$$\phi_3''' = \phi_3(\theta,\ \phi_2,\ \phi_3,\ \phi_2',\ \phi_3',\ \phi_2'',\ \phi_3'',\ \phi_2''') \tag{4}$$

When the $\{\ \theta,\ \phi_2,\ \phi_3,\ \phi_2',\ \phi_3',\ \phi_2'',\ \phi_3'',\ \phi_2'''\ \}$ is given with the initial values, we can solve (4) as a 3-order differential equation. Here we use one-step numerical computation (real-time differential iteration) to solve the trajectory of torso to realize on-line planning. In this paper, the Runge-Kutta 4-order method is adopted as the solving tool. Thereby, based on the real-time states, the torso trajectory planning in the next step is

$$\begin{cases} {}^{i+1}\phi_3 = {}^{i}\phi_3 + h/6 \cdot \left({}^{i}E_1 + 2\,{}^{i}E_2 + 2\,{}^{i}E_3 + {}^{i}E_4\right) \\ {}^{i+1}\phi_3' = {}^{i}\phi_3' + h/6 \cdot \left({}^{i}U_1 + 2\,{}^{i}U_2 + 2\,{}^{i}U_3 + {}^{i}U_4\right) \\ {}^{i+1}\phi_3'' = {}^{i}\phi_3'' + h/6 \cdot \left({}^{i}L_1 + 2\,{}^{i}L_2 + 2\,{}^{i}L_3 + {}^{i}L_4\right) \end{cases} \tag{5}$$

where i represents the present step, and the step length is $h = {}^{i}q_1 - {}^{i-1}q_1$. L, U, E are obtained through order reduction, $L(=\phi_3''') = U'$, $U(=\phi_3'') = E'$, $E = \phi_3'$.

3.3. *On-line gait planning*

The gait planning is looking for a periodic gait, and the gait planning is implemented as follows:

1) We start the waking gait search by choosing $\chi_{\text{ini}} = \left[\theta^+,\ \phi_2^+,\ \phi_3^+,\ \dot{\theta}^+,\ \phi_2'^+,\ \phi_3'^+\right]$. As the biped robot is walking on level ground, there is $\theta^+ = -\phi_2^+$ while both legs touch the ground.
2) By setting the numerical values for $\{p_2, p_3, p_4\}$, the swing leg trajectory is given. Then its derivations with respect to θ, $\{\ \phi_2', \phi_2'', \phi_2'''\ \}$, can be obtained.
3) The torso trajectory is on-line planned by (5).
4) Repeat 1) ~ 3) until that the swing leg touches the ground. Meanwhile, a vector is formed,

$$\chi^- = \left[\theta^-, \phi_2^-, \phi_3^-, \dot{\theta}^-, \phi_2'^-, \phi_3'^-\right] \tag{6}$$

   where (6) represents the states prior the impact.
5) The vector in (6) can be used to compute the states after impact. In this paper, the impact event is considered to happen in a short space of time, rather than immediately as an impulse. Thereby, we use force control during the impact procedure to control the velocity of joints, ensuring that the joint velocity after impact is close to the joint velocity at the gait starting.

6) Then, we check the gait search by the following law,

$$\left\| {}^1q - q^+ \right\| \le \delta, {}^1q = \left[{}^1\theta, {}^1q_1, {}^1q_2 \right]^T, q^+ = \left[\theta, q_2^+, q_3^+ \right] \tag{7}$$

where the nonnegative δ is equal to 0.001 in this paper. When the raw (7) holds, the search is stopped.

3.4. *Example on gait planning*

One solution of symmetrical gait (see Figure 2) is found with the initial values: χ_{ini} = [-0.2503, 0.2503, 0.3131, 1.2512, 0.3961, -0.237], where the unit of angle is *rad*. Meanwhile, the swing leg trajectory is determined by the coefficients $\{p_2=0.5, p_3=-0.2, p_4=-1.0\}$.

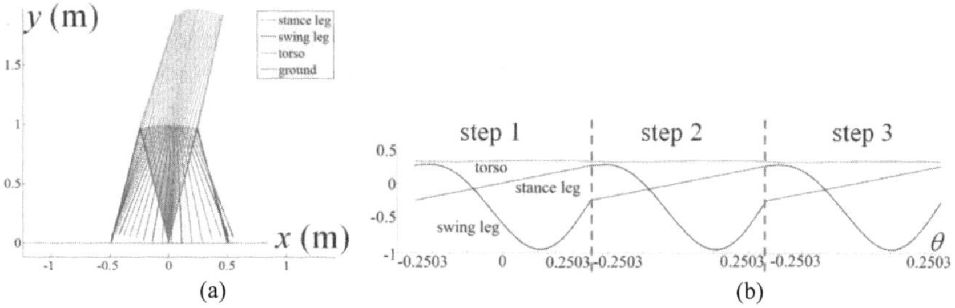

Figure 2. Solution of a periodic gait

As shown in Figure 2(a), it shows the trajectories of biped robot, where the evolution of θ is in the interval [-0.2503, 0.2503]. In Figure 2(b), at step 1, ϕ_2 starts and ends at 0.2503 and -0.2503 respectively; ϕ_3 starts and ends both at 0.3131. Moreover, the trajectories of step 2 and step 3 are the same as that of step 1. Hence, the designed gait in Figure 2 is a periodic gait.

4. Gait Stabilization

In this section, the controller stability is based on the periodic orbit and Poincare return map [1-4]. Thereby, we would design a nonlinear controller to realize an exponentially stable periodic orbit.

4.1. *Controller*

As the control algorithm in [8, 9], feedback PID loop, sliding mode control and force control are included in the controller. Feedback control ensures that the real trajectory is close to the command trajectory. Sliding mode control can make the control switch between motion control and force control. Force control is based on impedance control, which is a basic and necessary for realizing stabilization during impact procedure. Adaptive control based on online parameter estimation is used to rebuild accuracy model.

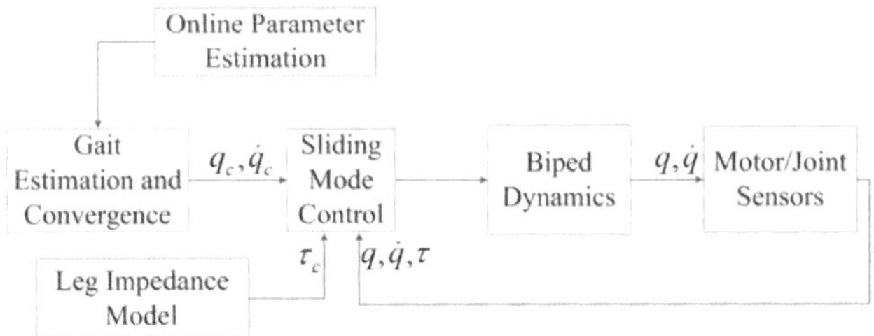

Figure 3. Control block for biped robot

4.2. *Stabilization tests*

First, the disturbance is assumed as periodic pulse force acting on three links with the value of 20*N*. In Figure 4(a), the red curve is the command curve, and the black curve is the tracking curve with disturbance. The comparison shows that the designed controller has strong ability on anti-disturbance.

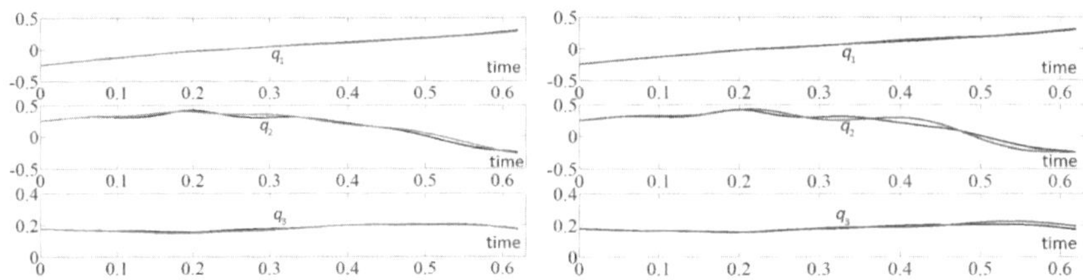

(a) (b)

Figure 4. Simulation results for tracking trajectories with disturbance and parametric variation

Second, it is assumed that all the physical parameters have 10% parametric variation with respect to their nominal values. The simulation also considers the disturbance in first test. Through on-line parameter estimation, the tracking trajectories can follow the desired trajectories well (see Figure 4(b)). It shows that the controller can maintain the trajectories in a vicinity of the desired nominal cycle.

5. Conclusions

In this paper, we presented gait planning and control algorithm for a three-link planar biped robot with two symmetric legs and upper torso. In this example, the robot has one actuator acting between the legs. Based on the differential iteration, the on-line periodic gait planning is designed. With the designed nonlinear controller, it is able to maintain the trajectories close to the desired trajectories.

References

1. E. Westervelt, J. Grizzle, C. Chevallereau, J. Choi, and B. Morris, Feedback Control of Dynamic Bipedal Robot Locomotion. CRC Press, Taylor and Francis Group (2007)
2. K. Sreenath, H. Park, I. Poulakakis, and J. Grizzle, *Int. J. Rob. Res*. **32**, 324-345 (2013)
3. J. Grizzle, G. Abba, and F. Plestan, *IEEE Transactions on Automatic Control* **46**, 51-64 (2001)
4. D.O. Morales, and P.X. La Hera, Design of energy efficient walking gaits for a three-link planar biped walker with two unactuated degrees of freedom, in *IEEE International Conference on Robotics and Automation* (2012)
5. S. Collins, and A. Ruina, A biped walking robot with efficient and human-like gait, in *IEEE International Conference on Robotics and Automation* (2005)
6. M. Hirose, and K. Ogawa, *Philosophical Transactions of the Royal Society*, **A(365)**, 11-18 (2007)
7. M. Vukobratovic, and B. Borovac, *International Journal of Humanoid Robot* **1(1)**, 157-173 (2004)
8. L. Liu, Y. Han, R. Lingarkar, N. Sinha, and M. Elbestawi On adaptive force motion control of constrained robots, in *Proceedings of IECON 1989* (1989)
9. L. Liu, W. Liang, Adaptive asymptotic stable biped locomotion, in *2014 33rd Chinese Control Conference* (2014)

INSPECTION PLATFORM FOR APPLICATIONS IN HUMANITARIAN DEMINING

H. MONTES[1, 2], L. MENA[1], R. FERNANDEZ[1], J. SARRIA[1] and M. ARMADA[1]

[1]*Field and Service Robotics Group, Centre for Automation and Robotics-CAR (CSIC-UPM), Spain;* [2]*Faculty of Electrical Engineering, Technological University of Panama, Panama,* hector.montes@ csic.es

The inspection platform that is described in this manuscript consists of a hexapod walking robot designed by the Centre for Automation and Robotics CSIC-UPM, Spain. This inspection platform will load a scanning manipulator arm which, in turn, carries a metal detector on its tool centre point. With the integration of both, hexapod robot and scanning manipulator, several test tasks about the search and localisation of antipersonnel mines will be carried out, within a controlled environment. The SCARA configuration of the hexapod robot legs will allow low energy consumption when the robot executes gaits on flat terrain or with reduced slope, due the decoupling of gravitational effects. This legged robot has a mass about 250 kg, and it can bear a high payload up to about 300 kg. Considering this load characteristic then the vibrational effects on the scanning manipulator will be reduced, when this carry out scanning tasks over the terrain.

1. Introduction

Detection and removal of landmines fields is a severe problem with amazing political, social and economic dimensions, which have been present during the last decades [1]. In this context, the international scientific community has shown a broad interest to solve this problem, which is addressed from different points of view and with different procedures. The anti-personnel mines, sub-munitions and unexploded ordnance (UXO) can remain active for several years (in some cases up to 50), they do not can discriminate between military and civilian people, killing or injuring indiscriminately to them, including children and humanitarian collaborators [2].

Currently, it is estimated that there are 110 million landmines buried in the ground, and more than 160 million are in stockpiles waiting to be planted or destroyed [3]. One solution to this problem could be the application of fully automatic systems or tele-operated systems to detect and remove the landmines. However, regardless of the latest advances in this field, this solution still seems far from possible.

In order to solve this problem, using the above mentioned systems, should be considered efficient sensors, detectors and positioning systems for detecting, locating and to identify the landmines. Besides, appropriate mobile vehicles must be developed in order to have on board the sensors for move them on infested fields of mines and, in this way, put off to human operator of a direct risk. Additionally, should be interesting to have a human-robot collaboration to carry out the humanitarian demining tasks, since this work is very complex and difficult to realize.

Many types of mobile vehicles that can take on board sensors to move them to a mine-infested area; e.g., wheeled vehicles, tracked vehicles and even legged robots; can carry out demining tasks efficiently and safely, relatively. Wheeled robots are the simplest and least costly [4-7]; tracked robots have excellent ability to travel on almost any terrain [1, 8, 9]; however, legged robots have a very interesting potential for this activity [10-15].

Legged robots for carry out humanitarian demining tasks have been developed, at least, in the last 15 years, and several prototypes of these robots have been tested experimentally. Some examples of these robotics platforms are TITAN VIII [12], AMRU-2 [11], RIMHO2 [10], COMET series [14], SILO6 [13, 15]. These walking robots are based on configurations of legs insect type. But have been developed other configurations of walking robots for demining, such as sliding frame systems [11, 16]. The need to develop legged robots to operate in certain situations where other humanitarian demining systems using wheels or tracks cannot operate properly or have low yield, has been investigated by several research centres (Roya Military Academy (RMA), Free University of Brussels, Chiba University, Tokyo Institute of Technology, The Spanish National Research Council (CSIC), other).

In this work the main characteristics of a hexapod robot designed and manufactured by the Centre for Automation and Robotics (CAR) CSIC-UPM are presented. The main objective of this hexapod walking robot is carry out tasks for localisation of antipersonnel mines, taking on board a scanning manipulator arm with a metal detector installed on its tool centre point. Primarily, Section 2 describes the mechanical design and system architecture of the hexapod robot. Afterwards, Section 3 presents some experimental results carried out on flat surface within a controlled environment. Finally, conclusions are given in Section 4.

2. General Configuration of the Hexapod Robot

The hexapod walking robot has been configured to carrying out tasks of humanitarian demining according to the requirements established in TIRAMISU Project*. This legged robot will carry on-board a scanning manipulator of 5 DoF,

---

* http://www.fp7-tiramisu.eu/

several sensors to help the scanning tasks, the industrial computer, electronics, control/power cards, Wi-Fi wireless communication system, DGPS, batteries, and other devices and accessories.

2.1. *Description of the mechanical system*

The hexapod robot legs have a SCARA configuration (RRP). This configuration decouples gravitationally the movement of the robot body (obviously, when it is walking on surfaces without slope). For this reason, this hexapod robot has an energetic autonomy relatively high, considering its mass (~250 kg) and the payload that it can to load (<300 kg). Therefore, the scanning manipulator arm with metal detector and other sensors and devices installed on-board of this hexapod robot shall not comprise a significant load to itself. Figure 1 shows the mechanical configuration of the hexapod robot. Table 1 shows the main characteristics of this walking inspection platform.

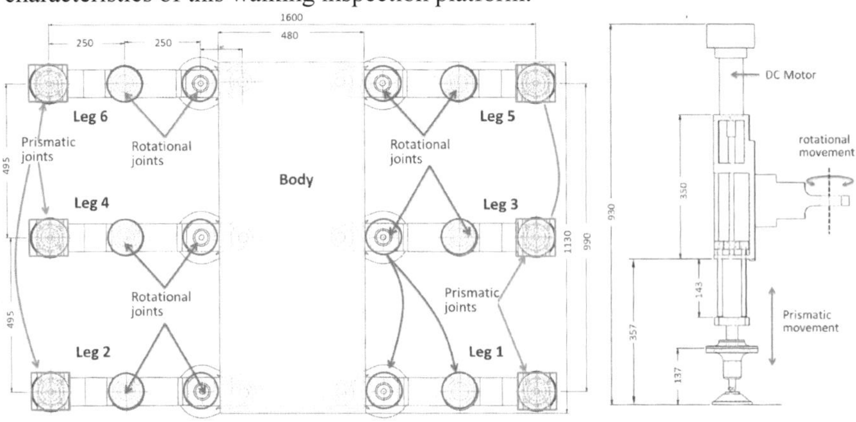

(a) (b)

Figure 1. Mechanical configuration of the hexapod robot (units in mm). (a) Top view of the robot; (b) Frontal view of the leg.

Table 1. Main mechanical characteristics of the hexapod walking robot.

Degrees of freedom		18
Stability		High
Robot mass		250 kg
Payload capacity		Up to 300 kg
Obstacle height to surpass		Up to 200 mm
Power supply		16-30 VDC, Typ. 24 VDC
Body size	Length	1130 mm
	Width	480 mm
Robot size	Max length	1130 mm
	Max width	1700 mm

The main robot gait is performed by means of an alternating tripod gait. However, other gait alternatives are being developed at the time of writing of this manuscript. The hexapod has three feet in contact with the ground all times, while the other legs are in transfer phase. In this gait two non-adjacent legs of one side and the central leg of the opposite side of the robot support the robotic platform, with high stability, while the other tripod is in the transfer phase [17].

2.2. *System architecture description*

The system architecture of the hexapod robot consist of an on-board computer, control cards, data acquisition boards, power cards, signals conditioner cards, positioning sensors, DC motors, Wi-Fi communication system, DGPS, batteries, and other devices and accessories. This control architecture provides a reliable starting point for developing several control strategies in order to carry out humanitarian demining tasks. Figure 2 shows the most important subsystems that comprise the control architecture assembled in the hexapod robot, and a perspective view of the robot [17].

The central idea is carry out stable gaits in order that scanning manipulator arm can perform suitable movements of its end effector, where is installed the metal detector head. In this regard, the inspection platform shall move to a fix position in order that the manipulator arm carries out the ground exploration. Standard strategies for the sweeping techniques will be considered. Some explanations related with these standards can be seen in [18]. According to the strategies of demining tasks (commanded by the operator), the displacement of the robot is proposed. These strategies are sent to the trajectory generator stage by mean of Wi-Fi communication, which uses the robot kinematics and the PID control to perform the controlled movement of the hexapod robot. Figure 3 shows the general control architecture of the hexapod robot.

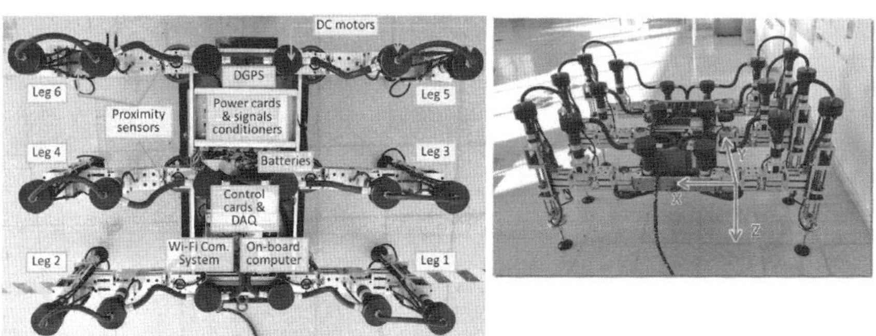

(a) (b)

Figure 2. (a) Subsystems of the control architecture of the hexapod robot; (b) Hexapod walking robot in CAR (CSIC-UPM) lab.

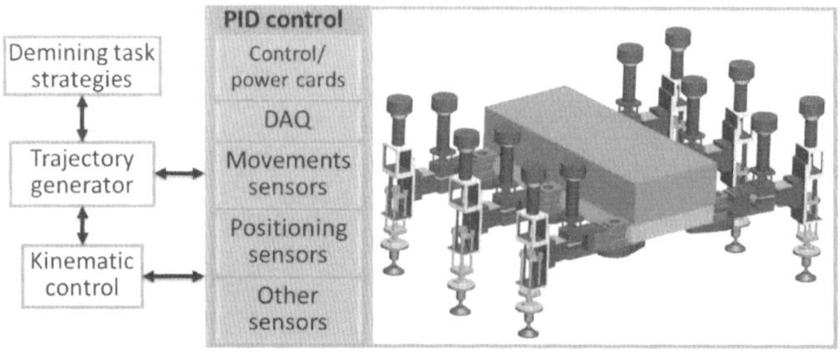

Figure 3. General control architecture of the hexapod robot.

3. Experimental Results

Several experimental tests with different gait patterns have been carried out in the CSIC lab. The hexapod robot has demonstrated high stability during the following of several trajectories. This robot has moved loads of more than 50 kg during some tests, and the additional power consumption has been insignificant with respect to the execution of gait patterns without extra payload. Therefore, will be feasible to carry on-board the scanning manipulator and other additional sensors, and this do will not contribute to a significant energy consumption.

Figure 4 shows some experimental results during the execution of a continuous gait using the alternating tripod mode. In order to see some details of the experimental results only two steps are presented in Figure 4. Figures 4(a) and 4(b) show the position of the shoulder and elbow joints of the leg 1 and leg 2, respectively. Each of these legs moves with different tripods. For this reason, these graphics represent the motion of each tripod, assuming that the corresponding legs of each tripod have similar motions. In Figure 4(b), backlash in elbow joint can be appreciated. Figure 3(c) shows the position of both legs along the Z axis, the alternating gait is perceived, besides a small unevenness on the soil is detected. Figure 4(d) show the lateral displacement of the extreme leg 1, the commanded straight line has an error about ±0.51mm, wherewith is considered reasonably right.

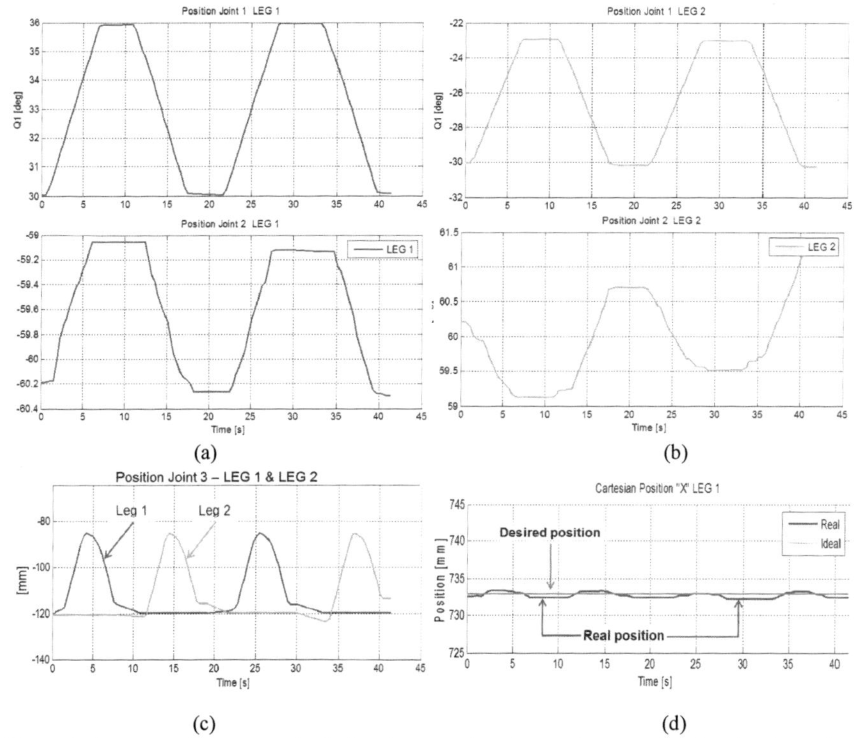

Figure 4. Some experimental results during the execution of a continuous gait using the alternating tripod mode.

The transfer and stance phases of the leg 1during two steps are shown in Figure 5. The small error along X axis can be observed in this figure, besides of the small error between each step. However, this error is negligible. When the robot performs a gait during long time, e.g., when it is performing some humanitarian demining task (landmine searching), the summation of this error can be compensated normally.

Figure 6 shows a photographic sequence of gait pattern of the hexapod robot. In this Figure the scanning manipulator arm is executing a search strategy.

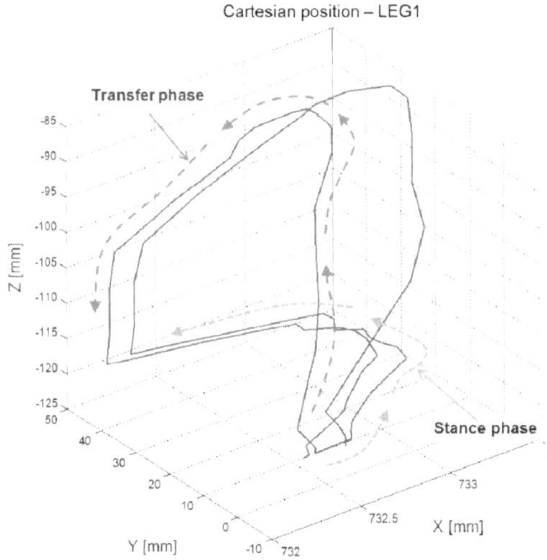

Figure 5. Phases of transfer and stance of the leg 1.

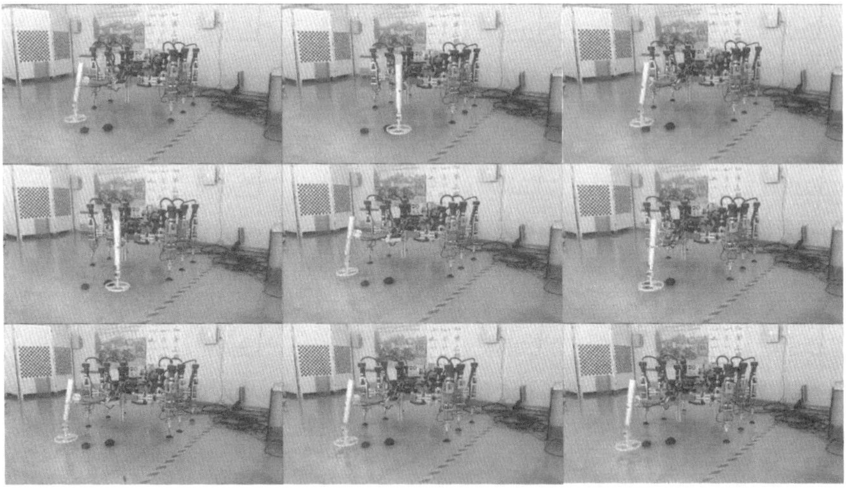

Figure 6. Photographic sequence (left to right and up to down) of a gait pattern of the hexapod robot.

4. Conclusions

In this paper an inspection platform (hexapod walking robot) for applications for humanitarian demining tasks has been presented. This platform will support on-board a scanning manipulator arm and other sensors to detect and localize anti-personnel mines, firstly, within a controlled environment. After several tests, it will attempt to use this inspection platform in a real environment with surrogate landmines.

Acknowledgements

The authors acknowledge funding of this work under TIRAMISU Project - Grant Agreement Nº 284747 of the 7FP, and partial funding from the RoboCity2030-III-CM project (Robótica aplicada a la mejora de la calidad de vida de los ciudadanos. Fase III; S2013/MIT-2748), funded by Programas de Actividades I+D en la Comunidad de Madrid and cofunded by Structural Funds of the EU. Dr. Héctor Montes also acknowledges support from Universidad Tecnológica de Panamá.

References

1. Y. Baudoin et al., CLAWAR, Year 2 Report: TASK 9, *WS-HUDEM* (1999).
2. *Landmine Monitor* (2010).
3. *Landmine Monitor* (2011).
4. Y. Baudoin and I. Doroftei, In Proc. *6th IARP WS-HUDEM*, 93 (2012).
5. Y. Baudoin, M. Habib, I. Doroftei, *Woodhead Publishing Limited*, (2011).
6. P. Kopacek and L. Silberbauer, 7th *IARP WS-HUDEM*, (2008).
7. E. Fukushima, M. Freese, T. Matsuzawa, T. Aibara, and S. Hirose, *Intl. J. Smart Sensing and Intelligent Sys.*, 1(3), 735 (2008).
8. K. Munsang et al., *IARP WS-HUDEM*, (2002).
9. D. Waterman, K. Nonami, R. Yuasa, S. Amano, S. Masunaga, H. Ono, *IARP WS-HUDEM*, (2005).
10. P. Gonzalez and M.A. Jimenez, *J. of Robotics Sys.*, 12(9): 599 (1995).
11. J.C. Habumuremyi et al., *CLAWAR'98*, (1998).
12. S. Hirose and K. Kato, In Proc. *CLAWAR'98*, 261 (1998).
13. P. Gonzalez de Santos, E. Garcia, J. Estremera, and M. Armada, M.A. *IARP WS-HUDEM*, (2002).
14. K. Nonami et al., *JSME Intl. Journal*, Series C46(3), 881 (2003).
15. P. Gonzalez de Santos, J. Cobano, E. Garcia, J. Estremera, M. Armada, *Mechatronics*, 17: 417 (2007).
16. L. Marques, M. Rachkov and A.T. Almeida, In Proc. *10th Mediterranean Conf. Control and Autom.*, (2002).
17. H. Montes, L. Mena, R. Fernández, J. Sarria, P. González de Santos, M. Armada, Proc. *RISE 2015, 8th IARP*, (2015).
18. R. Fernández, H. Montes, C. Salinas, P. González de Santos, M. Armada, Industrial Robot: An Intl. Journal, 39(5), 450 (2012).

QUADRUPED ROBOT MECHANISM DESIGN AND MOTION SIMULATION BASED ON SOLIDWORKS AND ADAMS*

PENGYANG XU, RUI SONG†, SHENGLEI MAO, XUEWEN RONG and YIBIN LI

School of Control Science and Engineering, Shandong University, Jinan, 250061, China

Abstract: Quadruped robot has the advantages of high speed, strong obstacle capability, high flexibility and so on. It has become a hot research field of robotics. The research of quadruped robot in our country is in a state of rapid development. The robot of large-scale high-performance and high stability is the goal of our research. But the design and research of small quadruped robots is also indispensable. At first we use SolidWorks to design the frame structure of the quadruped robot. Then we use simulation software ADAMS to achieve the quadruped robot gait of Trot. So we further determine the feasibility of the design.

Keywords: Quadruped robot; Mechanism design;3D printing; ADAMS simulation.

1. Introduction

According to the way of motion, mobile robots can be divided into wheeled robots, legged robots(single leg robots, double legs robots and multiple legs robots), tracked robots, crawling robots, swimming robots and so on. Because the legged robots have the advantages of flexible motion and good environmental adaptability and so on. Many countries have joined to the research of legged robots. Quadruped robots are similar with most animals of the nature. They conform to mechanics principles, so they have become a hot topic of research. Many domestic schools dedicated to the research of large quadruped robot. Such as Big Dog Robot SCalf of Shandong University[1]. It has been checked and accepted by Project 863. It can walk on pit road, on ice road and on snow ground. It also can resist side kick. However our country have not obtained satisfactory achievement in the small quadruped robot. Thus we

* This work is supported by the National Natural Science Foundation of China (61233014 and 61305130), the Shandong Provincial Natural Science Foundation (ZR2013FQ003 and ZR2013EEM027), National High Technology Research and Development Program of China (2015AA042201)

† Corresponding author: Rui Song, E-mail: rsong@sdu.edu.cn.

designed this bionic quadruped robot. The first generation of prototype has been achieved.

Because of the advantages of small, light, cheap, simple control and so on, small quadruped robots gain the interest of researchers. At present in the research field of bionic quadruped robot. Due to problems such as control algorithm, device connected, material, structure and so on. The quadruped robots that we designed always can't compare with real animals both from the appearance to performance. Therefore we designed a quadruped bionic robot with flexible bodies. This robot's calf adopts four connecting rod. We using servos to drive the upper leg and shoulder straightly. The torso combined with the spinal passive degrees of freedom[2]. So the bionic quadruped robot has good stability.

2. Mechanism design

We using SolidWorks to design the mechanism of the quadruped robot[3]. It includes the design of torso and legs. Torso adopts flexible methods. Designed as the mimic animal's spine. In this method the robot can walk steady on uneven pavement.

The design of leg is an important step of robot. The motion and stability of the robot depend on the leg to a great extent. Upper leg and calf adopt longitudinal joint. Shoulders adopt horizontal joint. This robot calf joints are driven by connecting rod[4]. The higher motor mount, the less torque of upper leg and shoulder motor are needed. There are various motion of the four connecting rod. It can achieve complex motion such as rotation, swing and so on. Therefore it widely used in all kinds of machinery and equipment.

The four connecting rod need enough motion space. So the removed part of upper leg that it don't need. At the same time it save material. In order to enhance the intensity of the leg, we add the stiffener to the place where the stress is great. So the leg avoid deformation under the great stress[5]. When the motor of the calf rotating. According to the motion relation between the four connecting rod, the calf will make the appropriate motion to simulate the reciprocating motion of the animals'. Because foot of this robot is not necessary, so we don't consider it's foot. As is shown in the picture. Figure 1 shows the four connecting rod. Figure 2 is the physical figure of the upper legs and calf. Figure 3 shows the location of the motor. A, B and C is the location of motor. Figure 4 shows the simple framework of the quadruped robot. Figure 5 is the whole physical picture of the quadruped robot.

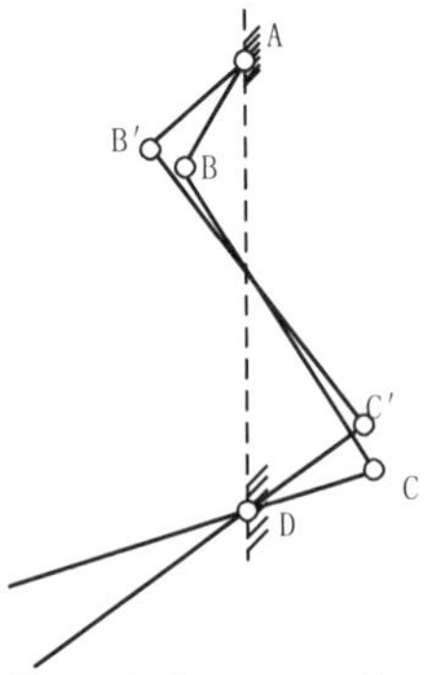

Figure 1: Four connecting rod

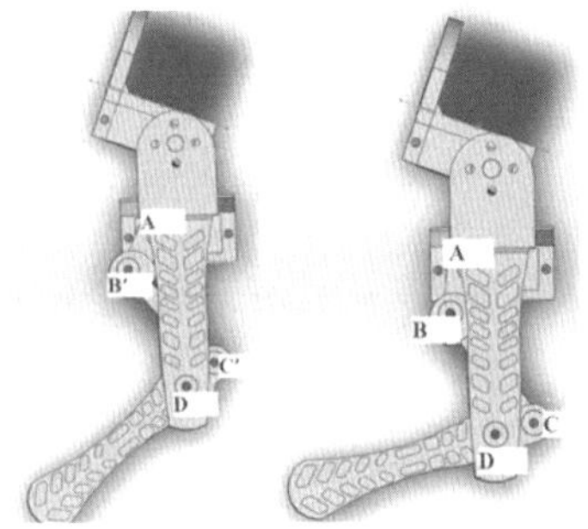

Figure 2: The physical figure of the legs

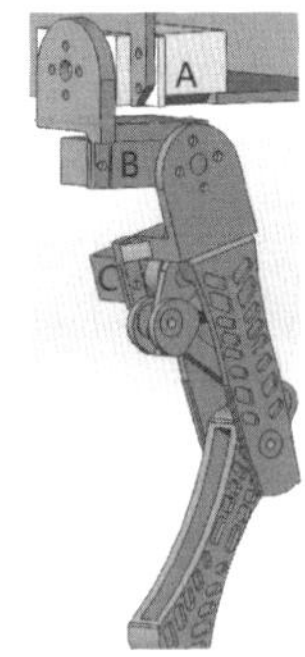

Figure 3:Location of motor

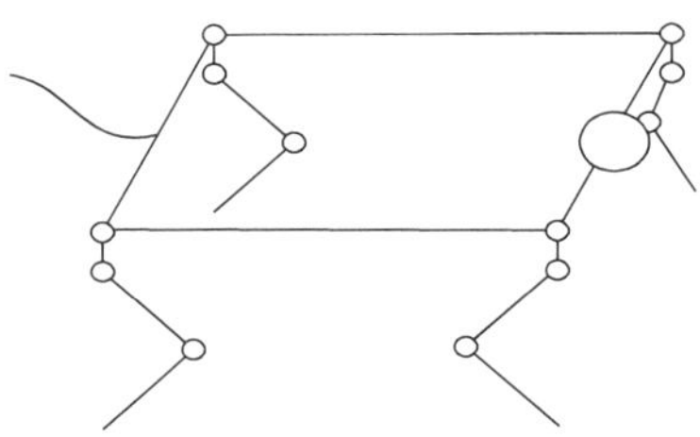

Figure 4: Simple framework of the robot

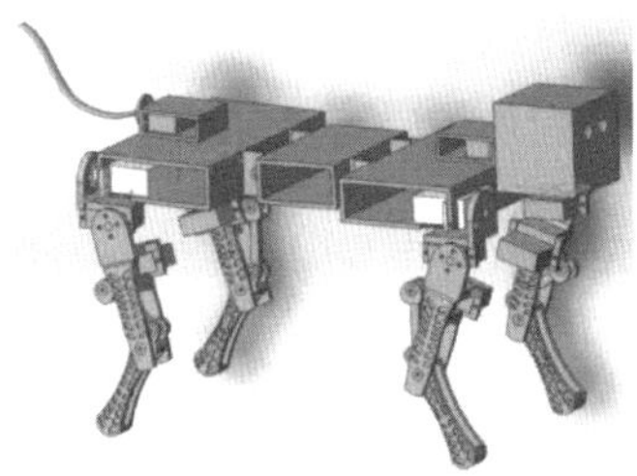

Figure 5: Whole physical picture of the robot.

The four connecting rod adopts the double rocker mechanism. As is shown in Figure 1, the point of A and D are fixed shaft. They can only rotating around the shaft. The point of B and C are mobile shaft. They can not only rotating around the shaft but also can move. The point A is the location of the motor shaft. AB and CD are all side link. AD is rack. BC is called connecting rod. AB and CD can't do the entire rotate. Rod of AB always rotates with the center of A. Because of the effect of connecting rod, the CD rod has the only way of motion with the center of D under the force of the BC rod. In the figure, the $A - B - C - D$ is the previous state of motion. the $A - B' - C' - D$ is the state of later motor rotation. The whole motion is to lift the leg.

The robot is 190mm high, with width of 160mm and length of 440mm. It's upper leg has length of 85mm and calf has length of 70mm. The material is ABS plastic, it's whole mass is about 1078g. The mass of leg is about 162g. The quadruped robot can use horizontal degrees of freedom to turn. It's conform to characteristics of natural animal. Meanwhile we add the head and tail to the body respectively. Therefore the quadruped robot can achieve the movements such as shake head, swing tail and so on. At the same time they have the effect of keeping balance.

We adopt 3D printing to produce the structure. 3D printing is a kind of rapid prototyping technology, which is based on the digital model file. It use adhesive material such as powder metal and plastic. This technology construct objects by the way of printing step by step[6]. With the 3D printing we can make the object that does not exist on the market.3D printing are flexible with high precision. Therefore we choose 3D printing to make robot framework. It is also a good combination of Solidworks and 3D printing technology.

3. System Simulation

The system need to make a whole simulation to make sure that the designed model has no problems. First we save the model that assembled in SolidWorks as the form of .x_t. Then we imported the .x_t file into ADAMS[7]. Next we need to define the Bodies, add the Connectors and Motions to the model. Because the motion simulation is proceed on the ground. So we need to add a plane to the system. And this plane must be fixedly connected with the ground. Then the effect is equivalent to the ground. After that we add contact force and friction to the four legs and the plane. At last we need to add the drive function to the leg joints to achieve the motion simulation.

As we all know, Robot's motion is achieved by the motor. Therefore motion simulation is achieved by making different motor rotate different angles respectively. The motors of shoulder, upper leg and calf cooperate with each other. It can achieve differences of gait. This simulation aims to verify whether the designed quadruped robot can achieve trot gait. The trot is the gait that diagonal legs swinging at the same time. It is a more practical dynamic walking way, which implementation is relatively simple[8].

When the quadruped robot walk with trot gait, the motion of left front upper leg is same with right back upper leg's. But the motion of left front calf is contrary to right back calf. Motion of left front upper leg is contrary to right front upper leg's. Motion of left front calf is also contrary to right front calf. After calculation, we decide to use STEP function as the drive signal. STEP function is a simple motor drive function. The function of STEP(time,0,0d,1,10d) express that the axis rotate 10 degree at the time of 0-1s[9]. The shoulder joint is to achieve the action of turn. So what it just to do is to rotate a suitable angle at the appropriate time. The upper leg joint is to achieve front and back swing. Thus we need a function that realize the circulation. We found that the trigonometric function $\mathrm{Y} = \mathrm{A}\sin(2\pi/\mathrm{T} \times \mathrm{t} + \varphi)$ can realize this function. At the same time, changing A can change the margin of swing and changing T can change the period of swing. We using $32 * \mathrm{PI}/360 * \sin(2 * PI/0.4 * time + PI)$ as the drive function of left front upper leg. Using $32 * \mathrm{PI}/360 * \sin(2 * PI/0.4 * time)$ as the drive function of right back upper leg. Using $32 * \mathrm{PI}/360 * \sin(2 * PI/0.4 * time + PI)$ as the drive function of right front upper leg. Using $32 * \mathrm{PI}/360 * \sin(2 * PI/0.4 * time + PI)$ as the drive function of left back upper leg. With this the robot can realize the basic movement. After research the dog's walking, we found that the motion of calf is basic same as the upper leg's. Thus we also use trigonometric function to drive the calf. So we using $32 * \mathrm{PI}/360 * \sin(2 * PI/0.4 * time)$ as the drive function of left front calf. Using $32 * \mathrm{PI}/360 * \sin(2 * PI/0.4 * time + PI)$ as the drive function of right back calf. Using $16 * \mathrm{PI}/360 * \sin(2 * PI/0.4 * time + PI)$ as the drive function of right front calf joint. Using $16 * \mathrm{PI}/360 * \sin(2 * PI/0.4 * time)$ as the drive function of left back calf.

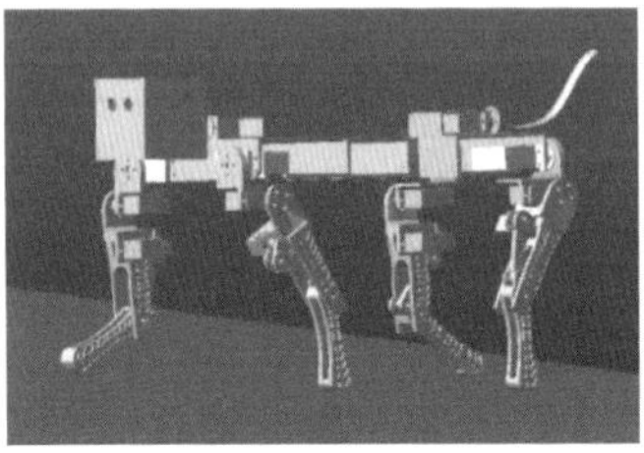

Figure 6: State 1

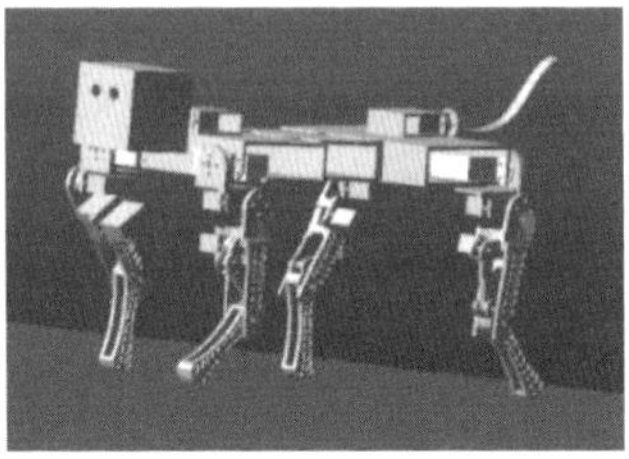

Figure 7: State 2

The simulation results are shown in figure. As is shown in figure 6, the quadruped robot's left front calf, right back calf is support phase. At the same time right front calf, left calf back is swing phase. Figure 7 shows another state. Two kinds of state changed to achieve the motion. Figure 8 is the motion curve of the robot. From the picture we can see the robot moves 450 mm at the time between of 1-5s. Figure 9 shows the vertical displacement curve of body when the robot walking. From figure 8 and figure 9 we can see the robot's motion is stable. Figure 10 and figure 11 show the swinging angle of upper leg and calf when the quadruped robot moving. From the figures we can see the robot's leg movement is a cycle when the robot is moving. Figure 12 and figure 13 are the torque curve of upper leg and calf motor need respectively. But there are spikes. This is because calf contact with ground periodically[10].

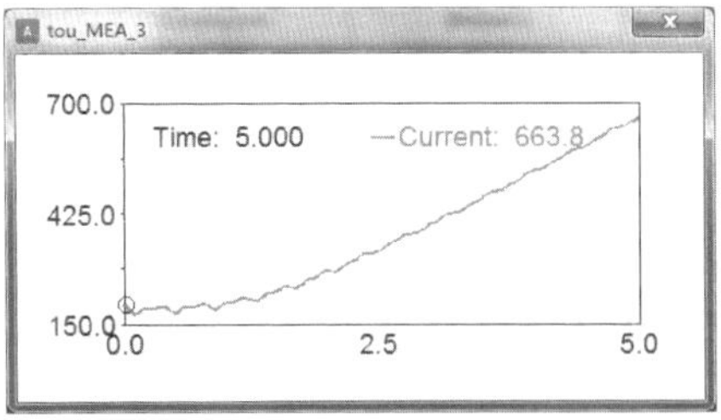

Figure 8: Motion curve

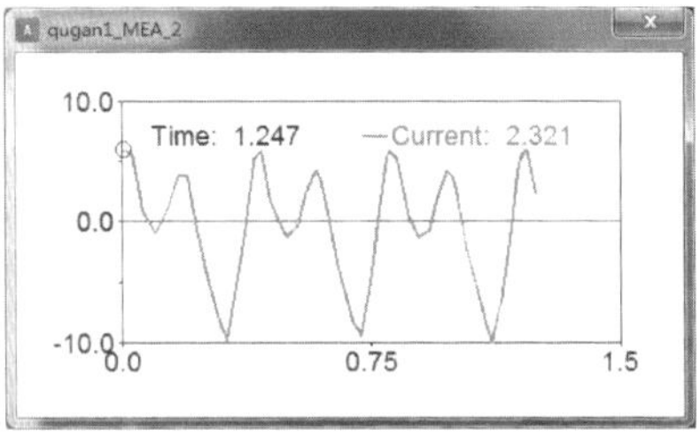

Figure 9: Vertical displacement curve of body

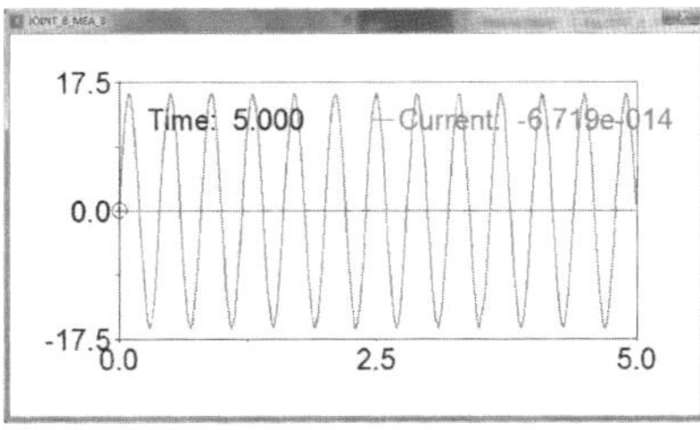

Figure 10: Swinging angle of upper leg

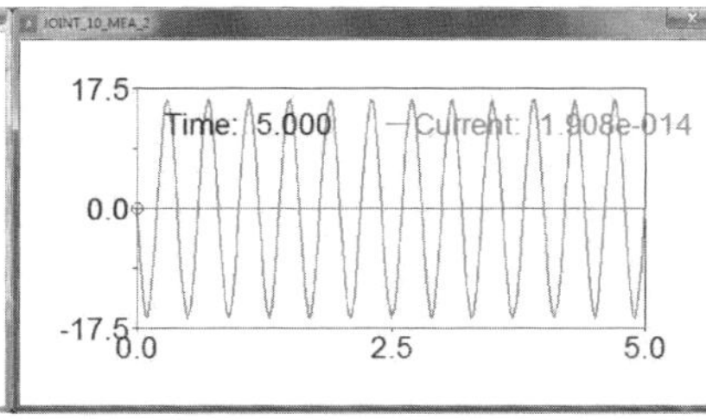

Figure 11: Swinging angle of calf

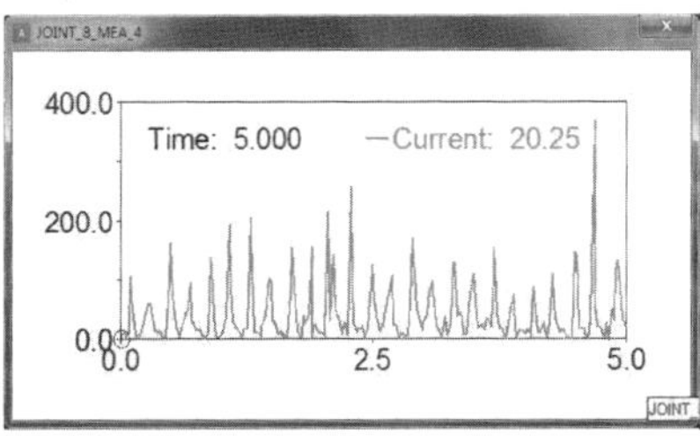

Figure 12: Torque curve of upper leg

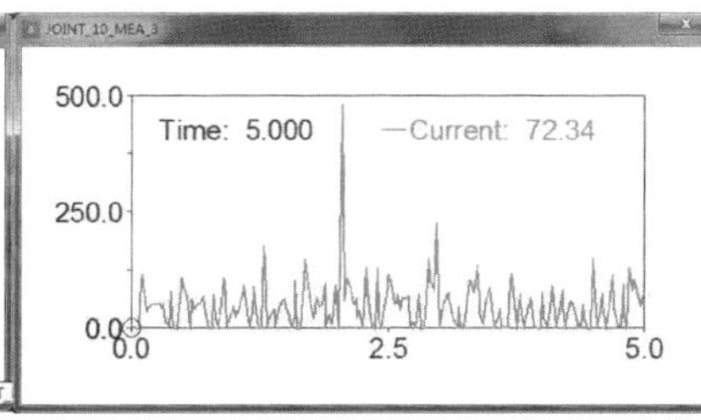

Figure 13: Torque curve of calf

4. Conclusion

At first, we designed mechanism of quadruped bionic robot. On the base of the traditional quadruped robot, we improved the structure of the legs and torso. Calf adopts connecting rod structure. With this method we can move the motor higher to reduce the torque that the motor need. Shoulder adopts transverse degree of freedom to achieve the turning function. The adding passive degrees of freedom to the trunk have the effect of buffer. It is better than rigid torso. Of course, we can also add the spring to the trunk to achieve the flexibility. The head and tail also have the effect of keeping balance. When the robot walking, it can shift the center of gravity by shaking it's head and wagging tail to improve the stability of the entire system.

We use trigonometric function to drive motor in this system. Through the simulation it can walk on the road stably. We can realize walk, pace, bound gait by changing drive function. At present the quadruped robot that we designed is the first generation of prototype. So there are some shortcomings such as the design of trunk still has differences with real animals. It can't make good imitation of reality quadruped's spine to achieve the effect of buffer. Now we are committed to the quadruped robot in the design of the new structure and gait.

References

1. Xuewen Rong. Mechanism Design and Kinematics Analysis of a Hydraulically Actuated Quadruped Robot SCalf [D]. Jinan: Shandong University,2013.

2. Bin Li, Xuewen Rong, Yibin Li. Review and Analysis of Quadruped Robots with Articulated Spine [D]. Jinan: Shandong University,2014.
3. Chaoxiang Chen. Advanced Assemble Tutorial of SolidWorks [M]. Beijing:China Machine Press, 2014.
4. Lianggui Pu, Guoding Chen, Liyan Wu. Mechanical Design [M]. Beijing: Higher Education Press , 2013.
5. Huizu Shan. Mechanics of Materials [M]. Beijing: Higher Education Press, 2006.
6. Woody• Lipson, Melba • Cuman. 3D printing - From imagination to reality [M]. Beijing: China Citic Press, 2013.
7. Weidong Guo. Virtual prototype technology and Applications Tutorial of ADAMS [M]. Beijing: Beihang University Press, 2008.
8. Xiuying Tang, Xiaodong Zhou, Kejian Nong. The Design of a Quadruped Walking Bionic Robot and Study on It's Gaits [J]. Agricultural Mechanization road with Chinese characteristics——Chinese Society for Agricultural Machinery 2008 Annual Conference Proceedings (The first volume), 2008.
9. Weili Lu, Xingao Bian, Jian Jiao. Gait analysis and motion control of quadruped crawling robot [J]. Mechatronics Engineering, 2012, 29(8).
10. Karssen J G, Haberland M, Wisse M, et al. The effects of swing-leg retraction on running performance: analysis, simulation, and experiment[J]. Robotica, 2014: 1-19.

A FREE GAIT PLANNING METHOD BASED ON THE FOOTHOLD SEARCH STRATEGY FOR QUADRUPED ROBOT*

SHUAISHUAI ZHANG, YIBIN LI, RUI SONG†, XUEWEN RONG, BIN LI

School of Control Science and Engineering, Shandong University, Jinan, Shandong 250061, China

Planning a robust free gait is an important thing to improve the terrain adaptability of quadruped robot. By using a robust free gait, quadruped robot can walk through the rough terrains with high complexity. The foothold search method is presented in this paper to search the optimum foothold for the swing foot, and the stability of the robot can be improved by using the proposed foothold search method. Furthermore, in order to ensure that quadruped robot can swing a foot forward with enough stability margin, just before the swing foot lift-off the ground, a four feet phase is added into the walking process of the robot to complete a body sway movement to enlarge the stability margin, and the Cubic Bezier curves are used to construct the COG trajectory. With the proposed free gait, the robot can walk through the rough terrains automatically and successfully. The correctness and effectiveness of the proposed method is verified via simulation.

1. Introduction

Compared with the wheeled and tracked robot, the quadruped robot can travel on the rough terrains. The quadruped robots are more stable than biped robots and have less complicated mechanism than hexapod robot [1].

The static gait can help quadruped robot to walk through the rough terrains. McGhee and Frank firstly gave a detailed analysis of the static gait of quadruped robot [2]. After that, a number of researchers gave different static gait planning methods [3, 4], including many which can help the quadruped robot walk through rough terrains with high complexity [5, 6]. When travelling on the rough terrains containing forbidden areas, quadruped robot generally use the non-periodic static gait called free gait [7].

In recent years, there are many free gait planning methods have been proposed for the quadruped robots. Shaoping Bai et al. proposed a free gait

* This work is supported by the National Natural Science Foundation of China (61233014 and 61305130), the Shandong Provincial Natural Science Foundation (ZR2013FQ003 and ZR2013EEM027), National High Technology Research and Development Program of China (2015AA042201).

† Corresponding author: Rui Song, E-mail: rsong@sdu.edu.cn.

called primary/secondary gait [7]. Yun Jung Lee and Zeungnam Bien presented a free crab gait planning method for quadruped robot by using a hierarchical strategy [8]. Prabir K. Pal and K. Jayarajan proposed of a free gait planning method for the straight-line motion for quadruped robot in a straight-line motion [9]. Hiroshi Igarashi et al. gave a free gait generation method for quadruped robot to unknown irregular terrain [10].

In the above free gait planning methods, there are no specific methods were proposed to search the foothold for the swing foot of the quadruped robot. In order to ensure that the swing foot can obtain an optimum foothold, a foothold search method is proposed in this paper. In addition, in order to enlarge the stability margin, before the swing foot lift-off the ground, according to the support pattern, the robot completes a sway motion in a four feet phase. With the free gait proposed in this paper, the robot can walk through the rough terrains containing forbidden areas, and the simulation results show the effectiveness and efficiency of the proposed free gait planning method.

2. The Model of the Quadruped Robot

The simulation model of the quadruped robot is shown in Figure 1. Each leg of the quadruped robot contains 3 degrees of freedom (DOF): a rolling rotary joint and two pitching rotary joints, as shown in Figure 1. The robot is 0.7m long and 0.35m wide. When the legs are fully extended, the robot is 0.75m high. The total weigh of the robot is 130kg and each leg weights 10kg. At the bottom of each foot, there is a touch sensor for the touchdown detection.

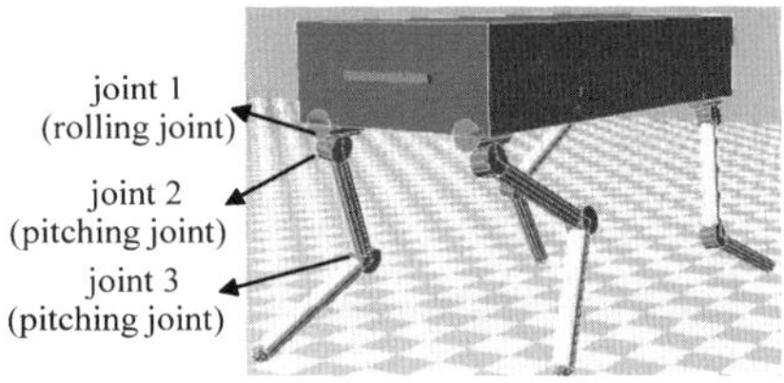

Figure1. The simulation model of the quadruped robot

3. The Foot Step Sequence and the Stability of the Quadruped Robot

3.1. *The foot step sequence of the quadruped robot*

The foot step sequence is not fixed when quadruped robot walking with the free gait. The robot should determine the foot step sequence according to the state of the robot during the walking process.

In order to ensure the robot can move forward constantly, the foot with the minimum kinematic margin should be selected to swing forward to restore the kinematic margin. When the swing foot of the robot touches the ground, all the four feet of the robot are on the ground, and the kinematic margins of the four feet can be obtained. Then, the foot with the minimum kinematic margin should be chosen to be the Following Swing Foot (FSF). Among the remaining three feet, the foot with the kinematic margin just lager than the kinematic margin of the FSF would swing forward after the FSF complete its swing movement, and this foot is named as AFSF (the swing foot after the FSF).

3.2. *The stability of the quadruped robot*

When walking on the rough terrains, the stability is the primary concern. The robot can obtain the largest stability margin if the projection of the center of gravity (COG) is located at the centroid of the support triangle [11]. In the free gait proposed in this paper, the projection of the COG is desired to be moved to the centroid of the support triangle just before the FSF lift-off the ground through a body sway movement, and the body sway movement is completed in a four feet support phase.

In this paper, the Cubic Bezier curves are used to construct the COG trajectory to guide the COG from the start point to the target point of the body sway movement. Taking the support partner shown in Figure 2 as an example, the calculation of the COG trajectory is shown as follow.

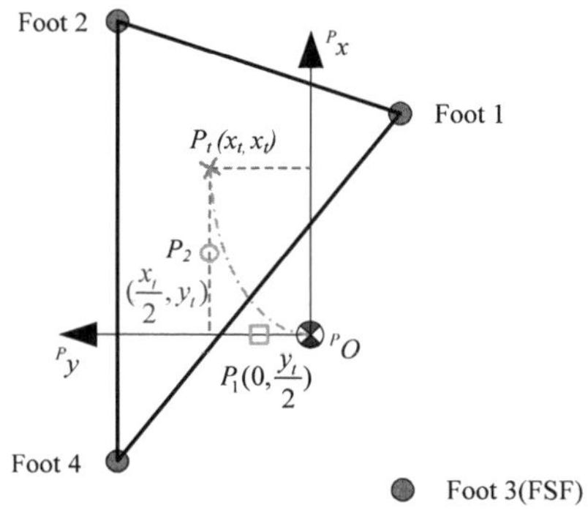

Figure 2. The target position of the sway movement

As shown in the Figure 2, the solid circles represent for the projection (on the ground surface) of the support feet, and the foot 3 is the FSF. The coordinate frame *{P}* is a planar frame with origin at the projection of the COG of the robot. The solid triangle shown in the Figure 2 is the support triangle of the robot when the FSF swings forward.

The start position of the sway movement is the origin of coordinate system *{P}*. The point $P_t(x_t, y_t)$, as the blue cross shown in the Figure 2, is the centroid of the support triangle, and it is the target position of the sway movement. The point $P_1(x_1, y_1)$ and the point $P_2(x_2, y_2)$ as shown in Figure 2 are the two control points of the Bezier curve, and the coordinate values of the point P_1 and P_2 can be respectively calculated using equations (1) and (2) as follow:

$$\begin{cases} x_1 = 0 \\ y_1 = \dfrac{y_t}{2} \end{cases} \tag{1}$$

$$\begin{cases} x_2 = \dfrac{x_t}{2} \\ y_2 = y_t \end{cases} \tag{2}$$

The COG trajectory constructed by the Bezier curves can be expressed as follow:

$$\begin{cases} x(t) = a_3 \cdot t^3 + a_2 \cdot t^2 + a_1 \cdot t + a_0 \\ y(t) = b_3 \cdot t^3 + b_2 \cdot t^2 + b_1 \cdot t + b_0 \end{cases}, \quad t \in [0,1] \tag{3}$$

Then, the coefficients of the equation (3) can be obtained by solving the simultaneous equations (4) and (5) respectively.

$$\begin{cases} a_0 = 0 \\ a_1 = 3 \cdot x_1 \\ a_2 = 3 \cdot x_2 - 6 \cdot x_1 \\ a_3 = x_t - 3 \cdot x_2 + 3 \cdot x_1 \end{cases} \tag{4}$$

$$\begin{cases} b_0 = 0 \\ b_1 = 3 \cdot y_1 \\ b_2 = 3 \cdot y_2 - 6 \cdot y_1 \\ b_3 = y_t - 3 \cdot y_2 + 3 \cdot y_1 \end{cases} \tag{5}$$

Substituting the equations (4) and (5) into equation (3), the COG trajectory in the four feet support phase can be obtained. Then, after the robot completes the body sway movement along the COG trajectory, the projection of the COG is located at the centroid of the support triangle. In this way, the quadruped robot can obtain the maximum value of the S_m when the robot swing a foot forward.

4. The Optimal Foothold Search Strategy

After the FSF is selected, the positions within the foothold search area of the FSF can be chosen as the footholds of the FSF, called the foothold candidates in this paper, would be obtained. Among the foothold candidates, an optimum foothold should be selected to be determined.

In order to make sure the quadruped robot can walk through the rough terrains successfully, the robot should obtain enough stability margin during the whole walking process. Enlarging the support triangle is an effective way to enlarge the stability margin of the robot. Therefore, among the foothold candidates, the foothold candidate which can help the robot to generate a biggest possible support triangle for the AFSF should be selected to be the foothold of the FSF, and the stability of the robot can be improved in this way.

Taking the robot state shown in Figure 3 as an example, the foothold search method proposed in paper is shown as following.

When the robot is in the state shown in Figure 3, the foot 4 is the FSF and the foot 2 is the AFSF. The dotted rectangles are the search areas of the feet. The red crosses represent for the example foothold candidates of the FSF.

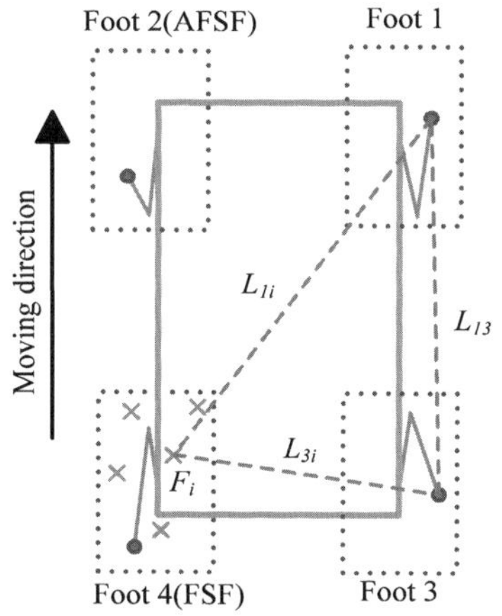

Figure 3. The example foothold candidates of the foot 4

Let F_i (x_i, y_i) be the *i-th* foothold candidate and its coordinate value which is expressed in the frame *{P}*. Let P_1 (x_1, y_1) and P_3 (x_3, y_3) be the leg-end positions of foot 1 and foot 3 which are respectively expressed in the coordinate frame *{P}*.

If the *i-th* foothold candidate is selected to be the foothold of the FSF. Let ΔS_T be the support triangle of the robot when the robot swings the AFSF forward.

According to the coordinates of corresponding points, the length of the L_{1i}, the L_{3i} and the L_{13} can be obtained as following:

$$\begin{cases} L_{1i} = \sqrt{(x_1 - x_i)^2 + (y_1 - y_i)^2} \\ L_{3i} = \sqrt{(x_3 - x_i)^2 + (y_3 - y_i)^2} \\ L_{13} = \sqrt{(x_1 - x_3)^2 + (y_1 - y_3)^2} \end{cases} \tag{6}$$

Then the perimeter of the ΔS_T can be calculated:

$$p_i = L_{1i} + L_{3i} + L_{13} \tag{7}$$

Finally, the area of the ΔS_T can be obtained:

$$S_i = \sqrt{p \cdot (p - L_{1i}) \cdot (p - L_{3i}) \cdot (p - L_{13})} \tag{8}$$

where S_i represents for the area of the ΔS_T when the *i-th* foothold candidate is selected to be the foothold of the FSF.

From the equation (6) to equation (8), it can be seen that the x_i and y_i are the variables in the equation (8), so the equation (8) can be rewrite as follow:

$$S_i = f(x_i, y_i) \tag{9}$$

According to the coordinate values of the foot candidates and the equation (9), the corresponding value of the S_i can be respectively obtained. Then the foothold candidate with the max value of the S_i would be selected to be the optimal foothold for the FSF.

5. Dynamics Simulation with Webots and Performance Analysis

Figure 4 shows the quadruped robot and the rough terrain in the simulation environment.

Figure 4. Overview of the rough terrain

There are some rectangular shaped objects which were randomly placed along the moving direction of the robot with different heights, as shown in Figure 4. During the walking process, the quadruped robot is only allowed to put its feet on these rectangular shaped objects, and the remaining areas are considered as forbidden areas.

Figure 5 shows a set of snap-shots when the robot was walking on the rough terrain shown in Figure 4. The quadruped robot can walk through the rough terrain shown in the Figure 4 automatically with the free gait proposed in this paper.

Figure 5. Sequence of snapshots of the robot walking process

Figure 6 shows the stability margin of the robot in the whole process of the robot walking through the rough terrain shown in Figure 4. From the Figure 6, it can be seen that the minimum stability margin during the whole process of the robot walking movement is more than zero, and the quadruped robot can obtain the enough stability margin when walk with the free gait which is proposed in this paper.

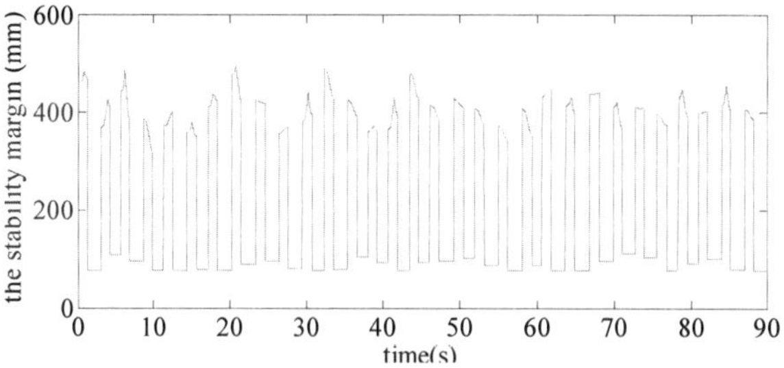

Figure 6. The value of the S_m in the whole walking process of the robot

6. Conclusion

In order to ensure the quadruped robot can walk through the rough terrains containing forbidden areas, a free gait generation method based on the foothold strategy is proposed in this paper. A foothold search method is presented to obtain an optimum foothold for the swing foot. With the foothold search method given in this paper, the stability margin of the quadruped robot is enlarged, and the terrain adaptability of the robot can be improved in this way.

From the simulation results, it can be seen that, with the free gait generation method proposed in this paper, the robot can walk through the rough terrain containing forbidden areas automatically. The simulation results show the effectiveness and the efficiency of the proposed free gait generation method.

References

1. Li Y, Li B, Ruan J, et al. "Research of mammal bionic quadruped robots: a review," in Proceedings of IEEE Conference on Robotics, Automation and Mechatronics (RAM2011), pp. 166-171. September 17 -19, (2011).
2. McGhee, Robert B., and Andrew A. Frank, "On the stability properties of quadruped creeping gaits," Mathematical Biosciences, vol.3, pp. 331-351, (1968).
3. Chen, Xuedong, et al, "Translational crawl and path tracking of a quadruped robot, Journal of Robotic Systems", vol. 19, no.12, pp.569-584, (2002).
4. Ma, Shugen, Takashi Tomiyama, and Hideyuki Wada, "Omnidirectional static walking of a quadruped robot, IEEE Transactions on Robotics," vol. 21, no. 2, pp. 152-161, (2005).
5. Pongas D, Mistry M, Schaal S. "A robust quadruped walking gait for traversing rough terrain," in Proceedings of IEEE International Conference on Robotics and Automation (ICRA), pp. 1474-1479, (2007).
6. V. Barasuol, J. Buchli, C. Semini, M. Frigerio, E. De Pieri, and D. Caldwell, "A reactive controller frame-work for quadrupedal locomotionon challenging terrain," IEEE International Conference on Robotics and Automation (ICRA), pp. 2554-2561, (2013).
7. Bai, Shaoping, Kin Huat Low, and Teresa Zielinska. "Quadruped free gait generation based on the primary/secondary gait," Robotica, vol.17, no. 4, pp. 405-412, (1999).
8. Lee, Yun Jung, and Zeungnam Bien, "A hierarchical strategy for planning crab gaits of a quadruped walking robot," Robotica, vol. 12, no. 01, pp. 23-31, (1994).
9. Pal, Prabir K., and K. Jayarajan, "Generation of free gait-a graph search approach," IEEE Transactions on Robotics and Automation vol.7, no.3, pp.299-305, (1991).
10. Igarashi H, Machida T, Harashima F, et al, "Free gait for quadruped robots with posture control," 9th IEEE International Workshop on Advanced Motion Control, pp. 433-438, (2006).
11. Rebula J R, Neuhaus P D, Bonnlander B V, et al. A controller for the littledog quadruped walking on rough terrain[C]//Robotics and Automation, 2007 IEEE International Conference on. IEEE, pp.1467-1473, (2007).

DRIVING COORDINATION BETWEEN THE KNEE AND ANKLE JOINTS DURING TIPTOE SUPPORT PHASE FOR BIPEDAL ROBOTS SPRINTING*

CHAO LI, RONG XIONG, QIU-GUO ZHU, WEI-NAN WU, JUN WU

State Key Laboratory of Industrial Control Technology, Zhejiang University Hangzhou 310027, China

This paper proposes to plan the bipedal robot sprinting using virtual forces control method owing to the toe joint between the support leg and ground is under-actuated. However, under the virtual forces control, the internal motion of the support leg becomes uncertain and the works of the joint torques may pit against each other during the tiptoe support phase in sprinting due to the added redundant toe joint. To solve this problem, a driving coordination between the knee and ankle joints method that balances the instantaneous works of the two joints by an angular velocity follower for the ankle joint is proposed. Simulations experiments verify the effectiveness of the proposed method.

1. Introduction

Making bipedal robots running like humans with higher speed is one of the key research issues in the field of bipedal robots. In 1986, Raibert developed the first planar bipedal running robot "Planar Biped" with linear legs and no foot, which completed running by alternate bounding with the left and right legs [1]. Since 2003, a variety of articulated bipedal running robots have been developed, such as MABEL [2] of Michigan University, ASIMO of Honda [3], HRP-2LR of AIST [4] and so on. Among them, some adopted legs with only hip and knee joints and no foot in the sagittal plane [2], while others used legs with added foot and ankle joint like humans in the sagittal plane [3][4].

It is commonly known that most bipedal robots with foot were landing in such a way that the heel and the tiptoe touch and leave the ground simultaneously all the time during their walking and running. However, Novacheck found that heel-to-tiptoe landing way is normally used in human's running, while only tiptoe is used for landing all the time in sprinting [5]. In addition, Lieberman analyzed the influence of different landing ways on the

* Research is supported by the National Natural Science Foundation of China (Grand No. 51405430 and Grand No. 61473258), and supported by the National 863 plan of China (No. 2012AA041703).

impact force and pointed out that the tiptoe landing way can smooth the change of the impact force [6]. According to the energy consumption data of different joints during human's running and sprinting [5], it turns out that the ankle consumes more energy than the knee in all scenarios. However, on most of the existing bipedal running robots, the fact is that the knee consumes more energy than the ankle. This is due to the restriction that the support foot must fully touch the ground, which greatly limits the performance of the ankle. To make full use of the ankle, we propose to ignore the restriction and adopt the tiptoe landing way in running (hereinafter referred to as sprinting).

For sprinting, the redundant Degree-of-Freedom (DoF) of the support leg during tiptoe support phase must be dealt with first. In this phase, the contact point between the foot and the ground can be modeled as a virtual toe joint, which is a redundant DoF. We need to introduce an extra constraint to solve this problem. How to define the constraint to make full use of the redundant DoF will be the key issue to achieve the bipedal sprinting.

2. Motion Planning of Bipedal Sprinting

Instead of adding an extra toe joint in the foot, we permit the foot rotate about the ankle and make only the tiptoe touch the ground in single support phase. Then, the planar bipedal robot model with foot is built and showed in Fig. 1(a). The robot is composed of a rigid body and two three-joint legs, which contain hip, knee and ankle joints. During sprinting, we assume the tiptoe and the ground is connected by a virtual and unactuated toe joint in the tiptoe support phase.

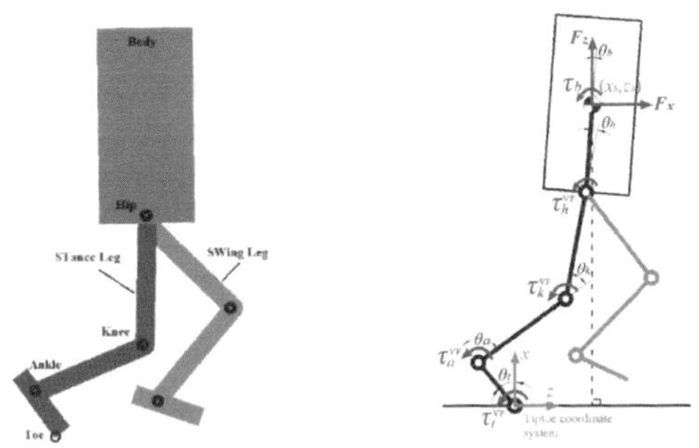

(a) (b)

Figure 1. Smplified model (a) and VFC model (b) of the planar biped robot.

The corresponding VFC model is showed in Fig.1(b). x_b and z_b are the horizontal, vertical positions of the Center of Mass (CoM) of the body, θ_b is the pitch angle of the body. θ_t , θ_a , θ_k , θ_h are the angles of the toe, ankle, knee and hip joints of the support leg respectively. F_x , F_z and τ_b are the virtual horizontal force, vertical force and torque imposed on CoM respectively. τ_t^{VF}, τ_a^{VF}, τ_k^{VF} and τ_h^{VF} are the equivalent torques imposed on the toe, ankle, knee and hip joints of the support leg respectively.

This paper studies how to generate joint torques of the support leg such that the planar bipedal robot do desired body motion efficiently and avoid that the torques of knee and ankle pit against each other. During tiptoe support phase, ignoring the effect of the swing leg, the forward kinematics of the support leg is:

$$\boldsymbol{q}_b = f_{\mathrm{FK}}\left(\boldsymbol{\theta}\right) \tag{1}$$

where $\boldsymbol{q}_b = \begin{bmatrix} x & z & \theta_b \end{bmatrix}^{\mathrm{T}}$, $\boldsymbol{\theta} = \begin{bmatrix} \theta_t & \theta_a & \theta_k & \theta_h \end{bmatrix}^{\mathrm{T}}$.

Given the virtual forces, the equivalent joint torques of the support leg can be easily calculated using the Force Jacobian Matrix [7][8],

$$\boldsymbol{\tau}^{\mathrm{VF}} = J\left(\boldsymbol{\theta}\right)^{\mathrm{T}} \boldsymbol{F} \tag{2}$$

where $\boldsymbol{F} = \begin{bmatrix} F_x & F_z & \tau_b \end{bmatrix}^{\mathrm{T}}$, $\boldsymbol{\tau}^{\mathrm{VF}} = \begin{bmatrix} \tau_t^{\mathrm{VF}} & \tau_a^{\mathrm{VF}} & \tau_k^{\mathrm{VF}} & \tau_h^{\mathrm{VF}} \end{bmatrix}^{\mathrm{T}}$.

During tiptoe support phase, there is no torque on the virtual toe joint. Setting $\tau_t^{\mathrm{VF}} = 0$ for the first line of Eq. (2), the constraint between virtual forces is got,

$$zF_x - xF_z + \tau_b = 0 \tag{3}$$

Due to the constraint of Eq. (3), we found that not all the virtual forces have the corresponding equivalent joint torques, which means, as long as any two of them are determined, the remained one can be calculated from Eq. (3). For sprinting, considering the stability of the bipedal robot and energy consumption, it is desired that the height and angle of the body keep unchanged during sprinting. Then, a PD Controller is designed to maintain the body height and angle,

$$\begin{cases} F_z = k_{zp}\left(z^d - z\right) + k_{zd}\left(\dot{z}^d - \dot{z}\right) + \mathrm{mg} \\ \tau_b = k_{bp}\left(\theta_b^d - \theta_b\right) + k_{bd}\left(\dot{\theta}_b^d - \dot{\theta}_b\right) \end{cases} \tag{4}$$

where z^d and θ_b^d are the desired height and angle of the body. $\dot{z}$ and $\dot{z}^d$ are the actual and desired vertical velocity of the body, $\dot{\theta}_b$ and $\dot{\theta}_b^d$ are the actual and desired angular velocity of the body. k_{zp}, k_{bp} and k_{zd} , k_{bd} are the parameters of the PD Controller. m is the body mass and g is the gravity acceleration.

Therefore, we specify F_z and τ_b solve for F_x ,

$$F_x = \frac{x}{z} F_z - \frac{1}{z} \tau_b \tag{5}$$

Plugging Eq. (4) and Eq. (5) back into Eq. (2), we get the equivalent joint torques of the support leg given the desired height and pitch angle of the body. The planar biped robot during tiptoe support phase is similar to linear inverted pendulum model [9], so the horizontal motion of the body can be controlled by properly choosing the position of the landing of the swing leg.

3. Driving Coordinate between the Knee and Ankle Joints

For the bipedal robot with only hip and knee joints, the above method is sufficient to be used to plan the motion of walking and running. However, when it comes to sprinting, the result will be totally different. Desired virtual forces can be determined by Eq. (4) and Eq. (5), thus, the accelerations of the body subjecting to virtual forces are:

$$\ddot{\boldsymbol{q}}_b = diag\left(m^{-1}, m^{-1}, I^{-1}\right)\left(\boldsymbol{F} - \begin{bmatrix} 0 & mg & 0 \end{bmatrix}^{\mathrm{T}}\right) \tag{6}$$

where $\ddot{\boldsymbol{q}}_b$ is the accelerations of the body, I is the rotational inertia of the body on the CoM.

Observing Eq. (6) we know that given the body mass, rotational inertia and the PD controller parameters, the body motion can be determined by its initial state of the phase and is independent of the motion of the support leg.

Taking the derivative of Eq. (1), we have the relationship between the velocities of the body and joints angular velocities of the support leg:

$$\dot{\boldsymbol{q}}_b = J(\boldsymbol{\theta})\dot{\boldsymbol{\theta}} \tag{7}$$

where $J(\boldsymbol{\theta})$ is the Jacobian Matrix of the support leg.

There are four variables involved in Eq. (7), as only three motion equations, there will be multiple solutions of the variables $\dot{\boldsymbol{\theta}}$, causing the motion of the support leg to be uncertain. From Eq. (2), we know that the equivalent joint torques is only associated with the current configuration of the support leg, thus virtual forces imposed on the body keep satisfying Eq. (4) and Eq. (5) regardless of the change of the configuration, and the body being controlled.

To solve the motion uncertainty of the support leg and avoid that the works of knee and ankle joints pit against each other, we attempt to coordinate the instantaneous work of the knee and ankle joints. Then, a condition of DCKA (Driving Coordination between the Knee and Ankle Joints) is introduced:

$$\frac{\tau_a \dot{\theta}_a}{\tau_k \dot{\theta}_k} = k \tag{8}$$

where $\tau_k \dot{\theta}_k$ and $\tau_a \dot{\theta}_a$ are the instantaneous work of the knee and ankle, k is the coefficient of DCKA and must be a positive value. Usually, k take the value of the ratio between the rated power of the knee and ankle joints.

If the relationship of the instantaneous work of the knee and ankle satisfy the condition of Eq. (8) all the time, there will be no internal energy loss of the knee and ankle joints ideally. Combining the Eq. (7) and Eq. (8), there are four variables in four equations, so the motion of the support leg is certain.

Given the actual angular velocity of the knee, we get the desired angular velocity of the ankle to satisfy the DCKA condition of Eq. (8) as follows:

$$\dot{\theta}_a^d = \frac{k \tau_k^{\mathrm{VF}} \dot{\theta}_k}{\tau_a^{\mathrm{VF}}} \tag{9}$$

where $\dot{\theta}_a^d$ is desired angular velocity of the ankle of the support leg.

In order to track the desired angular velocity of the ankle, an angular velocity follower for the ankle is designed as:

$$\tau_a' = k_d' \left(\dot{\theta}_a - \dot{\theta}_a^d \right) \tag{10}$$

where τ_a' represents the compensated torque on the ankle, and k_d' is the control parameter.

Thus, the final torques to be applied on the joints of the support leg is:

$$\boldsymbol{\tau}_{\mathrm{SU}} = \boldsymbol{\tau}^{\mathrm{VF}} + \begin{bmatrix} 0 & \tau_a' & 0 & 0 \end{bmatrix}^{\mathrm{T}} \tag{11}$$

4. Simulation Experiments

Due to the planar bipedal robot with SEA (series elastic actuator) joints being developed, some simulation experiments are taken to verify the effectiveness of the proposed methods. The mechanical parameters of the bipedal robot model is summarized in Table 1.

Based on the planar biped robot model, two simulation experiments are taken: the effectiveness of DCKA method in solving the motion uncertainty of the support leg, and then a simulation of a bipedal robot sprinting is carried out using virtual forces control and DCKA method. In our simulations, the desired height is 0.46m and the body pitch angle is 0rad, the control parameters are $k_{zp} = 50000$, $k_{zd} = 1000$, $k_{bp} = 3000$, $k_{bd} = 40$, $k_d' = 8$.

Table 1. Mechanical parameters of the planar bipedal robot model.

	Body	Thigh	Shin	Foot
Mass (kg)	20	1.0	1.0	0.5
Length (m)	0.4	0.25	0.25	0.1
Inertia (×10-4[kg.m2])	3333	43	43	7.83

4.1. *Comparative experiments using DCKA method and not*

Frist of all, an experiment not using DCKA constraint is taken, joint torques of the support leg are got only by Eq. (2), a simulation of the planar bipedal robot during tiptoe phase is taken as showed in Fig.2. Though the robot body can follow a desired motion under the virtual forces, the internal motion of the support leg is uncertain, which leads to an irregular movement. We can also observe that the support foot flips forward at 0.25s and the joints of the support leg are rotating freely.

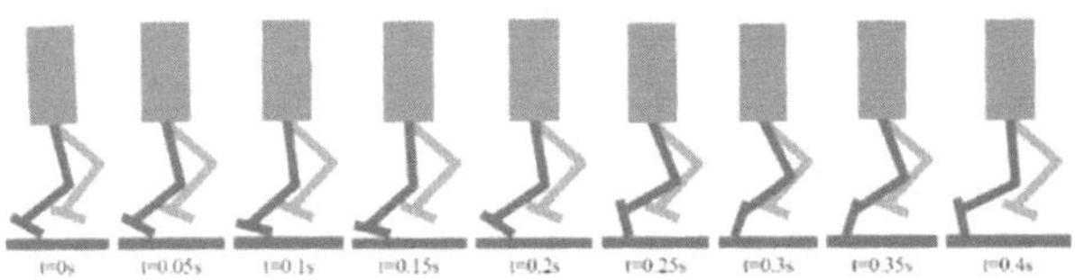

(a)

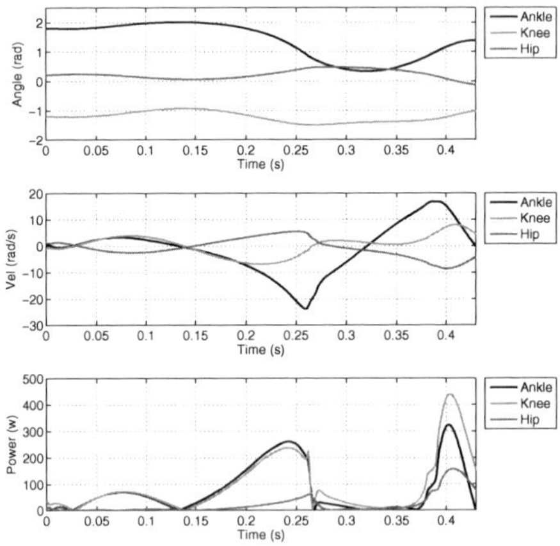

(b)

Figure 2. (a) The time elapsed image sequence of the bipedal robot during without using DCKA method. (b) The corresponding curves of the support leg.

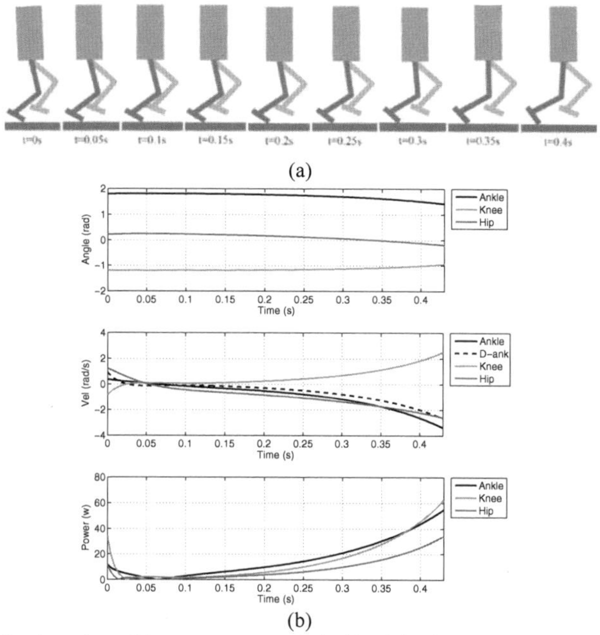

Figure 3. (a) The time elapsed image sequence of the bipedal robot using DCKA method. (b) The corresponding curves the support leg.

Then, we setup the simulation with the same initial and ending conditions as showed in Fig. 2. By setting the coefficient of DCKA is $k=1$ and applying torques on the support leg joints according to Eq. (11), we can obtain the same body motion during tiptoe support phase, together with the corresponding curves of the angles, angular velocities and powers of the ankle, knee and toe joints of the support leg, as shown in Fig.3. Compared with Fig.2, the motion of the support leg becomes smoother and the powers of the knee and ankle joints are both significantly reduced when DCKA method is applied.

4.2. *Simulation of the bipedal robot sprinting*

Setting the desired forward speed as 3m/s, the maximum power of the knee and ankle as both $250\text{w}(k=1)$, we carry out a simulation of the bipedal robot sprinting. Fig. 4 shows the curves of the forward speed, height and pitch angle of the body, where the solid line represents the actual value while the dashed line denotes the desired value. As shown in Fig. 4, the bipedal robot is able to sprint quite steadily, proving the validity of the DCKA method.

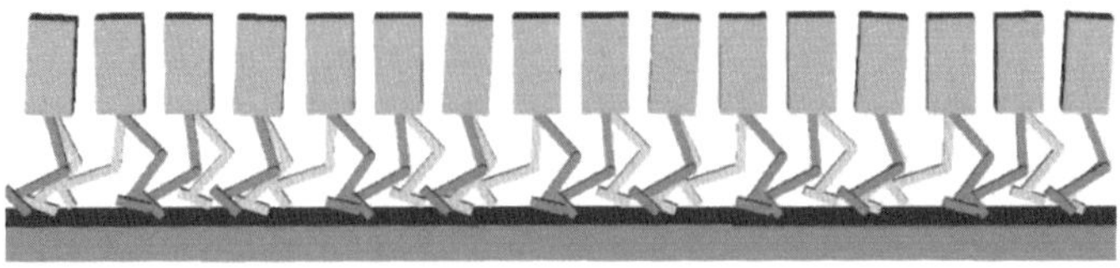

(a)

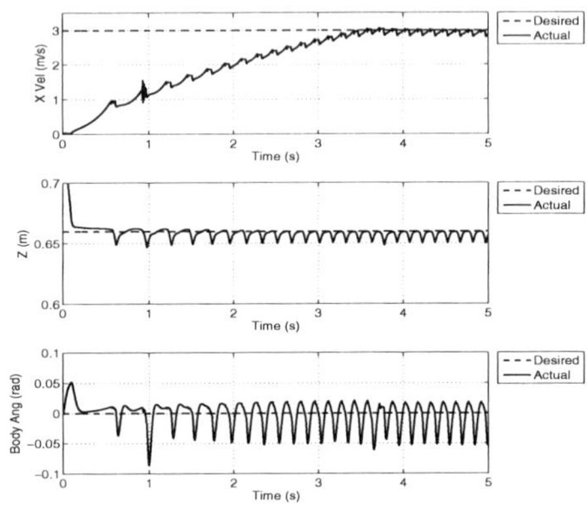

(b)

Figure 4. (a) The time elapsed image sequence of the bipedal robot sprinting. (b) The corresponding curves of the desired and actual body motion.

5. Conclusions

We have presented a control approach for bipedal sprinting using VFC and DCKA method where the internal motion uncertainty of the support leg is solved and the problem that the works of knee and ankle pit against each other is avoided. Simulation experiments have validated the effectiveness of the method. The ongoing work includes a series of physical experiments carried out on a planar bipedal robot with SEA joints, which is being developed in our laboratory, to verify our proposed method.

In addition, in this paper, we only focused on the support leg rather than the swing leg, which will be taken into account in the future. In sprinting, the angular velocities and powers of the joints of the swing leg also play an important role in improving the speed, how to optimize the motion of the redundant ankle joint of the swing leg to achieve higher speed is also critical.

References

1. Raibert M H. *Legged robots that balance*, Cambridge, MA: MIT press, 1986.
2. Grizzle J W, Hurst J, Morris B, et al. MABEL, a new robotic bipedal

walker and runner. *Proceeding of the American Control Conference (ACC)*, St Louis, USA, 2009, 2030-2036.
3. The Honda Humanoid Robot ASIMO, http://world.honda.com/ASIMO /technology/index.html
4. Kajita S, Nagasaki T, Kaneko K, et al. A running controller of humanoid bipedal HRP-2LR. *Proceeding of the IEEE International Conference on Robotics and Automation (ICRA)*, Barcelona, Spain, 2005, 616-622.
5. Novacheck T F. The biomechanics of running. *Gait & posture*, 1998, **7**, 77-95.
6. Lieberman D E, Venkadesan M, Werbel W A, et al. Foot strike patterns and collision forces in habitually barefoot versus shod runners. *Nature*, 2010, **463**, 531-535.
7. Pratt J, Dilworth P, Pratt G. Virtual model control of a bipedal walking robot. *Proceeding of the IEEE International Conference on Robotics and Automation (ICRA)*, Albuquerque, New Mexico, 1997, 193-198.
8. Craig J J. *Introduction to Robotics: Mechanics and Control*. 2nd edition. Boston, Addison-Wesley, 1989.
9. S. Kajita and K. Tani. Study of dynamic biped locomotion on rugged terrain-derivation and application of the linear inverted pendulum mode. *Proceeding of the IEEE International Conference on Robotics and Automation (ICRA)*, 1991, 1405-1411.

FOOT PLACEMENT CONTROL FOR BIPEDAL WALKING ON UNEVEN TERRAIN: AN ONLINE LINEAR REGRESSION ANALYSIS APPROACH

YANGWEI YOU*, ZHIBIN LI, NIKOS G. TSAGARAKIS and DARWIN G. CALDWELL

Department of Advanced Robotics, Istituto Italiano di Tecnologia, Genoa, Liguria 16163, Italy
**E-mail: yangwei.you@iit.it*

This paper presents a novel foot placement control algorithm for adaptive bipedal walking. In this method, the torso attitude and height are stabilized by synergic patterns so that the forward velocity and its change have a stable and nearly linear relation with the foot placement. Hence, our proposed online linear regression analysis well represents the local linear models by estimating continuously from measured data. Based on this estimation, an appropriate foot placement can be determined to control the forward velocity. Our simulation study successfully demonstrates the natural gait with accurate tracking of walking velocity, and the robustness of walking over uneven terrain.

Keywords: Foot placement control; Dynamic walking; Legged locomotion.

1. Introduction

To date, many humanoid robots have been built with impressive progresses in hardware and control.[1–5] This demands the software and algorithms to improve the feasibility and autonomy of those robots.[6] One important aspect for humanoid locomotion is to place the foot correctly,[7] which remarkably affects the stability. The work in[7,8] studied the relation between the foot placement and the system stability, and proposed foot placement indicator and estimator as the measure of balance.

Many methods for humanoid walking were developed based on the zero-moment point,[3,4,9,10] and the linear inverted pendulum model (LIPM) was often used for the foot placement which provides analytic solutions. However, the robots governed by LIPM move with unnatural bent-knee gaits.

Biological research reveals that the movement of legged animals can be represented better by a spring loaded inverted pendulum (SLIP) model.[11]

Raibert developed legged robots from one-legged hoppers to quadrupeds[12] and used the foot placement to regulate the forward velocity by a simple equation. However, to apply his controller, the physical model needs to be carefully designed to satisfy the assumption that the leg is sufficiently light compared to the body. Guocai raised a similar controller but he used only the ankle actuators to control the walking velocity.[13,14]

To address this issue, we propose a novel foot placement control based on the online linear regression analysis to identify the approximated linear relation between the foot placement, forward velocity and its change. Though the real relation is non-linear due to the multi-body dynamics, the foot placement can still be represented well by the local linear estimation which is being corrected continuously using the real measurements.

This paper is organized as follows. Section 2 presents the robot model and the pattern synergies. Section 3 elaborates the principles of the linear regression analysis for the foot placement control. Section 4 demonstrates the feasibility from our simulation study. We conclude in the last section.

2. Walking Pattern Generation

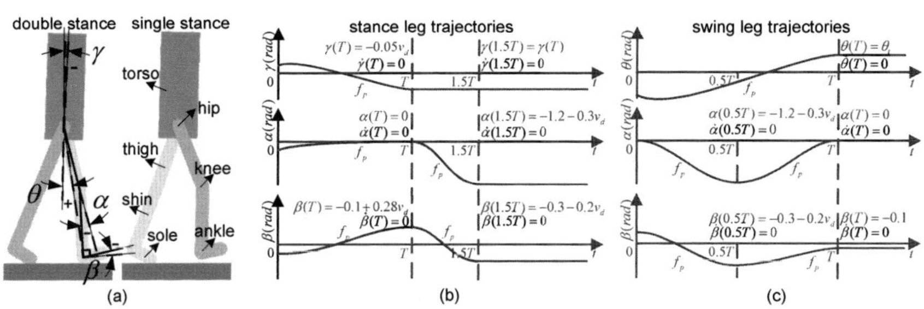

Fig. 1. Planar humanoid robot and its walking pattern.

To enable natural walking, the basic patterns of stance and swing legs are designed first, then the foot placement control can be implemented to adjust the specific trajectories for each step and keep the walking stability.

The simulated planar robot is comprised of 7 parts, one torso, two thighs, two shins and two soles, as shown in Fig. 1(a). Rotary actuators are mounted in the hip, knee and ankle joints. γ is the pitch angle of torso. θ is the swing angle, which is the angle between the vertical line and the line across the hip and ankle. θ_t is the touch-down angle, which is the swing angle at the moment of touch-down. α and β are the angles of knee and ankle joints respectively. The direction convention of these angles are defined in Fig. 1(a) with '+' and '-' signs. A 3rd-order polynomial is used for trajectory planning of a period from the time T_s to T_f to ensure the continuity of angle and angular velocity:

$$\begin{aligned}\phi(t) &= f_p(\phi(T_s), \dot{\phi}(T_s), \phi(T_f), \dot{\phi}(T_f), t) = \phi(T_s) + \dot{\phi}(T_s)t \\ &- \frac{3\phi(T_s) - 3\phi(T_f) + 2(T_f - T_s)\dot{\phi}(T_s) + (T_f - T_s)\dot{\phi}(T_f)}{(T_f - T_s)^2}t^2 \\ &+ \frac{2\phi(T_s) - 2\phi(T_f) + (T_f - T_s)\dot{\phi}(T_s) + (T_f - T_s)\dot{\phi}(T_f)}{(T_f - T_s)^3}t^3.\end{aligned} \tag{1}$$

where, $\phi(T_s)$, $\dot{\phi}(T_s)$, $\phi(T_f)$ and $\dot{\phi}(T_f)$ are respectively the initial angle and angular velocity and the final angle and angular velocity. For simplification, all the 3rd-order polynomials used below are referred as f_p.

The touch-down event of walking is detected when the swing leg really touches on the ground and the stance leg has supported for more than half of the expected stance time. At this moment, the trajectories of legs between this touch-down and the next one are planned based on the current state and expected forward velocity. Then, the stance leg is activated to swing while the swing leg starts to support. Although there is a short period of double support, this transition is not specially planned, but taken into account while planning the trajectories of single support phase. So the trajectory patterns of a leg can be classified according to the walking phases.

During stance, the hip actuator applies torque at the torso to track a planned trajectory of body pitch, which is related to forward velocity in order to imitate the leaning-forward behavior when people are walking. Meanwhile, actuators on knee and ankle are used to extend the leg to propel. Note that when the actual stance time is longer than the desired one and the swing leg still doesn't touch down, the stance leg will actively retract. The expected trajectories of these angles are piecewise three-order polynomials and detailed in Fig. 1(b), where T is the expected stance time.

During the swing phase, the swing angle is controlled by the actuator on hip joint to track the desired touch-down angle. Meanwhile, the leg is retracted first to have foot-ground clearance, and then extended by the knee and ankle actuators. When the actual swing time is bigger than expected and the touch down still does not occur, then knee and ankle joints will remain at the same position until touch-down happens. The swing angle is of particular interest, because it affects the touch-down angle θ_t and should be updated according to the current forward velocity. More details will be discussed in Section 3. The desired swing leg trajectories are also in Fig. 1(c).

3. Foot Placement Control

This section explains the foot placement control algorithm for adjusting the forward walking velocity. Previously, by observing the symmetry of running and the influence of foot placement, Raibert proposed a simple algorithm to control forward velocity as:[12]

$$x_f = \frac{\dot{x}T}{2} + k_{\dot{x}}(\dot{x} - \dot{x}_d), \tag{2}$$

where x_f is the foot placement, T is the duration of the stance phase, $\dot{x}$ is the current velocity, and $\dot{x}_d$ is the desired velocity. The item $\frac{\dot{x}T}{2}$ is the neutral point where the movement is symmetric and the average traveling speed remains unchanged after one step. The item $k_{\dot{x}}(\dot{x}-\dot{x}_d)$ is the feedback to displace the foot from the neutral point for stabilizing the forward speed against the errors and external disturbances.[12]

To some extent, Raibert's method works fine due to a good linear relationship among the foot placement, forward velocity and its change. To ensure this linearity, the mass ratio between the torso and leg of the robot has to be large, which limits the application on real humanoid robots with considerable mass of the legs.

Our idea is that though the mathematical function among the foot placement, forward velocity, and its change may appear as a nonlinear hyper surface, it can still be presented well by a set of local linear models. Therefore, to overcome the limitation of Raibert's method, we use the linear regression analysis on the real measurements to improve the prediction of foot placement by online updating the coefficients during walking.

Take the touch-down angle θ_t as the representative of foot placement. Regarding θ_t as the dependent variable, and forward velocity $\dot{x}$ and its change $(\dot{x} - \dot{x}_d)$ as the independent variables, their relationship can be estimated by the least square method as follows:

$$\begin{gathered}\theta_t = k_0 + k_1\dot{x} + k_2(\dot{x} - \dot{x}_d) \\ \begin{bmatrix} k_0 \; k_1 \; k_2 \end{bmatrix}^T = \begin{bmatrix} \boldsymbol{I}\cdot\boldsymbol{I}^T & \dot{\boldsymbol{x}}\cdot\boldsymbol{I}^T & \Delta\dot{\boldsymbol{x}}\cdot\boldsymbol{I}^T \\ \boldsymbol{I}\cdot\dot{\boldsymbol{x}}^T & \dot{\boldsymbol{x}}\cdot\dot{\boldsymbol{x}}^T & \Delta\dot{\boldsymbol{x}}\cdot\dot{\boldsymbol{x}}^T \\ \boldsymbol{I}\cdot\Delta\dot{\boldsymbol{x}}^T & \dot{\boldsymbol{x}}\cdot\Delta\dot{\boldsymbol{x}}^T & \Delta\dot{\boldsymbol{x}}\cdot\Delta\dot{\boldsymbol{x}}^T \end{bmatrix}^{-1} \cdot \begin{bmatrix} \boldsymbol{I}\cdot\boldsymbol{\theta}_t^T \\ \dot{\boldsymbol{x}}\cdot\boldsymbol{\theta}_t^T \\ \Delta\dot{\boldsymbol{x}}\cdot\boldsymbol{\theta}_t^T \end{bmatrix} \\ \boldsymbol{I} = [1\;1\cdots1]_{1\times n};\dot{\boldsymbol{x}} = [\dot{x}_1\;\dot{x}_2\cdots\dot{x}_n];\boldsymbol{\theta}_t = [\theta_{t1}\;\theta_{t2}\cdots\theta_{tn}]; \\ \Delta\dot{\boldsymbol{x}} = [\Delta\dot{x}_1\;\Delta\dot{x}_2\cdots\Delta\dot{x}_n];\Delta\dot{x}_i = \dot{x}_i - \dot{x}_{di}; i = 1,2\cdots n. \end{gathered} \tag{3}$$

where three coefficients (k_0, k_1, k_2) are used to approximate the linear relation, and compute the touch-down angle θ_t given the current velocity $\dot{x}$ and the desired velocity $\dot{x}_d$. Compared with (2), one more coefficient k_0

is added into the linear model to resolve a possible offset. All three coefficients (k_0, k_1, k_2) are updated based on the online linear regression analysis. For the regression analysis, n sets of data are collected online. Each data set consists of three elements: forward velocity just before touch-down $\dot{x}_i$, touch-down angle θ_t, and the forward velocity change after one step $\Delta\dot{x}_i$.

To ensure a good linear representation of the current local correlation, the latest data are used for the analysis. A queue that stores n sets of data is used, where the latest data is loaded while the oldest one is deleted continuously. To form an accurate local model, we avoid the singularity of data collection, by applying a data filter algorithm with thresholds $(\dot{x}^{th}, \theta_t^{th}, \Delta\dot{x}^{th})$ to determine whether the latest data is too similar with the existing ones.

The thresholds affect the steady-state accuracy and the robustness of forward velocity tracking. If they are too big, the sampled data are too far away from each other to form a precise local linear approximation. Whereas if too small, the data concentrate on a singular point thus cannot form a linear plane well. So the thresholds are the trade-off between steady-state accuracy and robustness. The number of data sets, n, also affects the control performance. If n is too big, the estimated linear model will refresh slowly and cannot response well to the system change. On the other hand, if too small, the estimated model changes too drastically and results in instability.

Given the predicted touch-down angle θ_t, the swing leg trajectory can be computed. The expected touch-down angle is calculated continuously and used to adjust the swing trajectory in response to the change of forward velocity. The swing angle is obtained as in (4), where θ_t is the expected touch-down angle calculated at the touch-down, and θ_t^{new} is the updated one during the stance phase, and T is the expected stance time.

$$\theta(t) = \begin{cases} f_p(\theta(0), \dot{\theta}(0), \theta_t, 0, t) + \theta_t^{new} - \theta_t, t \leq T; \\ \theta_t^{new}, t > T. \end{cases} \tag{4}$$

4. Simulation

We validated the performance of our proposed algorithm in the simulations with a humanoid as in Fig. 1(a). The mass of torso, thigh, shin and sole are 19.8, 2.8, 2.5 and 0.3 kg respectively, while their moments of inertia are 0.3, 0.02, 0.02, 3.7e-4 kgm^2, and the limb dimension of 0.4, 0.25, 0.25, 0.075 m. The motor PD controllers have the stiffness and damping of 287 Nm/rad and 5.7 Nms/rad for both knee and ankle. For the hip joint, the PD gains are 250 Nm/rad and 10 Nms/rad for controlling the torso attitude during stance, and are 100 Nm/rad and 2 Nms/rad for controlling the swing leg.

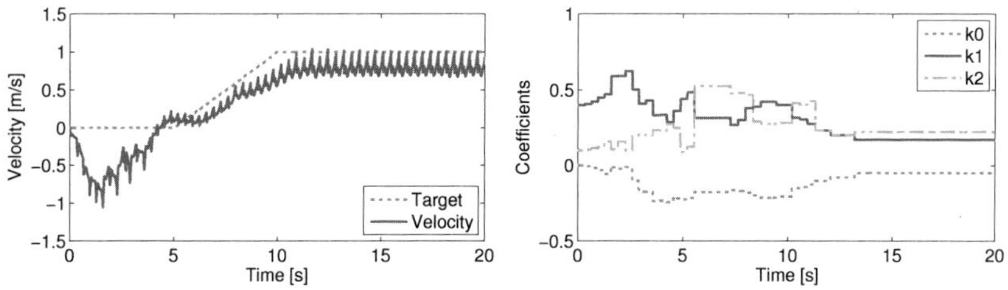

Fig. 2. Forward velocity tracking and online coefficient adaption of walking.

We used 6 sets of measured data for the online linear regression analysis. The thresholds $\dot{x}^{th}$, θ_t^{th}, $\Delta\dot{x}^{th}$ of the filter are 0.05 m/s, 0.02 rad, and 0.05 m/s. The initial data for the linear regression are listed in Table 1. They are chosen arbitrarily and cannot ensure stable walking without online update.

Table 1. Initial data for linear regression analysis.

Num.	$\dot{x}$	$\Delta\dot{x}$	θ_t	Num.	$\dot{x}$	$\Delta\dot{x}$	θ_t
1	1.000	0.000	0.400	4	1.000	0.000	0.400
2	0.000	1.000	0.100	5	0.000	1.000	0.100
3	0.000	0.000	0.000	6	0.000	0.000	0.000

The robot walked in place for 5 s to adjust the coefficients in (3), and then walked at 1 m/s within a transition of 5 s. At the beginning, the velocity fluctuated because the initial data for regression analysis was arbitrarily given. When the coefficients were adjusted well, then the walking velocity was tracked accurately. The online update of the coefficients and controlled walking velocity are shown in the Fig. 2. Then, the robot was controlled to walk at 1 m/s to cross an unknown step of 5 cm height. Fig. 3(a) shows the snapshots of successful walking over the step. The coefficients refreshing and walking velocity are shown in the Fig. 3(e) and 3(c).

Further, the robot was controlled to walk up and down a slope of 0.15 rad at 0.8 m/s. The slope was known for the robot, and the ankle angle $\beta(T)$ was revised according to the slope to ensure a good contact between the sole and slope. In practice, the slope can also be calculated according to the orientation of sole when it was fully in touch with the ground. Figure 3(b) is the snapshots of successful walking over the slope. The coefficients refreshing and walking velocity are shown in the Fig. 3(f) and 3(d).

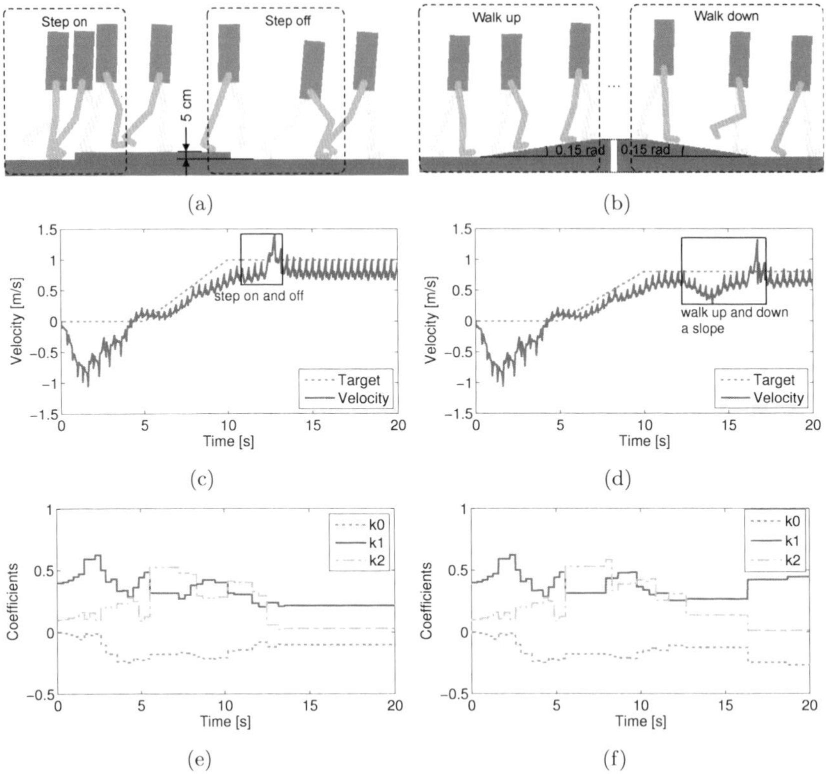

Fig. 3. Simulation of walking across uneven terrain.

5. Conclusion

We proposed a novel foot placement algorithm for online adaptive control of the bipedal walking. By controlling the torso attitude and height of the robot, the foot placement has a nearly linear relation with the forward velocity and its change. Therefore, our online regression analysis can estimate a good foot placement to regulate the walking velocity accurately, compared to the previous work.[12] The simulation study demonstrates the robustness of the algorithms by three successful validations: accurate tracking of forward velocity on the flat ground, blinding walking over a stair, ascending and descending known slopes. In the future, this method is expected to be implemented on a real 3D humanoid robot.

Acknowledgments

This work is supported by the FP7 EU project WALK-MAN (ICT 2013-10).

References

1. S. Kajita, K. Kaneko, F. Kaneiro, K. Harada, M. Morisawa, S. Nakaoka, K. Miura, K. Fujiwara, E. S. Neo, I. Hara *et al.*, Cybernetic human hrp-4c: A humanoid robot with human-like proportions, in *Robotics Research*, (Springer, 2011) pp. 301–314.
2. Z. Li, N. Tsagarakis and D. Caldwell, Walking trajectory generation for humanoid robots with compliant joints: Experimentation with coman humanoid, in *IEEE International Conference on Robotics and Automation*, 2012.
3. Z. Li, N. G. Tsagarakis and D. G. Caldwell, Walking pattern generation for a humanoid robot with compliant joints, *Autonomous Robots* **35**, 1 (2013).
4. C. G. A. Salman Faraji, Soha Pouya and A. J. Ijspeert, Versatile and Robust 3D Walking with a Simulated Humanoid Robot (Atlas) a Model Predictive Control Approach, *IEEE International Conference on Robotics and Automation* (2014).
5. N. Paine, J. S. Mehling, J. Holley, N. A. Radford, G. Johnson, C.-L. Fok and L. Sentis, Actuator Control for the NASA-JSC Valkyrie Humanoid Robot: A Decoupled Dynamics Approach for Torque Control of Series Elastic Robots, *Journal of Field Robotics* (2015).
6. S. Dalibard, A. El Khoury, F. Lamiraux, A. Nakhaei, M. Taïx and J.-P.Laumond, Dynamic walking and whole-body motion planning for humanoid robots: an integrated approach, *The International Journal of Robotics Research* (2013).
7. P. Van Zutven, D. Kostic and H. Nijmeijer, Foot placement for planar bipeds with point feet, in *IEEE International Conference on Robotics and Automation*, 2012.
8. D. L. Wight, E. G. Kubica and D. W. Wang, Introduction of the foot placement estimator: A dynamic measure of balance for bipedal robotics, *Journal of computational and nonlinear dynamics* **3** (2008).
9. S. Kajita, F. Kanehiro, K. Kaneko, K. Fujiwara, K. Harada, K. Yokoi and H. Hirukawa, Biped walking pattern generation by using preview control of zero-moment point, in *IEEE International Conference on Robotics and Automation*, 2003.
10. C. Chevallereau, D. Djoudi and J. W. Grizzle, Stable bipedal walking with foot rotation through direct regulation of the zero moment point, *IEEE Transactions on Robotics* **24**, 390 (2008).
11. H. Geyer, A. Seyfarth and R. Blickhan, Compliant leg behaviour explains basic dynamics of walking and running, *Proceedings of the Royal Society B: Biological Sciences* **273**, 2861 (2006).
12. M. H. Raibert, *Legged robots that balance* (MIT press Cambridge, MA, 1986).
13. G. Liu, M. Li, W. Guo and H. Cai, Control of a biped walking with dynamic balance, in *International Conference on Mechatronics and Automation*, *International Conference on Mechatronics and Automation* 2012.
14. G. Liu, F. Zha, M. Li, W. Guo, P. Wang and H. Cai, Control of a humanoid robot walking with dynamic balance, in *IEEE International Conference on Robotics and Biomimetics (ROBIO)*, 2013.

HYBRID POSITION/FORCE CONTROL OF THE PLANAR TROTTING QUADRUPED ROBOT

YIZHANG LIU, JIAN WANG, LIN LANG, HONGXU MA[†]

College of Mechatronics Engineering and Automation, National University of Defense Technology, Hunan Province, Changsha, China

This paper presents a hybrid position/force control algorithm to perform motion control of the quadruped robot based on a simplified planar model in sagittal plane which can describe the main dynamical characters of trotting gait. In order to realize the approximation of body motion decoupling, the vertical position and pitch of the body are controlled by using position servo and the horizontal position control is performed by using force control. And several constraints are considered to maintain the balance of the body during motion control, so as to increase the robustness of the robot while large interference are exerted. Simulation is carried out to validate the effectiveness of the proposed algorithm in Matlab/SimMechanics.

1. Introduction

Legged robots have many advantages over wheeled and tracked robots in some applications such as search and rescue in unknown rough terrain. Early quadruped robots were usually walking in static gait with controlling the center of mass from the support polygon formed by the legs in contact with the ground. However, the superiority of animals' agility refers to dynamically stable locomotion. Robots designed by Boston Dynamics are dynamically stable locomotion. Those robots could carry a 180kg load and follow a human leader anywhere he goes[1]. The performances of Boston Dynamics' robots are attractive and encouraging.

The dynamical systems approach has been proven to be successful in many robotic applications[2~4]. This approach is based on obtaining the dynamics of the systems. One can apply difference control methods directly or after transformation in obtained. The structures of those controllers are mainly based on computed torque drive method. And the performances of the controllers are mainly depended on the accuracy of modeling and generally care more about the internal forces of the system than physics constraints. Another available method is to design controller based on the kinematics[5,6],the motion of the robots mainly controlled by position controller and should be combined with online planning. Relatively, this method focuses only on the physics constraints and regardless the internal force. Therefore, a hybrid algorithm combine force control with position control is worth to be researched.

In this paper, an algorithm combine the two methods mentioned is presented, and several constraints are added to keep the body balance. The proposed algorithm realizes approximate decoupling of force control and position control, and could accommodate large pose interference.

2. Multi-Link Structure Modeling and Control Approach

Figure 1 shows the 3D quadruped system, whose body has six degrees of freedom and four springy legs. Each springy leg has three DOFs, including one translational joint and two rotational joints. The body attitude can be adjusted by those joints. When it is trotting, two feet are in swing phase while other two feet contacting the ground. The support legs' behavior is approximated to 3D spring damper multi-link structure systems, and the dynamics and kinematics of multi-link model can be easier built than the whole quadruped system. So we simplify the support phase together with the body to 3D multi-link structure system as Fig. 2, and trotting gait can be treated as changing of this systems. If we can control the multi-link model well, it would be easier to control the quadruped robot to trotting gait. Hence, it is quite important to discuss the control method of multi-link structure systems.

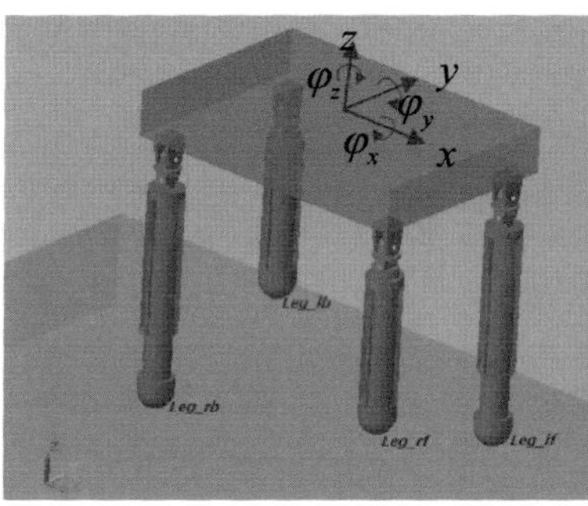

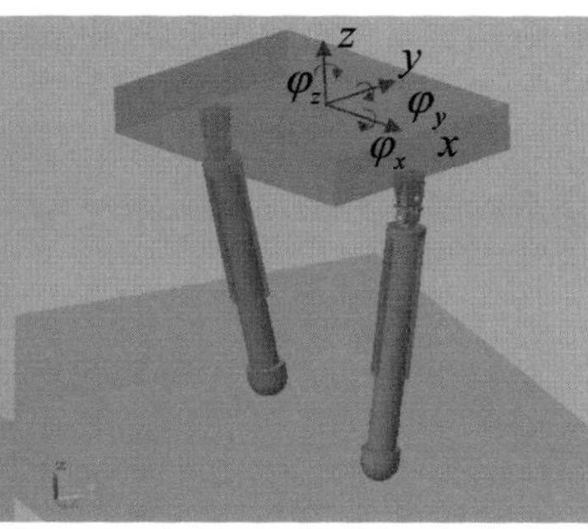

Fig. 1. Quadruped robot with simplified legs   Fig. 2. 3D multi-link mode

2.1. *Dynamic modelling*

Raibert had simplified the diagonal legs to a single leg with the concept of "virtual leg"[7]. However, if the diagonal legs are simplified to planar model as be shown in Fig. 3. All motion of the body in planar model could be controlled stable rather than two of them in supporting phase. The planar model was also used to analysis of virtual leg by Raibert. And it is an effective model for studying robots' dynamic behaviors. For the mass of the four-legged robot is mainly concentrated in the upper body, and the mass of legs are small enough to be ignored, when we build the dynamics of the simplified model, the mass and the inertia of the legs are ignored. The associated parameters are listed in Table 1.

Table 1.Variables and indices used

m	Body mass	x	Horizontal position	z	Vertical position
φ_y	Body pitch angle	g	Gravity	I	Body inertia
f_x	Horizontal force	f_z	Vertical force	θ	Incidence angle
l	Leg length	L	Half body length	τ	Torque on the hip
d or r	Desired or real	l_0	Leg rest length	a_l	Acceleration of leg length

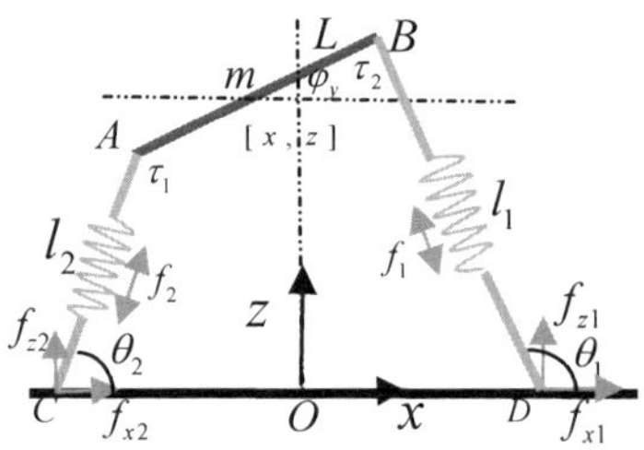

Fig. 3. Planar model of quadruped robot

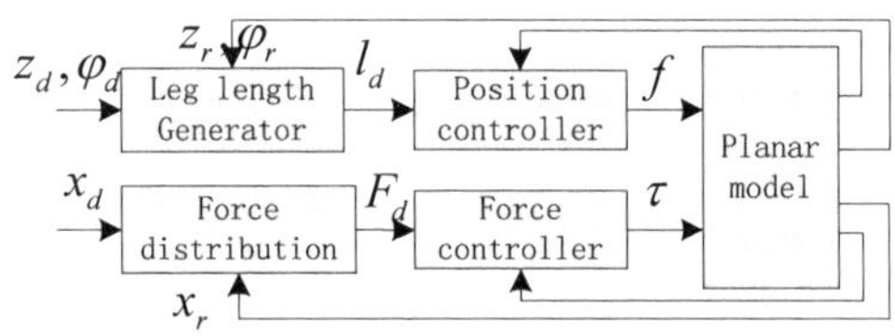

Fig. 4. simplified control diagram

The planar model have four different phases: flight phase, hind leg stance phase, front leg stance phase and double legs stance phase. The dynamic equations of double legs stance phase are listed as follows using Newton-Euler method and ignoring the mass and inertia of legs.

$$f_{x1} + f_{x2} = m\ddot{x} \tag{1}$$

$$f_{z1} + f_{z2} = mg + m\ddot{z} \tag{2}$$

$$\begin{aligned} I_{yy}\ddot{\varphi}_y = {} & f_{z2}L_2\cos(\theta_2) + f_{z1}L_1\cos(\theta_1) + (f_{z2} - f_{z1})L\cos(\varphi_y) \\ & - f_{x2}L_2\sin(\theta_2) - f_{x1}L_1\sin(\theta_1) + (f_{x2} - f_{x1})L\sin(\varphi_y) \end{aligned} \tag{3}$$

The kinematics can also be listed as follows:

$$\sin(\varphi_y) = \frac{l_2\sin(\theta_2) - l_1\sin(\theta_1)}{2L} \tag{4}$$

$$y = \frac{l_2\sin(\theta_2) + l_1\sin(\theta_1)}{2} \tag{5}$$

$$x = \frac{l_1\cos(\theta_1) + l_2\cos(\theta_2)}{2} \tag{6}$$

where θ_1, $\theta_2$are the incidence angles.

2.2. *Position control to pitch and vertical position*

Position control is appropriate when a manipulator is following a trajectory through space, and the high stiffness of position controller is attractive in robot control. The given trajectory must be continuously differentiable is one of the problems in using position control in robot control. So gait planning must be combined in using this method. The framework shown in Fig. 3 had proven to be effective in [8, 9]. The inner loop can be easily realized by position control method, and the main task is to design an appropriate leg length generator. The desired leg length can be easily obtained through leg inverse kinematics.

For planar model, equations (4) and (5) could be rewritten as:

$$\begin{bmatrix} l_1 \\ l_2 \end{bmatrix} = J_1 \begin{bmatrix} z \\ \sin(\varphi_y) \end{bmatrix} \tag{7}$$

where

$$J_1 = \begin{bmatrix} \frac{2}{\sin(\theta_1)} & -\frac{2L}{\sin(\theta_1)} \\ \frac{2}{\sin(\theta_2)} & \frac{2L}{\sin(\theta_2)} \end{bmatrix} \tag{8}$$

and J_l is the Jacobian that relates the body motion to the leg length, and θ_1, θ_2 are the incidence angles. J_l eliminate the effect caused by x. So it is approximate decoupling. Moreover it is obvious that $0<\theta_1<\pi$, $0<\theta_2<\pi$ when the robot is moving, so:

$$\begin{aligned} 0 < \sin(\theta_1) \le 1 \\ 0 < \sin(\theta_2) \le 1 \end{aligned} \tag{9}$$

And J_l is non-singular matrix and reversible.

By this approach, the vertical position and body pitch could be controlled by using simple position servo, and could carry larger load for the bigger stiffness of position control.

2.3. *Force control to horizontal position*

There are four control inputs in planar model; the change of legs length are used to control vertical position and body pitch, and the horizontal position could be controlled by the hip joint. In order to control the internal forces at the same time, the horizontal position is controlled by force controllers which are based on computed torque drive method.

As is shown in equation (1), there are two forces but only one objective, so another force distribution equation should be added. By adding this constraint, normal computed torque drive method could be used to control horizontal position.

To avoid the foot from sliding, the friction force should under the constraint of equation (12).

$$|f_{x1}| < \mu f_{z1}, |f_{x2}| < \mu f_{z2} \tag{10}$$

where μ is the friction coefficient.

Generally speaking, this constraint is used to limit the desired horizontal force to avoid the feet from sliding. SCI (slippage constraint index) was defined in [10] to represent the possibility of sliding as:

$$SCI = \max(\frac{F_x}{F_z}) \tag{11}$$

where F_x, F_y are the contact forces.

To make SCI as smaller as possible, the force distribution equation should be added as:

$$f_{x1d} / f_{x2d} = f_{z1} / f_{z2} \tag{12}$$

Combining equation (1) with(14), the desired horizontal force can be calculated as:

$$\begin{bmatrix} f_{x1d} \\ f_{x2d} \end{bmatrix} = \begin{bmatrix} f_{z1r} / (f_{z1r} + f_{z2r}) \\ f_{z2r} / (f_{z1r} + f_{z2r}) \end{bmatrix} \left(P_x(x_d - x_r) + D_x(\dot{x}_d - \dot{x}_r) \right) \tag{13}$$

The constraint of horizontal force ensures none of the supporting feet or both sliding at the same time when the robot is moving.

From Fig. 3 we can also list the dynamics relate to the torque of the hip:

$$\begin{bmatrix} \tau_1 \\ \tau_2 \end{bmatrix} = J_2 \begin{bmatrix} f_{x1} \\ f_{x2} \end{bmatrix} + G \tag{14}$$

where

$$J_2 = \begin{bmatrix} l_1 \sin(\theta_1) & 0 \\ 0 & l_2 \sin(\theta_2) \end{bmatrix}, \quad G = \begin{bmatrix} -f_{y1} l_1 \cos(\theta_1) \\ -f_{y2} l_2 \cos(\theta_2) \end{bmatrix} \tag{15}$$

and J_2 is the Jacobian that relates the horizontal force to the leg hip torque, G is the feed forward mainly caused by horizontal position, body pitch and gravity. So, the hip torque can be calculated as:

$$\begin{bmatrix} \tau_1 \\ \tau_2 \end{bmatrix} = J_2 \begin{bmatrix} P_{fx1}\left(f_{x1d} - f_{x1}\right) + I_{fx1}\left(\dot{f}_{x1d} - \dot{f}_{x1}\right) \\ P_{fx2}\left(f_{x2d} - f_{x2}\right) + I_{fx2}\left(\dot{f}_{x2d} - \dot{f}_{x2}\right) \end{bmatrix} + G \tag{16}$$

J_2 and G eliminate the effect caused by z and φ_y.

3. Prototype Simulations

To validate the effectiveness of the proposed algorithm, a planar model of 5-link structure is built in Simmechanics. The nonlinear spring-damper model presented in [12] is used to calculate the contact forces in the simulation. The parameters of the body and the 3D contact model listed in Table 2. To avoid slippage of standing feet, the maximum horizontal force is limited to be 0.3 times of the vertical force.

Table 2.Simulation environment

Body mass	50*kg*	Body size	0.6*0.37*0.1m^3
Mass of one leg	2*kg*	Fiction coefficient	0.35
Normal stiffness	8.5e6*N/m*	Normal damping coefficient	3.1e3 *Ns/m*
Tangential stiffness	12.75e6 *N/m*	Tangential damping coefficient	3.1e3 *Ns/m*

In the simulation, Some interference are added to examine the performance of the presented algorithm: During 0~0.5s, the planar model just standing on the ground; and then added an interference force of -100N to the body during 0.5~1s to examine the performance of this method in dealing with impacted by external force; when the simulation time was 1.5s,a step signal of 0.1m is added to the desired horizontal position to test the effect of force distribution method; in order to test the effect of the presented leg length Generator, a step signal of 10deg to the body pitch is added in 3s. The results are listed in the following figures.

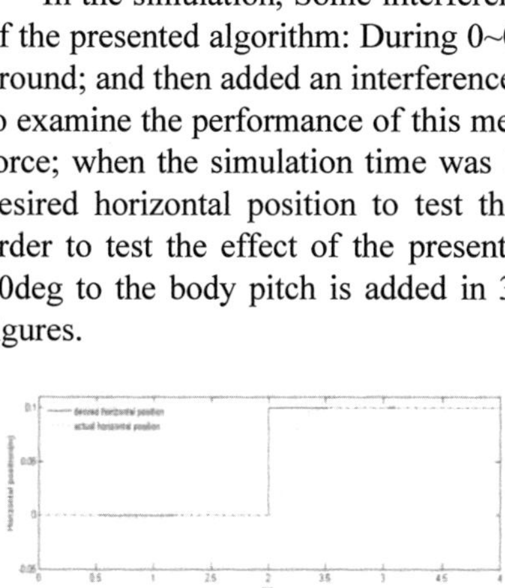

Fig. 6. Horizontal position

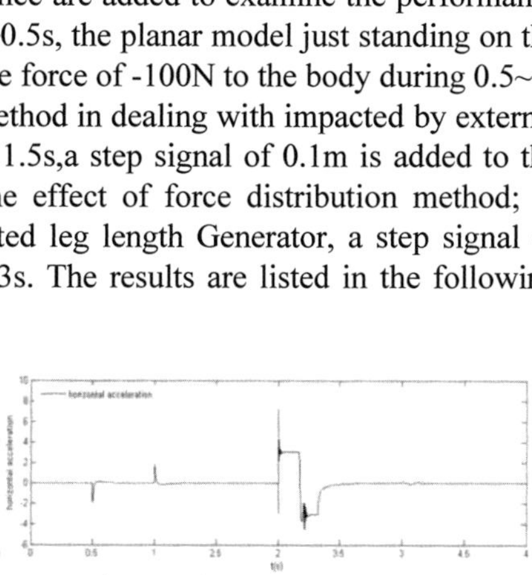

Fig. 7. Horizontal acceleration

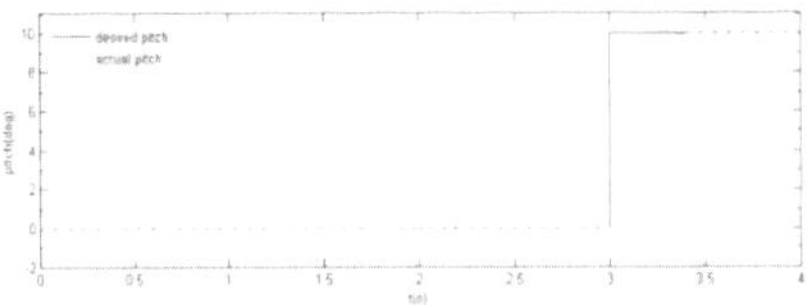

Fig. 8. Body pitch angle (deg)

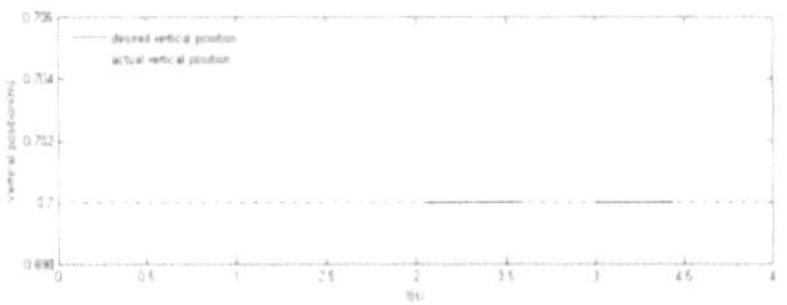

Fig. 9. Vertical position

It is obvious that, the added external force leads a wave of 2e-3m to the horizontal position, which dues to steady state error of the controller and the tangential stiffness of ground, but almost has no influence to the body pitch and vertical position. When we give a step signal to the horizontal position, the force controller could follow the desired position quickly, and the actual horizontal acceleration could be limited in $-3 \sim 3m/s^2$, owing to the constraint of friction constraint and the force distribution method. The peak in 2s caused by the mutation of desired acceleration, but almost had no influence to the feet for it was just an instant. Figure 7 has shown the effectiveness of position controller, we find the actual pitch could track the desired pitch quickly. And the change of vertical position mostly caused by the tangential stiffness of ground. In a word, by the presented approach, the horizontal position is approximate decoupling to the body pitch and its vertical position, and has strong robustness.

4. Conclusions

This paper presents a hybrid position/force control algorithm to perform motion control of the quadruped robot based on a simplified planar model in sagittal plane which can describe the main dynamical characters of trotting gait. And several constraints are added to maintain of the body balance during the body moving. The algorithm is better than simple position for the added constraint of internal forces and the added force distribute approach ensuring the SCI of the robot to be the minimum. Compare to simple force controller, the presented approach care more about the physics constraint of the robot, and the presented leg length generator have strong ability in dealing with trajectory that are not continuously differentiable. In the future, the proposed method will be combined with SLIP model to complete trotting gait in unknown rough terrains.

Acknowledgments

Supported by Hi-tech development program (Grant NO. 61473304) and National Science Foundation (Grant No. 61473304)

References

1. Jeremy Ma, Sara Susca, Robust Multi-Sensor, Day/Night 6-DOF Pose Estimation for a Dynamic Legged Vehicle in GPS-denied Environments.

2012 IEEE International Conference on Robotics and Automation. May 14-18 (2012).
2. Michele Focchi, Thiago Boaventura. Torque-control Based Compliant Actuation of a Quadruped Robot. The 12th IEEE International Workshop on Advanced Motion Control. March 25-27(2012).
3. G.Schöner, M.Dose, A dynamical systems approach to task-level system integration used to plan and control autonomous vehicle motion, Robotics and Autonomous Systems10(4) 253–267 (1992).
4. R. Blickhan, The spring-mass model for running and hopping, Journal of Biomechanics. 22(11–12) 1217–1227 (1989).
5. Ali Zamani, Mahdi Khorram. Dynamics and Stable Gait Planning of a Quadruped Robot. 2011 11th International Conference on Control, Automation and Systems, Oct. 26-29 (2011).
6. WANG Shuhai, JIN Yi. Gait Planning and Simulation of the Quadruped Robot Based on Virtual Prototype. Modeling, Identification and control, Wuhan, China, June 24-26 (2012).
7. Raibert Marc H. Legged robots that balance. J. Communications of the ACM. 29(6):499-514(1986).
8. Guoteng Zhang, Hui Chai. An Impact Recovery Approach for Quadruped Robot with Trotting Gait International Conference on Information and Automation Hailar, China (2014).
9. Barkan Ugurlu, Kana Kotaka Actively-Compliant Locomotion Control on Rough Terrain: Cyclic Jumping and Trotting Experiments on a Stiff-by-Nature Quadruped. 2013 IEEE International Conference on Robotics and Automation (ICRA) Karlsruhe, Germany (2013).
10. Fu Chenglong and Ken Chen.Theory and technology of Humanoid Robot. Tsinghua University Press, Peking China (2010).
11. Matt Zucker Learning and Optimization Methods for High Level Planning The Robotics Institute Carnegie Mellon University Pittsburgh, PA 15213-3890(2010).
12. Morteza Azad and Roy Featherstone. Modeling the contact between a rolling sphere and a compliant ground plane (2010).

EXPERIMENTAL STUDY ON THE FLEXIBLE STRAIGHT LEG FOR MOBILE ROBOTS

XUDONG WANG, ZHUO CHEN, TAO FANG, MIN XU, SHIWU ZHANG*

Dept. of Precision Machinery and Precision Instrumentation,
University of Science and Technology of China,
Hefei, Anhui, 230026, P. R. China.

There are lots of demands for the amphibious travelling style of mobile robots. This paper proposed a simple, elastic and useful leg for mobile robots. When moving on the ground, the leg could be deformed a curve-like one passively, forced by the ground reaction force. Moreover, the leg would recover straight, fin-like, and propel the robot in the water. The leg just needed to be fixed on the output shaft simply and its deformation could be realized without any additional mechanism. That made it could be widely applied to much more robots. To evaluate the movement performance of the leg, an experiment platform was designed and series of experiments were carried out. The results verified our assumption that the elastic leg would be useful for robot's travelling. The data obtained in the experiments would also help optimize this leg in the future.

1. Introduction

Nowadays, more and more mobile robots are applied into people's life. We can find their presences in many areas, such as scientific exploration and earthquake rescue. However, there are still less movement styles for robots movement, especially in the amphibious environment. Wheel robots move fast but have troubles in accommodating tough terrains. Traditional legged robots obtain good adaptability to rough terrains [1-7]. But they have not yet exhibited excellent ability in amphibious environments such as mud terrain. Thus, the locomotion mechanism through amphibious environments becomes important [8-10]. To solve that problem, arc-like legs were developed [8]. The arc-like legged robots combine the advantages of wheeled robots with traditional legged robots. They move faster than traditional legged robots and show better terrains adaptability than wheels. In the last few years, some robots equipped with arc-like legs have been designed. Among them, AmphiHex [11-12], was developed by our team. But there are still problems in ride comfort, trafficability characteristic, etc.

* **Corresponding author:** Shiwu Zhang
E-mail: swzhang@ustc.edu.cn

Obviously, with millions years of evolution, animals have shown their extraordinary movement performance. It is very important that the organs and tissues for locomotion are all elastic [13-14]. The elastic element could buffer the impact and store the energy through its own deformation. The shock absorber helps the smooth ride, which would protect the payload of the automobile [15-17]. Benefited from the bionic research, people put forward the spring-loaded inverted pendulum model [18-19] and other similar models [20]. In those researches, the elastic element was introduced into the design of legged robots.

Based on those researches, a novel leg would be proposed in this paper. This leg was made of elastic material, polyurethane, with flat and straight geometric profile. Then a leg with continuous distribution of elastic element was designed. The leg could show good ability to adapt the terrain. When it was straight, it would work as a fin and drive the robot swam in the water. When the robot landing, The elastic leg would be deformed curved passively, forced by the ground reaction force. Thus, the elastic leg could realize the function of amphibious locomotion only by itself without any additional structure. What's more, this new elastic leg may help solve the problems in ride comfort and trafficability characteristic with simple structure.

To verify the movement performance of this novel leg, an experiment platform has been set up. This platform could record the locomotion parameters of the movement unit, including output torque, propulsion force, the horizontal displacement and the fluctuation of the center of gravity in a single stride. The data would help evaluate the performance of the movement unit, and then verify the value of the legs or wheels. That would help validate the assumption and design. The results could also provide references for the further study. What is more, the digital imaging technology was also applied to describe the changes of the elastic leg and the motion track of the movement unit, which would provide data for the further theoretical study.

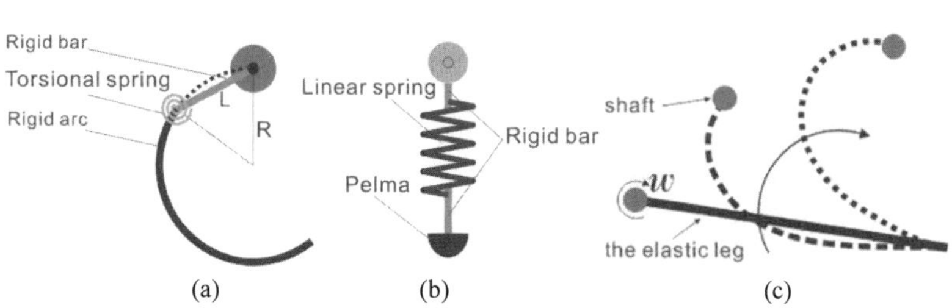

Figure 1. The elastic legs in mobile robots. (a) The arc-like leg based on the R-SLIP model [18-19]. (b) The straight leg with elastic element [20-21]. (c) The moving process of the elastic leg.

2. Material and Method

2.1. *Experimental system setup*

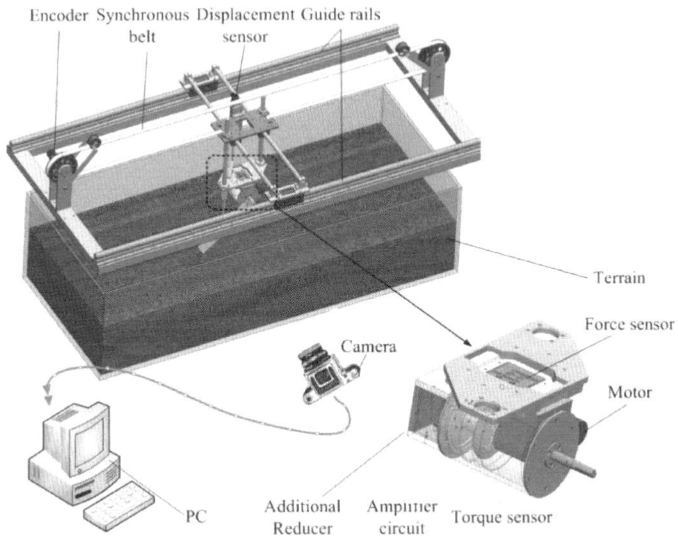

Figure 2. Single-legged experiment platform for the leg-terrain interaction.

In order to evaluate the movement performance of the elastic leg, the experiment platform was developed. The platform would simulate the movement through different terrains. The sensors, equipped in the platform, could measure and record the status of the motion unit, such as the fluctuation of the center of gravity of the motion unit, etc. The sliders helped the motion module move along the guide rails. Two synchronous pulleys were equipped in this platform and one was connected with the photoelectric encoder by the coupling. The encoder rotated synchronously with the synchronous pulley. The synchronous belt, connected with the motion unit, was used to transfer the displacement of the motion module to the angular displacement of the encoder. One torque sensor (range: ±10 Nm with accuracy: 0.5% F.S) connected the output shaft of gearbox with the elastic leg. It was shown in Fig.2 that, the signal of the torque would be amplified by the circuit, and then acquired by the data acquisition board. A pressure sensor (range: ±10 Kg with accuracy: 0.5% F.S) was applied to measure the propulsion force of the motion module. What's more, a displacement sensor (range: 175mm with accuracy: 0.5% F.S) was connected to the moving parts, measuring the displacement in the vertical direction, which was considered as the fluctuation of the center of gravity.

The data of all the sensors were acquired by the data acquisition board (PCI-1716L, 16 bits, 250Ks/s, Advantech). The motion module was driven by a DC motor (Maxon RE 30 DC Motor, 24 V, 60 W, with a 66:1 accessorial reduction gearbox). The details of the motion module were also shown in Fig. 2. Subject to the performance of the motor, an additional reducer (2:1) was assembled to enlarge the output torque of the motion module to provide enough propulsion torque. Besides, pictures, focused on the leg, would be taken by a high-speed digital camera (Casio, EX-F1) during the operation.

2.2. *Design of the leg*

It is observed that, legs play a very important role in the locomotion of legged robots. To a large extent, the leg determines the performance of the robot. People found that, many animals in nature have the Achilles tendons and ligaments which have character of non-linear elastic deformation. Tendons and ligaments serve as excellent elastic energy stores during animals' running gaits [13]. With the help of the stored energy, there would be less requirements for muscles when animals move through the tough terrains [13]. Inspired by that, the elastic leg was designed.

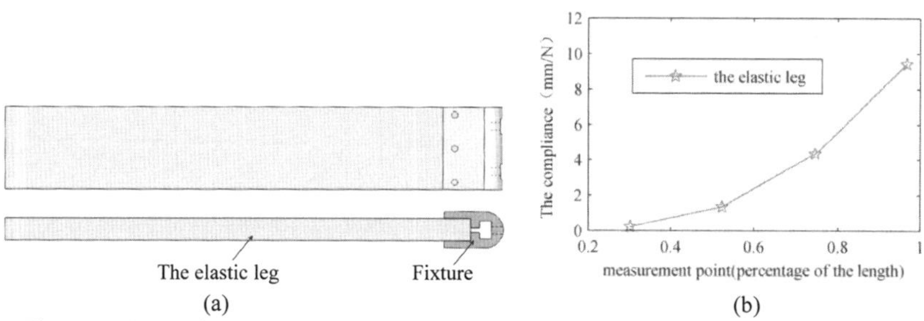

Figure 3. The elastic leg. (a) The structure of the elastic leg. (b) The stiffness of the elastic leg.

Showed in Fig. 3(a), the elastic leg was made of elastic material, polyurethane. The leg was designed straight for ability of amphibious locomotion. The leg would bend arc-like passively, rolling on the terrain, and would be convenient for flapping propulsion in the water. As an exploratory study, the leg was designed uniform thickness, 15mm. The length of this leg was 325mm, and the width was 54mm. The leg was connected to the motion unit by the fixture. The leg rotated with the motor and the leg would recoil and stretch during every circle. Then the stress state of this leg was similar to the cantilever beam with large deformation during rotation. That structure brought the elastic legs characteristic of variable elasticity in different position. The flexibility pattern of this elastic leg was shown in Fig. 3(b). The flexibility changes along the length of the leg and the elasticity becomes rigid from the tip to the root.

2.3. *The leg-terrain interaction experiments*

To verify the performance of the elastic leg, a series of experiments were conducted. The leg was applied to moving on the hard board. To test the load capacity of the elastic leg, this leg was loaded different weights, 1.4kg, 1.7kg, 2kg, 2.3kg. The leg was driven with angular speeds from 1 rad/s to 5rad/s and the data was obtained by the data acquisition board through sensors. And each experiment was repeated at least three times to exclude the unexpected cases. The procedure of the deformation of the leg were shown in Fig. 4(a).

3. Results and Discussion

It was mentioned that, to evaluate the value of this new design, we just select one leg for the experiment initially. The data results would be presented and discussed in this section. To avoid the odd data, at least three experiments for each parameter were conducted. Then the mean number was used to evaluate the movement performance of the elastic leg. The variance was calculated to analyze the distribution of the data and the error bars verified the validity of the experiments.

3.1. *The fluctuation of the center of gravity*

The ride comfort is an important quality for robots. As mentioned above, the hard road surface is a whole plank placed horizontally. In another word, there is no need for the movement unit consuming energy to lift itself. Besides, the violent fluctuation would also import impact to the payload. And it could bring damage to the internal components such as sensors, circuits. Thus, the fluctuation of the center of gravity should be the less the better. Obviously testified in Fig. 4(b), the elastic leg performed well in 2 rad/s to 5 rad/s. There was larger fluctuation in 1rad/s because the low angular speed offered enough time for the elastic leg deformed and lifted the movement unit up. When the angular speed was up to 5rad/s, the motion inertia shot up. The motion inertia in vertical direction caused the increase of the fluctuation. On the other hand, it could be observed that the payload just changed the amplitude of the fluctuation and exerted less influence on the regular of the fluctuation changing with angular speed. As a whole, the elastic leg may be a valuable movement style for robots especially in medium angular speed.

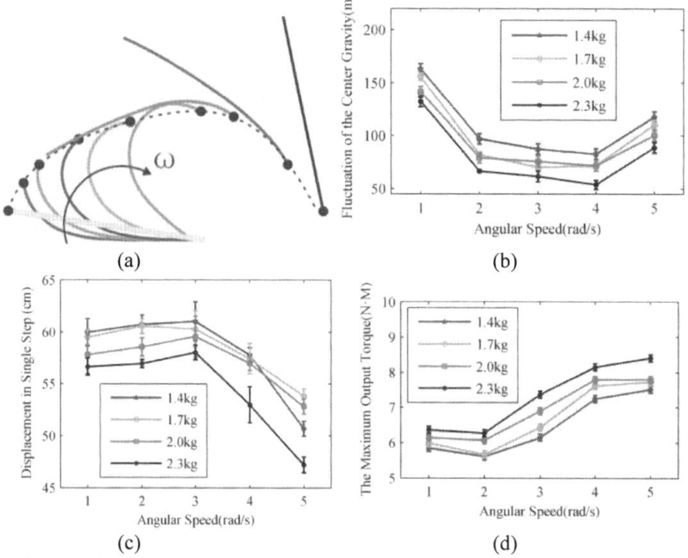

Figure 4. The experiment result. (a)The deformation process of the elastic leg. (b)The fluctuation of the center of gravity (c)The displacement in one single step. (d)The maximum output torque of the output shaft

3.2. *The displacement in single-step*

Figure 4(c) showed the single-step displacement of the elastic leg changing with angular speed and payload. It was indicated that, different payloads resulted in different amplitudes, the lighter, the larger. No matter what kind of payload, the displacement in single step increased first and then decreased with the increase of the angular speed. The reason might be that the effect of the horizontal inertia enhanced with the increase of angular speed. When the movement unit fell on the horizontal plank, the inertia in horizontal direction still kept the movement unit move forward for some distance. The effect became more visible when the angular speed increased. However, the slip between the leg and the plank also increased with the angular speed. That slip would weaken the ground reaction force. The slip would play the main role when mechanism was in high angular speed. As a result, the displacement in single step decreased. Those papers also show that within some speed limits, the leg rotates faster, the robot move a shorter distance.

3.3. *The maximum output torque*

The output torque of the motor is proportional to the current, thus the maximum output torque indicates the power demand. Higher torque means more powerful motor and more current. That puts forward higher requirements to the motor and the power supply. So it is more meaningful to research the extreme working state. For this reason, we choose the maximum output torque as the evaluation index. It need more power to transport the heavier payload, so the output torque of 2.3 kg shown in Fig. 4(d) was the highest. For the same reason, the maximum output torque increased with the angular speed. But the growth rate would change with different angular speed. That might be relative to the feature of the leg-land contact, which need more research.

4. Conclusions and Future Work

This paper put forward a new elastic leg for mobile robots. An experiment platform was set up and series of experiments were carried out. The practicability of the elastic leg was evaluated. The results showed that the leg was helpful for robot's travelling. Base on that, in the future work, more experiments need to be carried out. The movement performance of the leg in more different terrains such as mud would be evaluated. Besides, the design of the elastic leg still needs to be optimized.

Acknowledgment

This research has been financially supported by National Natural Science Foundation of China (51375468 and 61375095).

References

1. Murphy, Michael P., et al. "The littledog robot." The International Journal of Robotics Research (2010): 0278364910387457.
2. Zucker, Matt, et al. "Optimization and learning for rough terrain legged locomotion." The International Journal of Robotics Research 30.2 (2011): 175-191.
3. Klaassen, Bernhard, et al. "Biomimetic walking robot SCORPION: Control and modeling." Robotics and autonomous systems 41.2 (2002): 69-76.
4. Cordes, S., et al. "On the design of a four-legged walking machine." Proc. Int. Conf. on Autonomous Robotics. 1997.
5. Ilg, W., et al. "Bisam: From small mammals to a four legged walking machine." Proceedings of the Fifth International Conference on Simulation of Adaptive Behaviour. 1998.
6. Playter, R., M. Buehler, and M. Raibert. "BigDog." Defense and Security Symposium. International Society for Optics and Photonics, 2006.
7. Lu, Dongping, et al. "Design and development of a leg-wheel hybrid robot "HyTRo-I"." Intelligent Robots and Systems (IROS), 2013 IEEE/RSJ International Conference on. IEEE, 2013.

8. Saranli, Uluc, Martin Buehler, and Daniel E. Koditschek. “RHex: A simple and highly mobile hexapod robot.” The International Journal of Robotics Research 20.7 (2001): 616-631.
9. Boxerbaum, Alexander S., et al. “Design of an autonomous amphibious robot for surf zone operation: Part I mechanical design for multi-mode mobility.” Advanced Intelligent Mechatronics. Proceedings, 2005 IEEE/ASME International Conference on. IEEE, 2005.
10. Yu, Junzhi, et al. “Amphibious Pattern Design of a Robotic Fish with Wheel-propeller-fin Mechanisms.” Journal of Field Robotics 30.5 (2013): 702-716.
11. Liang, Xu, et al. “The AmphiHex: A novel amphibious robot with transformable leg-flipper composite propulsion mechanism.” Intelligent Robots and Systems (IROS), 2012 IEEE/RSJ International Conference on. IEEE, 2012.
12. Zhang, Shiwu, et al. “Initial Development of a Novel Amphibious Robot with Transformable Fin-Leg Composite Propulsion Mechanisms.” Journal of Bionic Engineering 10.4 (2013): 434-445.
13. Gray, James. How animals move. Cambridge University Press, 2013.
14. Aigner, Michael, and J. Heegard. “One-dimensional quasi-static continuum model of muscle contraction as a distributed control system.” Center for Turbulence Research, Ann. Res. Briefs (1999): 155-168.
15. Sharp, R. S., and D. A. Crolla. “Road vehicle suspension system design-a review.” Vehicle System Dynamics 16.3 (1987): 167-192.
16. Yue, C., T. Butsuen, and J. K. Hedrick. “Alternative control laws for automotive active suspensions.” American Control Conference, 1988. IEEE, 1988.
17. Bailey, J. R., J. K. Hedrick, and D. N. Wormley. “Rail vehicle suspension parameter identification.” Proceedings of the ASME Winter Annual Meeting. 1988.
18. Blickhan, Reinhard. ‘The spring-mass model for running and hopping.” Journal of biomechanics 22.11 (1989): 1217-1227.
19. Holmes, Philip, et al. “The dynamics of legged locomotion: Models, analyses, and challenges.” Siam Review 48.2 (2006): 207-304.
20. Huang, Ke-Jung, Chun-Kai Huang, and Pei-Chun Lin. ‘A simple running model with rolling contact and its role as a template for dynamic locomotion on a hexapod robot.” Bioinspiration & biomimetics 9.4 (2014): 046004.

DESIGN AND RESEARCH FOR TWO-DIMENSIONAL PASSIVE DYNAMIC WALKING ROBOT

JINJIAO XIE

Institute of computer application, Beijing Information Technology College
Beijing, 100018, China

In this paper, a two-dimensional passive walking robot is designed and processed, and dynamics model of the robot is established to perform the numerical simulation with the parameters of the prototype. Besides, with different parameters, the numerical simulation is performed to research the impact of parameter changes on gait. Finally, we do the prototype walking experiment.

1. Introduction

Biped robot has been an important part of study on robot. The traditional biped robots have similar structure and control mode: high torque drive system and high-gain feedback control; ZMP theory to guarantee stability of robot; trajectory tracking method to control the joints; real-time adjustment of robot posture to eliminate error. This kind of robots, which walk in accordance with planned gait, is usually called active walking robot. However, active walking robot has a stiff gait and complexity of control and drive system which leads to low energy efficiency. Therefore, in recent years, passive walking robot attracts increasing attention for its natural walking gait and low energy consumption.

Passive walking phenomenon was found very early when a walker toy that can naturally walk on a slope was invented. But until the 1980s, McGeer did the groundbreaking research on such walker that only depend on gravity and structural characteristics to realize biped walking and this kind of walking was named “passive walking” [1]. Along with the traditional active walking robot researched deeply, a lot of problems and bottleneck were encountered. It is realized that active robot’ walking way and human physiology have great gap, which is maybe the reason why it is difficult to improve the level of walking robot. If the active walking robot is in form to realize biped walking, passive walking robot is maybe closer to essence of biped walking.

After McGeer, many universities studied passive walking robot and developed a lot of prototype, including Biped by Cornell university[2], Toldder by MIT[3], Dennis by Delft University[4]. These researches proved that as long as reasonable drive motor is added instead of gravity, the robot developed on the basis of passive dynamics principle can realize stable walking in a horizontal ground and remain high energy efficiency. At the same time, it also proved that passive dynamics method is of great application value.

This paper designs and researches a two-dimensional passive walking robot. It is organized as follows. In Section 2, mechanical design of robot is introduced. In Section 3, we establish dynamic model of passive robot and simulate the walking process of robot designed in Section 2 through matlab. In Section 4, we analyze the influence of geometric parameters on gait. In Section 5, prototype test result is presented.

2. Mechanical Design

The robot can walk on slope. No drive, it walks only depending on gravity through swinging legs. It has four legs: two internal legs and two external legs. In order to limit the robot's motion in a two-dimensional plane, a set of legs is fixed together, which is regarded as one leg in motion and dynamics. Arc feet are fixed on legs. The robot is about 430mm tall and 250mm wide. Its three-dimensional model is as showed in Fig. 1.

3. Dynamics Modeling and Simulation of Passive Walking Robot

Passive Walking robot is of significance for its simple structure, high energy efficiency and walking ability in the case of only gravity without driver under the natural gait. Passive dynamic walking robot can produce a stable passive walking gait on an inclined plane. Therefore, we established mathematical model of passive walking, then did theoretical analysis and numerical simulation based on the model. This work is of great importance for passive walking robot structure design and controlling system design with a simple driver.

In this section, we used a similar compasses model structure, made some assumptions on dynamic walking process. Then we established the mathematical model of walking gait. Besides, we used Matlab numerical simulation method to simulate the gait model, solve the dynamic walking process and analyze the dynamic process of passive walking gait.

3.1. *Dynamics modeling*

A complete walk cycle of passive walking robot can be divided into two periods, swing period and collision period. Swing period is from the end of the collision period to the start of the next collision period; collision period is from foot contacting ground to foot leaving ground. We used the multi-body system dynamics modeling method to analyze the swing period and collision period.

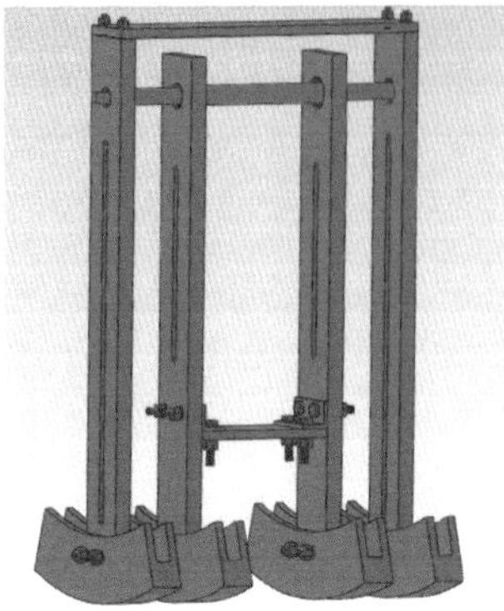

Figure 1. Model of robot

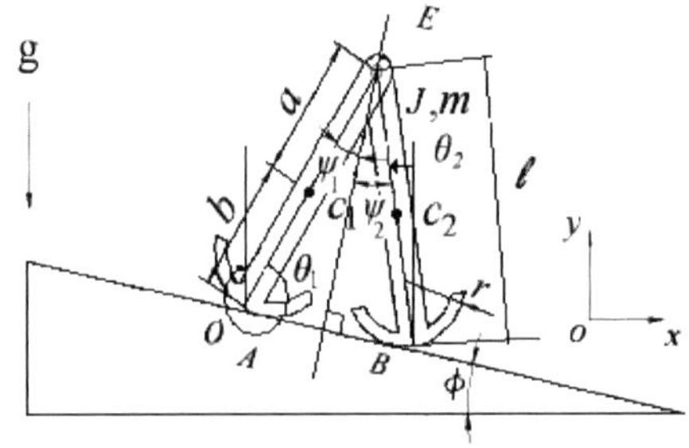

Figure 2. Two-dimensional walking robot simplified model

Lagrange method adopted, the dynamics model in the swing period is established. Ignoring air flow, friction, passive walking robot system can be transformed into a system composed of rod and mass. Then we can establish dynamics equations of robot' motion in swing period, in accordance with Lagrange dynamics equation.

Regarding θ_1, θ_2 as generalized coordinate, only gravity without other external force, we can establish the Lagrange equation, simultaneous equations composed by two equations, as follow:

$$\frac{d}{dt}\left(\frac{\partial L}{\partial \dot{\theta}_1}\right) - \frac{\partial \mathrm{L}}{\partial \theta_1} = 0 \tag{1}$$

In above equations, i = 1, 2;

$$\mathrm{L} = \mathrm{T} - \mathrm{V} \tag{2}$$

T — kinetic energy of the system;
V — potential energy of the system.

We obtain the dynamics equation of swing period as follows:

$$\begin{pmatrix} M_{11} & M_{12} \\ M_{21} & M_{22} \end{pmatrix} \begin{pmatrix} \dot{\theta}_1 \\ \dot{\theta}_2 \end{pmatrix} + \begin{pmatrix} N_{11} & N_{12} \\ N_{21} & 0 \end{pmatrix} \begin{pmatrix} \dot{\theta}_1 \\ \dot{\theta}_2 \end{pmatrix} + \begin{pmatrix} G_1 \\ G_2 \end{pmatrix} = 0. \tag{3}$$

In collision period, assuming that roles of supporting leg and swing leg are reversed after collision occurs, while collision is instantaneous and legs' angular velocity changes and legs' angle is consistent after collision occurs. Parameters with "-" indicate parameters before collision, parameters with "+" indicate parameters after collision.

Shown in Figure 2, supporting leg EA becomes the swing leg after the collision. In the collision process, EA is subject to three forces — supporting force at point A by plane, forward tensile force at point E by the body, while by itself the effect of gravity, but the three forces are limited values, and the collision process is extremely short, so they can be ignored in collision process. Therefore, according to the law of conservation of angular momentum, the dynamics model in collision period is established.

The equation of conservation of angular momentum, simultaneous equations composed of two equations, is written out as follows:

$$\begin{cases} L_B^- = L_B^+ \\ L_E^- = L_E^+ \end{cases}, \tag{4}$$

We obtain the dynamics model of collision period as follows:

$$\begin{cases} P^- \begin{pmatrix} \dot{\theta}_1^- \\ \dot{\theta}_2^- \end{pmatrix} = P^+ \begin{pmatrix} \dot{\theta}_1^+ \\ \dot{\theta}_2^+ \end{pmatrix} \\ P^- = \begin{pmatrix} P_{11}^- & 0 \\ P_{21}^- & P_{22}^- \end{pmatrix} \\ P^+ = \begin{pmatrix} P_{11}^+ & J + ma^2 \\ P_{21}^+ & P_{22}^+ \end{pmatrix} \end{cases} \tag{5}$$

3.2. *Numerical Simulation*

As long as the model parameters is reasonable, and initial conditions appropriate, the model established before can produce stable gait. According to the robot cited in Section 2 and simplified model as Fig. 2, we obtained the geometric parameters such as table1 and find fixed point: Ψ_1=0.31875, θ_1= -3.7238, θ_2= -2.8129. Simulation curve is made by matlab as shown in Fig. 3.

Table 1. Parameters of prototype.

Leg quality m(kg)	Leg length l(m)	Distance from leg centroid to hip a(m)	Radius of arc foot r(m)	Angle of inclined plane ϕ(rad)
1.8	0.375	0.18	0.2	0.04

Figure 3 indicates the passive walking robot's motion during 40 steps, in which blue lines represent EA leg, green lines represent EB leg, and * represent the angle and the angular velocity when the foot touching the ground. Figure 3 (a) provides the curve of angle between leg and normal direction varies with time. Figure 3 (b) provides the curve of angle velocity varies with time. Figure 3 (c) provides the curve of distance between foot and ground varies with time.

4. Analysis of Parameters' Influence on Gait

4.1. *Original data*

The original data is the prototype data (such as shown in Table 1), and the angle and angular velocity change with time as shown in Fig. 4, its speed changes with time as shown in Fig. 4(a), the walking distance changes with time as shown in Fig. 4(b). From these figures, the scope of cycle time T=0.6295s, speed v=0.75893m/s and step length λ=0.52761m.

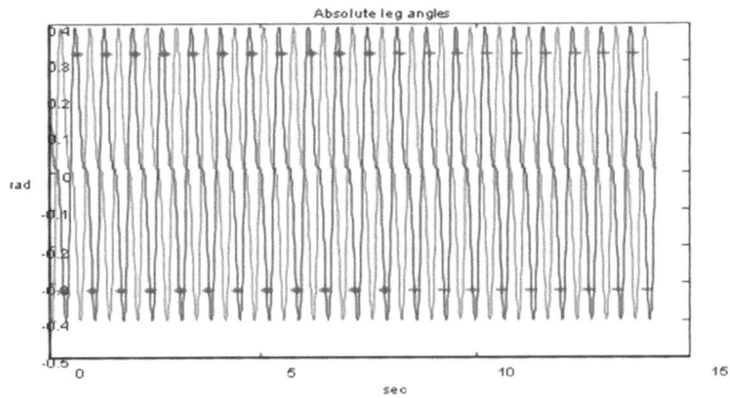

Figure 3(a). Curve of angle between leg and normal direction varies with time.

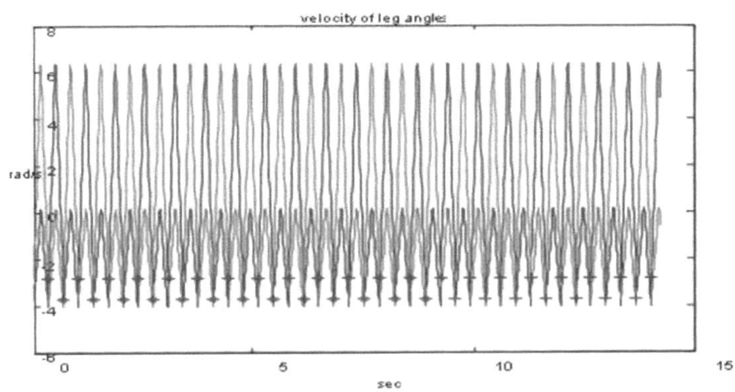

Figure 3(b). Curve of angle velocity varies with time.

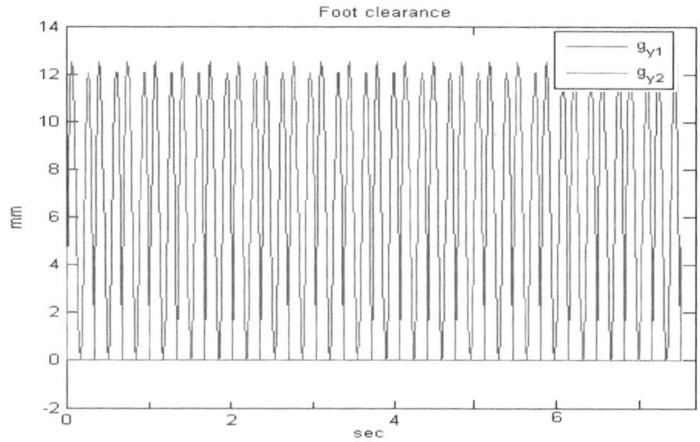

Figure 3(c). Curve of distance between foot and ground varies with time.

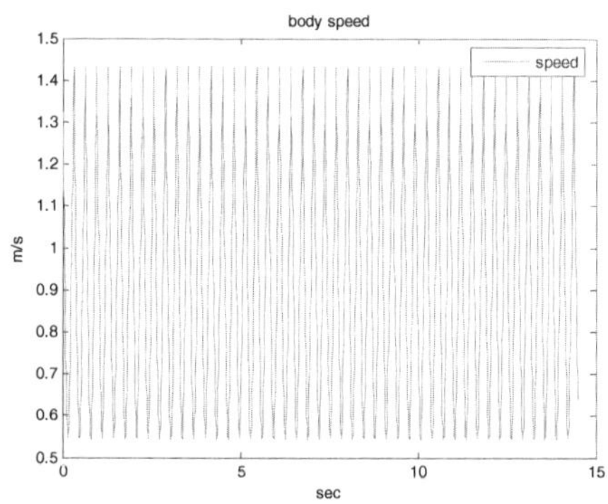

Figure 4(a). walking speed varies with time

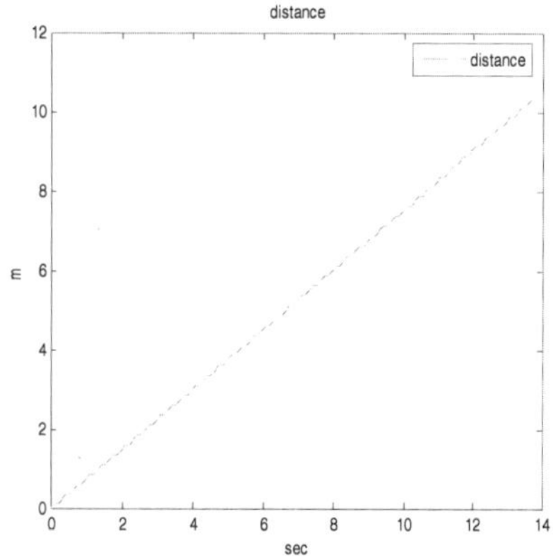

Figure 4(b). walking distance varies with time.

4.2. *The influence of leg's length on gait*

Changing length of legs, other parameters remaining unchanged, we can find the different fixed points and calculate angle and angular velocity in walking process through matlab. Fixed points, cycle time, walking speed and step length under different length are shown in Table 2.

From Table 2, it is known that increasing length of leg can make step length and speed significantly increased, cycle time significantly decreased.

Table 2. Gait parameter change with leg's length

Length of leg l(m)	Ψ_1	Fixed point $\dot{\theta}_1$	$\dot{\theta}_2$	Cycle time T(s)	Speed v(m/s)	Step length λ (m)
0.375	0.31875	-3.7238	-2.8129	0.6952	0.7589	0.5276
0.55	0.26338	-2.4583	-1.5179	0.6312	0.9426	0.5969
0.75	0.23873	-2.0396	-1.0586	0.6250	1.1333	0.7083
0.85	0.22128	-1.8011	-0.80166	0.6140	1.2830	0.7878

4.3. *The influence of distance between leg's centroid and hip joint on gait*

Changing distance from leg's centroid to hip joint, other parameters remaining unchanged, we can find the different fixed point and calculate angle and angular velocity in walking process through matlab. Fixed points, cycle time, walking speed and step length under different distance are shown in Table 3.

From Table 3, it is known that increasing distance from leg's centroid to hip joint can make step length and speed significantly decreased, cycle time increased.

Table 3. Gait parameter change with distance from leg's centroid to hip joint

Distance from leg's centroid to hip joint a(m)	Ψ_1	Fixed point $\dot{\theta}_1$	$\dot{\theta}_2$	Cycle time T(s)	Speed v(m/s)	Step length λ (m)
0.13	0.368	-3.627	-1.8507	0.5921	1.0494	0.6213
0.18	0.3188	-3.7238	-2.8129	0.6952	0.7589	0.5276
0.23	0.2722	-3.9447	-3.3203	0.6667	0.6311	0.4207
0.27	0.1919	-4.6187	-4.3001	0.6815	0.5060	0.34481

5. Prototype Experiment

According to the mechanical design in S ection 2, the theoretical analysis and simulation calculation is done in Section 3. But with the greatly simplified model, experiment is needed to test and verify the result. At the same time, the feasibility of passive walking, also need to test. We verify that theoretical model is reliable through the measurement of the actual walking step length and speed.

We processed an aluminium alloy prototype with wood foot, which can walk down on slope. In order to avoid scraping-ground, many 15 cm square blocks are placed on a piece of 2 m long board as the slope. The prototype can walk stably on the slope (0.04 rad, same with the simulation). The comparison of experiment results and simulation data is shown in table4. There is some error due to difference between simplified model and prototype and distinction between initial conditions and theoretic fixed point.

Table 4. Comparison of Prototype and Simulation

Data type	Step length (m)	Speed (m/s)
simulation	0.52761	0.75893
prototype	0.75	0.93

6. Conclusion

This paper established model of passive walking robot and designed and processed a prototype. Furthermore, numerical simulation and prototype experiment were done, which demonstrates feasibility of passive walking. Simulation with different parameters demonstrated length of leg and location of leg's centroid have a great influence on gait, which is of great value for mechanical design for passive robot.

References

1. McGeer,T, "Powered flight, child's play, silly wheels, and walking machines," Technical report, Simon Fraser University, Burnaby, British Columbia, Canada, 1989
2. Tedrake R, Zhang T W, Seung H S, "Stochastic policy gradient reinforcement learning on a simple 3D biped," presented at the IEEE/RSJ International Conference on Intelligent Robots and System, New York, USA:IEEE, 2004, pp. 2849-2854
3. Schuitema E, Hobbelen D G E, Jonker P P, "Using a controller based on reinforcement learning for a passive dynamic walking robot," presented at the IEEE/RAS International Conference on Humanoid Robots. NewYork, USA:IEEE,2005, pp. 232-237
4. Naila Moreira, "Easy Striders-New humanoids with efficient gaits change the robotic sland scape," Science News, 2005, 168(6), pp. 88.
5. Collins S, Ruina A, Tedrake R, et al, "Efficient bipedal robots based on passive-dynamic walkers," Science, 2005, 307, pp. 1082-1085

VELOCITY CONTROL FOR A MULTI-DOF TWO-WHEEL INVERTED PENDULUM ROBOT

XUESHAN GAO, YUBAI LIU, SHIGONG JIANG
Electromechanical college, Beijing Institute of Technology,
NO.5 Zhongguancun South Street
Beijing, China

FUQUAN DAI
Electromechanical college, Fujian Institute of Technology, NO.3 Xueyuan Street
Fujian, China

Two-wheel inverted pendulum (TWIP) robots attract public attention as environmentally-friendly products for personal transportation in recent years. This paper presents a multi DOF TWIP(MD-TWIP) robot to balance its structure as well as sliding and swinging mechanisms. To fulfill the trajectory planning task, we proposed a robust controller based on the sliding mode theory for equilibrium control. In addition, a velocity control strategy is also introduced based on the T-S fuzzy controller to achieved good results in robot motion. Results are verified in ADAMS and MATLAB platforms simultaneously. Furthermore, the proposed control strategy is implemented on a practical self-designed physical TWIP robot model.

1. Introduction

The two wheel inverted pendulum (TWIP) robots made the rapid advancement in the last decade and appeared in many areas of people's daily life such as assistant robot[1], telepresence robot[2], baggage transportation robot[3] and so on. Moreover, Segway[4] has made some successful products based on TWIP system which are applied in many fields as a reliable companion for transportation. The widely application of TWIP is due to its fundamental characteristic including as well as the compact structure, zero turning radius and no exhaust emission.

The inverter pendulum system also interests researchers as it is a complex nonlinear system. Consequently, some TWIP robots are proposed by going through system modeling, control schematic design and experiment on prototypes such as JOE[5] and WV-2R[6]. As the TWIP system operates around the equilibrium point, the nonlinear system can be transformed into linear model. Therefore, many researchers adopt linear control strategy[7-10] in the published literature. However, the linear controller manifests its weakness in robustness comparing with the nonlinear control approaches. Hence, the sliding mode control scheme has been applied to various robots, as one of the robust control algorithms. The Coupled sliding-mode control(SMC) approach of inverted pendulum is proposed and the SMC law is constructed by coupled sliding surface[11]. By applying the terminal sliding mode control (TSMC) on TWIP, it employs three switching sub controllers to achieve an optimal braking scheme[12]. Moreover, Fuzzy control logic[13, 14] provides a better user understanding to express and translate human linguistic for control strategies. Hence, to design better control plans and strategies, human knowledge and expertise in experience involves.

In this paper, a MD-TWIP is proposed with both translational and swing degrees of freedom to improve the stability and comfort as personal transporter. Moreover, a velocity control strategy based on T-S fuzzy controller is proposed and compared with two other methods.

The rest of the paper is outlined as follows. Section 2 provides the system model. The controller design is illustrated in Section 3. Simulation and experiments of velocity control strategies compression are contained in Section 4. Section 5 draws the conclusion.

2. System Modeling

Table 1. System Parameters

Parameter	Meaning
m_w , m_1 , m_2	Mass of the wheel, the robot body and the slider
I_w , I_b , I_s	Rotational inertia of the wheel, the body and the slider
l_1 , l_2	Distance from the wheel axle to the CoG and to the slider
r_w	The radius of the wheel
θ_1 , θ_w	The robot body pitch angle and the wheel rotational angle
α	The inclined angle of the slope
λ	The position of the slider
f	The control force on the slider
τ_w	The torque outputted by the balancing motor

The prototype of the MD-TWIP presented in this paper is shown in Figure 1. On the basis of the traditional TWIP system, two additional degrees of freedom are introduced. Therefore, by positioning the displacement of the slider on the guide rail, the velocity of the robot is controllable at the equilibrium state. Besides the sliding mechanism, the swinging structure realizes the improvement of robot stability in yaw steering motion.

Figure 1. The prototype of MD-TWIP

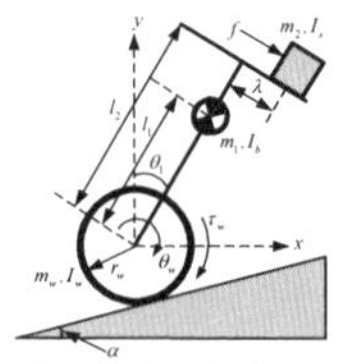

Figure 2. Model of MD-TWIP in rectilinear motion

In order to realize the velocity control strategy, the robot motion is decoupled into rectilinear motion shown in Figure 2. We can obtain the wheel coordinates, the center of gravity coordinates and the slider coordinates. By differentiating the aforementioned coordinates, we can calculate the kinetic energy of the three parts and the overall potential energy. Based on the overall potential energy and kinetic energy, the Lagrange function can be derived as Eq. (1):

$$L = T_w + T_1 + T_2 - U \tag{1}$$

The system dissipation function caused by friction is calculated according to three parts of friction: the friction between the wheels and the ground, the driving wheel transmission mechanism friction and the moving slider transmission mechanism friction.

$$E_D = \frac{1}{2} D_w \dot{\theta}_w^{\,2} + \frac{1}{2} D_1 \left(\dot{\theta}_w - \dot{\theta}_1\right)^2 + f_{c1}\left(\dot{\theta}_w - \dot{\theta}_1\right) + \frac{1}{2} D_2 \dot{\lambda}^2 + f_{c2}\dot{\lambda} \tag{2}$$

In rectilinear motion, the generalized coordinate:

$$\mathbf{q} = \begin{bmatrix} \theta_w & \theta_1 & \lambda \end{bmatrix}^T \tag{3}$$

The corresponding generalized force:

$$\mathbf{Q} = \begin{bmatrix} \tau_w & -\tau_w & f \end{bmatrix}^T \tag{4}$$

Based on Eq. (1), (2), (3) and (4), the dynamical equations of motion of the model can be obtained. It should be pointed out that if we assume the slope angle α is 0, the issue is degenerated into rectilinear motion on the flat ground.

As mention above, we adopt the slider mechanism to control the robot velocity which means the slider displacement λ is taken into account as one of the inputs of the system. Simultaneously, θ_w and θ_1 are regarded as the disturbance of the λ. In addition, $\dot{\theta}_w$, θ_1 and $\dot{\theta}_1$ are considered as the state variables and the robot remains balancing $\mathbf{x}^* = \begin{bmatrix} x_1 & x_2 & x_3 \end{bmatrix}^T = \begin{bmatrix} \dot{\theta}_w^* & 0 & 0 \end{bmatrix}^T$. Thus, the system is simplified as:

$$M_{11}\ddot{\theta}_w + M_{12}\ddot{\theta}_1 = \tau_w - G_1 sin\alpha - f_{c1} - D_1\dot{\theta}_w - D_w\dot{\theta}_w \tag{5}$$

$$M_{12}\ddot{\theta}_w + M_{22}\ddot{\theta}_1 = -\tau_w + f_{c1} + D_1\dot{\theta}_w + G_2\theta_1 + m_2 g\lambda \tag{6}$$

where

$$M_{11} = \left(m_w + m_1 + m_2\right) r_w^2 + I_w \tag{7}$$

$$M_{22} = m_1 l_1^2 + m_2 l_2^2 + I_b + I_s + m_2\lambda^2 \tag{8}$$

$$M_{12} = (m_1 l_1 + m_2 l_2) r_w \cos\left(\alpha + \theta_1\right) - m_2 r_w \lambda \sin\left(\alpha + \theta_1\right) \tag{9}$$

$$G_1 = \left(m_w + m_1 + m_2\right) r_w g \tag{10}$$

$$G_2 = \left(m_1 l_1 + m_2 l_2\right) g \tag{11}$$

3. Controller Design

3.1. *Equilibrium control with SMC*

In balancing control, the robot at the equilibrium point that $\theta_1=0$ and $\dot{\theta}_1=0$. Regarding the constraint $\lambda = 0$, $\ddot{\theta}_w$ and $\ddot{\theta}_1$ can be solved as:

$$\ddot{\theta}_w = \frac{F_{22}M_{12} - F_{11}M_{22} - \left(M_{12} + M_{22}\right)\tau_w}{M_{12}M_{21} - M_{11}M_{22}} \tag{12}$$

$$\ddot{\theta}_1 = \frac{F_{11}M_{21} - F_{22}M_{11} + \left(M_{21} + M_{11}\right)\tau_w}{M_{12}M_{21} - M_{11}M_{22}} \tag{13}$$

According to the characteristic of the TWIP system and the definition of variable structure controller sliding surface, we select the sliding surface as:

$$s = \theta_1 + c\dot{\theta}_1,\ c > 0 \tag{14}$$

Differentiating Eq. (14) yields:

$$\dot{s} = \dot{\theta}_1 + c\ddot{\theta}_1 \tag{15}$$

To force the system sliding along the sliding surface, three basic conditions should be satisfied: the existence and reachability of the sliding mode and the stability of the sliding motion. On the basis of the sliding mode condition discussed above, we select the derivative of the switch function $\dot{s}$ as :

$$\dot{s} = -\varepsilon\,\mathrm{sgn}\left(s\right),\ \varepsilon > 0 \tag{16}$$

Obviously, Eq. (13), (15) and (16) lead to τ_w :

$$\tau_w = \frac{cF_{22}M_{11} - cF_{11}M_{12} - \left(M_{12}M_{21} - M_{11}M_{22}\right)\left(\varepsilon sgn(s) + \dot{\theta}_1\right)}{c\left(M_{11} + M_{12}\right)} \tag{17}$$

3.2. *Velocity control with T-S FC*

A Takagi-Sugeno fuzzy controller (T-S FC) is proposed to control the velocity of the robot movement by the slider shifting. The inputs of the T-S FC are error signal E, changing rate of error signal EC and integration of error signal EI. The controller first fuzzifies the input signals and then fuzzy reasoning according to the fuzzy rules. Weighted sum method is adopt to calculate the output λ .

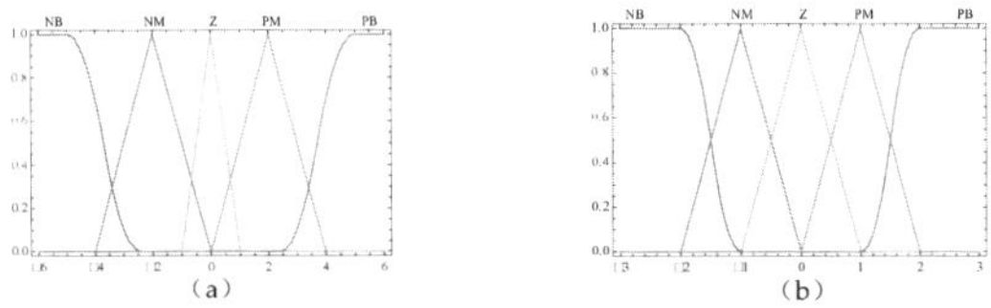

Figure 3. Error signal and error signal rate membership function

The input of error signal fuzzy sets on the domain $[-6,6]$ contains five elements (PB , PM , Z , NM , NB) and the membership function provided in Figure 3(a). Similarly, the changing rate of error signal fuzzy sets on domain $[-3,3]$ and the membership function is presented in Figure 3(b).

19 fuzzy rules are illustrated below:

R1: if E is PB and EC is PB then $\lambda = 0.35E + 0.3EC$

R2: if E is NB and EC is PB then $\lambda = 0.35E + 0.3EC$

R3: if E is PB and EC is NB then $\lambda = 0.35E + 0.3EC$

R4: if E is NB and EC is NB then $\lambda = 0.35E + 0.3EC$

R5: if E is PB and EC is not PB then $\lambda = 0.26E + 0.2EC + 0.001EI$

R6: if E is PB and EC is not NB then $\lambda = 0.26E + 0.2EC + 0.001EI$

R7: if E is NB and EC is not PB then $\lambda = 0.26E + 0.2EC + 0.001EI$

R8: if E is NB and EC is not NB then $\lambda = 0.26E + 0.2EC + 0.001EI$

R9: if E is PM and EC is PB then $\lambda = 0.3E + 0.22EC + 0.0032EI$

R10: if E is PM and EC is NB then $\lambda = 0.3E + 0.22EC + 0.0032EI$

R11: if E is NM and EC is PB then $\lambda = 0.3E + 0.22EC + 0.0032EI$

R12: if E is NM and EC is NB then $\lambda = 0.3E + 0.22EC + 0.0032EI$

R13: if E is PM and EC is PM then $\lambda = 0.28E + 0.26EC + 0.0038EI$

R14: if E is PM and EC is NM then $\lambda = 0.28E + 0.26EC + 0.0038EI$

R15: if E is NM and EC is PM then $\lambda = 0.28E + 0.26EC + 0.0038EI$

R16: if E is NM and EC is NM then $\lambda = 0.28E + 0.26EC + 0.0038EI$

$$\text{R17: if } E \text{ is } PM \text{ and } EC \text{ is } Z \text{ then } \lambda = 0.15E + 0.29EC + 0.0032EI$$
$$\text{R18: if } E \text{ is } NM \text{ and } EC \text{ is } Z \text{ then } \lambda = 0.15E + 0.29EC + 0.0032EI$$
$$\text{R19: if } E \text{ is } Z \text{ then } \lambda = 0.229E + 0.252EC + 0.0019EI$$

4. Simulation and Experiments

4.1. *Simulation on ADAMS- MATLAB/Simulink platform*

In this subsection, the simulation environment which combines ADAMS and Matlab/Simulink is employed to validate the control strategy proposed.

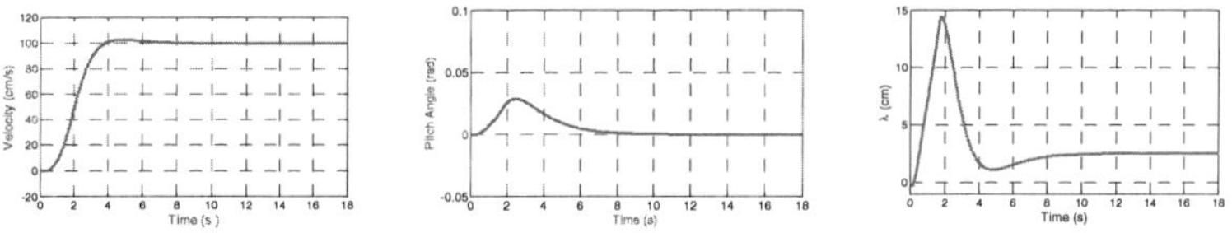

Figure 4. Robot velocity, pitch angle and slider displacement time responses under T-S FC for robot velocity control on the flat ground and the target velocity is 100cm/s.

Figure 4 demonstrates the velocity simulation under T-S FC. The velocity reaches the set point within 4 seconds and after only 2 second adjustment it keep the velocity 1m/s steadily. In the process of uniform motion, the pitch angle remains zero and the slider shifts to maintain the velocity due to the existence of friction. It conforms to the mathematical analysis of relationship between slider position and robot velocity in dynamic modelling.

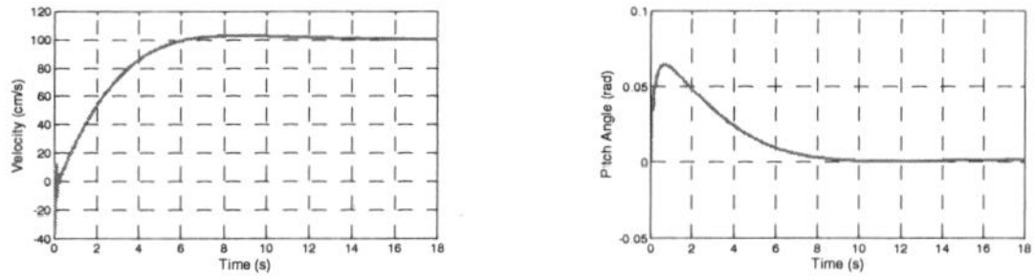

Figure 5 Angle control method pitch angle and slider displacement time responses for robot velocity control on the flat ground and the target velocity is 100cm/s.

Compared case 1 Angle control method

The traditional TWIP has to lean forward or backward continuously to control its velocity witch we can call it Angle control method. As we can see from Figure 5, the maximum pitch anlge is 0.06 rad during the acceleration process.

Compared Case 2 Equilibrium Setpoint method

As the existence of linear relationship between the slider displacement and the wheel velocity, we can design to move the slider to the specific position and therefore, the robot can reach the target velocity eventually. Even though the robot reaches the target velocity smoothly in Figure 6, it takes 30s to achieve the stable state which is much longer than the method we proposed.

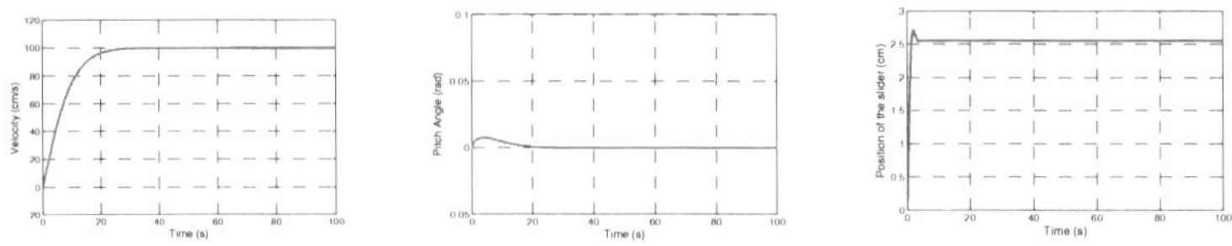

Figure 6. Equilibrium Setpoint method wheel velocity, pitch angle and slider displacement time responses for robot velocity control on the flat ground and the target velocity is 100cm/s.

4.2. *Implementation and experiments on MD-TWIP prototype*

The rectilinear motion strategy is that to reach the target velocity as soon as possible, hence, the slider is overdriven in the acceleration process. As the matter of fact, the robot can have a better acceleration effect rather than driving the slider to the set point directly. Figure 7 provides the corresponding velocity, pitch angle and slider displacement. As the figure indicates, the robot can reach the target velocity in about 4s and the pitch angle during the whole process is around zero.

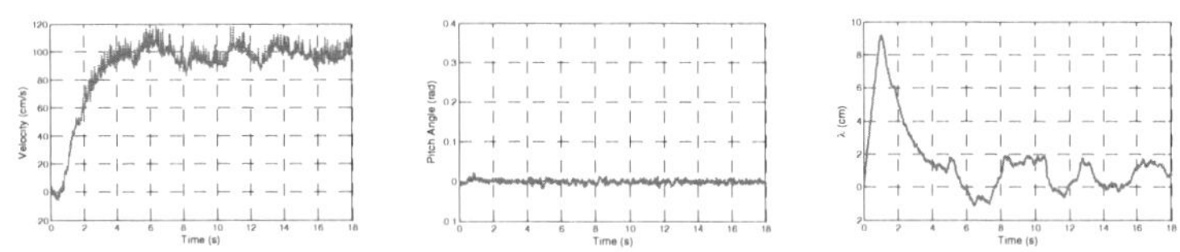

Figure 7. Experimental results of robot velocity, pitch angle and slider displacement under T-S FC for robot velocity control on the flat ground and the target velocity is 100cm/s

5. Conclusion

In this paper, an MD-TWIP is proposed with a movable slider and swinging configuration. The SMC and T-S FC are proposed for equilibrium and velocity control respectively. Simulations and experiment validate the model and the proposed velocity control approach achieved satisfied results and proved advantages over the other two methods.

References

1. L. Canete and T. Takahashi, Disturbance compensation in pushing, pulling, and lifting for load transporting control of a wheeled inverted pendulum type assistant robot using the extended state observer, in *IROS 2012*, 2012.
2. K. M. Tsui, M. Desai, H. A. Yanco, and C. Uhlik, Exploring use cases for telepresence robots, in *IEEE HRI*, 2011.
3. T. Takei, R. Imamura, and S. Yuta, Baggage Transportation and Navigation by a Wheeled Inverted Pendulum Mobile Robot, *J IEEE T IND ELECTRON* **56**, 3985 (2009).

4. H. G. Nguyen, J. Morrell, K. Mullens, A. Burmeister, S. Miles, N. Farrington, *et al.*, Segway robotic mobility platform, in *Mobile Robots XVII*, 2004.
5. F. zzGrasser, A. D'Arrigo, S. Colombi, and A. C. Rufer, JOE: a mobile, inverted pendulum, *J IEEE T IND ELECTRON* **49**, 107 (2002).
6. J. Solis, R. Nakadate, Y. Yoshimura, Y. Hama, and A. Takanishi, Development of the two-wheeled inverted pendulum type mobile robot WV-2R for educational purposes, in *IROS 2009* 2009.
7. K. Pathak, J. Franch, and S. K. Agrawal, Velocity and position control of a wheeled inverted pendulum by partial feedback linearization, *J IEEE T ROBOT* **21**, 505 (2005).
8. A. Shimada and N. Hatakeyama, High-Speed Motion Control of Wheeled Inverted Pendulum Robots, in *ICMA 2007*, 2007.
9. J. Akesson, A. Blomdell, and R. Braun, Design and Control of YAIP - An Inverted Pendulum on Two Wheels Robot, in *IEEE CCA/CACSD/ISIC 2006*, 2006.
10. K. Yeon Hoon, K. Soo Hyun, and K. Yoon Keun, Dynamic analysis of a nonholonomic two-wheeled inverted pendulum robot, *J. Intell. Robot. Syst* **44**, 25 (2005).
11. M.-S. Park and D. Chwa, Swing-up and stabilization control of inverted-pendulum systems via coupled sliding-mode control method, *J IEEE T IND ELECTRON* **56**, 3541 (2009).
12. D. Feng, J. Huang, Y. Wang, X. Gao, T. Matsuno, T. Fukuda, *et al.*, Optimal braking control for UW-Car using sliding mode, in *ROBIO 2009*, 2009.
13. C.-H. Chiu and C.-C. Chang, Design and development of Mamdani-like fuzzy control algorithm for a wheeled human-conveyance vehicle control, *J IEEE T IND ELECTRON* **59**, 4774 (2012).
14. J.-X. Xu, Z.-Q. Guo, and T. H. Lee, Design and implementation of a Takagi–Sugeno-type fuzzy logic controller on a two-wheeled mobile robot, *J IEEE T IND ELECTRON* **60**, 5717 (2013).

DYNAMICAL ANALYSIS OF LARGE DEFLECTION COMPLIANT LEG DURING TERRESTRIAL LOCOMOTION

TAO FANG, XUDONG WANG, ZHUO CHEN, MIN XU, SHIWU ZHANG*

Dept. of Precision Machinery and Precision Instrumentation
University of Science and Technology of China
Hefei, Anhui, 230026, China

An amphibious robot with straight compliant flipper-legs can conquer various amphibious environments. It can rotate its flipper legs and utilize the large deflection of the legs to locomote on rough terrains, and it can oscillate the straight flipper legs to propel underwater. This paper focus on the dynamics of the compliant straight flipper legs during terrestrial locomotion. Leg's motion is modeled dynamically using large deflection theory and simulated to investigate locomotion parameters including trajectory, velocity and propulsion efficiency. In order to validate the theoretical model of the locomotion dynamics, a single-leg experimental platform is set up to explore the locomotion performance of the flipper legs with various structural and kinematic parameters. The trajectories of the rotating axle of the leg during locomotion in simulation and experiment coincide approximately. The dynamical analysis in this paper for terrestrial locomotion facilitates the implementation of amphibious robots with compliant flipper legs.

1. Introduction

Years, more and more researchers are paying attention to bio-inspired amphibious robot. Scholars are trying to develop robots being able to adapt to both water and terrestrial environments. And it's better to perform well in efficiency and maneuverability. So far, amphibious robots have been developed into following categories: snake-like amphibious robot, wheel-propeller amphibious robot, flipper amphibious robot and so on. Snake-like amphibious robot like ACM-5 [1] and salamander-like [2, 3] robot has similar waved propulsion modus, which is highly efficient. Another amphibious robot AQUA series robot [4] transformed from R-Hex robot runs fast on land and swim gracefully. AQUA can also walk on the bottom of the water and run along the shore [5]. However, locomotion between water and ground needs to be artificially switched. Under water, it uses six flippers for propulsion, while on

* **Corresponding author:** Shiwu Zhang
E-mail: swzhang@ustc.edu.cn

the ground, it uses six compliant C-shaped legs. In AmphiHex-I, authors proposed a transformable flipper-leg mechanism to adapt complex amphibious environments [6, 7]. However, the active transformation of flipper leg increases complexity of the structure and controlling strategy. This paper introduces a simple compliant flipper leg to walk on terrain and swim underwater. The compliant flipper leg transforms into C-shape during terrestrial locomotion which is similar to the curved legs of R-Hex robot. To verify this conjecture, a series of polyurethane flipper legs are designed and a single-leg locomotion platform is set up to explore experimentally locomotion performance of the compliant legs in various structural and kinematic parameters.

In order to understand the dynamics of the terrestrial locomotion of compliant leg deeply, authors tried to model the dynamics theoretically. Through studying on the gait of C-shape leg using R-SLIP model, researchers found C-shape compliant leg guarantees R-Hex robot a stable gait [8]. With different stiffness of the leg, robot performs differently in stability, efficiency and robust [9]. Park et. al modeled dynamically for bio-Inspired Quadruped amphibious robot when it runs on the surface of water and land using Pseudo-Rigid-Body model [10]. Hai-Jun Su putted up a PRB 3R model in his research to simulate deflection of compliant cantilever beam [11]. Howell *etc.* significantly proposed a method to emulate large deflection of long thin beam, which could be summarized by three large deflection equations [12, 13]. These equations were then used to analyze long straight compliant flipper-leg's dynamical performance on land.

The paper mainly presents the dynamics of a complaint flipper leg for the locomotion of amphibious robot. We modeled dynamically of the locomotion process with large deflection. An experiment system was built for testing locomotion performance of the complaint leg. The experimental result validated the model and can facilitate the implementation of amphibious robots with compliant flipper legs.

2. Experiment System

To test flipper-leg's dynamical performance, an experiment system is built as showed in Fig. 1. The system consists of a horizon cylindrical slide way, two vertical cylindrical slide ways and driving unit circled in Fig. 1(b). Driving unit mainly contains a MAXON motor and a torque sensor, besides, mechanical drives, displacement sensor and some other parts are also belonged to payload. The red pointer shown in Fig. 1(b) on a disk paper drawn calibration line is used to measure rotated angle of the motor shaft. The DC motor is controlled precisely to change motor's rotation speed in clockwise direction. For exploring optimal properties for compliant legs, a series of legs are manufactured by varying thickness, hardness. The material is polyurethane, which is widely used as a super elastic material. Fig. 1(a) shows a simplified draft of real system. The coordinate is set in the center of the rotation tip of leg marked in the figure as follow. Snapshots of the motion during experiment are presented in Fig. 1(b).

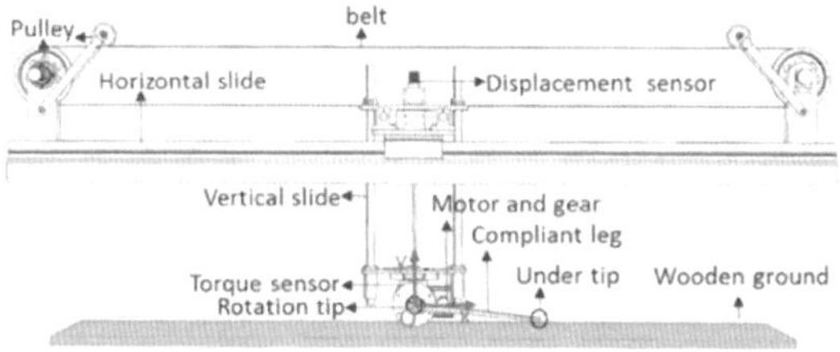

(a)

(b)

Fig. 1. (a) Draft of experimental system. (b) The snapshots of the locomotion of the compliant leg. The pointer is used to measure rotation angle of the motor, and the weight is used to change the payload.

3. Mechanical Model and Analysis

Since the bending of the leg is complex, the process of driving unit motion is divided into three phases. In the first phase, driving unit contacts with ground and keeps still. While the ground supporting force N_m is 0 N, driving unit starts moving and enters the second phase until the under tip of leg marked in Fig. 1 (a) separates from ground. In the third phase, the motion of the driving unit is similar to free falling with the assumption that the effect of the leg's oscillation on the driving unit is ignored.

3.1 First phase

Driving unit maintains still until F_y is large enough to force driving unit to leave ground. Calculating the amount of rotated angle Θ_0 after which the driving unit starting moving is presented below.

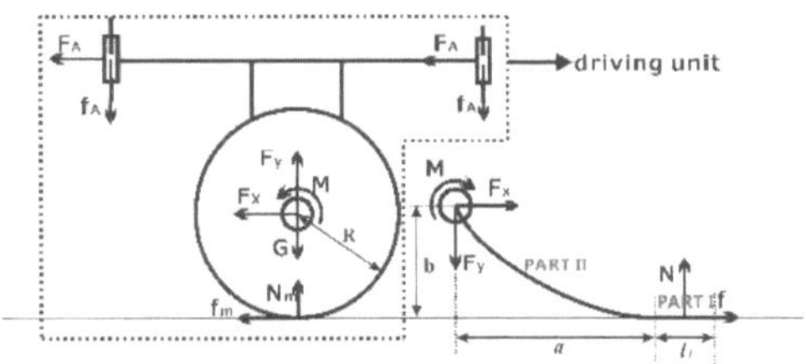

Fig. 2. Equilibrium diagram of the system before driving unit leaving ground (the 1st phase), showing separately driving unit and leg.

Since the compliant leg bends slowly, the motion is regarded as quasi-static so that the three large deflection equations can be adopted. F_x, F_y and M are respectively x-axial load, y-axial load and moment from motor. N_m and f_m are normal force and static friction between the bottom of the driving unit and ground. F_A and f_A are normal force and friction between leg and vertical slide way with friction coefficient μ_s. G stands for gravity of payload. In the coordinate built in Fig. 1(a), Mechanical equilibrium equations of driving unit is listed:

$$F_y + N_m = G + 2f_A. \tag{1}$$

$$2F_A h + f_m R = M. \tag{2}$$

$$f_m + 2F_A = F_x. \tag{3}$$

Then the leg is partitioned into two parts for mechanical calculation separately. The part I marked in Fig. 2 contacts with the ground. Friction f and normal force N between the leg and ground are relatively equal to F_x and F_y, Besides, moments are balance for leg:

$$N(a + 0.5l_1) + fb = M. \tag{4}$$

The friction coefficient between leg and ground is defined as μ_d. And $l1$ is length of part I. Three equations derived by L. L. Howell to calculate large deflection of the rotation tip are presented as follows [13]:

$$l_2\sqrt{F_o/2EI} = \frac{1}{2}\int_0^{\Theta_0} \frac{d\Theta}{\sqrt{[cos(\Theta_0 + \Phi) - cos(\Theta + \Phi)] + M_o{}^2/2EIF_o}}. \tag{5}$$

$$a = \frac{1}{2\sqrt{F_o/2EI}}\int_0^{\Theta_0} \frac{\cos(\Theta)\, d\Theta}{\sqrt{[\cos(\Theta_0 + \Phi) - \cos(\Theta + \Phi)] + M_o{}^2/2EIF_o}}. \tag{6}$$

$$b = \frac{1}{2\sqrt{F_o/2EI}}\int_0^{\Theta_0} \frac{\sin(\Theta)\, d\Theta}{\sqrt{[\cos(\Theta_0 + \Phi) - \cos(\Theta + \Phi)] + M_o{}^2/2EIF_o}}. \tag{7}$$

F_o denotes resultant force of F_x and F_y, and Φ is angle between resultant direction and x-axial direction. E is elastic modulus of the material and I denotes moment of inertial of leg's cross section. We calculated Θ_0 at different driving unit and thickness of leg. Through observation of the result, it's found that Θ_0 is nearly proportional to weight of the payload. The solution above is significant for designing gait and controlling of the robot accurately.

3.2 Second phase

It's found from experiment that at low rotation speed, the under tip of leg doesn't slip, which is the ideal status. In coordinate (showed in Fig. 1(a)), force balance equations are listed as follows:

$$F_x - 2F_A = m\ddot{x}. \tag{8}$$

$$-2f_A + F_y - mg = m\ddot{y}. \tag{9}$$

m is weight of payload, $\ddot{x}$ and $\ddot{y}$ are x axial and y axial accelerations. The motion of leg is regarded as a quasi-static process since it is slow. In this phase, vertical slide ways move in x-axial direction with payload, while in y-axial direction, it has no displacement. In x-axial direction, the acceleration can be formed as follow:

$$2F_A - 2f_1 - 2f_2 = m_2\ddot{x}. \tag{10}$$

m_2 is vertical slid ways' weight, N_1 and f_1 are force and friction between vertical and horizontal slid ways as shown below in Fig. 3. By coupling three large deflection equations, these equations can be solved entirely. However, the upper limit of integral (5) (6) (7) is an unknown variable Θ_0, besides, two order derivative is also hard to deal with. To solve these equations, an iterative method is adopted with initial conditions derived from the first phase.

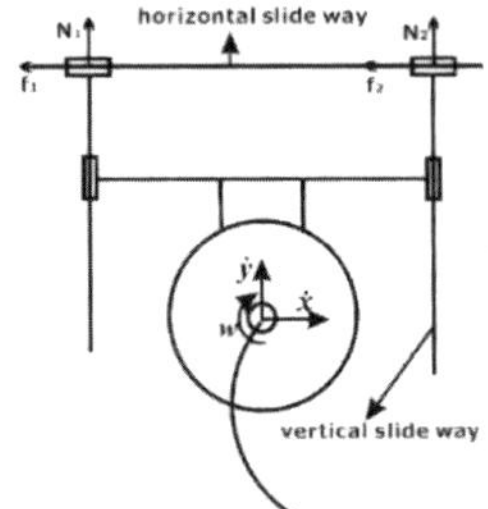

Fig. 3 Mechanical model after driving unit leaves the ground

Once friction f and normal forces N between leg and ground satisfied formula: $f/N \geq \mu_d$, the leg leaves ground entirely and gets into the third phase. The third phase is similar to a free falling process. In fact, due to existence of friction on slide way, it's not free falling. Nevertheless, the analysis of the third motion is simple so that we don't present it here.

4. Results and Discussions

4.1 Simulation result

Based on the mechanical analysis presented in Section 3, the trajectories of the body were calculated with respect to varying rotation speeds, as presented in Fig. 4(a).

The starting phases of the trajectories are similar low rotation speeds (< 2 rad/s). Then they separate. Generally, the step length is higher when the rotation speed becomes lower. And what's more, the starting phases of trajectories at a higher speed (> 2 rad/s) are higher than others, which were caused by slippage at higher speeds. The average horizontal speed of the driving unit reaches the highest when the rotation speed reaches 3 rad/s, as showed in Fig. 4(b).

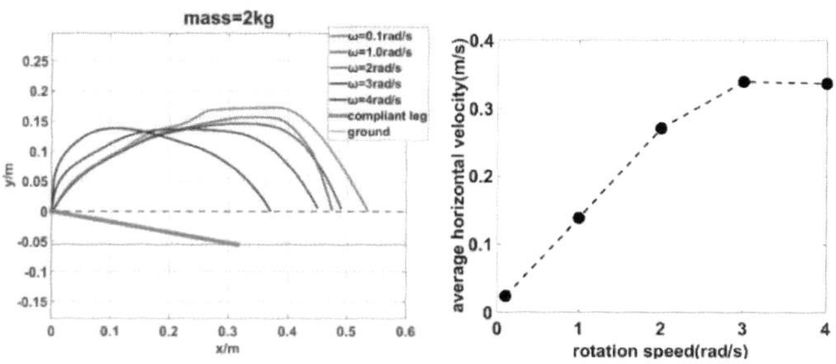

(a) (b)

Fig. 4. (a) Trajectories of rotation tip of leg at various rotation speeds from 0.1 rad/s to 4 rad/s. (b) Forward speed at various rotation speeds

Figure 5 shows a comparison between simulation trajectory and experimental trajectory at 0.5 rad/s rotation speed. The result indicates that in the beginning, the two trajectories are nearly the same. However, the later part of trajectory in simulation is higher than that in experiment. Error maybe caused by simplification of leg property. Polyurethane is not an ideal linear material. The elastic modulus changes to smaller with strain being larger.

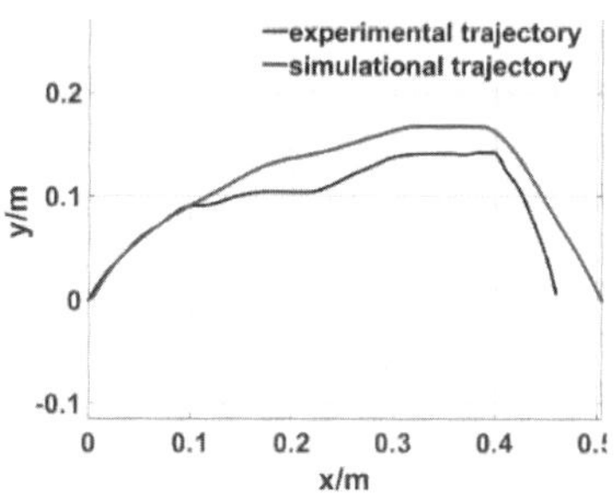

Fig. 5. Comparison between simulation trajectory and experimental trajectory. Rotation speed is 0.5 rad/s and the load is 2 kg. Shapes and features between both are the same.

4.2 Propulsion efficiency

Cost of transport (COT) is adopted to evaluate the power efficiency [14]. COT is defined as follow:

$$\text{COT} = \frac{P_i}{WV}. \tag{11}$$

In the formula above, P_i is input power of the system. W is gravity of moving parts, and V is forward speed. Apparently, COT is dimensionless, so that the propulsion efficiency in different systems could be compared. COT is calculated from the driving unit leaves ground to touching the ground as shown in Fig. 4(a) during a period of time T. Then, P_i is mean of product of the motor torque and rotation speed:

$$P_i = \frac{\omega}{T} \sum_T M(t) \vartriangle t_i. \tag{12}$$

Hence, COT in the simulation is finally formed as follow:

$$\text{COT} = \frac{\omega \sum_T M(t) \vartriangle t_i}{WVT}. \tag{13}$$

COT is calculated at several speeds and the result shows that COT becomes larger at higher rotation speed (see Fig. 6). Owing to increasing of motor's rotation speed, the time T of a gait would be shorter, and average $M(t)$ becomes larger. So though horizontal speed is smaller, taken together, COT is still higher at large rotation speed, which means the propulsion efficiency becomes lower. When the speed of motor rotation is brought up, much energy is released to air as heat. And at rotation speed higher than 2 rad/s, COT increases steeply due to large slippage.

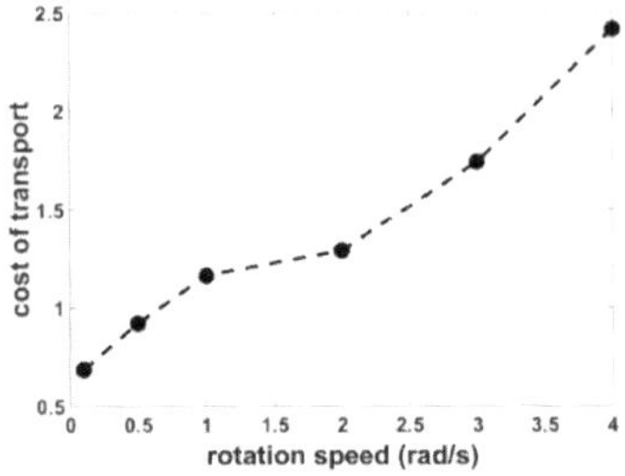

Fig. 6. Cost of transport at various roation speeds

5. Conclusions

The purpose of this paper is to present dynamics of compliant straight flipper leg. Through observing the locomotion performance on terrain of the leg made by polyurethane, authors found the driving unit forwards faster at a higher rotation speed of motor. However, when the rotation speed is too large, the velocity becomes lower due to slippage between leg and ground. To analyze propulsion efficiency, authors divide all the initially inputted power into several parts: expenditure in the transmission, energy released to air as heat and energy transported to mechanical energy. Cost of transport (COT) is adopted to quantify efficiency by varying rotation speeds and it's found that propulsion efficiency is higher at low rotation speed owing to more input energy is released to air and transformed to fuel.

6. References

[1] Hirose, S. Biologically Inspired Robots: Snake-Like Locomotors and Manipulators, Oxford University Press, Oxford (1993).

[2] Ijspeert, A. J., & Cabelguen, J. M. (2014). Control of Aquatic and Terrestrial Gaits in Salamander. *In Encyclopedia of Computational Neuroscience* (pp. 1-9). Springer New York.

[3] Ijspeert, A. J., Crespi, A., Ryczko, D., & Cabelguen, J. M. (2007). From swimming to walking with a salamander robot driven by a spinal cord model. *science*,315(5817), 1416-1420.

[4] Dudek, G., Giguere, P., Zacher, J., Milios, E., Liu, H., Zhang, P., & Ripsman, A. (2007). Aqua: An amphibious autonomous robot. *Computer*, (1), 46-53.

[5] website http://www.designboom.com/technology/aqua2-underwater-robot/

[6] Liang, X., Xu, M., Xu, L., Liu, P., Ren, X., Kong, Z., & Zhang, S. (2012, October). The amphihex: A novel amphibious robot with transformable leg-flipper composite propulsion mechanism. *In Intelligent Robots and Systems* (IROS), 2012 IEEE/RSJ International Conference on (pp. 3667-3672). IEEE.

[7] Yan, Q., Wang, L., Liu, B., Yang, J., & Zhang, S. (2012). A novel implementation of a flexible robotic fin actuated by shape memory alloy. *Journal of Bionic Engineering*, 9(2), 156-165.

[8] Ioannides, E., & Harris, T. A. (1985). A new fatigue life model for rolling bearings.*Journal of Tribology*, 107(3), 367-377.

[9] Galloway, K. C., Clark, J. E., & Koditschek, D. E. (2013). Variable stiffness legs for robust, efficient, and stable dynamic running. *Journal of Mechanisms and Robotics*, 5(1), 011009.

[10] Park, H. S., & Sitti, M. (2009, October). Compliant footpad design analysis for a bio-inspired quadruped amphibious robot. *In Intelligent Robots and Systems*, 2009. IROS 2009. IEEE/RSJ International Conference on (pp. 645-651). IEEE.

[11] Hai-Jun Su, (2009, May), A Pseudo-Rigid-Body 3R Model for Determining Large Deflection of Cantilever Beams Subject to Tip Loads, *Journal of Mechanisms and Robotics*, 1(2), 1-9

[12] Howell, Larry L. Compliant mechanisms. John Wiley & Sons, New York, 2001.

[13] Howell, L. L., & Midha, A. (1995). Parametric deflection approximations for end-loaded, large-deflection beams in compliant mechanisms. *Journal of Mechanical Design*, 117(1), 156-165.

[14] Tucker, V. A. (1975). The energetic cost of moving about: walking and running are extremely inefficient forms of locomotion. Much greater efficiency is achieved by birds, fish—and bicyclists. *American Scienti*st, 413-419.

LOCOMOTION PERFORMANCE OF AMPHIHEX –I ON ROUGH TERRAINS WITH FLEXIBLE FLIPPER LEGS

BIN ZHONG, ZELIN YU, YOUCHENG ZHOU, MIN XU, SHIWU ZHANG*

Dept. of Precision Machinery and Precision Instrumentation,
University of Science and Technology of China,
Hefei, Anhui, 230026, P. R. China.

Amphibious robots are attracting more and more attentions from researchers worldwide for their broad foreground in resource exploration, disaster rescue, and reconnaissance. AmphiHex-I with transformable flipper-leg composite propulsion mechanisms can adapt various terrestrial and water environments. In this paper, we explored the locomotion performance of AmphiHex-I on various terrains by dynamical simulation. The influence of the stiffness of the flipper legs was investigated comprehensively. The simulation result indicates that the propulsion efficiency of the flexible flipper leg is higher than that of the rigid curved leg when passing through stairs, the variant of center of gravity (CG) keeps lower when the robot adopt a flexible flipper leg on flat terrain, the average velocity of the robot with flexible flipper legs is higher when the initial phase difference is large. These results verify that the transformable flexible flipper-leg mechanisms enable AmphiHex-I to perform better in passing through complex terrestrial environments.

1. Introduction

Amphibians possess unique abilities to adapt to various environmental conditions and terrains. Recent years, some bionic amphibious robots have been developed. Snake-like robots are amphibious robots propelling on land and underwater by undulating their bodies [1]. Whegs is a hybrid propelling robot with a combination of legs and propellers that gives it good locomotion performance on rough terrain and underwater [2]. AmphiRobot-II can demonstrate fish-like swimming and wheeled crawling with both wheel and fin propulsion [3]. Moreover, Salamander Robot, a salamander-inspired amphibious robot can utilize body undulation and limb walking to demonstrate the ability to transit between terrestrial and aquatic locomotion [4, 5]. Among the various amphibious robots, the AUQA series robots stemming from the RHex, a cockcroach-inspired hexapod robot possessing an outstanding terrestrial

* **Corresponding author:** Shiwu Zhang
E-mail: swzhang@ustc.edu.cn

locomotion performance [6, 7] is most attractive. And the AQUA2 with replaceable flipper legs has been better developed to allow transition between locomotion modes autonomously [8]. There are likewise many other amphibious robots have been developed that we don't mention.

During the development process of amphibious robots, more and more attention has been given to analyze the influences of structures of legs on the locomotion performance. There are some interesting results. First of all, adjustable leg stiffness is necessary to close the performance gap between robots and animals [9]. Adjusting leg stiffness has great mechanical value in improving adaptability of the absolute elastic leg to achieve an efficient locomotion [10]. Variable stiffness elastic legs can improve the ability of robots to achieve a robust, efficient and stable dynamic running [11]. Moreover, the robot can even leap on a solid level substrate from a motionless state to some desired aerial apex condition with elastic legs [12]. And the Hexapod RHex shows great performance on stair climbing behavior with stair descending over various stair geometries [13].

We have proposed and developed an amphibious robot, AmphiHex-I, with autonomously transformable flipper-leg composite propulsion mechanisms to solve some problems of existing amphibious robots [14]. The flipper-leg composite propulsion mechanism is designed to transit between fin and leg, which allow AmphiHex-I to locomote underwater and muddy terrain perfectly. In this paper, we focus on studying the influences of the flexible flipper legs on the locomotion performance of AmphiHex-I on various rough terrains including rigid ground, slope, and stairs.

2. Flexible Flipper Leg Propulsion Mechanism

2.1. *Structure of flexible flipper leg*

Flexible flipper leg is composed of seven elements as showed in Figure. 1. The stiffness of the flexible flipper leg can be adjusted by varying the thickness of the steel plate. Angle α between two adjacent segments varies from 0° to 26° when the leg is loaded. If the cable of the flipper leg is pulled tightly, the flexible flipper can transform into rigid curved leg. In the study, we simplify the spring steel plate into six torsional springs, which connect every two adjacent segments. Figure.1 shows the transformation process of the flipper leg from flipper mode to rigid leg mode.

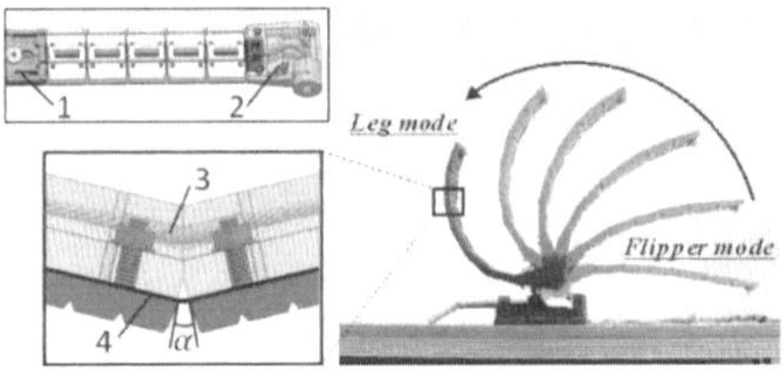

Figure 1. Structure of the flipper leg. 1. End segment 2.Head segment. 3. Cable. 4. Spring steel plate.

2.2. *Gaits of the AmphiHex-I*

In this section, we introduce two typical gaits used in the simulation and show the locomotion process of different gaits with flexible flipper legs and rigid legs.

2.2.1. *Tripod gait*

In this gait, as showed clearly in Figure.2, the front and the rear legs on one side keep the same movement with the middle leg on the other side. When one group of the three legs in the same gait are about to leave the ground, the other group of legs should touch the ground immediately, so the robot can keep balance. During the locomotion, two groups of legs keep a constant phase difference. The value of the phase difference Φ has a great influence on locomotion.

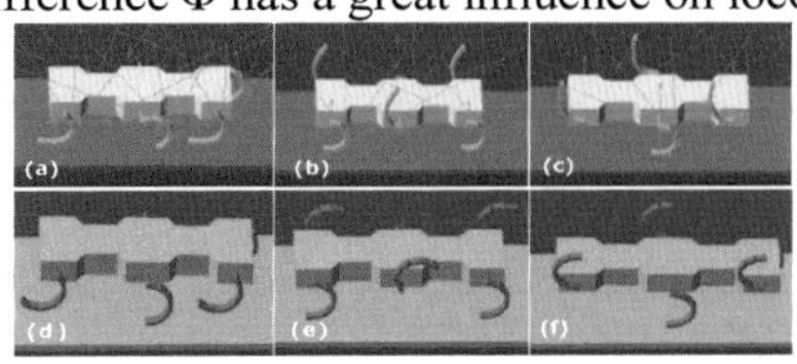

Figure 2. Description of the tripod gait. (a), (b), and (c) show the tripod gait with flexible flipper legs, (d), (e), and (f) show the tripod gait with rigid legs.

2.2.2. *Tetrapod gait*

In this gait, as showed in Figure.3, two front legs, two middle legs and two rear legs keep the same movement respectively. Different from the tripod gait, the initial phase differences between every group of legs and the perpendicular direction are preset, which are -55°, 0°, and 50°, respectively.

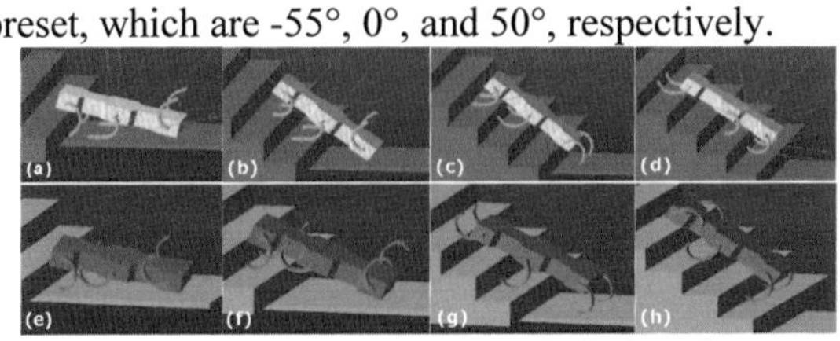

Figure 3. Description of the tetrapod gait. (a), (b), (c), and (d) show the tetrapod gait with flexible flipper legs, (e), (f), (g), and (h) show the tetrapod gait with rigid legs.

3. Locomotion Simulation

3.1. *Terrestrial environments*

In this simulation, we design three kinds of environments to explore the locomotion performance of AmphiHex-I, which include ground, slope, and stairs. The gradients of the slope consist of 20°, 25°, 30°, and 35°, respectively. The gradients of the stairs consist of 20°, 25°, 30°, 35°, 40°, and 45°, respectively.

Especially, in the stairs model, the height and the length of each step should be chosen carefully to make sure the robot can climb up. In the actual situation, the maximum height that AmphiHex-I can climb up is 25cm, so 15cm is chosen to be the standard height of every step in the simulation.

3.2. *Evaluation parameters*

We design four parameters to evaluate the locomotion performance of AmphiHex-I, which include the variant of center of gravity Δh, locomotion speed v, the output torque of legs *T*, and the locomotion efficiency η. $\Delta h, v$, and *T* can be exported directly in the simulation. While η needs to be calculated.

To calculate the locomotion efficiency, we need other three important parameters, including the value of the frictional force *F* from the ground, which can be got in Adams directly, the effective power of locomotion P_1 and the output power of motor P_2. P_1 and P_2 are expressed as following equations:

$$P_1 = F \cdot v \tag{1}$$

$$P_2 = T \cdot \omega \tag{2}$$

Where ω stands for the angular velocity of legs.

So, the equation to calculate the efficiency η is expressed as follow:

$$\eta = \frac{\int_0^{\tau} P_1 \cdot t}{\int_0^{\tau} P_2 \cdot t} = \frac{\int_0^{\tau} F \cdot v \cdot t}{\int_0^{\tau} T \cdot \omega \cdot t} \tag{3}$$

In the above equation, τ stands for the period of motion, which is set as 4 s in the tripod gait while 6 s in the tetrapod gait.

4. Results and Discussions

In this section, we discuss three different situations to study the locomotion performance of flexible flipper legs and rigid legs. The variables contain two different structures of the leg, different initial phase differences Φ of two gaits.

4.1. *Locomotion performance on flat ground*

In this section, we analyze the locomotion performance of AmphiHex-I with rigid legs and flexible legs from four parameters (Δh, locomotion speed, output torque and efficiency).

Figure 4 shows that Δh of both kinds of legs increase with increasing of the initial phase difference. And in the same gait, Δh with rigid legs is significantly higher than Δh with flexible flipper legs, which illustrates the flexible flipper legs have advantages in stable movement.

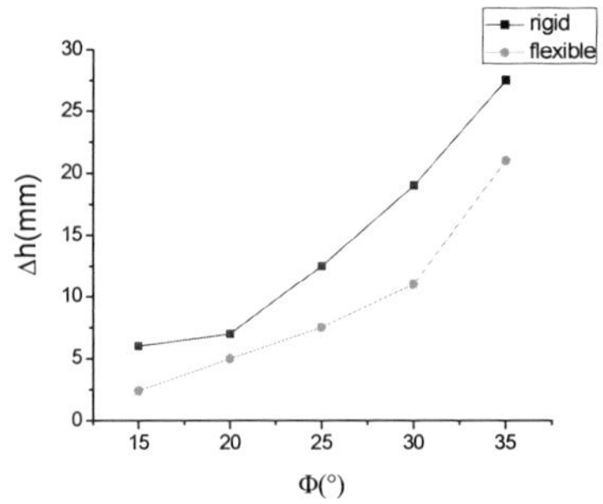

Figure 4. Variant of CG Δh.

Locomotion speed on flat ground is showed in Figure 5. Before 25°, the locomotion speed with flexible flipper legs and rigid legs is almost the same. After 25°, the locomotion speed with flexible flipper legs is higher, which illustrates the flexible flipper legs have advantages in high speed locomotion when the phase difference is large.

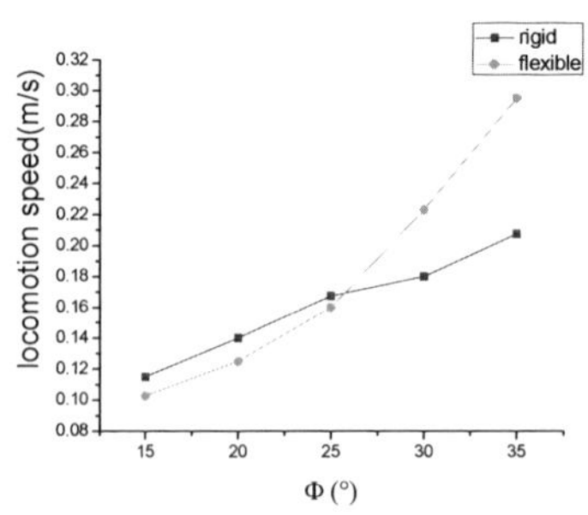

Figure 5. Locomotion speed of AmphiHex-I with rigid/flexible legs.

Figure 6 shows that the amplitude of the output torque of left middle leg with flexible flipper leg is always higher than the amplitude of the output torque with rigid leg in the same gait.

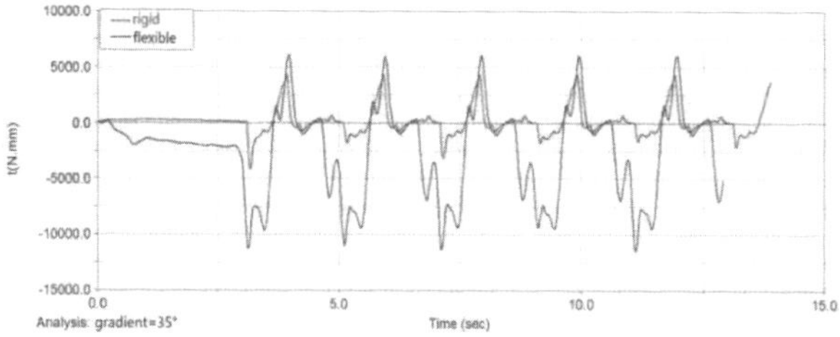

Figure 6. Output torque of the left middle leg.

According to the equation (3) and values of the parameters got from Adams, we make Table 1. Table 1 shows that both of the efficiency value with flexible flipper legs and rigid legs increase with the increasing of the phase difference before 25°. After 25°, the efficiency value of flexible flipper leg keeps almost the same, while the efficiency value of rigid leg is decreasing.

Table.1 Efficiency values of two kinds of legs.

η \ Φ	15°	20°	25°	30°	35°
Flexible flipper leg	10.36	19.16	35.24	37.07	36.8
Rigid curved leg	26.44	54.5	66.7	10.08	13.3

4.2. *Locomotion performance on climbing the slope*

In this section, we analyze the locomotion performance of AmphiHex-I with rigid legs and flexible legs from three parameters (locomotion speed, output torque and efficiency). For the result of Δh is similar to the situation on the ground, so we don't discuss in this section. Through numerous simulation experiments, we conclude that the maximum gradient the robot can climb up is 35°, whether the legs are flexible or rigid.

Figure 7 shows that the locomotion speed with flexible flipper legs is higher than the speed with rigid legs. The curves also illustrate that the flexible flipper leg give the robot the ability to move more stably.

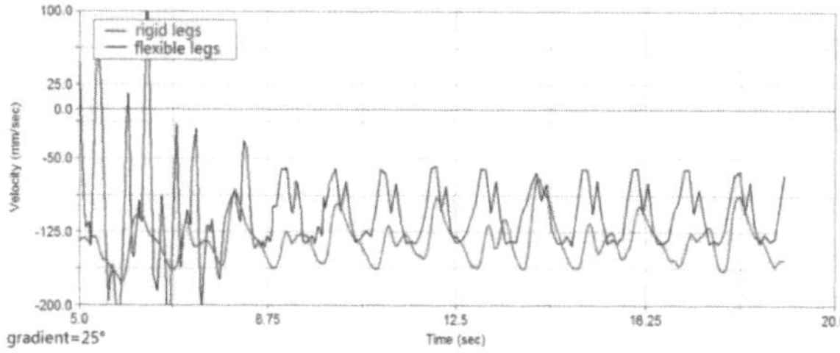

Figure 7. Locomotion speed of AmphiHex-I during climbing slope with 25° gradient.

Figure 8 shows that output torque of left middle leg with flexible flipper leg is higher than the output torque with rigid leg in the same gradient. However, the output torque with rigid leg is much more stable.

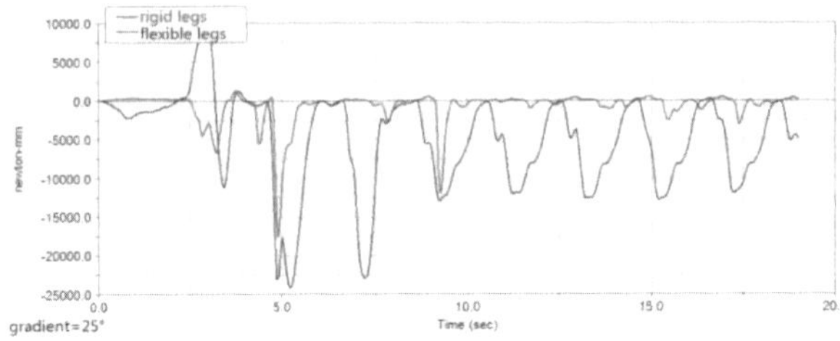

Figure 8. Output torque of the left middle leg during climbing slope with 25° gradient.

Table 2 shows that the efficiency values with flexible flipper legs are larger than the values with rigid legs. Moreover, compared with the results in Table 1, we find that values in Table 2 are much smaller. As we know, a majority of the energy converts into the gravitational potential energy, so the result is reasonable.

Table.2 Efficiency values with two kinds of legs in climbing slopes. θ stands for the gradient of the slope.

η \ θ	20°	25°	30°
Flexible flipper leg	19.4	15.6	9.6
Rigid curved leg	7.03	3.17	4.87

4.3. *Locomotion performance on stairs*

Through many simulation results, we conclude that the maximum gradient of the stairs the robot can climb up with flexible flipper legs is 45°, while it is 40° with rigid legs, which illustrates the flexible flipper leg has superiority in climbing stairs. And the results of the locomotion speed and output torque are similar to former results in 4.2. But locomotion with flexible legs is more stable during climbing the stairs.

5. Conclusions and Future Work

In this paper, we have presented a general description of structures of two kinds of legs, gaits of AmphiHex-I and some kinematic formulas of Adams simulation. Some preliminary theoretical analysis and simulation experiments are carried out to explore the effects of kinematic parameters to the locomotion performance of different legs. Some conclusions are derived from the simulation results: (1) the phase difference has a great influence on locomotion performance including the velocity, output torque and efficiency; (2) the flexible flipper legs have great advantages in locomotion performance on complex terrains. Future works are mainly focused on the following aspects: (1) the locomotion performance of the flipper legs composite propulsion mechanism working in various environments, including underwater, should be considered comprehensively. There are still lots of simulations to do; (2) the effects of the flipper leg's flexibility should also be studied and optimized; (3) we should find some better ways to calculate the efficiency under various environments.

Acknowledgment

This research has been financially supported by the National Natural Science Foundation of China (51375468).

References

1. Hirose S, Yamada H. Snake-Like robots machine design of biologically inspired robots. *IEEE Robotics & Automation Magazine*, 2009, **16**, 88-98.
2. Boxerbaum A S, Werk P, Quinn R D, Vaidyanathan R. Design of an autonomous amphibious robot for surf zone operation: Part I-Mechanical design for multi-mode mobility. *Proceedings of the 2005 IEEE International Conference on Advanced Intelligent Mechatronics*, Monterey, California, USA, 2005, 1459-1464.
3. Yu J, Ding R, Yang Q, Tan M, Wang W, Zhang J. On a bio-inspired amphibious robot capable of multimodal motion. *IEEE/ASME Transactions on Mechatronics*, 2012, **17**, 847-856.
4. Crespi A, Ijspeert A J. Online optimization of swimming and crawling in an amphibious snake robot. *IEEE Transactions on Robotics*, 2008, **24**, 75-87.
5. Ijspeert A J, Crespi A, Ryczko D, Cabelguen J M. From swimming to walking with a salamander robot driven by a spinal cord model. Science, 2007, **315**, 1416-1420.
6. Prahacs C, Saunders A, Smith M, McMordie D, Buehler M. Towards legged amphibious mobile robotics, *Proceedings of the Canadian Engineering Education Association*, 2004.
7. Dudek G, Giguere P, Prahacs C, Saunderson S, *et al.* An amphibious autonomous robot. Computer, 2007, **40**, 46-53.
8. AQUA2 amphibious robot is super cute and fast, less annoying than most pets because it has no head, [2010-07-08], *http://www.engadget.com/2010/07/08/aqua2-amphibious-robot-is-super-cute-and-fast-less-annoying-tha/*
9. Ayers J, Underwater Walking. *Arthropod Structure & Development*, 2004, **33**, 347-360.

10. Kevin C, Jonathan E, Mark Yim, Daniel E. Experimental investigations into the role of passive variable compliant legs for dynamic robotic locomotion. *Proceedings of the 2011 IEEE International Conference on Robotics and Automation*, Shanghai, China, 2011.
11. Kevin C, Jonathan E, Daniel E. Various stiffness legs for robust, efficient and stable dynamic running.[DOI: 10.1115/1.4007843].
12. Aaron M, D. E. Koditschek. Toward a vocabulary of legged leaping. *Proceedings of the 2013 IEEE International Conference on Robotics and Automation*, Karlsruhe, Germany, 2013.
13. D.Campbell, M.buehler. Stair descent in the simple hexapod 'RHex'. *http://www.cim.mcgill.ca/~arlweb*
14. Liang X, Xu M, Xu L, Liu P, Ren X, Kong Z, Yang J, Zhang S. The AmphiHex: a novel amphibious robot with transformable leg-flipper composite propulsion mechanism. *Proceedings of the 2012 IEEE International Conference on Intelligent Robots and Systems,* Algarve, Portugal, 2012, **124**, 265.

SECTION–8

MANIPULATION, INTELLIGENCE AND LEARNING FOR CLAWAR

RADIATION DOSING SOFTWARE CONTROL OF A ROBOT SYSTEM FOR THE ATLAS SCANNING FACILITY*

HECTOR MARIN-REYES

Department of Physics and Astronomy, The University of Sheffield, United Kingdom

RICHARD FRENCH

Department of Physics and Astronomy, The University of Sheffield, United Kingdom

Materials evaluation to establish component service lifetime in high radiation environments is often by means of irradiation. Achieving an accelerated lifetime "dose" of radiation drives better materials selection & system performance. The software control provides the ability to irradiate any area or single points of a sample using a proton beam from a Cyclotron. Samples cooled in a thermal chamber are moved at a constant velocity along a path or static point through the beam. The beam orientation is transverse to the sample movement, intersecting exactly with specific points on the sample. Accurate timing and positioning allow a controlled dose of radiation to the sample. Secondary dose monitoring is by means of measuring beam charge in a graphite Faraday cup though out the irradiation run. Confirming the level of induced radioactivity to the sample is by measuring the accumulated radioactive nuclei decay or half-life of a Ni foil mounted adjacent to the sample inside the thermal chamber. This scanning control of the beam delivers a highly uniform dose of radiation within a specified tolerance. Additionally, scanning path profiles can be adjusted prior the irradiation of samples. The radiation dosing software control of the Pre-configured XY-axis Cartesian Robot System is described in detail in this paper.

1. Introduction

Based at CERN, Geneva in Switzerland sited on the Large Hadron Collider (LHC) particle accelerator is the ATLAS experiment. ATLAS is a colossal, seven ton, sub atomic particle tracking detector, housed in a 10 story high underground experimental cavern comprising of many different sub detectors. Working at the core of ATLAS (the most complex machine in the world) is the Inner Detector, comprising of four separate Silicon Vertex Tracking Detectors. The Semiconductor Tracker (SCT), a silicon micro-strip tracker has 60 m^2 of

* This work is supported by STFC, UK-ATLAS Tracker Upgrade, EU FP7 AIDA & EU H2020 AIDA2

silicon distributed over 4 cylindrical layers and 18 planar discs with over 6 million implanted readout strips (6 million channels) coupled to the innermost Pixel & IBL detectors (with over 80 million 3D channels).

From the design phase, high radiation backgrounds have been predicted, later validated by real time measurements. The ATLAS SCT has a limited operational life due to radiation damage. The planned LHC upgrade (HL-LHC) will deliver higher energies (x10) requiring a new Inner Detector to have a service life of over 20 years. The ATLAS Upgrade Inner Tracker (ITk) [1], [4], is planned for 2022 to replace the SCT. Planned operation of the new ITk requires components be verified for operation in highly radioactive environments with robotic handling of replaceable detector elements.

The Sheffield ATLAS Group has recently commissioned a new proton irradiation facility for UK-ATLAS at The University of Birmingham to conduct radiation studies for proposed ITk detector components, passives & materials by providing a lifetime dose of radiation in a matter of hours or minutes. By using cutting edge robotic technology, mirroring high radiation levels found at HL-LHC, University of Sheffield staff apply their research experience on radiation effects, robotics and automatic control systems.

The irradiation facility (Commissioned in 2013) has (September 2014) tested over 211 samples successfully. Irradiation of samples is carried out in a thermally controlled cold chamber which moves continuously through a homogenous 10x10mm2 beam spot produced by a Scanditronix MC40 Cyclotron. This movement is provided by the pre-configured XY-Axis Cartesian Robot scanning system [2], see Figure 1.

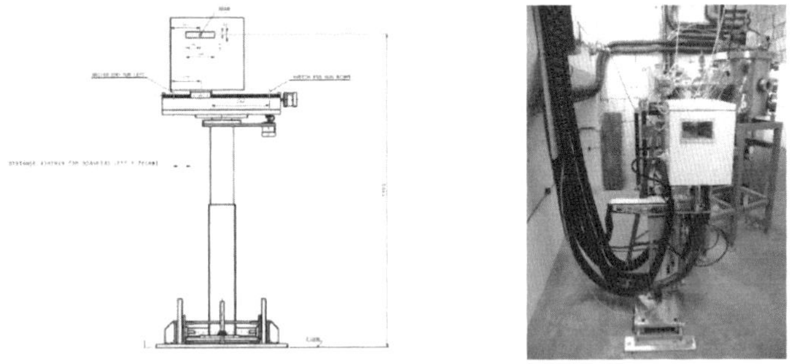

(a) (b)

Figure 1. (a) CERN – PS Original design c.1997, Joint construction with UK groups & CERN, (b) Pre-configured XY-Axis Cartesian Robot System c. 2012, Sheffield ATLAS group

2. LabVIEW Irradiation Program

Control of the scanning system is via NI cRIO Real-Time programmable controller with a LabVIEW GUI, see Figure 2. G Programming Language is a flowchart-like dataflow programming model. "Flow" of data through the application is represented graphically output of nodes connected with wires to the input of another.

Practical benefits of graphical programming focuses more on data and the operations performed, abstracting much of the administrative complexity of computer programming such as memory allocation and language syntax. New users need little familiarity with G code to flow charts or processes.

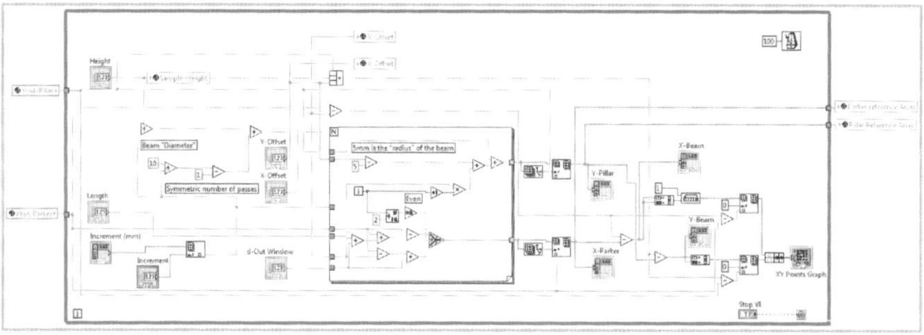

Figure 2. LabVIEW program to calculate irradiation path profile.

3. Sample Overheating Problem

Sample overheating is demonstrated by brief, significant, temperature increase during sample (silicon sensor) movement through the beam spot. This is due to the 27 MeV protons depositing 1.1W of energy in the silicon sensors.

Control system design previously considered a travel speed of 1mm/s in the X-axis. Irradiations performed at this speed achieved the desired fluence in a relatively short period of time (3-4 hours depending on the area size to be irradiated). Unfortunately new higher fluence demands have caused the operating conditions to change drastically. The irradiated samples started to show overheating, causing various unwanted phenomenon, such as annealing of silicon and melting of jigs, see Figure 3. This overheating problem caused the robotic system to operate out of its designed specification therefore various approaches to control irradiations have been proposed.

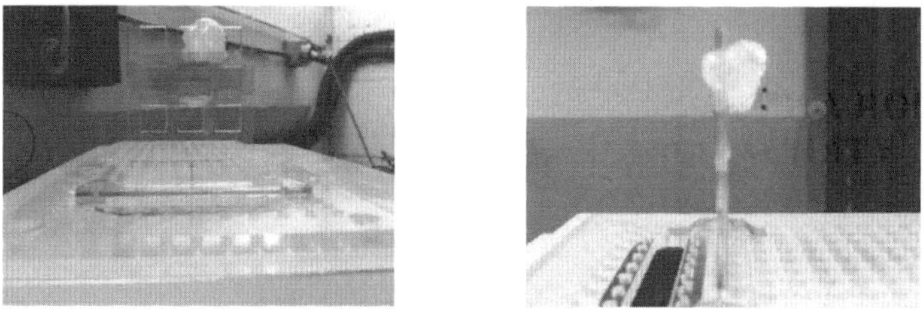

(a) (b)

Figure 3. Melted sample frame due to overheating: (a) Front view, (b) Side view.

4. Birmingham Facility Upgrade

Future prevention of overheating in subsequent irradiation tests required a temporary solution be implemented. Reduction of maximum beam current to 0.5 μA limits the potential to irradiate to high fluence as the scan time is doubled compared to using a beam current of 1 μA but allowed continued operations. Other improvements were made to the robotic platform in order to tackle the overheating problem [3].

4.1. *LN2 cooling system*

A prototype liquid nitrogen (LN2) dosing system using evaporative cooling was retrofitted to the thermal chamber. LN2 is dripped on to a heat sink located at the base of the cold box which evaporates to produce cold N2 gas directed across the samples by forced convection. Thermal chamber LN2 flow is adjusted automatically by the dosing system, which monitors the temperature, pumping in LN2 to achieve the required set point temperature of the cold box. The new prototype cooling system can achieve a stable temperature of −50 °C in 30 minutes.

4.2. *New thermal chamber*

A prototype thermal chamber has been made with a larger volume and window size allowing for larger samples to be irradiated in batches to cope with our reduced beam current and subsequent increased irradiation periods. More powerful fans have been installed to circulate the cold nitrogen gas. The box is purged with dry nitrogen before and after running.

4.3. *Redesign of carbon fibre frames*

Square recesses hold the 1 cm^2 silicon mini-sensors have been added to the frames (see Figure 4). By reduce the area of Polymide tape used to secure the samples allows a larger area of exposure to cold nitrogen therefore maintaining the required sub-0 °C temperatures on the sensors. Future redesign will eliminate Polymide tape by using a frame which sandwiches the sensor between two thin sheets of carbon fibre allowing maximum exposure to both faces of the sensor.

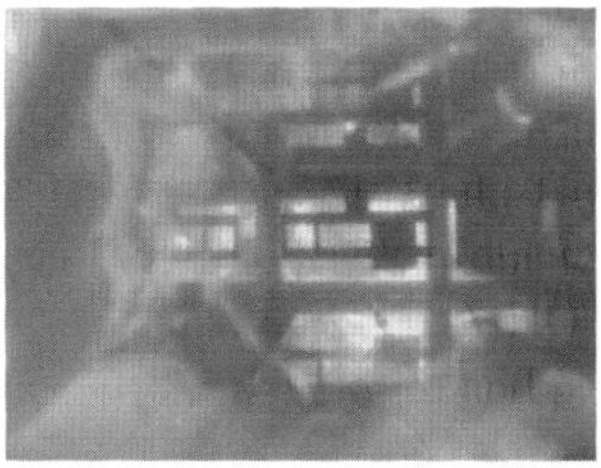

(a) (b)

Figure 4. (a) Previous carbon fibre frame (b) Back side inside the new cold box.

4.4. *Path profile irradiation control*

Reduction of sample exposure duration to the beam requires the increase of the scanning table horizontal speed from 1 mm s^{-1} up to a possible 10 mm s^{-1}. This therefore maintains the scan time. Whilst this does increase the frequency of sample heating cycles, each cycle is now faster, reducing the long periods which induce high temperatures therefore preventing annealing. With a greater scanning speed, large temperature spikes observed to the sensor are avoided reducing the overall maximum temperature recorded on the sensors.

5. Irradiation Program of AIDA & AIDA2

5.1. *Background*

Advanced European Infrastructures for Detectors at Accelerators (AIDA) is a European Union, Seventh Framework Programme supporting research infrastructures using a combination of collaborative project and transnational programmes of work conducted from February 2011 to January 2015.

A total of 691 researchers carried out 202 projects at European test beam and irradiation facilities (CERN, DESY, JSI, UCL, KIT) in the framework of the AIDA Transnational Access activities providing huge impact on detector R&D. The main area of Sheffield involvement with AIDA contributed to the improvement and equipment of irradiation and test beam lines at CERN (IRRAD). New beam line characterization, infrastructure installation, commissioning. IRRAD in the CERN PS East Area, uses Sheffield University robotic and thermal control technology that has been verified in our testing at the Birmingham Irradiation Facility. An online database, called IMHOTEP, on the properties of irradiated material & components was developed and populated with data.

AIDA2 is a European Union Horizon 2020 programme of work of a similar mode of operation to its predecessor, AIDA. From May 2015 the Birmingham Irradiation Facility will be further developed, characterisation infrastructure for high volume operation including production of refined thermal chambers and scanning robots will be installed and commissioned until April 2018. Transnational access will also be provided to the Birmingham Irradiation Facility as a result of EU funding.

5.2. *Carbon fiber continuous irradiation control*

ATLAS Carbon fibre frames, see Fig. 2, were used for sensor irradiations. Kapton tape is used to attach samples to the frame which then clips into a recessed mount on the underside of the thermal chamber lid. The sample position is measured relative to a reference point, 'Home position' ensuring precise sample location. This is essential to calculate the scan path, producing uniform irradiation. The scan path requires several input parameters from the user: the offset x (horizontal) and y (vertical) positions of the sample with respect to 'Home', the area to be scanned, the speed of the scan in the x-axis, and the beam current which is used to calculate the number of scans to achieve the required fluence [4]. Each run starts and ends with the 'Home' position and scans horizontally over the area with 0.5 cm vertical displacement between horizontal scans, see Figure 5.

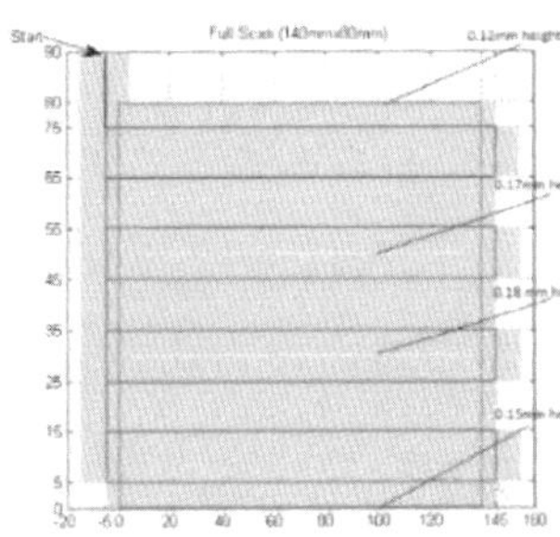

(a)

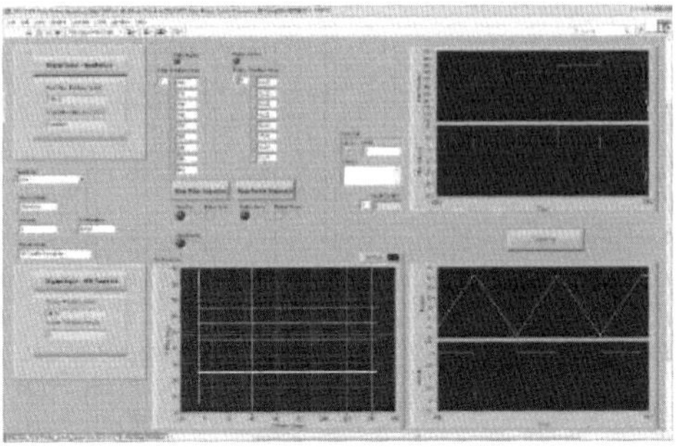

(b)

Figure 5. (a) Scanning path profile for a sample of 40x128 mm^2 with 10mm square beam spot, (b) Scanning path profile of the radiation dosing software [4].

5.3. *Continuous point by point fluence control*

Given the constraints of the horizontal and vertical speed of travel of the ATLAS thermal camber mounted to pre-configured XY-axis Cartesian robot system, it was still possible to achieve some irradiation which approximates to what the LHCb experimental VELO sensor required. The robot positioned the cold box to centre the beam on the required points, pause for corresponding time, and then continue to the next point it straightforward at lower intensity, see Figure 6.

x	y	r	Fluence
1	-1	0.71	1.96E+15
2	-1	1.58	4.25E+14
3	-1	2.55	1.71E+14
4	-1	3.54	9.21E+13
1	0.5	0.71	1.96E+15
2	0.5	1.58	4.25E+14
3	0.5	2.55	1.71E+14
4	0.5	3.54	9.21E+13
1	1.5	1.58	4.25E+14
2	1.5	2.12	2.43E+14
3	1.5	2.92	1.32E+14
4	1.5	3.81	8.00E+13
1	2.5	2.55	1.71E+14
2	2.5	2.92	1.32E+14
3	2.5	3.54	9.21E+13
4	2.5	4.30	6.34E+13
1	3.5	3.54	9.21E+13
2	3.5	3.81	8.00E+13
3	3.5	4.30	6.34E+13
4	3.5	4.95	4.86E+13

(a)

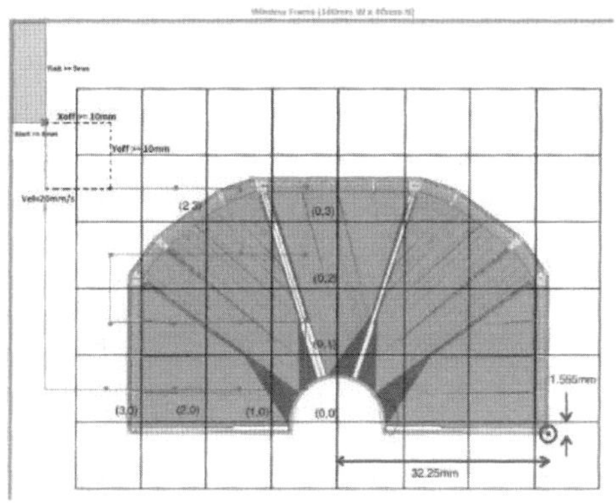

(b)

Figure 6. Birmingham-LHCb – VELO sample: (a) Fluence Map, (b) Irradiation path.

5.4. *Discrete point by point fluence control*

An irradiation run performed for The University of Oxford, ATLAS Group, involved irradiating at two points only (one immediately after the other one). Since the total cooling time will be reduced therefore, sensors will be dispatched on the same day of irradiation because the cooling time will be only of couple of hours. As the beam was much larger than the sample to be irradiated (beam spot area 10 mm^2, sample area less than 5 mm^2), there was no need to scan continuously - just to irradiate at one point for a certain period of time and then move the cold box to the last point location and wait until the dose has beam achieved.

Table 1. Irradiation Exposure times.

	Irradiation time	Chip Readout ON	Chip Readout OFF
Home Position (start)			
Position 1	1000 s		
Position 2	1000 s		
Homing Position		1800 s Cooling down	
Position 1	1000 s		
Position 2	1000 s		
Homing Position			1200 s Cooling down
Position 1	1000 s		
Position 2	1000 s		
Homing Position		1800 s Cooling down	
Position 1	1000 s		
Position 2	1000 s		
Homing Position		1800 s Cooling down	
Home Position (end)			

The beam current was set at 20 ηA, the tracking speed of the robot was set at 2 mm/s. The latter parameter value was not relevant as fixed points were irradiated only. The exposure times were manually tracked using 2 timers, see Table 1. The total irradiation time was thus 4000s per spot (10 mm^2). The beam current was repeatedly checked between runs as the aluminum shielding (plus boards etc.) prevented accurate reading of beam current during irradiation using the faraday cup. The temperature was stable at -48C throughout the irradiation and no issues were seen with the Norhof or the robotic system. Deciding how to un-mount the samples was tricky as they might have got frost up during transfer to the hot lab freezer and they were reasonably active at ~100 μS/hour.

Conclusion

Various irradiations have been performed for different UK groups (ATLAS, LHCb and Oxford). The radiation dosing software control of the Pre-configured XY-axis Cartesian robot system for the Atlas Scanning Facility in Birmingham has proven to be reconfigurable to each type of irradiation (continuous, point by point, timed controlled) thanks to the flexibility of the LabVIEW irradiation control program. Improvements on the robotic scanning system design have in principle overcome the overheating of the samples irradiated at higher fluence.

Acknowledgments

We are very grateful to the Birmingham ATLAS Group. We also thank to colleagues from the Sheffield ATLAS Group in our home university for their invaluable support.

References

1. P. Dervan, R. French, P. Hodgson, H. Marin-Reyes, K. Parker, J. Wilson, M. Baca, Upgrade to the Birmingham Irradiation Facility, **NIM A** In Press (2015).
2. H. Marin-Reyes, R. French, P. Hodgson, K. Parker, J. Wilson, P. Dervan, Pre-configured xy-axis Cartesian robot system for a new ATLAS scanning facility, Mobile Service Robotics: Proceedings of CLAWAR 2014,Poznan, Poland, World Scientific Publishing Co. (2014).
3. P. Dervan, R. French, P. Hodgson, H. Marin-Reyes, K. Parker, J. Wilson, M. Baca, Scanning facility to irradiate mechanical structures for the LHC upgrade programme, Proceedings of TIPP2014 conference, Amsterdam, The Netherlands, PoS - Proceedings of Science, **PoS(TIPP2014)419** (2014).
4. P. Dervan, R. French, P. Hodgson, H. Marin-Reyes, J. Wilson, The Birmingham Irradiation Facility, **NIM A 730** (2013).

AN AUTONOMOUS HAND-EYE CALIBRATION METHOD

YULI CHENG, YIFENG ZHANG, PENG WANG, RONG XIONG*

Department of Control Science and Engineering, Zhejiang University, Hangzhou 310027, P. R. China
**E-mail: rxiong@iipc.zju.edu.cn*

Hand-eye calibration to determine the relative pose between the robot and camera is a crucial problem for many tasks where the robot has to physically interact with the environment. This paper introduces an autonomous method for the calibration of eye-to-hand system, which is aimed at making the hand-eye calibration easy to operate in the industrial application. In this method, motion planning based on imaging model of camera is executed to make sure the calibration object mounted on robot end-effector in the range of camera vision. The poses of calibration object and robot end-effector are gathered in real time to calibrate and update the hand-eye relation until the result is convergent. Experiments have been conducted and experimental results show the effectiveness and accuracy of the proposed method.

Keywords: Autonomous hand-eye calibration;Motion space planning;Imaging model of camera

1. Introduction

Hand-eye calibration, which is aimed at determining the relative position and orientation between the robot and camera, is an important problem in robot applications. It has been widely used in visual servo, vision measurement, etc.

Mainly, there are two types of camera utilization for robot hand-eye system: eye-in-hand and eye-to-hand configuration. The camera is said as eye-in-hand when rigidly mounted on the end-effector. Conversely, the camera is said as eye-to-hand when it is mounted outside the robot and observes the robot and the workspace. This article focuses on the calibration problem of eye-to-hand system. Robot hand motion planning is performed in robot base coordinate system. So the relative position and orientation between camera and robot base must be known before we can guide robot hand motion through vision system. The most intuitive method is by means of measuring equipment, such as three coordinate measuring instrument. Though principle is direct, the measuring equipment is often expensive and the calibration process is time-consuming and needs manual operation.

In the modern intelligent factory, we want the idle robots to move to the busy assembly line to give assistance to improve the utilization rate of production equipment. In this application, we hope that the calibration is quick and executed automatically. To reduce the cost, the calibration tools would better be cheap. The classic solution is to employ a priori defined object, called the 'calibration object'.[1] Algebraically, the problem can be defined as a linear homogenous equation in the unknown pose matrix X, namely $AX = XB$, where A, B and X are all 4×4 homogeneous transformation matrices.

To solve the equation, the classic algorithms are proposed.[2–7] Such algorithms consist of two stages: initially the rotational axis of X is solved linearly according to the rotation axes of A and B, and the result is then used to solve the translation vector of X by a linear approach. While this kind of methods yield uncomplex and fast formulations, the rotational part estimation errors propagate to the translational part. Shiu and Ahmad[2] use least squares fitting of rotation, then translation. Tsai and Lenz[3] using angle-axis representation get closed-form solution. Chou and Kamel[5] propose method based on the singular value decomposition(SVD). Zhang and Shiu[9] apply nonlinear optimization for both parts, minimizing the Frobenius norms of homogeneous matrices of transformation errors. For online hand-eye calibration, Andreff et at.[14] propose a method whose derivation includes a new linear formulation of classical hand-eye calibration which is linear and numerically efficient.

The method proposed in this paper is mainly aimed at the automation of calibration for the eye-to-hand system and its contribution is mainly in the following two aspects.

(1) Robot hand motions are planned based on imaging model of camera to make sure the calibration object in the range of camera vision during calibration process.

(2) The whole automatic hand-eye calibration framework is designed to get convergent calibration results.

The remainder of this paper decomposes as follows. Section 2 is the formulation of the calibration problem for the eye-to-hand system. Section 3 describes the whole framework and gives the autonomous hand-eye calibration flow. The motion space planning algorithm based on imaging model of camera is presented in Section 4. Section 5 conducts some experiments on physical platform to validate the proposed method.

2. Problem Formulation

We employ a calibration object mounted on the robot end-effector and calibrate the eye-to-hand system through coordinate system transformation as shown in

Fig. 1, which shows the coordinate system transformations involved in the calibration process. The calibration object mounted on robot end-effector is moved from location i to $i+1$ in front of the camera. The key is that the coordinate frame transformations form a closed loop, and thus we have the following equation.

$$T_{b2c}{}^{(i)} = T_{r2c} \cdot T_{e2r}{}^{(i)} \cdot T_{b2e} \tag{1}$$

$$T_{b2c}{}^{(i+1)} = T_{r2c} \cdot T_{e2r}{}^{(i+1)} \cdot T_{b2e} \tag{2}$$

From Eq. (1) (2), we have the Eq. (3)

$$T_{b2c}{}^{(i+1)} \cdot (T_{b2c}{}^{(i)})^{-1} \cdot T_{r2c} = T_{r2c} \cdot T_{e2r}{}^{(i+1)} \cdot (T_{e2r}{}^{(i)})^{-1} \tag{3}$$

If $T_b{}^{(i)} = T_{b2c}{}^{(i+1)} \cdot (T_{b2c}{}^{(i)})^{-1}$ and $T_e{}^{(i)} = T_{e2r}{}^{(i+1)} \cdot (T_{e2r}{}^{(i)})^{-1}$, we can get $T_b{}^{(i)} \cdot T_{r2c} = T_{r2c} \cdot T_e{}^{(i)}$, which is the equation $AX = XB$, where B is the rigid motion of the robot end-effector and A is the corresponding calibration object motion.

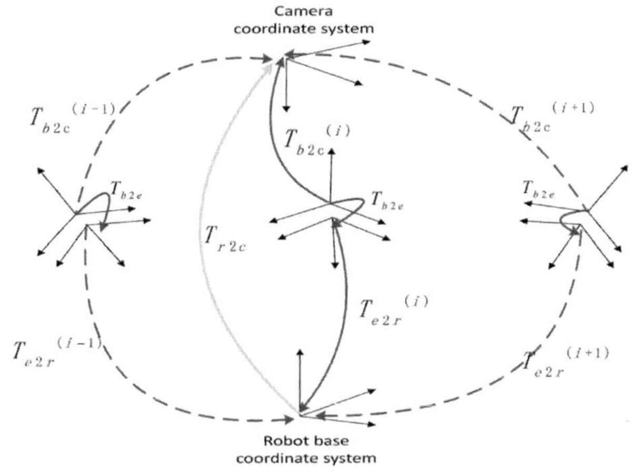

Fig. 1. Coordinate transformations involved in the calibration process.T_{b2c} represents the transformation from calibration object to camera. T_{b2e} represents the transformation from calibration object to end-effector. T_{e2r} represents the transformation from end-effector to robot base. T_{r2c} represents the transformation from robot base to camera.

3. Framework

We focused attention on the automation of calibration for eye-to-hand system. The framework of solution system has been designed, which is shown as Fig. 2. The whole framework composes of the implementation flow and the relevant modules.

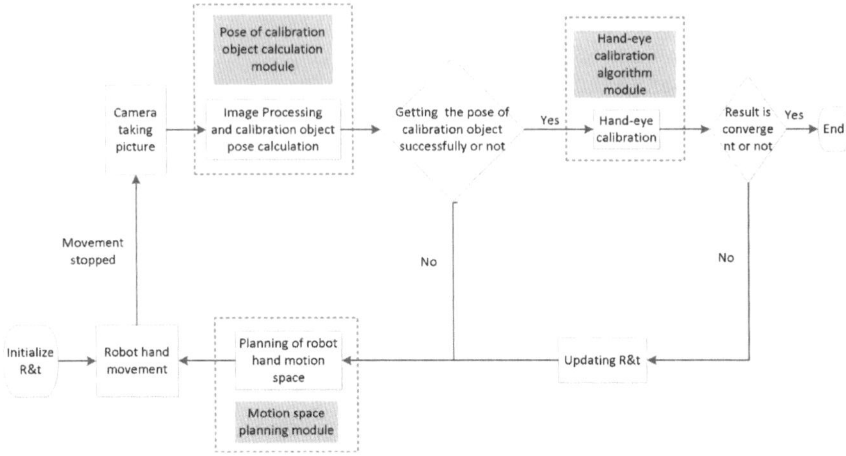

Fig. 2. The framework of autonomous hand-eye calibration system.

Before applying robot motion planning algorithm based on imaging model of camera, we need to know the rough transformational relation between camera and robot base coordinate. We set the rotation matrix from robot base coordinate system to camera coordinate system as R_{r2c}, and translation matrix as t_{r2c}. Whichever method adopted, at least two motions with non-parallel rotation axes must be used to determine the hand-eye transformation.

The details about the autonomous calibration system are described as follows. At first, we need to initialize the R_{r2c} and t_{r2c}. Manually operate robot to move the calibration object to the center of the field of vision. And then robot end-effector automatically performs two motions with non-parallel rotation axes. Two motions yield three stations of end-effector and calibration object. We can use the three groups of data to get two $AX = XB$ equations. To solve the equation, we get the ${R_{r2c}}^{(0)}$ and ${t_{r2c}}^{(0)}$ which are initial value of hand-eye relation.

Because of the noise existing in the process of calibration, there is a distance between initial value and ground truth. We just apply ${R_{r2c}}^{(0)}$ and ${t_{r2c}}^{(0)}$ to establishing the initial relation between robot base and camera. Then we can perform robot motion space planning in camera coordinate system though this is not so accurate. The function of motion space planning is realized by the motion space planning module, which is the key to improve the efficiency of calibration. Combined with the imaging model of camera, the method proposed in this paper gives a robot hand motion space planning solution. This method can limit the robot hand motion in the camera's view.

When the robot stops moving, we take picture of the still calibration object.

According to the current image, the position and orientation of calibration object can be calculated, which is realized by pose of calibration object calculation module. The process includes finding chessboard corners and refers the extrinsic camera parameter calibration method.[1] In such settings, we can get corresponding relation of pixel coordinate and 3D space coordinate of chessboard corners. With the known intrinsic camera parameter, we have the pose of calibration object.

If the pose of calibration object can be got successfully, we add the new pose of calibration object and the current pose of robot end-effector got from robot controller to the hand-eye calibration computing data, and perform calibration algorithm, updating $R_{r2c}{}^{(k)}$ and $t_{r2c}{}^{(k)}$. The module to realize this function is called hand-eye calibration algorithm module. In order to ensure the real-time of automatic online hand-eye calibration, we adopt the linear hand-eye calibration method in Ref.[3] instead of the time consuming iterative method.

Comparing $R_{r2c}{}^{(k)}$ and $t_{r2c}{}^{(k)}$ with the history calibration results, if the result has converged to set threshold or maximum number of calibration has been reached, we end the calibration. Otherwise, we set $R_{r2c}{}^{(k)}$ and $t_{r2c}{}^{(k)}$ as the new relation between robot base and camera and perform the next robot motion planning and calibration process.

In the process of calibration, $R_{r2c}{}^{(k)}$ and $t_{r2c}{}^{(k)}$ has been update, and more and more close to the ground truth. Except the manual operation in the initialization stage, the whole hand-eye calibration process is executed automatically. According to the requirement of calibration precision, the convergence threshold can be adjusted to balance the execution time and accuracy.

4. Relevant Algorithms

4.1. *Motion space planning algorithm based on imaging model*

In the past hand-eye calibration work, it can not ensure that the calibration object mounted rigidly on the robot gripper always appears in the range of vision during calibration process through robot motion planning, which results in the image used for camera extrinsic parameter calibration is sometimes invalid. This would require the robot hand move more extra stations or manual operation, reducing the efficiency of calibration.

To constrain the motion of calibration object mounted on robot end-effector in the range of camera view, we need to have a good knowledge of the imaging model of camera. The imaging model of camera is a so-called pinhole camera model. In this model, a scene view is formed by projecting 3D points into the image plane using a perspective transformation. The function of this model[17] can be written as Eq. (4)

$$s\begin{bmatrix} u \\ v \\ 1 \end{bmatrix} = \begin{bmatrix} f_x & 0 & c_x \\ 0 & f_y & c_y \\ 0 & 0 & 1 \end{bmatrix} \begin{bmatrix} x \\ y \\ z \end{bmatrix} \quad (4)$$

where (u,v) are the coordinates of the projection point in pixels, (x,y,z) are the coordinates of 3D point in the camera coordinate system. (c_x, c_y) is a principal point that is usually at the image center. f_x, f_y are the focal lengths expressed in pixel units. The upper boundary of u is $u_{\max}$, and the lower boundary is $u_{\min}$. The upper boundary of v is $v_{\max}$, and the lower boundary is $v_{\min}$. Therefore, the model of the range of camera vision can be expressed as Formula (5).

$$\begin{cases} f_x \cdot x - (u_{\max} - c_x) \cdot z < 0 \\ f_x \cdot x - (u_{\min} - c_x) \cdot z > 0 \\ f_y \cdot y - (v_{\max} - c_y) \cdot z < 0 \\ f_y \cdot y - (v_{\min} - c_y) \cdot z > 0 \end{cases} \quad (5)$$

This is a pyramid shape space and the optic center is the top point of the pyramid which is also the origin of camera coordinate system as shown in Fig. 3. By measuring, we can get the coordinates of boundary corner point in the end-effector coordinate system as q_1, q_2, q_3, q_4. Before ith motion executed, motion planning has been done and the transformation between end-effector coordinate system and robot base coordinate system is written as ${T_{r2c}}^{(k)} = \begin{bmatrix} {R_{r2c}}^{(k)} & {t_{r2c}}^{(k)} \\ 0 & 1 \end{bmatrix}$.

And then we can get the coordinates of boundary corner point in the robot base coordinate system as p_1, p_2, p_3, p_4, where $p_i = T_{r2c} \cdot T_{e2r} \cdot q_i$. Then judge whether p_1, p_2, p_3, p_4 all satisfy inequality constrain Formula (5). If the motion state satisfies all the conditions, the motion will be executed. Otherwise, give up this motion and plan next motion state.

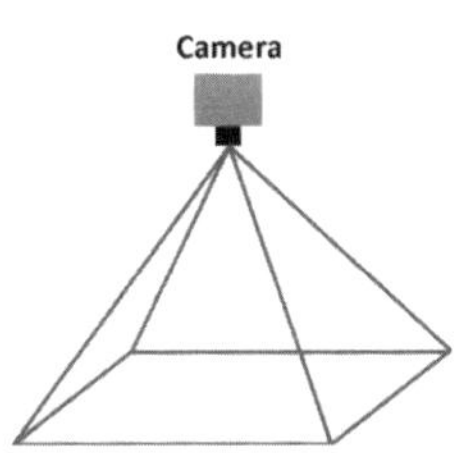

Fig. 3. Imaging model of camera.

We constrain the motion of calibration object in the range of camera vision as far as possible to improve the probability of getting valid calibration images.

Because the process of motion execution is more time-consuming than computing process, it can save time of the whole calibration process and improve the efficiency.

4.2. *Detailed hand-eye calibration algorithm*

Tsai[3] proposed a method to solve the equation $AX = XB$. The method is aimed at the eye-in-hand calibration problem. The method presented in this paper is for eye-to-hand system. In the section 2, we have deduced the mathematical model of eye-to-hand problem and found that it is still can be expressed as $AX = XB$, which is also can be described as the folllowing two equations.

$$R_b{}^{(i)} \cdot R_{r2c} = R_{r2c} \cdot R_e{}^{(i)} \tag{6}$$

$$R_b \cdot t_{r2c} + t_b{}^{(i)} = R_{r2c} \cdot t_e{}^{(i)} + t_{r2c} \tag{7}$$

Rotation matrix R can be represented as a rotation around axis $\left[n_1 \; n_2 \; n_3 \right]^T$, and the rotation angle is θ. We can use a modified version of Rodrigues formula,[3] and define P_r as $P_r = \frac{1}{2}\sin\theta(n_1, n_2, n_3)^t$. $Skew(P_r)$ is the skew-symmetric matrix generated by the 3D vector P_r. The transformation between R and P_r is expressed as Eq. (8).

$$R = (1 - \frac{|P_r|^2}{2})I + \frac{1}{2}(P_r P_r^t + \sqrt{4 - |P_r|^2} \cdot Skew(P_r)) \tag{8}$$

And Eq. (6) is equivalent to the Eq. (9).

$$Skew(P_b{}^{(i)} + P_e{}^{(i)}) P_{r2c}{}' = P_e{}^{(i)} - P_b{}^{(i)} \tag{9}$$

where

$$P_{r2c} = \frac{2P_{r2c}{}'}{\sqrt{1 + |P_{r2c}{}'|^2}} \tag{10}$$

Since $Skew(P_b{}^{(i)} + P_e{}^{(i)})$ is always singular, it takes at least two pairs of stations to solve for a unique solution for $P_{r2c}{}'$. This is a linear equation and when there are n pairs of motion stations, we can get n such equations and solving for $P_{r2c}{}'$ using linear least squares technique.

Combine Eq. (8) (10) and $P_{r2c}{}'$, we can get R_{r2c}. Then we use R_{r2c} to solve t_{r2c} in Eq. (7) with linear least squares technique in the same way. According to the method described above, we complete the calibration for eye-to-hand system using n pairs of stations. The whole solving process is linear and not iterative.

5. Experiments

To validate the method of autonomous hand-eye calibration designed in this paper, we conduct experiments on the physical platform. We demonstrate the experiment by an ABB robot (ABB IRB-120) and a camera(Flea3 FL3-FW-20S4C) whose resolution is 1624*1224. The experimental facility as shown in Fig. 4(a), includes a camera, an industrial robot, a calibration object and a PC to do processing. The calibration object is mounted on the robot end-effector and the camera is mounted outside the robot composing the eye-to-hand system. The calibration of camera intrinsic parameter has been done before this experiment. Using the experimental facility described above, we perform the autonomous calibration experiment for the eye-to-hand system.

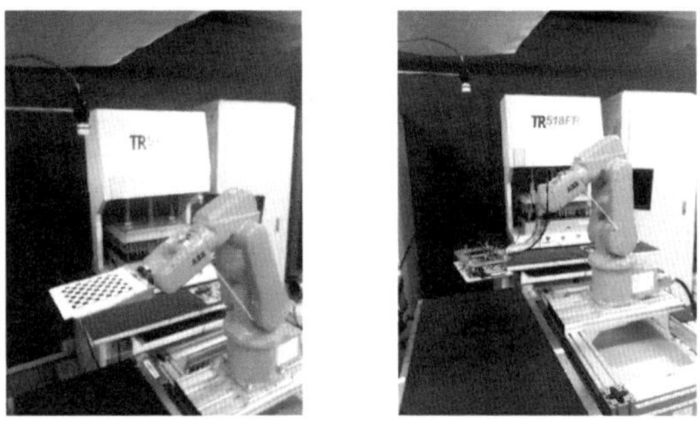

(a) Hand-eye calibration physical platform.

(b) Vision servo grab experiment platform.

Fig. 4. The physical platform.

The transformation relation between robot base and camera measured by three coordinate measuring instrument was used as ground truth to measure experiment errors. In this experiment, the maximum number of stations was set as 64, which means when robot have executed 64 valid motions, the hand-eye calibration program is stopped. As the experiment going on, there are more and more stations participating in the calibration computing. The rotation error and translation error are described as follows:

$$Err(R_{r2c}{}^{(k)}) = \left\| R_{r2c}{}^{(k)} - R_{r2c}{}' \right\|_1, \quad Err\left(t_{r2c}{}^{(k)}\right) = \left\| t_{r2c}{}^{(k)} - t_{r2c}{}' \right\|_1 \tag{11}$$

where $R_{r2c}{}^{(k)}$ and $t_{r2c}{}^{(k)}$ are the calibration results of k stations, $R_{r2c}{}'$ and $t_{r2c}{}'$

are the ground truth. The errors of rotation matrices and translation matrices are shown as Fig. 5.

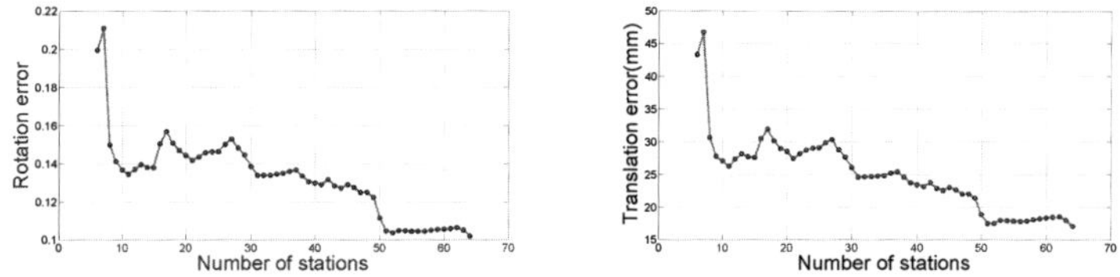

(a) Rotation error as a function of number of stations.

(b) Translation error as a function of number of stations.

Fig. 5. The convergence error curve.

From Fig. 5 we can see that with the increasing of number of stations, calibration error shows a trend of convergence. From Fig. 3, it can be seen that there will be more and more stations during the execution of the autonomous hand-eye calibration solution. Therefore, with the method proposed in this paper, a calibration result convergent to the set threshold can be got. At the same time, it can be seen that sufficient data help to overcome the interference of noise. And through setting convergent threshold or maximum number of stations, we can balance the time consumed and accuracy.

With the calibration result, we conducted the vision servo object grab experiment. And the grabbing goal was the circuit board positioned by the vision system. The experiment demonstrated that the circuit board can be grabbed from the assembly line and placed in the testing instrument successfully as shown in Fig. 4(b).

6. Conclusions

This paper introduced a method of autonomous calibration for the eye-to-hand system. To realize the automation of the whole calibration process, the motion planning based on imaging model of camera has been used to improve the possibility of obtaining valid calibration images. At the same time, a novel calibration flow has been designed to control the start and end of autonomous calibration. As the process going on, there will be more and more stations as the calibration computing data and the calibration result is updating all the time. Experimental results from the real setup show that the calibration error is convergent to the set threshold during the conducting of calibration. However, the setting of the maximum number of stations or the convergence threshold should be chosen by experience.

References

1. Zhang Z, A flexible new technique for camera calibration, in *IEEE Trans. on Pattern Analysis and Machine Intelligence*, 2000, 22(11): 1330-1334.
2. Shiu Y C, Ahmad S, Calibration of wrist-mounted robotic sensors by solving homogeneous transform equations of the form AX= XB, in *IEEE Trans. on Robotics and Automation*, 1989, 5(1): 16-29.
3. Tsai R Y, Lenz R K, A new technique for fully autonomous and efficient 3D robotics hand/eye calibration, in *IEEE Trans. on Robotics and Automation*, 1989, 5(3): 345-358.
4. Zhuang H, Roth Z S, Shiu Y C, et al, Comments on" Calibration of wrist-mounted robotic sensors by solving homogeneous transform equations of the form AX= XB", in *IEEE Trans. on Robotics and Automation*, 1991, 7(6): 877-878.
5. Chou J C K, Kamel M, Finding the position and orientation of a sensor on a robot manipulator using quaternions, in *The international journal of robotics research*, 1991, 10(3): 240-254.
6. Daniilidis K, Hand-eye calibration using dual quaternions, in *The International Journal of Robotics Research*, 1999, 18(3): 286-298.
7. Zuang H, Shiu Y C, A noise-tolerant algorithm for robotic hand-eye calibration with or without sensor orientation measurement, in *IEEE Trans. on Systems, Man and Cybernetics*, 1993, 23(4): 1168-1175.
8. Axelrod B, Huang W H, Improving hand-eye calibration for robotic grasping and manipulation, *Proceedings of Technologies for Practical Robot Applications (TePRA), 2012 IEEE International Conference on*, 2012:121 - 126.
9. Horaud R, Dornaika F, Hand-eye calibration, in *The international journal of robotics research*, 1995, 14(3): 195-210.
10. Heller J, Henrion D, Pajdla T, Hand-eye and robot-world calibration by global polynomial optimization, *Proceedings of Robotics and Automation (ICRA), 2014 IEEE International Conference*, 2014:3157 - 3164.
11. Chen H H. A screw motion approach to uniqueness analysis of head-eye geometry, *Proceedings of Computer Vision and Pattern Recognition*, 1991: 145-151.
12. Zhao Z, Liu Y, Hand-eye calibration based on screw motions, *Proceedings of Pattern Recognition, 2006. ICPR 2006. 18th International Conference*, 2006, 3: 1022-1026.
13. Strobl K H, Hirzinger G, Optimal hand-eye calibration, *Proceedings of Intelligent Robots and Systems, 2006 IEEE/RSJ International Conference*, 2006: 4647-4653.
14. Andreff N, Horaud R, Espiau B, On-line hand-eye calibration, *Proceedings of 3-D Digital Imaging and Modeling, Second International Conference*, 1999: 430-436.
15. Andreff N, Horaud R, Espiau B, Robot hand-eye calibration using structure-from-motion, in *The International Journal of Robotics Research*, 2001, 20(3): 228-248.
16. Angeles J, Soucy G, Ferrie F P, The online solution of the hand-eye problem, in *IEEE Trans. on Robotics and Automation*, 2000, 16(6): 720-731.
17. Bradski G, Kaehler A, Learning OpenCV: Computer vision with the OpenCV library. O'Reilly Media, Inc., 2008.

SLIDING MODE CONTROL – BASED INTERVAL TYPE-2 FUZZY LOGIC CONTROLLER FOR QUADCOPTER UAVs

AHMAD RIYAD FIRDAUS and M. O. TOKHI

Department of Automatic Control and Systems Engineering, The University of Sheffield, The United Kingdom

Quadcopter UAVs have several complex conditions including nonlinearity, verily coupled, and working area that tend to be subject to disturbances and uncertainties. Sliding Mode Control (SMC), which is one of well-known robust control approaches, is believed to be able in dealing with such issues. However, chattering phenomenon in SMC is still an issue that may lead to performance degradation. Interval Type-2 Fuzzy Logic Control which has the capability to accommodate uncertainties of the system, will play a vital role in eliminating this phenomenon. A combining of the two control methods is investigated in this paper, and the results obtained demonstrate good performance of the approach in dealing with nonlinearity, disturbance, and uncertainty as well as eliminating the chattering phenomenon of SMC.

1. Introduction

Quadcopter, which is lifted by 4 rotary blades connected to motors, has a complex system which may lead to setting up inaccurate model. Therefore, previous works have been conducted to address this issue, such as proportional integral derivative (PID) [1], linear quadratic regulator (LQR) [2], model-predictive control (MPC) [3]; and sliding mode control (SMC) [4],[5]. Among these, The SMC scheme shows better performance in terms of robustness, time response, and error tracking.

In addition to the capability in dealing with nonlinearity, SMC can handle any uncertainties and disturbances [6]. However, the chattering around sliding surface that may influence system performance, is still an issue to be solved. This work, Interval Type-2 Fuzzy Logic Control (IT2FLC) coupled with SMC, are proposed to improve the performance of the controller, such as eliminating chattering phenomenon, dealing with nonlinearity, and being insensitive to uncertainties and disturbances. Such hybrid control method has been employed for other systems, such as inverted pendulum [7], [8] and omnidirectional mobile robots [9] with different approach with the proposed method in this paper, and compared with type-1 fuzzy logic controller (T1FLC). They found that IT2FLC perform better T1FLC in dealing with chattering, uncertainty, and disturbance.

2. Quadcopter System Model

A quadcopter as can be seen in Figure 1 is established by two rotating clockwise rotors (2,4), and two rotating counterclockwise rotors (1,3), so that by increasing the rotors speed at the same rates will increase the altitude of the rotorcraft, and vice versa.

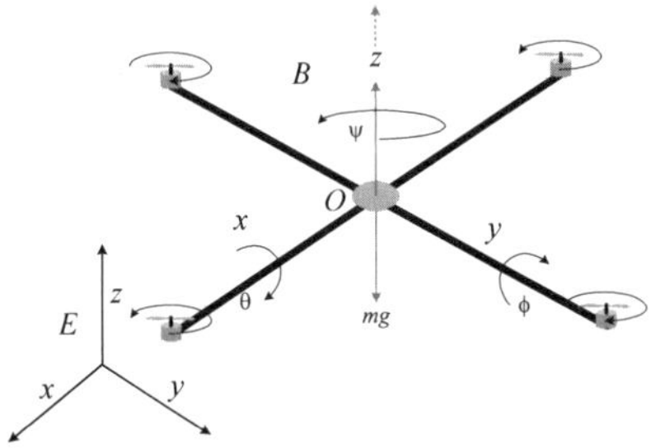

Figure 1. Quadcopter UAVs

By ignoring the total residue of angular velocity of the rotors, the dynamical model of quadcopter can be expressed as follows [5]:

$$\ddot{x} = \frac{1}{m}(c\phi s\theta c\psi + s\phi s\psi)u_1 - \frac{k_1\dot{x}}{m}$$

$$\ddot{y} = \frac{1}{m}(c\phi s\theta s\psi + s\phi c\psi)u_1 - \frac{k_2\dot{y}}{m}$$

$$\ddot{z} = \frac{1}{m}(c\phi c\theta)u_1 - g - \frac{k_3\dot{z}}{m} \tag{1}$$

$$\ddot{\phi} = \dot{\theta}\dot{\psi}\frac{I_y - I_z}{I_x} + \frac{l}{I_x}u_2 - \frac{k_4 l}{I_x}\dot{\phi}$$

$$\ddot{\theta} = \dot{\psi}\dot{\phi}\frac{I_z - I_x}{I_y} + \frac{l}{I_y}u_3 - \frac{k_5 l}{I_y}\dot{\theta}$$

$$\ddot{\psi} = \dot{\phi}\dot{\theta}\frac{I_x - I_y}{I_z} + \frac{1}{I_z}u_4 - \frac{k_6}{I_z}\dot{\psi} \tag{2}$$

where the angles (ϕ, θ, ψ) represent roll, pitch, and yaw, while s, and c represent sine, and cos respectively. The symbols of k_i, m, g, and l express the coefficients of drag (positive constants), the total mass, the gravity acceleration, and the distance of each rotor to center of mass of aircraft respectively, while I_x, I_y, I_z denote the inertia of rotorcraft in x, y, z axes respectively. Furthermore, the control inputs of quatcopter represented by u_1 for total thrust of rotorcraft in z-axis, and u_2, u_3, and u_4 for roll, pitch, and yaw respectively, are expressed as follows:

$$u_1 = (F_1 + F_2 + F_3 + F_4); u_2 = (-F_2 + F_4); u_3 = (-F_1 + F_3); u_4 = d(-F_1 + F_2 - F_3 + F_4)/b$$

where b, and d represent the lift coefficient and the scaling factor of force to moment.

3. Flight Control Design

In this work a hybrid control approach by exploiting the advantages of SMC and IT2FLC is proposed. Referring to the system dynamics of quadcopter in equations (1) and (2), this system is nonlinear and developed by two multiple subsystems including fully-actuated subsystem and under-actuated subsystem [5].

3.1. *Sliding mode control*

3.1.1. *Sliding surface design*

Generally, the sliding surface is represented by as:

$$\sigma(x,t) = Sx(t) \tag{3}$$

where S is a matrix of dimension m x n with positive constant elements. In a scalar form, "a time varying surface $\sigma(t)$" for second order can be defined as [6]:

$$\sigma(t) = se + \dot{e} \tag{4}$$

The variable s is strictly positive constant whereas e and $\dot{e}$ are tracking error and rate of tracking error respectively. Hence, the sliding manifolds for the fully-actuated subsystem (z-axis, and yaw/ϕ) are expressed by the following two equations:

$$\sigma_1 = s_z(z - z_d) + (\dot{z} - \dot{z}_d) \tag{5}$$

$$\sigma_4 = s_\psi(\psi - \psi_d) + (\dot{\psi} - \dot{\psi}_d) \tag{6}$$

For the under-actuated subsystem, the sliding surface design should consider the dependency of each state. In this case, the state x (translation in x-axis) depends on pitch angle (θ), whereas y (translation in y-axis) depends on roll angle (ϕ). Hence, the sliding manifolds for interdependent states can be seen as follows:

$$\sigma_3 = s_x(x - x_d) + (\dot{x} - \dot{x}_d) + s_\theta(\theta - \theta_d) + (\dot{\theta} - \dot{\theta}_d) \tag{7}$$

$$\sigma_2 = s_y(y - y_d) + (\dot{y} - \dot{y}_d) + s_\phi(\phi - \phi_d) + (\dot{\phi} - \dot{\phi}_d) \tag{8}$$

3.1.2. *SMC design*

Consider the general form of a dynamical system as:

$$\dot{x}(t) = f(x,t) + B(x,t)u(t) \tag{9}$$

The control signal (u) is developed by two signals, namely u_{eq} (equivalent control), and u_n (natural control). Therefore, the control law can be obtained by summing the two control signals as:

$$u = u_{eq} + u_n \tag{10}$$

From the equations (3), (9), and (10), the equivalent control (u_{eq}) can be expressed as [6]:

$$u_{eq} = -\left(S\hat{B}(x,t)\right)^{-1} S\hat{f}(x,t) \tag{11}$$

where the symbol (^) represents the best estimate of system parameters. Furthermore, maintaining the sliding condition, the natural control (u_n) can be expressed as [6]:

$$u_n = -k\left(S\hat{B}(x,t)\right)^{-1} sign(\sigma); \quad if\ (SB)\ invertible \tag{12}$$

3.2. *SMC-based IT2FLC design*

Generally, the architecture of SMC-based IT2FLC architecture is shown in Figure 2.

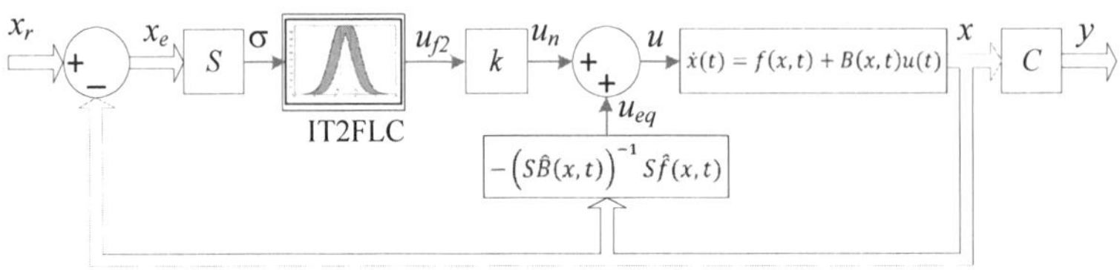

Figure 2. Architecture of IT2FLC

The basic idea of this method is to replace the switching function of natural control (u_n) with fuzzy logic system. Therefore, the proposed method tries to eliminate the chattering phenomenon during sliding mode condition in the sliding surface

3.3. *SMC-based IT2FLC for quadcopter*

From modelling and general form of dynamical system as expressed in equations (1), (2), and (9), the state vector X can be defined as follows:

$$x_1 = \phi;\ x_2 = \dot{\phi};\ x_3 = \theta;\ x_4 = \dot{\theta};\ x_5 = \psi;\ x_6 = \dot{\psi}$$

$$x_7 = z;\ x_8 = \dot{z};\ x_9 = x;\ x_{10} = \dot{x};\ x_{11} = y;\ x_{12} = \dot{y}$$

Therefore, $f(x,t)$ and $B(x,t)$ can be expressed as follows:

$$\dot{x}(t)=\begin{bmatrix} x_2 \\ x_4x_6\left((I_y-I_z)/I_x\right)-k_4lx_2/I_x \\ x_4 \\ x_2x_6\left((I_z-I_x)/I_y\right)-k_5lx_4/I_y \\ x_6 \\ x_2x_4\left(\left(I_x-I_y\right)/I_z\right)-k_6x_6/I_z \\ x_8 \\ -gk_3x_8/m \\ x_{10} \\ -k_1x_{10}/m \\ x_{12} \\ -k_2x_{12}/m \end{bmatrix}+\begin{bmatrix} 0 & 0 & 0 & 0 \\ 0 & l/I_x & 0 & 0 \\ 0 & 0 & 0 & 0 \\ 0 & 0 & l/I_y & 0 \\ 0 & 0 & 0 & 0 \\ 0 & 0 & 0 & C/I_z \\ 0 & 0 & 0 & 0 \\ cx_1cx_3/m & 0 & 0 & 0 \\ 0 & 0 & 0 & 0 \\ \frac{cx_1sx_3cx_5+sx_1sx_5}{m} & 0 & 0 & 0 \\ 0 & 0 & 0 & 0 \\ \frac{cx_1sx_3sx_5+sx_1cx_5}{m} & 0 & 0 & 0 \end{bmatrix}U \tag{13}$$

Furthermore, using equations (5) - (8) the sliding surface matrix is obtained as:

$$\sigma=\begin{bmatrix} 0\,0\,0\,0\,0\,0\,s_z\,1\,0\,0\,0\,0 \\ s_\phi\,1\,0\,0\,0\,0\,0\,0\,0\,0\,s_y\,1 \\ 0\,0\,s_\theta\,1\,0\,0\,0\,0\,s_x\,1\,0\,0 \\ 0\,0\,0\,0\,s_\psi\,1\,0\,0\,0\,0\,0\,0 \end{bmatrix}x+\begin{bmatrix} -s_zx_{7r} \\ -s_\phi x_{1r}-s_yx_{11r} \\ -s_\theta x_{3r}-s_xx_{9r} \\ -s_\psi x_{5r} \end{bmatrix} \tag{14}$$

The membership function of fuzzy input is built by the values of sliding surface (σ) as seen in Figure 3, while the fuzzy output is depicted in Figure 4. Each of these has 7 membership functions, namely: Negative Big (NB), Negative Medium (NM), Negative Small (NS), Zero (Z), Positive Small (PS), Positive Medium (PM), and Positive Big (PB).

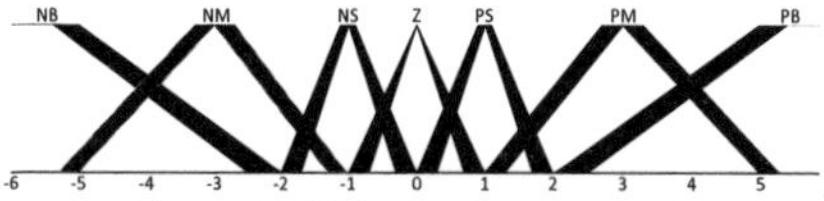

Figure 3. Sliding surface (σ) MFs

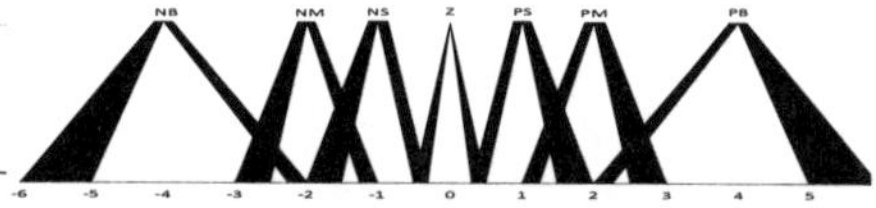
Figure 4. Fuzzy output MFs

Considering the fuzzy input and output, the rules-base to produce the desired natural control (u_n) can be set up as in Table 1.

Table 1. The Rules of IT2FLC-SMC

	σ (Sliding mode)						
	NB	NM	NS	Z	PS	PM	PB
Out$_{(fuzzy2)}$	PB	PM	PS	Z	NS	NM	NB

The natural control is obtained after type reduction and defuzzification process, and the control signal can be expressed as:

$$u_n=ku_{f2} \tag{15}$$

where k is a gain to normalize the output signal, while u_{f2} represents the crisp values obtained from IT2FLC. Thus, using equations (11) and (15), the overall control signal of SMC can be expressed as:

$$u = -\left(S\hat{B}(x,t)\right)^{-1} S\hat{f}(x,t) + ku_{f2} \tag{16}$$

4. Results and Discussion

In this sections a set of simulation exercises are presented to examine and demonstrate the performance of proposed control approach in controlling the altitude of the quadcopter. Settling times, percentages of overshoot, and steady-state errors are taken into account to evaluate the performance of the controller. Parameters mismatch between estimation and real plant, and external disturbance such as e.g. air drag have also been incorporated into the model to demonstrate the robustness of the proposed controller.

Table 2. Quadcopter Parameters

Variables	Values		Units
	Estimation	Plant	
m	1.1	1.1	Kg
l	0.21	0.21	m
I_x=I_y	1.22	1.22	Ns^2/rad
I_z	2.2	2.2	Ns^2/rad
k_i (i=1,2,3)	0.1	0.1	Ns/m
k_i (i=4,5,6)	0.12	0.12	Ns/m
g	9.81	9.81	m/s^2

The quadcopter parameters used in the simulation are shown in Table 2. Suitable values of sliding surface parameters (s) and gain (k) selected for desired performance of quadcopter were $s_x = 0.9$; $s_y = 0.9$; $s_z = 10$; $s_\phi = 4.5$; $s_\theta = 4.5$; $s_\psi = 10$; and $k = 70$. The simulation results obtained are shown in Figures 5–8.

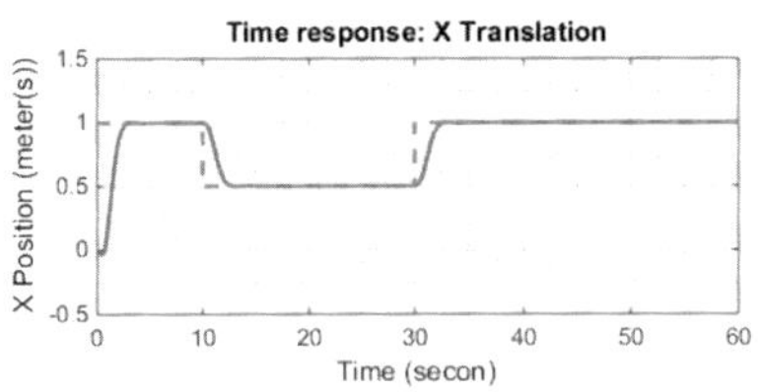

Figure 5. x-axis reponse

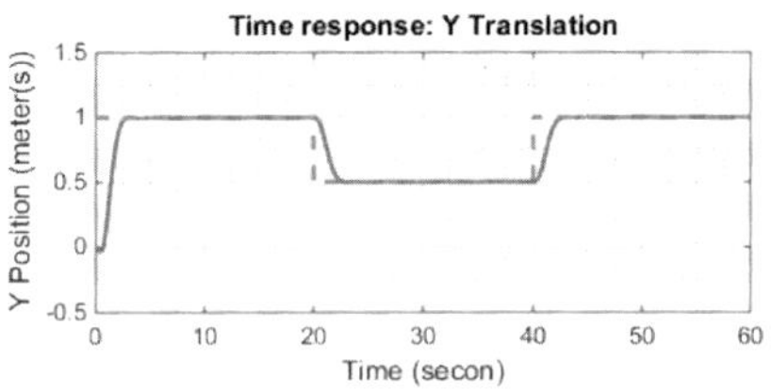

Figure 6. y-axis response

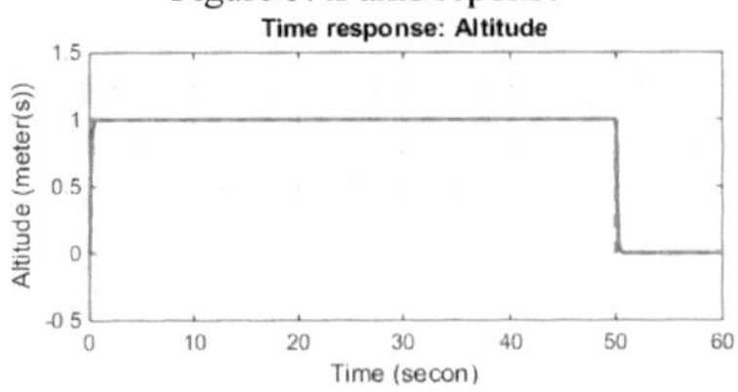

Figure 7. z-axis response

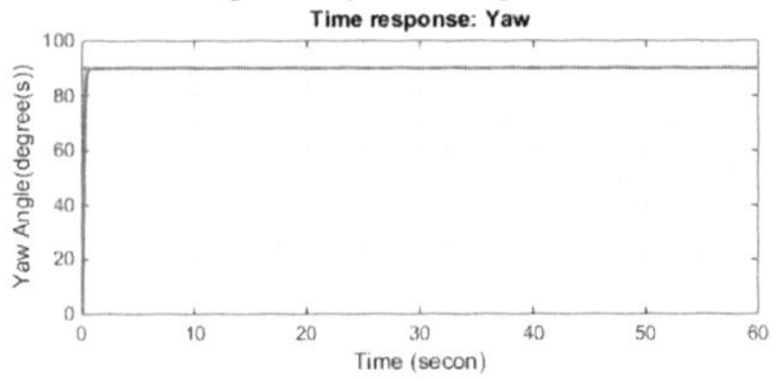

Figure 8. Yaw angle response

The figures show that the flight needed 2.74 seconds with 0.55% overshoot and near 0 steady-state error to reach a distance of 1 meter in x-axis, while in y-axis the flight took around 2.80 seconds with 0.45% overshoot and almost 0 steady-state error. Different achievement was demonstrated by the flight to track the reference in z-axis and yaw movements. The rotorcraft took 0.82 second to reach 1 meter altitude with no overshoot and steady-state error, while in yaw movement, it spent around 0.98 second to rotate 90° in horizontal plane with no overshoot and steady-state error. Moreover, changing the reference did not reduce the performance of the aircraft. In other words, the system states converged to the reference values in every condition. Refering to the figures, it is noted that there was no chattering around the surface in the sliding condition so that the trajectory can track the reference smoothly. This further demonstrated capability of the control method in eliminating the chattering phenomenon in the SMC scheme.

Although the air drag as an external disturbance has been incorporated in the model, simulation has accommodated parameters mismatch of the systems, and possible changes of inertia moment of quadcopter as seen in Table 3.

Table 3. Quadcopter Parameters

Variables	Values		Units
	Estimation	Plant	
I_x	1.22	**1.4**	Ns²/rad
I_y	1.22	**1**	Ns²/rad
I_z	2.2	**2**	Ns²/rad
k_i (i=1,2,3)	0.1	**0.15**	Ns/m
k_i (i=4,5,6)	0.12	**0.15**	Ns/m

According to the results shown in Figures 9 – 12, there were no noticable differences in the system responses. This means that the controller still maintained stability of the aircraft even in disturbed and uncertain condition. In summary, the natural control (u_n) produced by IT2FLC can keep the trajectory on the surface, so that the system becomes insensitive and robust in the presence of disturbances and uncertainties.

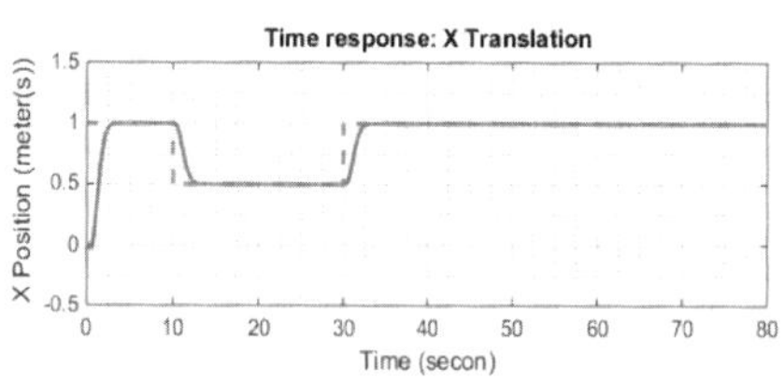

Figure 9. x-axis response

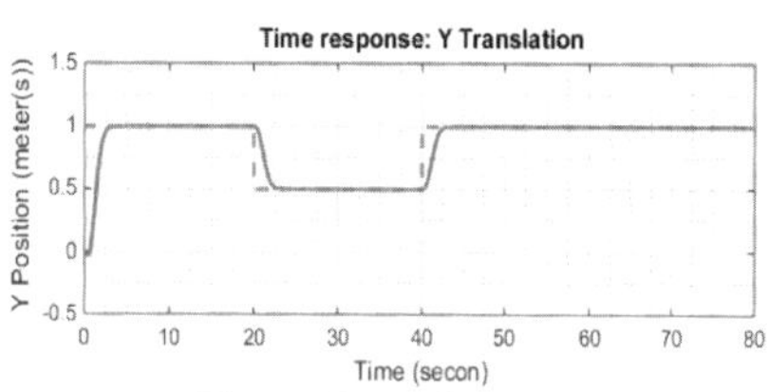

Figure 10. y-axis response

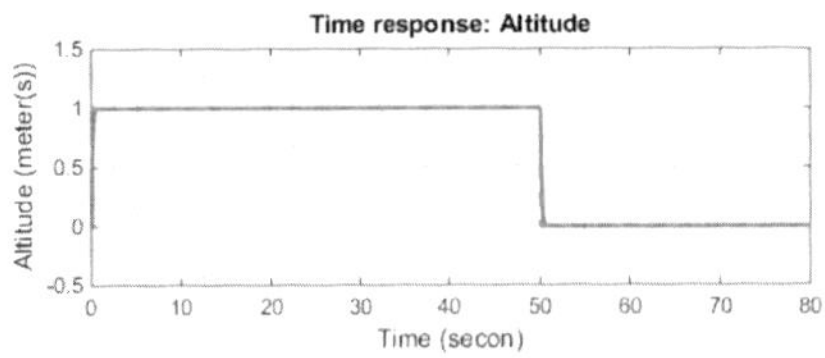

Figure 11. z-axis response

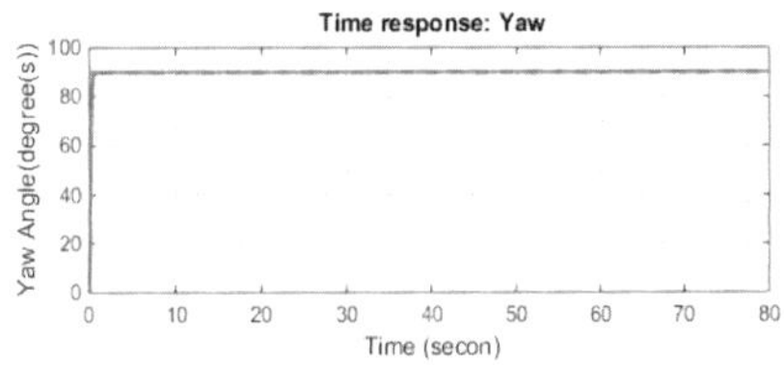

Figure 12. Yaw angle response

Finally, the proposed method was also compared with type-1 fuzzy logic controller (T1FLC) in term of eliminating the chattering phenomenon of SMC. T1FLC still cannot handle chattering phenomenon by appearing the oscillation in the control signals with the amplitude around 0.3, 0.3, and 7.65 for **x,y,** and **z** axes movement respectively.

5. Conclusion

A control approach incorporates design of sliding mode control for both fully and under-actuated subsystems, and the IT2FLC for eliminating the chattering as well as improving the quadcopter performance. Parameters mismatch and external disturbance have been incorporated into the system for robustness tests of the proposed approach. Comparative study has been conducted also between IT2FLC and T1FLC to address the issue in SMC. From the results achieved it can be concluded that: (1) SMC scheme can give a robustness method to deal with disturbances and nonlinearity by converging the states to the reference values with small overshoot and steady-state errors even if the references change suddenly, (2) In addition to dealing with parameters uncertainty, IT2FLC plays an important role in keeping the states trajectory in the sliding surface as well as eliminating the chattering around the surface, (3) IT2FLC outperform T1FLC in term of eliminating chattering phenomenon of SMC.

Finally, this proposed control system can be a promising method to be applied in real time application of quadcopter. Therefore, further research will focus on online tuning of the controller parameters, and experimental validations of the approach.

References

[1] D. Cabecinhas, R. Naldi, L. Marconi, C. Silvestre, and R. Cunha, "Robust Take-Off for a Quadrotor Vehicle," *IEEE Trans. Robot.*, vol. 28, no. 3, pp. 734–742, Jun. 2012.

[2] V. G. Adir and A. M. Stoica, "Integral LQR Control of a Star-Shaped Octorotor," *Incas Bull.*, vol. 4, no. 2, pp. 3–18, Jun. 2012.

[3] K. Alexis, G. Nikolakopoulos, and A. Tzes, "Model predictive quadrotor control: attitude, altitude and position experimental studies," *IET Control Theory Appl.*, vol. 6, no. 12, pp. 1812–1827, Aug. 2012.

[4] R. Zhang, Q. Quan, and K.-Y. Cai, "Attitude control of a quadrotor aircraft subject to a class of time-varying disturbances," *IET Control Theory Appl.*, vol. 5, no. 9, pp. 1140–1146, Jun. 2011.

[5] E.-H. Zheng, J.-J. Xiong, and J.-L. Luo, "Second order sliding mode control for a quadrotor UAV.," *ISA Trans.*, vol. 53, no. 4, pp. 1350–6, Jul. 2014.

[6] A. R. Firdaus and A. S. Rahman, "Genetic Algorithm of Sliding Mode Control Design for Manipulator Robot," *TELKOMNIKA*, vol. 10, no. 4, pp. 645–654, 2012.

[7] M. Y. Hsiao, T. H. S. Li, J. Z. Lee, C. H. Chao, and S. H. Tsai, "Design of interval type-2 fuzzy sliding-mode controller," *Information. Sciences*, vol. 178, pp. 1696–1716, 2008.

[8] Mardlijah, M. Abdillah, A. Jazidie, B. Widodo, and A. Santoso, "Performance enhancement of inverted pendulum system by using type 2 fuzzy sliding mode control (T2FSMC)," *Proc. 2011 Int. Conf. Electr. Eng. Informatics*, pp. 1–6, July 2011.

[9] M.Y. Hsiao and C.T. Wang, "Sliding-Mode Controller Design for Omnidirectional Mobile Robots," *International Conference on Advanced Robotics and Intelligent Systems*, pp. 80–85, May 2013.

ACQUISITION SLOPE SURFACE WALKING FOR HUMANOIDS VIA TRANSFER LEARNING

YI WANG, XIAOQIANG HAN, ZHAN LIU, DINGSHENG LUO* and XIHONG WU

Key Lab of Machine Perception (Ministry of Education)
Speech and Hearing Research Center, School of EECS, Peking University
Beijing, 100871, China
**E-mail: dsluo@cis.pku.edu.cn*

Reinforcement Learning (RL) is commonly used in learning new skills or adapting new situations for humanoid robots as task or environment changes. However, new skills acquisition through RL usually starts from a state of tabula rasa, which is a time-exhausted procedure and can not be suitable for real physical robot, especially when task is a complicated one. Thus, with the emphasis on skills learning efficiency, utilizing past knowledge or experience instead of from a tabula rasa becomes a hot topic. Knowledge transfer learning with RL has exhibited low sample complexity and then been heavily focused. In this research, underling the knowledge transfer learning on autonomous robot, the problem of autonomously modeling inter-task mapping between source task and target task is addressed, where a three-way Restricted Boltzmann Machine (RBM) is employed. To further decrease the sample complexity of transfer leaning on humanoid robots, biased sampling technique is proposed instead of random sampling. To evaluate the performance of the contribution of this research, experiments are performed on a physical robot, PKU-HR5.1, with task domain of bipedal walking on both flat and slope surfaces. Experimental results demonstrate the effectiveness and efficiency of the proposed approach.

Keywords: robot; acquire skills; RBM; TL; biased sampling.

1. Introduction

Learning with the interactions of environment directly on physical humanoid robot, Reinforcement Learning (RL) is an appealing paradigm[1].However, learning directly on humanoid robot is sample complexity sensitive due to the fact that the robot's dynamics are often not constant for motors' wear and tear. Thus, though RL is well defined and theoretically elegant, there are only a few successes on real robot with simply tabula rasa RL, such as [2][3]. Most are trained in simulation and then with the eventual learned behaviors (usually need an adaptation process) run on a physical robot[4][5]. Within RL framework, knowledge transfer on RL exhibits low sample complexity which is suitable for learning various skills directly on real robots.

The concept referring to reuse knowledge acquired from past learning is the major idea behind Transfer Learning (TL) in Reinforcement Learning[6], which further involves the learning of new skills related to new tasks in terms of past knowledge. The task learned previously is called the source task and the new task is called the target task. Transfer Learning in RL would speed up and even improve asymptotic performance of RL by transferring the knowledge from the source task to the target task. This would also reduce the RL's sample complexity for autonomous robot new skill acquisition. But TL need an explicit task mapping.Previous works assume the mappings are human-supplied for transfer in RL[7]. Also, some works rely on strong assumptions on certain similarity between task spaces, such as [8]. To fully achieve the capability of new skill acquiring autonomously (without the prior of mappings) under unexpected situations, in this research, we focus on how to learn the relationships between source and target tasks effectively. Considering that three-way RBM is a powerful but general framework that can be used to model similarity and describe an abstract common space for different tasks[9], it is employed for transfer knowledge between different MDP for RL.

Before training the inter-task mapping, samples of both source and target tasks should be collected, which is usually completed based on the criterion that both source and target tasks spaces are sampled randomly[7]. This would avoid model the mapping for only certain narrow areas of both source and target spaces. However, the random unbiased sampling would easily lead to high sample complexity, and encounter the curse of dimensionality problem, this may bring a disaster when learning is performed on physical robots. In order to let the mapping be tractable on real robots, we adopt a biased sampling technique that prefers sampling from the space near the optimal policy. Further, for the tasks of slope walking skill transferring, a phased sampling strategy is proposed based on the idea of biased sampling.

Three-way RBM that extend the topology of standard RBM has more powerful representation in capturing correlations between tasks, But the standard Contrastive Divergence learning [10] becomes intractable due to the problem of input-independent [11]. To address this issue, we use our new approximate inference method to train Three-way RBM, named Cyclic Contrastive Divergence (CCD)[12].

The remainder of the paper is organized as follows. In Section II, the transfer learning with main of our contributions including autonomous mapping learning, biased sampling are detailed. In Section III, experiments on PKU-HR5.1 real robot are performed, and results are discussed. In last section, conclusions of this research are given.

2. Transfer Learning via RBMS in RL

2.1. *Problem foundation*

Following that Reinforcement Learning is conventionally formalized as the Markov Decision Process (MDP), the task spaces of source and the target are also described as MDPs. Namely, source task $\mathcal{M}_1$ is $\mathcal{M}_1 =< S_1, A_1, \mathcal{T}_1, R_1, \gamma_1 >$ and target task $\mathcal{M}_2$ is $\mathcal{M}_2 =< S_2, A_2, \mathcal{T}_2, R_2, \gamma_2 >$, where for all $i \in \{1, 2\}$, S_i is either the continuous or discrete state space reached by the agent, A_i is the action space in which the agent can execute, and $\mathcal{T}_i$, R_i and γ_i correspond to the transition probability function, reward function and discount factor. As two tasks are different, the constituents of their MDPs vary respectively. More specifically, not only state or action space is different between source and target tasks, but also transition probability or reward can be unequal. Denote $\Omega_{\mathcal{M}}$ as a probability distribution over the space of task M whose formulation is MDP as previously noted. Therefore, the distribution $\Omega_{\mathcal{M}_1}$ of source task $\mathcal{M}_1$ and $\Omega_{\mathcal{M}_2}$ of target task $\mathcal{M}_2$ may be unequal due to the disparity between tasks. In orther words, they do not exhibit the same distribution, $\Omega_{\mathcal{M}_1} \neq \Omega_{\mathcal{M}_2}$. To transfer the knowledge autonomously, we denote the joint probability $p(\mathcal{V}_{\mathcal{M}_1}, \mathcal{V}_{\mathcal{M}_1})$ over $< \Omega_{\mathcal{M}_1}, \Omega_{\mathcal{M}_2} >$ which resembles the common supervised learning setting where training samples are drawn form a given distribution. Thus, a transfer learning algorithm takes the knowledge from source task, and by a functional mapping transfers these knowledge to target task.

2.2. *Inter-task modeling on three-way RBMs*

The primary concern of inter-task modeling is to develop a general and robust joint distribution model of inter-task autonomously. Restricted Boltzmann Machine, a probabilistic graphical model, describes probability distribution via a bipartite graph structure, which not only leads it easy in developing, understanding, and motivating probabilistic model, but also relaxes complex computations of marginal probability. Also, three-way RBMs can automatically build an inter-task mapping using source task and target task samples, because it can discover the latent similarities between the tasks, implicitly encoding an inter-task mapping. In particular, the multiplicative interactions of three-way RBM can let one task gate the parameters of representation model on another task. This means the hidden variables can encode the product of source task and target task.

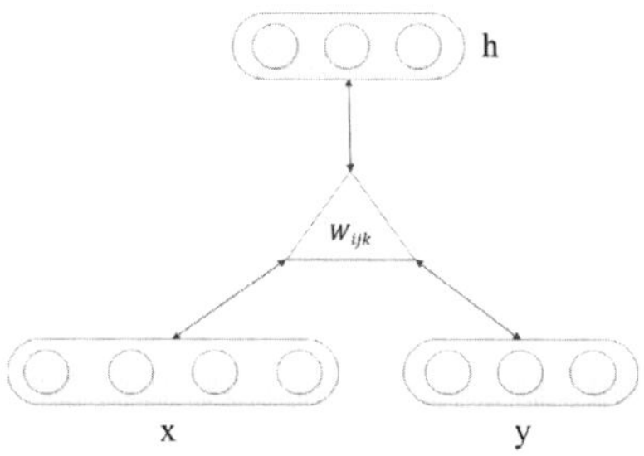

Fig. 1. Illustration of a three-way RBM

A three-way RBM consists of three layers as shown in Fig. 1: (1) a visible layer that takes the knowledge of source task as input, (2) a visible layer that takes the knowledge of source task as input and returns the transferred optimal policy as output, and (3) a hidden layer that encodes the inter-task mapping. These three layers are connected with a three-way interactions, multiplicative tensor. Formally, the energy function of three-way RBMs for modeling the joint distribution of $\Omega_{\mathcal{M}_1}$ and $\Omega_{\mathcal{M}_2}$ is

$$\begin{aligned} E(\mathbf{v}_1, \mathbf{v}_2, \mathbf{h}) = &-\frac{1}{2}\left(\frac{\mathbf{v}_1 - \hat{\mathbf{a}}_1}{\boldsymbol{\delta}_1}\right)^2 - \frac{1}{2}\left(\frac{\mathbf{v}_2 - \hat{\mathbf{a}}_2}{\boldsymbol{\delta}_2}\right)^2 \\ &- \mathbf{h}^T\hat{\boldsymbol{b}} - \mathbf{h}^T\mathbf{W}\left(\frac{\mathbf{v}_1^T\mathbf{v}_2}{\boldsymbol{\delta}_1^T\boldsymbol{\delta}_2}\right) \end{aligned} \tag{1}$$

where the visible source variable is $\mathbf{v}_1$ with a Gaussian $\mathcal{N}(\boldsymbol{\mu}_1, \boldsymbol{\delta}_1)$, the visible target variable is $\mathbf{v}_2$ with a Gaussian $\mathcal{N}(\boldsymbol{\mu}_2, \boldsymbol{\delta}_2)$, $\mathbf{h}$ is the hidden layer that consists of sigmoidal activations, and $\mathbf{W}$ is the three-way weight tensor connecting between each of the above layers. The input of visible source layer $\mathbf{v}_1$ is samples collected on $\Omega_{\mathcal{M}_1}$ and target source layer $\mathbf{v}_2$ is ones on $\Omega_{\mathcal{M}_2}$.

In order to reduce the computational complexity of training three-way RBMs, the three-way weight tensor is factored into sums of products[13], which makes the computational complexity reduce from $O(N^3)$ to $O(N^2)$.

Note that we define the joint distribution for both source and target task, rather than a conditional distribution of an output target space given input source space. This generative model builds the inter-task mapping of the joint distribution over source and target space that makes it possible to capture the mapping efficiently.

2.3. *Biased sampling*

When modeling the inter-task, $2 \times n$ samples $\mathcal{D}_1$ and $\mathcal{D}_2$ are needed. Normally, $< s^{(i)}, a^{(i)} >_1^n$ are generated randomly from the states and actions joint space $(\mathcal{S}, \mathcal{A})$ over both the source task and target task, where current state $s \in \mathcal{S}$ and action $a \in \mathcal{A}$. Through the transition model, the next new state $s'(s' \in \mathcal{S})$ is obtained. Then it gets the random samples $< s^{(i)}, a^{(i)}, s'^{(i)} >_1^n$ that describe the task's space. More specifically, samples on the source task space are $\mathcal{D}_1 = \{< s_1^{(i)}, a_1^{(i)}, s_1'^{(i)} >_1^n\}$ and target task $\mathcal{D}_2 = \{< s_2^{(i)}, a_2^{(i)}, s_2'^{(i)} >_1^n\}$.

However, on humanoid robot, it takes more training samples collected on the tasks and additional computational cost on learning the three-way RBM model. This results from the random sampling on both source and target task, which leads to limited representation of jointed space $\mathcal{S} \times \mathcal{A}$. Specifically, we assume that the space of $\mathcal{S} \times \mathcal{A}$ can be divided into two subspaces: successful space ($Space_{success}$) that describes the upright position of pole and the failed space ($Space_{failed}$) that represents the falling down position. Obviously, successful space only covers a bit of total space, that is $|Space_{failed}| \gg |Space_{success}|$[14]. Thus, by introducing expert knowledge, bias and selective sampling on $\mathcal{S} \times \mathcal{A}$ would reduce the sample complexity.

Furthermore, it is unrealistic to collect sample pairs of source space and target space respectively for the case of bipedal walking of humanoid robot. Because walking would be affected by humanoid robot's nonlinear dynamic property and the unconstant physical environment. Also, due to unbalance bipedal walking, there would be various falling postures, in other words, the transition state s' is not constant for a given $< s, a >$. However, the variance of transition state s' is small for the stable bipedal walking on humanoid robot.

2.4. *The process of transfer learning*

Transfer learning in RL has two main steps: (1) inter-task mapping: the inter-task mapping is learned through source and target task samples and (2) policy transfer: policy is transferred from the source to the target using as initialization for a RL algorithm.

After learning, the three-way RBM encodes an inter-task mapping between the source and the target task. This encoding is then used to transfer samples of transition triple $\mathcal{D}_1^*$ from optimal policy π_1^* of source task to the target task. The $\mathcal{D}_1^*$ is the optimal state transition triple sampled according to π_1^*. These triplets are passed through the visible source layer $\mathbf{v}_1$ of three-way RBM and are used to reconstruct initial target task samples $\mathcal{D}_2^0$ at the visible target layer $\mathbf{v}_2$, forming policy π_2^0. If the transferred policy π_2^0 on target task is close to the optimal policy π_2^*, then the transferred policy is expected to be an optimal or near-optimal behavior. They are then used as an initialization of RL algorithm, such as Sarsa[1] to learn an optimal policy π_2^* in the target task. The overall process of transfer learning with three-way RBMs is summarized in Algorithm 2.1.

Algorithm 2.1 Transfer learning with three-way RBMs

Input:

Samples $\mathcal{D}_1 = \{< s_1^{(i)}, a_1^{(i)}, s'^{(i)}_1 >\}_{i=1}^{n_1}$ that describe the source task space and its corresponding target task space samples $\mathcal{D}_2 = \{< s_2^{(i)}, a_2^{(i)}, s'^{(i)}_2 >\}_{i=1}^{n_2}$, optimal policy π_1^* on source task

Output:

Optimal policy π_2^* on target task

1: Obtain inter-task mapping $\rho_{transfer}$ by training a three-way RBM via CCD learning over $\mathcal{D}_1$ and $\mathcal{D}_2$
2: Sample source task according to π_1^* to attain $\mathcal{D}_1^*$
3: Transfer $\mathcal{D}_1^*$ to the target space using learned inter-task mapping and obtain $\mathcal{D}_2^0$ ($\mathcal{D}_2^0 = \rho_{transfer}(\mathcal{D}_1^*)$)
4: Use $\mathcal{D}_2^0$ or π_2^0 as an initialization of RL algorithm

---

3. Experiments and Results

To evaluate the efficiency of the proposed transfer learning, experiments of transferring walking ability on different slopes were performed.

3.1. *The PKU-HR5.1 robot platform*

The PKU-HR5.1 is the 5th version kid-size humanoid robot (with slightly renewal) designed by Peking University, China, which is 2.65kg weight and 47.8cm high. It has 20 DOF (degree of freedom), including 6 DOF for hip, 1 DOF for each knee, 2 DOF for each ankle, 2 DOF for head, 3 DOF for each arm. It uses 20 ROBTIS RX-28 servo actuators. All the servos are connected to the controller via RS485 with connection speed of 1Mbps. The FIT-PC2I equipped with inertial sensor is the main control system that placed within the torso of PKU-HR5.1. The inertial sensor contains one 3-axes gyroscope and one 3-axes accelerometer.

3.2. *Experimental setup*

Our experiments on the PKU-HR5.1 consider the task of walking ability transfer on different slope with various angles. The source task is the ability of walking on surface and the corresponding target task is slope bipedal walking.

- *Source Task:* The source task is bipedal walking on a flat surface whose stability margin is more larger than a slope[15]. As [16] discussed, the walking pattern on humanoid robot is largely dependent on the trajectory of Center of Mass (CoM) and the leg movement.
- *Target Task:* For the target task, it's still the PKU-HR5.1 system with the same stable walking goal. However, the walking plane is slope, making maintaining stable with optimal walking policy harder.

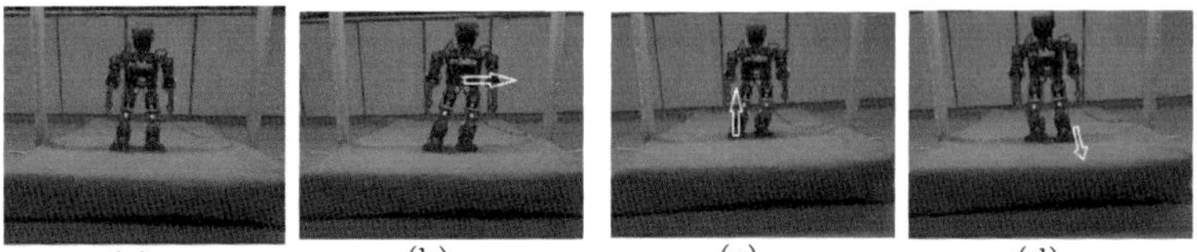

(a) (b) (c) (d)

Fig. 2. Four-phase biased sampling process on the slope surface for target task samples collection. (a) Standing; (b) Swinging; (c) Stepping Upward; (d) Stepping Forward.

As described in Section II., we propose a biased sampling method to sample near the optimal policy space for transfer learning on humanoid robot. This is inspired by the fact that bipedal walking can be regarded as synthesized movement of three different sub-movements: standing, swinging and stepping. The biased sampling contains four phases: (1) standing balance statically on the slope; (2) moving CoM left-right while standing; (3) stepping legs upwardly with small range while moving CoM; and (4) stepping forward like bipedal walking. Obviously, biased sampling avoids exaggerated moving in case of falling down with costly wear and tear. Fig. 2 shows the process of four-phase biased sampling.

After, sampling, we use proposed CCD learning to learn the inter-task mapping via a three-way RBM with 100 hidden units and 50 factors. The samples $\mathcal{D}_1$ and $\mathcal{D}_2$ are split into mini-batches with size 100. The CCD training process ran for 50,000 epochs for the convergence of reconstruction error.

3.3. *Experimental results*

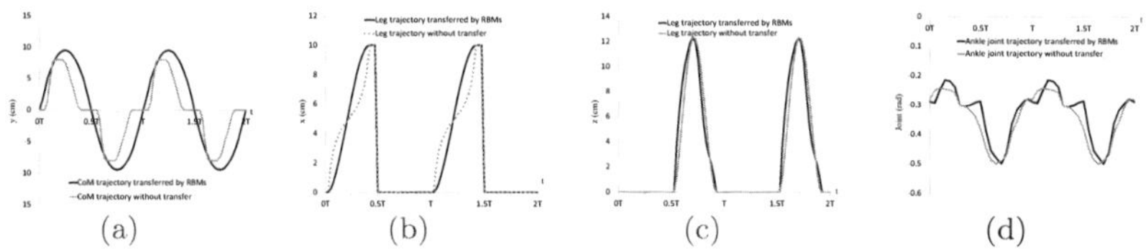

Fig. 3. Trajectory comparison on robot's CoM, leg and ankle joint between skill transfered with RBM and skill generated by dynamic model LIPM. (a) CoM Trajectory; (b) Leg movement in front-back direction; (c) Leg movement in up-down direction; (d) Trajectory of ankle joint.

On the policy transfer step, the target trajectories of CoM, leg movement and joint of ankle generated by the learned three-way RBM are shown in Fig. 3. The figure only shows one cycle of transferred movement trajectories of PKU-HR5.1 due to bipedal walk is reciprocating motion with cycle T. The CoM trajectory of left-right direction is shown in Fig. 3(a). CoM trajectory obtained by learned inter-task mapping on $10°$ slope is no different than the trajectory generated by Linear Inverted Pendulum Model (LIPM)[17]. Fig. 3(b) is the movement trajectory of front-back direction. Since the support leg of humanoid robot is fixed, Fig. 3(c) is the trajectory of up-down movement of swing leg as well. However, as Fig. 3(d) shows, though the overall trend of both transferred trajectory and LIPM is same, there are difference of amplitude and inflection point.

The comparison shows that decomposing the walking movement into three sub-movements and sampling with respect to the three sub-movements provide a better description of stable walking task's space. It also improves the accuracy of modeling inter-task mapping between tasks. Applying three-way RBM based transfer learning, humanoid robot learns to walk on the slope with few trail-and-error from tabula rasa. It can acquire new skills via generalizing knowledge to new situations.

4. Conclusion

Considering autonomous robot being demanded more powerful capacity so that to better serve in human society, the issue on autonomously acquiring new skills with the utilizing of past knowledge or experience is addressed in this research. And RBM based transfer learning under reinforcement learning paradigm is involved. Our purpose lies in two points, i.e., autonomous learning as well as efficient learning. The former means the skill learning is fulfilled without additional human intervention so that keep the autonomy of the robot; while the latter implies the skill acquisition process yields being performed on a real physical robot.

With the three-way RBM that powerful in capturing correlations between tasks, the framework of autonomously knowledge transferring in RL is structured, where the mapping between tasks (corresponding to skills) could be learned autonomously without human intervention. The idea of biased sampling technique is adopted to actively collect those samples that are more close to optimal policy, and a heuristic phased sampling strategy is proposed in a slope surface walking transfer problem.

Acknowledgment

The work is supported in part by National Basic Research Program (973 Program) of China (No. 2013CB329304), the National Natural Science Foundation of China (No. 90920302, No. 91120001, No.61121002), a "Twelfth Five-Year" National Science & Technology Support Program of China (No. 2012BAI12B01), the Seeding Grant for Medicine and Information Sciences of Peking University (No. 2014-MI-10) and the Key Program of National Social Science Foundation of China (No. 12&ZD119).

References

1. R.S. Sutton and A.G. Barto, "Reinforcement learning: an introduction," Cambridge, MA, USA:MIT Press, (1998).
2. J. Kober and J. Peters, "Policy search for motor primitives in robotics," Machine Learning, vol.84, no.1-2, pp.171-203, (2011).
3. A. Coates, P. Abbeel, and A.Y. Ng, "Learning for control from multiple demonstrations," in Proceedings of the 25th International Conference on Machine Learning, pp.144-151, (2011).
4. S. Barrett, M.E. Taylor, and P. Stone, "Transfer learning for reinforcement learning on a physical robot," in Proceedings of the 9th International Conference on Autonomous Agents and Multiagent Systems-Adaptive Learning Agents Workshop, (2010).

5. D. Luo, Y. Wang, and X. Wu, "Active online learning of the bipedal walking," in Proceedings of the IEEE-RAS International Conference on Humanoid Robots, pp.352-357, (2011).
6. M.E. Taylor and P. Stone, "Transfer learning for reinforcement learning domains: a survey," Journal of Machine Learning Research, vol.10, no.1, pp.1633-1685, (2009).
7. H.B. Ammar, D.C. Mocanu, M.E. Taylor, K. Driessens, K. Tuyls, and G. Weiss, "Automatically mapped transfer between reinforcement learning tasks via three-way restricted boltzmann machines," in Proceedings of the European Conference on Machine Learning and Principles and Practice of Knowledge Discovery in Databases, pp.449-464, (2013).
8. M.E. Taylor, P. Stone, and Y. Liu, "Transfer learning via inter-task mappings for temporal difference learning," Journal of Machine Learning Research, vol.8, no.1, pp.2125-2167, (2007).
9. J. Susskind, R. Memisevic, G.E. Hinton, and M. Pollefeys, "Modeling the joint density of two images under a variety of transformations," in Proceedings of the 24th IEEE Conference on Computer Vision and Pattern Recognition, pp.2793-2800, (2011).
10. G.E. Hinton, "Training products of experts by minimizing contrastive divergence," Neural Computation, vol.14, no.8, 1771-1800, (2002).
11. R. Memisevic, "Learning to relate images," IEEE Transactions on Pattern Analysis and Machine Intelligence, vol.35, no.8, pp.1829-1846, (2013).
12. D. Luo, Y. Wang, X. Han and X. Wu, "A Cyclic Contrastive Divergence Learning Algorithm for High-Order RBM," in Proceedings of the IEEE International Conference on Machine Learning and Applications, pp.80-86, (2014).
13. R. Marc' Aurelio, A. Krizhevsky, and G.E. Hinton, "Factored 3-way restricted boltzmann machines for modeling natural images," in Proceedings of the 13th International Conference on Artificial Intelligence and Statistics, pp.621-628, (2010).
14. D. Luo, Y. Wang, and X. Wu, "Discriminative apprenticeship learning with both preference and non-preference behavior," in Proceedings of the IEEE International Conference on Machine Learning and Applications, pp.315-320, (2013).
15. F. Ali, A.Z.H. Shukor, M.F. Miskon, M.K.M. Nor, and S.I.M. Salim, "3-D biped robot walking along slope with dual length linear inverted pendulum method (DLLIPM)," International Journal of Advanced Robotic Systems, vol.10, no.1, pp.377-389, (2013).
16. D. Luo, Y. Wang, and X. Wu, "Online learning of COM trajectory for humanoid robot locomotion," in Proceedings of the IEEE International Conference on Mechatronics and Automation, pp.1996-2001, (2012).
17. M. Friedmann, J. Kiener, S. Petters, and D. Thomas, "Versatile, high-quality motions and behavior control of humanoid soccer robots," IEEE-RAS International Conference on Humanoid Robots, pp.9-16, (2006).

SECTION–9

MEDICAL AND REHABILITATION ROBOTICS

A REAL-TIME GAIT PHASE DETECTION METHOD FOR PROSTHESIS CONTROL *

JUAN LI, XIAOHUA ZHOU, CHUNGUANG LI†, WEIDA LI, HONGMIAO ZHANG, HONG GU

School of Mechanical and Electric Engineering, Soochow University Suzhou 215021, China

Accurate and real-time gait phase detection is the fundamental question of intelligent prosthesis control. A good phase detection method can make the prosthesis control better and easier . Here we propose a real-time detecting method for gait phase and build a phase detection system for prosthesis. The system consists of 4 foot force sensors, two attitude sensors and an interaction interface program. Experiments were performed on three subjects with different walking velocities and different stride length. Gait phase could be detected correctly for all the subjects. The experimental results show that the proposed method is effective and accurate.

1. Introduction and Background

Artificial limbs, walking aids and rehabilitation robotics are the major means of recovery to stroke, paralysis and amputees. The control method of low limb robot is to guarantee its wearer walk in a more natural and safe way. Due to the typical movement states of gait characteristics, the control method based on the states has been widely concerned. Early in 1983, Bar divided one gait cycle into 10 states and proposed a control method of the finite state machine [1]. In 1988, Chitore put forward a method that established a digital controller to control the knee and ankle prosthesis based on finite state machine [2]. From 2010 to 2013, Borjian studied the feasibility of the state mapping control method to control the prosthetic knee on the basis of states [3]. Joshi divided the gait cycle into 4 typical phases based on knee joint curves, he realized the control of the knee joint by fuzzy adaptive control method [4]. Liu, Wang adopted the method of neural network to predict gait after the gait was divided into 6 states [5]. From the above researches, it can be seen that the gait phase detection is the base of the robot control method and plays a key role.

---

* This work is supported by National Natural Science Foundation of China (No.51475314 and No. 61203367).

† Corresponding author, Email:lichunguang@suda.edu.cn.

Existing gait phase detection systems are mostly consisted of a variety of sensors and their combination. Force Sensing Resistor(FSR) system, angular rate sensor system and accelerometer sensor system are typical. Early systems used pure mechanical micro-switches. Recent systems are based on threshold values obtained from one or more FSR transducers placed under the foot which can detect typical phases like heel strike, foot flat and heel off [6]. Now a wearable force plate system is applied to measure gait variability [7]. FSR system has become the standard of gait phase detection systems, however, it can not be used during swing phase because of its properties. Angular rate sensors (gyroscopes) provide measurements of angular displacement. A great advantage of gyroscopes is the fact that it is not affected by gravitation, but it has the problem of signal drift. Gait event detection on level ground and incline walking using a rate gyroscope has got better results [8]. The combination of angular rate sensor and FSR sensor can efficiently obtain the gait phase information. A real-time system composed of three FSR sensors combined with a gyroscope was investigated regarding detection of heel off and heel strike events [9],the detection rate was above 95% but it had a problem of time delay. With the development of MEMS technology, the accelerometer sensor has been used in the gait phase detection system already. This kind of inertial sensor acquires joint angle through integration of accelerations from thigh, shank and knee and brings integral error at the same time. An example is that a system showed a good off-line processing analysis ability through the positive and negative changes of sagittal acceleration [10]. Another system showed the capability of real-time detection of initial contact and final contact [11]. In addition, a combination of accelerometer sensor and angular rate sensor has achieved average error less than 7% [12].The combination of different sensors has become the trend of gait phase detection system [13].

The paper proposes a gait phase detection method which is based on simple principle and reduces calculation cost. This method has good real-time performance and strong adaptability to different subjects, different walking velocities and different stride length. The mentioned performance is required for prosthesis control in the future work.

2. Method

2.1. *Gait phase division*

Walking plays an important role in our daily life. As a typical rhythmic movement, force, torque and ankle of each joint show different characteristics in a gait cycle. One gait cycle can be divided into multiple phases based on these characteristics. The relationship of ankle torque and angle in a gait cycle was described in Figure 1(a). During the stance phase, ankle reflects three different characteristics in mechanics: when the heel touches the ground, ankle torque is small and remains unchanged basically; subsequently, ankle torque and angle value increase gradually, they are nonlinear until ankle angle achieves maximum dorsi-flexion; after that, ankle torque and angle decrease at the same time until the ankle torque reduces to zero, which is the end of the stance. During the swing phase, according to the knee angle curve shown in Figure 1(b), knee angle reflects two different characteristics: firstly the knee is in flexion with the angle increasing and the angular velocity is greater than 0; then the knee is in extension with the angle decreasing and the angular velocity is less than 0. According to these characteristics, the paper divides one cycle into 5 phases: CP (Controlled Plantar-Flexion), CD (Controlled Dorsi-Flexion),PP(Powered Plantar-Flexion), SF (Swing Flexion) and SE (Swing Extend). Each phase has the following features: CP begins at heel strike and ends at foot flat, ankle torque is proportional to ankle position; CD begins at foot flat and continues until the ankle reaches a state of maximum dorsi-flexion(heel off), the main function of the human ankle during CD is to store elastic energy; PP begins after CD and ends at the instant of toe-off, during PP the ankle is modeled as a torque source; SF begins at toe-off and ends at the knee flexion to the maximum, during SF the angular velocity is greater than 0; SE begins after SF and ends at heel-strike, during SE the angular velocity is less than 0[14].

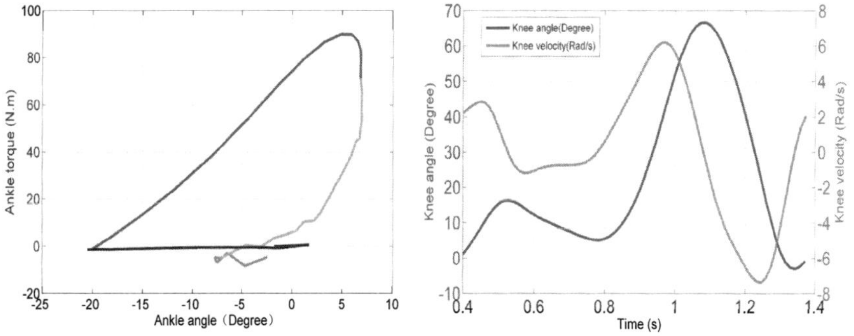

Figure 1. a) The relationship between ankle torque and angle in a gait cycle. b) Knee angle and velocity in a cycle

2.2. *Gait phase detection method*

FSR sensors (Interlink Electronics Company 402 type) are used to detect phases during stance and attitude sensors are used to detect phases during swing. CP, CD, PP are detected by FSR sensors and SF, SE are detected by the knee angle integrated with attitude sensor signal, as plotted in Figure 2.

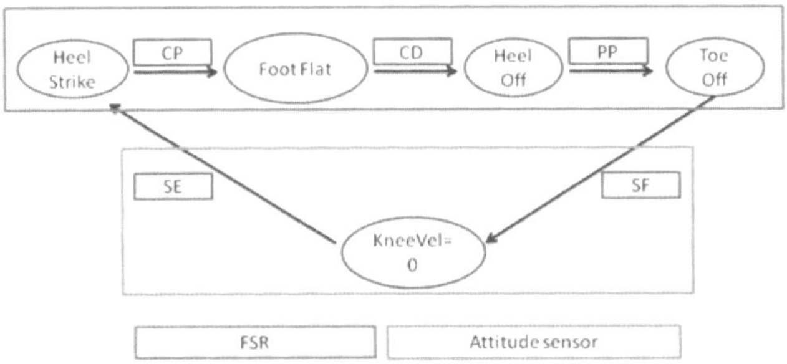

Figure 2. Gait phase detection method

The foot force system consists of four FSRs(two were placed underneath the heel and toe, another two were underneath the first and fourth heads of the metatarsal bones because of asymmetry during walking), Arduino Uno R3 board and its circuit, as described in Figure 3. The resistance of FSR changes with applied force. The signals are acquired by the board with working voltage 5V and 100Hz frequency. Its pressure signal ranges from 0 to 5V, corresponding to the digital signal from 0 to 1024.The number 350 is selected as threshold with several actual experiments. Phase information integrated is sent to the computer through a serial port in real time.

Angle measurement system is composed of two attitude sensors (Xsens Technology Company) as well as the motion capture module, as shown in Figure 4. Two attitude sensors are placed on the thigh and shank. The motion capture module collects the angle information with 100Hz frequency. Angle information is sent to the computer after integrated by internal DSP.

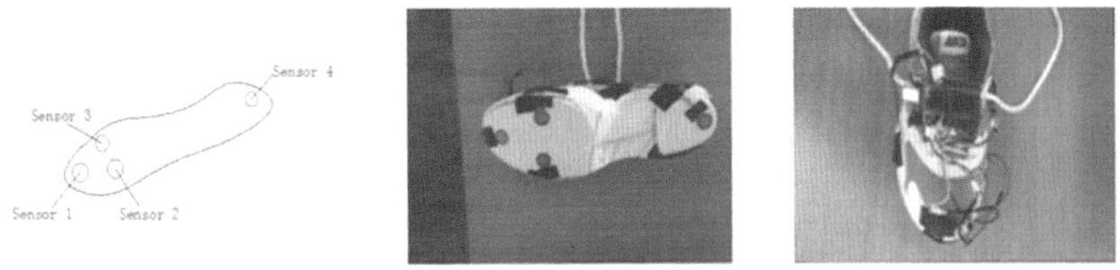

Figure 3. Foot force system

We used a gait phase detection truth table to describe the method (Table 1). 0 indicates the sensor signal is lower than the threshold, 1 indicates the sensor signal is higher than the threshold and X indicates the sensor signal has no influence on the result. Sensor index in the table is in accordance with Figure 3. For example, one of the sensors #1, #2, #3 signal is higher than the threshold and the sensor #4 signal is higher than the threshold, it means the gait is in CD phase.

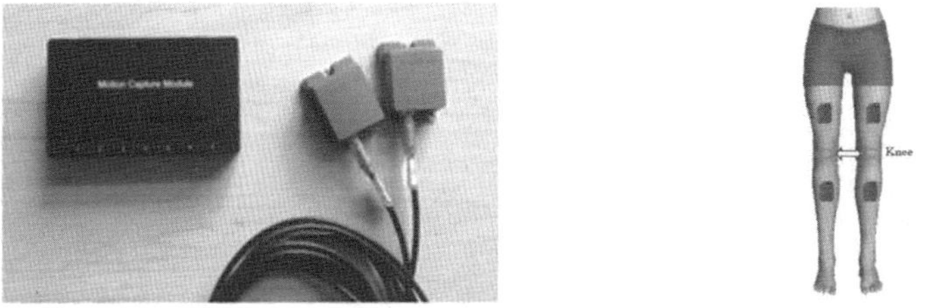

Figure 4. Angle measurement system

Table 1.Gait Phase Detection Truth Table

	Sensor1	Sensor2	Sensor3	Sensor4	KneeVel$\geq$0
CP	0	0	0	1	X
	1	X	X		
CD	X	1	X	1	X
	X	X	1		
	1	X	X		
PP	X	1	X	0	X
	X	X	1		
SF	0	0	0	0	1
SE	0	0	0	0	0

Human-computer interaction interface program is written based on MFC. Multithreading functions of MFC are used to combine the information from foot force system and angle measurement system. The program can complete the data real-time acquisition of the whole system, as shown in Figure 5. Figure 6 is the flow chart of the gait phase detection program. The program receives the foot force signal from the Arduino Uno R3 board firstly, shows the phase after a process of detection. If the threshold from every sensor is 0, it means swing phase has begun. The program will get angle information from angle measurement system and show the phase SF or SE based on angular velocity.

3. Experiment

To verify the proposed method, experiments were performed on 3 subjects (all healthy males) with different walking velocities and different stride lengths (Table 2).

Three subjects walked in three different cases:

1. Subject A walked horizontally at normal stride length but three different walking velocities: slow, normal and fast.
2. Subject A walked horizontally at normal velocities but three different stride length: small, normal and big.
3. Subject A,B,C walked horizontally at normal stride length and normal walking velocities.

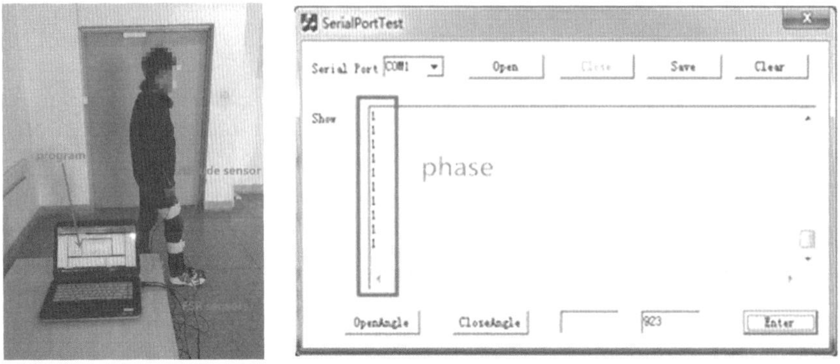

Figure 5. Gait phase detection system and interaction interface program

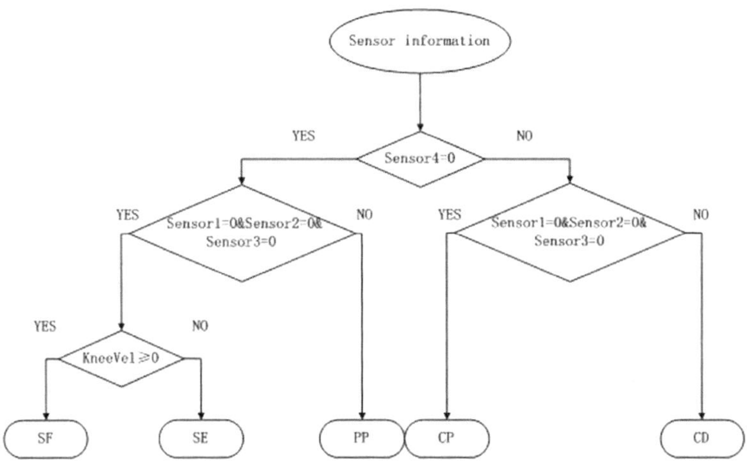

Figure 6. Flow chart of gait phase detection

Table 2 Walking velocities, stride length range in the experimental study

Speed range	0.55step/s~0.8step/s
Stride length range	65cm~85cm

The result in the situation of Subject A walking at a speed of 0.55step/s is described in Figure 7, which includes signals of FSR sensors, attitude sensor and phases. Figure 8 shows the relationship of gait percent and phase in the same situation. It can be seen that stance phase probably accounted for 60% of a gait cycle(the percentage of toe-off is 63%) and swing phase for 40%. This is consistent with the theoretical results. Also we can conclude from the results: different walking velocities and stride length has no effects on the percentage of stance and swing phase. The experimental results shows that the proposed gait phase detection method can adapt to different walking velocities, stride length and different subjects.

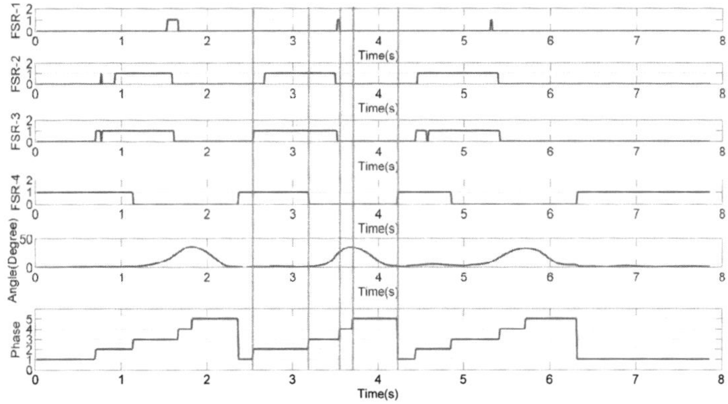

Figure 7. Gait phase detection at a speed of 0.55step/s

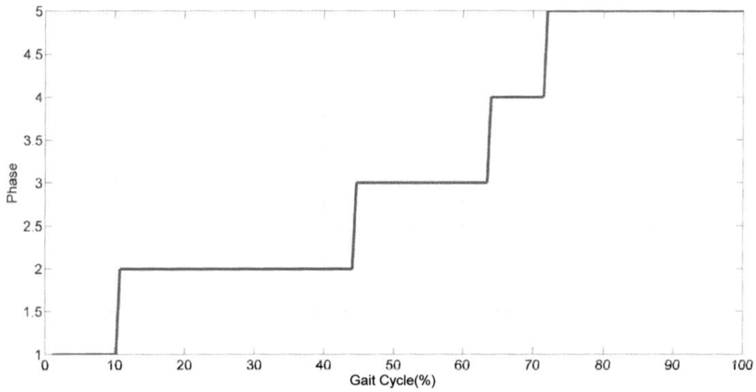

Figure 8. The relationship between gait percent and phase at a speed of 0.55step/s

4. Conclusion

A gait phase detection method that identified the phases CP, CD, PP, SF and SE was presented. Compared with other methods, this real-time system was based on simple principle and with limited calculation cost. And the experimental results also showed that the detection system is adaptive to different walking velocities, stride length and subjects on the flat floor. The accuracy of the gait phase detection system need to be explored. And our work in the future will focus on the application of the method in the real-time control of intelligent prosthesis.

References

1. Bar A, Ishai G, Meretsky P, et al. Adaptive microcomputer control of an artificial knee in level walking[J]. Journal of biomedical engineering, 1983, 5(2): 145-150.
2. Chitore D S, Rahmatallah S F, Albakry K S. Digital electronic controller for above knee leg prostheses[J]. International journal of electronics, 1988, 64(4): 649-656.
3. Borjian R, Khamesee M B, Melek W. Feasibility study on echo control of a prosthetic knee: sensors and wireless communication[J]. Microsystem technologies, 2010, 16(1-2): 257-265.
4. Joshi D, Mishra A, Anand S. ANFIS based knee angle prediction: An approach to design speed adaptive contra lateral controlled AK prosthesis[J]. Applied Soft Computing, 2011, 11(8): 4757-4765.
5. Liu Y, Wang C, Zheng D, et al. An ANN Based Approach for Gait Prediction of a Lower-Limb Exoskeleton with Plantar Pressure Sensors[M].Intelligent Robotics and Applications. Springer Berlin Heidelberg, 2013: 373-383.
6. Skelly M M, Chizeck H J. Real-time gait event detection for paraplegic FES walking[J]. Neural Systems and Rehabilitation Engineering, IEEE Transactions on, 2001, 9(1): 59-68.
7. Liu T, Inoue Y, Shibata K. A wearable ground reaction force sensor system and its application to the measurement of extrinsic gait variability. Sensors. 2010;10:10240–10255.
8. Catalfamo P, Ghoussayni S, Ewins D. Gait event detection on level ground and incline walking using a rate gyroscope. Sensors. 2010;10:5683–5702.
9. Pappas I P I, Popovic M R, Keller T, et al. A reliable gait phase detection system[J]. Neural Systems and Rehabilitation Engineering, IEEE Transactions on, 2001, 9(2): 113-125.
10. Mansfield A, Lyons G M. The use of accelerometry to detect heel contact events for use as a sensor in FES assisted walking[J]. Medical Engineering & Physics, 2003, 25(10): 879-885.
11. Gonzalez RC, Lopez AM, Rodriguez-Uria J, Alvarez D, Alvarez JC .Real-time gait event detection for normal subjects from lower trunk accelerations[J]. Gait & Posture, 2010,31: 322–325
12. Mayagoitia R E, Nene A V, Veltink P H. Accelerometer and rate gyroscope measurement of kinematics: an inexpensive alternative to optical motion analysis systems[J]. Journal of biomechanics, 2002, 35(4): 537-542.
13. Rueterbories J, Spaich E.G, Larsen B, Andersen O.K. Methods for gait event detection and analysis in ambulatory system. Med. Eng. Phys. 2010;32:545–552.
14. Au S, Berniker M, Herr H. Powered ankle-foot prosthesis to assist level-ground and stair-descent gaits[J]. Neural Networks, 2008, 21(4): 654-666.

SPEED-ADAPTIVE REFERENCE KNEE TRAJECTORY GENERATION BASED ON FOURIER FUNCTIONS[1]

JIAYUE WANG, JUAN LI[2], WEIDA LI, QIHUAN ZHU, HAIYAN HU, HONGMIAO ZHANG, CHUNGUANG LI, HAO GUO

School of Mechanical and Electric Engineering, Key Laboratory of Robotics and System of Jiangsu Province, Soochow University, Suzhou 215021, China

For the intelligent control of lower limb prostheses, kinematic reference trajectories are required. Due to individual difference and various preferred reference walking-speed, a general reference knee angle trajectory is difficult to adapt different people and their different walking speed. To solve this problem, gait kinematic information of thirty healthy subjects was collected at their preferred speeds and the relationship between kinematic models of knee joint and speeds was analyzed. Each knee angle trajectory was modeled based on Fourier functions, and the parameters of models were proved to be related to walking speed. Finally, kinematic models for knee angle trajectories were reconstructed based on regression equations. The reconstructed trajectories not only can adapt to the various walking-speeds, but also adapt to different people. The result of analysis showed that the reconstructed trajectories matched the measured motion trajectories well.

1. Introduction

To assist disabled people who survived from stroke or accident, many instruments are invented, for instance, exoskeleton robots and prostheses. Not only can they enable the people to walk again, but also enhance their life quality.

The control approach plays an important role in prostheses. During the past decade, many strategies were proposed, for instance, neural-network control [1,2,3,13,16,17], damping control [7,11,15], PID control [9], fuzzy control system [10] and control strategy using Electromyogram (EMG) [5,12]. Whatever the control approach we choose, a reference trajectory to define the motion is necessary. Generally, these reference trajectories are based on the pre-recorded data of unimpaired volunteers [6]. But these trajectories can't adapt to different people or speeds. Another method using neural-network to predict

---

[1] This work is supported by National Natural Science Foundation of China (No.51475314 and No. 61203367)

[2] Corresponding author: lijuan@suda.edu.cn

joint angle was proved to be successful [13,16,17], but the weakness of the neural-network is that it is hardly to be understood and the parameters of network do not have exact significance. A method that quantic-splines were fitted between the predicted key events to reconstruct a full gait cycle was presented [18]. The study researched gait kinematics at a wide range of walking-speeds. As previous researches showed, the gait pattern relates to walking-speed [4,8,14].

An approach to reconstruct kinematic trajectories that could adapt to different subjects and various speeds was proposed. Gait kinematic information of thirty healthy subjects was collected at their preferred speeds and the relationship between kinematic models of knee joint and speeds was analyzed. Through the relationship between parameters of fitting function of trajectory and speeds, a mathematical model for different walking-speed was reconstructed.

2. Method

2.1 *Subjects*

Thirty healthy volunteers in the age of 23 and 27 years, twenty males, ten females, weight 65kg±15, body-height 1.70m±0.14, participated in this study. All the volunteers had no neurologic, orthopedic or cardiopulmonary problems.

2.2 *Experimental Protocol*

Data was collected by a three-dimension motion analysis system made by Motion Analysis Corporation (Santa Rosa, CA, USA). Each volunteer was adhered with twenty nine reflective markers to acquire the joint angles using eight CCD cameras (Fig. 1) at the frequency of 100Hz during moving.

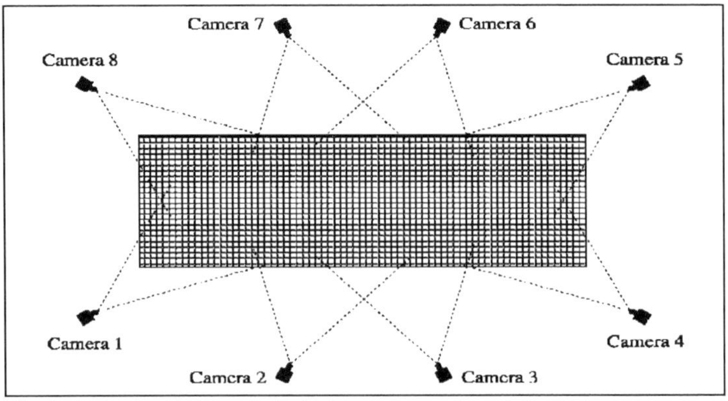

Figure 1. The top view of laboratory with eight cameras

The subjects were asked to walk in the laboratory at their preferred walking-speeds. The data was recorded for 4s during each trail to ensure at least a complete gait cycle. Before data analysis, every data was intercepted from the time when the extreme back of heel touches ground to that it touches ground again (Fig. 2).

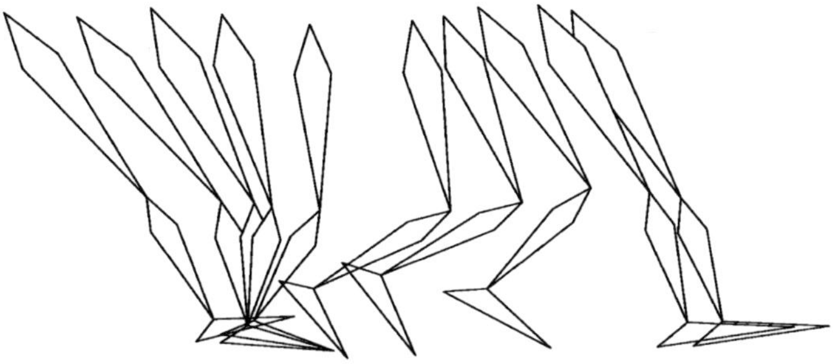

Figure 2. A complete gait cycle intercepted form data measured.

2.3 *Data analysis*

At first, a small quantity of abnormal data was rejected because of the drop of markers and abnormal gait. Each complete gait data was fitted with the curve fitting toolbox of MATLAB. Through several tests, it was found that the knee trajectories were fitted well by Fourier function with the R-square reaching 0.99. Hence, knee angle can be modeled as equation (1) based on Fourier function:

$$f(x) = a_0 + \sum_{n=1}^{m} (a_n \cos(n * \omega * x) + b_n \sin(n * \omega * x)) \tag{1}$$

Where, m=5 and other parameters ω, a_0, a_n and $b_n (n=1,2,3,4,5)$ would be calculated for each trajectories. In the model, $f(x)$ is knee angle at time x. The parameter ω has influence on the period of the model, and relates to walking-speed. The parameters a_n and b_n represent the amplitude of sine function and cosine function respectively. They both have relationship with walking-speeds, but the values of them are slightly different on account of different people.

Figure 3 depicts three knee joint trajectories of one subject with different speeds. As displayed in Fig. 3, along with the speed slowing down, the maximum of knee angle decreases, and the time of swing phase and stand phase become longer. It was also found that the parameter ω diminishes at the same time. In Figure 4, the left one depicts the relationship between ω and walking-speed for all data collected, while the other reflects the relationship for one subject. And the parameter ω was found to have a quadratic relationship with walking-speed as shown in Fig. 4.

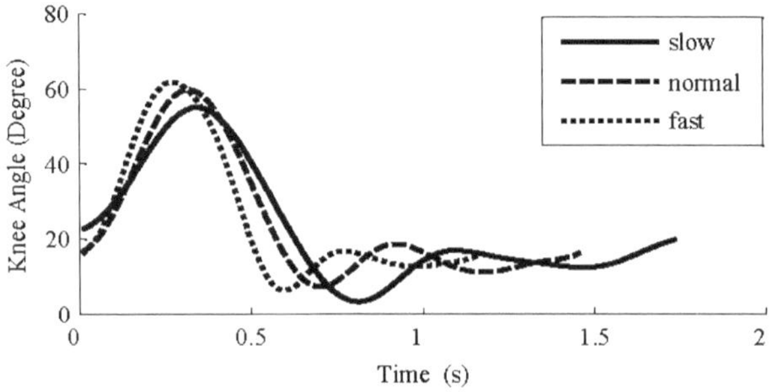

Figure 3. Three knee-angular trajectories for different speeds.

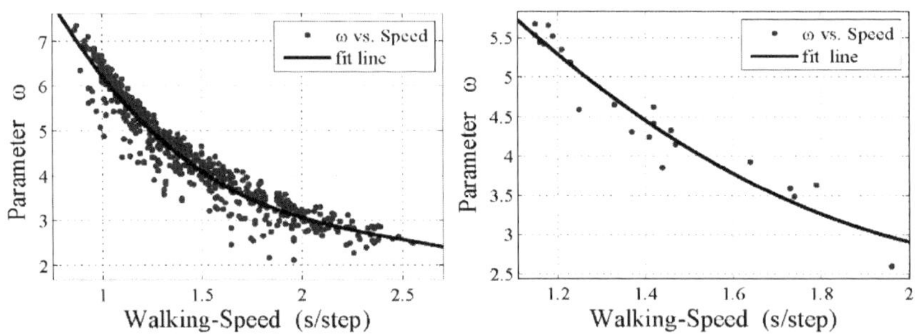

Figure 4. Fit lines of parameter ω and speeds for all data (left) and one subject (right).

The regression equations for parameters and speeds of one subject are provided in Table 1. In Table 1, v represents walking-speed, and all regression equations were obtained by the curve fitting toolbox. For different people, the regression equations have slightly differences, but have the same tendency. Hence, a method was proposed that the regression equations in Table 1 were took as reference, and regression equations for different people would be adjusted after the parameters compared to the reference.

Table 1. The regression equations between parameters and walking-speed (one subject).

Parameters	Relationships with Speed
ω	$6.059/(v - 0.02669)$
a_0	$-4.2701v^3+22.5078v^2-44.1732v+53.7053$
a_1	$36.7059v^3-172.3902v^2+278.1075v-147.0736$
b_1	$-12.5657v+35.9257$
a_2	$91.0892v^3-413.6768v^2+624.5107v-323.9828$
b_2	$54.4745v^3-270.5611v^2+448.3047v-241.362$
a_3	$-43.737v^3+194.5496v^2-292.3099v+147.7491$
b_3	$14.7561v^2-47.2701v+33.0124$
a_4	$-1.1314v^2+5.5805v-6.1215$
b_4	$1.1378v^3-11.2516v^2+25.4991v-17.1458$
a_5	$1.6162v^2-5.4151v+4.7962$
b_5	$5.8101v^2-17.6356v+12.5449$

2.4 *Model reconstruction*

2.4.1 *Reconstruct model for one subject*

The obtained regression equations above were used to reconstruct the reference trajectories of different speeds for the subject. Firstly, a set of parameters were calculated for the subject with a certain speed. Subsequently, based on model of equation (1), a complete gait trajectory was reconstructed. Results indicate that the reconstructed models matched the measured trajectories well. Figure 5 displays four knee-angular trajectories for different speeds. As shown in Fig. 5, the solid line represents knee trajectory measured in experiment, while the dash-dotted line is reconstructed curve based on model of equation (1).

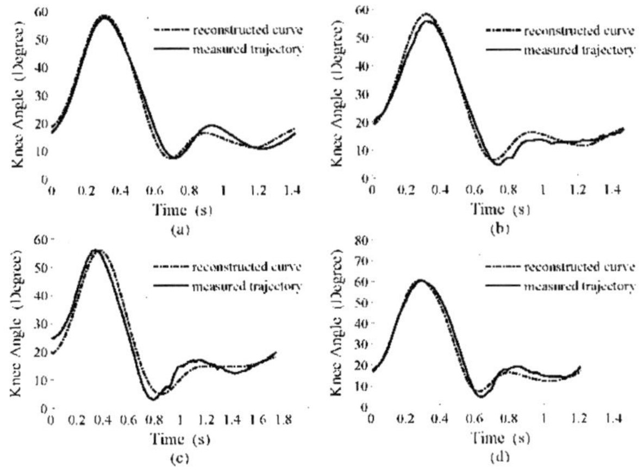

Figure 5. Fit lines with equation (1) for different speeds of one subject. The walking-speeds of (a), (b), (c) and (d) are 1.4s/step, 1.5s/step, 1.8s/step, and 1.2s/step respectively.

2.4.2 *Reconstruct model for others*

Because of individual difference, there are differences in the joint angle for different people. As shown in Fig. 6, regression models for two parameters that a_1 and b_1 of knee model based on equation (1) were displayed. And it was found that the tendency of angle change caused by speeds is similar. Therefore, in order to adapt to different people, a strategy, adjusting offsets based on reference model parameters, was proposed. Firstly, some knee angle trajectories of another subject were necessary for acquiring several groups of model parameters based on equation (1). Secondly, the series of regression equations in Table 1 were took as reference, and every equation was added with an offset ε_n (n=1,2,3,4,5…12) to obtain new equations for another subject. The offsets mean the difference between reference and other subject. And these equations can be written as $\omega'=\omega+\varepsilon_1$, $a_0'=a_0+\varepsilon_2$, $a_1'=a_1+\varepsilon_3$, $b_1'=b_1+\varepsilon_4$, $a_2'=a_2+\varepsilon_5$, $b_2'=b_2+\varepsilon_6$, $a_3'=a_3+\varepsilon_7$, $b_3'=b_3+\varepsilon_8$, $a_4'=a_4+\varepsilon_9$, $b_4'=b_4+\varepsilon_{10}$, $a_5'=a_5+\varepsilon_{11}$ and $b_5'=b_5+\varepsilon_{12}$.

Thirdly, In order to obtain all the offsets, model parameters of equation (1) for the other subject were compared with the parameters that calculated by the equations in Table 1 in the same walking-speeds. Subsequently, each offset ε_n was obtained from the average of differences of the comparison. Finally, a set of equations for the other subject were got, and knee angle trajectory would be reconstructed based on the new equations. Figure 7 displays four trajectories of the other subject using the method proposed.

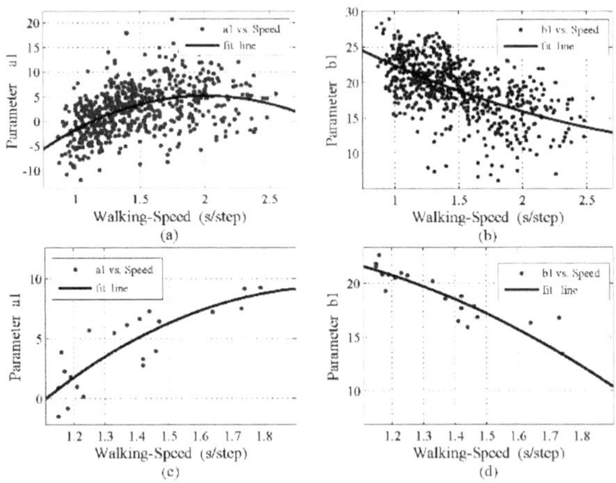

Figure 6. Fit lines between parameters of a_l and b_l with speeds. (a) and (b) are for all subjects, (c) and (d) are for one subject.

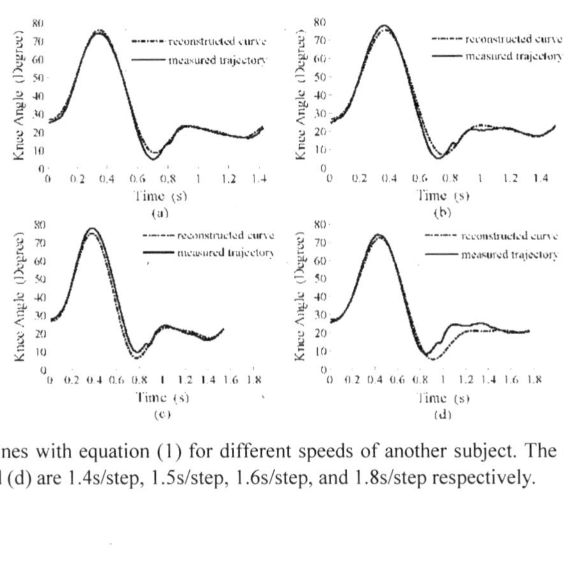

Figure 7. Fit lines with equation (1) for different speeds of another subject. The walking-speeds of (a), (b), (c) and (d) are 1.4s/step, 1.5s/step, 1.6s/step, and 1.8s/step respectively.

3. Results

To evaluate the performance of the reconstructed model, Root Mean Square Error (RMSE) was used. Additionally, the Correlation Coefficient (CC) was also used to quantify the similarity between the reconstructed and actual patterns. In Table 2, ten subjects were evaluated. The average RMSE was 5.35°, and the average CC was 0.95.

Table 2. The RMSEs and Correlation Coefficient (CC) for ten subjects.

	1	2	3	4	5	6	7	8	9	10
RMSE	5.263	5.977	4.317	6.638	5.005	3.164	6.470	5.091	4.115	7.449
CC	0.951	0.955	0.937	0.921	0.940	0.987	0.930	0.974	0.979	0.885

The results showed that parameters of the model of knee trajectories were related to walking-speed. The reconstructed patterns matched the measured data well. Hence, it was feasible to acquire patterns by adjusting equations based on reference model.

4. Conclusion

In this study, a method to reconstruct knee trajectories dependent on walking-speed is presented. And the strategy can also adapt to different people. It was found that the relationship between parameter ω and walking-speeds is obvious (Fig. 4), while other parameters distribute widely, but has similar tendency (Fig. 6). Several conditions may result in the phenomenon: (1) there are some difference between patterns of walking of different people, (2) body-height could be another reason, but several studies reflected body-height had little influence on the trajectory [4,18]. Therefore, the presented method to reconstruct reference trajectories based on reference model equations is feasible and results are good.

References

1. S. Jonic, T. Jankovic, V. Gajic, D. Popvic, Three machine learning techniques for automatic determination of rules to control locomotion. *Biomedical Engineering, IEEE Transactions on* **46**, 300-310 (1999).
2. V. D. Kalanovic, D. Popovic, N. T. Skaug, Feedback error learning neural network for trans-femoral prosthesis. *Rehabilitation Engineering, IEEE Transactions on* **8**, 71-80 (2000).
3. S. D. Prentice, A. E. Patla, D. A. Stacey, Artificial neural network model for the generation of muscle activation patterns for human locomotion. *Journal of Electromyography and Kinesiology* **11**, 19-30 (2001).
4. J. L. Lelas, G. J. Merriman, P. O. Riley, D. C. Kerrigan, Predicting peak kinematic and kinetic parameters from gait speed. *Gait & posture* **17**,

106-112 (2003).
5. J.-W. Lee, G.-K. Lee, Gait angle prediction for lower limb orthotics and prostheses using an EMG signal and neural networks. *International Journal of Control, Automation and Systems* **3**, 152-158 (2005).
6. A. O. Kapti, M. S. Yucenur, Design and control of an active artificial knee joint. *Mech Mach Theory* **41**, 1477-1485 (2006).
7. K. Fite, J. Mitchell, F. Sup, M. Goldfarb, in *Rehabilitation Robotics, 2007. ICORR 2007. IEEE 10th International Conference on.* (2007), pp. 902-905.
8. A. Ferrari *et al.*, Quantitative comparison of five current protocols in gait analysis. *Gait & posture* **28**, 207-216 (2008).
9. G. G. Scandaroli, G. Araujo Borges, A. F. da Rocha, F. A. de Oliveira Nascimento, in *Robotic Symposium, 2008. LARS '08. IEEE Latin American.* (2008), pp. 164-169.
10. R. Borjian, M. B. Khamesee, W. Melek, Feasibility study on echo control of a prosthetic knee: sensors and wireless communication. *Microsystem Technologies* **16**, 257-265 (2009).
11. D. Joshi *et al.*, in *Systems in Medicine and Biology (ICSMB), 2010 International Conference on.* (2010), pp. 143-147.
12. G. Yanli, X. Xiaoyun, C. Lingling, Y. Peng, in *IECON 2010 - 36th Annual Conference on IEEE Industrial Electronics Society.* (2010), pp. 1495-1499.
13. D. Joshi, A. Mishra, S. Anand, ANFIS based knee angle prediction: An approach to design speed adaptive contra lateral controlled AK prosthesis. *Applied Soft Computing* **11**, 4757-4765 (2011).
14. S. Seungmoon, H. Geyer, in *Robotics and Automation (ICRA), 2012 IEEE International Conference on.* (2012), pp. 511-516.
15. M. Liu, F. Zhang, P. Datseris, H. Huang, Improving Finite State Impedance Control of Active-Transfemoral Prosthesis Using Dempster-Shafer Based State Transition Rules. *Journal of Intelligent & Robotic Systems*, (2013).
16. Y. Liu, C. Wang, D. Zheng, G. Wang, X. Wu, in *Intelligent Robotics and Applications,* J. Lee, M. Lee, H. Liu, J.-H. Ryu, Eds. (Springer Berlin Heidelberg, 2013), vol. 8102, chap. 38, pp. 373-383.
17. M. M. Ardestani *et al.*, Human lower extremity joint moment prediction: A wavelet neural network approach. *Expert Systems with Applications* **41**, 4422-4433 (2014).
18. B. Koopman, E. H. van Asseldonk, H. van der Kooij, Speed-dependent reference joint trajectory generation for robotic gait support. *Journal of biomechanics* **47**, 1447-1458 (2014).

POWERED KNEE ORTHOSIS FOR PERFORMANCE OF ASSISTANCE AND REHABILITATION PURPOSES

M. SHYSH, A. SAFONOV

University of Magdeburg, Faculty of Engineering, Institute of Mobile Systems, P.B. 4120, 39106 Magdeburg, Germany

A. TELESH, U. SCHMUCKER

Fraunhofer IFF, Department Virtual Engineering, Sandtorstr. 22, 39106 Magdeburg, Germany

Such modern social issues as demographic ageing and relevance of living conditions enhancement, along with mobility maintenance of different population groups, makes powered orthosis and exoskeleton the core research and application issue. Being widely used, they play a key role for medication purposes, such as rehabilitation of patients suffering from stroke, spinal cord injuries or lower limb surgeries.

From the perspective of a healthy person, powered orthosis and exoskeleton can help increasing the body performance capabilities, allowing transferring a heavy weight, moving faster and further. In the long term, such devices can find an application in sport, tourism and industry.

This article describes a pilot version of a powered knee orthosis based on series elastic actuator (SEA) with flexible hydraulic transmission.

1. Introduction

This paper introduces a powered knee orthosis design that consists of force controlled electrohydraulic actuator, connected to hard-frame knee orthosis. The actuator used in the paper is an elaboration result of a research group "RobotsLab". This development was as well used in the construction of the robot "ROTTO" in 2007 and 2014 [1, 2].

In order to become widely used for assistance and therapeutic purposes during rehabilitation, wearable robotic devices (powered orthosis and exoskeletons) should possess the following main possibilities and dynamic characteristics:

- Lightweight construction: the device should be light, non-binding for human body dynamics, but robust and stable at the same time;

- Increasing Wearer‘s Performance: the device should increase human body capacities, allowing to perform the tasks which were unrealistic to perform without the device;
- Low Impedance: the device should not hinder natural smooth human actions. It should be as well compliant;
- Wearer’s Comfort and Safety: the device should be safe for the user and should not cause any physical discomfort during the usage. It must be handy and easy to put it on and off;
- Autonomy: the device should be independent, sustainable and it should have a longstanding energy storage capacity.

In this paper, we present a first prototype of the powered knee orthosis, which fulfills 4 out of the 5 listed requirements. Due to the fact that this is the first prototype, the question of autonomy is not in the scope of this paper. The weight of the construction may be further optimized using a lighter weight actuator.

2. Construction

As an actuating source of a powered knee orthosis, we used a force controlled electrohydraulic actuator. A choice in favor of electrohydraulic actuator is made due to the fact that classical hydraulic and pneumatic actuators possess a number of disadvantages, which make their application complicated in wearable robotic devices.

Among advantages of pneumatic actuators are such characteristics as small mass and fast trigging speed. Nevertheless, they have some disadvantages, mostly determined by the nature of the operation environment, i.e. air. Such negative attributes are complicated positioning, low smoothness and exactness of motion, speed instability under the load, as well as accumulation of energy in the air due to its high compressibility, leading to hitches and shocks.

In spite of the fact that classical hydraulic actuators generally develop a high power, they also have some significant drawbacks. Among those are large size and mass; necessity of compressors, pumps and pressure accumulators’ usage; modest dynamics.

Used electrohydraulic actuator makes a good fit drive for orthosis and exoskeleton. Its main virtue comparing to classical hydraulic and pneumatic actuators is absence of additional pressure sources (compressor, pump, etc.), low mass of actuating mechanism, construction simplicity and a high positioning dynamics [1].

The used actuator control system is flexible enough to carry out the realization of various control strategies. Due to this fact, described in this paper orthosis is able to fulfill various tasks, such as:

- Support and enhancement of a sit-to-stand movement;
- Support and enhancement of a knee flexion and/or extension;

- Continuous passive motion therapy;
- Controlled active motion therapy.

Figure 1 represents the construction of the developed powered knee orthosis.

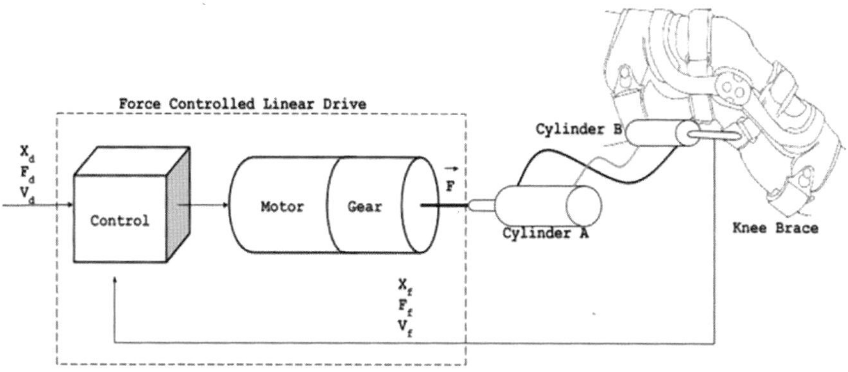

Figure 1. A construction of the powered knee orthosis

The mechanism consists of three main elements: a force controlled linear actuator, a hydraulic transmission and a hard-frame knee orthosis. The hydraulic transmission consists of two double acting differential cylinders, whose respective chambers are connected with elastic pipes. Shaft of the power cylinder connects to elements of orthosis frame, which allows performing knee joint bend and extension motions. It essentially diminishes the orthosis transferring weight in comparison to an electrical powered orthosis with electrical motor and reducer gear as an operating element [4, 5].

Series elastic actuator (SEA) can run in three modes: position control, force control and position control with a given force [1]. Current location of powered knee orthosis elements allows using it in different environment, as well as submergence. Distant dislocation of force sensory system makes it possible. Such construction increases robustness of device, makes it human-safe due to absence of electrical components, however, it brings inaccuracy in measurements, as a result of fluid friction in connecting pipes [1].

3. Experiments

Figure 2 shows the experimental prototype of the powered knee orthosis.
We used water as a working fluid and realized drive-control through the real-time interface MATLAB / XPC Target [1].

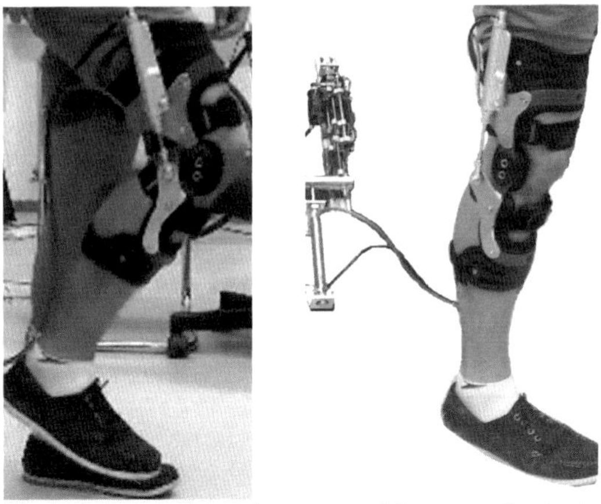

Figure 2. An experimental prototype of the powered orthosis

The aim of the experiments is to study a quality of meeting the requirements, such as low impedance, enhancement of human body capacities and dynamic properties of the device, as well as its applicability as a therapeutic device (CPM/CAM – Therapy).

The designed powered knee orthosis possesses a very low impedance and does not impede the natural motions of the person, who wears it. The measured herein force intensity corresponds to the force of a friction in cylinders and the knee joint and equals to 5 N (See Figure 3).

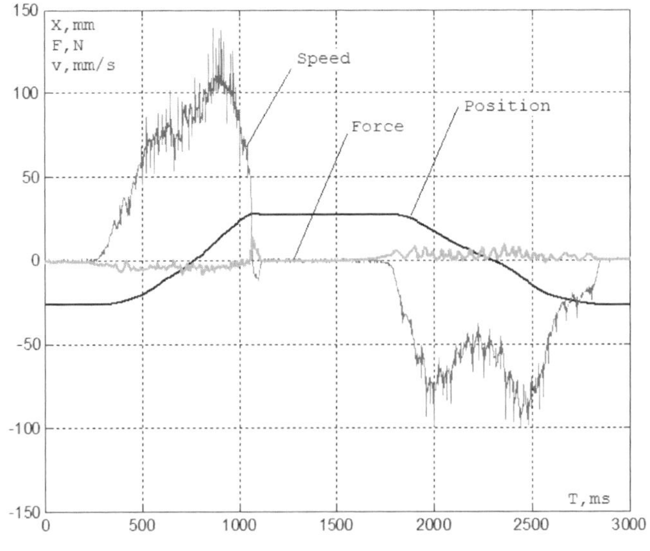

Figure 3. The low impedance experiment

The experiment with a reinforcement of knee joint bend and extension is as well embodied in this paper. Figure 4 shows the intensification of motion in both directions.

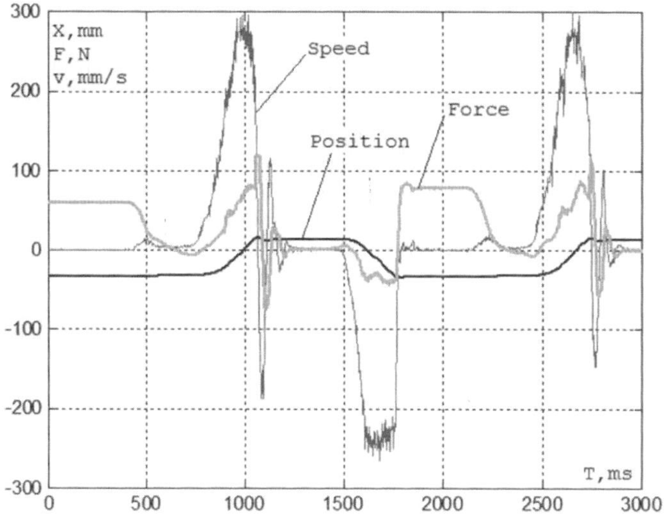

Figure 4.The knee flexion and extension enhancement

The maximal exerted by SEA force is equal to 80N; the maximum achievable speed is 300mm/s.

Another experiment that we conducted is an orthosis force generation, directed against a human leg movement. This approach is widely used in medicine, in controlled active motion therapy: a patient performs an exercise on the knee bend and extension using a training device with an assigned resistance. Figure 5 demonstrates the knee joint bend and extension with a given force. The drive realizes a constant impedance equal to 40N.

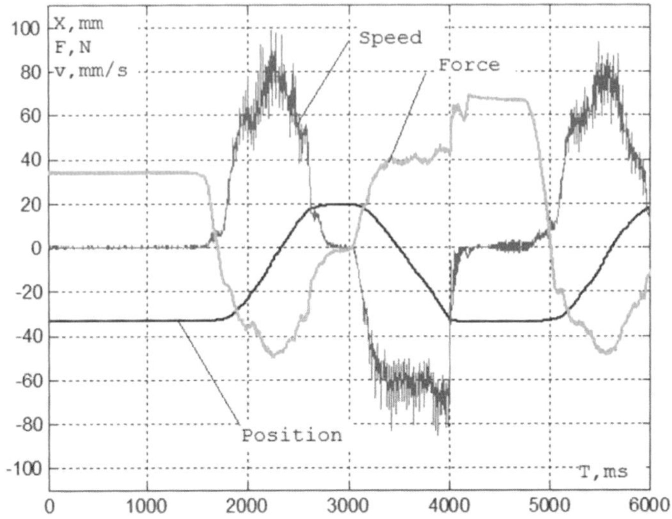

Figure 5. The knee flexion and extension with a given force

4. Conclusion

The designed powered knee orthosis is a low impedance wearable robotic device with a lightweight construction, capable to increase the human force, being convenient and human-safe. This device exhibits a flexible operating system and combines the advantages of the hydraulic system and the electric drive in terms of the control problem. Low impedance is approached by means of using the force controlled series elastic actuator with high compliance and back drivability. It is also obtained due to light weighted orthosis construction. The working element weights only 150g, while the knee brace weights 530g.

A remote force measurement significantly increases safety and reliability of the device. Due to the fact that the powered knee orthosis is built on the basis of the off-the-shelf knee brace, it is easy to wear, to put it on and off.

Even though the system is autonomous from a compressor and hydraulic valves, however it is dependent on electric power. This question requires a further research in the following papers. Among the other disadvantages of the designed powered knee orthosis are a scrubbing effect of liquid in cylinders and an emerging dead zone resulted from a remote force measurement. The magnitude of the dead zone is limited to ± 20N. We suggest using less thick liquids and measuring the force directly in the cylinder chambers in order to rectify the deficiencies.

References

1. A. Safonov, V. Bilous, M. Boiko, U. Schmucker, A. Telesh, Linear Electric Actuator with a Flexible Hydraulic Transmission. In: Mobile Service Robotics, Chapter 34, p. 289-296, 2014. doi:10.1142/9789814623353_0034.
2. A. Melnykov, M. Konyev, F. Palis, U. Schmucker, Linear elastic actuator of a biped robot “ROTTO”. In: Emerging trends in mobile robotics. - Singapore [u.a.]: World Scientific, ISBN 981-432797-2, S. 588-595, 2010 Kongress: CLAWAR; 13 (Nagoya) : 2010.08.31-09.03
3. J. E. Pratt et. al., The RoboKnee: an exoskeleton for enhancing strength and endurance during walking. In: Robotics and Automation, 2004. Proceedings. ICRA '04. 2004 IEEE International Conference on. Bd.3. 2004, 2430-2435 Vol.3. doi: 10.1109/ ROBOT.2004.1307425.
4. P. P. Pott et. al., Series Elastic Actuators for Man-Machine Cooperation. In: Innovative Small Drives and Micro-Motor Systems, 2013. 9. GMM/ETG Symposium. 2013, S. 1-5.
5. C. Souit et al., Design of a knee ankle robotic exoskeleton to induce perturbations during gait. Proc. Int. Congress of Mechanical Engineering (COBEM 2013), no. Cobem, (Ribeirão Preto, SP, Brasil). 2013.
6. M. Bortole et. al., A robotic exoskeleton for overground gait rehabilitation In: Robotics and Automation (ICRA), 2013 IEEE International Conference on. 2013, S. 3356-3361. doi: 10.1109/ ICRA.2013.6631045.
7. U. Onen et. al., Design and Actuator Selection of a Lower Extremity Exoskeleton. In: Mechatronics, IEEE/ASME Transactions on 19.2 (2014), S. 623-632. issn: 1083-4435. doi: 10.1109/ TMECH.2013.2250295.
8. A. Gams et. al., Effects of Robotic Knee Exoskeleton on Human Energy Expenditure. In: Biomedical Engineering, IEEE Transactions on 60.6 (2013), S. 1636-1644. issn: 0018-9294. doi: 10. 1109/TBME.2013.2240682.
9. A. Zoss, H. Kazerooni, A. Chu, Biomechanical design of the Berkeley lower extremity exoskeleton (BLEEX). In: Mechatronics, IEEE/ASME Transactions on 11.2 (2006), S. 128-138.issn: 1083-4435. doi: 10.1109/TMECH.2006.871087.

SECTION–10

MODELLING AND SIMULATION OF CLAWAR

WALL CLIMBING ROBOT MOTION WITH ADAPTIVE VACUUM CONTACT DEVICES

V. G. GRADETSKY, M. M. KNYAZKOV, A. A. KRYUKOVA, E. A. SEMYONOV and A. N. SUKHANOV

Institute for Problems in Mechanics of the Russian Academy of Sciences
101-1, Prospect Vernadskogo, 119526 Moscow, Russia
gradet@ipmnet.ru, sukhanov-artyom@yandex.ru

Simulation technique of studying adaptation motion of mobile wall climbing robot with sliding seal and two wheels is suggested. Such parameters as forces, pressure, velocities of contact device have been taken into account. The results of simulation present dynamic range of the robot motion with restriction on parameters. It is shown the limitation of characteristics of robot adaptive motion over complex surfaces. The obtaining results were used for the robot's prototype design.

Keywords: Mobile robots, adaptive motion, simulation, dynamic parameters, contact devices, sliding seal.

The research is supported by Russian Foundation for Basic Researches, Grant № 15-08-04117

1. Introduction

Adaptive motion realization is the most important task for satisfying reliability of wall climbing (WCR) robot's motion under the conditions when quality of surfaces is unknown for control system. Cracks, cavities, convexities, roughness and other irregularities are often met on the WCR's trajectory which may be appraised as sudden obstacles. WCR has to overcome such irregularities or obstacles while moving.

Many researches have been in progress to create adaptive motion for WCR including robots with vacuum contact devices. The spring actuated sensors were suggested to include in control loop to compensate loss of vacuum [1, 2, 9]. Control algorithms were developed to change the motion direction in the case of extreme condition [3-8]. Special sensors based on various physical phenomena such as ultrasonic, electromagnetic, optical etc. were applied [9].

At the same time process of contact with the surface was not sufficiently studied when automatic changing of vacuum degree for satisfaction of continuous robot's motion is produced.

The goal of our paper is to present some simulation results of adaptive WCR's motion with contact to surface via vacuum device.

Different parameters such as forces, pressure, velocities of robot have been taken into account. Limitations of characteristics of adaptive motion over complex surfaces are presented. The obtaining results can be used for design of WCR with sliding seal.

2. Methods and Experiments

There are two main types of locomotion system's adaptation of the WCR designed for moving over different surface: the mechanical one and software one. Adaptation of the drive system must permit the robot to reach the destination point regardless the surface quality and the changing quality of the environment. Changes in the surface of environment may include movable obstacles, other robots, and in certain cases robot may move from one surface to another. In such circumstances it is required to ensure reliable operation of the locomotion system and reliable contact with the surface.

Mechanical adaptation provides the ability to change the clearance of the robot with use of active and passive suspension elements. The active elements could be hydropneumatic systems or electromechanical systems. Such devices can significantly increase the adaptability of the robot's transport system for moving over the surface, but they seriously complicate the design of the robot and it control system. For the active adaptation elements certain sensors are required which can provide tracking of the surface and send data to the control system of the robot in real-time. The reliability of such a system will depend on the processing speed of data-signals from sensors and control signals goes to the active elements of adaptation.

Springs, elastic joints, low-pressure tires could be passive adaptation elements. Such adaptation of the robot system is not sufficiently complicates the construction of the robot neither requires control while robot's motion, but it cannot give a complete description of the robot's control system.

To improve the reliability of the locomotion system, it is necessary to satisfy the reliability of the robot's contact with the surface, particularly when the robot travels over complex surfaces, such as walls and ceilings, it is necessary to combine adaptation systems. In such cases servo systems are essential for motion, which could provide motion control of the robot in real-time. In addition, the combined adaptation system can obtain complete description of the environment around the robot, which can sufficiently facilitates the work of the operator with supervisory control mode.

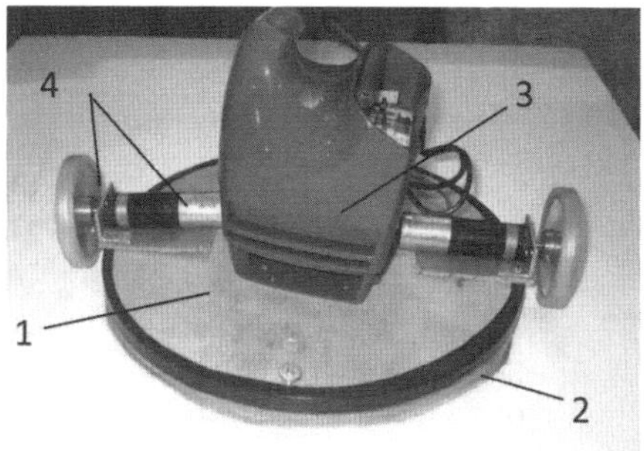

Fig. 1 Overview of WCR with a sliding seal

The design involves a mechanical system (1), a sliding seal (2), a vacuum generating device (3) and servo drives (4) (Fig. 1).

While moving, the WCR can meet cracks and irregularities of various depths. This leads to vacuum level drop, and detach of the robot from the wall. To solve this problem it is needed to apply an integrated structured information processing data received from the sensors: the information about position and speed of the wheel, the current sensor data and vacuum sensor data, and the vacuum level data in the area of contact of the robot with the surface. The design was implemented parallel operation of several control loops of WCR: vacuum level with speed of the air suction control loop and motion PI-control of the robot (Fig. 2).

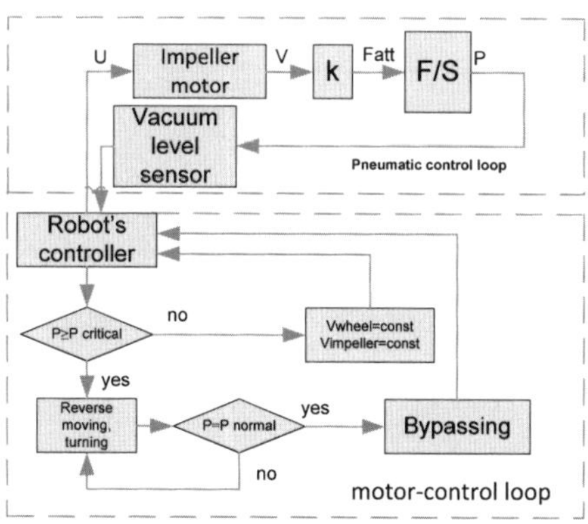

Fig. 2 Control loops of the WCR

Figure 2 shows a pneumatic robot system that monitors the pressure in the vacuum area. It is possible to measure the vacuum level between the robot and the surface using the air-pressure sensor. The data are sent to the controller where the information is compared with the experimentally identified reference values. Corrective actions are set for increasing the speed of rotation of the impeller blades. That is aimed for maintaining the desired vacuum level by increasing the supply voltage to the motor.

Another loop in the microcontroller is intended to overcome the obstacles such as cracks and irregularities in the surface, indicated at Figure 2 as motor-control (bottom). The level of vacuum in the contact area is compared with a critical level with a little dispersion. When the vacuum level is deeper than the critical value, it means that the robot should continue moving at a constant velocity in a predetermined direction. If the vacuum level is lesser than the critical value, then the robot's goes back with turning while the vacuum sensor does not provide with information about the right air-pressure. Then the robot bypasses the suspicious area and returns to the previous trajectory.

Figure 3 contains the mathematical model of the WCR's control system. It is based on P- and PI-control of impeller-motor and wheel-motors.

The electromechanical properties of motors with ironless rotors can be described by means of the following equations:

1. The power supply voltage U_0 is equal to the sum of the voltage drop produced by the current I in the resistance R_a of the rotor winding, and the voltage U_i induced in the rotor:

$$U_0 = I \cdot R_a + U_i \tag{1}$$

2. The voltage U_i induced in the rotor is proportional to the angular velocity ω of the rotor:

$$U_i = k_F \cdot \omega \tag{2}$$

3. The rotor torque M is proportional to the rotor current I via the torque factor k_T:

$$M = k_T \cdot I \tag{3}$$

By substituting the fundamental equations (2) and (3) into (1), it is possible to obtain the characteristics of torque/angular velocity for the DC-motor with an ironless rotor:

$$U_0 = M \cdot R_a + k_E \cdot \omega \tag{4}$$

The discrete transfer function of the close looped servo drive with three feedback loops is presented in (5).

$$\begin{cases} \bar{\Phi}_n(w) = \dfrac{(1+T_{sr}w)\cdot(1+\dfrac{T_s w}{2})}{1+A_1 w + A_2 w^2 + A_3 w^3 + A_4 w^4} \\ A_1 = (T_{sr} + \dfrac{T_s}{2} + \omega_{co}^{-1}) \\ A_2 = T_{sr}\cdot(\dfrac{T_s}{2} + \omega_{co}^{-1}) \\ A_3 = \dfrac{T_{sr}}{\omega_{co}\cdot\omega_{sco}} \\ A_4 = \dfrac{T_{sr}\cdot T_{aqs}}{\omega_{co}\cdot\omega_{sco}} \end{cases}, \quad (5)$$

where T_{sr} is time-constant of the speed controller, T_s is sampling time of digital system, ω_{co} is full cut-off frequency, ω_{sco} is speed cut-off frequency, $T_{eq.s}$ is equivalent time-constant, w is Laplace variable.

Software adaptation system provides a specific set of sensors tracking of changes on the surface relief in real-time and the critical level of vacuum inside the gripper. These sensors are mounted directly on the robot drives. Unlike the mechanical adaptation, algorithmic software does not change the specifications of the locomotion system. The data obtained from the sensors are sent to a control unit which changes algorithms according to the specified motion trajectory of the robot. Thus the robot avoids obstacles. If the obstacle is passable, the robot action algorithms have to set restrictions on the motion velocity or other parameters. This adaptation system does not require of serious complication of the robot design, but makes greater demands on the complexity of the control system of the robot.

The logical block (Subsystem on Fig. 3) gets data from the vacuum sensor and compares with predetermined value. Then it makes a decision for next actions. It is shown on the Figure 4.

WCR is moved over the surface in a predetermined direction at a predetermined speed of rotation of the wheels and the impeller blades of the air suction device. As soon as the robot runs into a crack, the value of the vacuum level is increased, thus the controller increases the power of the impeller-motor to increase the attaching force to the surface. In this parallel circuit in the drive wheels of the robot system are compared with the critical pressure value. If the pressure does not exceed a predetermined level, the robot continues to move along the previously selected path, equalizing the vacuum level by the speed of rotation of the impeller blades.

If the pressure in the vacuum area exceeds a predetermined reference value, the control system concludes that the crack is too great to overcome it, thus the robot moves backward and bypasses the crack.

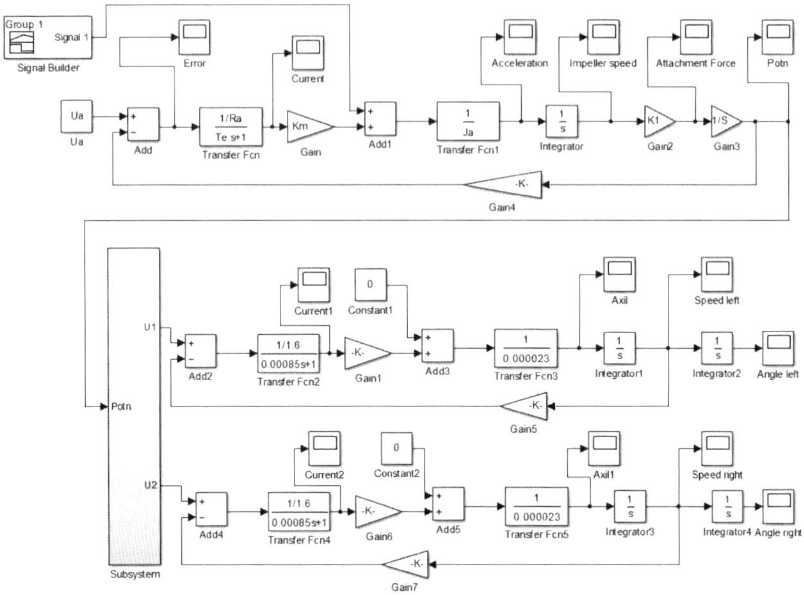

Fig. 3 The mathematical model of the robot's adaptive control system

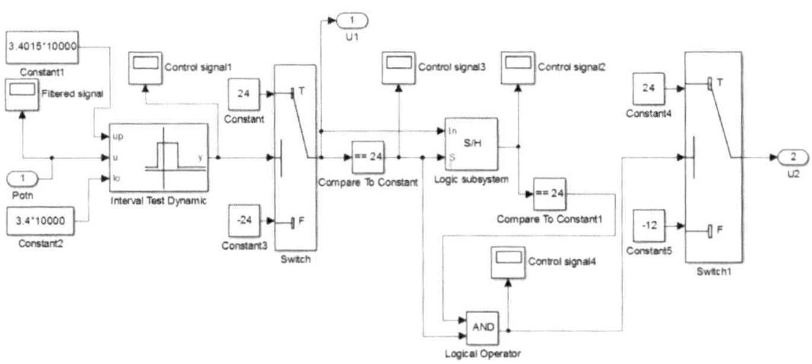

Fig. 4 The logic subsystem for wheel-drives control

The results of the experiment are shown on the figures 5, 6. We can see that when robot meets a crack in the surface it goes back and turns. After the controller gets the information about the equalizing the pressure in the vacuum area WCR bypasses the dangerous obstacle and returns to its trajectory.

Figure 5 describes the situation when the WCR loses air pressure in the vacuum chamber when robot is moving over a crack. It shows data sent to the controller's logical block.

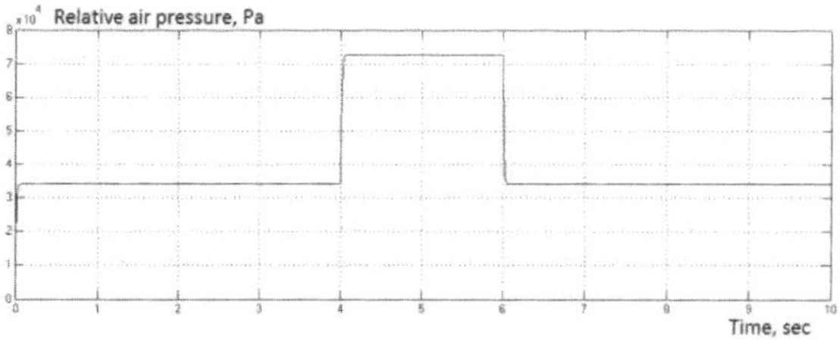

Fig. 5 Relative air pressure in the vacuum chamber of the robot

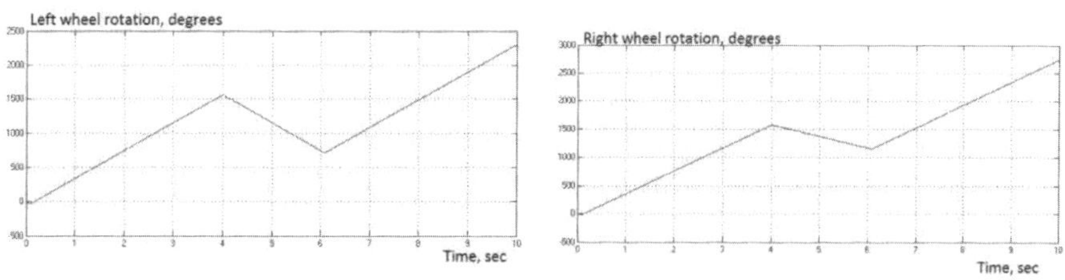

Fig. 6 Left wheel and right wheel rotation

Figure 6 shows the rotation process of robot's wheels while using the designed algorithm of control. They describe motion of the WCR in predetermined direction and reverse motion to avoid the area where robot can lose contact with the surface.

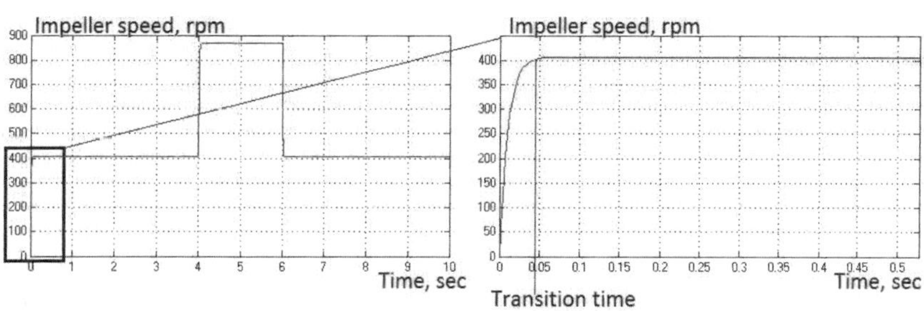

Fig. 7 Dynamics of the impeller speed and its transient process

Figure 7 shows dynamics of the rotation speed of the impeller, which provides changing of attachment force, and rotation of the WCR's wheels.

It is shown that transition time is no more than 0.05 sec.

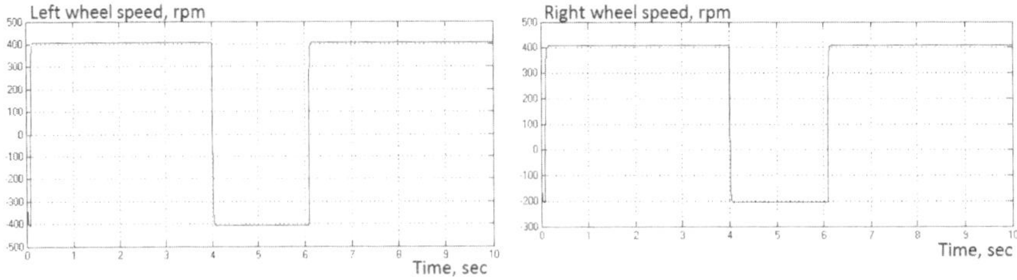

Fig. 8 Dynamics of The WCR's wheels rotation speed

Figure 8 describes dynamics of the rotation speed of the robot's wheels. The transition time is no more than 0.05 sec too. The obtained characteristics show the dynamic processes in the vacuum chamber of the robot and speed change of the impeller and wheels while moving over the undetermined surface. The transition processes are obtained with simulation of the WCR in MATLAB.

3. Conclusion

In this paper a simulation technique of studying adaptation motion of mobile wall climbing robot with sliding seal and two wheels is suggested. Such parameters as forces, pressure, velocities of contact device have been taken into account. The mathematical model of the WCR control system using MATLAB is designed. The results of simulation present dynamic range of the robot motion with restriction on parameters. It is shown the limitation of characteristics of robot adaptive motion over complex surfaces. The obtaining results can be used for the robot's prototype design.

References

1. Abe T. "Wall Climbing Robot for Inspection of concrete structure", Journal of Robotic Society of Japan, 1992, Vol.10, №5 pp. 590-593.
2. Fukuta T., Arai F. "A Study of Wall Surface Mobile Robot" ISME International Journal Ser. C. 1995, Vol.38, №2 pp. 292-299.
3. Longo D., Muscato G. "Adhesion techniques for climbing robots: State of the art and experimental consideration", 11th International Conference on Climbing and Walking Robots and the Support Technologies for Mobile Machines, CLAWAR'08. 08-10 September 2008, Coimbra, Portugal.
4. Luk B., Collie A., Billingsley J. "Robug II: An intelligent wall climbing robot", Robotics and Automation. Proc. Of IEEE International Conference 1991, Vol.3, pp. 2342-2347.
5. Shmid D., Berns K., Ohr J., "Analysis of sliding suction cups for negative pressure adhesion of a robot climbing on concrete walls", Proc. Of CLAWAR 2012 Internat. Conf. Adaptive Mobile Robotics, 23-26 July

2012, Baltimore, USA, pp. 813-820.

6. Illingworth L., Reinfeld D., "Vortex attractor for planar and non-planar surfaces", USA Patent 6.619.922, Sept. 16, 2003.
7. Gradetsky V., Knyazkov M., Fomin L., Chashchukhin V, "Mechanika miniatyurnykh robotov", M. Nauka, 2010, 271p.
8. Gradetsky V., Knyazkov M., Kriukova A., Semionov E., Sukhanov A., Fomin L., "Dynamics of vacuum contact devices of mobile robots", Mechatronics, Automatization and Control, 2014, №9, pp. 17-22, 9.
9. Gradetsky V., Veshnikov V., Kalinichenko S., Kravchuk L., "Control motion of mobile robots over arbitrarily oriented surfaces in the space", M., Nauka, 2001, 359p.

GAIT OPTIMIZATION BASED ON PHYSICS SIMULATION OF 3D ROBOT MODELS WITH A MODULAR ROBOTIC SIMULATION SYSTEM

DENNIS KRUPKE, NORMAN HENDRICH and JIANWEI ZHANG

Department of Informatics, Group TAMS, University of Hamburg, Hamburg, Germany
E-mail: {krupke, hendrich, zhang}@informatik.uni-hamburg.de

HOUXIANG ZHANG

Faculty of Maritime Technology and Operations
Aalesund University College, Norway
E-mail: hozh@hials.no

This paper presents a method for gait optimization of modular robotic locomotion. A physics-based simulation system for development and optimization of locomotion patterns is utilized. A graphical user interface (GUI) allows for interaction with the simulation at runtime and to supervise it. Long-term optimizations can be performed as background processes without GUI. Great benefit arises from the possibility to optimize even not well understood and complex control parameters whose correlations with the behaviour of the associated control algorithm and other parameters are not clear. The successful application of the system for control parameter optimization is shown in two different experiments. Each of these experiments is performed with two different optimization algorithms. The advantages and disadvantages of genetic algorithms versus classical reinforcement learning is discussed in the end.

Keywords: Modular robots; Gait optimization; 3D models; OpenRAVE; ODE.

1. Introduction

Even though the simplicity of modular robots and their locomotion can be be easily understood by humans, their algorithmic control remains a difficult task if stability and efficiency is desired. The efficiency of their movement patterns highly depends on many variable factors like the physical properties of a single module and the environmental characteristics. This paper describes a system that simplifies the process of locomotion development in modular robotics and allows for fast reconfiguration of set-

ups. Thus, similar optimizations can be performed very efficiently, after applying only small changes to the environment or to the robot, by reusing already configured set-ups.
Open source modular robot simulation systems with integrated gait optimization currently do not exist. There are general simulation systems like Gazebo,[1] OpenRAVE[2] or Webots that come with integrated 3D visualization and physics engines but there is neither modular robot specific support nor integrated mechanisms for gait optimization. Modular robotic specific solutions are normally not public and intended to be used just for one specific modular robot or modular robot series of single institutes. This work describes the use of a general tool for chain-like caterpillar or snake robots in gait optimization tasks.

2. Simulation System

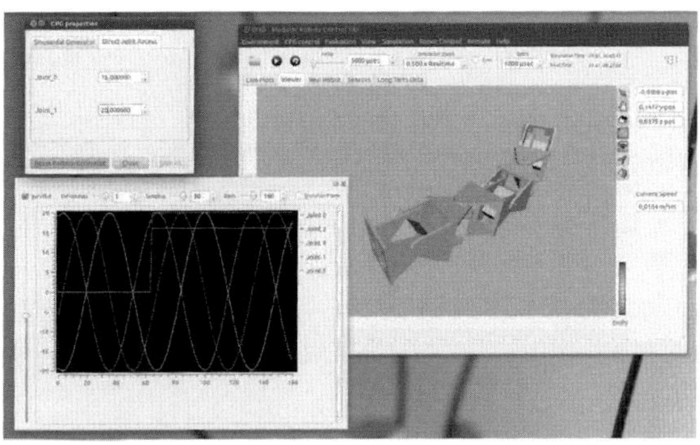

Fig. 1. The GUI of the control system visualizes the posture of the robot. In the plot window the desired angles of the joints are displayed versus real-time. The small window in the upper left corner is used to change important control parameters.

The simulation system is based on OpenRAVE[2] with some modular robotic specific extensions.[3] Comparisons of different locomotion strategies are simplified by using the same simulation or control configuration except the concrete composition of locomotion algorithms. Locomotion algorithms are implemented in dynamic libraries in order to support further extension of the system and to simplify the process of reconfiguration of a scene file. The control GUI allows for visualization of control signals and sensor readings at runtime. Different robot models can be integrated by implementing arbitrary robot control libraries and importing corresponding 3D models for the simulated robots. Models that have been created for 3D-printing can directly be used in the simulation.[4] Figure 1 shows the control GUI in a robot control task. By integrating optimization methods into the modular robot simulation and control system users can as well supervise the current process as identify configuration errors.

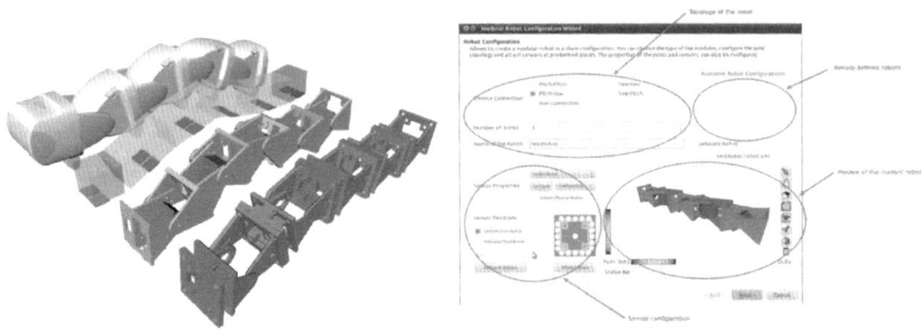

Fig. 2. Left: Four different robot models with similar structure were used in simulations for this work. A detailed model of the CUBO robot (robotechn), a commercial educational robot platform that can be reconfigured manually in very short time. A minimal model of a modular robot is used for first evaluations of new ideas. A 3D model of a prototype with tactile sensors based on YI.[5] The last one is a 3D-printable modular robot.[4] Right: The wizard page for robot configuration is suggested to be used for easy creation of a robot.

3. Optimization of Robot Gaits

In the following section two different experiments demonstrate the utility of the approach presented in this paper. First, a simple parameter is optimized with respect to locomotion speed. Here the parameter has an individual value per joint which increases the number of possible solutions in this optimization task with the number of joints. Second, a group of different parameters with different ranges is optimized. These parameters determine the behaviour of the robot depending on amount of contact with the floor. Since, snake-like locomotion is the result of applying forces to the ground in the oppositional direction of frictional forces, this algorithm tries to achieve the needed degree of contact with the ground. In this algorithm the parameters and their correlations among each other are not easily feasible for the human mind and automated optimization is needed in order to find good results.

For each experiment two different optimization methods are used to calculate the resulting parameter sets. The bio-inspired approach utilizes genetic algorithms (GA) and the classical approach of reinforcement learning makes

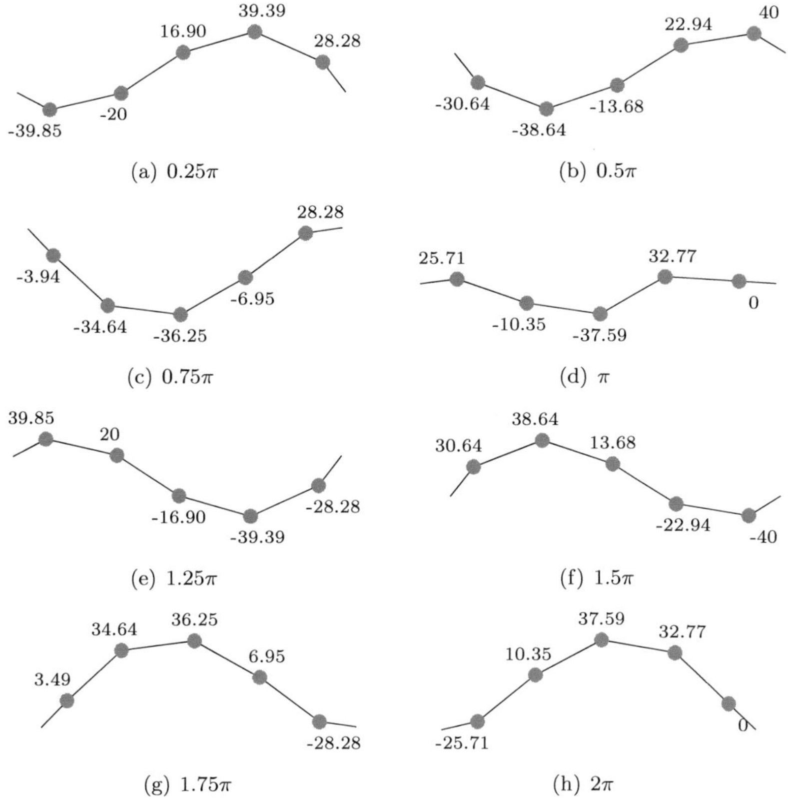

Fig. 3. Different stages of a sine-pattern with uniform amplitudes. Step (a) to (h) represent the whole locomotion cycle of $2*\pi$. The step-size between each stage is $0.25*\pi$. The movements of the robot result in displacement from left to right.

use of a modified version of the great deluge algorithm (GD). The GA is implemented with the GAlib[6] from the MIT which follows Goldberg's theories of genetic algorithms.[7] For the GD the algorithmic principle is extended to be more robust against local maxima by implementing an adaptive neighbourhood definition and retry credits. The longer no appropriate candidate is found the larger is the search radius. If still no candidate is available the search radius is reduced to the initial value. Both algorithms use the same weighting function for evaluation of passed simulations. The weighting function is based on the travelled distance in a fixed number of simulation steps, calculated by the Euclidean distance between the starting point and the position when the final simulation step is passed.

3.1. *Sine-wave optimization*

In the first experiment a pitch-pitch configured robot with five joints is simulated. The 3D models are YI modules[8] and the locomotion is generated by a chain of sinusoidal generators. As depicted in[5] sine functions allow for efficient locomotion with low computational effort and feature appropriate parameters to influence the dynamics of the resulting locomotion. In the experiment the influence of non-uniform amplitudes is analysed. The range of valid amplitudes is limited to the interval [0 40] while the phase parameters are set to a fixed reasonable value that generates smooth locomotion. Since the amplitude parameter is related to the stride length in walking models it is expected that the largest amplitudes would result in fastest locomotion. The locomotion pattern at different stages with fixed amplitudes of 40 is illustrated in figure 3.

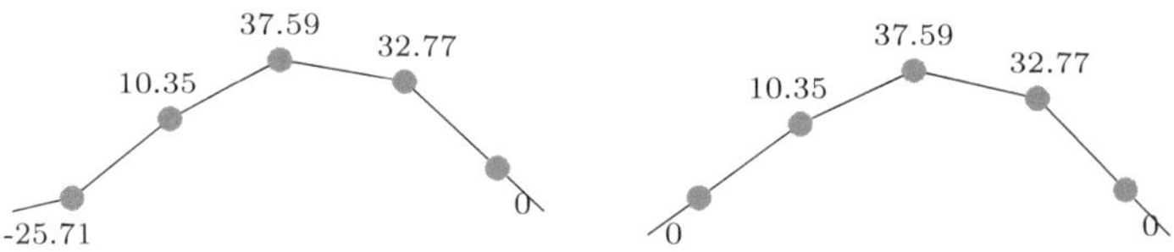

(a) Uniform amplitudes (b) Amplitude of last joint is zero

Fig. 4. Uniform and optimized sine-patterns in comparison: In 4(a) amplitudes are uniform. In 4(b) an optimized version of the same stage with individual amplitudes determines the posture of the robot. In simulations robots using amplitudes of version 4(b) had much greater displacement in the same amount of time.

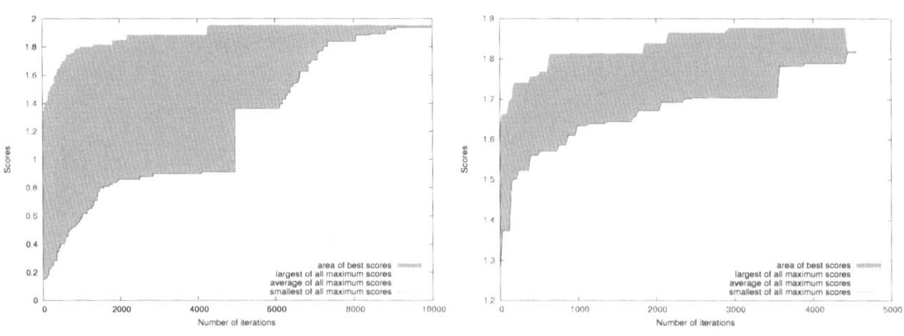

Fig. 5. Minimum, maximum and average scores of all amplitude optimizations. Left: Results of the repeated GD optimization. Right: Results of the GA optimization.

Figure 5 shows worst case and best case maximum scores as well as the average of all 30 trials. A maximum possible score is not found in every

trial. Single trials seem to get stuck in local maxima. Finally all three graphs end up in the same point because there was only one experiment that had so many repetitions until the program terminated.
The results show surprisingly that the most efficient locomotion results if the last joint is fixed. Figure 4 compares two postures where this effect has the largest influence. The posture based on the optimized locomotion reduces backward slippage by increasing the friction in the opposite direction of travel.

Table 1. Best 8 amplitude sets from GD-optimizations out of 30 trials

Amplitude	1	2	3	4	5	6	7	8
Joint 0	40	40	40	40	40	40	40	40
Joint 1	40	39.5	39.75	40	40	40	40	40
Joint 2	39.75	40	40	40	40	40	39.75	40
Joint 3	40	39.75	40	40	40	40	40	40
Joint 4	0	0	0	0.75	0	0	0	0
Distance [m]	1.958	1.956	1.954	1.953	1.950	1.949	1.949	1.947

Table 2. Best 8 amplitude sets taken from GA-optimizations out of 29 iterations

Amplitude	1	2	3	4	5	6	7	8
Joint 0	37.5	39.5	38.75	39.75	39.25	37	38.25	39.75
Joint 1	39.25	40	40	38.75	39.5	39.5	38.75	39
Joint 2	39.25	39	38	40	39.75	39.25	40	40
Joint 3	39.5	39.25	38.25	37.25	39.25	38.75	39.75	36.75
Joint 4	1	2.25	0.5	0.5	0.75	2.25	0	0.75
Distance [m]	1.878	1.874	1.865	1.846	1.839	1.828	1.822	1.820

3.2. *Optimization of adaptive locomotion*

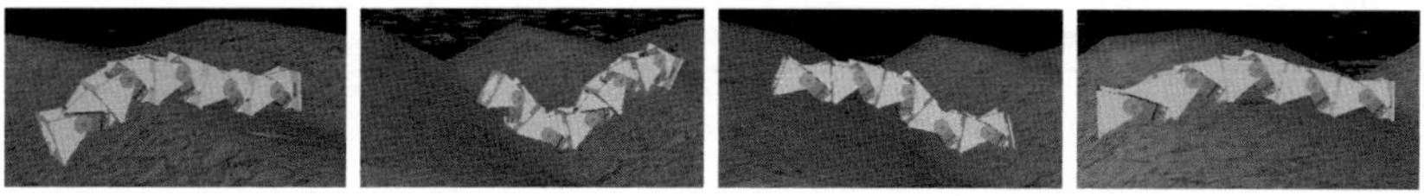

Fig. 6. Different stages while the robot passes an obstacle show how convex and concave shapes are handled in locomotion. The second image shows the most challenging part where very large offsets have to be applied to adjust the shape of the robot.

The task in this experiment is to pass an irregularly shaped object which is only possible with adaptive algorithms that make use of the perception of environmental properties. By analysing the current touch state between the robot and the ground the normal shape of the robot is adjusted[9] in order to pass the obstacle. Physically this is enabled by establishing the necessary amount of friction with the ground. Most of the parameters of this algorithm and their correlations among each other are not feasible for humans. In this experiment these parameters are automatically optimized with the goal to move as far as possible in a limited number of simulation steps. Figure 6 shows four different stages of one successful climbing trial after optimization with GD-algorithm.

Table 3. Parameter values of candidates from best 8 GD-optimizations out of 20

Parameter	1	2	3	4	5	6	7	8
Get-stuck	110	85	185	160	160	185	60	160
Recovering	0.33	0.74	0.56	0.55	0.56	0.13	0.37	0.18
Constant 1	11.8	3.3	2.7	6.6	2.6	1.4	0.6	2.4
Constant 2	21.6	1.9	0.7	1.6	2.3	0.8	0.7	1.3
Constant 3	11.8	9.7	5.1	6.5	2.7	2.2	2.5	1.9
Constant 4	0.6	1.3	2.4	5.34	2	1	16.2	1.4
Constant 5	20.2	1.1	1.2	0.9	0.4	0.9	1.5	0.1
Constant 6	1.5	5.9	1.7	6.8	0.5	0.6	0.3	0.9
Constant 7	2.36	5.1	5.76	6.5	4.1	2.9	4.3	2.7
Constant 8	11	1.5	1.2	1.7	2.3	0.9	0.9	0.6
Distance [m]	3.96	3.88	3.86	3.81	3.77	3.73	3.72	3.66

Table 4. Parameter values of candidates from best 8 GA-optimizations out of 20

Parameter	1	2	3	4	5	6	7	8
Get-stuck	175	125	100	125	175	175	125	150
Recovering	0.09	0.13	0.2	0.14	0.11	0.06	0.21	0.05
Constant 1	1	0.7	0.9	0.9	1	0.8	1	0.9
Constant 2	0.9	0.9	0.7	1	0.7	0.8	0.6	1
Constant 3	0.7	1	1	1	0.7	1	1	0.1
Constant 4	1	0.5	0.2	1	0.1	0.9	0.3	1
Constant 5	0.1	0.5	0.4	0.7	0.3	0.8	0.4	0.7
Constant 6	0.5	0.5	0.6	0.8	0.3	0.2	0.4	0.5
Constant 7	1	0.9	1	1	0.9	0.8	1	0.9
Constant 8	0.7	0.7	0.7	1	0.8	0.6	0.9	0.6
Distance [m]	3.19	3.11	3.08	3.07	3.05	3.02	3.02	3.02

4. Summary/Outlook

In this paper a locomotion optimization method is presented that is based on the interaction between a physics-engine and detailed 3D models of real robots. Two different optimization algorithms are compared in two different experiments. After all, both methods are capable of creating sufficient results in both experiments. GAs are more efficient at the beginning phase of the optimization but tend to converge very slowly. The GD-based approach is generally slower but achieves better results in the end. Future work should try to combine both methods to a more efficient algorithm that incorporates

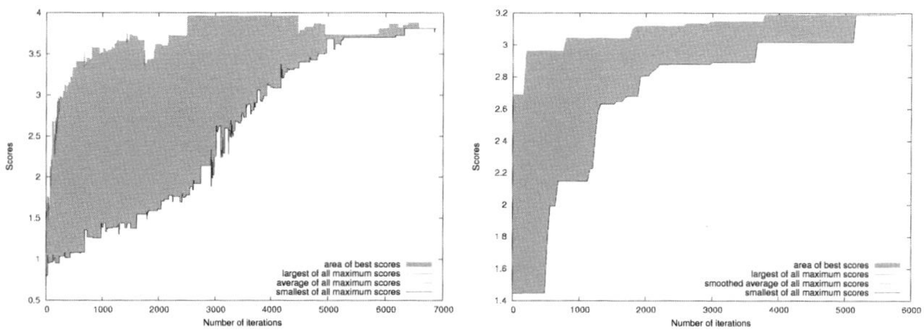

Fig. 7. Minimum, maximum and average of the resulting scores of repeated adaptive locomotion experiments. Left: Modified GD algorithm is used. Right: GA algorithm.

both advantages. The proposed system seems to be suitable for different optimization tasks. To increase the credibility of the method the results should be compared with measurements from real world experiments.

References

1. N. Koenig and A. Howard, Design and use paradigms for gazebo, an open-source multi-robot simulator, in *IEEE/RSJ International Conference on Intelligent Robots and Systems*, (Sendai, Japan, 2004).
2. R. Diankov and J. Kuffner, *Robotics Institute, Pittsburgh, PA, Tech. Rep. CMU-RI-TR-08-34* , p. 79 (2008).
3. D. Krupke, G. Li, J. Zhang, H. Zhang and H. P. Hildre, Flexible modular robotic environment for research and education, in *26th European Conference on Modelling and Simulation. ECMS2012*, 2012.
4. D. Krupke, F. Wasserfall, N. Hendrich and J. Zhang, *Printable Modular Robot: An Application Of Rapid Prototyping For Flexible Robot Design*, in *Mobile Service Robotics*, , ch. 33, pp. 277–285.
5. H. Zhang, J. González-Gómez and J. Zhang, A new application of modular robots on analysis of caterpillar-like locomotion, in *Mechatronics, 2009. ICM 2009. IEEE International Conference on*, april 2009.
6. M. Wall, Matthew's genetic algorithm library http://lancet.mit.edu/ga/, Accessed: 10/02/2015.
7. D. E. Goldberg, *The Design of Innovation: Lessons from and for Competent Genetic Algorithms* (Kluwer Academic Publishers, Norwell, MA, USA, 2002).
8. J. González-Gómez, H. Zhang, E. Boemo and J. Zhang, Locomotion Capabilities of a Modular Robot with Eight Pitch-Yaw-Connecting Modules, in *9th International Conference on Climbing and Walking Robots. CLAWAR06*, September 2006.
9. G. Li, Hierarchical control of limbless locomotion using a bio-inspired cpg model, PhD thesis, University of Hamburg, (Von-Melle-Park 3, 20146 Hamburg, 2013).

COMBINATION OF AFFINE DEFORMATION AND DYNAMIC MOVEMENT PRIMITIVE IN LEARNING HUMAN MOTION FOR REDUNDANT MANIPULATOR

JIN HU and RONG XIONG*

Institute of Cyber-Systems and Control, Zhejiang University, Hangzhou, 310027, China

* *Corresponding author, E-mail: rxiong@iipc.zju.edu.cn*

LfD(Learning from Demonstration) has the advantage of requiring no expert knowledge about the robot itself, which make large-scale application of robots possible. In this paper, we present a novel approach based on combination of affine deformation and DMP(dynamic movement primitive) to deal with two fundamental problems in LfD: data gathering and policy deriving. In the proposed approach, demonstrated motion data are gathered from optical-based motion capture system and DMP is used to sketch feature and derive policy. Combined with affine deformation, joint trajectory derived from the control policy can be refined so that the manipulator's physical constraints satisfied, the end point accuracy preserved and the execution time optimized. We verify the feasibility of our approach by reproducing a series of different motions with various trajectory profiles on a humanoid robot's arm basing on limited human demonstrations.

Keywords: affine deformation; DMP; LfD; redundant manipulator.

1. Introduction

Comparing with traditional methods,[1] motion planning based on LfD is advantageous in many ways. Learning drastically reduces the workload and complexity of robot programming. Human motion often contains prior knowledge such as environment information or preferred configuration. These information is important for specialized, complex motion, such as table tennis strike, and is hard to encode using traditional interpolation method.

The LfD learning problem can be segmented into two fundamental phases: gathering examples, and deriving policy from such examples.[2] Generally speaking, we need to solve two problems:

The first problem is what to learn. That means how to collect and record example motion data from demonstration. Precondition of successful LfD is that the recorded data form the teacher can be mapped to the robot and executed properly. The second problem is how to derive. That means how to sketch features of human motion and derive control policy from it. Human demonstration data are usually of high dimension and unstructured, which make traditional trajectory encoding method such as spline or polynomial interpolation inapplicable.

In this paper, we proposes a novel human motion learning framework for redundant manipulator based on combination of affine deformation and DMP. The main contribution of the framework includes: (1) apply affine deformation in the mapping process and use DMP to derive policy, which gives a new viewpoint to solve LfD problem; (2)refine trajectory reproduced by DMP system with affine deformation again, which can take physical constraints or other optimization criteria (such as execution time) into account.

We applied our approach on a 7-DOF redundant manipulator to learn typical table tennis strike motion from human. As a result, the humanoid robot can reproduce the strike motion at different target position with high similarity and efficiency.

The remainder of this paper is organized as follows. Related work is briefly reviewed in section 2. Section 3 describes the motion learning framework. Motion capture system structure and experimental results are demonstrated in section 4. Section 5 concludes this paper.

2. Related Work

In LfD, how to record and gather motion data is a fundamental problem. According to survey done by Argall et.al,[2]current methods can be concluded into two large categories: demonstration methods and imitation methods.

Demonstration methods are intuitive approaches. User may use joystick to teleoperate the learner system,[3] or drive its passive joint by hand.[4] These methods record execution data of each motor without embodiment mapping. But motor level demonstrations are difficult on redundant manipulators and data usually contain lots of noise.

Imitation methods try to get motion example directly from human. Typical approaches need human teacher wear motion capture device to measure body's state(position and time) when performing example action. Device have many types, such as optical based system,[5] exoskeleton motion capture systems.[6] Major problem is the size and structure difference between human and robot, which may lead to non-reachable position or trajectory after embodiment mapping.

Another fundamental problem of LfD is how to encode motion and derive control policy for execution. Existing approaches of policy deriving at end effector-level can be roughly divided into three categories: First, statistical model represented by Gaussian Mixture Model(GMM).[7] But trajectory may be discontinuous, thus causing undulation. Second, trajectory encoding using spline (e.g. B-spline).[8] For these method, choosing appropriate basis is difficult for a specific task. Third, trajectory encoding based on dynamic system.[9] Spring damping system is adopted to model motion trajectory and supervised learning method can be used to learn model parameter. However, it is difficult to consider physical limits of manipulator in such method.

Pham et.al[10] proposed affine deformation which deform trajectory in joint level. It make no use of exogenous model, e.g. spline function, Gaussian function or Markov property. Goal such as trajectory similarity or efficiency can be treated as criteria function. Constraints such as physical limits can be treated as restrictive conditions. Thus, problem of deformation can be transformed into optimization problem, which can be solved efficiently. In this paper, we combine end-effector level trajectory encoding for policy learning with joint level affine deformation for trajectory refinement, which provide new viewpoint to solve LfD problem.

3. Human Motion Learning Framework

The learning framework is illustrated in Figure 1. We first collect motion data using optical based systems. The captured data is processed to eliminate outlier point and noise. Then we deform the motion trajectory using affine deformation in order to satisfy physical constraints of robot and keep similarity. Result is projected into arm model to get corresponding end-effector trajectory. Next, we train these trajectories using DMP trainer to get their weight parameters. After learning, new trajectory is reproduced by transform system in task space giving new desired target or motion style and transformed into joint trajectory using inverse kinematic. Finally, affine deformation is applied again to generate executable trajectory to satisfy physical constraints or optimize other criteria(such as execution time). We have detailed discussion of the framework in 3.1 and 3.2.

3.1. *Human motion mapping into redundant manipulator*

In this section, we want to solve the problem of what to learn, specifically the embodiment mapping from human motion to robot arm. We use affine

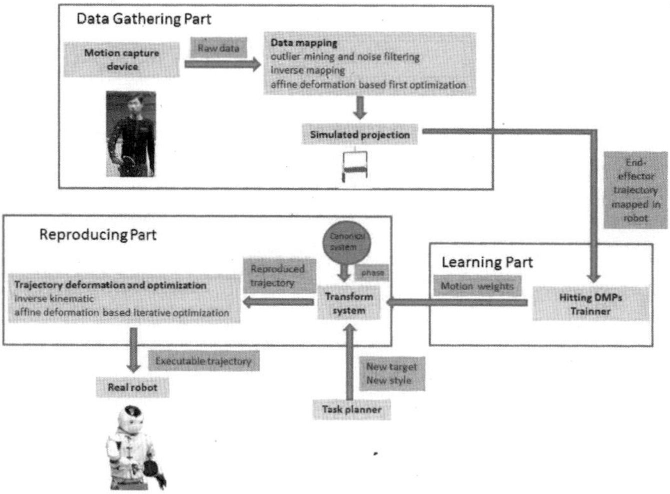

Fig. 1. Human motion learning framework using DMP and affine deformation

deformation to transform original human motion data gathered from optical based motion capture system into joint trajectory available for the robot.

Suppose we get captured joint position vector series $\boldsymbol{\theta}(t_i|i = 0, \cdots, n)$, then mapping approach has two goals:

1) Maintain high similarity between trajectory before and after mapping;

2) Satisfy physical limits of robot manipulator, such as position, velocity and acceleration constraints of each joint.

In affine deformation approach, deformation is calculated by linear mapping, i.e. $\boldsymbol{\theta}'(t_i) = \boldsymbol{\theta}(t_0) + \mathcal{M}(\boldsymbol{\theta}(t_i) - \boldsymbol{\theta}(t_0))$, where $\mathcal{M}$ is a linear matrix. Thus, similarity in the goal of the trajectory can be changed into similarity between $\mathbf{M}$ and identity matrix $\mathbf{I}$. As mentioned by Pham et.al,[10]one criteria of similarity between matrix is Frobenius distance. Using transformation matrix $\mathbf{M}$ in orthonormal basis $\mathcal{B}$, goal of similarity can be formulated as $\|\mathbf{M} - \mathbf{I}\| = \sqrt{\sum_{ij} m_{ij}^2} = \|\mathbf{m}\|_2$($\mathbf{m}$ is vector defined by rearrangement of non-zero terms in $\mathbf{M} - \mathbf{I}$)

Physical constraint can be viewed as inequality constraints, and formulated as

$$\begin{aligned} \mathbf{A}\boldsymbol{\theta}'(t_i) &\le \mathbf{b} \\ \mathbf{A}^v\dot{\boldsymbol{\theta}}'(t_i) &\le \mathbf{b}^v \\ \mathbf{A}^a\ddot{\boldsymbol{\theta}}'(t_i) &\le \mathbf{b}^a \end{aligned} \tag{1}$$

where $\boldsymbol{\theta}'(t_i)$, $\dot{\boldsymbol{\theta}}'(t_i)$ and $\ddot{\boldsymbol{\theta}}'(t_i)$ are deformed trajectory position, velocity and acceleration corresponding to position limits $\mathbf{b}$, velocity limit $\mathbf{b}^v$ and acceleration $\mathbf{b}^a$.

These two goal can be formulated as uniform quadratic optimization problem. The cost function is

$$\min \frac{1}{2} \|\mathbf{m}\|_2^2 \tag{2}$$

subject to inequality constraint

$$\begin{cases} \mathbf{APV}_i\mathbf{m} \le \mathbf{b} - \mathbf{A}\boldsymbol{\theta}(\mathbf{t_i}) \\ \mathbf{A}^v\mathbf{P}(\mathbf{V}_{j+1} - \mathbf{V}_j)\mathbf{m} - \mathbf{b}^v\Delta t \\ \le -\mathbf{A}^v\big(\boldsymbol{\theta}(t_{j+1}) - \boldsymbol{\theta}(t_j)\big) \\ \mathbf{A}^a\mathbf{P}(\mathbf{V}_{k+2} + \mathbf{V}_k - 2\mathbf{V}_{k+1})\mathbf{m} - \mathbf{b}^a\Delta t^2 \\ \le -\mathbf{A}^a\big(2\boldsymbol{\theta}(t_{k+1}) - \boldsymbol{\theta}(t_{k+2}) - \boldsymbol{\theta}(t_k)\big) \end{cases} \tag{3}$$

These inequalities derived from (1) using affine deformation equation (4) and forward difference method (5).

$$\boldsymbol{\theta}'(t_i) = \boldsymbol{\theta}(t_i) + \mathbf{PV_i m} \tag{4}$$

$$\begin{cases} \dot{\boldsymbol{\theta}}'(t_i) = \dfrac{\boldsymbol{\theta}'(t_{i+1}) - \boldsymbol{\theta}'(t_i)}{\Delta t} \\ \ddot{\boldsymbol{\theta}}'(t_i) = \dfrac{\dot{\boldsymbol{\theta}}'(t_{i+1}) - \dot{\boldsymbol{\theta}}'(t_i)}{\Delta t} \end{cases} \tag{5}$$

where Δt is the time interval between two captured points.

This problem can be solved effectively using standard constrained optimization techniques. We call the whole process *affine deformation based first optimization.*

3.2. *Trajectory learning and refinement for redundant manipulator*

In last section, human demonstration data has been transformed into motion of robot arm, which suit manipulator's physical limits. Then, these joint position series are casted into simulated 7-DOF arm with exactly the same structure of real manipulator.

The reason why we need to do the cast process is to reduce the side-effect of size difference between human and robot. After casting, trajectory of the end-effector is obtained, which act as input training example for policy deriving.

The gathered motion data are encoded by DMP system. Different from original DMP,[6] velocity information of target are important in robot reaching task, for example the table tennis playing tasks. So one modified type of DMP with moving target is applied in this section. Details are shown by J .Kober et.al.[11]

Locally weighted regression has been applied to train the parameters of w_i given motion trajectory. After getting weights, we can easily reproduce new trajectory giving different trajectory goals. New trajectory is transformed into joint trajectory using inverse kinematic. The problem of inverse kinematic is difficult to tackle when taking the joint limit into account. Affine deformation gives a solution.

We refine the reproduced joint trajectory through another optimization, noted as *affine deformation based iterative optimization.* The goal of this optimization is to minimize the difference in the end-effector trajectory while ensuring safety of the robot.

To do this, we define new cost function to obey dynamic constraints, as shown in (6)

$$C_1 = \min \int_0^T F_1 dt \tag{6}$$

In above function, $F_1 = [F(q_i) - F(q_i^o)]_x^2 + [F(q_i) - F(q_i^o)]_y^2 + [F(q_i) - F(q_i^o)]_z^2$, where function F represents robot forward kinematic function used to compute end-effector pose given joint position. q_i and q_i^o is joint position after and before refinement respectively. Similarity between end-effector trajectory before and after refinement is calculated from Euclidean distance criterion. Deformation matrix is computed from iteration optimization method at each time instant by minimizing Euclidean distance between these two trajectories.

Other constraints can also be added into optimization also. Here we take time optimization as example. Update cost function (6) by adding time criteria term $n\Delta t$, we have

$$\min_{w,\Delta t} w_1(C_1) + (1-w)(\frac{n\Delta t}{T}) \tag{7}$$

where $w \in [0,1]$ reflects the importance of different goals, T is original duration. Then, optimization is done follow the same rule described above.

4. Experiment

We demonstrate robot learning human table-tennis strike to verify our motion self-learning approach.

In previous work,[12] we have built two humanoid robots. Each equipped with two 7-DOF redundant arm to fulfill flexible table tennis playing and serving task.

As noted in section 3.1, we use optical based motion capture system "OptiTrack S250e" to record human motion trajectory. The system will observe and record 3D positions of visual markers every 8ms. 9 markers attached on teacher's body arm are carefully arranged, where two in main body, two in upper arm, two in forearm and three in paddle. After demonstration, we utilize Kalman filter to eliminate outlier point and moving-average filter to decrease noise of marker trajectories, thus, 3D position time series of nine markers are gathered as the example dataset for following learning. The human demonstration of table tennis strike is shown in figure 3(a).

Next, through inverse mapping we transform nine 3D marker's trajectories into teacher's joint trajectory. And then, affine deformation is used to map human joint trajectories to robot's joints regarding physical limits. The results are shown in figure 2.

Modified DMP with moving target is employed to learn the weight parameter of end-effector trajectory. We don't directly take the trajectory of wrist marker or racket marker as input for learning. Instead, joint trajectories from last step is projected to simulated arm model to derive end-effector trajectory. That will further reduce error caused by size difference. In the reproduction procedure, different striking movement will be generated by DMP according to new target point and be transformed into joint-space using inverse kinematic algorithm.

Before execution, joint trajectory generated from inverse kinematic is deformed again to optimize execution accuracy and execution time without breaking physical limits. The results of reproduction are shown in figure 4, 5, and result of time optimization is shown in figure 6.

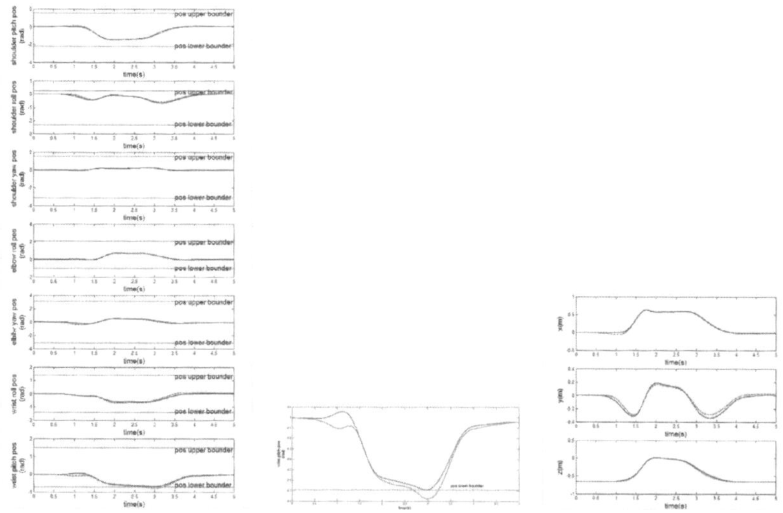

(a) 7 Joint position trajectory comparision (b) Enlarged view of wrist pitch joint trajectory (c) End effector trajectory comparision

Fig. 2. Result of *affine deformation based first optimization*(red dot line: before deformation, blue solid line: after deformation). In 2(a), 2(b) , we can see wrist pitch joint trajectory is deformed into safe area after it exceeds position lower bound.

(a) Human demonstration of table tennis forehand strike (b) Execution of original human motion mapped into manipulator

Fig. 3. Comparison of human demonstration and robot execution

(a) Reproduction of similiar move adding offset 0.2m in z axis (b) Reproduction of similiar move adding offset -0.2m in z axis

Fig. 4. Snapshot of strike move given different target

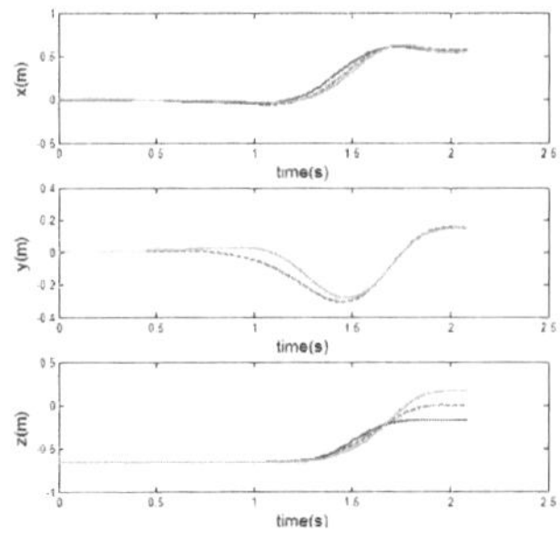

Fig. 5. End effector trajectory comparison given different target(red dot line: original end effector trajectory, blue solid line: trajectory with target adding offset -0.2m in z axis, green solid line represent trajectory: target adding offset 0.2m in z axis)

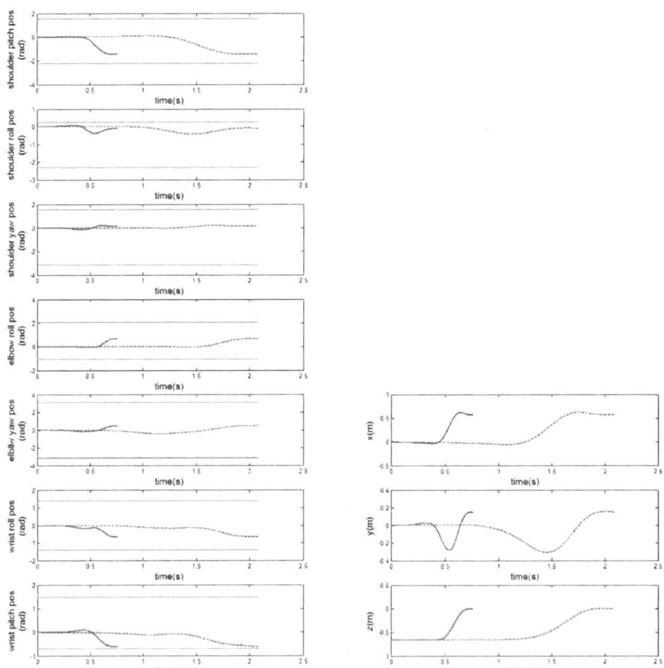

(a) Joint position trajectory comparision of time optimization

(b) End effector trajectory comparison of time optimization

Fig. 6. Result of time optimization(red dot line: before optimization, blue solid line: after optimization. Execution time saved: 1.3233s, but trajectories never go beyond the physical limits)

5. Conclusion

In this paper, we present a novel human motion learning framework for redundant manipulator based on combination of affine deformation and DMP. We show how existing techniques can be combined into a complete learning framework to make up drawback and improve the quality and execution efficiency of learned trajectories.

Our future work will focus on further improvement of robot's motion performance as a ping-pong player. It will include learning different type of human strike, which need motion segmentation and classification. In addition, a more complicated AI problem is to develop higher layer of decision system to choose and connect different motion primitive in longterm match.

References

1. L. Biagiotti and C. Melchiorri, *Trajectory planning for automatic machines and robots* 2008.
2. B. D. Argall, S. Chernova, M. Veloso and B. Browning, *Robotics and autonomous systems* **57**, 469 (2009).
3. P. K. Pook and D. H. Ballard, Recognizing teleoperated manipulations, in *Proceedings. IEEE International Conference on Robotics and Automation*, 1993.
4. A. G. Billard, S. Calinon and F. Guenter, *Robotics and Autonomous Systems* **54**, 370 (2006).
5. R. Amit and M. Matari, Learning movement sequences from demonstration, in *Proceedings. The 2nd International Conference on Development and Learning*, 2002.
6. P. Pastor, H. Hoffmann, T. Asfour and S. Schaal, Learning and generalization of motor skills by learning from demonstration, in *IEEE International Conference on Robotics and Automation*, 2009.
7. S. Calinon, F. Guenter and A. Billard, *IEEE Transactions on Systems, Man, and Cybernetics, Part B: Cybernetics,* **37**, 286 (2007).
8. A. Ude, C. G. Atkeson and M. Riley, Planning of joint trajectories for humanoid robots using b-spline wavelets, in *Proceedings. IEEE International Conference on Robotics and Automation*, 2000.
9. A. J. Ijspeert, J. Nakanishi, H. Hoffmann, P. Pastor and S. Schaal, *Neural computation* **25**, 328 (2013).
10. Q.-C. Pham and Y. Nakamura, Affine trajectory deformation for redundant manipulators., in *Robotics: Science and Systems*, 2012.
11. J. Kober and J. Peters, Movement templates for learning of hitting and batting, in *Learning Motor Skills*, 2014 pp. 69–82.
12. R. Xiong, Y. Liu and H. Zheng, A humanoid robot for table tennis playing, in *IEEE Workshop on Advanced Robotics and its Social Impacts (ARSO)*, 2011.

SECTION–11

PERCEPTION, LOCALIZATION AND RESCUE OPERATIONS

MULTI-SESSION SLAM OVER LOW DYNAMIC WORKSPACE USING RGBD SENSOR

YUE WANG[1], RONG XIONG[1*], SHOUDONG HUANG[2], JUN WU[1]

[1] *State Key Laboratory of Industrial Control Technology, Zhejiang University, Hangzhou, P.R. China.*

[2] *Centre for Autonomous Systems, School of Electrical, Mechanical and Mechatronics Systems, University of Technology, Sydney, Australia.*

[*] *Corresponding author: rxiong@iipc.zju.edu.cn*

Climbing robot, flying robot or human wearable devices usually execute daily tasks in a pre-defined workspace sharing with humans. Long-term operation for these robots posts three challenge: 3D pose estimation, cost limitation and unexpected low dynamics. To address these challenges, we propose a solution for performing multi-session SLAM using a RGBD sensor. The main model is a multi-session pose graph, which evolves over the multiple visits of the workspace. When the robot explores the new areas, its poses will be added to the graph. The poses in the graph will be pruned if their corresponding 3D point scans are out of date. Thus the scans corresponding to the poses kept in the current graph will always give a map of the latest environment. To detect the changes of the environment, an out-of-dated scans identification module is proposed. Pruning of the poses also decreases the computational burden in graph optimization. Experimental results using real world data acquired by a Kinect sensor show that the proposed framework is able to manage the map in date for low dynamic environments with a reduction in complexity and an acceptable error level compared to the method saving all poses.

Keywords: SLAM, low dynamic environment, RGBD sensor.

1. Introduction

Simultaneous localization and mapping (SLAM) has been a core technique enabling the autonomy of mobile platforms, such as robot car[1] and autonomous underwater vehicle.[2] Compared to these high cost and large devices, the climbing, flying, or human wearable devices began to raise the attention in recent years due to the high flexibility, small scale and low cost. These devices also call for SLAM to achieve the similar capacity of long-term operation. The main challenge for such a solution includes three aspects: (1) the climbing, flying, or human wearable devices have a motion pattern with more frequent change of orientation due to the disturbance from environment; (2) these devices aim on low cost, light-weighted and small scale, hence expensive or heavy sensors cannot be equipped; (3) these devices usually work in a pre-defined human sharable workspace with low dynamics, which means objects in the workspace may be moved, added or removed between multiple sessions.

To address the first two challenges, we apply the consumer-level RGBD sensor for perception, enabling the 3D pose estimation but also a dense environment map for subsequent navigation. The third challenge is caused by the change of objects (move, add, remove) across multiple sessions. This kind of dynamics calls for updating of the map, which is triggered by detecting the environment change from the map. In summary, the proposed solution can be applied for multi-session mapping in a low dynamic workspace with a low cost RGBD sensor.

The system consists of a multi-session SLAM component and a graph management component. The former built a SLAM graph while the latter,[3] out-of-dated scans identification as well as marginalization. We use a graph optimizer[4] to perform the optimization including pose to pose constraints in the current session and the integrated constraint from previous sessions. The out-of-dated scans identification module identifies the out-of-dated scans, of which the corresponding poses are pruned since they are no longer useful for map building and loop closure. The pruned poses are marginalized from the graph, forming an integrated constraint for the next session.[5] An illustrative example of the whole process at session t is shown in Fig. 1. To show the effectiveness of the proposed solution, we apply the system to a 10-session dataset collected by a hand hold Kinect sensor. The environment is an office desk with multiple objects being moved, added and removed. The dense map after 10 sessions is in the lower figure in Fig 2. The map without updating is obviously cluttered and confused.

The remainder of the paper is organized as follows: In Section 2, the out-of-dated scans identification module is presented in detail. In Section 3, we will demonstrate the experimental results using the real world datasets. The conclusion will be discussed in Section 4.

2. Out-of-dated Scans Identification

Our method detects the change in 2D depth image by comparing the real depth of scans from current session and synthesized depth using scans from previous session. First, the point cloud generated by scans from previous session are put into a volume, called previous volume (PV). Similarly, the

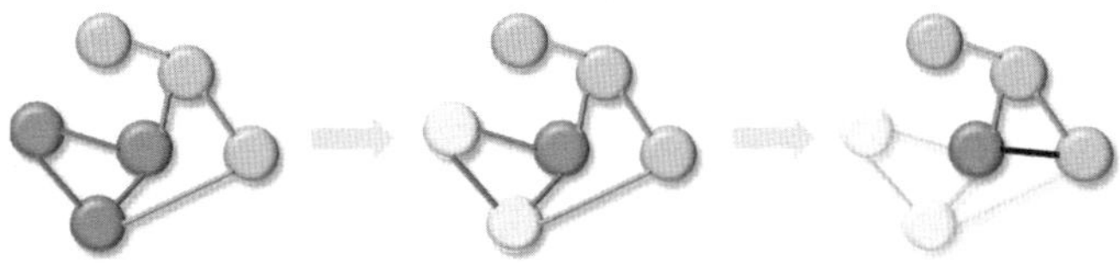

Fig. 1. An example of the procedures in the proposed algorithm. In the left graph, the blue nodes and edges indicate the final pose graph at session $t-1$, the green nodes and edges are poses and constraints obtained at session t, the whole graph is the initial pose graph at session t. In the middle graph, the yellow node is identified as out-of-dated node, which can only be the pose at previous sessions. In the right graph, the out-of-dated nodes are marginalized (translucent nodes), forming the final pose graph at session t, also the integrated constraint for the next session. The black edges are generated through marginalization.

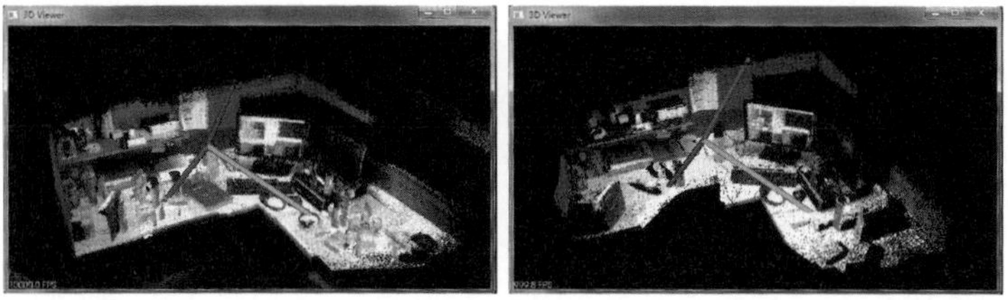

Fig. 2. A comparison of the reconstructed low dynamic workspace with objects inside being frequently added, removed and moved in point cloud after 10 sessions mapping using SLAM with all information as one session (left) and the proposed system considering the low dynamics (right).

point cloud generated by scans from current session are put into another volume, called current volume (CV). The two volumes should be in the same size. Then by comparing PV and CV, we have classifications for grids as below: Case 1, the grid in CV contains points while the corresponding one in PV, does not; Case 2, the grid in CV does not contain points while the corresponding one in PV, does; Case 3, the grid in CV and the corresponding one in PV both have points; Case 4, neither grid in CV nor the corresponding one in PV has point.

A grid here is a voxel in the volume, containing a cubic space. Note that it is not the same as that in occupancy grid mapping. It is only a container that saves a series of end points lying in its cubic region. Hence no ray casting is conducted. One can see that the change must happen in the grids belonging to Case 1 or Case 2. This is called Algorithm 1 which detects the change by simply using the similar rule in grid occupancy map based method, i.e. find the grid belonging to Case 1 or Case 2. The result as shown in the upper left figure in Fig. 3. The poor result is due to the

Fig. 3. Detection result using Algorithm 1 (left) Algorithm 2 (middle) and Algorithm 3 (right).

lack of the grid occupancy model, which solves the two problems implicitly: (1) there is no unknown status in our point cloud volume, so that the part not observed during the current session (previous sessions) is regarded equivalent to free status in occupancy grid map. Actually, such part should be treated as unknown status in occupancy grid map. (2) the point cloud acquired is of bad quality and no fusion mechanism as grid occupancy map can be applied. These two problems give some insights to promote the performance of change detection.

To solve the first problem, a camera projection model is applied explicitly to model unknown status. Given a point p in one scan, a synthesized depth from another pose can be computed as

$$u = KR^T(p - t) \tag{1}$$

where K is the camera intrinsic matrix, R and t is the pose. The third entry of u, $u(3)$, is the synthesized depth of p from the new pose. At the same time, we have the real depth d in the pixel $(u(1)/u(3), u(2)/u(3))$ of the depth image. If d is smaller than $u(3)$, then this point is occluded from this pose, thus no information is acquired (unknown). If d is larger than $u(3)$, then it means from this pose p, the point should have been observed but actually is not observed. The only reason is that this point is removed when measurement is taken from this pose, which gives the cues to the change in the environment, which is called Algorithm 2 with the result shown in Fig. 3.

The new result is much better, but some noisy points are also detected as dynamic, which is due to the second problem summarized. In grid occupancy map, the fusion mechanism can reduce the noise effectively. But in our case, there is no fusion mechanism. In addition, the quality of Kinect raw data is worse than that acquired by laser, especially it has holes. So we cannot assume that each grid is independent[6,7] and.[8] Note that the change in the environment is usually in the level of object, the change should also be detected in a higher level than the level of raw point. By applying connected component to the grids belonging to case 1(case 2), we can have clusters with each one having a series of neighboring grids belonging to the same cases. A cluster is now much more like an object and more robust to noise. By identifying whether a connected component is dynamic, a more robust algorithm is developed called Algorithm 3, of which the result is shown in Fig. 3.

Put all things together, the main steps of our method are: a traversal of all grids in the volume, connected components detection two times and applying measurement model in the level of connected component. The points fed to the measurement model are only small part of all points. If a 3D occupancy model is applied, the steps are: building the occupancy volume and a traversal of all grids in the occupancy volume. The building of occupancy volume takes time for ray casting of all pixels (has equal number as all points). Besides, ray casting is more time consuming than the simple matrix multiplication in (1). These two factors enable our method to be more efficient than using 3D occupancy model.

3. Experimental Results

In this section, we demonstrate the performance of the proposed algorithm using datasets collected from the real world. We show qualitative results using a small dataset with 2 sessions and quantitative comparison for a dataset with 10 sessions. The datasets are collected using a hand hold Kinect from a typical workspace with some objects being frequent operated. In experiments, the size of the grid is 0.05.

3.1. *2-session result*

In this experiment, the dynamics between two sessions including: a bottle and a mug are removed, a box is moved and a sitting person appears in the second session. The result of the out-of-dated scans identification is shown in Fig. 3, in which the red highlighted points indicate the removed bottle and mug, the moved boxes, and the newly appeared sitting person. The dense map generated by multi-session pose SLAM using all information without pruning is shown in the left image in Fig. 4. One can see that all objects appearing on the desk (bottle, mug and two duplicated boxes indicated by yellow circle). The map in the right image in Fig. 4 using the proposed method keeps track of the latest environment with a person sitting and two boxes removed in the 2nd session.

Fig. 4. The snapshots of the workspace environment in 1st and 2nd sessions (left two). The map reconstructed using all frames without considering the low dynamics cannot remove the objects which are not longer there in the 2nd session, while the proposed method can (right two).

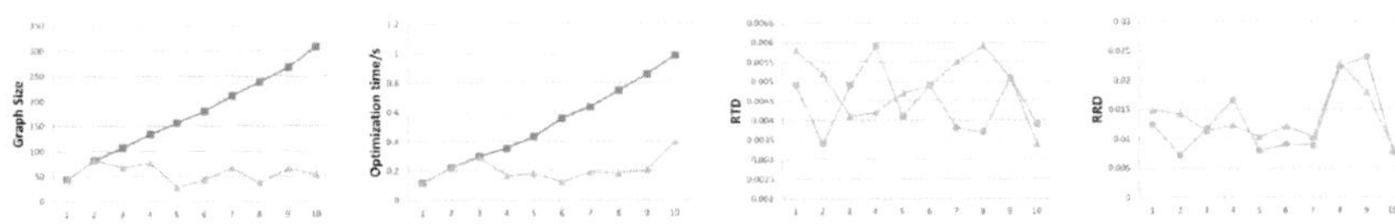

Fig. 5. Evolution of the graph size, the time for optimization, relative translational and rotational difference during sessions. Green is for Pruning, orange is for Pruning FEJ, Red is for No Pruning, x-axis indicates the session index.

3.2. *10-session result*

For quantitative comparison of the pose estimation quality, two schemes are tested in this experiment including:

- **No Pruning** multi-session pose graph SLAM using all information without any pruning (The best pose estimation one can achieve. So it is used as benchmark).
- **Pruning** proposed multi-session SLAM over low dynamic workspace.
- **Pruning FEJ** proposed multi-session SLAM over low dynamic workspace with first estimated Jacobian.[9]

To evaluate the performance, the last two schemes are compared with the first scheme using a 10-session SLAM in a low dynamic environment. The objects on the desk are added, removed or moved during sessions. The map reconstructed after 10 sessions by SLAM using all information is shown in Fig. 2, in which one can see that some objects appearing at different positions. So the map is hard to provide the accurate information of the environment, not appropriate for some grid or octree based localization and navigation techniques. The result after 10 sessions using the proposed framework is shown in Fig. 2, in which each object appears in its final position.

The size of the pose graph after each session saved in the system is shown in Fig. 5. The graph sizes in Pruning and Pruning FEJ are the same. Though the pose with out-of-dated scans provides geometric constraints, it also makes the graph grow continuously, leading to a higher complexity of both storage due to its saved RGBD scans, which, however, is useless. Pruning such poses releases the storage on one side. On the other side, the marginalization after optimization loses the possibility of re-linearization. To show the effects of both sides, we conduct two comparisons in sequel.

The computation time for optimizing the graph is shown in Fig. 5. Since the graph size is decreased, the optimization time of proposed system is reduced significantly. As the objects are frequently operated during sessions, the time complexity is almost constant. Note the evolution curve between session 1 and session 2, the trend of both methods are the same, because no dynamics occur in this interval.

To evaluate the accuracy of the pose estimates, the relative translational and rotational difference is used[10] as

$$RRD = \|\hat{R}_{i-1,i} - R_{i-1,i}\|_F, \; RTD = \|\hat{t}_{i-1,i} - t_{i-1,i}\|_2 \tag{2}$$

where $\hat{R}_{i-1,i}$ and $\hat{t}_{i-1,i}$ are the relative poses between the $i-1$th and ith pose using Pruning or Pruning FEJ, while $R_{i-1,i}$ and $t_{i-1,i}$, using No Pruning. The result is shown in Fig. 5. In this dataset, one can see that the difference is in the level of millimeters. Here we cannot state which one is more accurate, but it is clear that the error level during 10 sessions does not grow.

The difference between Pruning and Pruning FEJ comes from the value where Jacobian is estimated.[11] Theoretically speaking, Pruning FEJ will be better since it is more consistent. However, the experimental results shown in Fig. 5 does not demonstrate such a trend. The main reason we think is that: first the pose is optimized before it is marginalized. The linearized points are relatively accurate enough. Second a pose does not exist in the graph for a long time since the environment is dynamic, so the error may not be accumulated. As shown in this experiment, loss of possibility of re-linearization does not introduce serious error. Another error source is potential loop closures a pruned pose may have in the later sessions. This kind of error focuses on whether a pruned pose has loop closures in the future. It is possible that the pruned poses due to observing the dynamic does not have future loop closures. Thus missing potential loop closures does not have obvious effects on the result.

4. Conclusion and Future Works

In this paper, the problem of multi-session RGBD SLAM in low dynamic environment is studied in the context of pose graph SLAM. We propose a framework by equipping the multi-session SLAM system with an out-of-dated scans identification module to build a map over low dynamic workspace. Finally, the 2-session and 10-session experiments show the correctness and effectiveness of our method from two aspects of time complexity and accuracy.

Acknowledgement

This work is supported by the National Natural Science Foundation of China (grant number 61473258), the Program for Zhejiang Leading Team of S&T Innovation (grant number 2009R50014), and the Joint Centre for Robotics Research (JCRR) between Zhejiang University and the University of Technology, Sydney.

References

1. P. Newman, G. Sibley, M. Smith, M. Cummins, A. Harrison, C. Mei, I. Posner, R. Shade, D. Schroeter, L. Murphy *et al.*, *The International Journal of Robotics Research* **28**, 1406 (2009).
2. A. Kim and R. M. Eustice, *IEEE Transactions on Robotics* **29**, 719 (2013).
3. Y. Wang, R. Xiong and Q. Li, *The International Journal of Robotics and Automation* **28** (2013).
4. G. Grisetti, R. Kümmerle, C. Stachniss and W. Burgard, *Intelligent Transportation Systems Magazine, IEEE* **2**, 31 (2010).
5. Y. Wang, R. Xiong and S. Huang, *Advanced Robotics* , 1 (2015).
6. G. D. Tipaldi, D. Meyer-Delius and W. Burgard, *The International Journal of Robotics Research* **32**, 1662 (2013).
7. J. Saarinen, H. Andreasson and A. J. Lilienthal, Independent markov chain occupancy grid maps for representation of dynamic environment, in *Intelligent Robots and Systems, IEEE/RSJ International Conference on*, 2012.
8. A. Walcott-Bryant, M. Kaess, H. Johannsson and J. J. Leonard, Dynamic pose graph slam: Long-term mapping in low dynamic environments, in *Intelligent Robots and Systems, IEEE/RSJ International Conference on*, 2012.
9. G. P. Huang, A. I. Mourikis and S. I. Roumeliotis, A first-estimates jacobian ekf for improving slam consistency, in *Experimental Robotics*, 2009.
10. J. Sturm, N. Engelhard, F. Endres, W. Burgard and D. Cremers, "a benchmark for the evaluation of rgb-d slam systems, in *Intelligent Robots and Systems, IEEE/RSJ International Conference on*, 2012.
11. G. P. Huang, A. I. Mourikis and S. I. Roumeliotis, An observability-constrained sliding window filter for slam, in *Intelligent Robots and Systems, IEEE/RSJ International Conference on*, 2011.

GRAPH-SLAM BASED CALIBRATION OF AN EMBEDDED ASYNCHRONOUS MICROPHONE ARRAY FOR OUTDOOR ROBOTIC TARGET TRACKING

DAOBILIGE SU, JAIME VALLS MIRO and TERESA VIDAL-CALLEJA

Centre for Autonomous System (CAS), University of Technology, Sydney (UTS), Australia, daobilige.su@student.uts.edu.au,{jaime.vallsmiro, teresa.vidalcalleja}@uts.edu.au

This paper presents a strategy for sound source localisation using an embedded asynchronous microphone array for robotic target tracking application. Conventional microphone array technologies require a multi-channel A/D converter for inter-microphone synchronization making the technology relatively expensive. In our method, a synchronization free embedded asynchronous microphone array has released this requirement. The microphone array needs self calibration using graph-based SLAM method, which estimates starting time offset and clock difference/drift rate of each microphone channel using Gauss-Newton least square optimization. The proposed method is suitable for target tracking applications.

Keywords: Asynchronous microphone array; calibration; robotic; tracking; Graph-SLAM.

1. Introduction

Target localization and tracking is an important topic in robotics and industry automation.[1,2] There exist many successful applications which are based on laser range finder, camera and soundin the literature.[3] The application scenarios include for instance museum guidance[4] or hospital assistance.[5]

From the various solutions proposed for target tracking, the one based on sound source is of increasing interest as the acoustic waves can travel in all directions, can be detected at long distances from the sound source and beyond the line of sight.[6] While several commercial products exist (NI 9234 and RASP-24, they are generally too large in size to be practically integrated in generic mobile robotic platforms. Because of the dedicated hardware, they also tend to be relatively high in cost.[7]

Recent studies have started to relax this requirements.[7,8] Although those methods are able to compute microphone locations and the starting time offsets between different microphone channels, all of them made the assumption that the clock interval in each independent sound card dedicated to each channel is identical to that of other channels. In some circumstances that may be disregarded and the propose solutions will operate correctly, yet in other situations this error will accumulate over time to the point that the assumption is no longer valid. Sound cards, especially those in an affordable price range, have noticeable drifts in timing over just a few minutes. For embedded microphone array, since the size of the microphone array is relatively small, locations of microphones can be accurately fixed with the help of precise fabrication of its hardware frame.

The method proposed in this paper tackles these issues by calibrating the asynchronous microphone array using an optimization method based on pose-graph borrowed from the Simultaneous Localization and Mapping (SLAM) literature.[9] The formulation estimates the starting time offset and clock difference between each microphone using a probabilistic Gauss-Newton least square optimization. Once this is accurately established for the asynchronous microphone array, the results can be used online for target localization and tracking.

The rest of this paper is organised as follows. In section 2, the details of the proposed method is illustrated. In section 3, comprehensive simulations and experimental results are presented. Section 4 presents the conclusion and discussion about further work.

2. The Proposed Method

2.1. *System model*

In this method, the state vector of the SLAM framework is as follow.

$$X^T = (X_{mic}^T X_{src}^T) \tag{1}$$

where X_{mic} is the state vector of the microphone array and X_{src} is the state vector of the sound source. These two variables are described as below.

$$X_{mic}^T = (X_{mic\_1}^T \dots X_{mic\_N}^T) \tag{2}$$

$$X_{src}^T = (X_{src\_1}^T \dots X_{src\_K}^T) \tag{3}$$

where N is the total number of microphones and K is total number of sound source positions used in the calibration while it moves around the

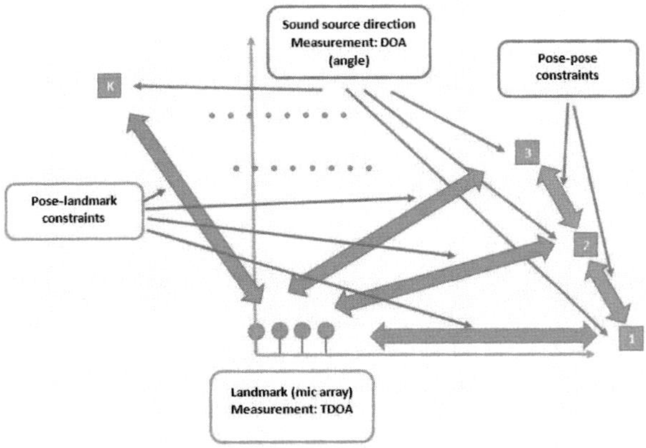

Fig. 1. Description of the poses, landmarks and constraints.

microphones. $X_{src\_k}(k = 1 \ldots K)$ is the angle of the sound source at time instance k. The individual state vector of each microphone $X_{mic\_n}(n = 1 \ldots N)$ is denoted as

$$X_{mic\_n}^{T} = (X_{mic\_n}^{\tau} X_{mic\_n}^{\delta})(n = 1 \ldots N) \tag{4}$$

where $X_{mic\_n}^{\tau}$ is the starting time offset and $X_{mic\_n}^{\delta}$ is the clock difference per second of microphone n. The first microphone is used as the reference and time offsets and clock differences of other microphones are computed relative to this microphone. So, $X_{mic\_1}^{\tau} = 0$ and $X_{mic\_1}^{\delta} = 0$.

The poses, landmarks and constraints of our method in the graph SLAM framework are shown in Fig. 1.

In our method, the trajectory of the sound source is a random walker like model in which the sound source angle of the next time instance is expected to be the same as that of previous time with the standard deviation of reasonable big value as defined below.

$$Z_{k-1,k}^{p-p} = 0(k = 1 \ldots K) \tag{5}$$

$$I_{k-1,k}^{p-p} = \frac{1}{\sigma_{p-p}^{2}}(k = 1 \ldots K) \tag{6}$$

where $Z_{k-1,k}^{p-p}$ and $I_{k-1,k}^{p-p}$ denote the measurement and information matrix between poses $k-1$ and k. σ_{p-p} is the standard deviation of the Gaussian distribution within which the DOA of next sound source should fall.

In pose-landmark constraints, the measurement is the TDOA values of the microphone array at each angle of sound source. The measurement of this pose-landmark constraint is defined as

$$Z_k^{p-l^T} = (TDOA_{mic\_2,mic\_1} \dots TDOA_{mic\_N,mic\_1}) \tag{7}$$

where $TDOA_{mic\_n,mic\_1}(n = 2 \dots N)$ is the TDOA between microphone n and microphone 1 as microphone 1 is used as reference. The information matrix for this pose-landmark constraint is

$$I_k^{p-l} = \frac{1}{\sigma_{p-l}^2}\mathbf{I} \tag{8}$$

where σ_{p-l} the standard deviation of Gaussian distribution within which the error of TDOA observation error should fall.

2.2. *Error functions and their Jacobian*

For each pose-pose constraint from sound source position $k-1$ to k, the computation of the error function is

$$e_{k-1,k}^{p-p} = (X_{src\_k} - X_{src\_k-1}) - Z_{k-1,k}^{p-p} = X_{src\_k} - X_{src\_k-1} \tag{9}$$

Therefore, the corresponding Jacobian matrix for this pose-pose constraint is as follow.

$$A_{k-1,k}^{p-p} = \frac{\partial e_{k-1,k}^{p-p}}{\partial X_{src\_k-1}} = -1 \tag{10}$$

$$B_{k-1,k}^{p-p} = \frac{\partial e_{k-1,k}^{p-p}}{\partial X_{src\_k}} = 1 \tag{11}$$

where A_{ij} and B_{ij} are Jacobian matrix of e_{ij} over the graph node X_i and X_j respectively.

In the pose-landmark constraint between microphone array and sound source location at time instance k, the error function is computed as below.

$$e_k^{l-p} = TDOA_{pre} - Z_k^{p-l}$$
$$= -\frac{1}{c}\begin{bmatrix} d_{mic} \\ \vdots \\ (N-1) * d_{mic} \end{bmatrix} cos(X_{src\_k}) + \begin{bmatrix} X_{mic\_2}^{\tau} \\ \vdots \\ X_{mic\_n}^{\tau} \end{bmatrix} + k\Delta t \begin{bmatrix} X_{mic\_2}^{\delta} \\ \vdots \\ X_{mic\_n}^{\delta} \end{bmatrix} - Z_k^{p-l} \tag{12}$$

where d_{mic} is the distance between adjacent two microphones, $TDOA_{pre}$ is the predicted TDOA value from the current state vector, c denotes the speed of sound and Δt refers to time interval between time instance k and $k+1$.

The Jacobian matrix correspond to this error function are computed as below.

$$A_k^{l-p} = \frac{\partial e_k^{l-p}}{\partial X_{mic}} = [\mathbf{0}\mathbf{J}_{mic\_2}^{l-p} \dots \mathbf{J}_{mic\_N}^{l-p}] \tag{13}$$

$$B_k^{l-p} = \frac{\partial e_k^{l-p}}{\partial X_{src\_k}} = \frac{1}{c} \begin{bmatrix} d_{mic} \sin(X_{src\_2}) \\ \vdots \\ (N-1) d_{mic} \sin(X_{src\_K}) \end{bmatrix} \tag{14}$$

where $\mathbf{J}_{mic\_n}^{l-p}$, $(n = 1 \dots N)$ is the Jacobian of e_k^{l-p} over state vector of microphone n. Since microphone 1 is used as reference, its state vector is constant value and the Jacobian of it is all zero. $\mathbf{J}_{mic\_n}^{l-p}$ is only nonzero at rule n and this nonzero rule is computed as

$$\mathbf{J}_{mic\_n}^{l-p}(n, :) = \begin{bmatrix} 1 \\ k\Delta t \end{bmatrix}^T \tag{15}$$

3. Simulations and Experimental Results

3.1. *Simulations results*

A Simulation of the proposed method with 4 microphones array is studied and the parameters used in the simulation are summarised in Table 1. The estimations of the starting time offset and the clock difference of each microphone channel are shown in Fig. 2. It can be seen from the figure that our method is able to estimate the values for these two variables with good accuracy.

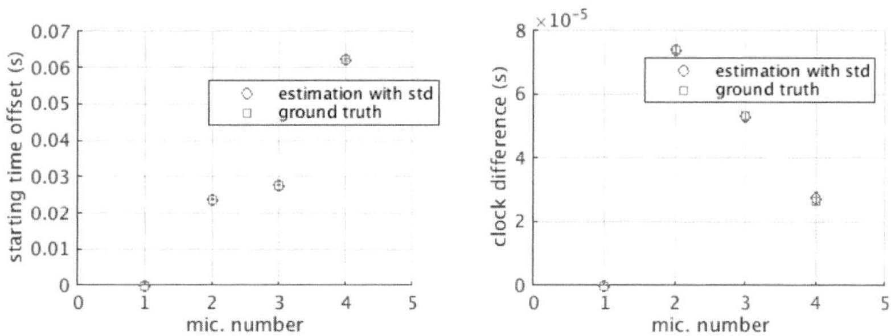

(a) starting time offset

(b) clock difference

Fig. 2. Simulation results compared to the ground truth values.

Table 1. Parameters setting in simulation

Parameters	Values
Number of microphones	4
Distance between microphones	0.05m
Maximum starting time offset	0.1s
Maximum clock difference	0.1ms
Observation (TDOA) noise STD	3.33ms
Sampling frequency	44.1 KHz
Random walker STD	50 (degree)
Adjacent sound source angle	1 (degree)
Number of calibration data	180
Speed of sound	340m/s
Maximum iterations	20
ϵ for ΔX	10

Secondly, the influence of the number of calibration data over the estimation accuracy is examined by simulations of various number of it with 20 Monte Carlo runs for each case. The results are shown in the Fig. 3. From the comparison of mean RMS errors plot, it is clear that less number of calibration data (smaller than 180) can dramatically degrade the estimation accuracy of both starting time offsets and clock differences.

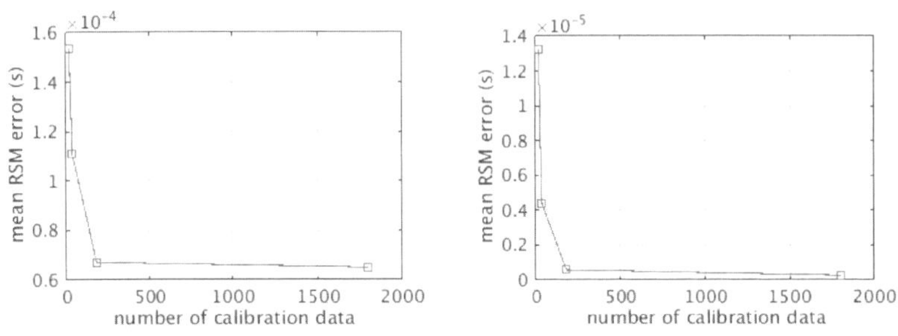

(a) starting time offset

(b) clock difference

Fig. 3. Mean RMS error w.r.t. number of calibration data.

3.2. *Experimental results*

Our experimental is performed in an indoor environment. In the experiment, we first collect the calibration data for estimation of starting time offsets and clock differences. After the calibration, the standard ESPRIT algorithm is used to find DOA estimation for sound source. The experimen-

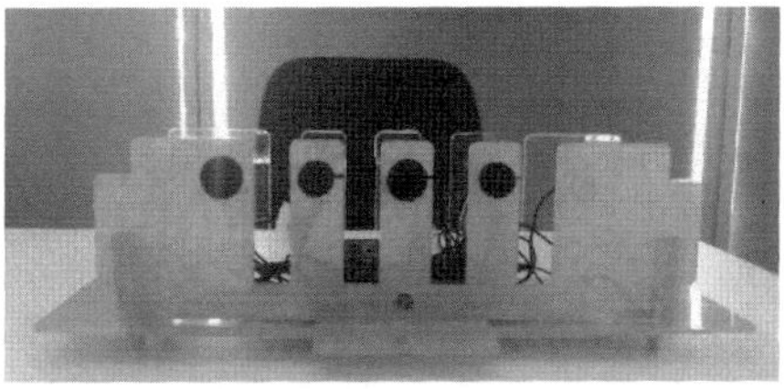

Fig. 4. Experimental setup of the asynchronous microphone array.

Table 2. Parameters setting in experiment

Parameters	Values
Number of microphones	4
Distance between microphones	0.05m
sound source	Samsung Galaxy S4 phone
sound wave	short time chirp
time interval of sound	0.5s
total duration of recording	1 min 14 s
ϵ for ΔX	10

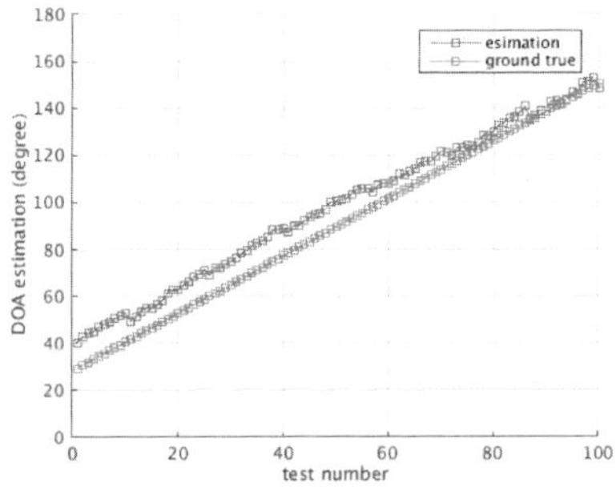

Fig. 5. DOA estimation results after the calibration.

tal setup of the microphone array is shown in Fig. 4. In the experiment, four microphones are individually sampled by their own USB sound cards. The extra parameters in the experiment are summarised in the Table. 2. Once calibrated, the DOA estimation results using ESPRIT algorithms w.r.t. ground truth values are shown in Fig. 5. The RMS error is 8.7961 degree.

4. Conclusion

In this paper, a pose graph optimization framework is used to calibrate a small size embedded asynchronous microphone array for estimating the starting time offsets and the clock differences between each microphone channel. The method relaxes the requirement of traditional microphone array based sound source DOA estimation methods which need a multi-channel analog-to-digital converter (ADC) to obtain synchronous reading of audio signals. Our method can be applied for human tracking application. The future work includes the study of 3D microphone array which estimate both azimuth and elevation angle of the sound source.

References

1. M. Kobilarov, G. Sukhatme, J. Hyams and P. Batavia, People tracking and following with mobile robot using an omnidirectional camera and a laser, in *IEEE International Conference on Robotics and Automation (ICRA 2006)*, 2006.
2. J. V. Miro, J. Poon and S. Huang, Low-cost visual tracking with an intelligent wheelchair for innovative assistive care, in *12th International Conference on Control Automation Robotics & Vision (ICARCV 2012)*, 2012.
3. Z. Chen and S. T. Birchfield, Person following with a mobile robot using binocular feature-based tracking, in *IEEE/RSJ International Conference on Intelligent Robots and Systems (IROS 2007)*, 2007.
4. S. Thrun, M. Bennewitz, W. Burgard, A. Cremers, F. Dellaert, D. Fox, D. Haehnel, C. Rosenberg, N. Roy, J. Schulte and D. Schulz, Minerva: A second generation mobile tour-guide robot, in *IEEE International Conference on Robotics and Automation (ICRA 1999)*, 1999.
5. J. F. Engelberger, *Robotica* **11**, 517 (1993).
6. M. Basiri, F. Schill, P. U. Lima and D. Floreano, Robust acoustic source localization of emergency signals from micro air vehicles, in *IEEE/RSJ International Conference on Intelligent Robots and Systems (IROS 2012)*, 2012.
7. H. Miura, T. Yoshida, K. Nakamura and K. Nakadai, SLAM-based online calibration of asynchronous microphone array for robot audition, in *IEEE/RSJ International Conference on Intelligent Robots and Systems (IROS 2011)*, 2011.
8. K. Hasegawa, N. Ono, S. Miyabe and S. Sagayama, *Latent Variable Analysis and Signal Separation* , 57 (2010).
9. G. Grisetti, R. Kummerle, C. Stachniss and W. Burgard, *IEEE Intelligent Transportation Systems Magazine* **2(4)**, 31 (2010).

INTERFACE FOR CORRECTING DISPARITY IN FMT VISION

KENICHI TOKUDA* and ATSUSHI HIRATA

Faculty of Systems Engineering, Wakayama University,
Sakaedani 930, Wakayama, 640-8510, Japan
**E-mail: kenichi.tokuda@ieee.org*
http://www.wakayama-u.ac.jp/ tokuda/

TETSUYA KINUGASA

Okayama Univrsity of Science,
1-1, Ridai-cho, Okayama, 700-0005, Japan
E-mail: kinugasa@mech.ous.ac.jp

TAKAFUMI HAJI

National Institute of Technology, Matsue Colleage,
14-4 Nishiikuma-cho, Matsue city, Shimane 690-0865, Japan
haji@matsue-ct.jp

HISANORI AMANO

National Research Institute of Fire and Disaster, Japan,
4-35-3 Jindaiji-higashimachi, Chofu, Tokyo, 182-0012, Japan
E-mail: amano@fri.go.jp

The covered area detection method, using the brightness change in an image and parallax correction, does not depend on a common field of vision. Camera images become important information when a rescue robot is operate by remote control, but mounting cameras is difficult on a rescue robot crawler that must get into cracks in rubble. We propose attaching cameras behind the crawler shoe. The biggest problem then, however, is that the shoe obstructs large parts of the camera image. To avoid this, we developed real-time image processing that complements the obstructed area through the use of two cameras. We then performed evaluation experiments to confirm the effectiveness of the proposed technique. The complement method can solve in a case of one parallax problem. However, several objects in a acquiring image occurs several parallaxes between objects. In this paper, we report a trial to solve the parallaxes problem by using a user interface software.

Keywords: crawler robot, tele-operation, image processing.

1. Introduction

FMT, Flexible mono-tread mobile,[1] is a rescue robot system, which is covered by crawler track except the both side surface (Fig. 1). In generally, to take the front and overview images need for remote control of tracked vehicles such as these robots, however, cameras mounted appropriately are difficult to position due to the crawler belt. The camera is therefore usually mounted on a part of the robot other than the drive, e.g., on the side of the vehicle. The Souryu-V[2] has cameras on the side and back of the head. Cameras protruding in narrow gaps such as those in rubble would keep such robots from moving safely in the surroundings. The Souryu-V's cameras protrude less and have a mechanism for reducing any force applied to the camera, but it is still difficult to eliminate contact with surroundings.

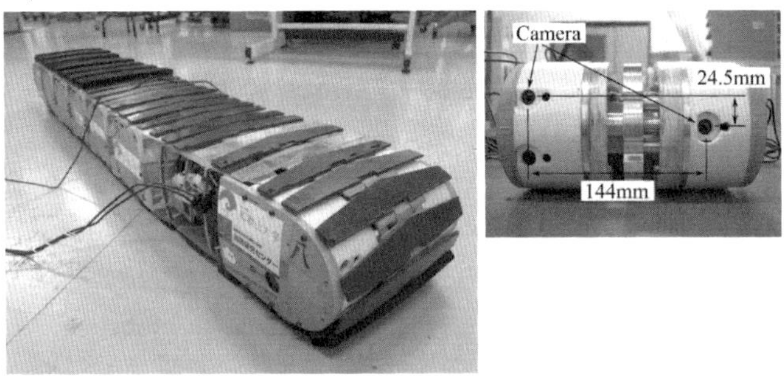

Fig. 1. FMT RT-04 overview

One way would be to install a camera inside the crawler belt, but the belt would block the camera's visual field as shown in Fig. 2. Specifically, the crawler shoe moves slowly and the visual field could be blocked completely during such motion, which could become a serious problem in remote control. To avoid this, the blocked image by the crawler shoes must be complemented in real time.

We propose using two cameras to complement images taken by a camera inside the FMT crawler belt but blocked by the crawler shoe. Our previous report[3] proposed a method to complement an image's patch by using brightness change of images from two cameras (Fig. 3). In the image processing method that we embedded succeed to complement some of the issues.[4] The complement method can solve in a case of one parallax problem. However, several objects in a acquiring image occurs several parallaxes between objects. There leaves parallax problem, we must improve the method to solve the problem that there are parallax objects in an image.

In this paper, we report a trial to solve the parallaxes problem by using a user interface software.

Fig. 2. Crawler shoe blocking

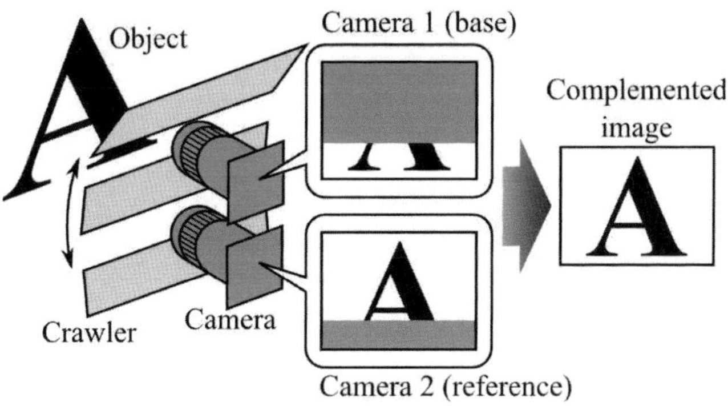

Fig. 3. Concept of image complement[4]

2. Complement Position and Estimation

We were able to complement a area covered by crawler-shoe with two cameras. However, when there are more than two big parallax objects in the image which the system acquired, this method does not work well. Therefore it is necessary to select one object among plural complement candidates, In this section, we show a complement candidate in user interface and algorithm to select by a human operator.

Domain recognition

(1) Select one object to complement a gap to assume parallax a cause.
(2) Using the coordinate pointed by an operator as a seed point, recognize the object area including the seed point.

(3) Area expansion method using the brightness level of the seed point to recognize an area.

Figure 3 shows the domain recognition flows.

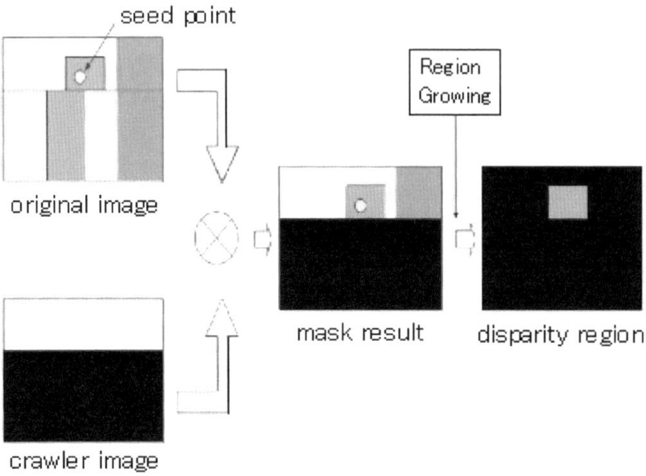

Fig. 4. Detecting region

Complement position decision

(1) The gap complementation for the position by SSD using the similar degree.
(2) Calculated complement position and revise a gap and constitute it as an image.

Figure 4 shows flows of the decision of the supplement position.

SSD is one of indexes to evaluate similarity between images. The index is given by following equality.

$$SSD = \sum_{i=0}^{M}\sum_{j=0}^{N} |I(x+i, y+j) - T(x+i, y+j)|^2$$

where $I(x, y), T(x, y)$ is brightness on the coordinate (x, y) in the each image.

Our proposal technique uses the SSD. We expand the object domain set to lower SSD area.

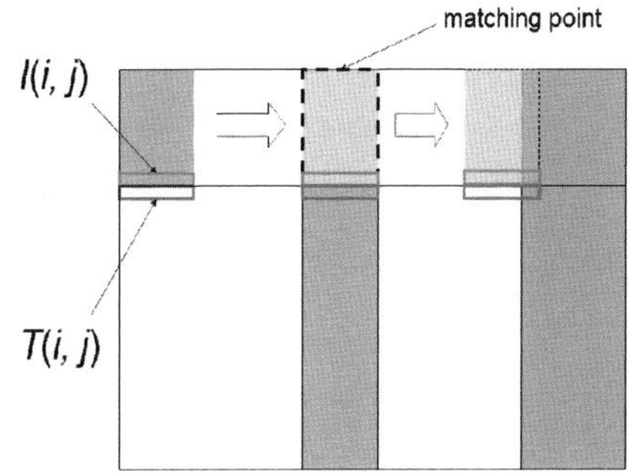

Fig. 5. Disparity estimation

3. Experiment

We used the device to show in the Fig. 6 to inspect whether propsed interface is effective. This device does structure in imitation of the face of the FMT. It is equipped with two USB cameras, and it is 142[mm] horizontally and 20[mm] in the vertical direction in the space between the camera. As for the OS, Windows 8.1, the CPU performed 1.8[GHz] of Intel(R) Core(TM) i3-3217U, the memory with a PC of 4[GB].

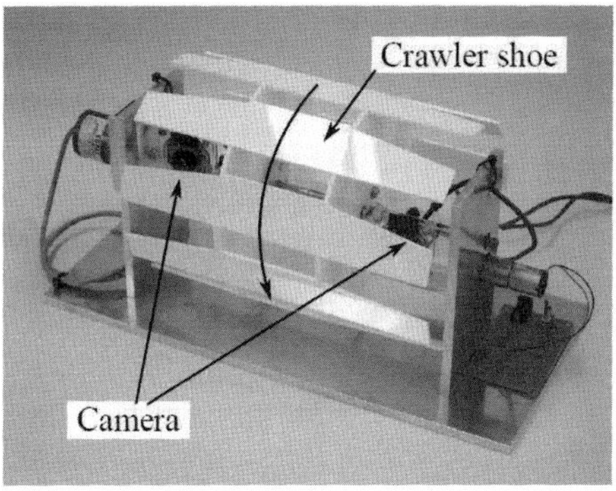

Fig. 6. Experimental device

3.1. *Disparity supplement*

The experimental environment is as follows, the distance from a camera to the wall is 1.0[m], to a yellow box is 1.5[m] and to the shelf is 2.0[m]. The size of this box is 155 × 340 × 150[mm]. We set it to match parallax with a shelf and complemented the parallax of a yellow box in the interface. We turned a device and continued complementing parallax and prevented a gap by the parallax from occurring (Fig. 7)

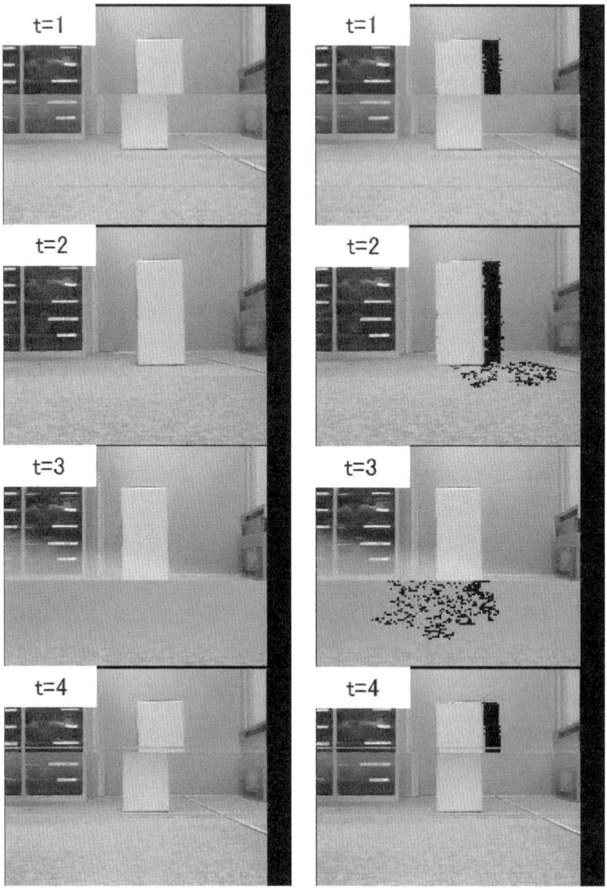

Fig. 7. Disparity correction

3.2. *Selected object correction*

Between a robot and a wall, we set three kinds of boxes which varied in the distance from a robot. Distance from a robot camera to the walls is 3 [m], By the user interface that we implemented to a PC, the robot operator chooses one correction object from plural objects. He can choose the object using the cursor of the mouse or a touch screen. The interface program recognized an object for a seed point at the point that a robot operator pointed to, and supplemented the parallax gaps. Figures 8–10 show results of complemented disparities.

Fig. 8. Disparity correct (target:Y)

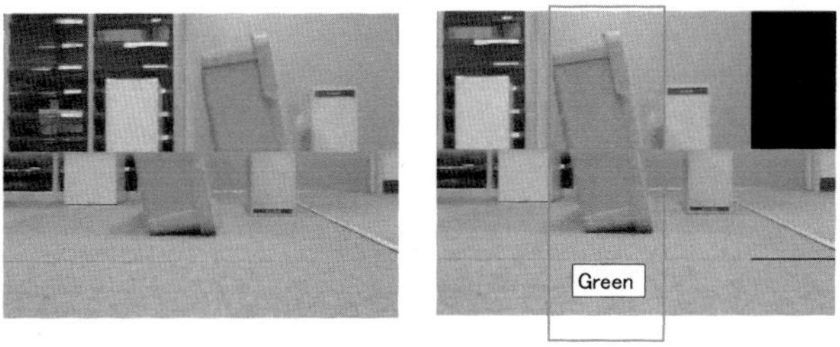

Fig. 9. Disparity correct (target:G)

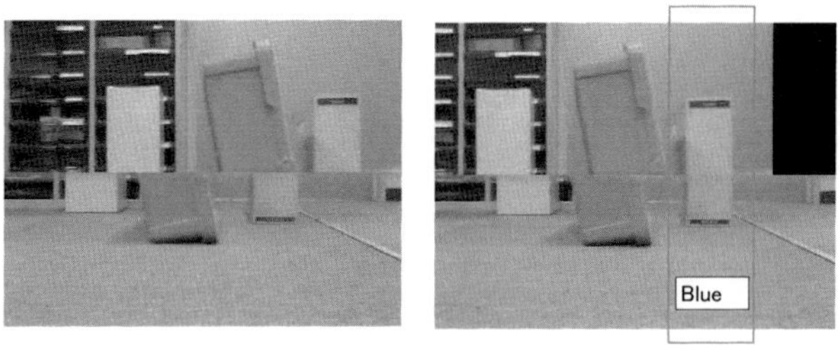

Fig. 10. Disparity correct (target:B)

4. Conclusion

In this paper, we proposed an interface implementation method for multiple disparities' correction. This interface is enable to correct one object with disparity gap pointed by a robot operator. When plural correct candidates exist, the operator can utilize this interface. Our experimental result showed that the interface implementation can solve the problem that a robot camera image includes multiple disparities. There leaves a gap correction's improvement precision as a future issue. Therefore we consider a geometric position of crawler-shoes.

Acknowledgements

A part of this work was supported by JSPS KAKENHI Grant Number 25242037.

References

1. T. Haji, T. Kinugasa, S. Araki, D. Hanada, K. Yoshida, H. Amano, R. Hayashi, K. Tokuda and M. Iribe, New Body Design for Flexible Mono-Tread Mobile Track: Layered Structure and Passive Retro-Flexion, *J. of Robotics and Mechatronics*, Vol.26 No.4, pp. 460-468, 2014.
2. M. Arai, Y. Tanaka, S. Hirose, H. Kuwahara, and S. Tsukui, De- velopment of Souryu-IV and Souryu-V: Serially connected crawler vehicles for in-rubble searching operations, *J. of Field Robotics*, Vol.25, No.1-2, pp. 31-65, 2008.
3. K. Tokuda, T. Hirayama, T.Kinugasa, T. Takafumi and H. Amano, COVERED AREA DETECTION BASED ON BRIGHTNESS CHANGE OF INNER CAMERA IMAGES FOR CRAWLER ROBOT in *Porc. of of the Sixteenth International Conference on Climbing and Walking Robots (CLAWAR2013)*, pp. 51-58, 2013
4. K. Tokuda, T. Hirayama, T. Kinugasa, T. Haji, H. Amano, and K. Yasuda, Complement Method for Obstructed Area on Images of Multiple Cameras Mounted Behind Crawler Shoe *J. of Robotics and Mechatronics*, Vol.27 No.2, pp. 146-155, 2015.

MECHANISM AND ANTI-EXPLOSION DESIGN OF AN OMNITREAD SERPENTINE ROBOT FOR SEARCHING IN COAL MINES*

GANGFENG LIU[1,2], JIHONG YAN[1], CHANGLE LI[1], ZHENFENG HAN[1], LEI ZHU[1], JIE ZHAO[1] and LIYI LI[2]

[1]*State Key Lab of Robotics and System, Harbin Institute of Technology, Harbin,150040, P. R. China*

[2] *School of Electrical Engineering and Automation, Harbin Institute of Technology, Harbin,150001, P. R. China*

Many disasters occurred in coal mines, but the explosive gas, less oxygen and hot air stopped the rescuers from doing their job. An omnitread serpentine robot –HITSR-I was developed to work in these areas. The purpose of this paper is to introduce the mechanism and anti-explosion design of the robot. With all these unique designs, the serpentine robot can search in hazardous areas. And the experiments have been carried out under coal mines.

1. Introduction

In recent years, a challenging study field of the robots, Search and Rescue Robots (SRRs) are more and more popular in assisting people in rescue operations. The SRRs can enter into the hazardous or narrow places in place of human to finish the search job and this can avoid or reduce the injury of the rescue people.

The SRRs must function in extremely hazardous and very complex disaster environments. Hence, it is very important to develop mobile mechanisms that enable robots to travel across the rubble or access to the interior of the rubble piles. Serpentine robots are ideally suited for search and rescue tasks in these environments because serpentine robots are slender, multi-segmented vehicles designed to provide greater mobility than conventional mobile robots. Serpentine robots are typically comprised of several rigid segments that are connected by muti-freedom joints. Many serpentine robots have been developed, such as follows.

---

* This work is supported by National High-tech R&D Program of China (No.2007AA041501, No.2014AA041601) and China Postdoctoral Science Foundation(No.2014M561338).

Takayama and Hirose developed the Souryu crawler. It consists of three segments connected to each other by a 2-DOF joint mechanism which is actuated by two lead screws (Takayama and Hirose, 2000; Masayuki et al., 2004). Osuka and Kitajima developed MOIRA, a serpentine omnitread robot which comprises four segments and the segments joints are actuated by pneumatic cylinders (Osuka and Kitajima, 2003; Haraguchi et al., 2005). KOHGA was developed by Kamegawa et al, which is equipped with two cameras. One is in the head, the other is in the tail that can be pointed forward (Miyanaka et al., 2007). Caleb III was designed in Korea which consists of three body segments connected in serial (Sang-Jin et al., 2007). Omni Tread OT8 and OT4 have been developed In the University of Michigan. OT8 consists of five segments and can fit through a hole of 8 inches in diameter. OT4 consists of seven segments and can fit through a hole of 4 inches in diameter. It has bellow-based joint actuators and the bellows are more compact and do not require any space in the segments (Granosik and Borenstein, 2005; Hansen, 2007; Borenstein and Borrell, 2008; Borenstein, 2010; Hopkins, 2009).

All these robots surely have high mobility, but they did not meet the demands in coal mine rescue. We need special design in mobility in the coal mine, and anti-explosion structure to ensure the safety during the robot's work.

2. Philosophy of the Design

In our design, we must ensure the high mobility and reliability of the robot in the coal mine.

First, to get to the searching destination, the robot must have the ability to overcome channels, cross obstacles and travel through cavity. Thus we have to design the special movement and driving mechanism for the robot.

Secondly, gas or other explosive gases exist in some disaster fields. Robot may ignite the surrounding explosive air (in normal operation and under defined abnormal operating conditions) and this is forbidden. The robot must be designed specially for working in those places.

To get over the problems, we have proposed a new serpentine robot HITSR-I, shown in Figure 1.

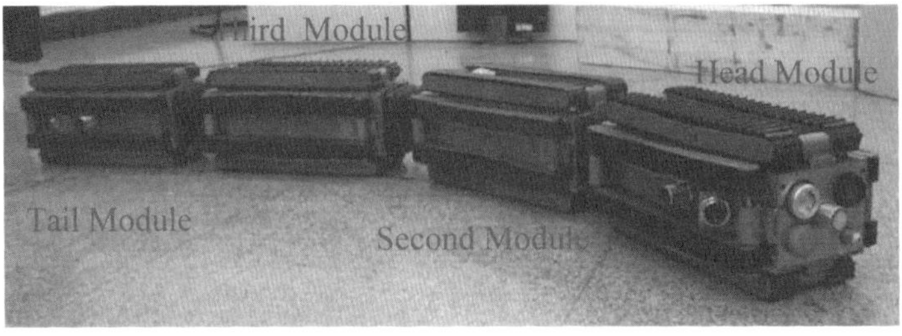

Figure 1. Serpentine search and rescue robot HITSR-I

The robot HITSR-I comprises four segments which are linked to each other by 2-DOF joints. Moving tracks on all four sides of each segment guarantee propulsion even when the robot rolls over. The joints are actuated by two brushless DC motors which produce sufficient torque to lift one or two segments up and overcome obstacles.

There are pressed CO_2 containers inside the robot. Fully sealed up structure shells keep the inside pressure higher than the outside. These designs make the robot meet the requirements for Class 1, Div. 1 & 2 and Zone 1 & 2 (Han Zheng and Zhu Yongsheng, 2003) hazardous locations. The robot is equipped with thermometer, pressure, camera, gas sensors etc., which help the robot operator survey the scene of the disaster and collect relevant data far away.

3. Mechanism Design

3.1. *The overall configuration*

To reach the goal of mobility and reliability, we design the robot to be a serpentine robot composed of tracked segments. The content and function of each segment are shown in Figure 2.

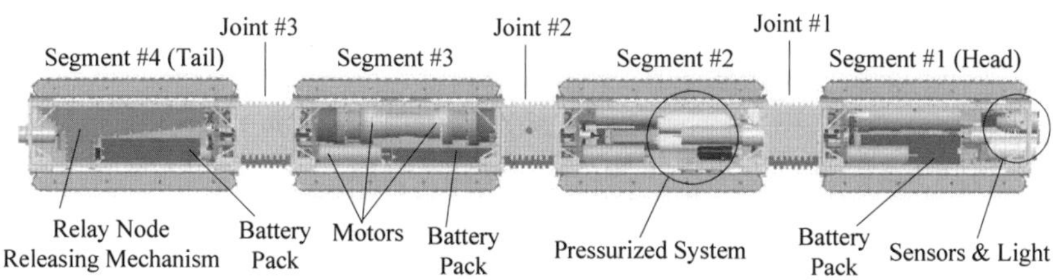

Figure 2. The overall configuration of the robot

The sensors and the main control board are in the head segment. It is equipped with five ultrasonic distance sensors, an attitude transducer sensor, a CCD camera and other gas sensors. The purged and pressurized system is in the second segment. The function of the third segment is to enhance the capability of climbing obstacles. The forth segment is the tail which contains communication nodes. There are also five ultrasonic distance sensors in this segment. The overall length of the robot is 1.7m and the total weight is 50Kg.

3.2. *All around tracks*

To enhance the mobility, all around tracks were designed for every segment. The omni-track driving mechanism is shown as Figure 3. All the crawlers attached in every sides of each segment are driven by the track driving motor through four synchronous belts simultaneously.

There are panels in the opposite sides of each tread for dust resistance shown in Figure 4. With this design, the robot can work much better in sandy terrain. The belt driving pulley is driven by the synchronous belt which is driven by the omni-track driving mechanism of Figure 3. The tracks are supported by the pulley array. Sealed type bearings and mechanical seal are used on the driving shaft.

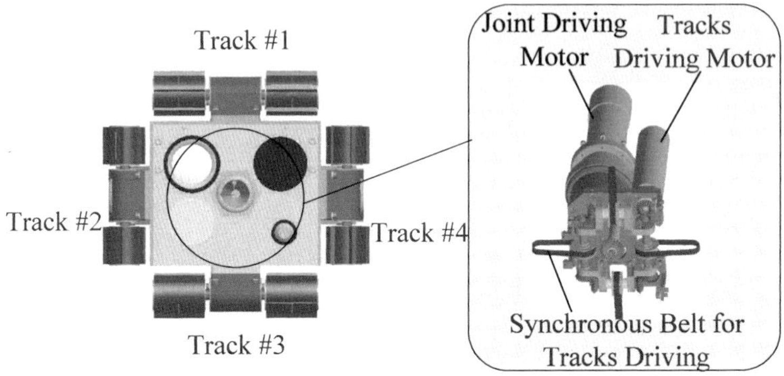

Figure 3. Omni-track driving Mechanism

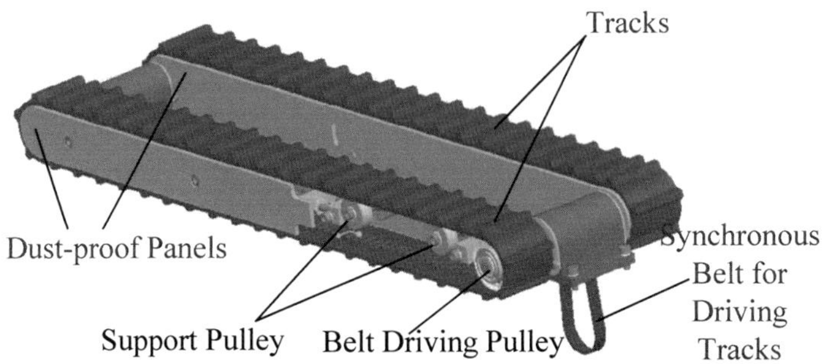

Figure 4. Dust-proof track mechanisms

3.3. *Driving and joint mechanism*

Between any two segments is one universal joint. It is made of two forks and a ball bearing-mounted gimbal connecting the two forks. The joint mechanism is shown in Figure 5.

The joint has two DOFs which are pitching direction and yawing direction. The two DOFs are driven by the motors in the two tandem segments respectively. The joint can be actuated up to 45° in the four principal directions (up, down and side to side). The two tandem segments are connected through a sealed pipe which turns the segments into a whole space, and wiring lines pass through the pipe.

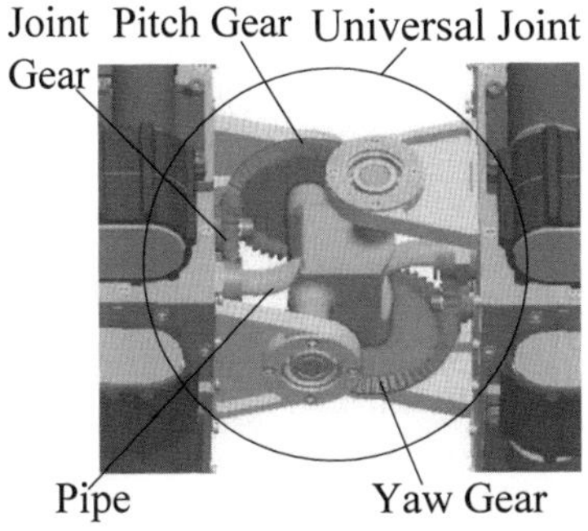

Figure 5. The 2-DOF joint mechanism

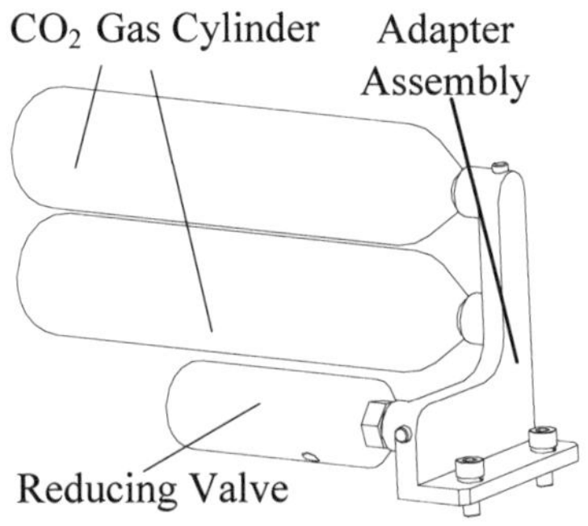

Figure 6. Gas supplying mechanism

4. Anti-explosion Design

Purged and pressurized systems provide a safe and acceptable method of protection in hazardous locations worldwide. The equipment used in the hazardous places should meet the certification requirements for the country where it is used. For example, in North America there are NEC/NFPA standards, the ATEX Directive 94/9/EC is for the European countries and IEC 60079 Standards are for worldwide. In China, GB3836: Electrical apparatus for explosive gas atmospheres should be met for the design of our robot. A gas supplying mechanism was designed to keep the robot under certain pressure, as Figure 6 showed.

The gas supplying mechanism consists of gas containers, sealed shell and the pressure control system. The O shaped rubber rings are fixed inside the end cover of the shells. And mechanical seal is applied on the shaft end. Material with fairly high hardness and wear resistance is selected as the dynamic seal element. The air inside the robot segments all runs into one another through pipes between segments. The gas containers release CO_2 to keep the pressure inside 200±30Pa, higher than outside. Failure to maintain pressure within the enclosure and adequate exhaust flow from enclosure vent is detected by an indicator.

Finally, the pressurized system endurance test was completed. In our experiments, we equipped the robot with two new pressed gas containers and keep the pressure inside 250Pa higher than the outside atmosphere. The inner pressure can be kept not less than 70% in 65 minutes, under which the robot will work safely.

5. Experiments and Discussion

In this section, experiments were set up to verify the robot's performance in some of the cases including gap traversing, crossing bridge and blocks, steering. Also field tests were carried out in the Shuangyashan Coal Mine Company.

5.1. *Performance experiments*

The robot traversed a gap 60 cm wide, shown in Figure 7. This is 35% of its length. It can even traverse a little wider gap based on geometry and center of the mass.

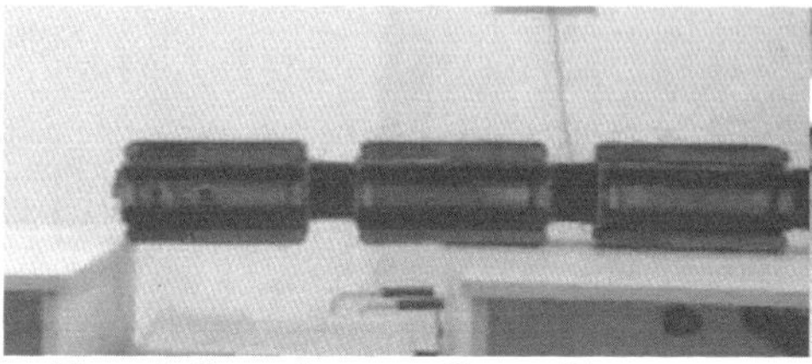

Figure 7. Traversing a gap 60 cm wide.

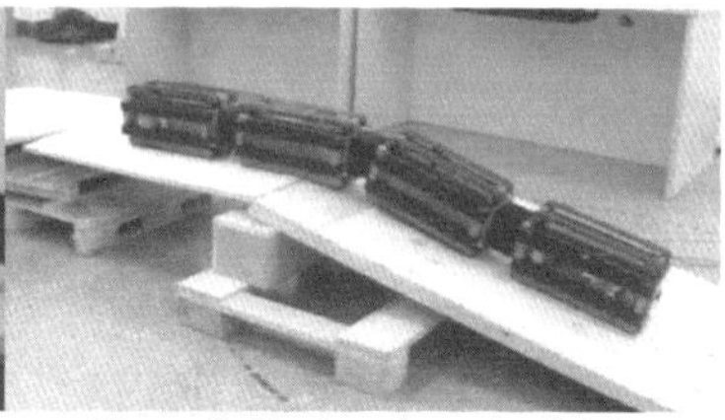

Figure 8. Getting through a bridge

We built a simple bridge with a slope of 20°shown in Figure 8. When the motors of the lift joints were released, the lift joints became passive joints and they could adapt the broken terrain.

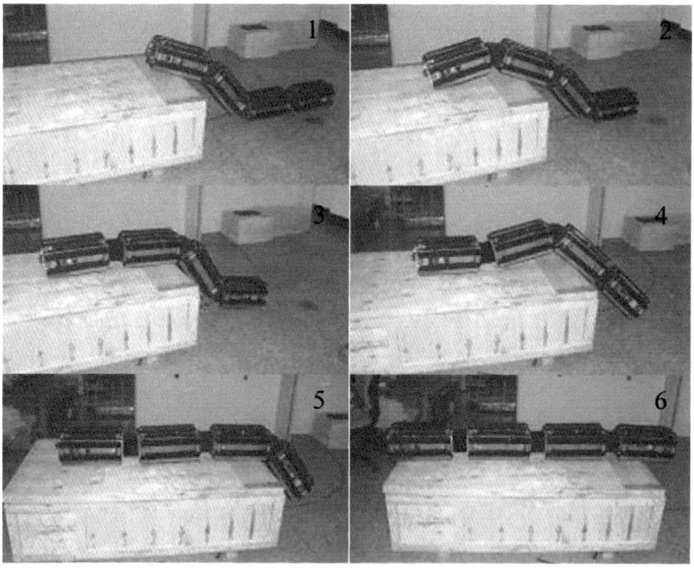

Figure 9. Climbing over block 40 cm high

Figure 9 shows the experiment of climbing over a block of 40 cm high. The robot can climb over it by lifting its segments.

The steering experiment was also conducted. The results show that the minimum turning radius on the laboratory flow is 0.4 m by turning all of its articulate joints to 45°. If only turning the first and second joints, its minimum turning radius is 0.6 m, shown in Figure 10.

Figure 10. Steer on flat floor

5.2. *Field test of the robot*

In order to verify the integral performance of the robot in coal mines, the robot was operated in Shuangyashan Coal Mine Company to carry out a searching task.

Firstly, the robot was used in the training tunnel of the company, shown in Figure 11. In this test, the robot finished the tasks including gap traversing, holes passing through, blocks crossing over, and steering.

Figure 11. The scene of working in the training tunnel

Secondly, in Xinan Coal Mine of the company, the robot was used to carry out searching task, as Figure 12 showed. In this test, the robot travelled across the gravel over 100 m, found the leaking gas bottle and fulfilled the task.

All of the tests show that the serpentine tracked robot has special advantages in search jobs and verify the correct philosophy and effective mechanism in this paper.

Figure 12. The scene of robot searching in coal mine

6. Conclusion

Facing the highly unstructured environment after disasters, this paper introduced a new omnitread serpentine robot- the HITSR-I, which was used in coal mines. The philosophy of the design, the mechanism for obstacle-surmounting and anti-explosion were described. Experiments were carried out to validate the performance of the robot. Finally the field test demonstrates the effectiveness and safety of the robot. Key features of this HITSR-I robot are:

(1) Body surrounded by tracks on all sides. This design makes the robot indifferent to roll-overs, which are bound to happen when the robot travels over rugged terrain. The four segments are linked to each other by 2-DOFs active joints. So the robot can climb over obstacles that are much higher than the robot itself.

(2) Purged and pressurized systems are developed for the robot. This design makes the robot have the ability of working in the environment with explosive gas, and even water.

(3) There are panels in the opposite sides of each tread for dust resistance. So the robot can work well in the sandy terrain.

For the next generation HITSR-II robot, we will improve the mechanism design to make it smaller and more flexible.

Acknowledgments

This work is supported by National High-tech R&D Program of China (No.2007AA041501, No.2014AA041601) and China Postdoctoral Science Foundation (No.2014M561338).

References

1. Takayama, T. and Hirose, S., Development of Souryu-I connected crawler vehicle for inspection of narrow and winding space. *Industrial Electronics*

Society, 2000. IECON 2000. 26th Annual Conference of the IEEE pp. 143-148(2000).

2. Masayuki, A., Takayama, T. and Hirose, S., Development of "Souryu-III": connected crawler vehicle for inspection inside narrow and winding spaces. *Intelligent Robots and Systems*, 2004. (IROS 2004). Proceedings. 2004 IEEE/RSJ International Conference on pp. 52-57(2004).
3. Osuka, K. and Kitajima, H., Development of mobile inspection robot for rescue activities: MOIRA. *Intelligent Robots and Systems*, 2003. (IROS 2003). Proceedings. 2003 IEEE/RSJ International Conference on pp. 3373-3377(2003).
4. Haraguchi, R., Osuka, K., Makita, S. and Tadokoro, S., The development of the mobile inspection robot for rescue activity, MOIRA2. *Advanced Robotics*, 2005. ICAR '05. Proceedings, 12th International Conference on pp. 498-505(2005).
5. Miyanaka, H., Wada, N., Kamegawa, T., Sato, N., Tsukui, S., Igarashi, H. and Matsuno, F., Development of an unit type robot KOHGA2 with stuck avoidance ability. *Robotics and Automation*, 2007 IEEE International Conference on pp. 3877-3882(2007).
6. Sang-Jin, O., Hyok-Jo, K., Jihong, L. and Hoon, C. Mathematical modeling for omni-tread type snake robot. *Control, Automation and Systems*, 2007. ICCAS '07. International Conference on pp. 1445-1449(2007).
7. Borenstein, J. and Borrell, A., The OmniTread OT-4 serpentine robot. *Robotics and Automation*, 2008. ICRA 2008. IEEE International Conference on pp. 1766-1767(2008).
8. Borenstein, J., The OmniTread serpentine robot: design and field performance (2010).
9. Han Zheng and Zhu Yongsheng, Selection of Electric Equipment in Explosive Gas Atmosphere. *Electric Switchgear*(2): 28-32 (2003).
10. J.K. Hopkins, A Survey of Snake-Inspired Robot Designs, *Bionispiration and Biomimetics*, 4(2):021001(2009).

SECTION–12

PLANNING AND CONTROL

LIDAR-BASED NAVIGATION-LEVEL PATH PLANNING FOR FIELD-CAPABLE LEGGED ROBOTS

I. HAVOUTIS*, D. G. CALDWELL and C. SEMINI

Dynamic Legged Systems Lab,
Department of Advanced Robotics, Istituto Italiano di Tecnologia,
Genova, 16163, Italy
**E-mail: ioannis.havoutis@iit.it*

In this paper we present a navigation framework for field deployment for the hydraulically actuated quadruped, HyQ.[1] Our framework uses a lidar sensor to perceive obstacles and free space around the robot and plans a path through traversable areas to given goals. A path following procedure generates velocity and angular rate commands that are sent to a locomotion controller that produces the trotting gait pattern, following the desired path to the goal.

Keywords: Legged Locomotion, Navigation, Quadruped, LIDAR.

1. Motivation

Legged robots are able to access a wide range of terrain types of varying difficulty. In unstructured environments, where only a number of discrete footholds are possible (disaster sites, forests, etc.), legged quadrupedal systems typically use walking gaits.[2,3] In environments where smooth, continuous support is available (flats, fields, roads, etc.), where exact foot placement is not crucial for the success of the behavior, legged systems use a variety of more dynamic gaits, e.g. trotting, pacing, galloping.[4,5]

Legged robots operating in the field are often directly controlled by a human. In many use cases, such systems can operate in an autonomous mode, provided that adequate perception and navigation capabilities are added. For example, such *locomotion autonomy* can enable a legged robot working in a construction site to reach material pick-up and drop-off checkpoints without human supervision. To this end we develop a navigation and locomotion framework that is capable of perceiving obstacles and free space around the robot using a lidar sensor, then plans and reliably follows paths to given goals using a trotting locomotion controller.

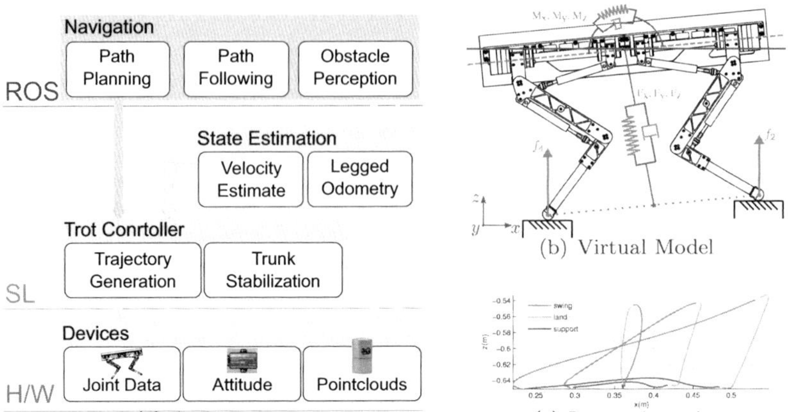

(a) Architecture

(b) Virtual Model

(c) Leg trajectories

Fig. 1. (a): Overall architecture of the approach. (b): The virtual elements used to calculate forces and moments that stabilize the quadruped's trunk. (c): Trotting trajectories that correspond to the foot of the left front leg in the robot's body reference frame. The different colors denote the three gait cycle phases while the three trajectories correspond to trotting in place, trotting forward with a velocity of 0.5m/s and 1.0m/s.

1.1. *Related work*

A large body of literature in robotics is devoted to quadrupedal locomotion. Early examples include the work of Raibert[6] on hydraulically and pneumatically actuated legged machines. Recent examples include Hawker and Buehler,[7] and Remy et al.,[8] focusing on passively compliant elements and light-weight leg structures. Close to our research stands Boston Dynamics' BigDog though little is still known about its control structure or experimental evaluation. In previous work we experimented with a feedback/feedforward control structure similar to the one presented here, coupled with a CPG-based trajectory generation procedure.[5] We have also experimented with an active compliance control structure that emulates a passive telescopic leg behavior.[4]

Navigation-level path planning has received relatively small attention from the legged robots community as many of the applications concern close distances and often indoor usage. Such cases are often best suited with a variety of stereo and RGB-D sensors or combinations thereof. On the other hand many of the teams of the DARPA Urban Challenge made extensive use of lidar sensors and similar path planning approaches[9,10] as such systems are tailored toward field deployment.

2. Locomotion & Navigation Framework

An overview of our framework is presented in Fig.1(a). Our approach is divided into a navigation and a locomotion level, which in practice are handled with two separate onboard computers. The details of the locomotion and navigation levels follow.

2.1. *Locomotion*

We chose a trotting gait as we are interested in locomotion speeds ranging from trotting on the spot to fast-paced trot (1m/s). The trot is a symmetrical gait in which diagonally opposite legs swing in unison. This allows a desirable division of the total force that the leg actuators should be able to produce when supporting the weight of the robot, when thrusting, and when receiving impact forces at touchdown. Our locomotion approach divides the control procedure into two subsystems. One generates the trajectories that the two leg pairs execute given external user input, while the other calculates a feed-forward trunk stabilizing input according to the current state of the system. In the interest of space we briefly describe these subsystems below.

2.1.1. *Trajectory generation*

While trotting, the four legs are naturally divided into two pairs. Each pair follows an identical foot trajectory. This trajectory depends on the commanded forward velocity, turning rate, step and swing height, and the gait cycle period. The step height dictates the body height in the body reference frame during the support phase, and the swing height defines the leg swing apex. The gait cycle period is used to calculate the timing of the leg trajectories as this is intuitively further subdivided in three distinct phases; swing, land and support (Fig. 1(c)). We set the step height to -0.65m and the swing height to -0.55m, according to the dimensions of HyQ.[1] In our experiments we use a gait cycle period of 0.5s ($2Hz$), that is suitable for velocities up to 1m/s.

2.1.2. *Trunk stabilization*

To stabilize the trunk of the robot we follow a virtual model control approach.[2] We calculate virtual forces and moments according to a reference state and the current state of the system (Fig. 1(b)) as computed by an onboard state-estimation algorithm. The virtual forces and moments are then transformed to forces f_i that the feet in contact need to apply. We use a least-squares optimization that provides the least norm solution, i.e. the minimum force solution. The feet forces are subsequently mapped to feed-forward torques (τ_{ff}) for the joint actuators of the legs that are in stance. This is done by: $\tau_{ff} = J^T\mathbf{f}$, where $\mathbf{f}$ is the vector of (linear) forces that each foot in stance needs to apply to emulate the virtual model behavior, and J is the Jacobian of the legs that are in contact.

2.2. *Navigation*

The navigation framework deals with all the procedures that are required for the quadruped to navigate autonomously in a large-scale unstructured environment. In brief we separate this into four procedures that are needed to achieve the navigation task. Namely these are: *obstacle perception*, *path planning* and *path following*, and are explained in detail in the following sections.

2.2.1. *Obstacle perception*

The lidar (Velodyne, HDL-32E) produces about 700.000 points per second, with an 10Hz frame-rate. This produces a pointcloud $\mathbf{P} = [p_i, \dots p_n]^T$ by 'scanning' the geometry of the environment. We then need to classify the points to two sets, the *obstacles set*, $\mathcal{O} = \{p_j\}$, and the *free-space set*, $\mathcal{F} = \{p_k\}$. We are most interested in a robust detection of the points belonging to the obstacle set as this is important for our navigation needs, i.e. these signify the areas that we need to avoid in the path planning step that follows.

We start by discretizing the area around the robot into a square grid where the edge of the square is $L = 10m$, while the resolution is set to $\ell = 0.1m$, i.e. each grid-cell is $0.1m^2$. We then consider all the points that belong to the point-cloud, $\mathbf{P}$, and accordingly assign them to their respective grid cells, $g_{i,j}$, according to their position. During this we keep track of the extrema points, in our case the minimum and maximum height points, that belong to each volume that a grid-cell defines along the vertical.

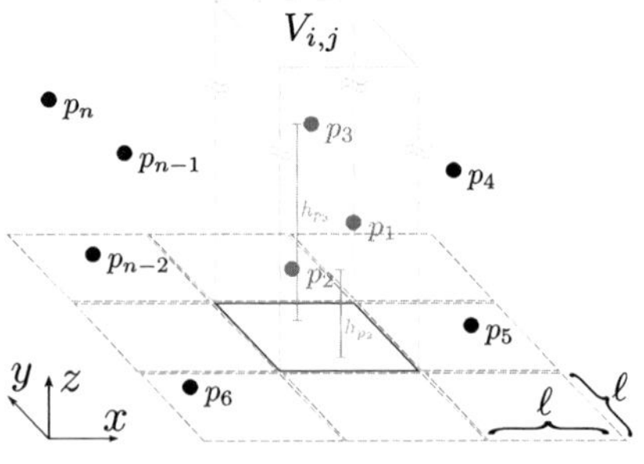

Fig. 2. Obstacle perception procedure. See Sec. 2.2.1 for details.

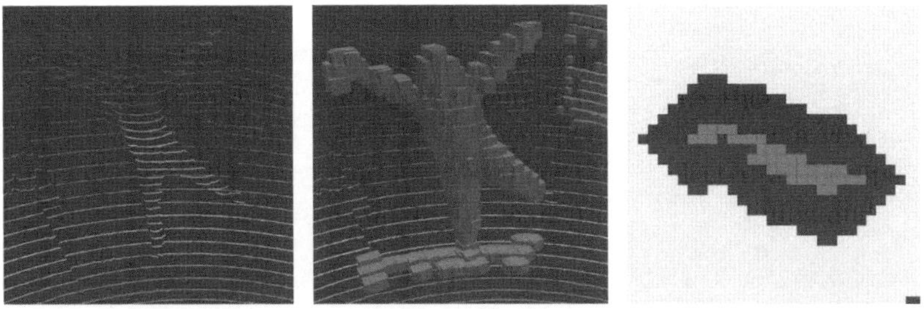

(a) Pointcloud. (b) Obstacle points. (c) Occupancy grid.

Fig. 3. The obstacle segmentation pipeline. (a) The pointcloud scan as published by the lidar. (b) The points that belong to the obstacle set $\mathcal{O}$. Note that the obstacle points are visualised as red cubes. (c) The (inflated) occupancy grid. The red cells are the direct projection of the obstacle points to the horizontal grid. The obstacle cells are then inflated according to the geometry of the robot, here represented by the grey grid-cells.

If the height difference between these two points is greater than a set obstacle threshold, $t_{\mathcal{O}}$, then all the points that belong to this volume (column) are added to the obstacle set.

For example consider the pointcloud in Fig.2. The points p_1,p_2 and p_3 belong to the grid-cell $g_{i,j}$ and accordingly to the volume, $V_{i,j}$, that this defines. The difference between the highest (p_3) and the lowest (p_1) points of the volume $V_{i,j}$ is then considered and accordingly all the points that belong to the volume in question are added to the obstacle set $\mathcal{O}$ while the rest are added to the free-space set $\mathcal{F}$. The obstacle threshold, $t_{\mathcal{O}}$, is typically set at $0.25m$. A real-world example of the obstacle segmentation procedure is available in Fig. 3(a) and 3(b). The former presents the raw pointcloud as scanned, while the latter is the result of the obstacle perception procedure, where obstacle points are visualised as red cubes. Note that with this procedure there can be grid-cells that are not assigned any points. This grows with the distance from the sensor as the distance between two consecutive 'scan-rings' increases and is by definition true for areas that are 'shadowed' (occluded) by any other geometry present.

According to $\mathcal{O}$ we populate an occupancy grid $G_{\mathcal{O}}$ that is communicated to the path planning step. Alternatively one can choose a voxel grid representation if 3D path planning is required. We selected an occupancy grid representation as it is adequate for navigation-level planning and drastically reduces the communication overhead.

2.2.2. *Path planning*

The obstacle perception step produces an occupancy grid that contains all the perceived obstacle areas within the defined grid. To simplify the path

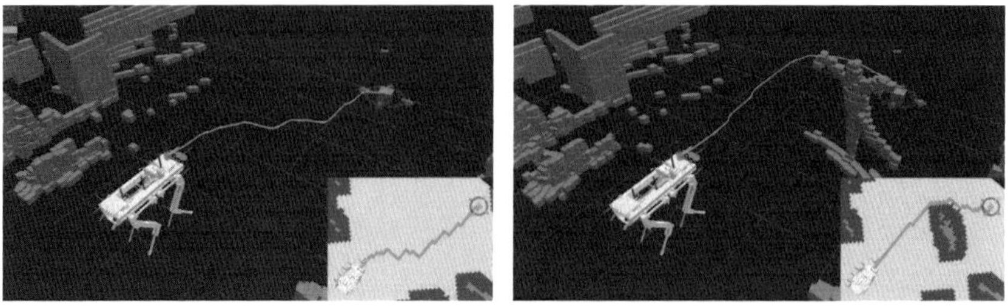

Fig. 4. Examples of the A* path planning results with the procedure described in Sec.2.2.2. The planned paths in 3D with the obstacle set points $p_j \in \mathcal{O}$ visualized as red cubes. *Inset plots*: Top-down view of the A* planned path to the current goal. The light-grey path on the grid is the path of cell-centers that is returned by the A* search algorithm. The orange line is the smoothed path that the quadruped follows.

planning problem we use the obstacle inflation approach. Accordingly all obstacle cells are 'inflated' according to an inflation factor that is dependent on the robot's geometry. In our trials we use an inflation factor of $0.5m$, according to the robot's width, which in practice works well since the quadruped has a preferred direction of motion. An omnidirectional locomotion approach would require a more elaborate obstacle inflation step that would take the possible orientation of the robot into account.

Once the inflated grid has been computed we use the A* search algorithm to find the shortest path from the current position of the robot to the goal position. The goal position is provided as user input and is expressed in the odomety (global) frame. For our current A* implementation we use a four-way connectivity and the Manhattan distance metric, while we use the Euclidean distance as a heuristic function. The produced path, $g_{i,j}^{1...n}$, is a sequence of cells in the inflated occupancy grid that is then transformed to the sequence of cell center-points.

The choice of four-way connectivity and Manhattan distance for the shortest path computation serves to keep the computational load to a minimum but in general produces non-smooth paths. We use a gradient decent approach to smooth the resulting path. This works iteratively by computing a smoothed version, $g_{sm}^{1...n}$, of the original path according to a smoothing coefficient. This is run until convergence, which in our experimental trials was on average below 10 iterations.

2.2.3. *Path following*

Next the planned path is converted to a meaningful command for the quadruped to follow. For this we use a 'carrot' path follower which operates as follows. The given path points, ranging from the quadrupeds current lo-

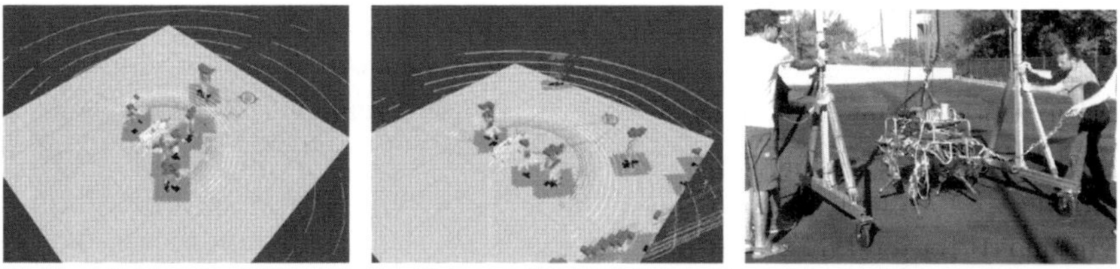

Fig. 5. *Left* and *middle*: Views of the world model that the robot builds-up. Occupancy grid and planned path to goal. *Right*: Picture of the HyQ trotting at the parking lot.

cation to the goal position, are sequentially examined. The path point that is d_c (length of the carrot stick) away from the robot is chosen to compute a twist command, i.e. a linear and angular velocity $t_c \in R^6$, towards this intermediate goal. The command is computed in a *PD* fashion and is constrained to forward linear velocity and angular yaw rate. A control logic that produces turning on-the-spot is also implemented for cases where the goal is beyond a reachable turning angle. Once no 'carrot' goal exists on the planned path, i.e. $\forall g_{sm} \nexists g^i_{sm} \geq d_c$, the goal has been reached.

3. Experimental Trials

We ran a number of experimental trials with the robot navigating outdoors (Fig.5). The site of trials was a parking lot where adequate space was available to evaluate the quadruped's behavior. On average we maintained a relatively cautious upper boundary for the velocity and the angular rate of the locomotion control which limited the speed at which goals were achieved. The quadruped never collided with obstacles or people moving around it. Nonetheless, a cautious inflation factor can limit the ability of the robot to navigate through narrow passages, for example single doors. On the other hand we do not expect to encounter such geometries in the field.

4. Conclusion

In this paper we presented a navigation framework, for legged robots operating in the field, that can provide locomotion autonomy. We showed how a lidar sensor can be used to create adequate perceptual input to enable navigation-level planning. We showed how lidar data can be used to classify obstacle and free space and subsequently use this world model to compute paths that avoid static and moving obstacles and reach goals that are tens of meters away in an outdoor environment. This approach makes use of a robust trotting controller to handle the robot's locomotion behavior. This way reliable path following was achieved by communicating desired velocity and angular rate commands.

In future work we aim to enable long and short-range map building and use semantic information to reason about path suitability. In addition we are developing a supervision system that will be able to choose between different locomotion modes, e.g. walking, trotting, galloping, based on the characteristics of the environment at hand.

Acknowledgments

This research is funded by the Fondazione Istituto Italiano di Tecnologia.

References

1. C. Semini, N. G. Tsagarakis, E. Guglielmino, M. Focchi, F. Cannella and D. G. Caldwell, Design of HyQ – a hydraulically and electrically actuated quadruped robot, in *Journal of Systems and Control Engineering*, 2011.
2. A. W. Winkler, I. Havoutis, S. Bazeille, J. Ortiz, M. Focchi, D. G. Caldwell and C. Semini, Path planning with force-based foothold adaptation and virtual model control for torque controlled quadruped robots, in *IEEE International Conference on Robotics and Automation (ICRA)*, 2014.
3. A. W. Winkler, C. Mastalli, I. Havoutis, M. Focchi, D. G. Caldwell and C. Semini, Planning and Execution of Dynamic Whole-Body Locomotion for a Hydraulic Quadruped on Challenging Terrain, in *IEEE International Conference on Robotics and Automation (ICRA)*, 2015.
4. I. Havoutis, C. Semini, J. Buchli and D. G. Caldwell, Quadrupedal trotting with active compliance, in *IEEE International Conference on Mechatronics (ICM)*, 2013.
5. V. Barasuol, J. Buchli, C. Semini, M. Frigerio, E. R. De Pieri and D. G. Caldwell, A reactive controller framework for quadrupedal locomotion on challenging terrain, in *IEEE International Conference on Robotics and Automation (ICRA)*, 2013.
6. M. H. Raibert, *Legged robots that balance* (The MIT Press, Cambridge, MA, USA, 1986).
7. G. Hawker and M. Buehler, Quadruped trotting with passive knees: design, control, and experiments, in *2000 IEEE International Conference on Robotics and Automation (ICRA'00)*, 2000.
8. C. Remy, M. Hutter and R. Siegwart, Passive dynamic walking with quadrupeds - extensions towards 3d, in *Robotics and Automation (ICRA), 2010 IEEE International Conference on*, 2010.
9. D. Dolgov, S. Thrun, M. Montemerlo and J. Diebel, Practical search techniques in path planning for autonomous driving, in *Proceedings of the First International Symposium on Search Techniques in Artificial Intelligence and Robotics (STAIR-08)*, (AAAI, Chicago, USA, June 2008).
10. D. Ferguson, T. M. Howard and M. Likhachev, Motion planning in urban environments, in *Journal of Field Robotics*, (11-12) (Wiley Subscription Services, Inc., A Wiley Company, 2008).

A STRATEGY FOR FINDING THE MOST RELIABLE PATH IN UNCERTAIN ENVIRONMENTS*

BOTAO ZHANG
Institute of Automation, Hangzhou Dianzi University, Hangzhou, Zhejiang Province, 310018, China

QIANG LU
Institute of Automation, Hangzhou Dianzi University, Hangzhou, Zhejiang Province, 310018, China

JIAN WANG
Department of Control Systems and Informatics, ITMO National Research University, St. Petersburg, 197101, Russia

Abstract: A strategy for finding the most reliable path is proposed to deal with target search problem in this paper. When a robot chases a target in Semi-structured environments with uncertainty and hazards, different roads may lead to disparate risks. A kind of reliability based topological map is designed, in which uncertainties such as road condition and threats are considered. And then, a motion planning method is presented to minimize the expected-risk in carrying out a target search task. Further simulation and physical-world experiments are presented to demonstrate the feasibility and validity of this approach.

Keywords: probabilistic robotics; target search; motion planning.

1. Introduction

Target search refers to search for one or several targets in an environment with one or more robots, which involves machine vision[1], motion planning[2], artificial Intelligence and so on[3]. Planning a path for robot target search, as one of the most challenging problems, has attracted much attention over the past several years. Many novel planning methods for various applications have been presented to handle target search problems, such as home service, search and rescue, plant exploration and security patrol[3,4].

* This work is supported in part by the Natural Science Foundation of Zhejiang Province (No. LQ14F030012, No. LQ13F030014) and the National Natural Science Foundation of China (No.61375104).

Most of these existing methods can be classified into two classes: deterministic target search methods and probabilistic target search methods. Previously, studies of motion planning for target search mainly focused on deterministic approaches. In these studies, the most important criterion to evaluate the quality of exploration strategy is time. In recent years, researchers find that many factors can contribute to uncertainty of target search tasks, such as unpredictable environments, limited sensors, algorithmic approximations, etc., robots are supposed to be able to deal with enormous uncertainties in the physical world[5]. Hereby, probabilistic target search methods are put forward. Sarmiento et al. proved that a path for target search that minimize the expected-time does not necessarily minimize the distance[2], and proposed a locally optimal continuous search methods to minimize the expected-time for target search. Sebastian pointed out that the safety of a robot can be threatened by narrow corridors, which can increase the uncertainty[6], and the terrain can influence a robot by the traversability of road and the localization of the robot[7].

This study aims to find a strategy for planning the most reliable path in environments with uncertainty, and then employs it to improve the reliability in target search. A kind of reliability based topological map is designed, in which uncertainties such as road condition and threats are considered. And then, the proposed path planning method is applied to search the above reliability based topological map.

2. A Strategy for Finding the Most Reliable Path

2.1. *Target search with the most reliable path*

With no uncertainties, a robot can maintain a constant speed, and the most important criterion to evaluate the quality of exploration strategy is time. Under this situation, a shorter road is superior to any longer one. However, in the physical-world, narrow corridors or terrible terrains could slow a robot down and might bring damage to a robot. In some cases, the most reliable road may not be the shortest one. Therefore, uncertain factors should be considered in path planning for a mobile robot.

Problem Definition: For a map $G(J,O)$, composed of a vertex set $O=\{O_k \mid O_k \in G, k=1,2,3,...,m\}$ and a edge set $J=\{J_k \mid J_k \in G, k=1,2,3,...,n\}$, the reliable probability $p_{O_iO_j}$ is to estimate the reliability between O_i and O_j. If there are x factors having influence on reliable probability $p_{O_iO_j}$, and sub-reliable probabilities of these factors can be represented as $p^1_{O_iO_j}, p^2_{O_iO_j}, p^3_{O_iO_j}, \cdots, p^x_{O_iO_j}$. $p_{O_iO_j}$ is calculated by

$$p_{O_iO_j} = f(p^1_{O_iO_j}, p^2_{O_iO_j}, p^3_{O_iO_j}, \cdots, p^x_{O_iO_j}) = \prod_{k=1}^{x} p^k_{O_iO_j}. \quad (1)$$

The global objective function for finding the most reliable path is

$$f(C) = \prod_{i=1}^{n-1} p_{L_iL_{i+1}}, \quad (2)$$

where $C = \{L_1, L_2, ..., L_n\}$ is a sequence of vertexes for target search.

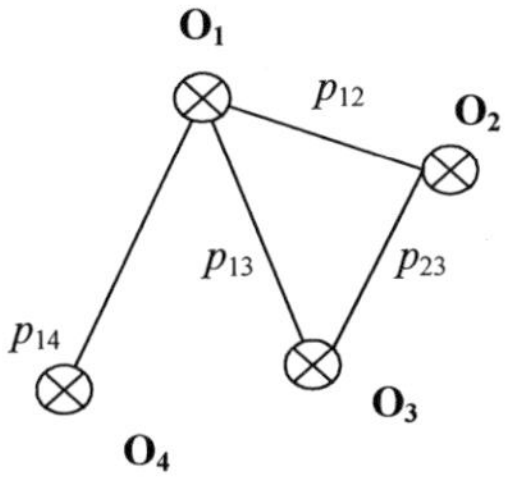

Fig. 1 Reliability based topological map

Figure 1 is a toy-like reliability based topological map $G_1(J,O)$, we will illustrate the reliable matrix with this map. The reliable matrix of $G_1(J,O)$ is

$$\hat{D}_1 = \begin{bmatrix} \hat{p}_{11} & \hat{p}_{13} & \hat{p}_{14} & \hat{p}_{15} \\ \hat{p}_{21} & \hat{p}_{22} & \hat{p}_{23} & \hat{p}_{24} \\ \hat{p}_{31} & a_{32} & \hat{p}_{33} & \hat{p}_{34} \\ \hat{p}_{41} & \hat{p}_{42} & \hat{p}_{43} & \hat{p}_{44} \end{bmatrix}, \quad \text{where, } \hat{p}_{ij} = \begin{cases} -\ln(p_{O_iO_i}) & p_{O_iO_i} \in (0,1) \\ \text{inf} & p_{O_iO_i} = 0 \\ 0 & p_{O_iO_i} = 1 \end{cases}. \quad (3)$$

There are many kinds of search tasks, the proposed method will only be applied to a special one. For a semi-structured environments ***E*** with some small real-time changes, for example, a building with several people goes around, a set of observation points $O = \{O_k \mid O_k \in E, k = 1,2,3,...,n\}$ at which robot can sense the environment is given as input. The location of target is stochastic and updates at an interval. Once the sensor network captures the target and sends its location, the robot should plan the most reliable path to the target.

2.2. *Sequence planning for the most reliable path*

In graph theory, the Floyd-Warshall algorithm is a well-known and well-studied algorithm to search for the sequence of vertexes and edges with minimum weight in directed and undirected graph[8]. In this study, it will be improved to solve the most reliable path problem.

The major differences between the Floyd-Warshall algorithm and path planning algorithm needed in this kind of target search are as follows.

(1) The Floyd-Warshall algorithm only presents the theoretical minimum path; it has nothing to do with the real feasibility. However, in the robot search task, if something blocked the set course, the robot needs to plan another path to the goal. (2) In the Floyd-Warshall algorithm, any weight of edge is constant. However, the reliable information of environments should be updated whenever necessary. (3) There are many influential factors for reliability of roads in target search, varying from robot to robot, and task to task. Therefore we have to choose an appropriate $p^{x}_{O_iO_j}$, and use formula (1) to calculate the reliability of road.

The following formula is to transform the probability matrix D_1 from workspace into the reliable matrix $\hat{D}_1$ based on the Floyd-Warshall. Take Fig.1 as an example:

$$\hat{D}_1 = \begin{bmatrix} p_{O_1O_1} & p_{O_1O_2} & p_{O_1O_3} & p_{O_1O_4} \\ p_{O_2O_1} & p_{O_2O_2} & p_{O_2O_3} & p_{O_2O_4} \\ p_{O_3O_1} & p_{O_3O_2} & p_{O_3O_3} & p_{O_3O_4} \\ p_{O_4O_1} & p_{O_4O_2} & p_{O_4O_3} & p_{O_4O_4} \end{bmatrix} \Rightarrow \hat{D}_1 = \begin{bmatrix} \hat{p}_{11} & \hat{p}_{12} & \hat{p}_{13} & \hat{p}_{14} \\ \hat{p}_{21} & \hat{p}_{22} & \hat{p}_{23} & \hat{p}_{24} \\ \hat{p}_{31} & a_{32} & \hat{p}_{33} & \hat{p}_{34} \\ \hat{p}_{41} & \hat{p}_{42} & \hat{p}_{43} & \hat{p}_{44} \end{bmatrix}, \quad (4)$$

The Sequence Planning algorithm is as follows.

Step 1: Input the probability matrix D_1 according to the workspace;

Step 2: Compute the reliable matrix $\hat{D}_1$ using formula (4);

Step 3: Sequence planning with the Floyd-Warshall algorithm;

Step 4: If the update time is up or the robot is on a blocked road, then, update the reliable information and location of target, and repeat setp1.

The procedure figure of the strategy for finding the most reliable path in environments with uncertainty is shown in Fig. 2.

3. Experiment Results

3.1. *Simulations*

In simulations, the robot is set to move at a speed of 0.5m/s, S_0 denotes the initial location of the mobile robot. In environment I, the narrow corridor is a threat for the mobile robot. The shortest path is generated by the Dijkstra and

APF, shown in Fig. 3(a). The most reliable path planned by the proposed algorithm is shown in Fig. 3(b). From results in Table1, we can see that the most reliable path is not the shortest path. This path, the robot might spend more time to carry out the task, however, it takes less risk.

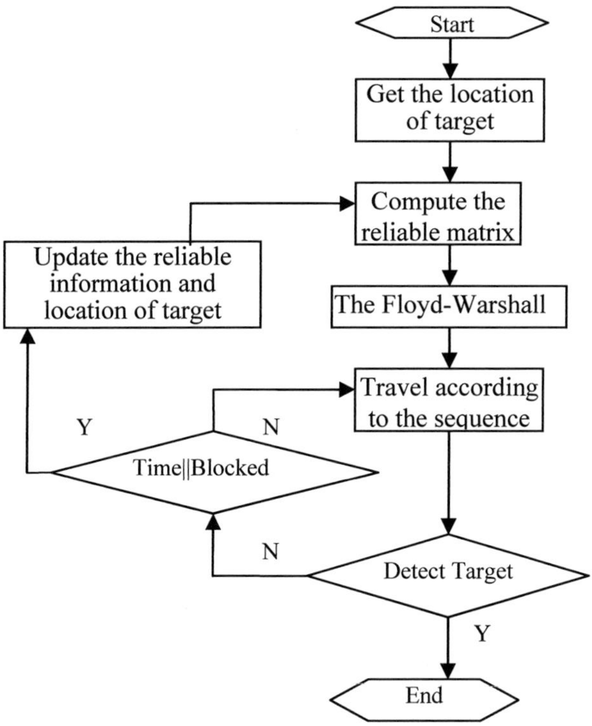

Fig. 2 Procedure figure of the most reliable path algorithm

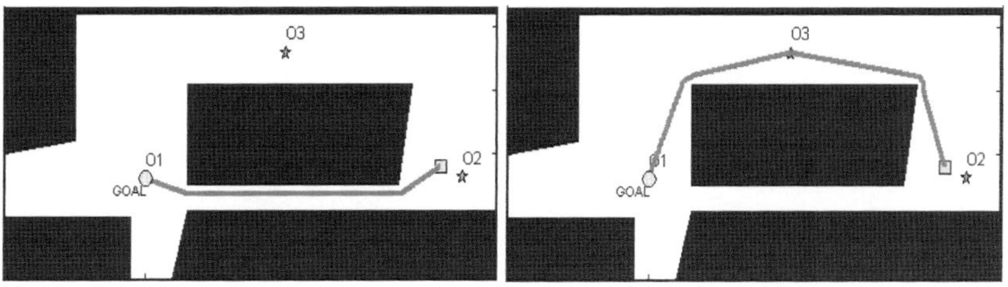

(a) (b)

Fig. 3 The Dijkstra and the proposed method in environment I

Table 1: Comparison of the shortest and the most reliable path in environment I

	The shortest path	The most reliable path
Reliability	17.2 %	43.0%
Time (s)	21.99	32.84
Average velocity (m/s)	0.50	0.50

There are ten observation points with no narrow corridor in environment II. Each edge between two points has different reliability. A mobile dangerous hunter is put into the workspace, which can cause damage to the robot. The shortest path is generated by the Dijkstra and APF, shown in Fig. 4(a). Table 2 shows that the time to complete the target search task with the Dijkstra and APF algorithm is shorter than that of the most reliable path approach in this particular environment. In most cases, the robot would choose the path in Fig. 4(b), however, if it encounters the hunter, the route will be re-planned, as shown in Fig. 5.

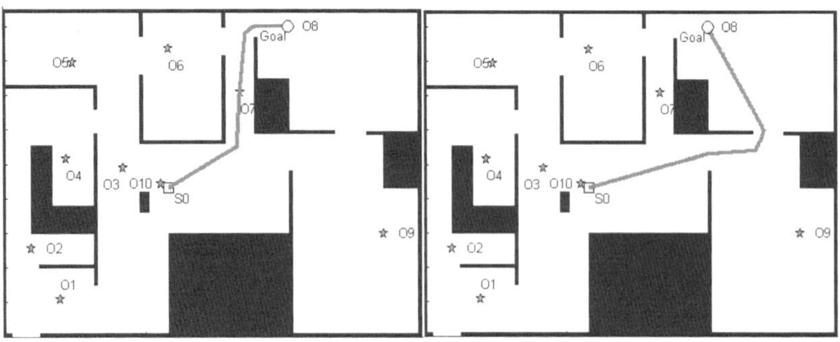

(a) The Dijkstra method (b) The most reliable path method

Fig. 4. Simulations in environment II

Table 2: Comparison of the shortest and the most reliable path in environment II

	The shortest path	The most reliable path	The most reliable path (encounter danger)
Reliability	11.7%	79.3%	62.4%
Time (s)	75.5	103.0	189.6
Average velocity (m/s)	0.50	0.50	0.5

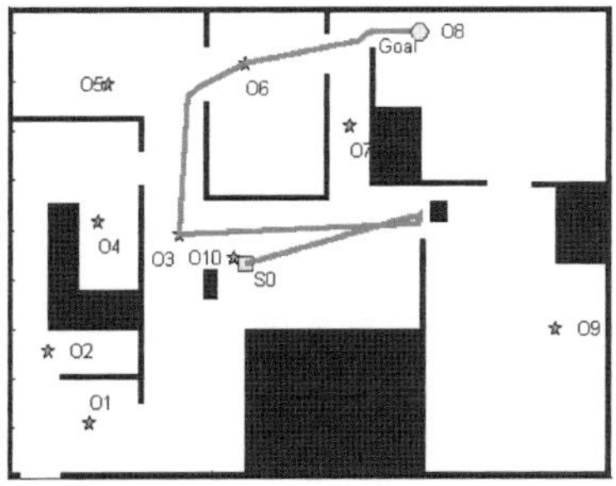

Fig. 5 The most reliable path in environment II (encounter danger)

3.2. *Physical-world experiment*

This semi physical-world is built in the laboratory, in which, the mobile robot can not keep a constant speed for various reasons, such as friction of the ground, turnings at corners. The robot also needs time to detect obstacles and the target. In following experiments, the robot named Powerbot, equipped with vision systems, segmented bump, ranging sonar and laser, was applied to look for a table tennis ball. The CamShift algorithm is used to recognize the object with a given hue, this algorithm could track table tennis ball with histogram.

Table 3: Comparison of the shortest and the most reliable path

	The shortest path	The most reliable path
Reliability	30.4%	65.5.0%
Time (s)	31	37
Average velocity (m/s)	0.194	0.105

There are two paths between the robot and the ball, the shorter one is narrow and with many obstacles. Five independent runs are carried out with each method to test the average of time cost and reliability. Compared with the shortest path, the most reliable path method generates a road a bit longer, but takes less time to track. When the robot travels along the shortest path, it spends too much time in preventing itself from colliding obstacles.

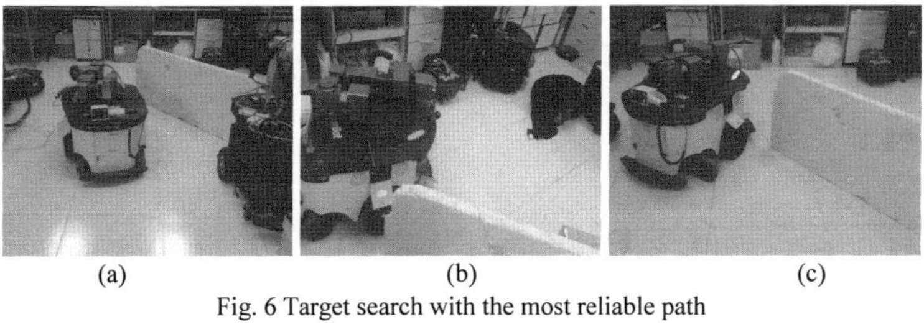

(a) (b) (c)

Fig. 6 Target search with the most reliable path

4. Conclusions

A strategy for finding the most reliable path in environments with uncertainty has been proposed in this study. As a conservative method, it is especially suitable for the situation in which it is difficult to bear the damage of the robot. In addition, this approach is suitable in time abundant search tasks. Simulation results and physical-world experiments have proved that the proposed method performs well on finding the most reliable path. However, this generalization need to be further supplemented and verified by more replicated studies so that they can be more applicable.

References

[1] Aydemir, A., Pronobis, A., Göbelbecker, M. and Jensfelt, P. Active Visual Object Search in Unknown Environments Using Uncertain Semantics [J]. IEEE Transactions on Robotics, 2013, 29 (4): 986-1002.

[2] Sarmiento, A., Murrieta-Cid, R. and Hutchinson, S. An Efficient Motion Strategy to Compute Expected-Time Locally Optimal Continuous Search Paths in Known Environments [J]. Journal of Advanced Robotics. 2009, 23 (12-13): 1533-1560.

[3] Anderson G. T., Tunstel E. W. and Wilson E. W. Robot system to search for signs of life on mars [J]. IEEE Transactions on Aerospace and Electronic Systems. 2007, 22 (12): 23-30.

[4] Shinji K., Mineo F. and Takashi O. Emergency response by robots to Fukushima-Daiichi accident: summary and lessons learned [J]. Industrial Robot. 2012, 39 (5): 428-435.

[5] He, R., Brunskill E. and Roy N. Efficient Planning under Uncertainty with Macro-actions [J]. Journal of Artificial Intelligence Research. 2011, Volume 40: 523-570.

[6] Thrun S., Burgard W. and Fox D. Probabilistic Robotics[M], The MIT Press, 2005.

[7] Zhu X, Qiu C, Minor M A. Terrain Inclination based Three-dimensional Localization for Mobile Robot in Outdoor Environments[J]. Journal of Field Robotics, 2014, 31(3):477-492.

[8] Hougardy, S. The Floyd–Warshall algorithm on graphs with negative cycles [J], Information Processing Letters. 2010, 110 (8-9): 279–281.

MOTION CONTROL BASED ON VELOCITY RESOLUTION OF A TROTTING QUADRUPED ROBOT

LIN LANG, JIAN WANG, HONGXU MA, QING WEI†

College of Mechatronics Engineering and Automation, National University of Defense Technology, Hunan Province, Changsha, China

This paper proposes a control strategy based on velocity resolution for velocity and orientation tracking of a trotting quadruped robot. In trotting gait, the pair of diagonal stance legs and body of robot can be approximately modeled as a virtual planar-seven-link mechanism and a SLIP model by projecting them into two orthogonal planes. To reduce tracking error of velocity, the foot placements are computed based on SLIP model. As there is misalignment between the centroid of body and the geometric center of body, the core of the trotting quadruped robot is switching control in the option of this paper. Based on Lyapunov theory, definition of stable region and sufficient condition of convergence of trotting quadruped robots are proposed. Finally, an experiment is performed to validate the effectiveness of proposed strategy.

1. Introduction

Trotting quadruped robots have been investigated by many researchers in the past few decades. During recent years, several successful quadruped robots have been developed, for example, Big Dog [1,2], AlphaDog [3], Cheetah [4], which were developed by Boston Dynamics, have strong acclimatization and robustness. However, not much information about the control algorithms is given out. LittleDog [5], HyQ [6] are two other famous prototypes. Detailed description of state of the art on quadruped robots in China is presented in literature [7].

A number of algorithms have been proposed to solve the planning and control problems of trotting quadruped robots. Raibert M proposes a velocity controller based on SLIP model [8]. However, SLIP model ignores the coupling relationship between position and orientation of body. So, another models and methods are needed. Victor Barasuol and Jonas Buchli, propose a reactive controller framework for robust quadruped locomotion based on a task space CPG, get significant simulations and experimental results on the hydraulically

† This work has been funded by the National Natural Science Foundation of China (No. 61473304) and National Hi-tech Research and Development Program of China (Great NO. 2015AA042202).

actuated quadruped robot HyQ. However, it is difficult to analyze the stability of gaits [9]. Reinforcement learning method, as a model-free algorithm, is widely used in motion control of the quadruped robots. By maximizing the long-term rewards sum, it approximates to the real system to get the optimal strategy. However, there are many problems in the transferring from the simulation experiment to the actual system, the efficiency of the algorithm and so on [10].

In trot gait, two pairs of diagonal legs are periodically switching between stance phase and swing phase. There have been significant theoretical results about switching control and the conditions of its stability [11], most of them are based on Lyapunov theory. However, there is little literature that applies switching control to stability analysis of trotting quadruped robots.

The main contributions of this work are:

Firstly, the dynamic model of a trotting quadruped robot is described by a planar-seven-link mechanism and a SLIP model.

Secondly, the orientation stability of a trotting quadruped robot is analyzed by using switching control theory.

This article is organized as follows. Based on velocity resolution, the dynamic model and velocity control strategy of a quadruped robot are deduced in section 2. And section 3 proposes a sufficient condition of orientation convergence of trotting quadruped robots by using switching control theory. In section 4, experiments are performed to validate the effectiveness of proposed strategy and algorithms, respectively. At last, the conclusion is drawn.

2. Velocity Control Strategy Based on Velocity Resolution

2.1. *Dynamic model based on velocity resolution*

A pair of diagonal legs is always simultaneously driven during trotting. In the following analysis, Left Rear (LR) leg and Right Front (RF) leg are in swing phase, while Right Rear (RR) leg and Left Front (LF) leg are in stance phase.

Two projection planes are defined:

V-plane is a plane which satisfies two conditions:

(1). Pass through stance feet.

(2). Parallel to the gravity direction.

S-plane is a plane satisfying following conditions:

(1). Perpendicular to V-plane.

(2). Pass through the mid-point of stance feet.

Projected to V-plane and S-plane, a trotting quadruped robot can be simplified as a virtual planar-seven-link mechanism and a SLIP model, shown in Fig. 1.

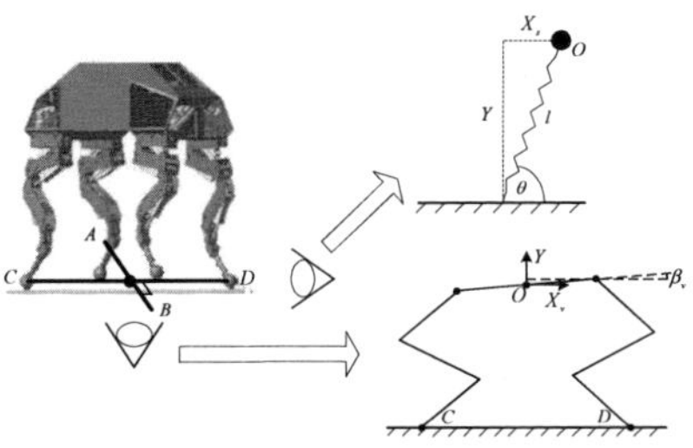

Figure 1. Approximate model of a trotting quadruped robot.

Where O is the centroid of body, X_s is horizontal position of body in S-plane, X_v is horizontal position of body in V-plane, Y is vertical position of body, β_v is the pitch in V-plane, l is the leg length, θ is the leg angle.

The dynamic model and control strategy about planar-seven-link mechanism is presented to make $[X_v \;\; Y \;\; \beta_v]^T$ tracking $[X_{vd} \;\; Y_d \;\; \beta_{vd}]^T$ in literature [12]. In S-plane, dynamic model of SLIP is following:

$$\ddot{l} = -\frac{k}{M}(l - l_{TD}) - g\sin\theta + l\dot{\theta}^2 \tag{1}$$

$$\ddot{\theta} = -2\frac{\dot{l}}{l}\dot{\theta} - \frac{g}{l}\cos\theta \tag{2}$$

In this paper, using sectionalized linearization method, we get approximated analytic solution of SLIP model:

$$l = \tilde{l} + a_l \cos(\omega_l t + b_l) \tag{3}$$

Where $\omega_l = \sqrt{\frac{k}{M} - \dot{\theta}_{TD}^{\;2}}$, $\tilde{l} = \frac{k l_{TD} - Mg\sin\theta_{TD}}{M\omega_l^2}$, M is the mass of the body, k is the spring stiffness of SLIP, TD is touchdown mark, g is gravitation acceleration. a_l and b_l can further be determined by initial conditions: $b_l = a\tan\left(-\frac{\dot{l}_{TD}}{\omega_l\left(l_{TD} - \tilde{l}\right)}\right)$, $a_l = \frac{\dot{l}_{TD}}{-\omega_l \sin b_l}$.

The approximated analytic solution of θ is:

$$\theta = \frac{1}{l}(u_0 + \varepsilon u_1) \tag{4}$$

$$u_0 = c_1 e^{\lambda t} + c_2 e^{-\lambda t} \tag{5}$$

$$u_1 = c_1 n e^{\lambda t}\left(\cos\psi - \frac{2\lambda}{\omega_l}\sin\psi\right) + c_2 n e^{-\lambda t}\left(\cos\psi + \frac{2\lambda}{\omega_l}\sin\psi\right) \tag{6}$$

Where ε, λ, ψ, n are parameters of sectionalized linearization, c_1 and c_2 are determined by θ_{TD} and $\dot{\theta}_{TD}$.

The velocity of body in S-plane is:

$$V_{xs} = \dot{X}_s = \dot{l}\cos\theta - l\sin\theta\dot{\theta} \tag{7}$$

$$V_y = \dot{Y}_s = \dot{l}\sin\theta + l\cos\theta\dot{\theta} \tag{8}$$

2.2. *Control strategy based on velocity resolution*

The velocity of quadruped robot is synthesized by components of V-plane and S-plane, shown in Fig. 2. ϕ is the angle between stance line and horizontal velocity.

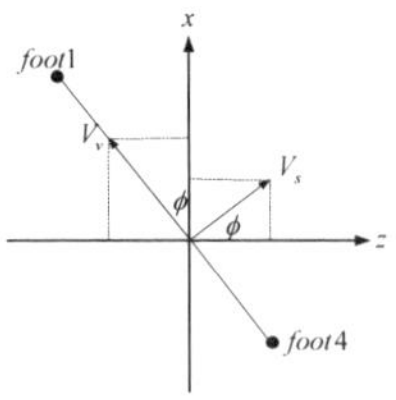

Figure 2. Diagram of velocity resolution.

Based on Fig. 2, the desired velocity and height in S-plane and V-plane is

$$\begin{aligned} V_{sxd} &= V_{xd}\sin\phi + V_{zd}\cos\phi \\ V_{vxd} &= \frac{V_s\cos\phi - V_{zd}}{\sin\phi} \end{aligned} \tag{9}$$

$$\begin{aligned} Y_{sd} &= Y_{vd} = Y_d \\ Y_s &= Y_v = Y \end{aligned} \tag{10}$$

where V_{xd} and V_{zd} are the desired velocity of body along x and z direction. As V_{vx} and Y_v are controlled completely in V-plane, when V_{vxd} and Y_{vd} are calculated to match V_{sxd} and Y_{sd} based on equation (9) and (10), the velocity and height of body are only related to V_{sx} and Y_s. Based on the approximated analytic solution of dynamics model of SLIP, V_{sx} and Y_s are determined by initial conditions. Initial conditions include X_{sTD}, V_{sxTD}, Y_{sTD}, V_{syTD} and $\dot{\theta}_{TD}$, which are determined by the motion of robot in pervious stance cycle.

Define the tracking error as equation (11), LO is liftoff mark.

$$e = \left|\frac{V_{sxdLO} - V_{sxLO}}{V_{sxdLO}}\right| + \left|\frac{Y_{sLO} - Y_{sdLO}}{Y_{sdLO}}\right| \tag{11}$$

Based on equation (9-11), traversal search θ_{TD} value from region $(90^{\circ}, 103^{\circ}]$, making tracking error minimum. In this paper, the step-length is 0.5°.

3. Posture Control Strategy

In trot gait, for a point-foot quadruped robot, the stance region is a line at any time. So it is statically unstable. The robot gets dynamical stability by two pairs of diagonal legs periodically switching between stance phase and swing phase. Anytime in trotting cycle, β_v is controllable in V-plane, which is nonlinear combination of pitch and roll of body. The definition of orientation stable region and sufficient condition of convergence are proposed in this section.

3.1. *Dynamic model of orientation angle*

In a trotting gait cycle, the switching process of different stance phases and the definitions of symbols are shown in Fig. 3.

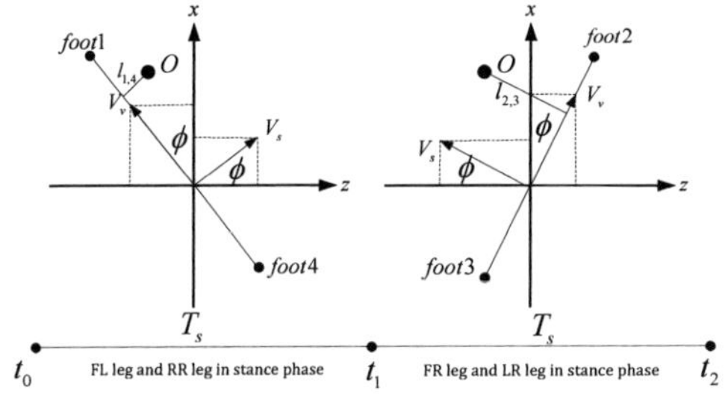

Figure 3. Diagram of switch process.

β_v and β_s are calculated based on geometric constraint of mechanical structure, where β_s is the pitch in S-plane.

When LF leg and RR leg are in stance phase, $\beta_v = \arcsin\dfrac{\sin\beta\cos\gamma + \tan\phi\sin\gamma}{\sqrt{1+\tan^2\phi}}$, $\beta_s = \arcsin\dfrac{\sin\beta\cos\gamma\tan\phi - \sin\gamma}{\sqrt{1+\tan^2\phi}}$.

When RF leg and LR leg are in stance phase, $\beta_v = \arcsin\dfrac{\sin\beta\cos\gamma - \tan\phi\sin\gamma}{\sqrt{1+\tan^2\phi}}$, $\beta_s = \arcsin\dfrac{\sin\beta\cos\gamma\tan\phi + \sin\gamma}{\sqrt{1+\tan^2\phi}}$.

The dynamic model about β_s in S-plane is:

$$J\ddot{\beta}_s = \tau \tag{12}$$

where J is moment of inertia of body, τ is gravity torque.

3.2. *Stability analysis*

The orientation of trotting quadruped robot is stable, if following condition is satisfied:

$$\begin{aligned} |\beta| &< K_\beta \\ |\gamma| &< K_\gamma \end{aligned} \tag{13}$$

Where $K_\beta > 0$ and $K_\gamma > 0$ are the boundaries of the stable region.

Theorem 3.1 The orientation of trotting quadruped robot is stable, if in every gait cycle, following condition is satisfied:

$$\beta_v = \begin{cases} \beta_v(t_0)\sin\left(\pi/2 - \pi(t-t_0)/2T_s\right) & \tau \in [t_0, t_1) \\ \beta_v(t_1)\sin\left(\pi/2 - \pi(t-t_1)/2T_s\right) & \tau \in [t_1, t_2) \end{cases} \tag{14}$$

$$\left(\frac{1+\tan^2\phi}{(1+\tan\phi)^2}\right)\left|\left(\sin\left(-\frac{\tau_{2,3}T_s^2}{2J}\right) + \frac{\tan\phi A_{TS}\sqrt{1-B_{TS}{}^2} + B_{TS}}{(1+\tan\phi)^2}\right)\right| < \left(\sin^2(\beta(t_0)) + \sin^2(\gamma(t_0))\right) \tag{15}$$

Where $\tau_{2,3} = Mgl_{2,3}$, $\tau_{1,4} = Mgl_{1,4}$, $A_{TS} = \tan\phi \dfrac{\sin\left(-\frac{\tau_{1,4}}{2J}T_s^2 + \beta_s(t_0)\right)}{1+\tan\phi}$,

$B_{TS} = \dfrac{\sin\left(-\frac{\tau_{1,4}}{2J}T_s^2 + \beta_s(t_0)\right)}{1+\tan\phi}$, $\beta_s(t_0) = \arcsin\dfrac{\tan\phi\sin\beta(t_0)\cos\gamma(t_0) - \sin\gamma(t_0)}{\sqrt{1+\tan^2\phi}}$.

Proof. Considering the following Lyapunov function candidate:

$$V = \sin^2\beta + \sin^2\gamma \tag{16}$$

Synthesizing equation (12-15), we can get:

$$V(t_2) = \sin^2\beta(t_2) + \sin^2\gamma(t_2) = \left(\frac{1+\tan^2\phi}{(1+\tan\phi)^2}\right)\sin^2\left(-\frac{\tau_{2,3}}{2J}T_s^2 + \beta_s(t_1)\right)$$

$$< \left(\frac{1+\tan^2\phi}{(1+\tan\phi)^2}\right)\left|\sin\left(-\frac{\tau_{2,3}}{2J}T_s^2 + \beta_s(t_1)\right)\right|$$

$$< \left(\frac{1+\tan^2 \phi}{(1+\tan \phi)^2} \right) \left| \sin\left(-\frac{\tau_{2,3}}{2J} T_s^2 \right) + \sin\left(\beta_s\left(t_1\right)\right) \right| \tag{17}$$

If in every gait cycle, equation (15) is satisfied, we can get $V(t_2) < V(t_0)$. So the orientation of trotting quadruped robot is stable. ■

4. Experiment

In order to validate the effectiveness of proposed pose control strategy, we conducted a moving forward experiment with trot gait. The stance cycle is 0.3s. The stable region of orientation is $|\beta| \le 1.5\text{deg}$, $|\gamma| \le 1.5\text{deg}$. The initial state is $\beta = 4.5\text{deg}$, $\gamma = 0\text{deg}$, $V_x = 0m/s$, $Z = 0m/s$, the objective state is $\beta = 0\text{deg}$, $\gamma = 0\text{deg}$, $V_x = 1m/s$, $Z = 0m/s$. Figure 4(a) shows the outlines of the robot in the smooth, and the other results of experiment are shown in Fig. 4(b-f).

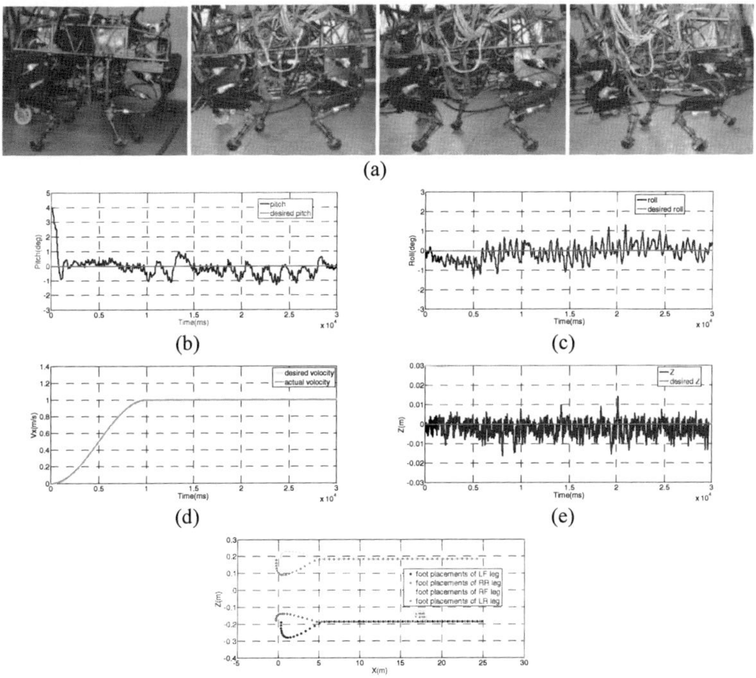

(f)

Figure 4 results of moving forward experiment with trot gait (b) pitch; (c) roll; (d) velocity of body along x axis; (e) position of body along z axis; (f) foot placements.

From Fig. 4, we can see that the pitch and roll converge to the stable region in 1 sec. The chattering in the curves of orientation variable is caused by the switching between stance phase and swing phase. Because at the time of switching, the definitions of S-plane and V-plane change instantaneously, the gravity torques change too, the cycle of chattering is equal to the gait cycle (0.6s). Furthermore, the boundary of stable region is the critical condition of equation (15), if the pitch or roll is not in the stable region, theorem 3.1 is satisfied. The actual position of body along z axis is between -0.015m and 0.015m. From Fig. 4(d), It is shown that the desired velocity is tracked accurately. Figure 4(f) illustrates that the foot placements in the process of accelerating are closer to the stance line than them in the process of uniform speed. So the effectiveness of proposed motion control strategy based on velocity resolution is validated.

5. Conclusions

In this paper, the stance legs are approximated as a virtual planar-seven-link mechanism and a SLIP model. Approximate analytic solution of SLIP model is described using sectionalized linearization method. And a velocity control strategy is proposed to make the desired velocity of body being tracked by adjusting the foot placements of swing legs. Then a sufficient condition of convergence of trotting quadruped robots is presented based on switching control theory.

References

1. Playter R, Buehler M, Raibert M, Bigdog. Unmand Systems Technology VIII, 6230(1/2): 1-6. (2006) .
2. Raibert M, Blankespoor K, Nelson G et al, Bigdog, the rough-terrain quadruped robot, International Federation of Automation Control, Vol.17, no. 1. (2008).
3. http://spectum.ieee.org/automation/robotics/military-robots/ls3-alphadog-robot-begins-outdoors-assessment.
4. http://spectum.ieee.org/automation/robotics/military-robots/boston-dynamics-building-fast-running-robot-cheetah-agile-humanoid.
5. Mrinal Kalakrishnan, Jonas Buchli, Peter Pastor et al, Fast, Robust Quadruped Locomotion over Challenging Terrain. IEEE International Conference on Robotics and Automation. (2010).
6. Barkan Ugurlu, Ioannis Havoutis, Claudio Semini et al, Dynamic Trot-Walking with the Hydraulic Quadruped Robot-HyQ: Analytical Trajectory Generation and Active Compliance Control. IEEE International Conference on Robotics and Automation. (2013).
7. Tan Min, Wang Shuo, Research Progress on Robotics. ACTA AUTOMATICA SINICA. (2013).

8. M. H. Raibert, Legged Robot that Balance. The Robotics Institute Carnegie-Mellon University. (1986).
9. Victor Barasuol, Jonas Buchli, Claudio Semini et al, A Reactive Controller Framework for quadruped Locomotion on Challenging Terrain. IEEE International Conference on Robotics and Automation. (2013).
10. Nicolas Zeitlin, Reinforcement learning methods to enable automatic tuning of legged robots. Electrical Engineering and Computer Sciences University of California at Berkeley Technical Report. (2012).
11. Joao P. Hespanha, Daniel Liberzon, Eduardo D. Sontag, Nonlinear observability and an invariance principle for switched systems. Proceedings of the 41st IEEE Conference on Decision and Control. (2002).
12. Lin Lang, Jian Wang, Force Control Based on Sliding Mode of a Trotting Point-foot Quadruped Robot. International Journal of Robotics and Automation. (2015).

OPTIMAL TRAJECTORY PLANNING WITH MINIMUM-ACCELERATION FOR HUMANOID MANIPULATOR BY USING DIFFERENTIAL EVOLUTION

REN ZIWU, ZHENG SHUAIYIN, WANG ZHENHUA*, SUN RONGCHUAN, YU SHUMEI and LI CHUNGUANG

Robotics and Microsystems Centre, Soochow University, Suzhou, 215021, P. R. China
**E-mail: zwren@suda.edu.cn*

In order to alleviate the counteractive effect on the body of humanoid biped robot while the humanoid manipulator operates under rapid continuously reaction environment (e.g. baseball playing, ping-pong playing, etc.), a trajectory planning method with minimum-acceleration based on differential evolution (DE) is presented in this paper for humanoid manipulator with seven degrees of freedom (DOFs). In this method, quadrinomial polynomial is used to describe the segments of the operation process and return process of the manipulator movement at joint space. Through optimizing the angular acceleration of the trajectory at the target point moment using DE, minimum-acceleration trajectory with continuous profile property is obtained, which can reduce the disturbance against the humanoid biped robot to a certain extent. Simulation experiment results show the effectiveness of this method for the trajectory planning problem studied.

Keywords: Trajectory planning; Minimum-acceleration trajectory; Humanoid manipulator; Differential evolution.

1. Introduction

Trajectory planning problem of the humanoid manipulator is a hot research topic in humanoid biped robot field, whose purpose is to serve the human better.[1] As to the trajectory planning of robot, it can be divided into two methods according to different planning space, one is Cartesian space planning and the other is joint space planning. Cartesian space planning involves large inverse kinematics computation, so its real-time performance is relatively poor.[2]

Due to the rod shape and mechanism design, each joint of the humanoid manipulator has its angle limitation. Also the manipulator's mechatronics system is limited by the size of its actuators, so it cannot produce large velocities and accelerations. Over the last couple of decades many researchers make efforts to address these constraints problems. Chan and Dubey use weighted least-norm solution to avoid joint limits for redundant joint manipulators, which guarantees joint limit avoidance and minimizes unnecessary self-motion comparison with the gradient projection method.[3] Xiang et al. present general weighted least-norm (GWLN) method for the control of redundant manipulators, experiment on the 7-DOF redundant manipulator illustrates that it can both guarantee the obstacle free and not violate the joint limits in contrast to the directional gradient projection method.[4] Garg and Kumar use genetic algorithm (GA) techniques for robot arm to identify the optimal trajectory based on minimum joint torque requirements, in which polynomial of 4th degree is adopted in time for trajectory representation to joint space variables.[5] Wang et al. presented a novel approach based on a type of simplified motion planning (SMP) to generate online trajectory for manipulator systems, which can find minimum-acceleration trajectory planning (MATP) to optimize the arm motion.[6]

The motion trajectory of the humanoid manipulator should be planned within all the kinematic constraints (angle, angular velocity, angular acceleration etc), otherwise the trajectory can not be achieved. In addition, under the rapid continuously reaction environment (e.g. baseball playing, ping-pong playing, etc.), due to the dynamic coupling between the humanoid manipulator and the body of humanoid biped robot (or unfixed base), the humanoid manipulator would produce great counteractive effect on the body, which not only disturbs the balance and stability of the robot body but also influences the trajectory precision of the manipulator end-effector significantly. Hence, how to alleviate the counteractive effect on the body of humanoid biped robot should be considered when planning the trajectory of the humanoid manipulator.

Corresponding to the counteractive effect on the humanoid body, the acceleration continuity and the acceleration value of the manipulator trajectory are essential factors that influences the disturbance to humanoid body's state of equilibrium.[6] Generally, trajectory optimization is achieved by minimizing a suitable performance index within all the kinematic constraints. On account of this point, the acceleration of the trajectory has been considered as performance index for humanoid manipulator operating under rapid continuously reaction environment, and a trajectory planning method based on differential evolution (DE) is presented in this paper. In this method, quadrinomial polynomial is used to describe the segments of the operation process and return process of the manipulator movement at joint space. Through optimizing the angular acceleration of the trajectory at the target point moment using DE, minimum-acceleration trajectory with continuous profile property is obtained, which can reduce the disturbance of the manipulator against the body to a certain extent. Simulation experiment results in 7-DOF humanoid manipulator show this effectiveness of the method for the trajectory planning problem studied.

2. Problem Description

As to the 7-DOF manipulator of the humanoid biped robot, supposing that the initial state of the static point is denoted by $\boldsymbol{\theta}_0 = [\theta_{10}, \theta_{20}, \theta_{30}, \theta_{40}, \theta_{50}, \theta_{60}, \theta_{70}]$, the state of the target point at t_{f1} moment is denoted by $\boldsymbol{\theta}_f = [\theta_{1f}, \theta_{2f}, \theta_{3f}, \theta_{4f}, \theta_{5f}, \theta_{6f}, \theta_{7f}]$, and the joint velocity of the target point at t_{f1} moment is denoted by $\dot{\boldsymbol{\theta}}_f = \left[\dot{\theta}_{1f}, \dot{\theta}_{2f}, \dot{\theta}_{3f}, \dot{\theta}_{4f}, \dot{\theta}_{5f}, \dot{\theta}_{6f}, \dot{\theta}_{7f}\right]$, then the trajectory planning problem of the humanoid manipulator for high-speed objective can be transformed into a trajectory with continuous acceleration properties optimum problem from all candidate trajectories of the system, which could be expressed by the following mathematical description:

$$\boldsymbol{\theta}^*(t) = \arg\min\max\left\{\left|\ddot{\boldsymbol{\theta}}_i(t)\right|, t \in [0, t_{f1} + t_{f2}], i = 1, 2, \cdots, 7\right\} \tag{1}$$

subject to

$$\begin{array}{ll} \boldsymbol{\theta}^*(0) = \boldsymbol{\theta}_0 & \boldsymbol{\theta}^*(t_{f1}) = \boldsymbol{\theta}_f \\ \dot{\boldsymbol{\theta}}^*(0) = \mathbf{0} & \dot{\boldsymbol{\theta}}^*(t_{f1}) = \dot{\boldsymbol{\theta}}_f \end{array} \quad \text{(Operation process)} \tag{2}$$

$$\begin{array}{ll} \boldsymbol{\theta}^*(t_{f1}) = \boldsymbol{\theta}_f & \boldsymbol{\theta}^*(t_{f1} + t_{f2}) = \boldsymbol{\theta}_0 \\ \dot{\boldsymbol{\theta}}^*(t_{f1}) = \dot{\boldsymbol{\theta}}_f & \dot{\boldsymbol{\theta}}^*(t_{f1} + t_{f2}) = \mathbf{0} \end{array} \quad \text{(Return process)} \tag{3}$$

$$\lim_{t \to t_0} \ddot{\theta}_i(t) = \ddot{\theta}_i(t_0) \quad \forall t_0 \in [0, t_{f1} + t_{f2}], \quad i = 1, 2, \cdots, 7 \tag{4}$$

where $\boldsymbol{\theta}^*(t)$ is the selected optimum trajectory, t_{f1} is execution time of the operation process, t_{f2} is execution time of the return process, formula (2) and (3) respectively correspond to the boundary conditions of operation process and return process, formula (4) indicates the trajectory demand of continuous acceleration properties. In addition, the selected optimum trajectory should meet the following kinematic constraints conditions of the manipulator:

$$\begin{array}{ll} \theta_i^{\min} \le \theta_i^*(t) \le \theta_i^{\max} & \forall t \in [0, t_{f1} + t_{f2}] \\ \dot{\theta}_i^{\min} \le \dot{\theta}_i^*(t) \le \dot{\theta}_i^{\max} & \forall t \in [0, t_{f1} + t_{f2}] \\ \ddot{\theta}_i^{\min} \le \ddot{\theta}_i^*(t) \le \ddot{\theta}_i^{\max} & \forall t \in [0, t_{f1} + t_{f2}] \end{array} \tag{5}$$

where $i = 1, 2, ..., 7$, $\theta_i^{\min}$ and $\theta_i^{\max}$ are respectively the lower bound and the upper bound of the joint i angle, $\dot{\theta}_i^{\min}$ and $\dot{\theta}_i^{\max}$ are respectively the lower bound and the upper bound of the joint i angular velocity, $\ddot{\theta}_i^{\min}$ and $\ddot{\theta}_i^{\max}$ are respectively the lower bound and the upper bound of the joint i angular acceleration.

3. Motion Planning Strategy

As shown in figure 1, when the 7-DOF humanoid manipulator operates ping-pong playing, the whole motion trajectory is constituted by 2 segments which represents operation process and return process respectively. Because each segment is point-to-point trajectory planning problem, it can be achieved at joint space. According to the boundary conditions of each process and the trajectory demand of continuous acceleration profile, since cubic polynomial interpolating the trajectory would produce the discontinuous points of acceleration profile at t_{f1} moment,[7] a quadrinomial polynomial can be used to interpolate the joint trajectory of the operation process and return process.

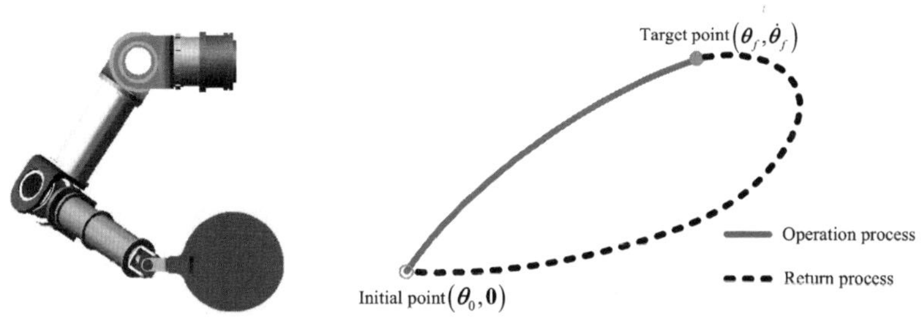

Fig. 1. Operation process and return process of the trajectory for ping-pong playing

At stage of operation process, the manipulator moves from the initial static point to the target point at t_{f1} moment. Hence, the boundary conditions of joint angle meet $\boldsymbol{\theta}(0) = \boldsymbol{\theta}_0$, $\boldsymbol{\theta}(t_{f1}) = \boldsymbol{\theta}_f$, and the boundary conditions of joint angular velocity meet $\dot{\boldsymbol{\theta}}(0) = \mathbf{0}$, $\dot{\boldsymbol{\theta}}(t_{f1}) = \dot{\boldsymbol{\theta}}_f$. Supposing that the joint angular acceleration at t_{f1} moment is $\ddot{\boldsymbol{\theta}}(t_{f1}) = \ddot{\boldsymbol{\theta}}_f$, then the operation process can be described by a quadrinomial polynomial as:

$$\begin{aligned} &\boldsymbol{\theta}_i(t) = a_{i0} + a_{i1}t + a_{i2}t^2 + a_{i3}t^3 + a_{i4}t^4 \\ &t \in [0, t_{f1}] \quad i = 1, 2, \cdots, 7 \end{aligned} \tag{6}$$

where $a_{i0}, a_{i1}, \cdots, a_{i4}$ are constants, and the boundary constraint conditions of operation process are given as:

$$\begin{aligned}
&\theta_i(0) = a_{i0}\\
&\theta_i(t_{f1}) = a_{i0} + a_{i1}t_{f1} + a_{i2}t_{f1}^2 + a_{i3}t_{f1}^3 + a_{i4}t_{f1}^4\\
&\dot{\theta}_i(0) = a_{i1}\\
&\dot{\theta}_i(t_{f1}) = a_{i1} + 2a_{i2}t_{f1} + 3a_{i3}t_{f1}^2 + 4a_{i4}t_{f1}^3\\
&\ddot{\theta}_i(t_{f1}) = 2a_{i2} + 6a_{i3}t_{f1} + 12a_{i4}t_{f1}^2
\end{aligned} \tag{7}$$

Hence, the five unknowns can be solved as

$$\begin{aligned}
a_{i0} &= \theta_{i0} \qquad a_{i1} = 0\\
a_{i2} &= \tfrac{6}{t_{f1}^2}\left(\theta_{if} - \theta_{i0}\right) - \tfrac{3}{t_{f1}}\dot{\theta}_{if} + \tfrac{\ddot{\theta}_{if}}{2}\\
a_{i3} &= -\tfrac{8}{t_{f1}^3}\left(\theta_{if} - \theta_{i0}\right) + \tfrac{5}{t_{f1}^2}\dot{\theta}_{if} - \tfrac{\ddot{\theta}_{if}}{t_{f1}}\\
a_{i4} &= \tfrac{3}{t_{f1}^4}\left(\theta_{if} - \theta_{i0}\right) - \tfrac{2}{t_{f1}^3}\dot{\theta}_{if} + \tfrac{\ddot{\theta}_{if}}{2t_{f1}^2}
\end{aligned} \tag{8}$$

At stage of return process, the manipulator would return to the initial point at the $t_{f1} + t_{f2}$ moment, and the angular velocity of each joint returns to zero. Therefore, the joint angle boundary conditions of this stage satisfy $\boldsymbol{\theta}(t_{f1}) = \boldsymbol{\theta}_f$, $\boldsymbol{\theta}(t_{f1}+t_{f2}) = \boldsymbol{\theta}_0$, the boundary conditions of the joint angular velocity satisfy $\dot{\boldsymbol{\theta}}(t_{f1}) = \dot{\boldsymbol{\theta}}_f$, $\dot{\boldsymbol{\theta}}(t_{f1} + t_{f2}) = 0$. Assuming the joint angular acceleration at t_{f1} moment is $\ddot{\boldsymbol{\theta}}(t_{f1}) = \ddot{\boldsymbol{\theta}}_f$, then the quadrinomial polynomial is adopted to describe the return process as:[8]

$$\begin{aligned}
&\boldsymbol{\theta}_i(t) = b_{i0} + b_{i1}\left(t - t_{f1}\right) + b_{i2}(t - t_{f1})^2 + b_{i3}(t - t_{f1})^3 + b_{i4}(t - t_{f1})^4\\
&t \in \left[t_{f1}, t_{f1} + t_{f2}\right], \quad i = 1, 2, \cdots, 7
\end{aligned} \tag{9}$$

where $b_{i0}, b_{i1}, \cdots, b_{i4}$ are constants, and the boundary constraint conditions of return process are given as:

$$\begin{aligned}
&\theta_i(t_{f1}) = b_{i0}\\
&\theta_i(t_{f1} + t_{f2}) = b_{i0} + b_{i1}t_{f2} + b_{i2}t_{f2}^2 + b_{i3}t_{f2}^3 + b_{i4}t_{f2}^4\\
&\dot{\theta}_i(t_{f1}) = b_{i1}\\
&\dot{\theta}_i(t_{f1} + t_{f2}) = b_{i1} + 2b_{i2}t_{f2} + 3b_{i3}t_{f2}^2 + 4b_{i4}t_{f2}^3\\
&\ddot{\theta}_i(t_{f1}) = 2b_{i2}
\end{aligned} \tag{10}$$

From these, we can obtain that the five unknown constants are:

$$\begin{aligned}
b_{i0} &= \theta_{if} \quad b_{i1} = \dot{\theta}_{if} \quad b_{i2} = \tfrac{\ddot{\theta}_{if}}{2}\\
b_{i3} &= \left(4\theta_{i0} - 4\theta_{if} - 3\dot{\theta}_{if}t_{f2} - \ddot{\theta}_{if}t_{f2}^2\right)\Big/t_{f2}^3\\
b_{i4} &= \left(-3\theta_{i0} + 3\theta_{if} + 2\dot{\theta}_{if}t_{f2} + \ddot{\theta}_{if}\tfrac{t_{f2}^2}{2}\right)\Big/t_{f2}^4
\end{aligned} \tag{11}$$

Based on the above, the parameters to be determined for planning the trajectory are the joint angular acceleration $\ddot{\boldsymbol{\theta}}_f$ at t_{f1} moment. For the

7-DOF manipulator case, there are seven parameters to be determined through optimization.

4. Optimizing the Trajectory Using DE

The rapid movement of the humanoid manipulator disturbs the balance and stability of the humanoid biped robot, which influences conversely the trajectory precision of the end-effector. Hence, a optimum trajectory with minimum-acceleration and continuous profile property should be selected to alleviate the counteractive effect on the robot.

As mentioned above, for the 7-DOF humanoid manipulator there are 7 parameters should be optimized, i.e. $\left[\ddot{\theta}_{1f}, \ddot{\theta}_{2f}, \ddot{\theta}_{3f}, \ddot{\theta}_{4f}, \ddot{\theta}_{5f}, \ddot{\theta}_{6f}, \ddot{\theta}_{7f}\right]$, where $\ddot{\theta}_{if}$ $(i = 1, 2, \cdots, 7)$ is the angular acceleration of the i joint at t_{f1} moment. Differential evolution (DE)[9] is a population-based intelligent search approach which solves optimization problem through individuals cooperation and competition. The flowchart of DE for selecting motion trajectory of 7-DOF humanoid manipulator is shown as Figure 2.

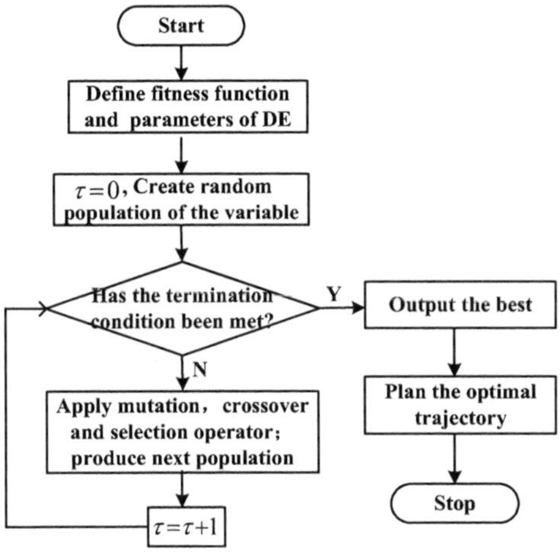

Fig. 2. Flowchart of DE for optimizing motion trajectory

In order to plan a optimum trajectory of the humanoid manipulator using the DE, according to the formula (1), the fitness function of DE algorithm is defined as the following:

$$\min f(\boldsymbol{K}) = \max\left\{\left|\ddot{\boldsymbol{\theta}}_i(t)\right|, t \in [0, t_{f1} + t_{f2}], i = 1, 2, \cdots, 7\right\} \tag{12}$$

where K represents individual of the DE. If the trajectory violates the kinematic constraints of the humanoid manipulator, i.e.

$$\begin{cases} \theta_i(t) \notin \left[\theta_i^{\min}, \theta_i^{\max}\right] \\ \dot{\theta}_i(t) \notin \left[\dot{\theta}_i^{\min}, \dot{\theta}_i^{\max}\right] \\ \ddot{\theta}_i(t) \notin \left[\ddot{\theta}_i^{\min}, \ddot{\theta}_i^{\max}\right] \end{cases} \quad (13)$$
$$\forall t \in [0, t_{f1} + t_{f2}], \quad i = 1, 2, \cdots, 7$$

penalties measure $f(\boldsymbol{K}) = C$ is employed to punish the corresponding fitness value, where $C > 0$ is a big positive constant.

5. Simulation

In order to validate the performance of the proposed method, a humanoid biped robot with a 7-DOF manipulator for Ping-Pong playing is employed as example. The physical model and joint model of the manipulator are respectively shown in Figure 3. Detailed introduction for this task can be referred to the reference 7. Also, this proposed planning method will be compared with the global minimum-jerk trajectory planning (MJTP).

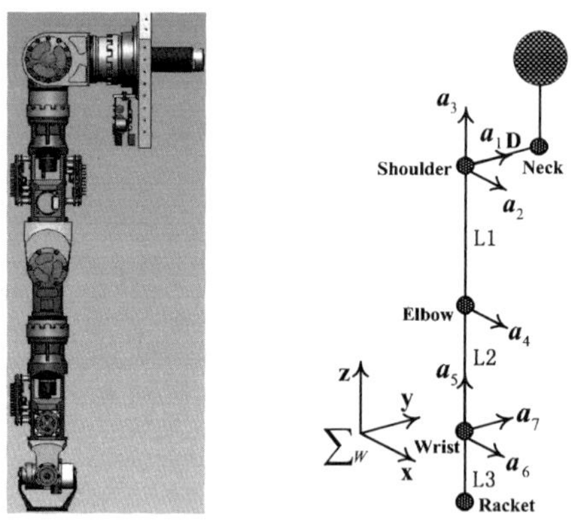

Fig. 3. Physical model (left) and joint model (right) of the 7-DOF manipulator

According to the humanoid manipulator shown in Figure 3, the angle range, maximum angular velocity and maximum angular acceleration of each joint are listed in the following Table 1-2:

The movement of the humanoid manipulator for Ping-Pong playing includes operation process and return process. Supposing the boundary con-

Table 1. Joint angle range of 7-DOF humanoid manipulator (°).

Range	Joint angle						
	θ_1	θ_2	θ_3	θ_4	θ_5	θ_6	θ_7
Upper	-126	-133	-180	-60	-180	-80	-42
Lower	90	15	90	120	180	80	85

Table 2. Maximum angular velocity/acceleration of 7-DOF humanoid manipulator.

Maximum	$\dot{\theta}_1$	$\dot{\theta}_2$	$\dot{\theta}_3$	$\dot{\theta}_4$	$\dot{\theta}_5$	$\dot{\theta}_6$	$\dot{\theta}_7$
velocity (rad/s)	13	18	12	20	11	10	4.5
Maximum	$\ddot{\theta}_1$	$\ddot{\theta}_2$	$\ddot{\theta}_3$	$\ddot{\theta}_4$	$\ddot{\theta}_5$	$\ddot{\theta}_6$	$\ddot{\theta}_7$
velocity (rad/s^2)	200	150	300	200	120	200	100

ditions and execution times of these two process are as following:

$$\begin{aligned}
&\boldsymbol{\theta}_0 = [0.0200, -0.5909, -0.8551, 1.5930, 1.5708, 0.8441, -0.5763] \\
&\boldsymbol{\theta}_f = [-0.3438, -0.3220, -0.8813, 1.4156, 1.2132, 0.9866, -0.1244] \\
&\dot{\boldsymbol{\theta}}_0 = [0, 0, 0, 0, 0, 0, 0] \\
&\dot{\boldsymbol{\theta}}_f = [-4.7659, 2.1737, -0.1073, -6.7378, 0.7667, 2.3225, -2.0398] \\
&t_{f1} = 0.3 \quad t_{f2} = 0.5
\end{aligned} \tag{14}$$

In order to obtain a optimum trajectory with minimum-acceleration and continuous profile property, DE is used to determine the angular acceleration of each joint at t_{f1} moment. In DE, the search range of each variable is determined according to the Table 2, scaling factor in mutation operator $F = 0.75$, crossover probability $CR = 0.90$, and the maximum evolution iteration T=60. The fitness function of DE is defined as formula (12), thereinto, penalties coefficient C is 10^4 when the trajectory is out of the kinematic constraints. Through trajectory optimization, optimal angular accelerations of each joint at t_{f1} moment are gotten, which is described as:

$$\begin{aligned}
\ddot{\boldsymbol{\theta}}_f &= \left[\ddot{\theta}_{1f}, \ddot{\theta}_{2f}, \ddot{\theta}_{3f}, \ddot{\theta}_{4f}, \ddot{\theta}_{5f}, \ddot{\theta}_{6f}, \ddot{\theta}_{7f}\right] \\
&= [-6.1861, -43.1335, 59.8396, -10.8677, 36.0457, -40.0019, -24.4951]
\end{aligned} \tag{15}$$

the corresponding fitness value is $f(\boldsymbol{K})$=100.2365. From this value, we can conclude that the selected trajectory meets all kinematic constraints of the humanoid manipulator. Figure 4 shows the angle, angular velocity and angular acceleration versus time curve of each joint for the selected optimum trajectory. From this figure, it can be seen that the angular acceleration curve of the joint does not exist breaking point, which can alleviate the counteractive effect on the humanoid body to a certain extent.

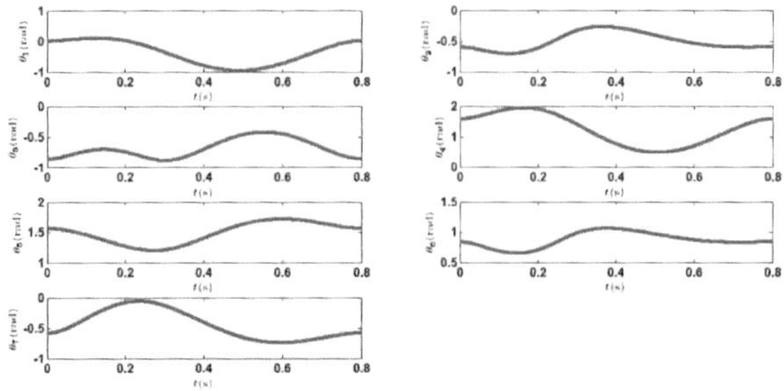

(a) Angle versus time for each joint

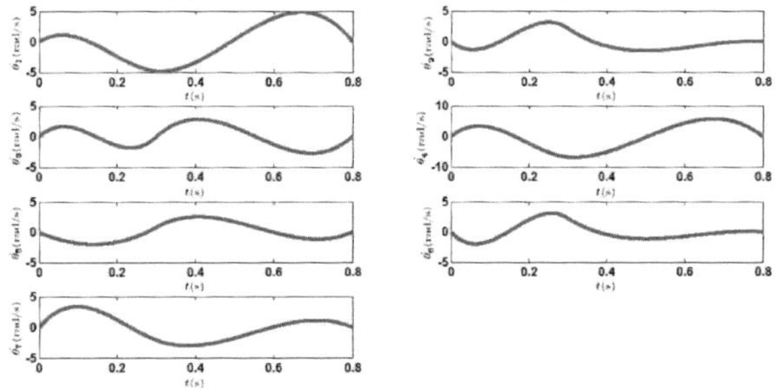

(b) Velocity versus time for each joint

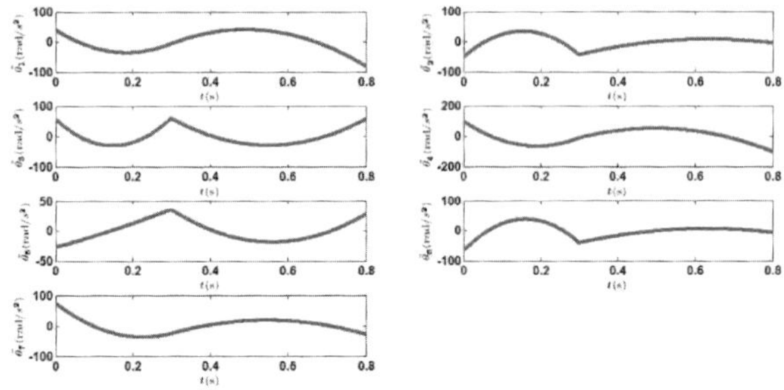

(c) Acceleration versus time for each joint

Fig. 4. Angle, velocity and acceleration curve of each joint for the 7-DOF manipulator

For this same task, the global minimum-jerk trajectory planning (MJTP)[10] with quadrinomial polynomial is also used to achieve this motion, and compare with the method proposed in this paper. The Table 3 respectively lists the maximum acceleration value of the trajectory optimized by these two methods. From this Table we can see that the maximum acceleration value of the trajectory optimized by the proposed method in this paper is smaller than that of MJTP method, which indicates that the method in this paper is favorable for alleviating disturbance of the humanoid manipulator against the humanoid biped robot. The trajectory planning method in this paper is more feasible and effective.

Table 3. Maximum acceleration of trajectory using different methods.

	This paper's Method	MJTP
Maximum acceleration	100.2365 (rad/s^2)	122.3047 (rad/s^2)

6. Conclusion

Trajectory planning for the 7-DOF humanoid manipulator under rapid continuously reaction environment is presented. In this method, quadrinomial polynomial is used to describe the whole process, and minimum-acceleration trajectory with continuous profile property is obtained through DE. Simulation results show the effectiveness of the method. In future, this method can be extended by integrating example joint force, torques etc.parameters to reduce the disturbance against the humanoid biped robot.

Acknowledgment

The authors acknowledge the support of the National Natural Science Foundation of China (Grant No. 61273340, 61203367).

References

1. Y. Tian, X. P. Chen, D. Y. Jia, F. Meng and Q. Wang, *Robot* **33**, 332 (2011).
2. J. J. Craig, *Introduction to Robotics: Mechanics and Control, Third Edition* (China Machine Press, Beijing, 2009).
3. T. F. Chan and R. V. Y. Dubey, *IEEE Transations on Robotics and Automation* **11**, 286 (1995).
4. J. Xiang, C. W. Zhong and W. Wei, *IEEE Transactions on Robotics* **26**, 660 (2010).
5. D. P. Grag and M. Kumar, *Engineering Applications of Artificial Intelligence* **15**, 241 (2002).
6. Y. S. Wang, L. Sun, L. Zhou and J. T. Liu, *Acta Automatica Sinica* **40**, 1328 (2014).
7. Z. W. Ren, Q. G. Zhu and R. Xiong, *Acta Automatica Sinica* (in press).
8. S. G. Yue, D. Henrich, W. L. Xu and S. K. Tso, *Robotica* **20**, 269 (2002).
9. R. Storn and K. Price, *Journal of Global Optimization* **11**, 341 (1997).
10. P. F. Huang, K. Chen, J. P. Yuan and Y. S. Xu, Motion trajectory planning of space manipulator for joint jerk minimization, in *Proceedings of the IEEE International Conference on Mechatronics and Automation*, (Harbin, China, 2007).

A SIMPLE MODELING METHOD AND TRAJECTORY PLANNING FOR A CAR-LIKE CLIMBING ROBOT USED TO STRIP COATING FROM THE OUTER SURFACE OF PIPES UNDERWATER

HANSONG WANG, CANJUN YANG, XUELEI DENG AND JINCHANG FAN

The State Key Lab. of Fluid Power Transmission and Control, Zhejiang University, Hangzhou, China

Abstract—The paper presents an analysis modeling and trajectory planning for a car-like climbing robot used for stripping coatings from the outer surface of underwater pipes used in oil platform offshore. The simple modeling method we proposed is based on the mapping relationship between modeling in plane and in cylinder, and using the method we designed trajectories for the robot to travel all over the outer surface of the pipes completely and efficiently. Simulation and experimental tests were carried out in order to verify the satisfaction of the analysis modeling and to validate performance of the control system. Although the method worked well in most cases, more factors and conditions need to be taken into consideration in the control in the future works.

Keywords: Car-like Climbing Robot, Kinematic Model, Mapping, Trajectory

1. Introduction

Recently, climbing robotic technology has been applied rapidly in several industries such as stripping paint from vessels, inspections of petrochemical, nuclear and other power plants and so on, which usually are hazardous tasks. With the rapid development of offshore oil exploration technology, regular coating cleaning and rust removal from pipes on the platform for oil exploration has become an urgent need to meet, for coating and rust on the pipes will greatly reduce the service life of platform and traditional artificial cleaning method is not only dangerous but also inefficient with high cost. Fig. 1 shows the pipes with coating and rust, Fig. 2 shows the traditional artificial cleaning by divers.

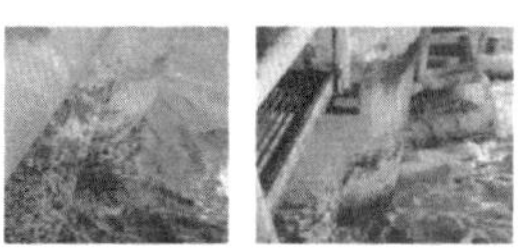

Fig. 1 The pipes

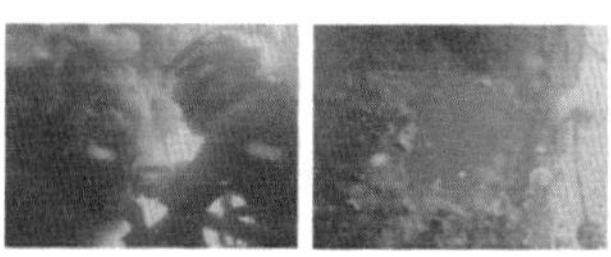

Fig. 2 The traditional artificial cleaning

In the paper, we present a car-like climbing robot to solve this problem. And the article aims to find a useful modeling method for the robot moving on cylinder and plan advisable trajectories for safe and efficient navigation.

2. Robot Design and The Modeling Method

2.1. *Overview of the climbing robot*

As show in Fig. 3, the car-like climbing robot (called CKCR for short) has four magnetic wheels which make it adhesive to the wall of pipe, the two rear wheels and two front wheels are respectively driven synchronously, and the orientation angle is determined by the front wheels. The cleaning equipment using cavitation technology is in front of the robot. Energy, signal and high-pressure water needed for cleaning are provided by umbilical cable. And there are two passive DOFs for adaptability.

2.2. *D-H coordinate system of CKCR*

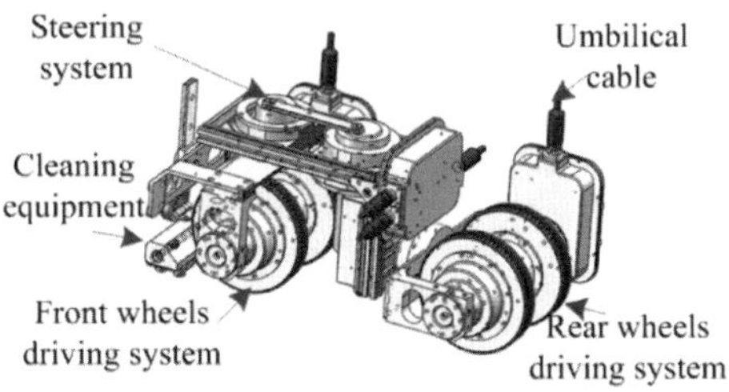

Fig. 3 Overview of CKCR

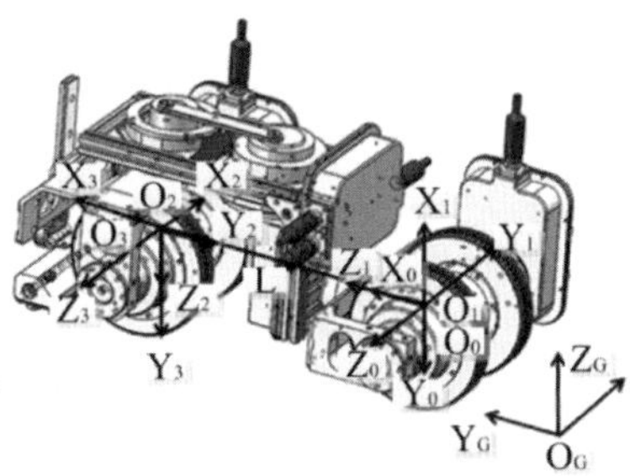

Fig. 4 D-H coordinate system of CKCR

We chose the center of the rear wheels (O_o) as datum of CKCR, and build the D-H coordinate system of CKCR as shown in Fig. 4. And the robot's kinematics analysis is made in the geodetic reference coordinate system (XG, YG, ZG). The related parameters of D-H coordinate system can be found in Table. 1, in which θ_b is angle between axis X_0 and axis X_G,which is shown in Fig. 5, θ p, θ t, θ o are the pitch angle, twist angle, orientation angle of corresponding DOFs of CKCR respectively, and the definition of other parameters can be understood in the Eq. (1).

Table.1 parameters of D-H coordinate system

Joints	x_i	y_i	z_i	α_i	β_i	θ_i	Range of θ_i
0	x_b	y_b	z_b	−90°	0°	θ_b	(0°, 360°)
1	0	0	0	−90°	0°	θ_p	(0°, 360°)
2	0	0	L	0°	−90°	θ_t	(-45°, 45°)
3	0	0	0	90°	0°	θ_o	(-30°, 30°)

We can get transformation matrixes of adjacent coordinate system

$$ {}^{G}_{0}T = Trans(z, z_b)Trans(x, x_b)Trans(y, y_b)Rot(z, \theta_b)Rot(y, \beta_b) \tag{1} $$

Similarly, we can get ${}^{0}_{1}T$, ${}^{1}_{2}T$, ${}^{2}_{3}T$. And ${}^{G}_{3}T = {}^{G}_{0}T\,{}^{0}_{1}T\,{}^{1}_{2}T\,{}^{2}_{3}T$.

2.3. *Kinematic model in plane*

The two passive DOFs won't work in plane, and θ_p=90°, θ_t=0°. Here I make an assumption: any continuous derivable curve of the CKCR's trajectory can be seen as being built up by arcs with different curvatures. So it is foundational to study on the circular trajectory planning. Taking its structure into consideration, we simplified the CR as a two-wheel car. As shown in Fig. 6. When S_1 (the trajectory of the rear wheels' center O_0) is round, S_2 (the trajectory of the front wheels' center O_3) is also round, and their centers of circle are the same. According to the geometric relationship showed in Fig. 6, we can get that:

$$ \theta_o = \arctan(L / r_0) \tag{2} $$

$$ r_3 = \sqrt{r_0^2 + L^2} \tag{3} $$

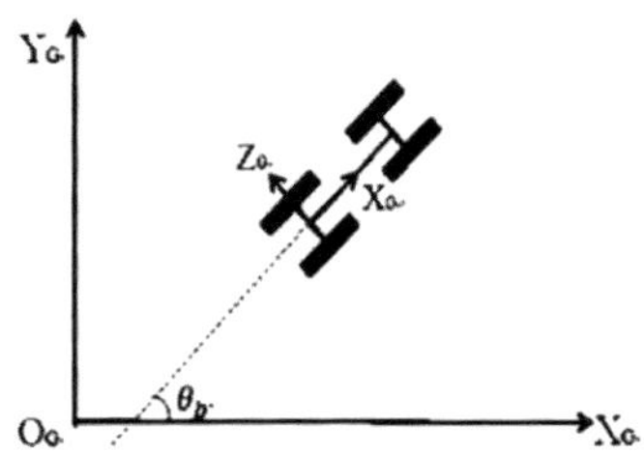

Fig. 5 The definition of θ_b

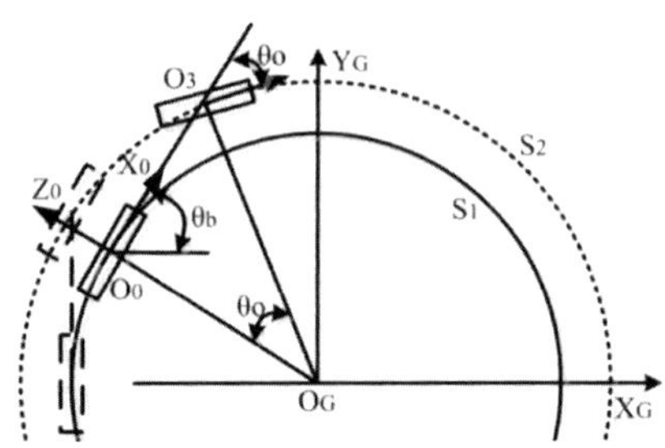

Fig. 6 Analysis of circular trajectory of O_0

We can conclude that when r_0 (the radius of the circular trajectory of the center of rear wheels) is identified, we can get θ_o (the orientation angle) and r_3 (the radius of the circular trajectory of the center of front wheels). The rotation speeds of the front and rear wheels (ω_f, ω_b respectively) are proportional to the radius of the trajectory: $\omega_f / \omega_b = r_0 / r_3$.

According to the theory, we can get the change of θ_o for any continuous derivable curve trajectory. In the following, we took the curve $y = -0.0015x^2$ as an example. The corresponding parameters are shown in Fig. 7. And the trajectory of O_0, O_3 are shown in Fig. 8.

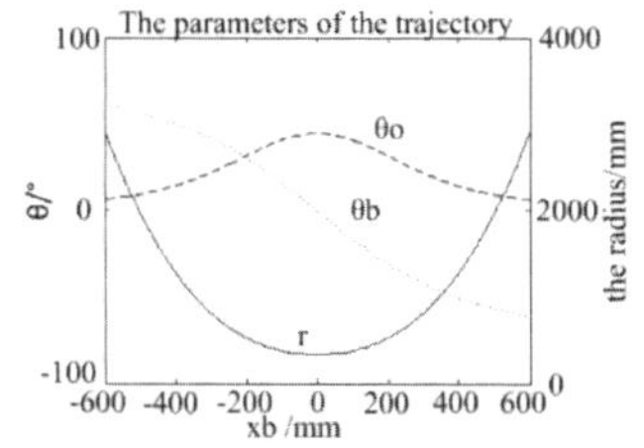

Fig. 7 The corresponding parameters of the curve

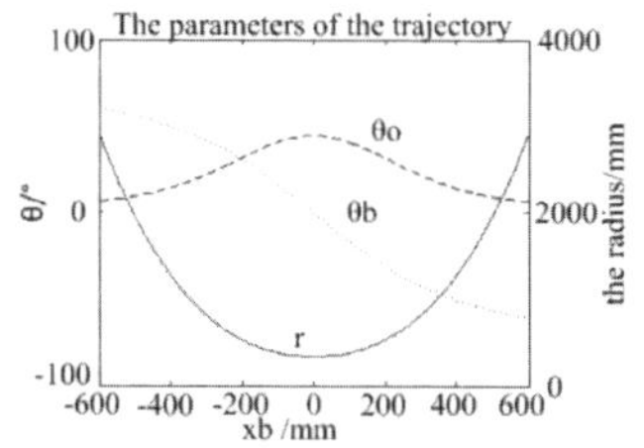

Fig. 8 The trajectory of O_0, O_3

2.4. *The mapping relationship between modeling in plane and cylinder*

On the cylinder with a small diameter, CKCR's two passive DOFs work, and the contact points between wheels and surface change as CKCR's position and posture change, which make the situation complex. Fortunately, cylinder can be expanded to a plane, which makes it possible to build up the relationship between modeling in cylinder and in plane. There are two important factors need to be considered in the mapping relationship. One is the change of the equivalent distance between O_0, O_3, since L is the related chord in cylinder, but it will be the related arc after cylinder being expanded. Another is the change of the contact points between wheels and surface, which usually stay the same in plane, the difference will affect the speeds of the front and rear wheels.

2.4.1. *The mapping relationship of position*

The change of the equivalent distance between O_0 and O_3 is shown in Fig. 9. The cylinder is not the outer surface of pipes, but a supposed surface where O_0, O_3 are, and it's radius (R) is the sum of radius of pipe (R_c) and wheel (r), R=Rc+r. Before expansion, $|O_0O_3|$ the equivalent distance between O_0 and O_3 is L. After expansion, the equivalent distance between O_0 and O_3 is $|O_0O_3'|$, and $|O_0O_3'| = \widehat{O_0O_3} = L'$, here $\widehat{O_0O_3}$ is arc of O_0O_3. Point C is projection of point O_3 in the plane which is parallel to XZ plane and contains point O_0. To simplify the modeling, we set point B on axis Z. And θ_b turns into $\theta_b{}'$. It is obvious that the coordinate of point O_0 is (0, 0, R), set β as the central angle of arc $\widehat{O_0C}$, then we get: $\beta = L'cos\theta_b'/R$. The coordinate of point C is $(R\sin\beta, 0, R\cos\beta)$, and the coordinate of point O_3 is $(R\sin\beta, L'\sin\theta_b', R\cos\beta)$. According to $|O_0O_3| = L$, we can get:

$$2R^2\cos\frac{L'\cos\theta_b'}{R} - L'^2\sin^2\theta_b' = 2R^2 - L^2 \tag{4}$$

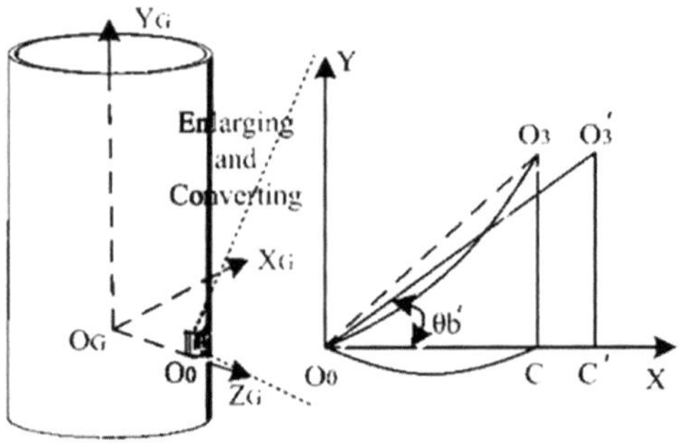

Fig. 9 The mapping relationship of position

And $\theta_o = \arctan(L' / r_0)$, $r_3 = \sqrt{r_0^2 + L'^2}$. Using the method, when we know the trajectory in plane, we can get related control parameters of CKCR in cylinder. We chose the same curve mentioned in section 3.2 to analyze. The trajectories of O_0, O_3 are shown in Fig. 10, and related parameters are shown in Fig. 11.

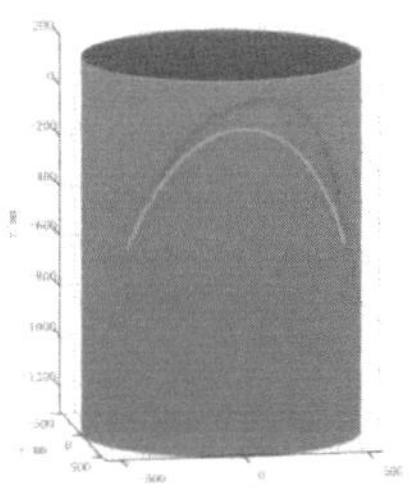

Fig. 10 The trajectories of O_0 O_3

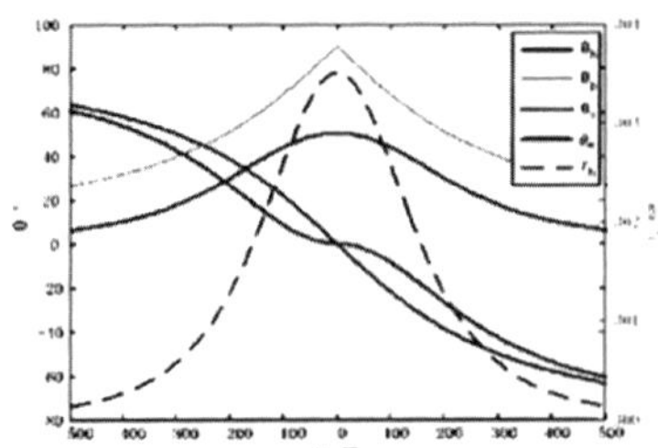

Fig.11 Related parameters

2.4.2. *The mapping relationship of speed*

We simplified the CR as a two-wheel bike, and we can found that when it moved on the cylinder the vectors $\overrightarrow{O_0C_0}$ $\overrightarrow{O_3C_3}$ are towards Y_G axis (axis of pipe), where points C_0 C_3 are the contact points of front and rear wheels respectively. We calculated s (the ratio of speed in cylinder and speed in plane) based on the geometric phenomenon. We can get that

$$s = \frac{v_c}{v_p} = \frac{\widehat{C_0C_3}}{\widehat{O_0O_3}} \tag{5}$$

Here v_c is the speed in cylinder, Vp is the speed in plane. When CKCR moves along the axis of pipe, X_0 is parallel to Y_G, and s=1. When CKCR moves along circumferential direction of pipe, X_0 is vertical to Y_G, and s=Rc/R, ranging from 1.08 to 1.245 according to the actual size. When diameter of pipe is 1676mm, s=1.089, and the speed of CKCR is low, usually 1~2mm/s for full cleaning, so the difference of speed on the surface and plane can be ignored.

3. Trajectory Planning

In this section, we presented three common trajectories for the robot moving on the surface of pipes and simply compared their possibility, safety and efficiency. A reasonable trajectory must meet some conditions, for example robot must move on the surface cleaned before ensuring enough adherence and cable used for the transmission of energy and signal can't circle the pipes during motion. The three trajectories are exhibited in the flowing.

3.1. *Trajectory 1*

Generally speaking, the trajectory as shown in Fig. 12 is the first choice for a climbing robot without umbilical cable to travel all over the outer surface of pipes. The trajectory is a helix around the pipe, and it can be seen as a straight line in plane with the mapping method as shown in Fig. 12 .In order to make a complete cleaning, the trajectory is of point O_0(the center of rear wheels) should meet the constraints: $H \leq W$, more specifically, $\theta \leq \arctan\{W/(2\pi R)\}$. W is CKCR's cleaning width, R is the radius of pipe, H and θ are shown in Fig. 12.

It's a simple and efficient trajectory. However, umbilical cable will circle the pipes, which make it can't be used in the situation only when robot is without umbilical cable.

3.2. *Trajectory 2*

The second trajectory is common as exhibited in Fig. 13. Robot moves along the direction of axis of pipes up and down regularly. There are several constraints should be considered as presented in the following.

- $r \geq r_{min}$. Here r is CKCR's turning radius, r_{min} is the minimum turning radius.$r_{min} = L/\tan\theta_{max}$. Here $r_{min} = 2.85W$ (L=330mm,W=200mm).
- $2\alpha k = n + 1$. Here $n = \text{ceil}(2\pi R/W)$, to simplify the situation, set $r = \alpha W$, α=3, 3.5, 4, 4.5……and k=1,2,3,4……

3.3. *Trajectory 3*

The third trajectory is shown in Fig. 14. Robot moves along the circumferential direction of pipes forward and back only one circle regularly. To improve cleaning efficiency, CKCR can move back faster without cleaning. To make the control of turning route simple, we choose a circular trajectory in which the steering angle stay the same as shown in Fig. 15. We can get that: $H \leq W$ and $r_f - r_b = (r_f - r_b)cos\theta + H$.

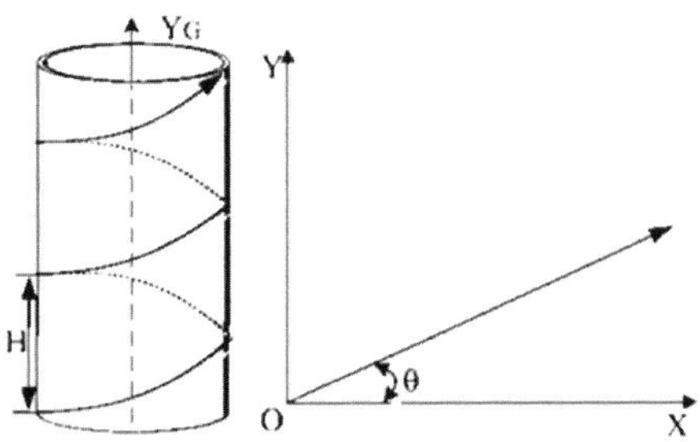

Fig. 12 Trajectory 1

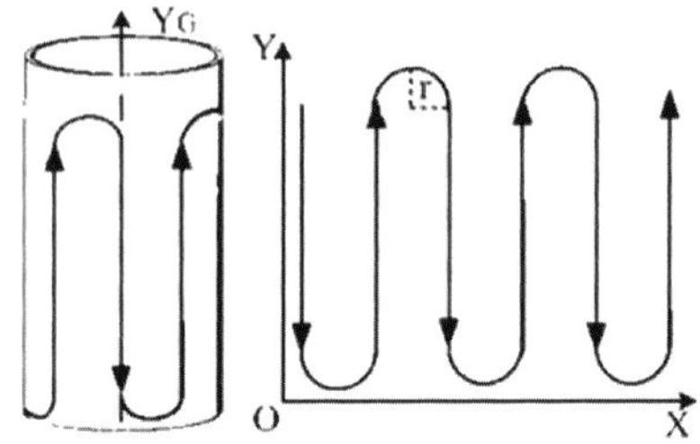

Fig. 13 Trajectory 2

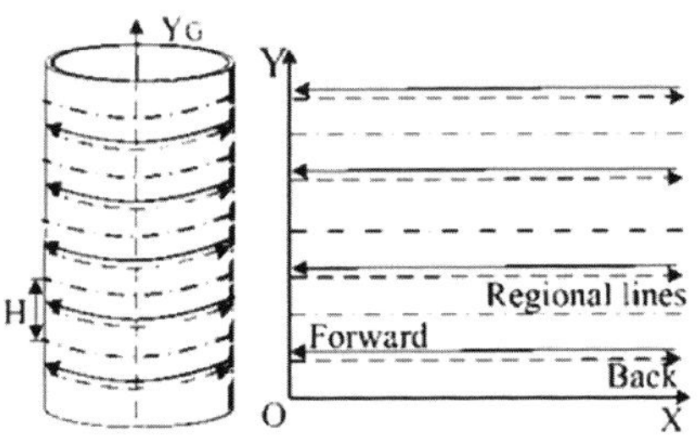

Fig. 14 Trajectory 3

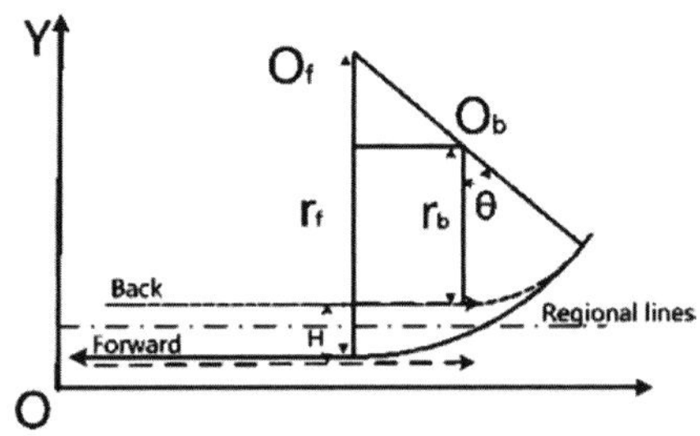

Fig. 15 Turning route of trajectory

3.4. *Comparison of the three trajectories*

Table.2 Comparison of the three trajectories:

trajectory	possibility	controllability	efficiency	locating	the changing rate of Cable's length
1	not	simple	high	easy	low
2	yes	complex	low	difficult	high
3	yes	simple	low	easy	low

After analyzing the three trajectories, we made a comparison, the result is exhibited in Table.2. It's obvious that trajectory 3 is the best choice.

4. Simulated and Experimental Results

In the experiment, we put CKCR on the surface of the pipe (R=1676mm) under underwater, using a ROV with a camera System to detect CKCR's motion. We tried to control CKCR to move on a curve simulated in section 3. The change of controlling parameter θ_0 is exhibited in Fig. 11, and we used the speed control model to control the driving motors in which we can roughly ignore the effect of gravity. Figure 16 shows the pictures of CKCR's motion. We can see that CKCR moves along the curve substantially.

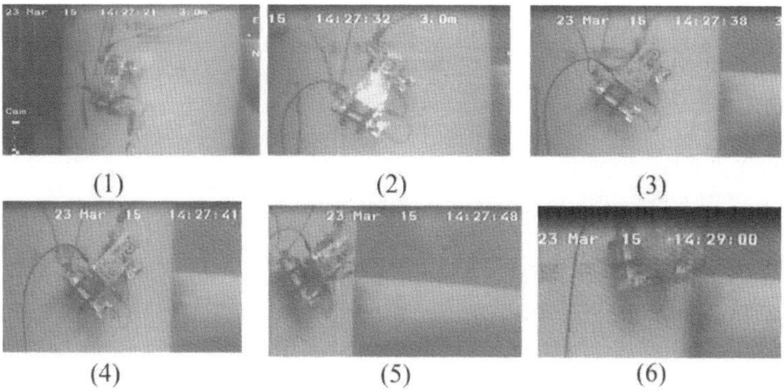

Fig. 16 Pictures of CKCR's motion

5. Conclusion

This paper presents a climbing robot to be used to strip coatings from the outer surface of underwater pipes used in oil platform offshore. A modeling method using the mapping relationship between modeling in plane and in cylinder is proposed. And three trajectories for the robot to travel all over the outer surface of the pipes completely are exhibited. Then we carried out an experiment to detect CKCR's motion. However, there are many works need to be done in the future.

Future works will focus on the development of the kinematic model using mapping technology, dynamic model to compensate gravity and effect of waves and other predictable attack underwater, kinematic and dynamic controllers, positioning systems and navigation/mapping system. The experiment system which can well evaluate the trajectory of robot should also be carried out.

References

1. Ross B, Bares J, Fromme C. A semi-autonomous robot for stripping paint from large vessels[J]. The international journal of robotics research, 2003, 22(7-8): 617-626.
2. Espinoza R V, de Oliveira A S, de Arruda L V R, et al. Navigation's Stabilization System of a Magnetic Adherence-Based Climbing Robot[J]. Journal of Intelligent & Robotic Systems, 2014: 1-17.
3. Kim J H, Lee J C, Choi Y R. LAROB: Laser-Guided Underwater Mobile Robot for Reactor Vessel Inspection[J]. Mechatronics, IEEE/ASME Transactions on, 2014, 19(4): 1216-1225.
4. Lamiraux F, Lammond J P. Smooth motion planning for car-like vehicles[J]. Robotics and Automation, IEEE Transactions on, 2001, 17(4): 498-501.

5. Chu B, Jung K, Han C S, et al. A survey of climbing robots: Locomotion and adhesion[J]. International journal of precision engineering and manufacturing, 2010, 11(4): 633-647.
6. Haynes G C, Khripin A, Lynch G, et al. Rapid pole climbing with a quadrupedal robot[C]//Robotics and Automation, 2009. ICRA'09. IEEE International Conference on. IEEE, 2009: 2767-2772.
7. Provancher W R, Jensen-Segal S I, Fehlberg M A. ROCR: An energy-efficient dynamic wall-climbing robot[J]. Mechatronics, IEEE/ASME Transactions on, 2011, 16(5): 897-906.
8. Fierro R, Lewis F L. Control of a nonholonomic mobile robot: backstepping kinematics into dynamics[C]//Decision and Control, 1995., Proceedings of the 34th IEEE Conference on. IEEE, 1995, 4: 3805-3810.
9. Olsson H, Åström K J, De Wit C C, et al. Friction models and friction compensation[J]. European journal of control, 1998, 4(3): 176-195.

SECTION–13
UNDERWATER AND SEA ROBOTICS

TOWARDS DEEP-SEA MONITORING WITH SMIS - EXPERIMENTAL TRIALS OF DEEP-SEA ACOUSTIC LOCALIZATION

SERGEJ NEUMANN, DAVID OERTEL, HEINZ WÖRN

Karlsruhe Institute of Technology – Intelligent Process Control and Robotics (IAR-IPR), 76131 Karlsruhe, Germany
{sergej.neumann, david.oertel, woern}@kit.edu

MARTIN KUROWSKI, DETLEF DEWITZ

University of Rostock – Institute of Automation, 18055 Rostock, Germany
{martin.kurowski, detlef.dewitz}@uni-rostock.de

JOANNA J. WANIEK, DAVID KAISER, ROBERT MARS

Leibniz Institute for Baltic Sea Research Warnemünde (IOW), 18119 Rostock, Germany,
{joanna.waniek, david.kaiser, robert.mars}@io-warnemuende.de

Successful deep-sea exploration (up to 6000m) performed by autonomous underwater vehicles (AUVs) depends heavily on accurate AUV localization. In the absence of GPS underwater, this requires acoustic communication, localization and geo-referencing underwater participants, e.g. by a GPS-positioned surface vehicle, which is necessary to periodically correct drift errors from dead-reckoning. Within the SMIS project this is supposed to be achieved by a team of AUVs assisted by an unmanned surface vehicle (USV) and a sea-bed-station. In order to derive constraints for the team behaviour, the quality and stability of the acoustic localization needs to be examined thoroughly in advance. This delivers various constraints for later at-sea operation, e.g. the maximum distance between AUV and USV and the maximum horizontal distance to the USV for a given AUV depth. Additionally, heuristic estimates for reproducibility of long-range position measurements can be derived which indicate imminent errors based on acoustic localization only. This paper reports associated tests and their results using USBL modems in deep-sea trials performed in the Atlantic ocean.

Keywords: deep-sea communication, acoustic localization, USBL, position estimation.

1. Introduction

Worldwide increasing activities for exploring and surveying the seabed can be observed in the maritime industry mainly due to rising demands on winning raw materials from the deep-sea. In this regard, high-capacity monitoring systems have to be developed. Within the research project SMIS - Subsea Monitoring via Intelligent Swarms[a] different unmanned underwater and surface vehicles will be designed and combined to an innovative system for efficient monitoring of large-scale deep-sea (up to depth of $6000m$) areas by a team of robots including an unmanned surface vehicle (USV), autonomous underwater vehicles (AUVs) and a seabed station (SBS)[1] for recharging AUVs.

A key problem of long-term underwater operations is the navigation task of the underwater vehicles which requires dead-reckoning from Attitude Heading Reference Systems (AHRS), Doppler Velocity Logs (DVL) and depth sensors. The resulting drift error needs compensation e.g. by acoustic localization and e.g. (Extended) Kalman Filter[2,3] or Particle Filter[4,5] for position estimation.

In SMIS, acoustic localization fixes for AUVs are provided by both USV and SBS via ultra-short baseline (USBL). Additionally, the USV serves as (satellite) communication link to the operator even in heavy-sea states.[6] USBL is used here due to compact device sizes and easy deployment (except for SBS[7,8]) despite a lower localization accuracy compared to LBL and SBL[9] systems. Please refer to[10] for a detailed review of underwater localization and navigation methods.

2. Problem Formulation

To ensure reliable communication and localization, the experimental trials and this study was threefold: (a) investigate the positioning characteristics via USBL, (b) examine the maximum comm. and loc. range at deep sea, (c) analyse the acoustic noise introduced at sea level, in order to find a suitable design of an unmanned surface vehicle (USV). The positioning characteristics was obtained using a stochastic approach, where many localization fixes from different experiments where analysed. For gathering the required data, the low noise acoustic drifter Emmi was designed. The involved partners of the acoustic communication trials are shown in Fig. 1.

[a]Project SMIS is supported by the German Federal Ministry of Economic Affairs and Energy (BMWi) under the registration number FKZ 03SX348.

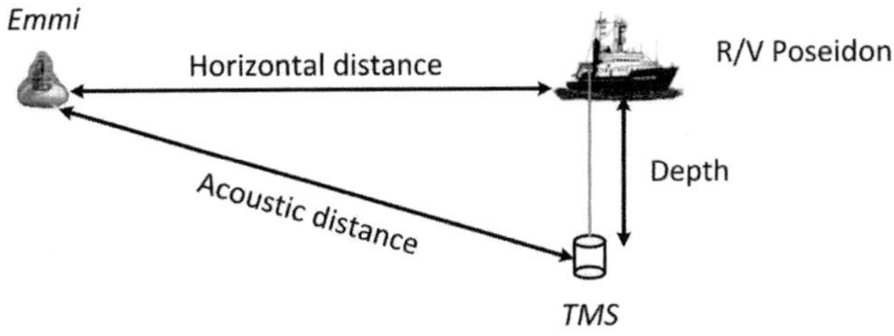

Fig. 1. Involved partners of the acoustic trials

3. Experimental Set-up

For all acoustic measurements EvoLogics USBL S2C 7/17 modems have been used, mounted on the three participants: R/V, TMS and Emmi.

The TMS from Enitech was designed as an interface device between a ship and a ROV. It handles data transfer and power supply through a ship's cable with length of $6000m$.

Emmi itself consists of a surface marker buoy also used as a lifting body and an underwater acoustic part. The lifting body has a weight of about $120kg$, a diameter of $1200mm$, a height of $600mm$, and a buoyancy of about $250kg$. The buoy is equipped with an AHRS including a differential GPS, radio beacon and WiFi for on-ship operation. The overall dimensions in mm are $L750 \times H500 \times B500$. The total weight is about 77kg. A USBL modem is mounted on a chain/rope with configurations defined in Table 1. Fig. 2 shows the overall set-up of the acoustic surface drifter Emmi schematically.

Table 1. Emmi configurations

Configuration	Elements	Length (o.a.)
short	Buoy, connection ropes and acoustic part	$2.3m$
long	Buoy, $5m$ rope, $20m$ rope, connection ropes and acoustic part	$27.3m$
long-c	Buoy, $5m$ rope, chain, $20m$ rope, chain, connection ropes and acoustic part	$28.9m$

3.1. *Experiment conditions*

There were several limitations within the test set-up which had to be compensated for. (a) In the absence of appropriate sensors on the TMS, the exact position and orientation of TMS was not known and had to be estimated by R/V GPS and cable length. (b) All deployed acoustic USBL modems are equipped with a low budget 3DOF orientation tracker with strong error drift. (c) Due to the relatively small size of Emmi, the position of the acoustic underwater part is strongly effected by the waves.

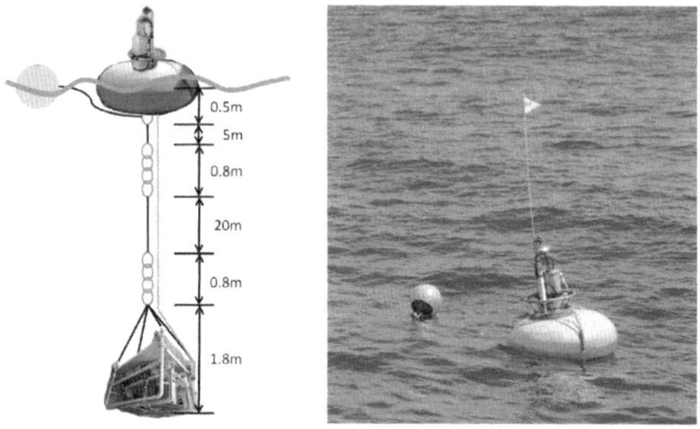

Fig. 2. Set-up of the acoustic drifter Emmi

4. Experiments and Results

4.1. *Experiments conducted*

Multiple experiments were conducted on several days between $28th$ of May to $11th$ of June 2014 with set-up from Section 3. In most of the six test days considered here, the focus was on localizing Emmi using the modem mounted on the TMS to emulate a deep-sea AUV localizing itself through a USV. The test days in short: (Day 1) First acoustic test between R/V and TMS up to approx. $4km$ depth. Emmi was released in *long-c* configuration. (Day 2) Emmi configuration changed to *short* for day $2-5$. Weather: Bft $2-3$, waves ca. $1m$. TMS was lowered to almost $5km$ depth. Horizontal distance R/V to Emmi up to $2.6km$. (Day 3) Same as day 2 with stronger wind of Bft 5 and horizontal distance R/V to Emmi up to $2.8km$. (Day 4) Very long range acoustic test. Wind Bft $3-4$. TMS depth up to $5km$ with R/V to Emmi horizontal distance up to $6.5km$ resulting in a slant range of ca. $8.2km$ between Emmi and TMS. (Day 5) Acoustic Baffle added to Emmi. Wind Bft 4. Horizontal distance up to $8.9km$ resulting in a slant range (Emmi to TMS) of ca. $10.2km$. (Day 6) Wind Bft $3-4$. Horizontal distance up to $8.2km$ resulting in slant range up to $9.6km$.

Besides that, multiple conductivity, temperature and depth (CTD) measurements to estimate sound velocity where taken.

4.2. *Preliminary considerations*

When analysing the characteristic of a sensor, one has to consider the different sources of error that can occur. Four major error sources have been identified for the acoustic localization with USBL modems.

- Orientation errors introduces by the integrated orientation tracker
- Limitations in the resolution of round-trip time and angle of arrival
- Assumption of a constant sound velocity
- Multipath arrivals through reflections and refraction effects[11]

A good time resolution is crucial for slant range accuracy. Evologics S2C 7/17 USBL modems are equipped with a high precision clock, which are supposed to enable a slant range resolution of $0.01m$. The angle of arrival resolution depends on topology and distance between installed hydrophones and is stated as $0.1deg$. For a USBL fix in $5000m$ distance this would result in a position error of $\pm 4.4m$. Due to the small base line, this is a rather big error when compared to other acoustic localization methods like LBL.

4.3. *Noise measurement for Emmi*

On most test days, noise measurements within $1-2$ minutes were performed to check the if environment noise (wind, waves etc.) should be avoided by lowering the Emmi's acoustic modem. Weather conditions can be considered generally good for all test days, see section 4.1.

The deployed acoustic modems use a Received Signal Strength Indicator (RSSI) value to indicate the received signal power during communication or the local noise strength in so-called noise mode.

Within the manufacturer's manual, RSSI is reported as given in dB re $1V$ with typical values between $-110dB$ and$-90dB$ where greater values indicate more noise with a manufacturer's recommendation to keep this value below $-80dB$.

The results can be found in Table 2. Taking these measurements was triggered by examining the long-c configuration through a submerged camera which recorded heavy noise by metallic chain elements resulting in poor communication quality.

In another test it was checked how much the ship's noise affects the noise measurements. For this test, the ship passed Emmi at a distance of about $180m$ ($90m$) with a velocity of $8kn$ where noise measurements raised up to a value of -90 (-80). The consequence of this for the USV design can be found in the conclusion, Section 5.1.

Table 2. Emmi noise RSSI, sample statistics in dB re $1V$

day	conf.	min	mean	max	std deviation
1	long-c	-	-	-	-
2	short	-102.00	-100.31	-80.00	3.69
3	short	-98.00	-89.57	-71.00	4.87
4	short	-102.00	-99.49	-65.00	7.24
5	short, Baffle	-103.00	-101.78	-86.00	2.56
6	long, Baffle	-103.00	-101.47	-83.00	3.81

4.4. *Long-range localization*

Due to the shortcomings mentioned in Section 3, no ground truth of TMS is available. Thereby an estimation of the acoustic localization accuracy is infeasible. However the precision, meaning the reproducibility of the measurement, can be estimated for short periods of time (e.g. $< 30s$) in which the participants are assumed to be motionless. This assumption is reasonable with respect to the considered distance of several km.

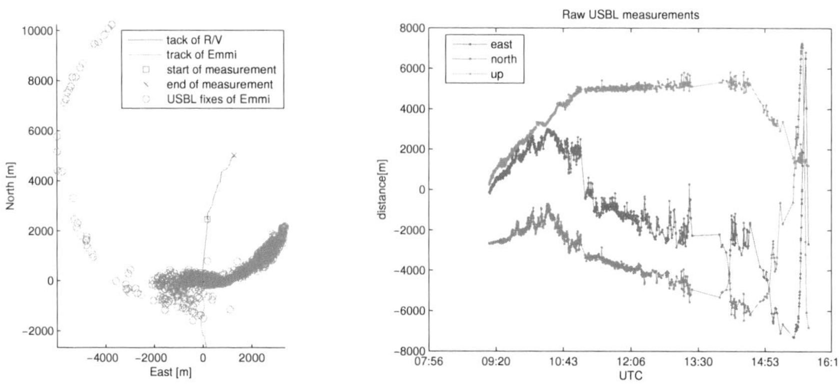

Fig. 3. Submerged *TMS* localizing Emmi while departing from each other. Position fixes (left) vs. raw USBL data (right)

Figure 3 shows measurements made on day 4. In this experiment, Emmi was launched first and the research vessel drove north for $\sim 3km$ at $\sim 1kn$ to increase the distance before submerging TMS to $5000m$ while Emmi drifted south. The (right) figure show the USBL fixes that where transformed from the local sensor frame into a east-north-up coordinate system with orientations measured by the mentioned orientation tracker. In Fig. 3 it is apparent, that the measurements don't match with the actual track of Emmi due to orientation drift errors. In order to extract useful information about the localization precision from the noisy data, a combination of filter functions was applied, i.e. transformation into a spherical coordinate system followed by a moving average filter with time constrains. To account for the changing propagation time and occasional absences of USBL fixes, the window size for the moving average filter was chosen to be proportional to propagation time $p(i)$ of a sample i. Defining

$$I =: \{1, \ldots, n\} \tag{1}$$

as the indices of the sample and $t(i)$ as the time stamp of the sample i, we define the moving window borders for $t(i)$ as

$$t_k(i) = t(i) + ws_{ma}, \tag{2}$$

$$t_j(i) = t(i) - ws_{ma}, \tag{3}$$

where ws_{ma} denotes the window size of the moving average and equals

$$ws_{ma}(i) = p(i) w_{tf}. \tag{4}$$

w_{tf} is the window time factor that can be adjusted to enlarge or reduce the window size. With this we get to the moving average

$$x_{ma}(i) = \frac{1}{k(i) - j(i)} \sum_{l=j(i)}^{k(i)} x(l) \tag{5}$$

where x is the measured localization fix in spherical coordinate system and j and k are defined as

$$k(i) := max \{\tau \in I | t(\tau) \leq t_k(i)\} \tag{6}$$

$$j(i) := min \{\tau \in I | t(\tau) \geq t_j(i)\} \tag{7}$$

This results in a smoother, less noisy dataset which will be considered as an expectation value for the windowed standard deviation filter. Similar to eq. 5 the window size for the standard deviation ws_{std} was time constrained with

$$b(i) := max \{\tau \in I | t(\tau) \leq t(i) + ws_{std}\} \text{ and} \tag{8}$$

$$a(i) := min \{\tau \in I | t(\tau) \geq t(i) - ws_{std}\}. \tag{9}$$

The stadard deviation for the localization precision was estimated by

$$\sigma(i) = \sqrt{\frac{1}{b(i) - a(i) - 1} \sum_{h=a(i)}^{b(i)} (x(h) - x_{ma}(i))^2}. \tag{10}$$

The underlying assumption is, that during a short period of time there are no big movements between both USBL transceivers. Many different window sizes have been tried experimentally for ws_{ma} and ws_{std} with similar result which can be seen when comparing Fig. 5 and Fig. 4. Figure 5 shows the so obtained standard deviation for the days 3 and 4. The mean of the calculated deviations vary only slightly for each day, as can be seen in Table 3. As the results suggest, typical localization errors lie between 1.3 and 2.5 degree for azimuth and between 0.6 and 1.2 degree for elevation angles. For the slant range a typical error between $0.5m$ and $2m$ was obtained. In Cartesian space this would result in an absolute position error, e.g. between $125m$ to $250m$ for a slant range of $5000m$. The results can give only a rough impression of the localization precision of acoustic USBL measurements. Although the time constrained moving average filtered parts of the errors made by the orientation tracker, it is assumed that some of that influence remains in the data. If that assumption is correct, the localization precision could be improved with a high-class orientation tracker.

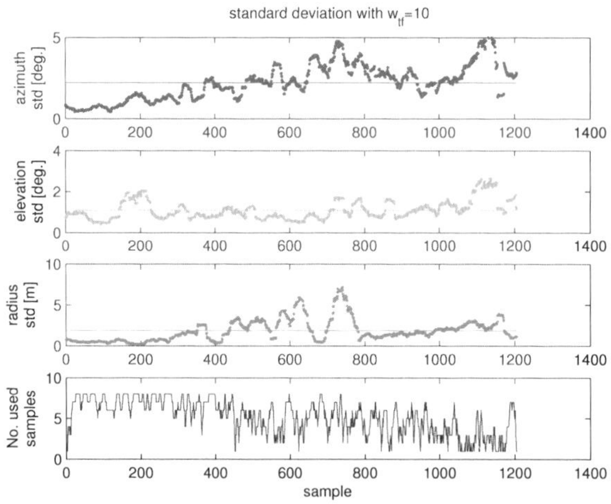

Fig. 4. Standard deviation of USBL fixes and number of used samples for moving average with $w_{tf} = 10$ for Day 4

4.5. *Long range communication*

The results from section 4.4 are based on small round-trip test-packages sent via the acoustic link. An overview of the general mean success ratio

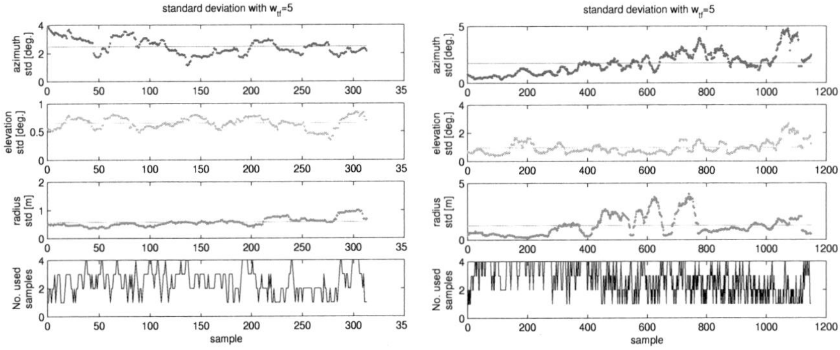

Fig. 5. Standard deviation of USBL fixes and number of used samples for moving average with $w_{tf} = 5$ for Day 3 (left) and Day 4 (right)

Table 3. mean of standard deviation by day

Day	azimuth[deg]	elevation[deg]	radius[m]
3	2.477	0.653	0.580
4	1.765	0.962	1.260
5	1.258	1.230	1.995
6	1.319	1.163	1.847

(i.e. TMS successfully received acknowledgement by Emmi) for different slant-ranges can be found in Fig. 6. The success-ratio dropped considerably for distances $\geq 8400m$ and communication was practically impossible for slant-ranges $\geq 9500m$.

5. Conclusion and Future Work

5.1. *Conclusion*

For comparable sea-conditions, the following heuristic constraints for operations with the SMIS system were derived.

In terms of environment noise, the USBL modem attached to the USV does not need to be at depth of more than $2.3m$. To avoid interference with ship's noise, a distance of at least $100m$ should be kept to the operating ship(s).

As a rough upper bound for AUV-USV localization reproducibility, the data collected suggests positioning uncertainties below $125m - 250m$ for a slant-range of $5000m$.

Package loss examinations suggest slant-range between AUV and USV

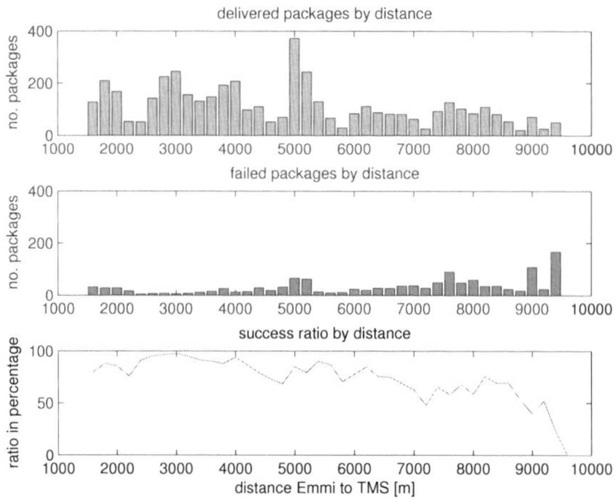

Fig. 6. Communication between TMS and Emmi: Number of packages delivered successfully vs. lost packages vs. overall success-ratio. Data merged from all relevant test days. Distance partitioned into buckets of $200m$.

should not exceed $8400m - 9500m$ providing mission constraints for SMIS team operation.

Constraints derived in this paper will have a large influence on the parameters and methods used for both control algorithms and mission planning of the underwater robotic team in SMIS.

5.2. *Future work*

In order to measure absolute USBL positioning capabilities, further tests are required with fixed modem positions or known ground-truth for position and orientation. This can also serve as basis for acoustic ray tracing examinations. Additionally, localization characteristics need to be observed for faster moving AUVs, where the motion and acoustic noise introduced by the thrusters can impair the positioning precision.

An additional sea trial is scheduled for May 2015, where further experiments will be conducted.

Acknowledgment

The SMIS team thanks the BMWi for funding this work through the SMIS project under registration number FKZ 03SX348. Also, the SMIS team would like to thank the crew of the R/V Poseidon as well as the colleagues of the company EvoLogics for their professional and friendly support during the experimental trials.

References

1. F. Boeck, M. Golz, S. Ritz and G. Holbach, Smis - subsea monitoring via intelligent swarms, design challenges of an autonomous seabed station, in *ASME 2014 33rd International Conference on Ocean, Offshore and Arctic Engineering Volume 8B: Ocean Engineering*, (San Francisco, California, USA, 2014).
2. A. Alcocer, P. Oliveira and A. Pascoal, Underwater acoustic positioning systems based on buoys with gps, in *Proceedings of the Eighth European Conference on Underwater Acoustics*, 2006.
3. M. Mandt, K. Gade and B. Jalving, Integrating dgps-usbl position measurements with inertial navigation in the hugin 3000 auv, in *Proceedings of the 8th Saint Petersburg International Conference on Integrated Navigation Systems, Saint Petersburg, Russia*, 2001.
4. M. F. Fallon, G. Papadopoulos, J. J. Leonard and N. M. Patrikalakis, *The International Journal of Robotics Research* , p. 0278364910380760 (2010).
5. P. Rigby, O. Pizarro and S. B. Williams, Towards geo-referenced auv navigation through fusion of usbl and dvl measurements, in *OCEANS 2006*, 2006.
6. S. Ritz, G. Holbach, F. Boeck and M. Golz, *Jahrbuch der Schiffbautechnischen Gesellschaft* **Volume 108**, p. in Process (2014).
7. H.-H. Chen, *Ocean Engineering* **35**, 1448 (2008).
8. J. Opderbecke, At-sea calibration of a usbl underwater vehicle positioning system, in *OCEANS'97. MTS/IEEE Conference Proceedings*, 1997.
9. K. Vickery, Acoustic positioning systems. a practical overview of current systems, in *Proceedings of the 1998 Workshop on Autonomous Underwater Vehicles*, (Cambridge, MA, USA, 1998).
10. L. Paull, S. Saeedi, M. Seto and H. Li, *Oceanic Engineering, IEEE Journal of* **39**, 131 (2014).
11. K. G. Kebkal and R. Bannasch, *The Journal of the Acoustical Society of America* **112**, 2043 (2002).

DESIGN AND IMPLEMENTATION OF A ROBOTIC FISH BASED ON REVOLUTE WINGTIPS[1]

ZHAO GONG, YUERI CAI, HONGWEI MA, YONG CAO and SHUSHENG BI

Robotics Institute, Beijing University of Aeronautics and Astronautics, Beijing, China

Pectoral fin motion discipline of the cownose ray as a nature prototype is analyzed. According to the analysis results, a bionic pectoral fin with multiple fin rays and one revolute wingtip is designed. The angle of attack distribution principle of the bionic pectoral fin is also analyzed. The analysis results show that the angle of attack of pectoral fin reaches its maximum at the wingtip. In order to improve the deformation ability of the tip of rigid fin ray, a novel revolute joint is designed and applied to the bionic pectoral fin. The simulation results show that the maximum angle of attack of the revolute wingtip can reach about 35°.The relationship between the pitching angle of fin tip and the average thrust/lift generated by pectoral fins is analyzed. In the thrust/lift experiments, with the help of the revolute joint, the maximal average thrust of 3.3N is achieved and the maximal absolute value of average lift is less than 2N. In the swimming experiments, a maximum forward speed of 0.7BL/s (body length/s) is reached. Under the same motion parameters, the maximum speed decreases to 0.35BL/s when the revolute joint is locked.

1. Introduction

Natural selection has ensured that the mechanical systems evolved in fish are efficient with regard to the habitat [1]. The fishes possess advantages of high efficiency, remarkable maneuverability and less noise than conventional underwater vehicles. Their remarkable swimming abilities inspire people to improve the propulsion performance of UUVs [2, 3].

Most fishes generate thrust by bending their bodies into a backward-moving propulsive wave that extends to its caudal fin, and this type of swimming is classified under body and/or caudal fin (BCF) locomotion. Other fish have developed alternative swimming mechanisms that involve the utility of their median and pectoral fins, termed median and/or paired fin (MPF) locomotion [4].

Compared with BCF mode, MPF mode has more advantages in

* This work is supported by National Natural Science Foundation of China (No.51205011), Fundamental Research Funds for the Central Universities (YWF-14-RSC-107) and Specialized Research Fund for the Doctoral Program of Higher Education (No.20111102130004).

maneuverability and stability [5-7]. Cownose ray, as one species of fishes utilizing MPF modes, has high maneuverability and stability and is studied by many researchers [8-10]. Robotic fish prototypes mimicking cownose ray usually have flat and large pectoral fins [8] which improve the stability in yaw, pitch, and roll directions.

In this paper, a robotic fish propelled by oscillating pectoral fins is designed. A novel revolute joint is designed to improve the propulsive performance of the pectoral fin. The mechanical structure and control system of the robotic fish are illustrated in Section 2. The pectoral fins are designed based on the analysis of cownose ray. The design of the revolute joint is the emphasis of Section 3. A series of simulations and swimming experiments are carried out in Section 4 and the conclusions and future work are given in Section 5.

2. Mechanical Structure of Robo-Ray VI

As shown in Fig. 1, cownose ray is a kind of cartilaginous fish, which is characterized by streamlined flat body, high aspect ratio pectoral fins, and wing-like plan view. For convenience of description, spanwise and chordwise directions are defined respectively, and the coordinate system is set, as shown in Fig. 1. Spanwise direction is parallel with the line connecting two fin tips. The direction from head to tail, which is perpendicular to spanwise, is defined as chordwise direction. The origin is set on the root of the leading edge of oscillating pectoral fin where there is no obvious deformation during linear cruising. X-axis is parallel with spanwise direction. Y-axis is parallel with chordwise direction. Z-axis is determined by the right-hand rule.

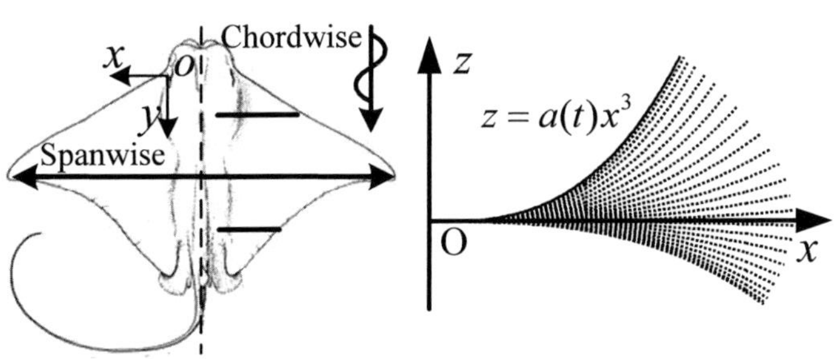

Fig. 1. Top view of cownose ray and the cubic function curve [11]. A series of curves show the different oscillating positions of the pectoral fin during a whole oscillating period.

The structure design of fin rays is a key to the mechanical design of the robotic fish. The main function of fin rays is to propel the fin with mimicking the oscillating motion of the nature sample. According to [11], the chordwise profile of cownose ray's pectoral fin shown in Fig. 1 can be described approximately by cubic function:

$$z = a(t)x^3. \tag{1}$$

Where $a(t)$ is a coefficient associated with time.

To fit the function curve, one-stage and two-stage slide–rocker mechanisms are employed. By optimizing the length of each linkage, the shape difference between the mechanism profile and the cubic function is within an acceptable range.

The mechanism schematics of the robotic fish are illustrated in Fig. 2, and the revolute joints are driven by the springs when the fin rays oscillate at a certain phase bias.

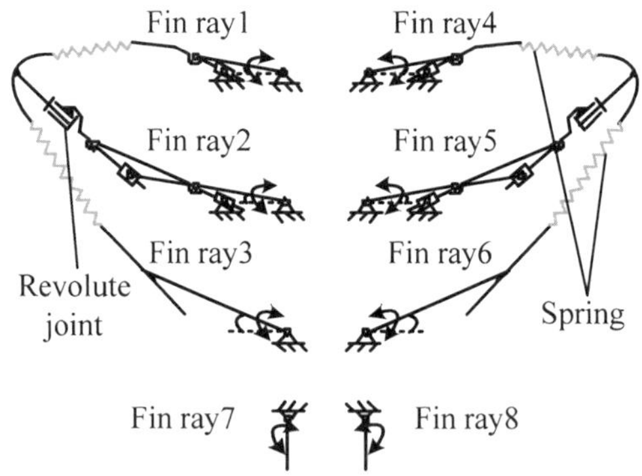

Fig. 2. Mechanism schematics of the robotic fish. The right pectoral fin and the left pectoral fin contain three fin rays respectively. Tail fins contain two fin rays. Revolute joints in fin ray2 and fin ray4 are linked by spring with their front and back fin rays respectively.

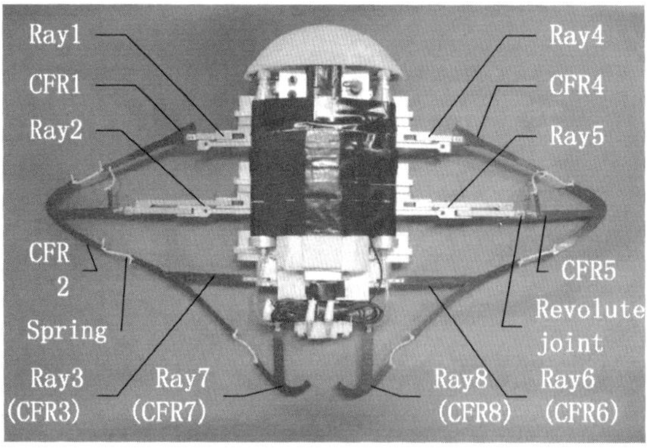

Fig. 3. Bionic prototype. Rubber hoses are utilized as springs to drive the revolute joints.

The prototype of Robo-Ray VI is shown in Fig. 3. Eight carbon fiber rods (CFR for short) are employed respectively in eight fin rays, one for each. CFR1 in fin ray1 and CFR2 in fin ray2 are respectively fixed on the aluminum rocker of each fin ray. Fin ray3, fin ray6, fin ray7, and fin ray8 consist of carbon fiber robs only. CFR2 in fin ray2 and CFR5 in fin ray5 are respectively connected to the rockers through revolute joints and are the tips of the pectoral fins mentioned in this paper. The triangular profile of CFR2 and CFR5 is obtained by mimicking the projection shape of cownose ray's pectoral fins.

Each fin ray of pectoral fins is driven by a high torque servo (37kgf.cm at 7.4V). O-ring and X-ring sealing techniques are employed to make the prototype waterproof. The prototype is covered by permeable skin when it swims in water.

3. Design of Revolute Joint

According to [12], the linear velocity and the angle of attack of oscillating foil are the two key factors for thrust generation. For the oscillating rigid fin rays, the linear velocity of a certain point is proportional to the distance from the revolving shaft. The linear velocity reaches its maximum at the fin tip.

For the robotic fish with multiple fin rays, the angle of attack is generated by adjacent fin rays oscillating at a certain phase difference. To investigate the distribution of the angle of attack when pectoral fins oscillate, the coordinate system is set as same as the one in Fig. 1. As shown in Fig. 4, at the same spanwise position x_0, the vertical positions of three fin rays are z_1, z_2, and z_3 respectively, which satisfy

$$\begin{cases} z_1 = a(t)x^3 \\ z_2 = a(t-\Delta t)x^3 \\ z_3 = a(t-2\Delta t)x^3. \end{cases} \tag{2}$$

Where Δt means three fin rays oscillating at a certain phase difference. So the angle of attack α_1 can be expressed:

$$\alpha_1 = \arctan\frac{z_2 - z_1}{l} = \arctan\frac{a(t-\Delta t)x^3 - a(t)x^3}{l}. \tag{3}$$

This is an increasing function, and α_1 increases with x increasing. Angle of attack α_1 reaches its maximum at the fin tip. Therefore, the mechanical structure of the fin tips must have the pitching motion or deformation ability. So the revolute joint is designed to achieve the fin tip pitching motion ability.

The revolute block of the revolute joint is fixed on the rocker, and the rotation shaft is fixed on CFR2 or CFR5. All the movements except the rotation on spin axis are restrained by the revolute joint. The revolute joints are driven by the pulling force produced by the springs during fin rays' oscillating. If the CFR2 and CFR5 are locked with the rockers, the tip deformation of pectoral fins will be discontinuous. In contrast, if the CFR2 and CFR5 can revolve freely, the angle of attack in the fin tip will be generated more easily.

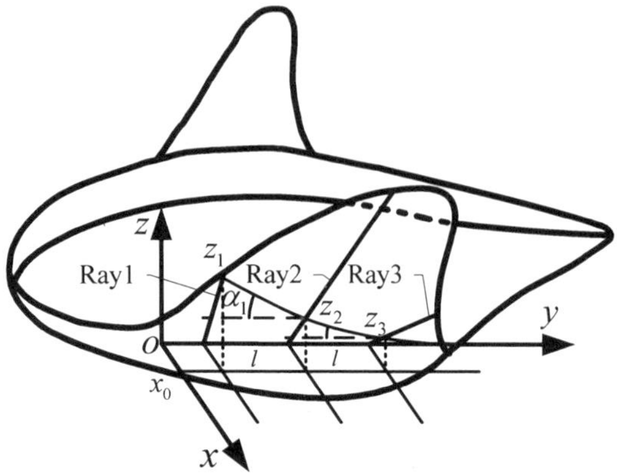

Fig. 4. Coordinate system definition and the deformation of the oscillating fins.

4. Simulations and Experiments

4.1. *Kinematics simulations*

In order to analyze rotating characteristics of the CFR2 and CFR5 in the fin tips, the kinematics simulations are conducted by using Automatic Dynamic Analysis of Mechanical Systems (ADAMS). 3D model of the robotic fish is established in SolidWorks, and is imported into ADAMS through X_T format interface. Constraints and motions are added in ADAMS. Pitching angle of CFR2 on the fin tip is measured by orientating measure function. Pitching angle of CFR5 is the same as pitching angle of CFR2 due to the symmetry.

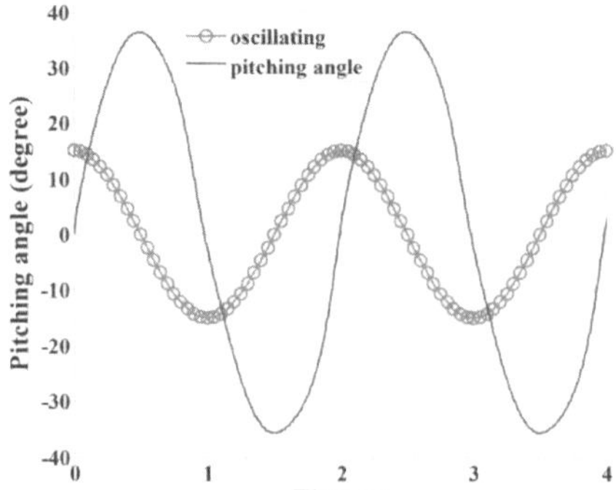

Fig. 5. The relationship between pitching angle/oscillating angle and time. The pitching angle follows sinusoidal law, and phase difference between pitching motion of CFR2 and oscillating motion of fin ray2 is 90°.

The oscillating amplitude of each fin ray is 16°, the phase difference is -50°, and the oscillating frequency is 0.5Hz. Phase difference is the phase of fin ray j minus the phase of fin ray i (j=i+1, i=1, 2, 4, 5). Fig. 5 shows the relationship between pitching angle/oscillating angle of the CFR2 and time. According to [12], when the phase difference between pitching motion and heaving motion is 90-100°, swinging hydrofoil can obtain optimal propulsion performance. So the motion of the revolute joint is feasible. Fig. 6 and Fig. 7 show the maximum pitching angle of CFR2 at different motion parameters.

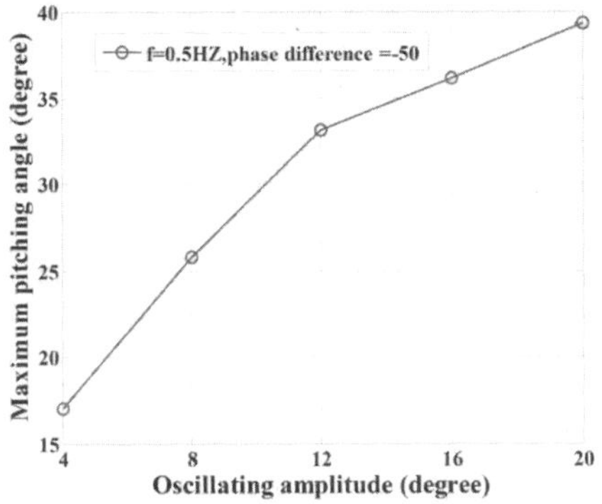

Fig. 6. The relationship between oscillating amplitude and pitching angle. When f = 0.5Hz and phase difference= -50°, the pitching angle increases from 17° to 39.3° with the oscillating amplitude increased from 4° to 20°.

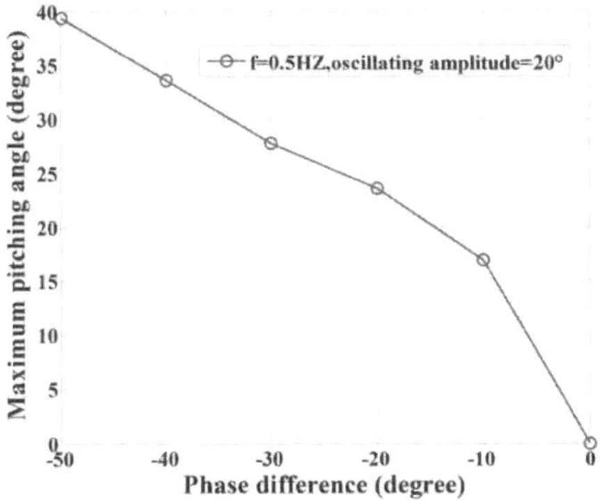

Fig. 7. The relationship between phase difference and pitching angle. When f = 0.5Hz and oscillating amplitude = 20°, the pitching angle decreases from 39.3° to 0° with the phase difference increased from -50° to 0°.

Fig. 6 and Fig. 7 also show that the oscillating amplitude and the phase difference are the two factors which have positive effects on the maximum pitching angle of CFR2 and CFR5. The effects of pitching angle on average thrust and lift is investigated in the following experiments.

4.2. *Experiments*

Thrust/Lift experiments and swimming experiments of the robotic fish are carried out in the swimming pool. Maximum average thrust of 3.3N is achieved and 3D locomotion including cruising, pivot turning, descent/ascent, and rolling motion is achieved.

4.2.1. *Thrust/Lift experiments*

The average thrust and lift are measured at the same motion parameters set in the kinematics simulations.

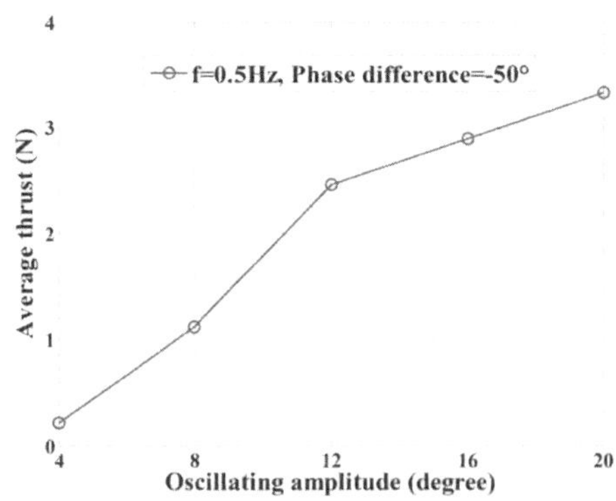

Fig. 8. The relationship between oscillating amplitude and average thrust. When f=0.5Hz and phase difference= -50°, the average thrust increases from 0.3N to 3.3N with the oscillating amplitude increased from 4° to 20°.

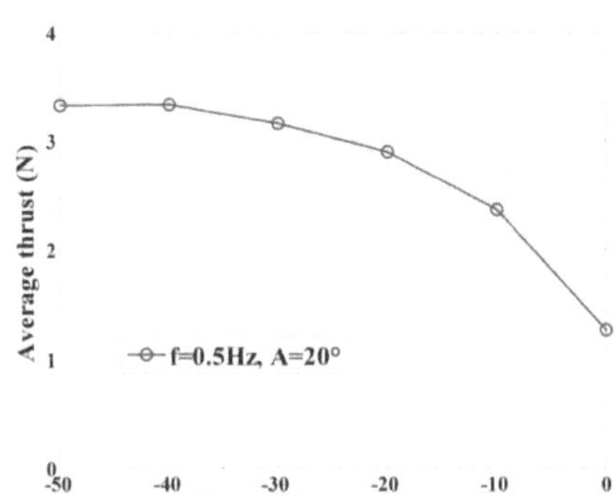

Fig. 9. The relationship between phase difference and average thrust. When f=0.5Hz and oscillating amplitude = 20°, the average thrust decreases from 3.3N to 1.3N with the phase difference decreased from -50° to 0°.

The influence of oscillating amplitude on average thrust is shown in Fig. 8. The influence of phase difference on average thrust is shown in Fig. 9. From the comprehensive analysis of the simulation and experimental results, it is seen that the changing rule of the average thrust is approximately consistent with the changing rule of the maximum pitching angle under the same motion parameters. Therefore, it is concluded that the pitching motion of fin tip is a key factor for the generation of thrust.

The influence of amplitude and phase difference on average lift is shown in Fig. 10 and Fig. 11 respectively. Average lift is near zero, and the influence of amplitude and phase difference is little, which shows that revolute joints keeps spatial symmetry during fin rays' oscillating.

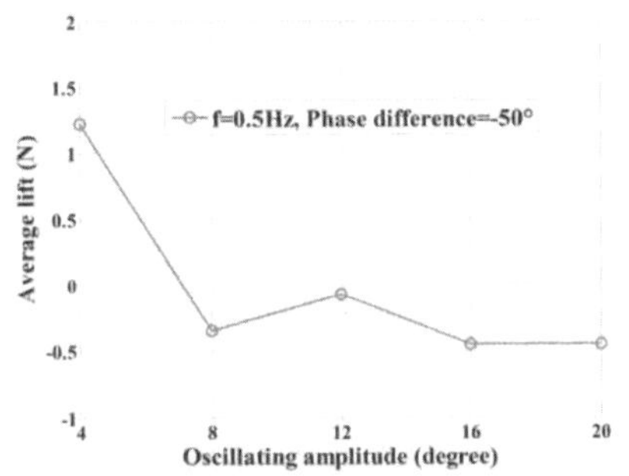

Fig. 10. The relationship between amplitude and average lift.

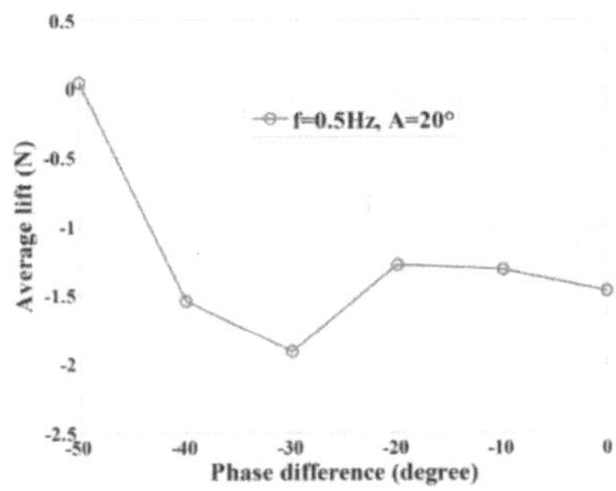

Fig. 11. The relationship between phase difference and average lift

4.2.2. *Swimming Experiments*

The maximum swimming velocity of 0.4m/s (0.7 times of body length per second) is measured at f = 0.5Hz, phase difference = -50°, amplitude = 20°. The maximum velocity decreases to 0.2m/s when the revolute joint is locked.

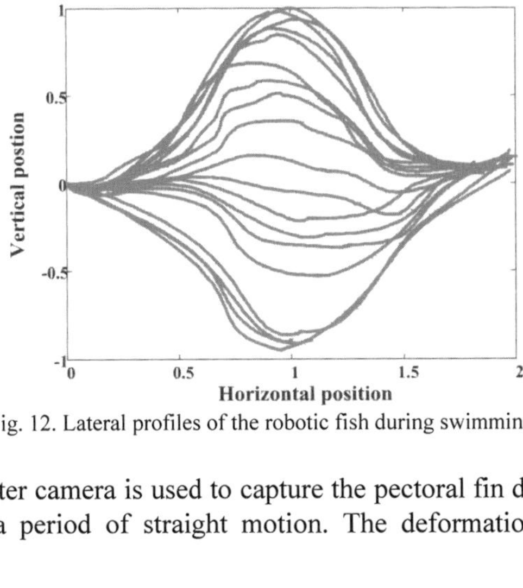

Fig. 12. Lateral profiles of the robotic fish during swimming.

An underwater camera is used to capture the pectoral fin deformation of the robotic fish in a period of straight motion. The deformation process of the pectoral fin is shown in Fig. 12. The deformation of pectoral fin tips is smooth and continuous, which indicates that the design of the revolute joints is feasible.

5. Conclusions and Future Works

In this paper, a robotic fish propelled by oscillating pectoral fins is developed. A novel revolute joint on fin tip is designed and applied to improve the pitching deformation ability of the rigid fin rays. A series of simulations and experiments are carried out to verify the effectiveness of the design of the bionic pectoral fin, especially the revolute joint. The simulation results show that the pitching angle of the revolute joint reaches its maximum 39.3° at amplitude = 20° and phase difference = -50°. The average thrust reaches its maximum 3.3N at the same parameters, and the robotic fish can achieve a maximum forward speed of 0.4m/s (about 0.7BL/s), and if the revolute joint is locked, the maximum speed will decrease by 50%.

Further studies of the bionic fin, such as research on the effect of the flexibility distribution of the fin rays will be carried out. Optimization of the revolute fin tips, including the material, the structure and size, will also be considered to improve the swimming performance of the robotic fish.

References

[1] Domenici P,Blake R.The kinematics and performance of fish fast-start swimming[J].Journal of Experimental Biology,1997, 200(8): 1165-1178.

[2] Sfakiotakis M,Lane D M,Davies J B C.Review of fish swimming modes for aquatic locomotion[J].Oceanic Engineering,IEEE Journal of,1999,24(2):237-252.

[3] Triantafyllou M S,Triantafyllou G S,Yue D K P.Hydrodynamics of fishlike swimming[J].Annual review of fluid mechanics,2000, 32(1): 33-53.

[4] P.Webb.Form and Function in Fish Swimming[J].Scientific American, 1984,251(1):58-68.

[5] Rosenberger L J.Pectoral fin locomotion in batoid fishes: undulation versus oscillation[J].Journal of Experimental Biology, 2001, 204(2):379-394.

[6] Blake R W. Fish functional design and swimming performance[J]. Journal of fish biology,2004, 65(5):1193-1222.

[7] Suzuki H,Kato N,Suzumori K.Load characteristics of mechanical pectoral fin[J].Experiments in Fluids,2008,44(5):759-771.

[8] Cai Y, Bi S, Zheng L. Design and experiments of a robotic fish imitating cow-nosed ray[J]. Journal of Bionic Engineering, 2010, 7(2): 120-126.

[9] Xu Y C, Zong G H, Bi S S, Gao J. Initial development of a flapping propelled unmanned underwater vehicle. Proceedings of the 2007 IEEE International Conference on Robotics and Biomimetics, Sanya, China, 2007, 524–529.

[10] Yang S B, Qiu J, Han X Y. Kinematics modeling and experiments of pectoral oscillation propulsion robotic fish. Journal of Bionic Engineering, 2009, 6, 174–179.

[11] Zheng L, Bi S, Cai Y, et al. Design and optimization of a robotic fish mimicking cow-nosed ray[C]// Proceedings of the 2010 IEEE International Conference on Robotics and Biomimetics.New York:IEEE,2010: 1075-1080.

[12] Read D. A., Hover F. S., Triantafyllou M. S. Forces on oscillating foils for propulsion and maneuvering[J]. Journal of Fluids and Structures, 2003, 17(1): 163-183

DESIGN AND CFD ANALYSIS FOR A BIOMIMETIC DOLPHIN-LIKE UNDERWATER GLIDER

ZHENGXING WU*, JUNZHI YU, ZONGSHUAI SU, JUN YUAN, and MIN TAN

State Key Laboratory of Management and Control for Complex Systems,
Institute of Automation, Chinese Academy of Sciences, Beijing, 100190, China.
**E-mail: zhengxing.wu@ia.ac.cn*

This paper presents an innovative design concept for a biomimetic dolphin-like underwater glider. As an excellent combination, it offers the advantages of both robotic dolphins and underwater gliders to realize high-maneuverability, high-speed and long-distance motions. As the first step, a skilled and simple dolphin-like prototype with only gliding capability is developed. The hydrodynamic analysis in the glider using Computational Fluid Dynamics (CFD) method is executed to explore the key hydrodynamic coefficients of dolphin-like glider including lift, drag and pitching moment, and also to analyze the dynamic and static pressure distribution. Finally, experimental results have shown that the dolphin-like glider could successfully glide depending on the pitching torques only from buoyancy-driven system and controllable fins without traditional internal movable masses.

Keywords: underwater glider; robotic dolphin; biomimetic robots; CFD.

1. Introduction

Underwater gliders have caught much attention in the last three decades.[1–3] As an excellent autonomous underwater vehicle (AUV), the underwater glider has such astonishing performance in characteristics of long distance, extended duration and low cost which enormously promote the significant and effective applications in huge oceans.

Since the first underwater glider was inspired by Stommel in 1989,[4] various underwater gliders have been developed and successfully applied in scientific research and ocean observatory, like Slocum,[5] Seaglider[6] and Spray.[7] Generally, these gliders always orientate the applications in deep oceans, and could be able to operate in over 1000 m depth and last up to nearly 360 days.[8] Traditional design style brings underwater gliders so many advantages in deep oceans, but reduces their maneuverability, such as fast turn in minor radius. In some complex aquatic environments, high maneuverability is vital for underwater gliders to complete certain missions. Consequently, an excellent underwater glider with both great endurance and high maneuverability should be developed.

This paper provides an innovative design concept for a dolphin-like glider to improve the maneuverability. As an excellent combination of robotic dolphins and underwater gliders, the dolphin-like underwater glider can not only perform a fast and flexible locomotion in dolphin flapping style, but also glide for a long distance depending on its buoyancy-driven system. Besides, owing to its controllable pectoral fins and flattened fluke in horizonal plane, the dolphin-like glider could quickly obtain enough pitching torques to adjust its gliding attitude even without traditional internal moveable masses. Our previous work mainly focused on the mechanical design and locomotion control for robotic dolphins.[9,10] Therefore, we have more confidence in the implementation of dolphin-like motion. By comparison, we more care about the realization of the gliding part. Consequently, a dolphin-like glider prototype is first developed to testify the glide motion. For a better space utilization rate and drag reduction, the dolphin-like glider adopts a well profile from killer whale. The hydrodynamic performance in gliding motion is analyzed by Computational Fluid Dynamics (CFD) method, and key hydrodynamic coefficients including lift, drag and moment are also provided for future dynamic analysis. Finally, the experiments testified that the dolphin-like glider could successfully glide upwards and downwards depending on the pitching torques only from buoyancy-driven system and controllable pectoral fins and flattened fluke, even without traditional internal moveable masses.

The rest of this paper is organized as follows. The mechanical design for the simple dolphin-like glider is described in Section 2. The gliding performance analysis using CFD method is detailed in Section 3. Experimental results and analyses are further offered in Section 4. Finally, conclusion and future work are summarized in Section 5.

2. Mechanical Design of the Dolphin-like Glider

As mentioned above, we have confidence in the mechanical design and motion control for a robotic dolphin, and here only focus on the realization of the gliding motion. So a simple testify prototype for gliding motion is developed. Notice that traditional internal moveable masses are specially removed to highlight that the dolphin-like glider could obtain enough pitching torques from buoyancy-driven system, controllable pectoral fins and flattened fluke.

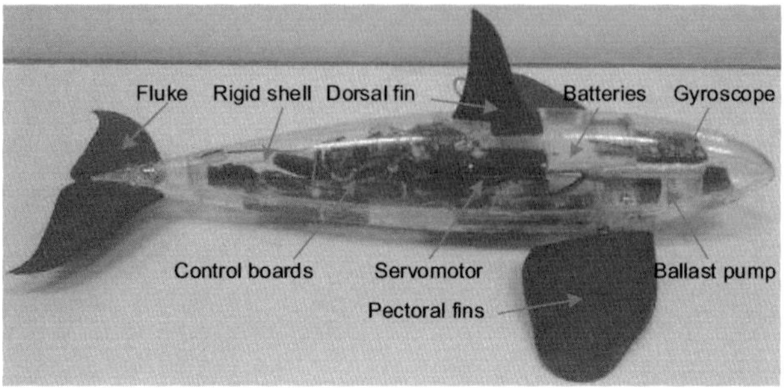

Fig. 1. Mechanical design of the dolphin-like glider.

The mechanical design of the dolphin-like underwater glider is schematically shown in Fig. 1. Generally, the dolphin-like glider is 0.368 m long and weighs 0.752 kg. A rigid well-streamlined body modeled after killer whale is adopted for a better space utilization rate and lift-drag ratio. Notice that pectoral fins and flatten fluke are specially amplified 1.5 times around the center of the mass for larger pitching torques. As shown in Fig. 2, the simple

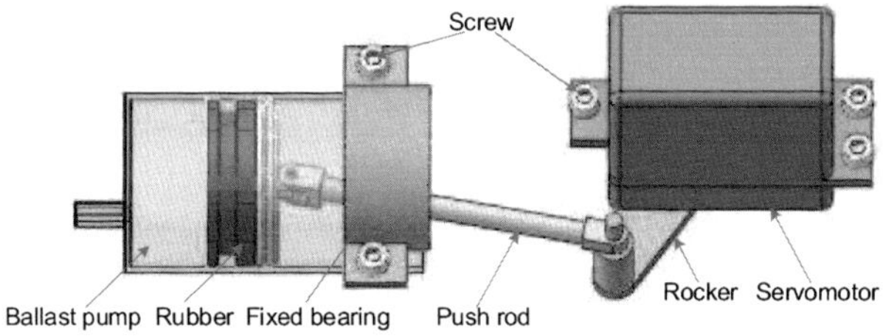

Fig. 2. Mechanical design of the buoyancy-driven system.

and skilled buoyancy-driven system is composed of an injecting syringe, a waterproof rubber, a digital servomotor and a push-pull structure with an aluminium push rod and a copper rocker. When the servomotor working, the copper rocker is turned to drive the aluminium push rod to make the rubber move back and forth in the injecting syringe. Meanwhile, the water in injecting syringe changes the buoyancy of the glider to provide the pitching torques for attitude adjustment.

3. CFD Simulation and Analysis

In this section, CFD methods are employed to analyze the gliding performance of the dolphin-like glider. Because the fins could manually controlled, we need separately compute hydrodynamic coefficients of lifts, drags and pitching moments on dolphin body, pectoral fins and fluke, which could be applied for the hydrodynamic analysis in glide motion. For an accurate and convenient CFD simulation, the commercial software ANSYS 15.0 is employed.[11] ICEM CFD software is adopted as the pre-processing tool to build a mesh for the glider which forms the finite flow domain. Fluent is applied to simulate the flow and pressure distribution around the glider when it is gliding in water.

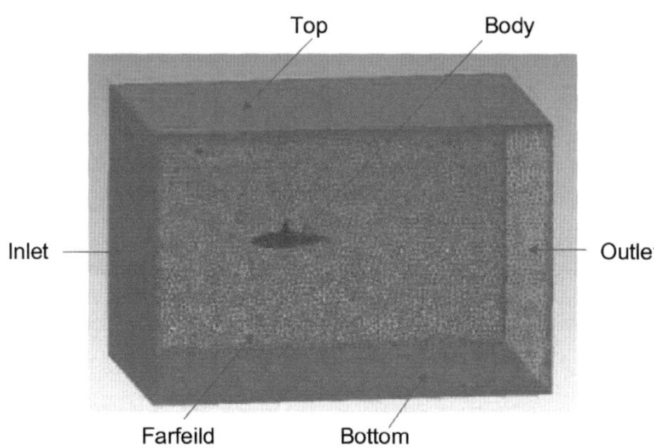

Fig. 3. Boundary conditions for the dolphin-like glider.

In order to gain great adaptability and high quality, an unstructured tetrahedron mesh is formed to describe the flow domain, as shown in Fig. 3. As for the boundary conditions of the CFD simulation for the dolphin's body, the Inlet, two times body length from the nose of the glider, is set as velocity-inlet with $v = 0.1$ m/s, and the Outlet, three times body length from the fluke of the glider, is set as outflow. In order to avoid reflected effect, the Top and Bottom are also set as velocity-inlet with $v = 0.1$ m/s. Other boundaries like Farfeild and Int-fluid are all set as no-flip walls, and the glider surface boundary is also set as no-slip wall. Meanwhile, for a better simulation results, seven prismatic layers are stacked onto the surface mesh. The CFD simulations about the pectoral fins and flattened fluke adopt the similar boundary conditions. In addition, the fluid is supposed as an incompressible and steady one, and $k-\omega$ SST (Shear-Stress-Transport) turbulence model with low-Re corrections is adopted in Fluent simulation.

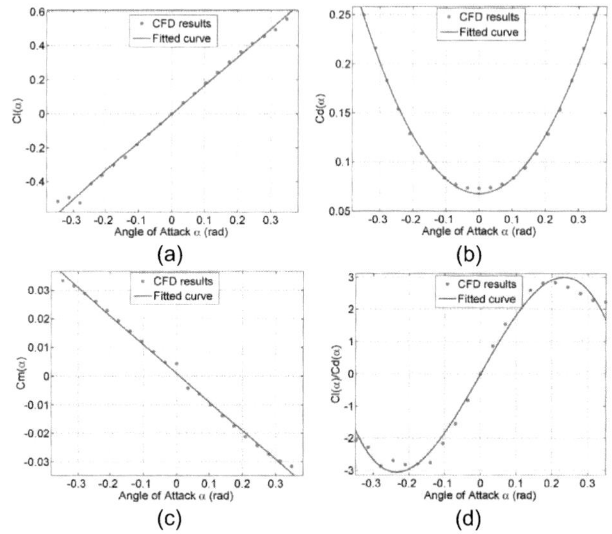

Fig. 4. Hydrodynamic coefficients over AoA of the pectoral fins. (a) Lift coefficient; (b) Drag coefficient; (c) Pitching moment coefficient; (d) Ratio of the lift and drag.

At first, the hydrodynamic coefficients including lift, drag and pitching moment coefficients of the dolphin's body, pectoral fins and fluke have been analyzed for various angles of attack (AoA) α. For example, Fig. 4 shows the coefficients over AoA of the pectoral fins. Notice that the reference area adopts the front one. As conventionally modeled, lift and pitching moment coefficients are linear relationship with the α while drag coefficient is in a quadratic form. The detailed expression is shown in Eq.(1). The hydrodynamic coefficients of the body and fluke have the similar expressions listed as Eq.(2) and Eq.(3).

$$\begin{cases} C_{d_p}(\alpha) &= 1.481\alpha^2 - 0.000357\alpha + 0.6758 \\ C_{l_p}(\alpha) &= 1.667\alpha - 0.003487 \\ C_{m_p}(\alpha) &= -0.09995\alpha + 0.0009239 \end{cases} \tag{1}$$

$$\begin{cases} C_{d_b}(\alpha) &= 1.946\alpha^2 - 0.01843\alpha + 0.2492 \\ C_{l_b}(\alpha) &= 1.406\alpha - 0.01649 \\ C_{m_b}(\alpha) &= -0.1784\alpha + 0.001153 \end{cases} \tag{2}$$

$$\begin{cases} C_{d_f}(\alpha) &= 1.344\alpha^2 - 0.002419\alpha + 0.09103 \\ C_{l_f}(\alpha) &= 1.601\alpha \\ C_{m_f}(\alpha) &= 0.3446\alpha - 0.000524 \end{cases} \quad (3)$$

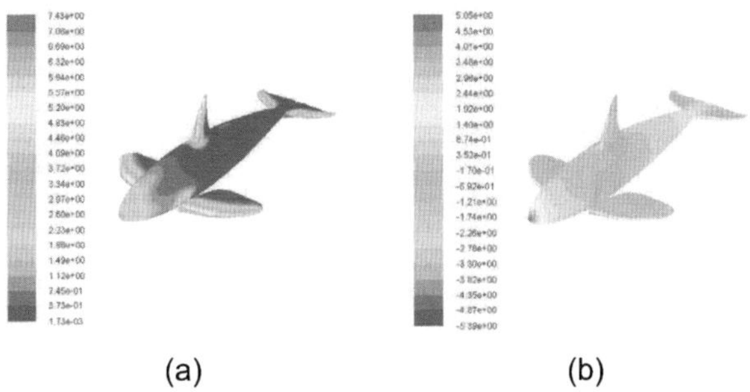

Fig. 5. Pressure contour of the dolphin-like glider. (a) Dynamic pressure; (b) Static pressure.

The analysis in the dynamic and static pressure distribution around the dolphin-like glider are also executed. Fig. 5 separately displays these pressure distribution around the glider. It shows that the highest pressure is at the tip of the glider's nose and the windward side of every fin or fluke, due to the interreaction between the fluid and the dolphin-like glider. These pressures on the rest of the glider surface is both lower because of the smooth flow.

4. Experiments and Discussion

To evaluate the performance of the dolphin-like glider, extensive experiments have been carried out.

The first experiment was designed to testify the downward gliding motion. At the beginning, the glider absorbed about 4.6 g water to keep balance in the surface of the water. The turn angle β between pectoral fins and body was manually changed to explore how the angle β affected the glide speed. When receiving the descending command, the glider began to absorb the other 4.6 g water to glide downwards. According to the experimental results, the glider gained the highest horizontal speed up to 0.096 m/s, when $\beta = 10°$.

The second experiment was carried out to testify the upward gliding motion. Similar with downward glide, the upward gliding motion could be successfully realized through draining away 4.6 g water, as shown in Fig. 6. The glider obtain the highest horizontal speed up to 0.063 m/s.

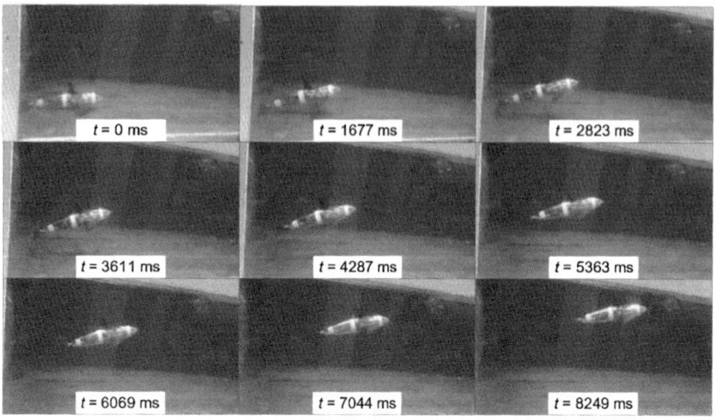

Fig. 6. Snapshot sequence of upward gliding motion.

Traditional underwater gliders usually employ the internal moveable masses to regulate the gliding attitude. Due to moving back and forth, the internal masses often occupy large space that leads to a low space utilization rate. Comparatively, the dolphin-like glider provided in this paper could obtain enough pitching torques from both buoyancy-driven system and controllable fins including pectoral fins and flattened fluke. Moreover, the fluke often provides a considerable pitching moment because of the relative larger moment arm. In this situation, the buoyancy-driven system only need a little volume for water, about $\pm 0.6\%$ of the whole displacement. The volume of the buoyancy-driven system and turn angle of the controllable fins could also be used as controlled input variables for an expected accurate attitude. Moreover, flexible pectoral fins and flattened fluke could bring a quick response to the attitude adjustment.

5. Conclusions and Future Work

In this paper, we have provided a novel design concept of a dolphin-like glider. As an excellent combination, the dolphin-like glider could not only quickly swimming forward in dolphin style, but also quietly glide for a long distance. In order to testify and analyze the gliding motion without the traditional internal moveable masses, a dolphin-like glider prototype has been developed. With the help of CFD simulation, the hydrodynamic characteristics in gliding motion has been analyzed and important hydrodynamic coefficients over the angle of attack and pressure distribution have also been provided for the following dynamic analysis. Experimental results verify that the dolphin-like glider could successfully glide upwards and downwards relying on the pitching torques only from buoyancy-driven system and controllable pectoral fins and flattened fluke, even without traditional internal moveable masses.

The ongoing and future work will focus on the mechanical design and motion control for a real dolphin-like underwater glider possessing both dolphin-like swimming and quiet gliding motion.

Acknowledgments

This work is supported by the National Natural Science Foundation of China (Nos. 61375102, 61273326, 61333016) and the Beijing Natural Science Foundation (No. 3141002).

References

1. D. L. Rudnick, R. E. Davis, C. C. Erikesn, and et al, Underwater gliders for ocean research, *Marine Technology Society Journal*, **38**(1): 48–59, 2004.
2. G. Griffiths, C. Jones, J. Ferguson, and et al, Undersea gliders, *Journal of Ocean Technology*, **2**(2): 64–75, 2007.
3. S. Zhang, J. Yu, A. Zhang, and F. Zhang, Spiraling motion of underwater gliders: modeling, analysis, and experimental results, *Ocean Engineering*, **60**: 1–13, 2013.
4. H. Stommel, The slocum mission, *Oceanography*, **2**(1): 22–25, 1989.
5. D. C. Webb, P. J. Simonetti, and C. P. Jones, SLOCUM: an underwater glider propelled by environmnetal energy, *IEEE Journal of Oceanic Engneering*, **26**(4): 447–452, 2001.
6. C. C. Eriksen, T. J. Osse, R. D. Light, and et al., Seaglider: a long-range autonomous underwater vehicle for oceanographic reaearch, *IEEE Journal Oceanic Engneering*, **26**(4): 424–436, 2001.
7. J. Sherman, R. E. Davis, W. B. Owens, and J. Valdes, The autonomous underwater glider "Spray", *IEEE Journal Oceanic Engneering*, **26**(4): 437–446, 2001.
8. Teledyne Webb Research, http://www.webbresearch.com/slocumglider.aspx
9. J. Yu, M. Wang, M. Tan and J. Zhang, Three-Dimensional swimming, *IEEE Robotics and Autonomous Magazine*, **18**(4): 47–58, 2011.
10. J. Yu, Z. Su, M. Wang, M. Tan, and J. Zhang, Control of yaw and pitch maneuvers of a multilink dolphin robot, *IEEE Transactions on Robotics*, **28**(2): 318–329, 2012.
11. ANSYS Inc., http://www.ansys.com/

MECHANICAL DESIGN OF A TWO-JOINT ROBOTIC FISH*

CHENG ZHANG, JUNZHI YU, MIN TAN
State Key Laboratory of Management and Control for Complex Systems
Institute of Automation, Chinese Academy of Sciences, Beijing, 100190, China

In this paper, we present the mechanical design of a novel robotic fish capable of fast swimming and yet with less joints. Prior to designing, the principle and goals for design are analyzed, in which the size, central pattern generator (CPG) model, and oscillatory frequency of robotic fish play an important role. Then, the mechanical structure of the two-joint robotic fish is designed. The first joint of the robotic fish is driven by a gear motor. The gear motor is linked to an eccentric wheel which rotates in a free slide. The second joint is driven by a connecting rod to make the two joints form an angle between them, allowing free direction adjustment and flexibility. After that, the mechatronic design and CPG-based control are described. Underwater tests are performed on the robotic fish, validating the effectiveness of the proposed design scheme. Particularly, the robotic fish reached a maximum speed of 0.7 body lengths per second, which is expected to be much faster after improvement.

1. Introduction

After long time evolution, fish has developed great ability of fast swimming, moving flexibly, and performing complicated movements, which attracts many researchers to investigate the characteristics of fast swimming and high maneuverability of fish. After coming into the 21th century, the development of oceanic engineering and related enabling technologies like robotics and automation empowers human beings to harness marine resources and to protect the marine environment. Robotic fish does not have a propeller and thereby creates less noise. Often powered by electric energy, the robotic fish cause minimal disturbance to the environment. More importantly, robotic fish share the characteristics of fast-swimming and high maneuverability[1].

As for the applications of robotic fish, it can detect the water pollution[2]. It is also of great help in oceanic exploration and archaeology. Besides, as robotic fish looks and acts like real fish, it can get close to real fishes and will not scary them away, which can aid in biological research.

* This work is supported by the National Natural Science Foundation of China (Nos. 61375102, 61273326, 61333016) and the Beijing Natural Science Foundation (No. 3141002).

The mechanical parts have a great influence on the performance of the robotic fish. In principle, different propulsive mechanisms, sizes, and morphologies will result in different performance, corresponding to different levels of task execution. In this paper, we aim to develop a new robotic fish capable of fast swimming and yet with less joints.

2. Goals and Principle for Design of Robotic Fish

One of the most important factors for designing robotic fish is the speed. Although there are many reasons that influence the speed of robotic fish, the following three requirements are the most significant.

1. CPG control model for generating fishlike swimming.
2. Oscillation frequency of the tail of robotic fish.
3. Power of the motor and mass of the robotic fish.

As for Requirement 1, there are many researches on CPG model. The CPG model for robotic fish is like the sinusoidal curve, or like part of it. For example, Wu's[3] and Wang's[4] robotic fish which are governed by CPG models reach good swimming performance.

Concerning Requirement 2, SPC-II has a maximum frequency of about 6 Hz in the air[5]. This robotic fish achieved a relatively high swimming speed by means of both well-built CPG model and high frequency[6]. Theoretically, the robotic fish which uses servo motors to provide the propulsive force cannot reach high frequency because the fastest rotating speed is about 0.09 s/60° for servo motors. That is, $\omega = 60°/0.09 \approx 667°/s$. If the oscillation amplitude of the tail is 60°, servo motor will rotate $60° \times 4 = 240°$ in one oscillation period. Thus, the oscillation period becomes $T = 240°/667° \approx 0.36$ s. The oscillation frequency of the tail corresponds to $f = 1/T = 1/0.36 \approx 2.78$ Hz. The realistic frequency will be lower while moving through the water, maximally around 2–2.5 Hz. Note that even the faster servo motor will not reach 4 Hz in the water. Therefore, the speed is much limited. In such a situation, we decide to design the robotic fish using gear motor instead of servo motor.

With regard to Requirement 3, the power of the motor is larger and the mass of the robotic fish is lighter (i.e., the size is smaller), robotic fish will swim faster.

3. Mechanical Design of the Robotic Fish

The mechanical structure of the robotic fish is shown in Figure 1, the total length of the robotic fish is about 310 mm to make sure the fish is small and light enough. The motor rotates in one direction. The motor drives an eccentric wheel to make the two joints and the fish tail oscillate back and forth to produce sufficient propulsive forces.

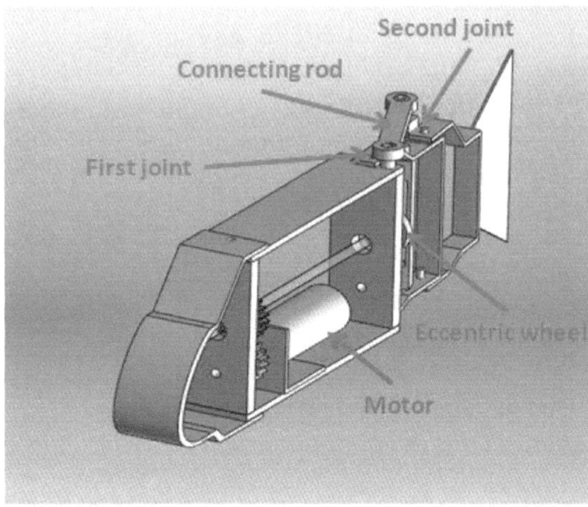

Figure 1. Mechanical structure of the two-joint robotic fish.

3.1. *Mechanical design of the first joint*

As shown in Figure 2, the first joint adopts the eccentric wheel structure. When the eccentric wheel rotates in one direction, the first joint of the robotic fish will reach an angle. Besides, the eccentric wheel still has two other holes on it to make the robotic fish have many choices to change the oscillation amplitude. As shown in Figure 3, the slide for the eccentric wheel to rotate along is designed as a free one, allowing rotation around the axis itself, which reduces much resistance of the transmission mechanism.

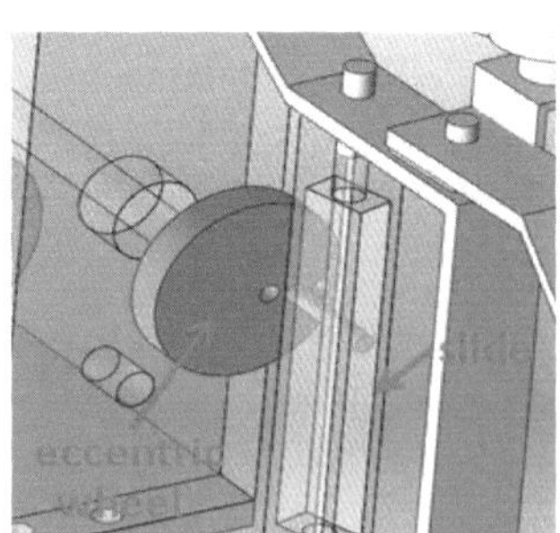

Figure 2. Mechanical structure of the first joint.

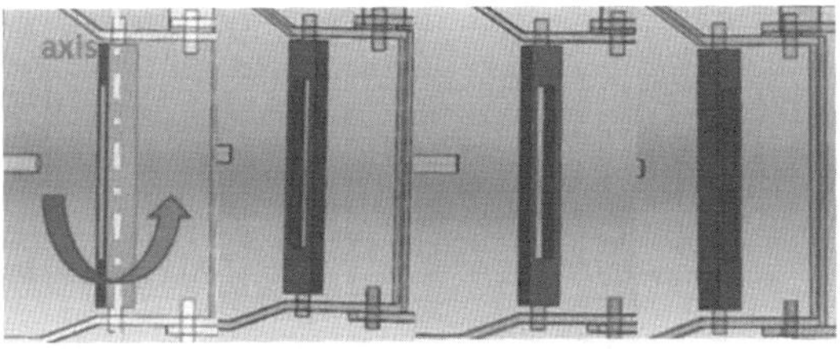

Figure 3. Mechanical structure of the free slide.

3.2. *Mechanical design of the second joint*

The second joint is driven by a connecting rod. The angle between two joints is varied as the connecting rod moves to a different position. The moment when the connecting rod moves to a definite position is shown in Figure 4(a), where the angle between the two joints is defined as α.

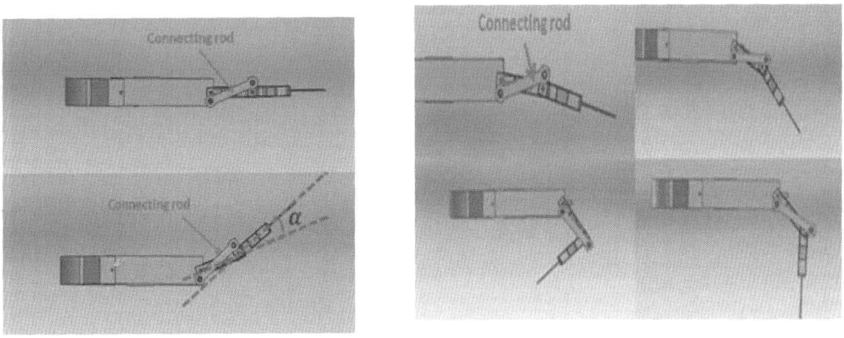

(a) (b)

Figure 4. Mechanical structure of the second joint moving to an angle.

By adjusting the diameter of the eccentric wheel, with the aid of the connecting rod, the angle between two joints will change, which endows the the ability of maneuverability as shown in Figure 4(b).

4. Swimming Mode and CPG Design

4.1. *State of straight swimming*

The robotic fish tail follows the CPG model as a sinusoidal curve. As shown in Figure 5, the two-joint fish tail theoretically tracks the red line. Note that, for real fish, the amplitude of the head is less than that of the tail.

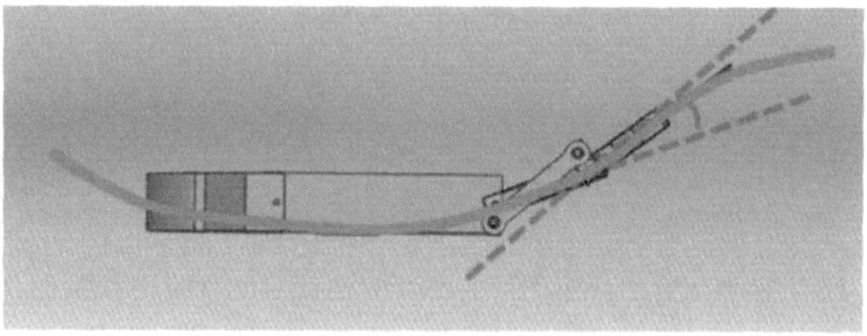

Figure 5. CPG model on robotic fish

Mathematically, one sinusoid corresponds to only one intrinsic amplitude. To determine a proper CPG model, we need a combination of different sinusoidal curves which have different amplitudes. Therefore, we determine the CPG model as the following formula:

$$CPG = \sum_{j=1}^{m} a_i \sin(b_i x + c_i) \quad (1)$$

where m is the number of joints, a_i, b_i, and c_i are constants. As the number of the joints for the developed robotic fish is two, the CPG module is designed as:

$$CPG = a_1 \sin(b_1 x + c_1) + a_2 \sin(b_2 x + c_2) \quad (2)$$

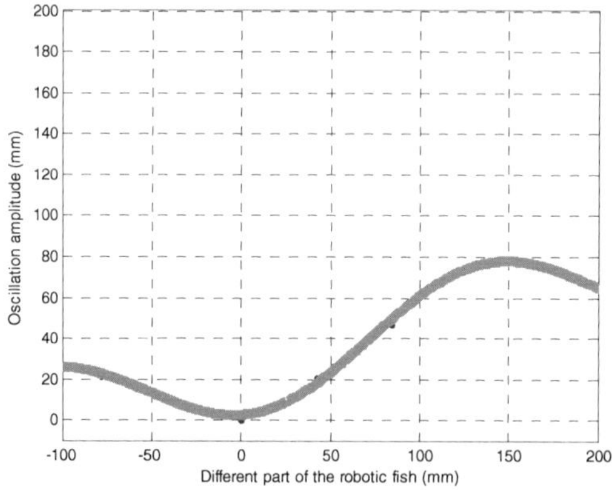

Figure 6. Simulation of CPG model.

After simulation according to the real size of the robotic fish, the result is shown in Figure 6.

The simulation formula is derived as follows:

$$CPG = 278.1\sin(0.0007544x + 0.09097) + 23.22\sin(0.02517x - 1.825) \quad (3)$$

Then, we can use this simulation result to predict the swimming patterns and postures of the robotic fish.

4.2. *State of performing a turn*

Then, we design the control method to make a turn. The basic turning idea for our robotic fish is that the driving motor in the tail rotates in one direction for a short period of time and then rotates in the opposite direction for a short period of time. As demonstrated in Figure 7(a), when the motor rotates, the gears will rotate which drive the axle rotating. The axle is linked to the eccentric wheel, which drives the robotic joint. The axle rotates in one direction for a short period of time and reverse for the same time, which helps the robotic fish make a turn. When to repeat this process, the tail of the robotic fish can repeat the process in Figure 7(b).

In detail, as shown in Figure 7(b), the tail firstly stays in the horizontal position. Then the motor rotates in one direction, and the tail moves to an angle. After that, the motor rotates in the opposite direction and the tail returns to the horizontal position. Apparently, when the motor rotates fast, the oscillation frequency of the tail increases. Repeating this process for a couple of times, the robotic fish can achieve a turn.

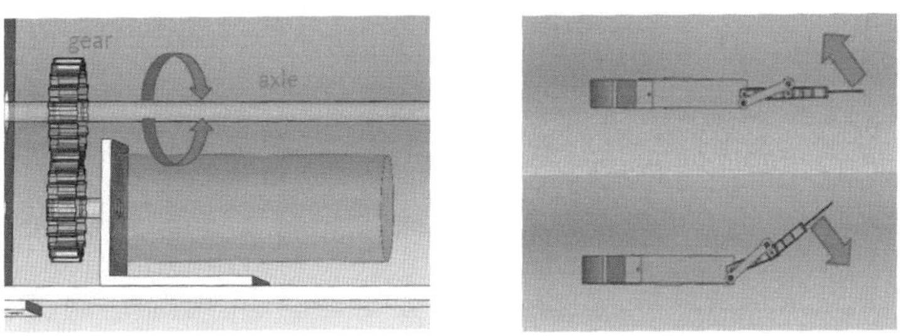

(a) (b)

Figure 7. Schematic diagram of achieving a turn.

5. Design of Hardware System

The electric system of the robotic fish includes battery (11.1 V lithium battery), the board of controller (Cortex-M4), the board of driver module (L298N), the communication module (RF200), and the gear motor. The whole electric system is illustrated in Figure 8.

After completion of the mechatronic design, the robotic fish is constructed and tested. The characteristic parameters of the robotic prototype are shown in Table 1.

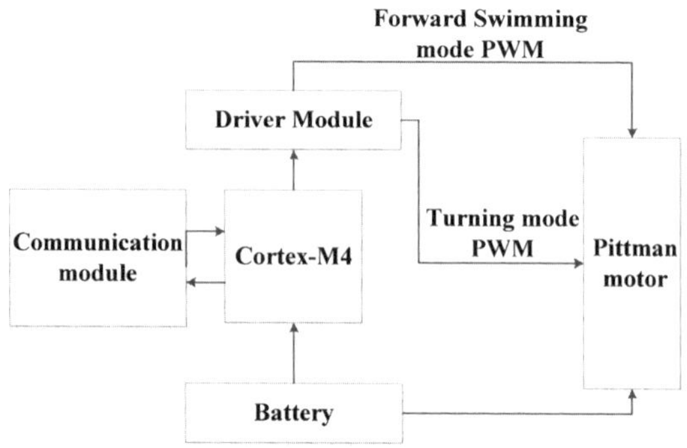

Figure 8. Block diagram of the electric system for the robotic fish.

Table 1. Technical parameters of the robotic fish.

Size	310×77.5×28 mm^3
Weight	863 g
Number of joints	2
Duration time	1.5 h
Microcontroller	Cortex-M4
Battery	11.1 V Lithium battery
Communication module	RF200
Driver module	L298N
Gear motor	Pittman motor

6. Underwater Tests

To validate the proposed design scheme, underwater tests are performed on the developed robotic fish with waterproofing treatment. When the oscillation frequency rises, the swimming speed of the robotic fish increases. Remarkably, the robotic fish reached a maximum speed of 0.22 m/s, corresponding to 0.7 body lengths per second (BL/s). A snapshot of steady swimming over a period is shown in Figure 9.

7. Conclusion

In this paper, we have developed a new robotic fish capable of fast swimming and yet with only one motor. Specifically, a two-joint mechanical configuration is adopted based on an eccentric wheel to drive the first joint, while the second joint is driven by the connection rod. The simplified CPG model is proposed and implemented to yield relatively flexible and stable swimming. Underwater tests on the real robot verify the effectiveness of the proposed design scheme. In particular, the robotic fish achieved a maximum speed of 0.7 BL/s. However, the power of the gear motor still needs improvement. If the power of the gear motor increases, the rotating speed and torque will also improve much. In future work, the propulsive speed and maneuverability of the single motor driven robotic fish are expected to be greatly enhanced after improvement on hydrodynamic optimization and motor configuration.

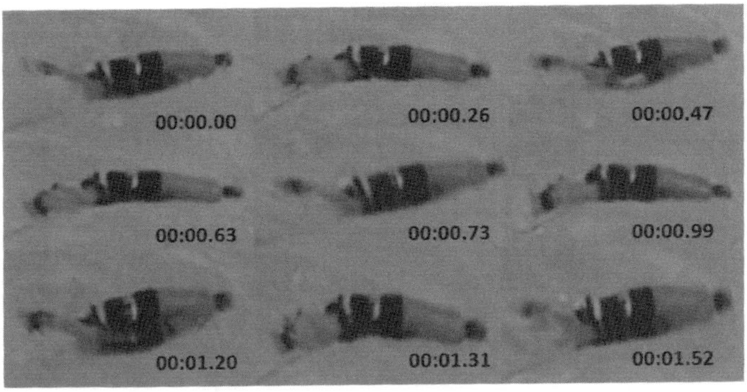

Figure 9. Snapshot of underwater tests on the robotic fish.

References

1. Z. Su, J. Yu, M. Tan, and J. Zhang. Implementing flexible and fast turning maneuvers of a multijoint robotic fish. *IEEE/ASME Trans. Mechatronics*, 19(1), 329-338, 2014.
2. C. Wei and J. Yu, Mechanical design and implementation of a robotic dolphin with 3-DOF flippers. *Proc. 30th Chinese Contr. Conf.*, 4078-4082, 2011.
3. Z. Wu, J. Yu, and M. Tan, CPG parameter search for a biomimetic robotic fish based on particle swarm optimization. *Proc. 2012 IEEE Int. Conf. Robot. Biomim.*, 563-568, 2012.
4. J. Yu, M. Wang, Z. Su, M. Tan, and J. Zhang, Dynamic modeling and its application for a CPG-coupled robotic fish, *Proc. 2011 IEEE Int. Conf. Robot. Autom.*, 159-164, 2011.
5. J. Liang, T. Wang, and L. Wen, Development of a two-joint robotic fish for real-world exploration, *J. Field Robot.*, 28 (1), 70-79, 2011.
6. R. J. Clapham and H. Hu, iSplash-I: High performance swimming motion of a carangiform robotic fish with full-body coordination, *Proc. 2014 IEEE Int. Conf. Robot. Autom.*, 322-327, 2014.

A NOVEL HYDRAULIC MECHANISM FOR BIO-INSPIRED UNDULATING ROBOT: MODELING AND MORPHOLOGICAL ANALYSIS *

HAIJUN XU, TIANJIANG HU, XIANG ZHANG and LEI ZHANG
College of Mechatronics and Automation, National University of Defense Technology, Changsha, Hunan, 410073, China

Hydraulic-driven Bionic Undulating Robot (HBUR) is regarded as one of fish-inspired underwater robots, whose bionic fin rays are actuated by a specially designed hydraulic system, and aims to promote the flexible performance of bionic motion and hydrodynamics. To understand in-depth the undulating mechanism of the bionic hydraulic system, so as to develop a more advanced bionic structure and valve system, this paper concentrates on model developing and morphology analyzing of undulating motions on HBUR. First of all, mathematical models for HBUR are setup with considerations on mechanism of bionic valve, as well as on bionic structure of hydraulic swing actuator (HSA). With some experiments, these analytic models were verified. Then, undulating morphologies and its hydrodynamics were investigated and discussed, to evaluate the actuating performances of the bionic hydraulic system. Results show that the specially designed valve could actuate HSA swinging and HBUR undulating as expected, under a much more simplified mechanical structure compared to the servo-valve approach. Characteristics of morphologies and hydrodynamics indicate that HBUR will be very promising for implementing a more progressive undulating robot underwater.

1. Introduction

Inspired by the special undulating shape and excellent hydrodynamics of fish fin, several bionic robots have been developed for generating undulating motion[1-3]. For instance, [4-6] presented two kinds of cam-based bionic undulating robot, which are constructed with rigid transmission structure, and may lead to structure and motion failure when underwater obstacles occurred.

To improve bionic performances, such as motion flexibility and structure safety, shape memory alloy and parallel actuator based technologies have been applied for developing new variants, and have attracted a lot developments[7-10].

* This work is supported by the Natural Science Foundation of China (NSFC), with the granted NO.: 51105365 and 61005077.

However, related experiments revealed that some downsides such as the complicated mechanism and costly control system were still hard to overcome.

Based on the understanding above, bio-hydraulic mechanism has been proposed for the construction of a hydraulic-driven bionic undulating robot (HBUR)[11]. For an in-depth understanding and investigation, its undulating morphologies and the hydraulic mechanism needs to be modeled and discussed.

2. Mechanical Implementations and Modeling of HBUR

2.1. *Construction of bionic undulating structure*

Figure 1 shows the fish "Gymnarchus niloticus" and its undulating fin consisted of several flexible fin rays[12-14]. To imitate the undulating motion, a bionic undulating robot is presented in Fig. 2 by arranging several Hydraulic Swing Actuators (HSAs) linearly on a bracket. When the HSAs are driven to swing periodically and sequentially, undulating motion will be formed.

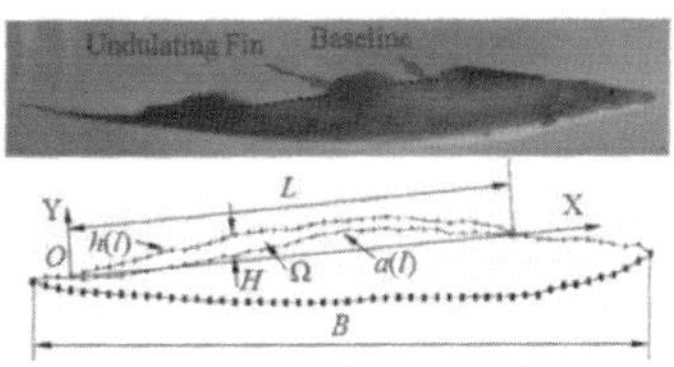

Figure 1. Undulating fin on fish back.

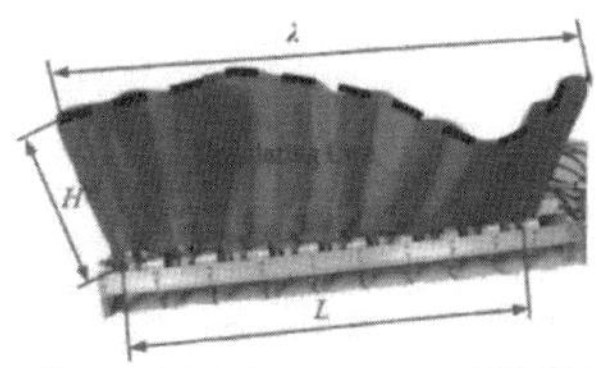

Figure 2. Main structure of HBUR.

In Fig. 1, L, H, B, $a(l)$, $h(l)$ denote the baseline length along X axis, maximum height of fish fin, body length, baseline shape and height of fin ray respectively, where l is a variable along X axis and has a range of $0 \le l \le L$. For engineering convenience, bionic fin rays fitted on the HSAs are selected to have the same length, which means that $h(l) = H$, as is shown in Fig. 2.

2.2. *Implementation and modeling of hydraulic swing actuator*

To imitate the swing motion, a typical HSA is proposed in Fig. 3.

Defining $\theta_i > 0$ when pole is on the right of Y_i axis, q_i and $\dot{\theta}_i$ get positive value if it swing anticlockwise. To model the motion of HSA, flow pressure p_{Li} and flow rate q_{Li} are introduced by $p_{Li} = p_l - p_r$ and $q_{Li} = (q_l + q_r)/2$. Applying Newton's second law, the dynamic motion gives

$$J\ddot{\theta}_i = p_{Li}A_p\frac{R+r}{2} - 2k_sR\theta_i - \mu\dot{\theta}_i - T_l, \quad -\theta_{max} \le \theta_i \le \theta_{max} \tag{1}$$

where, J is rotatory inertia of pole and $A_p = (R-r)BA$ is side sectional, k_s is spring stiffness, μ is damping factor and T_l is load torque on pole.

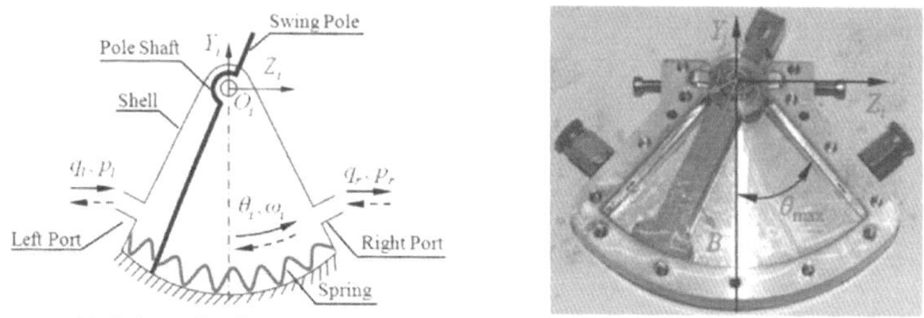

(a) Schematic diagram (b) Engineering structure

Figure 3. Mechanism and implementation of hydraulic swing actuator

With respect to the structure parameters in Fig. 3(a), flow continuity equation gives the relationship of θ_i and q_{Li} as

$$q_{Li} = \frac{R+r}{2} A_p \dot{\theta}_i + k_{lp} p_{Li} + \frac{V_{Li}}{\beta} \dot{p}_i \tag{2}$$

where, k_{lp} is leakage coefficient around pole, β is bulk modulus of oil, V_{Li} is volume inside HSA and taken into consideration, and can be calculated by

$$V_{Li}(t) = \begin{cases} \dfrac{R^2 - r^2}{2} \cdot B \cdot (\theta_{\max} + \theta_i) + V_{\min} & \dot{\theta}_i \geq 0 \\ \dfrac{R^2 - r^2}{2} \cdot B \cdot (\theta_{\max} - \theta_i) + V_{\min} & \dot{\theta}_i < 0 \end{cases} \tag{3}$$

Usually, oil compressibility can be neglected if the flow pressure is low and the chamber has small scale. Then Eqs. (2) and (3) can be rearranged as

$$\dot{\theta}_i(t) = \frac{q_{Li} - k_{lp} p_{Li}}{A_p \dfrac{R+r}{2}} = \frac{2(q_{Li} - k_{lp} p_{Li})}{(R^2 - r^2)B} \tag{4}$$

2.3. *Mechanism and modeling of hydraulic valve*

The traditional method is to drive every HSA with a servo valve. However, structure challenge will occur, and control system complexity will be unacceptable if a consistent undulating motion profile is expected[15-18].

Inspired by the co-swing motions among fish fin rays, a novel valve named plate-shaped valve was developed in [19] and tested in Fig. 4. Mechanism of the plate-shaped valve can be demonstrated by Fig. 5.

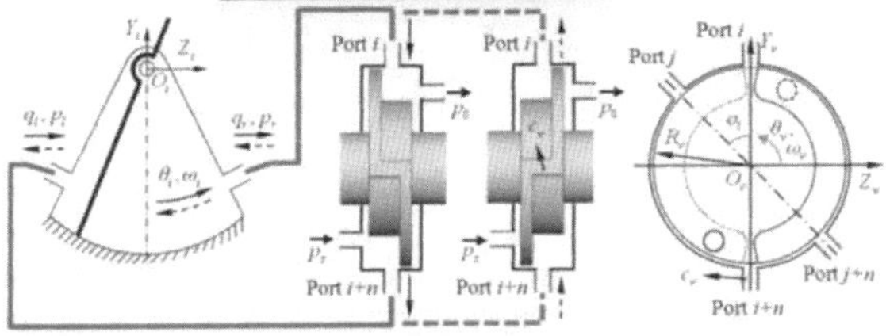

Figure 4. Plate-shaped valve. (a) (b) (c)

Figure 5. Mechanism of plate-shaped valve.

When flange with width c_v on spool is crossing a working port, it will cover the port partially. Especially as is demonstrated in Fig. 5(c), the side view of Fig. 5(a) and (b), ports i and $i+n$ are covered completely. According to Fig. 5, opening of port i in one revolution of spool rotation can be depicted by

$$x_{vi}(\theta_v)=\begin{cases} R_v\theta_v & 0\le\theta_v\le c_v/R_v \\ c_v & c_v/R_v<\theta_v\le\pi-c_v/R_v \\ \pi R_v-R_v\theta_v & \pi-c_v/R_v<\theta_v\le\pi+c_v/R_v \\ -c_v & \pi+c_v/R_v<\theta_v\le 2\pi-c_v/R_v \\ R_v\theta_v-2\pi R_v & 2\pi-c_v/R_v<\theta_v<2\pi \end{cases} \tag{5}$$

where, R_v is spool diameter, c_v is the maximum opening of working port. Apparently, opening of port $i+n$ is opposite to that of port i. Usually $\theta_v=\omega_v t$ is met. Negative value of x_{vi} indicates that port i is connected to the return port.

As shown in Fig. 5(c), while considering another pair of ports j and $j+n$, they will be covered by the flange of spool following the some rule as ports i and $i+n$, but with a fixed time delay caused by the position angle φ_l. Particularly for port j, its opening can be expressed as

$$x_{vj}(\theta_v)=x_{vi}(\theta_v-\varphi_l) \tag{6}$$

Due to the same covering mechanism of port, mathematical models of servo valve could be utilized for modeling plate-shaped valve. Supposing all the working ports are zero-lapped, when spool achieves a displacement x_{vi} at moment t, flow rate q_l can be specified by

$$q_{Li} = C_d k_A x_{vi} \sqrt{\frac{2(p_S - |p_{Li}|)}{\rho}} \tag{7}$$

where, C_d is port discharge coefficient, k_A is port section area coefficient, p_S is flow supply pressure and p_L is flow load pressure, ρ is oil mass density.

Especially, if supply pressure p_S keeps stable, Eq. (7) can be simplified into the following form by applying Taylor extraction method.

$$q_{Li} = k_q x_v - k_c p_{Li} \tag{8}$$

where, k_q is flow rate coefficient, k_c is load coefficient. Eq. (8) indicates that flow rate will be determined by the load pressure and spool position.

Applying Laplace Transfer approach on Eqs. (1), (4) and (8), then

$$\begin{cases} \theta_i(s) = \dfrac{A_p \dfrac{R+r}{2(k_{lp}+k_c)} k_q}{Js^2 + [(\mu + k_T) + A_p^2 \dfrac{R+r}{2(k_{lp}+k_c)}]s + 2k_s R} x_{vi}(s) \\ q_{Li} = \dfrac{k_q k_{lp}}{k_c + k_{lp}} x_v + \dfrac{A_p k_c}{k_c + k_{lp}} \dot{\theta}_i \end{cases} \tag{9}$$

With Eqs.(5) and (9), angular responses of hydraulic swing actuator actuated by valve ports at $\varphi_l = 0$, $\pi/4$ and $\pi/2$ are calculated and revealed in Fig.6, while spool rotating at $f_v = 2\text{Hz}$. Fig.7 gives the experimental results of HSAs, under the angular locations $\varphi_l = 0$, $2\pi/9$ and $4\pi/9$.

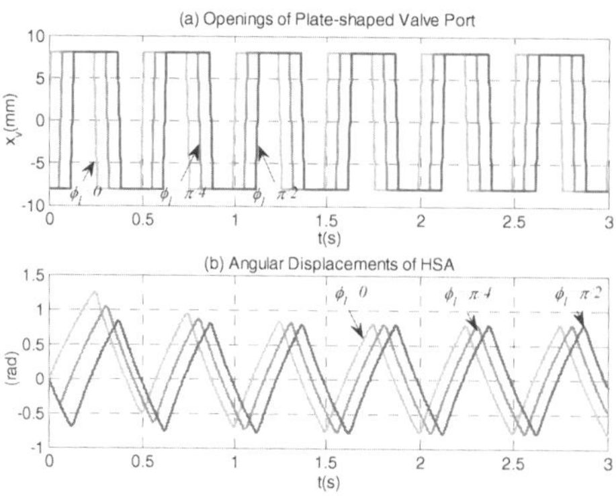

Figure 6. Calculating results of HSA motions. motions.

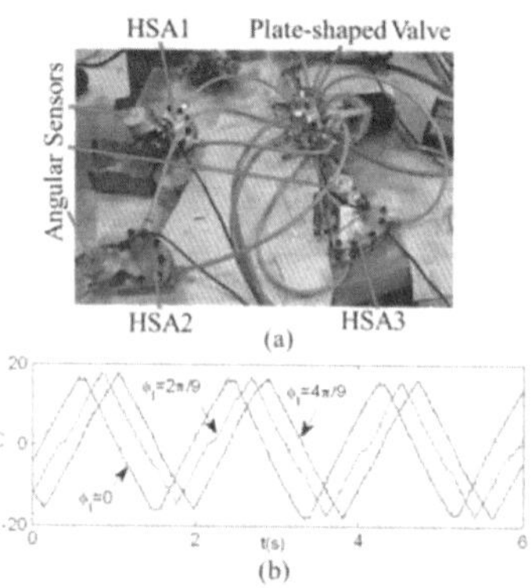

Figure 7. Experimental data of HSA

It can be observed from Fig. 6(a) and (b) that, when spool rotates at a constant angular speed, openings of ports around sleeve are controlled to change in trapezoidal waveform, and HSAs are actuated to swing under a process like sawtooth profile. Since working ports have different angular locations around sleeve, when covered by the same spool, they drive HSAs following the same swinging waveform but under fixed swing phases, in accordance to port locations. Fig.7(b) shows similar characteristics compared to the calculating results, indicating that the mathematical models for hydraulic system is valid.

3. Morphologies and Hydrodynamics of Bio-Undulations

3.1. *Undulating morphologies actuated by plate-shaped valve*

To generate a regular undulating motion, φ_l of working ports on plate-shaped valve should be specified according to the undulating motion and the structure parameters. When n number of HSAs are comprised together for a HBUR, swing phase of each HSA can be specified by

$$\varphi_l = \frac{2\pi\Delta d}{\lambda}(i-1) \tag{10}$$

where, $\Delta d = L/(n-1)$ is the distance between two adjacent HSAs.

With respect to physiological structure and motion of fish fin, its smooth undulating shape Ω can be described by ruled-surface equations as[20]

$$\begin{aligned} \Omega : \boldsymbol{r}(l,s,t) &= \begin{bmatrix} x & y & z \end{bmatrix}^T \\ &= a(l) + sh(l)c(l,t) \\ &= \begin{bmatrix} l \\ sh(l)\cos\theta(l,t) \\ sh(l)\cos\theta(l,t) \end{bmatrix} \end{aligned} \tag{11}$$

where, $\theta(l,t)$ is angular displacement of fin ray located at $x = l$. s is a variable indicating an arbitrary point on the specified fin ray with height $sh(l)$.

Eqs. (6), (9), (10) and (11) are used for calculating the morphologies, with $f_v = 2\text{Hz}$, $c_v = 8mm$, $\lambda = 480mm$ and $240mm$. Results are shown in Fig. 8.

Figure 8 reveals that undulating profile can be formed when all the HSAs are actuated by the plate-shaped valve, and different wavelength of undulating morphologies can be achieved if port locations are set with different values.

3.2. *Morphological evolution and hydrodynamics against B*

Figure 9 exhibits a key snapshot of the durative undulating motion of fish fin, the oscillating profile projection of fin upper edge is called undulating line, and designated with maximum amplitude A and wavelength λ.

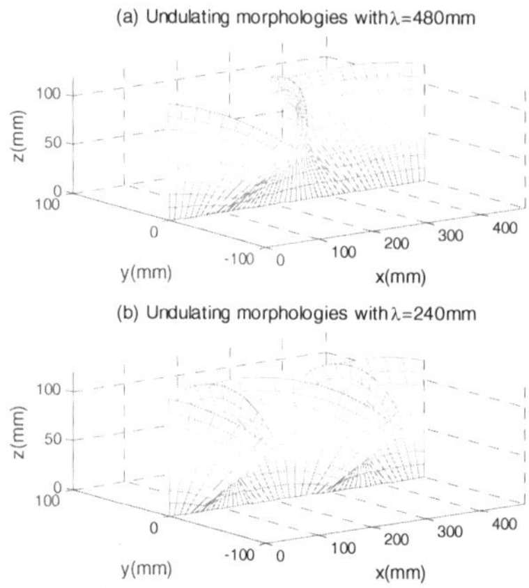

Figure 8. Morphological results for HBUR.

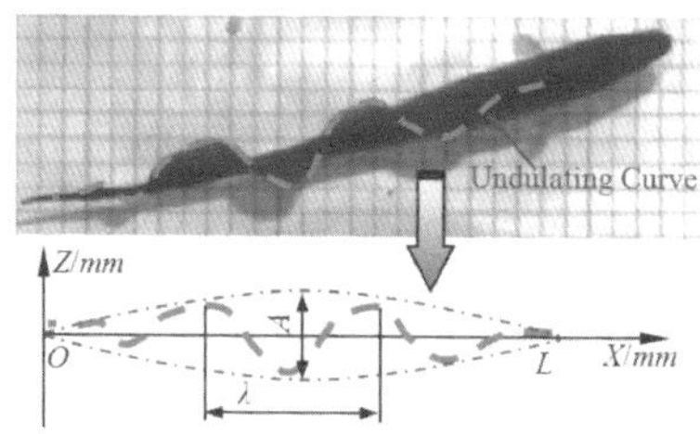

Figure 9. Oscillating profile and projection of fin.

Oscillating features of undulating line can be understood as that at any moment, it shows a spindled envelop shape with amplitude decreases gradually from the middle to the two sides, indicating that swing ranges of different fin rays decline in the same way.

According to Eq. (11), the undulating line can be delineated by

$$\begin{bmatrix} x & y & z \end{bmatrix} = \begin{bmatrix} l & 0 & h(l)\sin\theta(l,t) \end{bmatrix} \tag{12}$$

Since Eq. (9) indicates that the swing motion is related not only to the spool displacement but also to the structure parameters of HSA, especially R r and B. So as to keep the spindled envelop shape, a potential approach is matching the thicknesses of HSAs with certain values, so that swing motions of HSAs will follow the same process but under certain ranges.

Without loss of generality, thickness of pole is designed as

$$B(x) = \frac{B_{\min}}{c_B - \dfrac{4(1-c_B)}{L^2}x(x-L)} \tag{13}$$

where, $c_B = B_{\min}/B_{\max}$ is the ratio between the maximum and minimum thicknesses of poles inside HSAs.

To evaluate Eq. (13), undulating lines of HBUR are generated by Eqs. (9) to (12), with $\lambda = 160mm$, $c_B = 0.25$, $c_B = 1$, and displayed in Fig. 10(b), while Fig. 10(a) gives the ideal morphologies. In addition, hydrodynamics of the morphologies in Fig. 10(b) are investigated with CFD platform[21-23], and hydro-forces are revealed in Fig. 11.

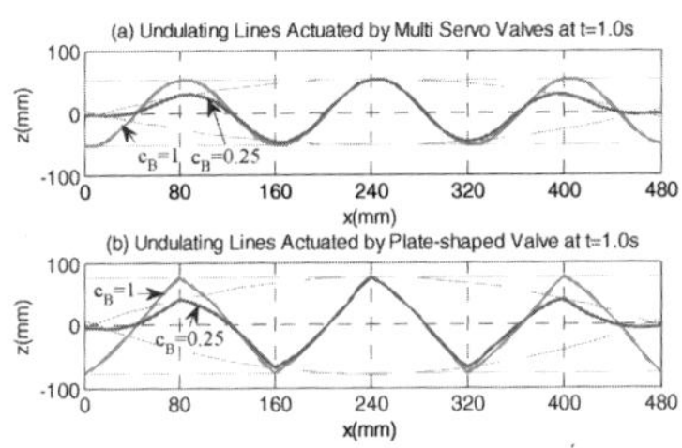

Figure 10. Undulating lines against different *B*.

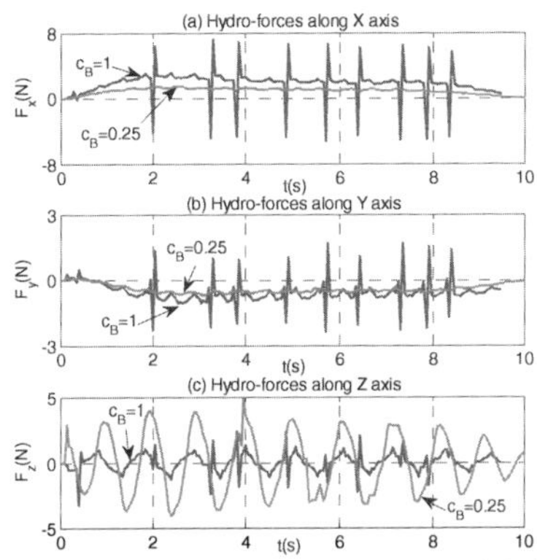

Figure 11. Hydrodynamics of undulating morphologies.

As are depicted in Fig. 10, when $c_B = 1$ is met, the ideal undulating line as well as the one actuated by plate-shaped valve are both enveloped by two parallel lines, indicating that all the HSAs are swinging under the same range, because of the same thickness they have. However, while $c_B = 0.25$ is applied, the undulating lines are reshaped and enveloped by two spindled curves, for the reason that swing ranges of HSAs and the undulating morphologies have changed.

Moreover, Fig. 11(a) and (b) reveal that hydro-force along x axis is smaller when undulating morphology shows a spindled envelop shape, and become less fluctuant. Although periodical value of hydro-forces along *z* axis is zero, the fluctuation of hydro-forces have been enlarged in Fig. 11(c), because the spindled envelop shape has made the morphology symmetry along *z* axis worse.

4. Conclusions

Mechanism and morphologies of a hydraulic-driven bionic undulating robot have been modeled and discussed in the paper, and conclusions are as follows:

(1) Plate-shaped valve has the potential for developing a more progressive HBUR system, because it can actuate the HBUR and generating a consistent undulating motion, under an extremely simplified valve structure.

(2) The mathematical models in the paper are valid for describing the dynamic motions and morphologies of HBUR, and are effective for studying further the kinematics and hydrodynamics of the undulating motions.

(3) Spindled envelop shape of undulating morphologies can make the bionic motion of HBUR closer to fish fin motion, in this way hydro-forces will decline in value but become more stable.

Researches such as the flexible motion characteristics, are actively under development.

References

1. L. J. Shang, S. Wang, M. Tan, et al. *Conf. Mod., Sim., and Ide*. **2**, 912 (2009).
2. M. M. Rahman, S. Sugimori, H. Miki, et al. *J. of Bio. Eng*. **10**, 265(2010).
3. F. F. Liu, C. J. Yang, Y. Q. Xie, et al. *IEEE Conf. on Rob., Auto., and Mec*. 718(2008).
4. H. J. Xu, C. Y. Pan, H. B. Xie. *Mach. Manu*. **43**, 21(2005).
5. K. H. Low, A. Willy. *Rob. and Bio*.. 45 (2006).
6. M. M. Rahman, Y. Toda, H. Miki, et al. *J. of Bio. Eng*.. **8**, 25(2011).
7. Z. B. Yang, Y. K. Wang, F. Gao, et al. *Conf. on Elec. & Mech. Eng. and Info. Tech*. 1404(2011).
8. Cole, J. Nicholas, P. Currie. *Bio. Arch*.. **2**, 98(2012).
9. A. Ghanbari, M. M. Fakhrabadi, M. Siahmansouri, et al. *Conf. on Eng. Sys. D. and A*.. 639(2010).
10. M. Sfakiotakis, D. M Lane, B. C. Davies, et al. *Conf. on Rob. and Auto*.. 2356(2001).
11. H. J. Xu, C. Y. Pan, D. B. Zhang, H. B. Xie. *Mech. Sci. and Tec. for Aero. Eng*.. **29**, 1243(2010).
12. F. Li, T. J. Hu, G. M. Wang, L. C. Shen. *J. of Bio. Eng*. **2**, 115(2005).
13. V. L. George, L. Jeanette, S. Ryan, et al. *Mar. Tech. Soci. J*.. **45**, 41(2011).
14. U. K. Muller, D. B. Van, L. W. Van, et al. *J. of Exp. Bio*.. **211**, 196(2008).
15. M. Y. Salloom, Z. Samad. Int. *J. of Ad. Manu. Tec*.. **58**, 279(2012).
16. F. L. Quilumba, L. K. Lee, W. J. Lee, A. *IEEE Trans. on In. App*.. **50**, 1313(2014).
17. R. Ramakrishnan, S. S. Hiremath, M. Singaperumal, et al. *Energy*. **67**, 422-(2014).
18. S. Z. Yang, L. J. Xing. *Valve*. **5**, 33(2006).
19. H. J. Xu, C. Y. Pan, Q. Li, F. D. Gao. *Ad. in Mech. D*.. **1**, 348(2011).
20. L. J. Shang, S. Wang, M. Tan, et al. *Robotica*. **30**, 913(2012).
21. Y. H. Zhang, J. H. He. *J. of Rob. and Mech*.. **24**, 568(2012).
22. M. A. Rapo, H. S. Jiang, M. A. Grosenbaugh, et al. *J. of Exp. Bio*.. **212**, 1494(2009).
23. M. I. Lamas, J. D. Rodríguez, C. G. Rodríguez, et al. *Pol. Mari. R*.. 18, 10(2011).

AUTHOR INDEX